CW01430327

Forgotten Books

ISBN 978-0-331-01372-6
PIBN 11107902

FB &c Ltd, Dalton House, 60 Windsor Avenue, London, SW19 2RR.
Company number 08720141. Registered in England and Wales.

For support please visit www.forgottenbooks.com

Form A. [PROPOSAL FORM FOR MEMBERSHIP.]

The Iron and Steel Institute.

ESTABLISHED 1869.
INCORPORATED BY ROYAL CHARTER 1899.

Mr. [a] .... (a) Name in full.

Address ....

Business or Profession ....

being of the required age, and desirous of becoming a Member of The Iron and Steel Institute, we, the undersigned, from our own personal knowledge, do hereby recommend him for election.

His qualifications are [b] .... (b) State qualifications fully.

....

....

....

Witness our hands, this..............day of ....

....
....
....
} Names of three Members required.

Address to which communications are to be sent..................

....

Business address, if different (for record only).......................

....

It would be a convenience if the Candidate's Card were sent at the same time.

In accordance with Bye-law No. 22, Members who are elected at a General Meeting in the Autumn in any year are required to pay only half the Subscription for that year.

[Associateship Proposal Form overleaf.]

Form E. [PROPOSAL FORM FOR ASSOCIATESHIP.]

The Iron and Steel Institute.

ESTABLISHED 1869.
INCORPORATED BY ROYAL CHARTER 1899.

Mr. .... Name in full.

Address .... Candidates over 24 years old (30 years for members of certain local technical societies) are not eligible as Associates.

Date of Birth...

Present Occupation...

being desirous of becoming an Associate of The Iron and Steel Institute, we, the undersigned, from our personal knowledge, do hereby recommend him for election.

Particulars of metallurgical or engineering training:

(a) Courses in metallurgy at.. State name of University, College, Technical School, Laboratory or Works.

(b) Pupil or apprentice with..

....

Witness our hands, this..............day of..................................

.... Names of three Members required.

....

....

Address to which communications are to be sent:

....

....

Business address, if different (for record only):

....

....

I hereby certify that Mr.. To be signed by Secretary of local technical society (if any) of which applicant is a member.

is a member of..

....Secretary.

It would be a convenience if the Candidate's Card were sent at the same time.

[Membership Proposal Form overleaf.]

Form A (1).

PROPOSAL FORM FOR JOINT MEMBERSHIP OF THE IRON AND STEEL INSTITUTE and THE INSTITUTE OF METALS.

The Iron and Steel Institute.

ESTABLISHED 1869.
INCORPORATED BY ROYAL CHARTER 1899.

Applicable to those who are 30 years of age and over (or, in the case of applicants to join The Institute of Metals, to those who are 26 years of age and over).*

(*a*) I am a member of the †Iron and Steel Institute / Institute of Metals and wish to join the †Iron and Steel Institute / Institute of Metals.

(*b*) I †am submitting / have submitted a proposal form in order to join the †Iron and Steel Institute / Institute of Metals and wish also to join the other Institute.

Signature ....

Address ....

....

Date ....

For Office use only.

I certify that the above applicant is a †Member / Associate of the †Iron and Steel Institute / Institute of Metals and that his subscription (if any due) has been paid to date.

Signature of Secretary of the †Iron and Steel Institute / Institute of Metals.

....

Date ....

† Words not applicable to be crossed out.

[*Application Form for those under these ages, overleaf.]

Form E (1).

PROPOSAL FORM FOR JOINT MEMBERSHIP OF THE IRON AND STEEL INSTITUTE and THE INSTITUTE OF METALS.

The Iron and Steel Institute.

ESTABLISHED 1869.
INCORPORATED BY ROYAL CHARTER 1899.

Applicable to those who are under 30 years of age (or, in the case of applicants to join The Institute of Metals, to those who are under 26 years of age).*

(c) I was born on..19......

(d) I am a †Member / Associate of the Iron and Steel Institute and wish to join the Institute of Metals.

(e) I am a †Member / Student Member of the Institute of Metals and wish to join the Iron and Steel Institute.

(f) I †am submitting / have submitted a proposal form in order to join the †Iron and Steel Institute / Institute of Metals and wish also to join the other Institute.

Signature ....

Address ....

....

Date ....

For Office use only.

I certify that the above applicant is a †Member / Associate / Student Member of the †Iron and Steel Institute / Institute of Metals and that his subscription (if any due) has been paid to date.

Signature of Secretary of the †Iron and Steel Institute. / Institute of Metals

....

Date ....

† Words not applicable to be crossed out.

[*Application Form for those over these ages, overleaf.]

BRITISH STANDARDISED STEEL SAMPLES

The Council of The Iron and Steel Institute announce that by agreement of the Director of The National Physical Laboratory, the standardised steel samples issued jointly by The Iron and Steel Institute and The National Physical Laboratory can now be supplied post free at the rate of 10*s*. 6*d*. per bottle containing 50 grammes, together with certificate of standardisation.

The Standards at present available are :

No. 1.—Sulphur (S = 0·027%).
No. 2.—Sulphur (S = 0·071%).
No. 3.—Phosphorus (P = 0·029%).
No. 5.—Carbon (C = 0·65%) Acid O.-H. Steel.
No. 6.—Carbon (C = 0·10%) Basic O.-H. Steel.
No. 8.—Carbon (C = 0·27%) Acid O.-H. Steel.
No. 9.—Carbon (C = 1·09%) Acid O.-H. Steel.
No. 11.—Manganese (Mn = 0·69%) Acid O.-H. Steel.
No. 12.—Cast Iron Standard Sample (Si = 2·22%, P = 1·14%, Mn = 0·50%, S = 0·075%).

They can be obtained, by sending order with remittance, direct from :

THE NATIONAL PHYSICAL LABORATORY,
(Metallurgy Dept.), Teddington, Middlesex.

These samples can also be obtained from all Chemists dealing regularly in Laboratory Ware.

THOMAS SWINDEN, Esq., D.Met., Member of Council.
Bessemer Gold Medallist, 1941.

Frontispiece.

No. 1 1941

THE JOURNAL

OF THE

IRON AND STEEL INSTITUTE

Vol. CXLIII.

EDITOR

K. HEADLAM-MORLEY

SECRETARY

ASSISTANT EDITOR

A. E. CHATTIN, B.Sc. (Hons. Met.), A.I.C.

ASSISTANT SECRETARY

PUBLISHED AT THE OFFICES OF THE INSTITUTE
4 GROSVENOR GARDENS, LONDON, S.W.1.
1941

LIST OF THE COUNCIL AND OFFICERS.

President.

John Craig, C.B.E.

Past-Presidents.

C. P. Eugène Schneider, D.Sc.
Sir Francis Samuelson, Bt.
Sir Frederick Mills, Bt., D.L., M.P.
Sir W. Peter Rylands, Bt., J.P.
Frank W. Harbord, C.B.E.
Benjamin Talbot.
Col. Sir W. Charles Wright, Bt., K.B.E., C.B.
William R. Lysaght, C.B.E.
The Rt. Hon. the Earl of Dudley, M.C.

Hon. Treasurer.

James Henderson.

Vice-Presidents.

Fred Clements.
Cecil H. Desch, D.Sc., Ph.D., F.R.S.
Arthur Dorman.
William H. Hatfield, D.Met., F.R.S.
Capt. Robert S. Hilton.
J. S. Hollings.
Sir William Larke, K.B.E.
Cyril E. Lloyd.

Hon. Vice-Presidents.

The Rt. Hon. Lord Airedale of Gledhow.
Charles John Bagley.
Sir Maurice H. L. Bell, Bt., C.M.G.
Carl A. F. Benedicks, Ph.D.
Alfonso de Churruca.
James A. Farrell.
Sir Edward J. George.
Léon Greiner.
Léon Guillet, D.Sc.
John E. James.
The Hon. R. D. Kitson, D.S.O., M.C.
Aloyse Meyer.
Ing. Dr. Vladislav Sýkora.
Thomas Turner, M.Sc.

Members of Council.

William J. Brooke.
Peter Boswell Brown.
Capt. H. Leighton Davies, C.B.E.
Charles Alfred Edwards, D.Sc., F.R.S.
Ian Frederick Lettsom Elliot.
Edmund John Fox.
J. Sinclair Kerr.
Gustavus Henry Latham, J.P.
Edward Fulton Law.
The Hon. R. G. Lyttelton.
Andrew McCance, D.Sc.
J. R. Menzies-Wilson.
Thomas Swinden, D.Met.
Arthur B. Winder.

Hon. Members of Council.

Director, British Iron and Steel Federation, Sir Wm. Larke, K.B.E.
President of the Cleveland Institution of Engineers, L. F. Wright.
President of the Lincolnshire Iron and Steel Institute, A. Robinson.
President of the Newport and District Metallurgical Society, G. H. Latham.
President of the Staffordshire Iron and Steel Institute, G. R. Bashforth.
President of the Swansea Technical College Metallurgical Society, O. J. Thomas.
President of the West of Scotland Iron and Steel Institute, Col. Alan Stein, M.C., T.D., D.L.

Secretary.—K. Headlam-Morley.

Hon. Secretary.—G. C. Lloyd.

Assistant Secretary.—A. E. Chattin, B.Sc. (Hons. Met.), A.I.C.

Librarian.—R. Elsdon.

Offices.—4, Grosvenor Gardens, London, S.W. 1.

Telegraphic Address.—"Irosamente, Sowest, London."

Telephone.—SLOane 0061.

PREFACE.

The present volume contains eight papers presented at the Annual General Meeting of the Institute held in London on May 1st, 1941; of these, one was sponsored by the Corrosion Committee, one was submitted by the Stresses in Moulds Panel of the Ingot Moulds Sub-Committee and one was a Report prepared by the Oxygen Sub-Committee, the latter two bodies being sub-committees of the Committee on the Heterogeneity of Steel Ingots. The discussion and correspondence on these papers are also included, together with the authors' replies, except in one case where, owing to illness, the author was unable to draft his reply in time for it to be printed here; it is hoped that the author's recovery will be rapid and complete, and that he will be able to comment on the correspondence before the No. II. volume of the *Journal* for this year goes to press. In addition, the replies of two authors resident overseas to the discussion of their papers presented at the Autumn Meeting in 1940, which were not received in time for inclusion with the papers themselves, will be found in this book.

Section I. of this volume contains the above-mentioned material and also the Minutes of Proceedings of the Meeting, including the speeches made at the Members' Luncheon which preceded it, together with the Report of Council and Statement of Accounts for 1940.

Section II. is devoted to a survey of literature on the manufacture and properties of iron and steel and kindred subjects, and consists of a collection of abstracts of articles from the Transactions and Proceedings of scientific societies and from the technical press. This Section also contains reviews of recent books and bibliographies of literature dealing with the manufacture and properties of iron and steel. The matter included in this Section has already appeared in the *Bulletin of The Iron and Steel Institute*, which is issued monthly.

In front of the title page is inserted a list of British Standardised Steel Samples issued jointly by The Iron and Steel Institute and The National Physical Laboratory, showing where and on what terms the samples are available. Proposal forms for Membership and Associateship and also for Joint Membership of this Institute and the Institute of Metals will likewise be found in the same place.

4, Grosvenor Gardens,
London, S.W.1.
August, 1941.

CONTENTS.

LIST OF PLATES.

Abbreviations used in Text.

Å. Ångström unit(s).
A.C. air-cooled; alternating current.
A.H. air-hardened.
amp. ampère(s).
amp.hr. ampère-hour(s).
atm. atmosphere(s) (pressure).
Bé. Baumé (scale).
b.h.p. brake horse-power.
B.o.T. Board of Trade.
B.Th.U. British thermal unit(s).
C. centigrade (scale).
cal. calory; calories.
c.c. cubic centimetre(s).
c.d. current density.
c.g.s. centimetre-gramme-second unit(s).
cm. centimetre(s).
coeff. coefficient.
const. constant(s).
c.p. candle-power.
cu. cubic.
cwt. hundredweight(s).
D.C. direct current.
dia. diameter(s).
dm. decimetre(s).
e.m.f. electromotive force.
F. Fahrenheit (scale).
ft. foot; feet.
ft.lb. foot-pound(s).
g. gramme(s).
gal. gallon(s).
H.F. high-frequency.
h.p. horse-power.
h.p.hr. horse-power-hour(s).
hr. hour(s).
in. inch; inches.
in.lb. inch-pound(s).
K. absolute temperature (Kelvin scale).
kg. kilogramme(s).
kg.cal. kilogramme-calory; -calories.
kg.m. kilogramme-metre(s).
km. kilometre(s).
kVA. kilovolt-ampère(s).
kW. kilowatt(s).
kWh. kilowatt-hour(s).
lb. pound(s).
L.F. low-frequency.
m. metre(s).
m.amp. milliampère(s).
mV. millivolt(s).
max. maximum.
mg. milligramme(s).
min. minimum; minute(s).
ml. millilitre(s).
mm. millimetre(s).
m.m.f. magnetomotive force.
N. normal (solution).
N.T.P. normal temperature and pressure.
O.H. open-hearth; oil-hardened.
O.Q. oil-quenched.
oz. ounce(s).
p.d. potential difference.
*p*H hydrogen-ion concentration.
r.p.m. revolutions per minute.
sec. second(s).
sp. gr. specific gravity.
sq. square.
T. tempered.
temp. temperature.
V. volt(s).
VA. volt-ampère(s).
Wh. watt-hour(s).
W.Q. water-quenched.
yd. yard(s).
° degree(s).

SECTION I.

MINUTES OF PROCEEDINGS
AND PAPERS OF
THE IRON AND STEEL INSTITUTE.

ANNUAL MEETING
1941

MINUTES OF PROCEEDINGS

AND

PAPERS AND DISCUSSIONS

AT THE

ANNUAL MEETING, 1941.

THE SEVENTY-SECOND ANNUAL MEETING OF THE IRON AND STEEL INSTITUTE was held at Grosvenor House, Park Lane, London, W. 1, on Thursday, May 1st, 1941. The Technical Session at 3.0 P.M. was preceded by a Luncheon; the PRESIDENT (Mr. John Craig, C.B.E., D.L.) presided over both functions.

MEMBERS' LUNCHEON.

Following the precedent of the previous year, the Annual Dinner was replaced by a Luncheon, which took place at Grosvenor House, Park Lane, London, W. 1, at 1 P.M. Nearly five hundred and fifty Members and their friends, including Ladies, were present; in addition to those more closely connected with the iron and steel industry, the following were among those who accepted invitations: The Rt. Hon. R. G. Menzies, P.C., K.C., LL.D. (Prime Minister and Minister for Co-ordination of Defence, Australia); His Excellency M. Bjorn Prytz (Swedish Minister) and Mme. Prytz; His Excellency M. C. Simopoulos (Greek Minister) and Mme. Simopoulos; His Excellency M. Ivan Soubrotitch (Yugo-Slav Minister) and Mme. Soubrotitch; His Excellency M. Joseph Bech (Minister of Foreign Affairs of the Grand Duchy of Luxemburg); The Rt. Hon. Lord Hankey, P.C., G.C.B., G.C.M.G. (Chancellor of the Duchy of Lancaster), and Lady Hankey; Sir Andrew Duncan, G.B.E., M.A., LL.D. (Minister of Supply); The Rt. Hon. S. M. Bruce, C.H., M.C. (High Commissioner for Australia), and Mrs. Bruce; Mr. W. J. Jordan, J.P. (High Commissioner for New Zealand), and Mrs. Jordan; Col. J. J. Llewellin, C.B.E., M.C., T.D., M.P. (Parliamentary Secretary, Ministry of Aircraft Production); Sir Rupert Howorth, K.C.M.G. (Clerk of the Privy Council), and Lady Howorth; Field-Marshal The Rt. Hon. Lord Milne, G.C.B., G.C.M.G., D.S.O., and Lady Milne; Marshal of the Royal Air Force The Rt. Hon. The Viscount Trenchard, G.C.B., G.C.V.O., D.S.O., and Viscountess Trenchard; Lt.-Gen. Sir John Brown, K.C.B., D.S.O., C.B.E.

(Director-General of the Territorial Army); Engineer-Vice-Admiral Sir George Preece, K.C.B. (Chief Engineer of the Fleet), and Lady Preece; Sir E. V. Appleton, F.R.S. (Secretary of the Department of Scientific and Industrial Research), and Lady Appleton; Sir Henry Dale, C.B.E., F.R.S. (President of the Royal Society), and Lady Dale; Sir William Bragg, O.M., K.B.E., F.R.S. (Director of the Royal Institution); Sir Leopold Savile, K.C.B. (President of The Institution of Civil Engineers), and Lady Savile; Mr. W. A. Stanier (President of the Institution of Mechanical Engineers) and Mrs. Stanier; Councillor L. Eaton Smith, J.P. (Mayor of the City of Westminster); and Brig.-Gen. Sir Harold Hartley, C.B.E., F.R.S. (Vice-President of the London, Midland and Scottish Railway Company), and Lady Hartley.

Presentation of the Bessemer Gold Medal for 1941 *to Dr. T. Swinden.*

The loyal toast having been honoured, the PRESIDENT (Mr. John Craig) presented the Bessemer Gold Medal for 1941 to Dr. Thomas Swinden (Member of Council; Director of Research, The United Steel Companies, Ltd., Stocksbridge, near Sheffield). In doing so he said that one of the pleasures of the presidency of the Institute was that the holder of that office was entrusted with the presentation of the Bessemer Gold Medal. That was a great pleasure and a great honour. The Council had determined with complete unanimity that the Medal for 1941 should be presented to Dr. Swinden. (*Applause.*) The award was officially declared to be " in recognition of the value of Dr. Swinden's original investigations into the metallurgy of steel and of his eminent services to the organisation and direction of research in the steel industry." Those who knew Dr. Swinden well realised that the Council acted not only wisely but well in selecting him for such a high honour.

Dr. Swinden's work in connection with tungsten steel and his close study of molybdenum in steel were well known. His most important recent work was an investigation into the properties of cold-rolled steel and the control of grain size, and that work had been described in papers published in the Institute's *Journal.* Dr. Swinden's work in improving the quality of the products of the iron and steel industry had not been confined to the theoretical study of the problems of the industry. He had built up one of the finest research organisations in the country. He took a leading part in the direction of organised research for the industry and was Chairman or Vice-Chairman of numerous Research Committees. He had also rendered great service to the country by giving advice which had proved of the utmost value to various Controls.

If he might add something of more personal interest, he would say that if there was one town which had a thorough knowledge of the steel trade that town was Sheffield, yet even in that enlightened town Dr. Swinden had shed further light, and he thought that those engaged in the industry in Sheffield would be very willing to join in doing honour to Dr. Swinden on the present occasion.

For reasons which would be appreciated by most people, the Medal which he was to hand to Dr. Swinden was not a gold medal. He thought it would have been most appropriate to make it of Sheffield steel, but that had not been done, perhaps because it was felt that Scotland or some other area would object to the use of an advertisement in that form. (*Laughter.*) Therefore a neutral metal had been used, but the gold would be forthcoming at a later date. In the meantime he would ask Dr. Swinden to receive a medal which was meant to be gold and which was gold in spirit.

The President then handed the Medal to Dr. Swinden amid loud applause.

Dr. T. Swinden, in acknowledging the award, said that in normal times the recipient of the Bessemer Gold Medal availed himself of the opportunity afforded by its presentation to say something of his work and aspirations, but that was not possible in the present circumstances. He was permitted, however, to say how deeply sensible he was of the very great honour which had been conferred upon him and to thank the President for the all too kind terms in which he had referred to the work for which the Council had thought him worthy of the award. The fact that the medal was a token medal might in fact be of considerable historic interest in the future. His pleasure in receiving the award had been very much enhanced by the letters that he had received, and he would like to take the present opportunity of saying how deeply he appreciated them. In common with many previous recipients, he regarded the award not only as a personal one but also as a recognition of the efforts of those for whom and with whom he had had the honour and pleasure of working for so many years.

In conclusion, he could only say that he regarded the award as an incentive to further service to the Institute and the industry and that he would do his utmost to render that further service.

A Prime Minister once wrote: "Rational, industrious, useful beings are divided into two classes: those whose work is work and whose pleasure is pleasure, and those whose pleasure and work are one. Fortune has favoured those who belong to the second class." In that respect he counted himself fortunate. (*Applause.*)

The Toasts.

The Right Hon. R. G. Menzies, P.C., K.C., LL.D. (Prime Minister and Minister for Co-ordination of Defence, Australia), in proposing the toast of

"*The Iron and Steel Institute and Industries,*"

said: I must first apologise for coming here so late, but the fact is that there is a War Cabinet in this country which, in my experience of it, observes the most irregular hours, and if you have any complaints to make about my late arrival I hope you will address them in writing to the Prime Minister of this country. (*Laughter.*) When I arrived I had a very hurried lunch, for which I shall no doubt suffer later on. (*Laughter.*) I fell into conversation with the President, and the moment I heard him and saw his name I said to myself: "Is it not marvellous? Whenever you find yourself in Great Britain in any really profitable industry a Scotsman is there." (*Laughter.*) Finding that I was between a Craig and a Duncan, I said to the President: "Are there any non-Scotsmen connected with this Institute?" I wanted to know, because I wished to find out whether I should have to talk down to my audience or up to it. (*Laughter.*) The President replied: "Oh yes; in fact, apart from Duncan and myself, I do not think there are any Scotsmen connected with it." Then I looked at the list of those who are supposed to be here and I found that the only one who can claim to be an Englishman is the Earl of Dudley—and his seat is empty. So I address myself to you as brother Scots, and I ask you to do what is most pleasing to the Scottish heart—to have a drink in a good cause; and the good cause will be the toast of The Iron and Steel Institute and Industries.

It would be very undesirable if a guest of yours, arriving late, kept you too late as a result, and so I shall endeavour to say something on this vast topic of The Iron and Steel Institute and the iron and steel industries within as brief a compass as possible.

There was a time, many centuries ago, when wars were fought by small bodies of men, and when communities went on their ordinary way, sometimes hardly conscious that a war was being fought; but, as time has passed, as technique has developed and as scruples have become less, war has spread its net further and further, until at this time we find ourselves engaged in a war that has been called a "total war," in which no-one who is within range of any instrument of death can regard himself or herself as being on anything other than active service. The people of this country, for example, are without exception at war and in war, and more and more as time goes on they realise, just as we who are further away in Australia and people on the American Continent begin to realise, that to win a total war you must have a total organisation of every resource that you have. More than that, the longer this war goes on the more we begin to realise its most outstanding characteristic. There has been, I suppose, no other war in history in which so many conquests have taken place in so short a time. It is true that Napoleon found his way ultimately across the whole of Europe, but it took him years to do it. Hitler finds himself to-day straddling most of Europe, and at all material times it has taken him only a matter of weeks to achieve each conquest. This war has been going on for over a year and a half, and, although

Europe has never seen such changes and such conquests in such a period before, the total casualties are a mere fraction of those that most people expected when this war began.

Now, all those eloquent facts have demonstrated—and I hope that we have realised it—that this is a war in which the factory, the organisation, the machine are much more important than at any time before in the world's history. I do not know whether there are any of us left in the British world who still want to talk in terms of numbers, who still want to think in terms of divisions, who still want to work out mathematical calculations about the total man-power; but, if there are, let us abandon that habit while there is yet time. We have to estimate our strength in this war by asking ourselves just what industrial power, translated into terms of machines of war, we can put into the hands of our soldiers. (*Applause.*)

We are a slow-moving people. I do not say that with pride, but with regret. We have too much the disposition to think of to-day in terms of yesterday. Only a few years ago, it took us some time to realise that the industrial efficiency of the individual was very largely conditioned by the number of horse-power that he had at his elbow, and in this war we must realise clearly, and pattern the whole of our policy upon it, that the efficacy of a British Army, whether from the United Kingdom, from Australia, from New Zealand, from South Africa or from Canada, will depend not only on the courage and skill of the men who compose it but also, and mightily, upon the number of tanks, vehicles and guns with which we have been able to equip them. I venture to say, therefore, that, properly understood, this is above all other wars an *industrial* war; and the work being done by the heavy industries of the British world is therefore seen to be the most vital work, so vital that I am not sure that everyone in the iron and steel industry should not be regarded as being on active service. I am not sure that you, Mr. President, should not at once be made a General! I am not sure that we should not have some outward symbol by which we recognise not merely that the iron and steel industries of the British world are making iron and steel, but that in the most literal sense the iron and steel industries have gone to war.

Speaking on behalf of one Government in the British Empire, I want to say that nothing can ever adequately express the debt of Governments and the debt of peoples to the great iron and steel industries not only of Great Britain but of the British world. It is only a few years ago that I used to find myself in London, sometimes in this very room, in which vast dinners used to be given and vast speeches made—some of them, I regret to say, by myself—and I remember very well that in those days we argued about some strange things. I used to meet such friends as I had in the iron and steel industry of Great Britain, and they would say to me, almost sadly: "What about you people in Australia? I am told that you grow very good wheat; I am told that your wool is excellent; I am told that your butter is superb. But why do you, with your small population, set about making iron and steel? Why do not you leave it to us?" (*Laughter.*) In the last few weeks that I have been here I have had many grim and humorous reflections on that question, because I have said to myself: "If we had stuck to making butter, would Lord Woolton be giving it to you to-day? If we had stuck to growing wheat and meat, would it be on your tables in some form or another to-day? Would the shipping problem be any less, and would the Battle of the Atlantic be less important if we had done those things?" Of course not; but Australia would have been less important in this war. Australia, instead of being, as she is, one of the really great manufacturers of munitions in this war, would be to-day on the list of those for whom my friend Sir Andrew Duncan has to provide. He would be saying: "I want so many guns and such and such munitions for Great Britain, so much for the Middle East, and I must also provide for Australia." Instead of that, he could now tell you that he is able to say: "What are my sources of supply? One of them, honourably placed, is Australia." (*Applause.*)

I do not say that because I want to stir up any echoes of those past debates, but I say it because I want to add this. My country could not at this moment be progressing as she is in all forms of engineering production for war if it were not for the iron and steel industry of Australia. Never a week goes by in my office in Australia when I do not feel or express gratitude for the wisdom of those men who decided that for the security of Australia, to enable her to play her full part should any trouble occur in the world, there should be laid down firmly and broadly the foundations of heavy production. The results, of course, have been beyond anyone's expectations. We not only see iron and steel in this country as the foundation of the defence of Great Britain, we not only see iron and steel in America as the foundation of an ultimate and crushing victory, but we see iron and steel in Australia as the basis upon which the greatest effort of our lifetimes is being put forth. (*Applause.*)

It is very frequently possible to make speeches and pleasant remarks about matters which are relatively superficial. It is quite possible to lay a just emphasis upon a small matter. But it is quite impossible for any guest of yours to lay undue or excessive emphasis upon the importance at this time of the industry which you represent. When some historian of the future looks back on the records of this war, and when in a freer world and a happier time he writes the story of what has occurred in these years, he will perhaps remember some things which appear at the moment to be transient. He will remember some great speeches made by some great men in this country; he will remember some magnificent decisions taken by Parliament in this country. He will remember —nobody would ever dream of forgetting—the superb courage and spirit shown by the men who fight for this country and by the women who fight for this country. He will remember all these things; but I believe that above all he will say that in this war the real foundation of success was laid by those who did not think that they were on active service, who had no uniform, who worked in grimy overalls, juggling with large masses of molten metal; because, he will say, this was the beginning of an age in which the industries of a country were seen to be among the great fighting services of that country. When that has been recorded, I believe that the position of the iron and steel industry and the relation of large-scale industry to the conduct of the affairs of the State will have been seen in a new light, and that we shall go forward into a period when politics will not be regarded as the eccentricity of a few people, but will be seen as a sphere of activity in which the great men of industry will be recognised as great men of their country and great men of affairs. In that sense I am not at all sure that our brilliant and distinguished friend Sir Andrew Duncan, for whose work in this war I have intense admiration, will not be regarded as the founder of a sort of industrial dynasty in Government circles. If he is, I shall be delighted, because when the business man, the great industrialist and the politician are no longer seen as point and counter-point but as people who ought to coincide and to have a joint responsibility, and who may very well be the same people, I believe that we are going to develop an enormous political power of a kind that we have not previously realised. (*Applause.*)

It is a great pleasure and a great honour to submit to you the toast of The Iron and Steel Institute and the Iron and Steel Industries. (*Applause.*)

The President (Mr. John Craig, C.B.E., D.L.), who responded, said: It is a great honour to us at this Luncheon that our Institute should be referred to in the way that it has been by the Prime Minister of Australia. His presence in this country has been to everyone most stimulating and fortifying, and has been a great source of encouragement to all the people of the homeland; it has been a great encouragement to us that someone should come from the Dominions so fortified by ability, enthusiasm, energy and knowledge to help us in this day of trial, and many have gained inspiration from his presence and from his speeches and have been helped by the conscious power that lies behind him and the great country which he represents. (*Applause.*) We welcome him here to-day not only as a great statesman from a great country but likewise as coming from a country which is worthily maintaining the highest traditions of our industry.

I can recall the arguments of a few years ago, when those in our industry in the old country were unduly perturbed and made anxious by witnessing the development of the industry in Australia; and I had intended in what I had planned to say to recall those days and to express thankfulness that, after all, the industry in Australia had developed, and developed on such a sound foundation. Nature has been extraordinarily kind to Australia, giving her wonderful iron ore deposits and all the other requisites for the making of steel, thus surely indicating that some day she was bound to make steel. It is fortunate that her great industry fell into the hands of the people who in fact direct it. Many of us are acquainted with them and are fully conversant with the work that they have done. They planned on broad lines, and laid the foundations of the industry in the best possible way; and to-day we are very grateful to the enlightened men who developed the industry there on no small lines. We are all well aware, and almost envious, of the excellent plant which exists in Australia, and we admire the men who lead it and who have taken advantage of all the knowledge which can be gathered in America, on the Continent and in our own country; because prior to the war not many months passed without someone coming from Australia on purpose to see the latest developments, and, if they were good, very soon they were copied. We are grateful to them for stepping in and for being so helpful in this great crisis.

It is good to hear the industry spoken of as it has been. I have been in this room at a time when to speak of The Iron and Steel Institute and the iron and steel industries was almost to apologise for appearing on the platform. They were regarded as having had their day, and they received no support from anyone. The only statement by Mr. Menzies to which I can take exception is that he has called this "a profitable trade." (*Laughter.*) I suppose that it ought to be, and perhaps we can take it that he believes that it ought to be, and that is an asset in itself. I have lived in days when it was otherwise, and now, when it is profitable, the Chancellor of the Exchequer is making quite sure that he will be the only profiteer. (*Laughter.*)

As regards the work of the Institute, perhaps to-day there is a better outlook. In the old days it was the skilful speculator who was regarded as the distinguished steelmaker, the man who sized up the markets and bought and sold. Personally, I always thought that steelmaking should be steel making, and based on the provision of an article for national use. The Institute struggled in those days to uphold the technical and scientific side of the industry, and did so through difficult years; but it has now come into its own. We are now realising more than ever that it is the scientific side of steelmaking which proves of the greatest benefit in time of crisis; it is not how cheaply we can make steel but how good we can make it which counts to-day. If ever there was a time when good steel was required it is to-day, and I am thankful to believe that the steel which is being provided for our Services, perhaps not so quickly as some would like, is reliable, and they can trust their lives to it. I think it has been proved that the steel with which the Services have been provided will not take second place to steel made in any other part of the world. The technical members of the industry have been greatly encouraged by Mr. Menzies, who has indicated, I think, that those who are engaged in the making of steel are very much more important than those who are engaged in buying and selling. I am happy to think that encouragement is given to those who have a technical outlook on life.

As far as the Institute is concerned, the great work which it did in pre-war years lay in giving to young men a consciousness of the value of research, a feeling that they were never to be satisfied with what was being done, that they were ever to be critical of what their fathers had done and never to accept what their grandfathers had done as limiting what they could do with the present, but that they must themselves bring a new mind to bear and harness the new knowledge constantly placed at their disposal. We are to-day rendering to the State a service which would have been impossible if we had continued to conduct our business on what might be termed commercial lines. I trust that the younger men will realise that steelmaking is a very young business, and that what we know about it is infinitesimal compared with what remains to be known. A great many things are being discovered to-day which are quite revolutionary. I am old enough to realise that steel is being produced to-day of a quality and in a manner which not very many years ago a good many of us believed to be quite impossible.

The Institute is encouraged by this gathering, and greatly encouraged by the speech which Mr. Menzies has made, with his intimate knowledge of what is happening elsewhere. We may have to turn our eyes far away to learn something new, but I am sure that Australia is most willing to co-operate and to put at our disposal anything which her people have learned in steelmaking.

That should have been the end of my speech, and the next toast should have been proposed by my friend Lord Dudley. Unfortunately he has been detained on urgent duties in connection with the important post which he holds of Civil Defence Commissioner for the Midlands, and I have been asked in his place to propose the toast of

"The Guests."

It is a great honour to be permitted to propose this toast. Our guests are a very distinguished company. It is true that, as has already been remarked to me since I entered this room, they are perhaps the most distinguished company of guests that has ever been present at a function of The Iron and Steel Institute. We have with us the Swedish Minister, and the Ministers of our gallant allies, Greece and Yugo-Slavia. (*Applause.*) We have some members of the Cabinet, including a member of the War Cabinet, who has addressed us. We have a number of guests from America, and we cannot mention America without saying how profoundly grateful we are for all that she has done and is doing to help us. (*Applause.*) We were always far enough away to be quite friendly with the Americans in the past so far as steelmaking is concerned (*laughter*), and to-day we are greatly their debtors for what they are doing, especially in the steel world, and we welcome them to our table this afternoon. There are present the High Commissioners of Australia and New Zealand and representatives of Canada, South Africa and India as well as of our gallant ally Poland. We have also with us members of the learned societies, and those on whom we look with great favour in times of peace because they consume the articles that we produce; and we always hope that they will develop their learning along the lines of using more steel! We hope that they will succeed in their efforts to help civilisation by the development of steel production. There are present also representatives of the purely scientific societies, including our old friend Sir William Bragg (*applause*), who a year ago made a memorable speech here. We recognise with gratitude that even pure science can be very helpful to us, because, so far as some of the problems which confront the practical men are concerned, it is the pure scientist who is opening the door which may lead to their solution.

We take the presence of all our guests this afternoon as an encouragement. Many of them are working long hours and strenuously, but I like to think that they are here to-day because they wish to say to us in the iron and steel industry that they recognise that we are engaged on important work, and they would like to encourage us to do more. Our guests are thus full of encouragement for us, and on your behalf as well as my own I give them a most warm welcome and thank them for being here to-day.

I have very great pleasure in coupling this toast with the name of our good friend from Luxemburg, M. Bech. Many of us have very tender memories of frequent visits to his country, where we were treated most hospitably and where we recognised that the industry was in very capable hands. We have many members in that country, and many friends, and we wish them well. We have members in many countries, and we welcome those of them who are here to-day. We have friends in enemy countries who perhaps wish us well, in spite of all. I have great pleasure in asking you to drink the toast of our Guests, coupled with the name of M. Bech. (*Applause.*)

His Excellency M. Joseph Bech, Minister of Foreign Affairs of the Grand Duchy of Luxemburg, who responded, said: It is an honour and a great pleasure to reply on behalf of the guests. I thank your President for the cordial way in which he has proposed this toast, and I regret that I am unable to improvise the words of sincere gratitude in which I should like to express my thanks for the words that he addressed to my country. I can only say " Thank you " from the bottom of my heart. After saying this, I need hardly add that I certainly do not owe the pleasant duty of speaking on behalf of so many distinguished people to my insufficient knowledge of English; probably I owe the privilege to two Members of this Institute, Thomas and Gilchrist, the inventors of the basic converter process of steel manufacture, which made it possible for the industry of Luxemburg to rank among the largest iron and steel producers in Europe. I wish here to pay homage to these English metallurgists, whose invention definitely established the prosperity of my country.

A few days ago I visited an antique shop near Piccadilly, and came upon a beautiful old picture with the Latin inscription " Omnis salus in ferro "—all safety lies in iron. It represented a suit of armour, gauntlets and shield and sword, laid ready for the knight to wear. At first sight this inscription—which might well be a motto for this Institute—seemed, if I might say so, somewhat ironical (*laughter*), for the place was surrounded by the devastation caused by German iron; and yet that device is true, and to-day is truer than ever, for our safety, our salvation lie in iron. As one of your Ministers said recently, superiority of soul and spirit is not sufficient to fight a better equipped foe. Never in history were there knights of higher spirit fighting for a nobler cause than the soldiers, sailors and airmen of Great Britain, her Dominions and her Allies (*applause*), and at this moment, in the presence of Mr. Menzies, the eminent Prime Minister of Australia, and of the High Commissioner for New Zealand and the Greek and Yugo-Slav Ministers, how could I fail to pay special tribute to the heroism of their soldiers? (*Applause.*) The spirit is there indeed, and, as for the weapons, you, the men of the iron and steel industries, with the magnificent and almost unlimited help of the United States, are laying them ready for victory and salvation. (*Applause.*) It is not for me to gauge the immense productivity of this country and of the United States, but your great Prime Minister has told us that Great Britain and the United States together are producing more steel than the rest of the world combined; and who better than Sir Andrew Duncan can value the greatness of your war effort and the debt which this country and all of us owe to the high intelligence of the industrial leaders and the skill of the British workers, and who more than you can value Sir Andrew's presence in the Government? (*Applause.*)

In his recent splendid speech, the Prime Minister spoke of a certain bad man and of a tiger. This man is in fact riding a tiger, and, as the old Chinese proverb says, " he who rides a tiger cannot get off." Hitler's measureless and wanton aggressive policy will surely drive him on and on to the fatal end which awaits all conquerors and oppressors of peoples. (*Applause.*) Until that fate overtakes him, we may have to face occasional set-backs. In this connection, let me tell you of an age-old ritual which we have in Luxemburg, and which historians believe to be of English origin. To me it seems symbolic of the ebb and flow of the tides of war, and of the indomitable tenacity of the British people. On Whit-Tuesday every year a procession takes place, and has done so for the past twelve hundred years, to the tomb of our patron saint, St. Willibrord, an English missionary who came from Northumberland to christianise the Ardennes. This procession is to-day world-famous. Some twenty thousand pilgrims proceed in a most peculiar manner to the church where the saint lies buried. Singing and praying, they wend their arduous way, dancing three steps forwards and two backwards, and so on for hours; but, in spite of all the backward steps, the faithful and tenacious pilgrims have never yet failed finally to reach the shrine. So it will be in the present struggle; even if occasionally we have to take two, three or even four steps backwards, it cannot affect the final outcome of the war. (*Applause.*) Under the leadership of Great Britain, humanity will reach the sanctuary of freedom and justice. England will once again win the last battle. (*Applause.*)

In conclusion, I should like to recall that when I had the honour and pleasure of welcoming the Institute to Luxemburg a few years ago, I expressed the hope that you would come back soon. I did not then expect that I should be coming to you! You will realise the feelings with which in the present circumstances I repeat the invitation; you cannot come too soon to a free and independent Luxemburg! (*Applause.*)

I am sure of the approval of the guests when I thank the President and the Institute for their very kind hospitality. In the case of those of us who, like myself, have found refuge and hospitality in this country, our gratitude to you and to England is all the more profound. I am happy to propose the health of Mr. Craig, your President. We all know that his eminent qualities of leadership are equal to the onerous burdens and great responsibilities that war-time brings to his high office. I ask you to join with me in wishing him continued health and prosperity in all his undertakings. (*Applause.*)

The President, after expressing his appreciation of the compliment, went on to say that there was present the former Secretary of the Iron and Steel Institute, Mr. G. C. Lloyd, who had celebrated his eightieth birthday the day before. All the Members would join in wishing him, in spite of his age, many happy years to come. (*Applause.*)

Technical Session.

The Session for the transaction of formal business and the presentation and discussion of papers, also held at Grosvenor House, opened at 3 P.M., the President (Mr. John Craig, C.B.E., D.L.) being in the Chair.

The Minutes of the previous Meeting, held in Sheffield on November 12th, 1940, were taken as read and signed.

Obituary.

The President (Mr. John Craig) said that his first duty was a sad one, namely, to refer to the death of Mr. Albert Peech and Mr. Charles J. Walsh. Mr. Peech was a Vice-President of the Institute and Mr. Walsh a Member of the Council. They were both well known to the Members of the Institute and would be greatly missed, and he was sure that the members would wish their condolences to be sent to Mrs. Peech and Mrs. Walsh. He would arrange for that to be done and also for a note expressing the sympathy of the Institute to be sent to the Company of which both Mr. Peech and Mr. Walsh had been Directors. Mr. Peech had linked up the Members with the past and Mr. Walsh with the present, and their deaths were a great loss to the industry.

At the request of the President, the Members stood in silence for a few moments.

Ballot for the Election of New Members and Associates.

Dr. L. Northcott (Teddington, Middlesex) and Dr. A. L. Norbury (Cleobury Mortimer, near Kidderminster) were appointed scrutineers of the ballot, and they reported in due course that the following sixty-three candidates for membership and forty-five for associateship had been duly elected.

Members.

Baillie, William	Birmingham.
Bales, Sidney Hartley, M.Sc., F.I.C.	Cambridge.
Bannister, John Boynes	Scunthorpe, Lincs.
Beswick, Wilfrid Ritzema	Gerrards Cross, Bucks.
Billimoria, Lim Motabhai	Jamshedpur, India.
Bootes, Roland Kenneth	Wolverhampton.
Bootman, Hedley	Bradford, Yorks.
Bryan, Andrew Meikle, J.P., B.Sc., M.I.Min.E.	Edinburgh.
Buchanan, William	Darlington.
Buchanan, William Gray	London.
Buvers, Harry	Rotherham.

. . . . Rot
ENGLAND, HAROLD HAYWOOD . Rot
FAVEL, MAURICE . . . . Joh.
GLADWIN, WILLIAM MORLEY . . Shei
GOW, CHARLES MILNE . . . New
tr.
GRIFFITHS, GEORGE HENRY . . Rotl
HAMILTON, ANDREW NOBLE . . New
tra
HARRISON, JOHN WILLIAM . . Darli
HEWITT, WILLIAM VIGORS . . Newc
tra
JONES, ROBERT . . . Shott
LAZENBY, THOMAS . . Bentl
LEVER, ERNEST HARRY . . Lond
LEWIS, JOHN . . . . . Newp
McLENNAN, IAN MUNRO, B.Eng. . Newca
trali
MANTERFIELD, DAVID . . . Sheffie
METHLEY, BERNARD W., F.I.C. . Rothe
MITCHELL, THOMAS EDWIN . . Scunth
MUKHERJEE, KALI KINKAR . . Kulti,
NADEN, JOHN WILLIAM RIVINGTON. Cheste
OKEDEN, RICHARD GODFREY CHRISTIAN PARRY Tirrikil
trali
PARKER, JAMES FRANCIS . . Prestw
bridg
PEARSON, STANLEY WRIGHT . . Rother
PINCHES, ELWYN . . . . Hemel
PLUCK, JOHN EDWARD . . . Rother
RAMBUSH, NEILS EDWARD . . Stockt
RICHARDS, JACK CHRISTIAN, B.Eng., B.A. Newcas
trali
ROBERTS, FRANCIS GEORGE . . Newcas
trali
ROBERTSON, ALEXANDER ROLLO . Sheffiel
ROLLASON, ERNEST

TRUZZELL, JOHN EVERSFIELD .	Johannesburg, S. Africa.
TUDOR-EVANS, A. G. . . .	Manchester.
TURNER, HAROLD J. . . .	Rotherham.
VAIDYE, DIGAMBAR GOPAL, B.Sc. (Bombay)	Jamshedpur, India.
VERNON, WILLIAM HAROLD JUGGINS, D.Sc., Ph.D. (Lond.), B.Sc. (Birm.), D.I.C., F.I.C. . .	Teddington, Middlesex.
WALDEN, ROLAND HILL . .	Rotherham.
WALTON, FRANK	Sheffield.
WARDELL, VINCENT ANDREW. .	Wollongong, N.S.W., Australia.
WATKIN, EVAN RAE SINCLAIR .	Scunthorpe, Lincs.
WEDDELL, RALPH EDWARD GOLDSBROUGH	Tipton, Staffs.
WOOD, PHILIP W. A. . . .	Sydney, N.S.W., Australia.
WRAGGE, WILLIAM BENJAMIN, B.Sc. (Tech. Hons.), A.I.C. . . .	Cheadle Hulme, Cheshire.

ASSOCIATES.

AIYER, KRISHNA VAIDYANATHAN .	Cambridge.
AYRES, HAROLD STANLEY . .	Scunthorpe, Lincs.
BINNER, STANLEY . . .	Slough, Bucks.
BIRKHEAD, MAURICE . . .	Deepcar, near Sheffield.
BOSE, ANJAB KUMAR, B.Met. (Hons.)	Sheffield.
BRADBURY, ERNEST JAMES, B.Eng. (Hons.)	Birmingham.
BURNS, BRIAN D. . . .	Teddington, Middlesex.
CHURCH, ROBERT STANLEY . .	Kempston, Beds.
CLARK, WALTER	Scunthorpe, Lincs.
COCKCROFT, ROBIN ARTHUR MIDGLEY, B.Sc.	Scunthorpe, Lincs.
COTTON, RAYMOND FREDERICK .	Birmingham.
CRESSWELL, ROBERT ARTHUR .	Birmingham.
DAWSON, CLEMENT . . .	Scunthorpe, Lincs.
EBORALL, RICHARD JOHN LANE .	Cambridge.
EDWARDS, ALAN ROWLAND . .	Ivanhoe, Victoria, Australia.
ELLIS, JAMES HAROLD . . .	Kettering, Northants.
ENGLISH, ALAN	Scunthorpe, Lincs.
FOX, MICHAEL JOHN . . .	Birmingham.
GILMORE, JAMES	Scunthorpe, Lincs.
GRAY, PETER ROBERT . . .	Gainsborough, Lincs.
HOLDEN, HARRY ASHTON . .	Birmingham.
HOSKINS, HENRY GUILDFORD .	Farnborough, near Banbury, Oxon.
JACKSON, GEORGE HENRY . .	London.
KUJUNDZIC, NEBOJSA . . .	Leeds.
LEE, HSUN, B.Eng., Ph.D. . .	Sheffield.
McADAM, GEORGE DOUGLAS . .	Leeds.
MARINKOV, CEDOMIR B., B.Sc. .	Leeds.
MAYER, SIMON ERNEST . . .	London.
METCALFE, GORDON JOLLIFFE .	Farnborough, Hants.

NIXON, ERIC WILLIAM . . .	Brigg, Lincs.
OLIVER, CHARLES	Scunthorpe, Lincs.
OWEN, WALTER S. . . .	Liverpool.
PARKER, ROBERT HADFIELD .	Scunthorpe, Lincs.
PHAFF, HERMAN JACK . . .	Sheffield.
REID, JAMES HENRY MICHAEL ALLAN	Thornton Heath, Surrey.
SHEPHERD, NEAL DERWENT . .	Beckennet, Cumberland.
STUBBINS, CHARLES . . .	Scunthorpe, Lincs.
THOMAS, GORDON DAVID . .	Birmingham.
THOMAS, IVOR HAYDN . . .	Dowlais, Glam.
TOWNDROW, RONALD PHILIP, M.Sc.	Scunthorpe, Lincs.
ULUBAY, ALI	Swansea.
WALKER, JOHN LESLIE . . .	Manchester.
WARD, GEORGE	Scunthorpe, Lincs.
WILSON, DONALD VERNON . .	Birtley, Co. Durham.
WRIGHT, DAVID ARTHUR FELTON .	Smethwick, Staffs.

Presentation of the Report of Council and Statement of Accounts for 1940.

The PRESIDENT moved that the Report of Council and Statement of Accounts for 1940 (*see* pp. 21 P to 47 P) be adopted.

The HON. TREASURER (Mr. James Henderson) seconded the motion and it was carried unanimously.

Election of Vice-Presidents and Members of Council.

The SECRETARY (Mr. K. Headlam-Morley) said that, in accordance with Bye-Law No. 10, the names of the following Vice-Presidents and Members of Council had been announced at the Autumn Meeting in 1940 as being due to retire at the Annual General Meeting in 1941 :

Vice-Presidents : Mr. Arthur Dorman, Mr. J. S. Hollings and Mr. C. E. Lloyd.

Members of Council : Captain H. Leighton Davies, Mr. I. F. L. Elliot, Principal C. A. Edwards, F.R.S., Mr. E. F. Law and the Hon. R. G. Lyttelton.

No other Members having been nominated up to one month previous to the present Annual Meeting, the retiring Vice-Presidents and Members of Council were now presented for re-election. (*Agreed.*)

Williams Prize for 1940.

The SECRETARY announced that the Council had awarded the Williams Prize for 1940 to Mr. B. YANESKE for his paper on "The Manufacture of Steel by the Perrin Process."

Andrew Carnegie Research Grants.

The SECRETARY said that grants had been made by the Council in 1940 to the following candidates :

M. BALICKI (University College, Swansea).—£100 in aid of a research on the cold-hardening of steel and the verification of the recrystallisation theory put forward by Professor Krupkoffski and the candidate.

H. MORROGH (British Cast Iron Research Association, Birmingham).—£100 in aid of a research on sulphide inclusions produced by the neutralisation of sulphur in cast iron by aluminium, copper, nickel, chromium and molybdenum.

He also stated that it had been decided that no award of the Andrew Carnegie Medal should be made for the year 1940.

Complete List of Papers Presented at the Annual General Meeting, 1941.

H. MORROGH : "The Polishing of Cast-Iron Micro-Specimens and the Metallography of Graphite Flakes."

H. MORROGH : "The Metallography of Inclusions in Cast Irons and Pig Irons."

L. NORTHCOTT : "The Influence of Turbulence upon the Structure and Properties of Steel Ingots."

T. F. RUSSELL : "The Thermal Relations between Ingot and Mould." (Paper No. 4/1940 of the Committee on the Heterogeneity of Steel Ingots (submitted by the Stresses in Moulds Panel of the Ingot Moulds Sub-Committee)).

E. S. TAYLERSON : "Atmospheric Corrosion Tests on Copper-Bearing and other Irons and Steels in the United States." (Paper No. 4/1940 of the Corrosion Committee (communicated by Dr. W. H. Hatfield, F.R.S.)).

E. M. TRENT : "The Formation and Properties of Martensite on the Surface of Rope Wire."

"Intercrystalline Cracking in Boiler Plates." A Report from THE NATIONAL PHYSICAL LABORATORY.

Part I.—Introduction. By C. H. DESCH, F.R.S.

Part II.—Prolonged-Stress Tests on Iron and Steel Specimens Immersed in Hot Sodium Hydroxide Solutions. By C. H. M. JENKINS and FRANK ADCOCK.

Part III.—Exposure of Iron and Steel Specimens to Sodium Hydroxide Solutions at High Temperature and Pressure. By FRANK ADCOCK (assisted by A. J. COOK).

Part IV.—Strain-Etch Markings in Boiler-Plate Material of Acid Open-Hearth Manufacture. By FRANK ADCOCK and C. H. M. JENKINS.

Part V.—Some Experiments on the Behaviour of Specimens of Boiler Plate and Boiler Joints Subjected to Slow Cycles of Repeated Bending Stresses while Immersed in a Boiling Aqueous Solution." By H. J. GOUGH, F.R.S., and H. V. POLLARD.

 [*Continued overleaf.*

Complete List of Papers Presented at the Annual General Meeting, 1941 (continued).

"Third Report of the Oxygen Sub-Committee." (Paper No. 5/1941 of the Committee on the Heterogeneity of Steel Ingots (submitted by the Oxygen Sub-Committee)).

Section I.—Introduction.

Section II.—The Vacuum Fusion and Vacuum Heating Methods.

Part A.—The Determination of Oxygen, Nitrogen and Hydrogen in Steel. A Survey of the Vacuum Fusion Method, with a Note on the Solid Solubility of Oxygen in High-Purity Iron. By H. A. Sloman.

Part B.—The Fractional Vacuum Fusion Method for the Separation of Oxides and Gases in Steel. Further Practice and Typical Results. By T. Swinden, W. W. Stevenson and G. E. Speight.

Part C.—The Determination of Hydrogen by Vacuum Heating. By W. C. Newell.

Section III.—The Aluminium Reduction Method.

Part A.—The Development and Comparison of Two Procedures for the Aluminium Reduction Method for Determining Oxygen in Steel. General Summary Showing the Applicability of the Aluminium Reduction Method. By N. Gray and M. C. Sanders.

Part B.—A Description of the Aluminium Reduction Method as Operated at Stocksbridge. By W. W. Stevenson and G. E. Speight (under the direction of T. Swinden).

Part C.—The Determination of Total Oxygen in Pig Iron by the Aluminium Reduction Method. By E. Taylor-Austin.

Section IV.—The Chlorine Method.

Part A.—Description of the Procedure now Adopted for the Chlorine Method. By E. W. Colbeck, S. W. Craven and W. Murray.

Part B.—General Summary Showing the Applicability and Utility of the Chlorine Method. By E. W. Colbeck, S. W. Craven and W. Murray.

Section V.—The Alcoholic Iodine Method.

Part A.—Present Position, Limitations and Possibilities of the Alcoholic Iodine Method, with a Note on Factors Affecting the Presence of Phosphorus in the Residue. By T. E. Rooney.

Part B.—A Simplification of the Alcoholic Iodine Method for the Determination of Oxide Residues in Steel. By W. W. Stevenson and G. E. Speight (under the direction of T. Swinden).

Section VI.—The Aqueous Iodine Method.

Part A.—Recent Developments in the Determination of Oxide Inclusions in Pig Iron by the Modified Aqueous Iodine Method. By E. Taylor-Austin.

Part B.—The Present Position of the Determination of Oxide Inclusions in Pig Iron and Cast Iron. By J. G. Pearce.

Section VII.—The Hydrogen Reduction Method. Submitted by W. W. Stevenson.

Section VIII.—The Analysis of Non-Metallic Residues Extracted by the Alcoholic Iodine Method. By G. E. Speight.

Section IX.—Reports on Materials Examined.

Part A.—An Examination of the Oxygen Content of a Basic Bessemer Rimming Steel. By T. Swinden and W. W. Stevenson.

Part B.—A Note on the Examination of a Series of Carbon Steels. Submitted by W. R. Maddocks.

Section X.—General Summary. By T. Swinden (Chairman of the Oxygen Sub-Committee).

Section XI.—Work in Hand and Future Programme.

Presentation of Papers.

The following papers were presented for verbal discussion (a list of all the papers included in the Meeting programme will be found on pp. 17 P and 18 P):

"Third Report of the Oxygen Sub-Committee." (Paper No. 5/1941 of the Committee on the Heterogeneity of Steel Ingots (submitted by the Oxygen Sub-Committee)).

"The Polishing of Cast-Iron Micro-Specimens and the Metallography of Graphite Flakes," by H. MORROGH.

"The Metallography of Inclusions in Cast Irons and Pig Irons," by H. MORROGH.

The last two papers were discussed jointly.

The PRESIDENT apologised to Dr. L. NORTHCOTT that time did not permit the presentation of his paper on "The Influence of Turbulence upon the Structure and Properties of Steel Ingots," which had been provisionally included in the list of papers to be discussed at the Meeting.

The proceedings then terminated.

REPORT OF COUNCIL.

THE Annual Report on the proceedings and work of The Iron and Steel Institute during the year 1940, and the Statement of Accounts for the same period, are submitted by the Council for the approval of Members at this the Seventy-Second Annual General Meeting. For the convenience of Members, some sections of the Report include information up to March 31, 1941.

The activities of the Institute have been maintained, but they have been adjusted to meet war-time conditions. Both the Annual and Autumn Meetings were held, and, though the programmes were necessarily curtailed, they were well attended and the discussions were satisfactory. Publication of the *Journal* and *Bulletin* of abstracts has been continued. Considerable use has been made of the Library and Information Departments by the Services, Government Departments and the Industry. Research activities carried on in conjunction with the Iron and Steel Industrial Research Council have been extended and research programmes adjusted to include problems of urgent importance.

Four members of the staff, Messrs. H. Davison, H. G. Hale, L. G. Poole and J. C. Robinson, are serving in the Army; other members of the staff have undertaken voluntary duties as Home Guards, Air Raid Wardens and in other civilian services. The Secretary was given permission to join the Chrome Ore, Magnesite and Wolfram Control of the Ministry of Supply as Deputy Controller.

ROLL OF THE INSTITUTE.

The membership of the Institute at December 31, 1940, was two thousand six hundred and sixty-six, a decrease of thirty-eight from the record figure of the previous year. In addition, the names of ninety-seven Members resident on the Continent whose subscriptions had not been paid before the end of the year have been placed on a suspense list. These include fifty Members resident in Belgium, Czechoslovakia, France, Norway, Poland and Rumania and forty-five Members and two Associates resident in Finland, Yugoslavia, Spain and Sweden.

Four Honorary Members, one Life Member and seventy-three ordinary Members of German or Italian nationality have ceased to be Members of the Institute, in accordance with Bye-Law No. 36.

Details of the membership compared with those of the two preceding years are as follows :

	31/12/'38.	31/12/'39.	31/12/'40.
Patron . . .	1	1	1
Honorary Members .	13	15	11
Life Members . .	69	68	67
Ordinary Members .	2341	2407	2339
Associates . . .	206	213	248
	2630	2704	2666

The above total membership figures for each year include eight Members whose names are retained in the List of Members in an honorary capacity by order of the Council.

Variations in the membership since the formation of the Institute in 1869 are shown in Fig. 1.

The decrease in membership is due to the extension of the war and consequent inability of Members resident in many countries in Europe to remit subscriptions. A further reduction, not exceeding one hundred and sixty, in the number of Members resident on the Continent must be anticipated, and it is proposed that those whose subscriptions have not been paid before the end of the year shall also be placed on the Suspense List as from January 1, 1941.

During the year under review one hundred and fifty-six new Members and sixty-one new Associates were elected. These include ninety-seven Members and twenty-five Associates nominated by Companies which subscribe to the Special Subscriptions Fund, and twenty-six Members and twenty-seven Associates who joined under the scheme of collaboration with the Institute of Metals. Two Associates were elected under the special arrangement with the American Institute of Mining and Metallurgical Engineers and one under the arrangement with Jernkontoret. Nineteen Associates were transferred to Membership and five former Members were reinstated. Thirty-nine Members and two Associates resigned and the deaths of thirty-one Members were reported.

Obituary.

The Council regret to record the deaths of the following twenty-six Members which occurred during the year 1940:

Anslow, Frank (Glasgow)	March 17th.
Armstrong, Charles Edward (Millom, Cumberland)	February 13th.
Best, Frederick (Bickley, Kent) . .	August 12th.
Carpenter, Sir Harold, F.R.S. (Swansea) (Past-President)	September 14th.
Crompton, Colonel Rookes Evelyn Bell, C.B., F.R.S. (Ripon)	February 15th.

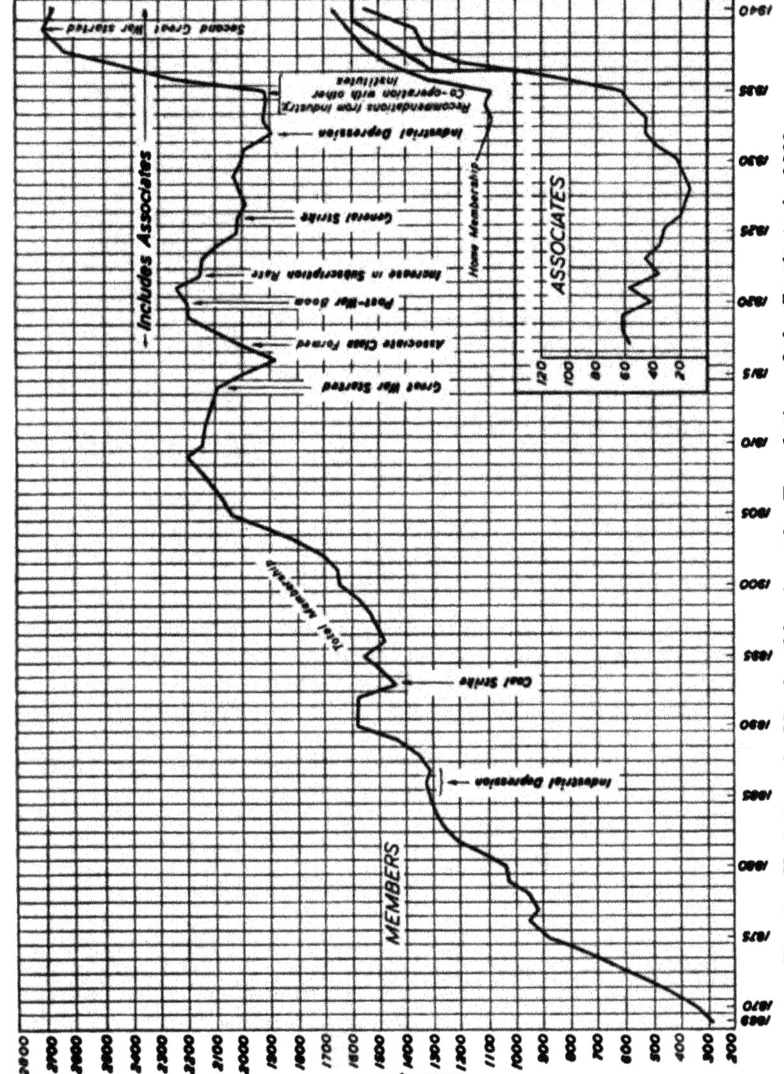

FIG. 1.—Variations in Membership since the Foundation of the Institute in 1869.

GILLOTT, JOSEPH PERCIVAL (Hornsea, Yorks.)	February 19th.
GJERS, JOHN (Middlesbrough) . .	July 2nd.
HADFIELD, Sir ROBERT, Bart., F.R.S. (London) (Past-President) . .	September 30th.
HINGLEY, GERALD BERTRAM (Dudley) .	May 20th.
HOLGATE, THOMAS EDWARD (Darwen) .	February 7th.
LEWIS, A. CECIL (Newport, Mon.) . .	January 27th.
LYSAGHT, DANIEL CONNOR (Chepstow, Mon.)	May 15th.
MACARTNEY, WALTER CREALOCK (Chesterfield)	March 9th.
MERZ, Dr. CHARLES H. (London) . .	October 15th.
MOORWOOD, JOHN MARTIN (Sheffield) .	April 8th.
PEACH, WILLIAM ISAAC (Chesterfield) .	January 17th.
PICCIOLI, ARTURO (Portovecchio di Piombino (Leghorn), Italy)	October 10th.
RAINE, ALFRED JOHNSON (Newcastle-on-Tyne)	March 1st.
SANDBERG, OSCAR FRIDOLF ALEXANDER, O.B.E. (London)	February 15th.
SCHUSTER, LEONARD WALTON (Manchester)	July 6th.
SMITH, FRED, C.B.E. (Leeds) . . .	May 3rd.
STRICK, GEORGE HENRY (Cheltenham) .	May 3rd.
THOMAS, HUBERT SPENCE (London) (Hon. Vice-President)	January 16th.
TOY, HARRY BROOKS (Nunthorpe, Yorks.)	February 26th.
WALBER, ALEXANDER CORRIN (London) .	July 14th.
WOLSTENHOLME, GEORGE ETHELBERT (Sheffield)	July 3rd.

The deaths of the following two Members took place earlier than 1940, but were not previously reported :

BENSON, ROBERT SEYMOUR (Bishopsteignton, S. Devon)	1938.
SETTERWELL, HUGO (Stockholm, Sweden)	1939.

Sir Robert Hadfield, F.R.S., was the Senior Past-President of the Institute, Sir Harold Carpenter, F.R.S., a Past-President and Mr. Spence Thomas an Honorary Vice-President; Mr. Toy had been an Honorary Member of Council during his Presidency of the Cleveland Institution of Engineers. Sir Harold Carpenter and Sir Robert Hadfield were Bessemer Gold Medallists. Mr. Strick, Colonel Crompton, Mr. Benson, Sir Robert Hadfield and Mr. Holgate had been Members of the Institute for more than fifty years, having been elected in 1876, 1877, 1881, 1885 and 1887, respectively.

Obituary notices have been printed in the No. II. volume of the *Journal* for 1940.

The Council regret to have to record the death of Mr. A. O. Peech (a director of The United Steel Companies, Ltd., and a Vice-President of the Institute) and of Mr. C. J. Walsh (managing director of the same Company and a Member of Council), which took place on March 22 and on February 22, 1941, respectively.

FINANCE.

(*The Statement of Accounts for* 1940 *is attached to this Report.*)

The Balance Sheet shows that the Institute's reserves were strengthened during the year, and the Income and Expenditure Account that the war had had little adverse effect on current finances, largely owing to substantial economies. A reduction in several sources of income must, however, be anticipated in future. Both accounts show the importance of the House Fund and of Industrial Subscriptions in assuring a satisfactory financial position.

General Fund.—Balance Sheet.

On the asset side of the Balance Sheet the total of sundry debtors has increased by £751 (from £1713 to £2464) and of payments in advance by £205 (from £695 to £900); cash at bank and in hand has increased by £2324 (from £517 to £2841). Expenditure on air-raid precautions (£327) charged against the House Fund in the accounts for 1939 has been written off. In accordance with usual practice, subscriptions in arrears, the stock of *Journals* and office furniture and the Library have not been valued.

Liabilities show a reduction of £291 (from £939 to £648) in the total of sundry creditors and a reduction of £241 (from £512 to £271) in subscriptions and *Journal* sales paid in advance.

Substantial additions have been made to Suspense Account, including the transfer of the balance of the House Fund (£661), and the balance of the Income and Expenditure Account (£530) has been transferred to Reserve Account.

General Fund.—Income and Expenditure Account.

While income, apart from special contributions from industrial subscriptions, decreased by £417 (from £14,224 to £13,807), expenditure was reduced by £1,555 (from £18,279 to £16,724). Income from special contributions increased by £867 (from £4166 to £5033), resulting, after transference to Reserve and other Accounts, in a final excess of income over expenditure for the year of £530 (in place of £228 excess of expenditure over income during the previous year). Income from subscriptions was less by £421 and from sales of publications by £141; further reductions under both these headings must be anticipated as the result of the extension of the

Rotherham Forge and Rolling Mills Co., Ltd.
Round Oak Steel Works, Ltd.
Joseph Sankey & Sons, Ltd.
Simon-Carves, Ltd.
Walter Somers, Ltd.
South African Iron and Steel Industrial Corpn., Ltd.
South Durham Steel and Iron Co., Ltd.
South Wales Siemens Steel Association.
Spear and Jackson, Ltd.
Stanton Ironworks Co., Ltd.
Staveley Coal and Iron Co., Ltd.
Steel Company of Canada, Ltd.
Steetley Lime and Basic Co., Ltd.
John G. Stein & Co., Ltd.
Stewarts and Lloyds, Ltd.
John Summers & Sons, Ltd.
Tata, Ltd.
Richard Thomas & Co., Ltd.
Tinsley Rolling Mills Co., Ltd.
Union Steel Corporation (of South Africa), Ltd.
United Steel Companies, Ltd.
Upper Forest and Worcester Steel and Tinplate Works, Ltd.
Vickers, Ltd.
Thos. W. Ward, Ltd.
Wellman Smith Owen Engineering Corporation, Ltd.
Whitehead Iron and Steel Co., Ltd.
Woodall-Duckham Vertical Retort and Oven Construction Co. (1920), Ltd.

Changes on the Council.

During the year the following changes on the Council were made: Mr. John Craig, C.B.E., succeeded The Right Hon. the Earl of Dudley, M.C., as President at the Annual Meeting. Sir Edward J. George [1] was nominated an Honorary Vice-President. Mr. Fred Clements and Captain R. S. Hilton were nominated Vice-Presidents, and Mr. G. H. Latham and Mr. C. J. Walsh [1] were elected Members of Council. As reported above, Mr. Walsh died on February 22, 1941, and Mr. A. O. Peech on March 22, 1941.

Colonel Alan Stein, M.C., T.D., D.L., and Mr. A. Robinson became Honorary Members of Council in succession to Mr. P. M. Ritchie and Mr. N. Nisbet for the periods of their Presidency of the West of Scotland Iron and Steel Institute and of the Lincolnshire Iron and Steel Institute, respectively.

In accordance with Bye-Law No. 10, the names of the following Vice-Presidents and Members of Council were announced at the

[1] Recorded in the Report of Council for 1939.

Autumn Meeting as being due to retire at the Annual General Meeting in 1941 :

Vice-Presidents : Mr. Arthur Dorman, Mr. J. S. Hollings and Mr. C. E. Lloyd.

Members of Council : Captain H. Leighton Davies, Mr. I. F. L. Elliot, Principal C. A. Edwards, F.R.S., Mr. E. F. Law and the Hon. R. G. Lyttelton.

No other Members having been nominated up to one month previous to this Annual Meeting, the retiring Members are presented for re-election.

Honours Conferred on Members.

(*To December 31st*, 1940.)

The Council offer their warm congratulations to Members of the Institute on honours and appointments received during 1940.

Sir Andrew Duncan, G.B.E., M.P., who on the outbreak of war had been appointed Controller of Iron and Steel, joined the Cabinet as President of the Board of Trade, was summoned to the Privy Council and later became Minister of Supply; he was succeeded as Controller of Iron and Steel by Col. Sir Charles Wright, Bt., K.B.E., C.B.

Many Members of the Institute hold important positions in the Iron and Steel and other Controls of the Ministry of Supply or are serving on technical and advisory Committees.

Mr. Axel Hultgren received the Robert Woolston Hunt Award from the American Institute of Mining and Metallurgical Engineers, and Mr. F. M. Beckett the Elliott Cresson Medal from the Franklin Institute. Mr. W. J. Dawson was awarded the E. J. Fox Gold Medal of the Institute of British Foundrymen, and Dr. P. D. Merica the Platinum Medal of the Institute of Metals. The Japan Institute of Metals presented the Honda Medal and Prize to the late Sir Harold Carpenter, F.R.S.; he was the second recipient of this award and the first foreigner to receive it. Mr. R. Lowe was awarded the Gold Medal of the Junior Institution of Engineers for 1939 and was also appointed a Member of Council of that body.

Professor P. Chevenard was appointed President of the Société Française de Physique for the duration of the war. Mr. W. Wilkinson Wood was elected Master Cutler of the Cutlers' Company in Hallamshire for the year 1940–41; he had already held this position in 1924, and the present is the first occasion on which an ex-Master Cutler has taken office a second time after a lapse of years. The President (Mr. John Craig, C.B.E.) was made an Honorary Member of the Chemical, Metallurgical and Mining Society of South Africa. Dr. E. Gregory was elected a Member of Council of the Institute of Chemistry of Great Britain and Ireland. Mr. G. R. Bashforth was re-elected President of the Staffordshire Iron and Steel Institute and Mr. H. H. L. Lockley became President of the Sheffield Metallurgical Association; Colonel Alan Stein, M.C., and Mr. A. Robinson were elected Presidents of the West of Scotland and the Lincolnshire Iron and Steel Institutes, respectively. Fellowship of the City and Guilds of London Institute was conferred on Dr. C. H. Desch, F.R.S. Dr. W. H. Hatfield, F.R.S., was appointed a Member of Council of the University of Sheffield, and Dr. R. J. Sarjant was elected President of the Sheffield and District Cambrian Society. Mr. E. A. Williams became Sheriff for Carmarthenshire.

Dr. C. W. H. von Eckermann became Professor at Stockholms Högskola Bergsingenjoren (Stockholm Technical College of Mining Engineers). Captain H. Leighton Davies was elected Chairman of the Welsh Plate and Sheet Manufacturers' Association; Major J. M. Bevan and Mr. T. O. Lewis became Vice-Chairmen of that body; Captain Leighton Davies was also appointed Chairman of the Swansea Tinplate Conference, with Major J. M. Bevan and Mr. F. S. Padbury as Senior and Vice Chairmen respectively, and also became President of the Swansea Metal Exchange. Mr. J. Sinclair Kerr was made District Officer of No. 4 Lancashire District and Representative of the St. John's Ambulance Brigade on the East Lancashire Joint County Committee of the British Red Cross and the St. John's Ambulance Associations.

Bessemer Gold Medal.

The Bessemer Gold Medal for 1940 was awarded to Dr. Andrew McCance in recognition of his original investigations into the application of the laws of physical chemistry to the open-hearth process and his critical examination of the thermo-dynamics of the process; of his valuable scientific papers; and of his services in the application of science to the iron and steel industry in Scotland.

The Council have decided to award the Bessemer Gold Medal for 1941 to Dr. T. Swinden in recognition of the value of his original investigations in the metallurgy of steel and of his eminent services in the organisation and direction of research in the steel industry.

Andrew Carnegie Medal.

No award was made of the Andrew Carnegie Medal during 1940.

Williams Prize.

The Williams Prize of £100 for 1940 was awarded to Mr. B. Yaneske for his paper on "The Manufacture of Steel by the Perrin Process." The paper is printed in the No. II. volume of the *Journal of The Iron and Steel Institute* for 1940.

Ablett Prize.

No papers were submitted in 1940 in competition of the Prize offered by Captain C. A. Ablett, O.B.E., B.Sc., M.Inst.C.E.,

Managing Director of the Cooper Roller Bearings Co., Ltd. The Council have gratefully accepted Captain Ablett's offer to renew the Prize, value £50, for 1941; by agreement with him, the conditions governing the award have been altered and are now as follows:

(*a*) Competing papers shall deal with a subject connected with engineering in iron or steel works.

(*b*) Competing authors shall be British subjects employed in the iron and steel industries of Great Britain or the British Empire. The competition shall be open to both Members and Non-Members of The Iron and Steel Institute.

(*c*) The decision of the Council of The Iron and Steel Institute shall be final as to whether a paper is eligible and on all matters arising out of the award of the Prize.

Competing papers should be marked "Ablett Prize Paper," and should reach the Secretary of the Institute not later than May 31, 1941. The successful paper will be presented at the Autumn Meeting and printed in the Institute's *Journal*.

Andrew Carnegie Research Scholarships.

Grants were made by the Council in 1940 to the following candidates:

M. Balicki (University College, Swansea).—£100 in aid of a research on cold-hardening of steel and verification of the recrystallisation theory put forward by Professor Krupkoffski and the candidate.

H. Morrogh (British Cast Iron Research Association).—£100 in aid of a research on sulphide inclusions produced by the neutralisation of sulphur in cast iron by aluminium, copper, nickel, chromium and molybdenum.

The Worshipful Company of Blacksmiths.

No recommendation for admission to the Worshipful Company of Blacksmiths was made for the year 1940.

Meetings.

Annual Meeting.

The Annual General Meeting of the Institute was held on Thursday and Friday, May 2 and 3, 1940, at the Offices of the Institute, 4, Grosvenor Gardens, London, S.W.1. The Retiring President, The Right Honourable the Earl of Dudley, M.C., was unable to be present, and, in his absence, Mr. James Henderson, Honorary Treasurer, opened the proceedings and inducted the President-Elect, Mr. John Craig, C.B.E., into the Chair. Mr. Craig gave his Presidential Address, after which nine papers were presented for discussion, including four submitted under the auspices of Joint Research Committees of the Institute and the British Iron and Steel Federation, and one report on a research carried out with the aid of a grant from the Andrew Carnegie Research Fund.

Members' Luncheon.

In place of the Annual Dinner a Luncheon was held at Grosvenor House, Park Lane, London, W.1, on Friday, May 3, 1940, at 1 P.M. About six hundred Members and their friends were present. The following proposed or replied to toasts:

Sir William Bragg, O.M., K.B.E., F.R.S. (President of the Royal Society).
Mr. John Craig, C.B.E. (President).
Rear-Admiral (now Vice-Admiral) B. A. Fraser, C.B., O.B.E. (Third Sea Lord and Controller of the Royal Navy).

Autumn Meeting.

The Autumn Meeting was held in Sheffield on Tuesday, November 12, 1940; by arrangement with the Presidents and Councils of the Sheffield Society of Engineers and Metallurgists and the Sheffield Metallurgical Association, it was a Joint Meeting with these two Societies. No visits to works or excursions were arranged and invitations were not issued to Ladies.

Following a Luncheon, at which about two hundred and seventy-five Members and Guests were present, two sessions for the discussion of papers were held under the Chairmanship of the President (Mr. John Craig, C.B.E.) during the afternoon and evening, with a short interval for tea. Eleven papers were included in the programme (of which five were presented and discussed), and of these three were submitted by Joint Research Committees of the Institute and the British Iron and Steel Federation, and one was a report on a research carried out with the aid of a grant from the Andrew Carnegie Research Fund.

A Committee of Arrangements, under the Chairmanship of Dr. W. H. Hatfield, F.R.S. (Vice-President), had been formed to advise on the preparations for the Meeting, and the Council desire to acknowledge their indebtedness to this Committee for their valuable assistance.

Joint Meetings.

(*For the period from* 30*th April*, 1940, *to* 31*st March*, 1941.)

In addition to the Autumn Meeting in Sheffield referred to above, the following Joint Meetings were held during the winter months with the Societies named:

Tuesday, 15th October, 1940 : The Lincolnshire Iron and Steel Institute.

Place and Time : The Modern School, Cole Street, Scunthorpe, at 7.30 P.M.
Chairman : Mr. N. Nisbet, President of the Lincolnshire Iron and Steel Institute.

Paper :
"Anti-Piping Compounds and their Influence on Major Segregation in Steel Ingots," by Dr. E. Gregory.

Monday, 16th December, 1940 : The Cleveland Institution of Engineers.

Place and Time : The Cleveland Scientific and Technical Institute, Corporation Road, Middlesbrough, at 6.30 P.M.
Chairman : Mr. L. F. Wright, President of the Cleveland Institution of Engineers.

Paper :
"The Manufacture of Steel by the Perrin Process," by B. Yaneske. (In the absence of the author the paper was presented by Mr. R. Mather, to whom the thanks of the Council are due.)

The Meetings were well attended and the discussions interesting. The Council wish to record their appreciation to the Presidents, Councils and Secretaries of the Local Societies, as well as to the authors of the papers, for their contributions to the success of the Meetings.

Owing to conditions arising out of the war it was not possible to hold as many Joint Meetings as usual. The Council, however, record with pleasure that the friendly relations previously reported were continued with the following Societies :

Cleveland Institution of Engineers.
Ebbw Vale Metallurgical Society.
Lincolnshire Iron and Steel Institute.
Manchester Metallurgical Society.
Newport and District Metallurgical Society.
Sheffield Metallurgical Association.
Sheffield Society of Engineers and Metallurgists.
Staffordshire Iron and Steel Institute.
Swansea Technical College Metallurgical Society.
West of Scotland Iron and Steel Institute.

Relations With Other Societies and Technical Institutions.

As far as the circumstances of the war permitted, friendly relations with other Scientific Societies and Technical Institutions in Great Britain, the Dominions and the neutral countries, were maintained. The Council particularly value the many signs of friendship which have been received from the American Iron and Steel Institute and the American Institute of Mining and Metallurgical Engineers.

Collaboration with the Institute of Metals has been continued on the same close footing as in former years; there has been an increase of forty-six in the number of Members and Associates who are Members or Student Members of the Institute of Metals and these now number six hundred and eighty-two.

Publications.

Papers and Committee and Andrew Carnegie Research Reports.—Two volumes of the *Journal* were published during the year. These contained Mr. Craig's Presidential Address and the following seven Committee Reports, two Andrew Carnegie Research Reports and eleven papers:

John Craig: "Presidential Address."

A. J. Bradley, F.R.S., W. L. Bragg, F.R.S., and C. Sykes: "Researches into the Structure of Alloys."

G. P. Contractor and F. C. Thompson: "The Damping Capacity of Steel and Its Measurement." (Paper No. 1/1940 of the Alloy Steels Research Committee).

C. A. Edwards, F.R.S., D. L. Phillips and H. N. Jones: "The Influence of some Special Elements upon the Strain-Ageing and Yield-Point Characteristics of Low-Carbon Steels."

U. R. Evans: "Report on Corrosion Research Work at Cambridge University Interrupted by the Outbreak of War." (Paper No. 1/1940 of the Corrosion Committee).

E. Gregory: "Anti-Piping Compounds and their Influence on Major Segregation in Steel Ingots." (Paper No. 1/1940 of the Committee on the Heterogeneity of Steel Ingots).

W. H. Hatfield, F.R.S., and G. W. Giles: "Non-Metallic Inclusions in Steel. Quantitative Evaluation.—Part I." (Paper No. 3/1940 of the Committee on the Heterogeneity of Steel Ingots (submitted by the Inclusions Sub-Committee)).

Wm. A. Haven: "The Manufacture of Pig Iron in America."

E. A. Jenkinson: "The Iron-Plating of Specimens for Microscopical Examination."

H. Lepp: "The Oxygen/Hydrogen/Molten-Iron System."

H. Lipson and N. J. Petch: "The Crystal Structure of Cementite, Fe_3C."

H. Lipson and A. J. C. Wilson: "Some Properties of Alloy Equilibrium Diagrams Derived from the Principle of Lowest Free Energy."

Bo O. W. L. Ljunggren: "Method of Sclero-Grating Employed for the Study of Grain Boundaries and of Nitrided Cases; Grain Structures Revealed by Cutting." (Andrew Carnegie Research Report).

P. Lloyd and E. A. C. Chamberlain: "Corrosion of Steels by Molten Nitrates."

W. C. Newell: "The Estimation of Hydrogen in Steel and other Metals." (Paper No. 2/1940 of the Committee on the Heterogeneity of Steel Ingots (submitted by the Oxygen Sub-Committee)).

L. Reeve: "The Corrosion of Mild and Copper-Bearing Steel Panels in Iron-Ore Wagons." (Paper No. 3/1940 of the Corrosion Committee (communicated by Dr. T. Swinden)).

O. A. Saunders and H. Ford: "Heat Transfer in the Flow of Gas through a Bed of Solid Particles."

T. Swinden and W. W. Stevenson: "An Accelerated Spray Test for the Determination of the Relative Atmospheric Corrodibility of Ferrous Materials." (Paper No. 2/1940 of the Corrosion Committee (submitted by the Laboratory Research Sub-Committee)).

A. A. TIMMINS: "The Decomposition of Pearlite in Grey Cast Iron." (Andrew Carnegie Research Report).
R. R. F. WALTON: "The Practical Side of Blast-Furnace Management, with Especial Reference to South African Conditions."
B. YANESKE: "The Manufacture of Steel by the Perrin Process."

The monthly *Bulletin of The Iron and Steel Institute* was also published during the year; as usual it was reprinted as Section II. of the *Journals* issued for the corresponding periods. The Bulletin is supplied free of charge to Members on application; the subscription rate to non-Members is 30*s*. per annum ($6 to members of the American Iron and Steel Institute, the American Institute of Mining and Metallurgical Engineers and the American Society for Metals).

JOINT LIBRARY AND INFORMATION DEPARTMENT.

Joint Library.

The number of volumes sent out on loan during 1940 was 4270, compared with 4150 in 1939, requests for books having been received from many companies, research establishments and Government departments as well as from Members of the Institute and of the Institute of Metals.

Numerous text-books have been presented to the Library or purchased, and the Council wish to take this opportunity of thanking those authors and publishers from whom presentations have been received. A list of the additions made to the Library is issued quarterly, and copies will be sent to Members on request.

The majority of the articles abstracted in the *Bulletin of The Iron and Steel Institute* are filed in the Library, and are available for loan. The original articles can be purchased for Members on request, and photographic copies can be supplied under certain conditions.

Collaboration with the Institution of Civil Engineers and the Science Library.

The valuable collections of scientific works included in the Science Library and the Library of The Institution of Civil Engineers are available for consultation or loan under certain conditions. Members who wish to avail themselves of these facilities should communicate with the Librarian of the Joint Library, 4, Grosvenor Gardens, London, S.W.1.

Information Department.

The Information Department has dealt with an increased number of enquiries, and bibliographies have been compiled at the request of Members. Members are invited to avail themselves of the bibliographical and information services.

The following translations were prepared, and copies are available for consultation in the Library or for purchase, at a nominal price :

No. 19.—" Investigation of the Working of the No. 3 Blast-Furnace at the Zaporzhstal Works," by I. Kozlovich. (Translated from the Russian.)

No. 20.—" The Solubility of Nitrogen in Steel," by Lennart Eklund. (Translated from the Swedish.)

No. 21.—" A Note on the Solidification of Steel Ingots," by R. Hohage and R. Schäfer. (Translated from the German.)

No. 22.—" New Trend in the Methods of Determining Oxygen in Alloy Steels," by G. Ya. Veynberg. (Translated from the Russian.)

No. 23.—" ' Stakhanov ' Methods of Rapid Open-Hearth Melting at the Stalin Kuznetskiy Works," by V. Savostin. (Translated from the Russian.)

No. 24.—" ' Stakhanov ' Methods of Operating the No. 3 Blast-Furnace at the Zaporzhstal Works," by D. Pospelov. (Translated from the Russian.)

No. 25.—" Lead-Bearing Steels," by L. Guillet. (Translated from the French.)

No. 26.—" Technical Cohesive Strength," by W. Kuntze. (Translated from the German.)

No. 27.—" A Method of Melting Steel in a Basic Electric-Arc Furnace," by K. Matsuyama, T. Iki and K. Muramoto. (Translated from the Japanese.)

Research.

Collaboration with the Iron and Steel Industrial Research Council was continued on the same basis as in former years.

The following is a list of the Joint Committees of the Institute and the Iron and Steel Industrial Research Council and of their Sub-Committees; the number of meetings recorded during 1940 was 59 (56 in 1939) :

Alloy Steels Research Committee : Chairman, Dr. W. H. Hatfield, F.R.S. Established June, 1934. Meetings held during 1940 : five.

Sub-Committee A, Thermal Treatment : Chairman, Mr. P. B. Henshaw. Established January, 1936. Meetings held during 1940 : four.

Hair-Line Crack Sub-Committee : Chairman, Dr. W. H. Hatfield, F.R.S. Established July, 1938. Meetings held during 1940 : six.

Special Aero-Components Sub-Committee : Chairman, Dr. W. H. Hatfield, F.R.S. Established July, 1940. Meetings held during 1940 : (not recorded).

Corrosion Committee : Chairman, Dr. W. H. Hatfield, F.R.S. Established July, 1928. Meetings held during 1940 : five.

Laboratory Research Sub-Committee : Chairman, Dr. U. R. Evans. Established June, 1930. No meetings held during 1940; activities carried on by correspondence.

Protective Coatings Sub-Committee : Chairman, Mr. T. M. Herbert. Established January, 1936. Meetings held during 1940 : one.

Marine Corrosion Sub-Committee : Chairman, Dr. G. D. Bengough, F.R.S. Re-formed November, 1938. Meetings held during 1940 : two.

Sub-Committee on Low-Alloy Steels: Established June, 1938. No meetings held during 1940; activities carried on by correspondence.

Sub-Committee on the Corrosion of Buried Metals (working in collaboration with the Committee on Soil Corrosion of Metals and Cement Products of the Institution of Civil Engineers). Established October, 1937. No meetings held during 1940; activities carried on by correspondence.

Heterogeneity of Steel Ingots Committee: Chairman, Dr. W. H. Hatfield, F.R.S. Established May, 1924. Meetings held during 1940: five.

Ingot Moulds Sub-Committee (joint with the Open-Hearth Committee of the Iron and Steel Industrial Research Council): Chairman, Mr. R. H. Myers. Established November, 1934. No meetings held during 1940; activities carried on by correspondence. (Stresses in Moulds Panel, one.)

Joint Conference on the Physical Chemistry of Steelmaking (formed jointly by the Committee on the Heterogeneity of Steel Ingots and the Open-Hearth Committee of the Iron and Steel Industrial Research Council): Chairman, Dr. T. Swinden. Established September, 1938. No meetings held during 1940.

Liquid Steel Temperature Sub-Committee: Chairman, Mr. E. W. Elcock. Established March, 1929. Meetings held during 1940: one.

Oxygen Sub-Committee: Chairman, Dr. T. Swinden. Established January, 1936. Meetings held during 1940: two. (Chemists' Panel, three.)

Inclusions Sub-Committee: Chairman, Dr. W. H. Hatfield, F.R.S. Established November, 1936. Meetings held during 1940: six.

Standard Methods of Analysis Sub-Committee: Chairman, Dr. E. Gregory. Established, September, 1939. Meetings held during 1940: seven.

Steel Castings Research Committee: Chairman, Mr. W. J. Dawson. Established November, 1934. Meetings held during 1940: six.

Moulding Materials Sub-Committee: Chairman, Mr. W. J. Rees. Established March, 1936. Meetings held during 1940: one.

Foundry Practice Sub-Committee: Chairman, Dr. C. J. Dadswell. Established May, 1938. Meetings held during 1940: four.

Appointment of Representatives.

The following is a list of the Institute's representatives on various governing bodies and committees for the year 1940; it has been brought up to date to March 31, 1941:

British Association, Fuel Economy Committee: *appointment open.*

British Cast Iron Research Association: Professor T. Turner.

British Corporation Register of Shipping and Aircraft, Technical Committee: Dr. A. McCance.

British Electrical and Allied Industries Research Association,
Sub-Committee J/E, Joint Committee, Steels for High Temperatures: Dr. W. H. Hatfield, F.R.S., Dr. T. Swinden.
Sub-Committee J, Earthing to Water Mains: Dr. J. C. Hudson.

British Iron and Steel Federation, Statistical Committee: Mr. K. Headlam-Morley.

British Refractories Research Association, Council: Mr. W. J. Brooke.

British Standards Institution,
Chemical Engineering Divisional Council: Mr. E. F. Law.
Engineering Divisional Council E/-: The Hon. R. G. Lyttelton, Mr. J. Sinclair Kerr; *one appointment open.*
Sub-Committee M33/7, Protective Glasses for Welders and Industrial Purposes: Dr. C. H. Desch, F.R.S.
Technical Committee CEB/1, Cement: Mr. F. W. Harbord, C.B.E.
Technical Committee CEB/6/1, Concrete Blocks: Mr. F. W. Harbord, C.B.E.
Technical Committee CH/17, Symbols used in Diagrams of Chemical Engineering Plant: Mr. A. E. Chattin.
Technical Committee EL/28, Fans: Mr. A. F. Webber.
Iron and Steel Industry Committee IS/-: Dr. T. Swinden.
Technical Committee IS/1, Co-ordination of Iron and Steel Specifications: Dr. T. Swinden.
Technical Committee IS/6, Steel Castings: Dr. R. H. Greaves.
Technical Committee IS/8, Creep Properties: Dr. W. H. Hatfield, F.R.S.
Technical Committee IS/15, Iron and Steel for Shipbuilding: Sir Edward J. George.
Technical Committee IS/17, Cast Iron Columns for Street Lighting: Mr. J. G. Pearce.
Technical Committee IS/35, Cast Iron: *appointment open.*
Technical Committee IS/35/3, Malleable Steel Castings: Mr. C. H. Kain.
Technical Committee ME/22, Marking and Colouring of Foundrymen's Patterns: Mr. F. W. Lewis.
Technical Committee ME/23, Brinell Hardness Testing: Dr. W. H. Hatfield, F.R.S.
Technical Committee ME/25, Testing of Thin Metal Sheet and Strip: Dr. T. Swinden.
Technical Committee ME/32, Engineering Symbols and Abbreviations: Dr. T. Swinden.
Solid Fuel Industry Committee, SF/-: Mr. A. F. Webber.
Technical Committee SF/1, Nomenclature and Definitions: Mr. A. F. Webber.
Technical Committee SF/2, Underfed Screw Type Stokers: Mr. A. F. Webber.

Technical Committee SF/4, Heating Stoves: Mr. A. F. Webber.

Technical Committee on Metallic Finishes: Mr. F. C. Platt.

Units and Technical Data Co-ordinating Committee: Sir Wm. Larke, K.B.E.

City and Guilds of London Institute, Advisory Committee on Metallurgy: Mr. E. C. Greig.

Constantine College, Advisory Committee: Mr. E. W. Jackson.

Empire Council of Mining and Metallurgical Institutions: Mr. F. W. Harbord, C.B.E., Mr. K. Headlam-Morley.

Engineering Public Relations Committee,

Main Committee: Mr. James Henderson.

Executive Committee: Mr. K. Headlam-Morley.

Hong-Kong University, Home Committee: *appointment open.*

Imperial College of Science and Technology, Board of Governors: Mr. James Henderson.

Imperial Institute, Mineral Resources Department, Iron and Ferro-Alloy Metals Committee: Mr. K. Headlam-Morley.

Institute of Fuel, Council: *appointment open.*

Institute of Welding, Council and Representative of Patron Institution: Mr. K. Headlam-Morley.

Institution of Mechanical Engineers, Research Committee on High-Duty Cast Irons for General Engineering Purposes: Dr. J. E. Hurst.

Iron and Steel Industrial Research Council: Mr. F. W. Harbord, C.B.E., Dr. W. H. Hatfield, F.R.S., Mr. K. Headlam-Morley.

Joint Committee on Materials and their Testing: Mr. K. Headlam-Morley.

Liverpool University, Court of Governors: Sir W. Peter Rylands, Bt.

Lloyd's Register of Shipping, Technical Committee: Mr. James Henderson, Mr. P. Baxter.

Mechanisation Board (Army Council): Mr. F. W. Harbord, .B.E.

National Physical Laboratory, General Board: Dr. T. Swinden, Dr. W. H. Hatfield, F.R.S.

Ramsey Memorial Laboratory, Advisory Committee: Mr. F. W. Harbord, C.B.E.

Royal School of Mines, Advisory Board: Mr. F. W. Harbord, C.B.E.

Royal Society, General Board for Administering Government Grants for Scientific Investigations: The President.

School of Metalliferous Mining (Cornwall), Board of Governors: Mr. J. S. Hollings.

Science Museum, Advisory Council: *appointment open.*

Sheffield University, Court of Governors: Mr. A. B. Winder.

[The Statement of Accounts for 1940 will be found in the following pages.]

THE IRON AND

BALANCE SHEET,

LIABILITIES.	£ s. d.	£ s. d.	£ s. d.
Sundry Creditors :—			
Office Rent		300 0 0	
Telephone Calls		27 4 9	
Travelling Expenses		6 0 4	
Autumn Meeting		5 16 0	
Library Books		1 0 4	
Cleaning, Heating, Lighting and Water		140 19 11	
Journal Printing		6 1 10	
Bulletin		119 18 10	
Grants		10 0 0	
Yearly Index		8 6	
Subscriptions in Suspense		30 3 9	
			647 14 3
Subscriptions in Advance :—			
Home Members		34 7 9	
Overseas Members		37 6 9	
Associates		5 0 0	
			76 14 6
Journal Sales :—			
Amount in advance, 1941		14 18 8	
Received on account of Vol. II., 1940		179 3 6	
			194 2 2
Suspense Account as at 1st January, 1940 :—	£ s. d.		
Reserve for 10-year Index	250 0 0		
Add Further transfer, 1940	50 0 0		
		300 0 0	
Entrance Fees	1,911 14 0		
Add Further transfer, 1940	290 19 6		
		2,202 13 6	
Library Account		500 0 0	
Repairs and Decorations	500 0 0		
Transfer re 4, Grosvenor Gardens per contra ...	661 0 5		
		1,161 0 5	
			4,163 13 11
Life Composition Fund			3,259 14 8
Iron and Steel Institute :—			
Capital as per last Balance Sheet		23,882 2 4	
Accumulated Excess of Income over Expenditure	£ s. d. 1,855 5 1		
Add Excess of Income over Expenditure for the year	530 8 9		
		2,385 13 10	
			26,267 16 2
			£34,609 15 8

I have examined the above Balance Sheet

STEEL INSTITUTE.

31st DECEMBER, 1940.

ASSETS.

	£ s. d.	£ s. d.	£ s. d.
Sundry Debtors :—			
Subscriptions in arrear			Not valued.
Amount due from Carnegie Scholarship Fund		249 3 3	
Telephone Deposit		1 0 0	
Travelling Expenses		3 11 6	
Telephone		8 4 7	
Printing and Stationery		4 8	
Salaries		10 0 0	
Sundries		1 5 0	
Income Tax recoverable		1,566 19 3	
Rent Receivable		275 0 0	
Publishing Expenses :—			
Printing	14 3 0		
Abstracts	28 17 6		
Reviews	3 1 6		
Advance Copies	122 5 0		
		168 7 0	
British Iron and Steel Corporation, Ltd., re distribution of Basic Refractories		145 16 8	
International Association for Testing Materials		34 7 11	
			2,463 19 10
Payments in Advance :—			
Journal Sales		627 6 11	
Insurance		53 9 8	
Library Books		27 16 8	
Staff Superannuation Fund		173 15 6	
Rates		12 9 11	
Cleaning, Heating, Lighting and Water		4 16 10	
			899 15 6
Stock of Journals			Not valued.
Office Furniture and Library			Not valued.
Cash at Bank and in Hand :—			
Secretary's Account		178 1 4	
Deposit Account		1,050 17 5	
Deposit Post Office Savings Account		722 5 9	
General Account		808 4 1	
Cash at Office		81 6 11	
			2,840 15 6
Investments at cost per Schedule :—			
General Fund			24,904 15 4
(The Market value of these Investments at 31st December, 1940, was £27,102 13*s.* 11*d.*)			
Research Committee :—			
Amount advanced			240 14 10
Life Composition Fund :—			
Investments at cost per Schedule			3,259 14 8
(The Market value of these Investments at 31st December, 1940, was £3,202 0*s.* 4*d.*)			
4, Grosvenor Gardens :—			
Structural Alterations and Decorations, Furnishing, Legal Charges and Incidental Expenses as at 1st January, 1940		14,170 18 0	
	£ s. d.		
Less Subscriptions received	14,505 3 0		
A.R.P. Expenditure now written off ...	326 15 5		
		14,831 18 5	
		661 0 5	
			£34,609 15 8

of the Institute and certify it to be correct.

(*Signed*) W. B. Keen,
Chartered Accountant.

INCOME AND EXPENDITURE ACCOUNT

1939 £	1939 £	INCOME.	£	s.	d.	£	s.	d.	£	s.	d.
139		Entrance Fees	87	5	6						
138		Do. Companies' Nominations	203	14	0						
	277					290	19	6			
	277	*Less* Transfer to Reserve Fund				290	19	6	—		
		Annual Subscriptions :—									
4,383		Members, Home Current				4,180	1	0			
134		Do. Companies' Nominations				303	13	0			
173		Do. Arrears				169	5	6			
	4,690								4,652	19	6
1,803		Members, Overseas Current				1,412	17	6			
4		Do. Companies' Nominations				2	12	6			
124		Do. Arrears				122	14	3			
	1,931								1,538	4	3
187		Associates, Current				187	15	0			
11		Do. Companies' Nominations... ...				26	5	0			
9		Do. Arrears				1	11	6			
	207								215	11	6
		Sales of Publications :—									
1,651		Journals, etc.				1,534	4	3			
111		Bulletin				87	2	4			
	1,762								1,621	6	7
		Interest on Investments (Gross) :—									
1,059		General Fund				1,052	17	0			
139		Life Composition Fund				123	3	2			
16		Bessemer Medal Fund				16	0	0			
	1,214								1,192	0	2
	6	Interest on Deposit Account							12	5	1
	18	Sundry Receipts							29	2	6
		Institute of Metals :—									
750		Rent Receivable				750	0	0			
350		Contribution to Joint Library	375	0	0						
		Less Refund	25	0	0						
						350	0	0			
	1,100								1,100	0	0
		Iron and Steel Industrial Research Council :—									
750		Grant for Bulletin				750	0	0			
2,000		Grant for Secretarial Services				2,000	0	0			
500		Grant for Information Service				500	0	0			
	3,250								3,250	0	0
		Carnegie Research Fund :—									
	50	Transfer in respect of Grants							37	10	0
	—	British Iron and Steel Corporation, Ltd., re distribution of Basic Refractories							158	6	8
	4,054	Balance, being Excess of Expenditure over Income carried down							2,916	15	10
	£18,279								£16.724	2	1

£	£		£	s.	d.	£	s.	d.
		Special Subscriptions :—						
3,292		Contributions receivable during 1940 ...	3,462	3	6			
874		Income Tax recovered	1,571	5	7			
	4,106					5,033	9	1
	227	Balance, being Excess of Expenditure over Income				—		

FOR THE YEAR ENDED 31st DECEMBER, 1940.

1939 £	1939 £	EXPENDITURE.	£ s. d.	£ s. d.	£ s. d.
6,747		Salaries (including Pensions and Overtime) ...		7,191 13 6	
—		War Bonus		254 18 0	
				7,446 11 6	
		Less Contributions from Ministry of Supply and British Iron and Steel Corporation, Ltd.	552 15 8		
150		Do. Carnegie Research Fund	150 0 0		
				702 15 8	
	6,597				6,743 15 10
	48	National Insurance			51 3 10
	341	Staff Superannuation Fund			286 7 5
	1,200	Office Rent			1,200 0 0
	24	Repairs and Decorations and A.R.P.			438 16 11
	771	Cleaning, Heating, Lighting and Water ...			804 11 0
	369	Library Books, Binding, etc.			432 16 2
	245	Office Furniture			106 3 1
	143	Annual Meeting			142 11 6
	162	Autumn Meeting			81 7 11
		Publishing Expenses:—			
2,254		Journal: Printing and Paper		1,384 13 5	
22		Translations and Reviews ...		15 10 0	
211		Postage		153 5 11	
315		Advance Copies: Printing		228 2 2	
130		Postage		66 5 11	
575		Bulletin: Printing		381 5 3	
77		Postage		93 3 9	
142		List of Members		—	
	3,726				2,322 6 5
	707	Stationery and Printing			402 0 7
	400	Postage and Receipt Stamps			330 4 6
	242	Travelling and Entertainment Expenses ...			183 16 6
	64	Insurance			45 15 8
	31	Auditors' Fees for 1939			43 10 0
	136	Telephone Rental and Calls			198 0 0
	229	Office Disbursements and Sundry Expenses ...			175 18 5
	45	Translations Service			51 6 4
	77	Bank Interest and Charges			—
	22	Bessemer Medal			23 17 6
	—	Blast Furnace Symposium			14 2 6
		Grants:—			
50		British Electrical and Allied Industries Research Association		50 0 0	
25		British Refractories Research Association ...		25 0 0	
11		Joint Committee on Materials and their Testing		10 10 0	
50		British Standards Institution		50 0 0	
—		Mellor Memorial		10 0 0	
25		Institution of Mechanical Engineers, Research on High Duty Cast Iron		—	
2		Engineering Public Relations Fund ...		—	
10		British Foundry School		—	
21		Sheffield University Foundry Courses ...		—	
6		British Standards Institution, International Meeting, Helsinki		—	
	200				145 10 0
	2,500	Iron and Steel Industrial Research Council Grant			2,500 0 0
	£18,279				£16,724 2 1

£	£		£ s. d.	£ s. d.	£ s. d.
	4,055	Balance brought down			2,916 15 10
		Transfer Reserve and Suspense Account:—			
—		Repairs and Decorations		500 0 0	
50		10-year Index		50 0 0	
—		Library Account		500 0 0	
	50				1,050 0 0
		Companies' Nominations:—			
139		Entrance Fees		203 14 0	
		Subscriptions Account:—	£ s. d.		
134		Home Members	303 13 0		
4		Overseas Members ...	2 12 6		
11		Associates	26 5 0		
				332 10 6	
	288				536 4 6

WILLIAMS PRIZE FUND.

BALANCE SHEET, 31st DECEMBER, 1940.

LIABILITIES.	£ s. d.	£ s. d.	£ s. d.	ASSETS.	£ s. d.	£ s. d.
Capital Value representing the Market Value of £3,000 3½% Conversion Loan at 21st September, 1926, when Fund was inaugurated		2,220 0 0		£3,452 15s. 7d. 3½% Conversion Loan (at cost)	2,670 0 0	
Income and Expenditure Account:—				Cash at Bank	101 3 6	
Balance as at 1st January, 1940	530 6 8					2,771 3 6
Add Excess of Income over Expenditure for the year to 31st December, 1940 ...	20 16 10					
		551 3 6				
			2,771 3 6			
			£2,771 3 6			£2,771 3 6

INCOME AND EXPENDITURE ACCOUNT FOR THE YEAR ENDED 31st DECEMBER, 1940.

INCOME.	£ s. d.	EXPENDITURE.	£ s. d.
Interest on 3½% Conversion Loan	120 16 10	Award	100 0 0
		Excess of Income over Expenditure for the year ...	20 16 10
	£120 16 10		£120 16 10

ANDREW CARNEGIE RESEARCH FUND.

BALANCE SHEET, 31st DECEMBER, 1940.

LIABILITIES.

	£ s. d.	£ s. d.	£ s. d.
Sundry Creditors :—			
Grants		250 0 0	
Amount due to General Fund		249 3 3	
Auditors' Fees		10 10 0	
M. Tibor Ver's Account		5 12 1	
			515 5 4
Amount of Original Fund ($100,000) ...		21,241 5 6	
Add Amounts since Capitalised		2,912 9 6	
		24,153 15 0	
Add Surplus Income as at 1st January, 1940 ...	1,829 0 4		
Excess of Income over Expenditure for the year	579 14 11		
		2,408 15 3	
			26,562 10 3
			£27,077 15 7

ASSETS.

	£ s. d.	£ s. d.
Investments at cost per Schedule		26,149 18 10
(The Market value of these Investments at 31st December, 1940, was £30,559 10*s.* 4*d.*)		
Cash at Bank :—		
On Current Account	272 4 1	
On Deposit Account	502 2 1	
Post Office Savings Account	23 10 11	
		797 17 1
Income Tax Recoverable		129 19 8
		£27,077 15 7

INCOME AND EXPENDITURE ACCOUNT FOR THE YEAR ENDED 31st DECEMBER, 1940.

INCOME.

	£ s. d.
Interest on Investments (Gross)	1,179 0 10
Interest on Deposit	2 18 5
	£1,181 19 3

EXPENDITURE.

	£ s. d.	£ s. d.	£ s. d.
Scholarship Grants		225 0 0	
Do. Iron and Steel Industrial Research Council	75 0 0		
Less Refund by General Fund of the Iron and Steel Institute ...	37 10 0		
		37 10 0	
			262 10 0
Printing Reports			177 12 10
Contribution to General Fund on account of Salaries			150 0 0
Sundries			1 6
Audit Fee			10 10 0
Medal			1 10 0
Balance, being Excess of Income over Expenditure for the year			579 14 11
			£1,181 19 3

THE IRON AND STEEL INSTITUTE.

SCHEDULE OF INVESTMENTS AT 31st DECEMBER, 1940.

SHOWING NOMINAL VALUES, COST VALUES AND PRESENT MARKET VALUES.

Nominal Value. £	s.	d.	Nature of Security	Market Value, 31st December, 1940. £	s.	d.	Cost Value. £	s.	d.
			GENERAL FUND OF THE INSTITUTE.						
2,197	1	0	3½% War Stock ...	2,262	19	2	2,161	0	8
1,324	7	4	3½% Conversion Loan ...	1,370	14	4	1,253	14	0
447	0	0	Southern Railway 4% Debenture Stock	467	2	3	449	2	4
1,872	0	0	London and North Eastern Railway 4% 2nd Guaranteed Stock	1,141	18	4	2,149	13	3
2,241	0	0	Do. do. 4% 1st do.	1,680	15	0	2,431	14	5
2,649	4	0	2½% Consolidated Stock	2,039	14	7	1,800	0	0
1,500	0	0	Buenos Ayres Great Southern Railway 4% Debenture Stock	585	0	0	1,594	12	9
2,954	1	0	4% Consolidated Stock	3,298	3	11	3,175	18	1
437	0	0	London and North Eastern Railway 3% Debenture Stock	299	6	10	376	12	5
12,242	19	3	4% Funding Loan 1960–90 (transferred from Special Purposes Fund)	13,956	19	6	9,512	7	5
				£27,102	13	11	£24,904	15	4
			LIFE COMPOSITION FUND.						
587	13	10	3½% Conversion Loan ...	608	5	2	513	2	3
50	10	0	3% Local Loans	45	3	11	42	0	0
1,330	0	0	London and North Eastern Railway 3% Debenture Stock	911	1	0	1,254	17	6
594	0	0	London Passenger Transport Board 4½% "A" Stock (July 1933)	647	9	2	638	6	1
205	15	6	3½% War Loan	212	19	6	211	8	10
681	13	0	4% Funding Loan 1960–90	777	1	7	600	0	0
				£3,202	0	4	£3,259	14	8

Nominal Value.	Nature of Security.	Market Value, 31st December, 1940.	Cost Value.
	ANDREW CARNEGIE RESEARCH FUND.		
£ s. d.		£ s. d.	£ s. d.
6,897 14 11	3½% War Stock...	7,104 13 7	6,895 16 6
800 0 0	Do.	817 18 5	794 2 0
2,693 12 0	3½% Conversion Loan ...	2,787 17 6	2,006 19 0
227 9 4	3% Local Loans	203 11 7	202 10 0
2,300 0 0	India 3½% Stock, 1931	2,208 0 0	1,600 2 9
2,560 0 0	India 3% Stock, 1948 ...	2,124 19 3	1,525 3 3
2,250 0 0	North Eastern Electric Supply Company 3½% Consolidated Debenture Stock ...	2,036 5 0	2,261 10 0
1,500 0 0	Great Western Railway 4% Debenture Stock	1,592 5 0	1,204 5 3
2,000 0 0	London Midland and Scottish Railway 4% Debenture Stock	1,990 0 0	1,693 0 6
1,312 0 0	London and North Eastern Railway 4% 1st Guaranteed Stock ...	984 0 0	880 11 11
5,000 0 0	Do. do. 4% Debenture Stock	4,600 0 0	3,540 8 5
6,000 0 0	Do. do. 3% do.	4,110 0 0	3,545 9 3
		£30,559 10 4	£26,149 18 10
	WILLIAMS PRIZE FUND.		
3,452 15 7	3½% Conversion Loan ...	£3,573 12 6	£2,670 0 0
	BESSEMER MEDAL FUND.		
	Trustees :—Lord Airedale and Sir Francis Samuelson.		
400 0 0	London Midland and Scottish Railway 4% Debenture Stock	£398 0 0	

(*Signed*) JAMES HENDERSON,
Hon. Treasurer.

(*Signed*) K. HEADLAM-MORLEY,
Secretary.

I have examined the foregoing Balance Sheets and Income and Expenditure Accounts with the Books and Vouchers of the Institute and certify them to be correct. I have also verified the Balances at the Bankers and the Securities for the Investments shown above.

23 QUEEN VICTORIA STREET,
LONDON, E.C. 4.
31*st March*, 1941.

(*Signed*) W. B. KEEN,
Chartered Accountant.

THE INFLUENCE OF TURBULENCE ON THE STRUCTURE AND PROPERTIES OF STEEL INGOTS.*

By L. NORTHCOTT, D.Sc., Ph.D., F.I.C. (Research Department, Woolwich).

(Figs. 20 to 42 — Plates I. to XI.)

Synopsis.

Seven steel ingots were cast by different methods which were selected as offering different conditions of turbulence of the molten metal in the mould. The methods employed involved : (1) Bottom-casting, (2) top-casting, a single stream down the centre of the mould, (3) the use of a multi-hole tundish, (4) a sand mould with a single stream, (5) a single stream near one side of the mould, (6) top-casting with a single stream, stirring with a poker after casting, and (7) casting in a sloping mould. The macrostructure, chemical segregation and mechanical properties of the ingots were studied.

Further information on the influence of the casting method upon turbulence was obtained from a number of small composite non-ferrous alloy ingots, in the preparation of which a red alloy was poured first and was immediately followed by a white alloy of similar density and melting point. Knowledge of the distribution of the stream in the mould was then obtained by examining the distribution of the differently coloured alloys in the ingot.

Differences in the structures of the steel ingots are discussed from the point of view of the influence of turbulence, and theories on the mechanism of the solidification of steel are put forward.

Introduction.

It is known that in the preparation of an ingot or casting of any metallic alloy, variations in the casting technique may lead to wide variations in the crystal structure. In steel-ingot manufacture, the factors in the actual filling of the mould to which most attention has been paid—that is, apart from questions affecting the composition or degree of killing of the liquid steel—have been the casting temperature, the speed of casting and the shape and size of the ingot.

Experience in the casting of other metals in addition to steel has shown that another factor, namely, turbulence, may exert a pronounced effect on both the structure and properties. When discussing the solidification of a metal it is customary to consider first of all a mould filled with liquid metal at rest; such a state of affairs rarely occurs, however, and probably not at all in the normal method of steel-ingot manufacture. Descriptions have been published from time to time of steel ingots cast by somewhat unorthodox methods, in which the turbulent conditions were likely to be different from those existing in normal casting practice, and

* Received October 7, 1940.

certain features in the structure of these and of ordinary steel ingots have been explained as being due to the turbulence of the liquid metal in the mould, but knowledge of the actual conditions of turbulence and of the influence of this factor is extremely limited. It was, therefore, considered desirable to investigate the problem. The application of a non-turbulent method of casting to the preparation of small steel ingots has been recently described.[1] One of the chief advantages of that method was the ability to cast at such a low temperature that coarse crystal structures and segregation were readily avoided. In view, however, of the probable application of such a process being limited to the smaller sizes of ingots, the problem of turbulence still remained to be investigated. This has now been done by preparing a series of otherwise strictly comparable ingots in which the only variable was the method of filling the mould. The influence of thus changing the conditions of turbulence was determined by examining the structure and properties of the ingots so prepared. The results of this examination are described in Section I. in the present paper. The pa er also contains, in Section II., a description of a series of small composite alloy ingots cast under conditions simulating those of the large steel ingots and prepared in order to obtain information on the influence of the casting conditions upon turbulence. Finally, Section III. is composed of a critical discussion of the experimental results, and arguments are advanced to explain certain of the effects observed.

I.—Steel Ingots.

Casting Conditions.

In order to limit the variables in casting to turbulence alone, it was considered desirable that all the ingots should be cast from the same melt, that casting temperature and speed of pouring should be constant, and that moulds of the same shape and size should be used. The steel was made in a 7-ton electric furnace and all the ingots were cast from the same ladle of metal. A separate tundish for each ingot ensured a constant head of metal. In all, seven 15-cwt. ingots were prepared, all top-poured except the first, six of them into ordinary cast iron moulds and one into a sand mould, in the following order :

QXS : Bottom-cast.
QXT : Top-cast, single stream down centre of mould.
QXU : Top-cast through a 16-hole tundish.
QXW : Top-cast, single stream down centre into a sand mould.
QXX : Top-cast, single stream near side of mould.
QXY : Top-cast, as *QXT*, but stirred with poker after casting.
QXZ : Top-cast into tilted mould with single stream striking lower side of mould.

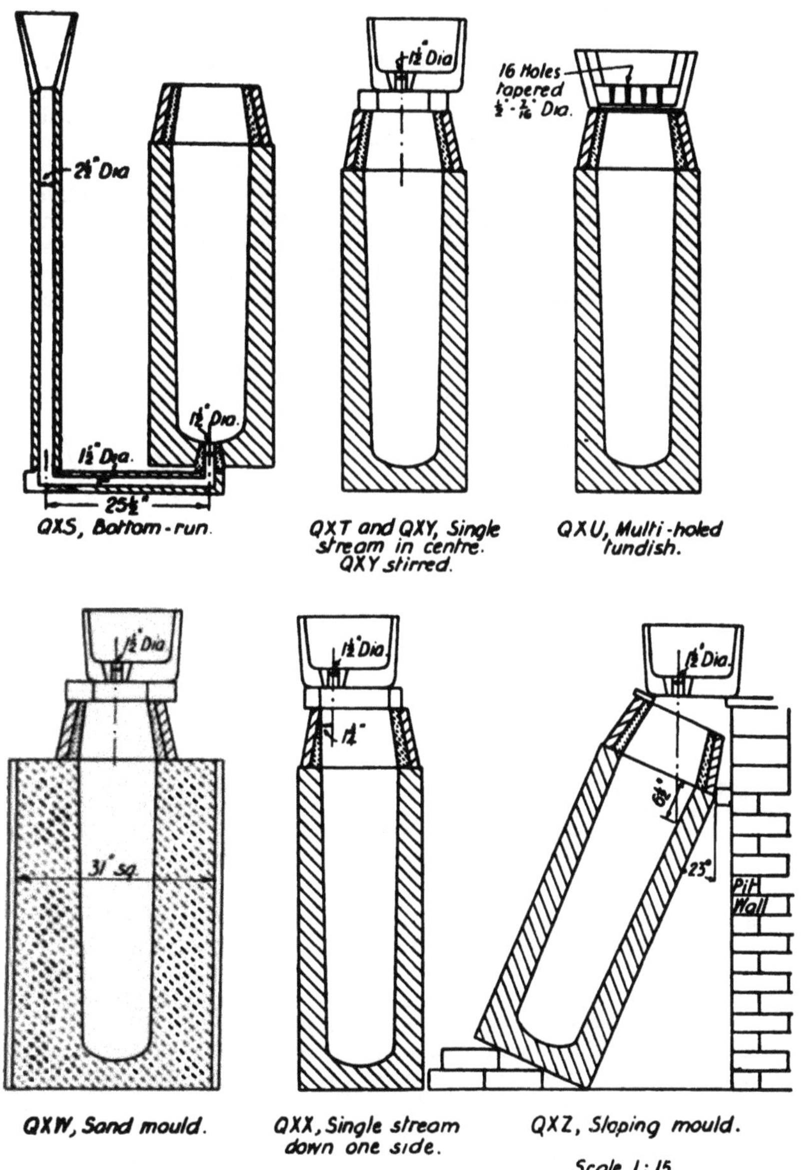

Fig. 1.—Arrangement of the 11-in. Square Moulds for Casting the Ingots.

Two other ingots were cast from the same melt after the seven described above, but have not been included in this examination.

The arrangement of the moulds is illustrated in Fig. 1, and what were assumed to be the conditions of turbulence are shown diagrammatically in Fig. 2. Standard 11-in. square ingot moulds were used, except, of course, for the sand mould. The runner bricks for the bottom-cast ingot were standard, but it was arranged for only one ingot to be poured. The second ingot, *QXT*, was representative of normal top-casting practice, except for the use of the tundish. The multi-hole tundish used for the third ingot, *QXU*, had 16 holes tapering from $\frac{1}{2}$ to $\frac{7}{16}$ in. in dia. and was made up from a mixture of firebrick, alundum and fireclay, and fired at 1100° C. before use.

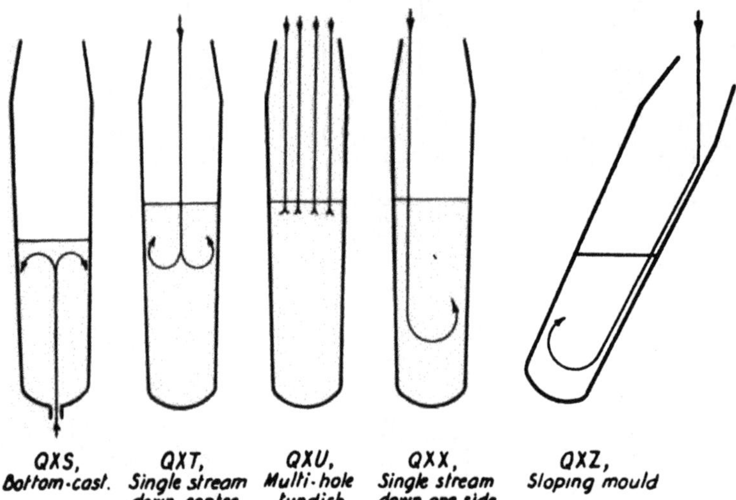

FIG. 2.—Assumed Motion of Liquid in Moulds.

This tundish, which appeared undamaged by the casting operation, enabled the mould to be filled by 16 small streams in a proximately the same time as the standard tundish with a single large hole. The mould for the fourth ingot, *QXW*, was made using an ingot as pattern with a 3-in. facing of rough compo on ordinary foundry backing sand. The mould was painted with the usual foundry black paint made from crucible pots and, after drying, with tar. In the casting of the fifth ingot, *QXX*, the stream occasionally impinged on the side of the feeder head, but the surface of the finished ingot showed no evidence that the stream had run down the face of the mould. The tundish was removed from the sixth mould, ingot *QXY*, immediately after casting and the metal stirred continuously with long $\frac{1}{2}$-in. dia. iron pokers for 6 min., except for stoppages to renew the poker, and again for 10 sec. after a further 2 min. Although the head was still open after 12 min. from the commencement of stirring, solidification in the ingot was so far advanced that further stirring was impossible. The stirred ingot was included to determine the influence of stirring upon both the crystal structure and the segregation. Experiments by Heggie[2] and Northcott[3] with materials having strongly directional crystallising properties had shown that the production of an equi-axial structure was only possible when the solidifying mass was stirred. The last mould was tilted at 23° to the vertical so that the metal stream struck the lower side of the mould along the centre line at a distance of 6 in. below the head. There was no welding on of the

TABLE I.—*Melting and Casting Particulars.*

Charge.	Carbon steel scrap		7 tons 13 cwt.
Slag.	With charge:	Lime	3 cwt.
	Fed later, after part removal of first slag:	Lime	4 cwt.
		Spar	½ cwt.
Additions.	In furnace:	Ferrosilicon (45%)	90 lb.
		Ferro-phosphorus (26·3%)	25 lb.
		High-carbon ferro-manganese	26 lb.
		Low-carbon ferro-manganese	71 lb.
		Aluminium	8 lb.
	In ladle:	Aluminium	2 lb.
		Ferrous sulphide	28 lb.
Apparent temperature at tapping.	Metal		1510–1505° C.
	Slag		1540° C.
Held in ladle.	3 min. 40 sec.		
Ladle nozzle.	1¼ in. in dia.		
Tundish nozzle.	1½ in. in dia.		
All ingots settled well in the mould.			

Casting Temperatures and Times.

Ingot Mark.	Type of Ingot.	Apparent Temp. From Ladle. ° C.	Apparent Temp. From Tundish. ° C.	Time. Sec.
QXS	Bottom-cast.	1430–1435	...	42
QXT	Single stream down centre.	1435–1440	1430	54
QXU	Multi-hole tundish.	1450–1460*	...	50
QXW	Sand mould.	1435	1430	45
QXX	Single stream down one side.	1430–1435	1425–1430	65
QXY	Poker-stirred.	1430	1420–1425	50
QXZ	Sloping mould.	1430	1425–1430	54

* Apparent temperature probably 10–15° too high owing to increased radiation from larger tundish.

stream, but the mould face was slightly scored where the stream touched. On those ingots in which the direction of casting was important stamp marks were applied for purposes of identification.

To facilitate the examination for heterogeneity, a medium-carbon steel of fairly high sulphur and phosphorus content was selected. The casting particulars are summarised in Table I.

Method of Ingot Sectioning.

After comparing the surface quality of the different ingots, the feeder heads were cut off and then an axial slice, 1 in. thick, was sawn off and the axial face ground and polished. A sulphur print was obtained and the axial face then etched to reveal the macrostructure. Using a drill ¾ in. in dia., samples for chemical analysis were taken at the positions adopted by the Committee on the Heterogeneity of Steel Ingots * for determining heterogeneity. Transverse bars, 1 in. deep and the full width of the ingot, were cut at distances of 6 in., 2 ft. and 3 ft. 6 in. from the base to enable more detailed examination to be made of material representing the lower, middle and upper part of each ingot. Tensile test-pieces were machined from a position just below the *A* and *F* analysis positions, with the axis of the test-pieces parallel with the length of the ingot, and others at right angles from just above the *A* and *F* positions. Information on the sulphur distribution, macrostructure, microstructure, chemical segregation and mechanical properties was thus obtained from the 1-in.-thick axial slice of each ingot. As will be described later, additional samples were obtained from certain of the ingots for a more complete examination to be made.

Ingot Examination.

Surface Quality.

No great differences were observed in the quality of the surfaces as between the different chill-cast ingots. The first two ingots, *QXS* and *QXT*, probably possessed the best surfaces although some pitting and general roughness were noticed in the lower quarters of both ingots. The multi-hole-tundish ingot, *QXU*, also had a good surface, apart from some pitting and streakiness where streams touched at two places. Ingot *QXW*, cast in the sand mould, showed several patches of roughness with a shallow type of under-cutting cavity and possessed the poorest surfaces of all, but the roughness appeared to be shallow and likely to be easily removed by machining. The surfaces of ingot *QXX* showed no evidence of the stream having been near one face. Ingot *QXZ*, cast in the tilted mould, showed a slightly convex oval patch of roughness and porosity, about 9 in. long and 5 in. wide, starting at 6 in. from the base of the feeder

* *See* Fifth Report on the Heterogeneity of Steel Ingots, *Iron and Steel Institute*, 1933, *Special Report No.* 4, Fig. 2, p. 25.

head, as evidence where the stream struck the lower face of the mould. The lower half of this face also showed some roughness along the centre line, but the other faces appeared satisfactory.

*Macrostructure.**

It will be most convenient to consider the structure of each ingot in turn:

Ingot QXS, *Bottom-Cast.*—The composition of this ingot was remarkably uniform throughout, there being comparatively little evidence of either Λ- or V-segregation, nor was there any obvious central porosity. Only at the top of the ingot, for a distance of about 8 in. below the head, was the primary crystal structure at all coarse. (*See* Fig. 20.)

Ingot QXT, *Single Stream down Centre.*—This ingot also appeared quite sound, but there was slightly more segregation of the Λ- and V- type than in *QXS*. (*See* Fig. 21.)

Ingot QXU, *Multi-Hole Tundish.*—There was a strong similarity between this ingot, represented in Fig. 22, and the bottom-cast one, except for the coarser columnar structure at the bottom of *QXU*; V-segregation was less than in the top-cast *QXT* ingot.

Ingot QXW, *Sand-Cast.*—Very pronounced piping, coarse Λ-segregation and crystal structure of a coarse open type characterised this ingot, which is shown in Fig. 23. The lower part of the pipe contained slag. The fact that the pipe extended down into the body of the ingot for about one quarter of the ingot length is an indication of the greater degree of solidification which takes place in a chill mould as compared with a sand mould by the time pouring is completed, since in the former there is then less liquid to solidify and consequently less shrinkage to be made good by feeding from the head. It is conceivable, too, that in a chill-casting the first-formed layer of compact chill crystals forming the surface is stronger and acts as a skin preventing penetration of liquid from the inside, whereas in a sand-casting of the size and cross-section of the ingot being examined it may be possible for the hydrostatic pressure of the liquid to enable it to break through the comparatively loose first-formed crystals and penetrate between these and the mould from which they have shrunk. Support for this view is offered by the appearance of the surfaces of this ingot, and by micro-examination of etched sections. The point has been mentioned since it prevents an accurate estimation of the relative rates of solidification in sand and chill moulds being made from the measurement of the internal piping.

Ingot QXX, *Stream down One Side.*—This ingot is shown in Figs. 24 and 25, the latter revealing the "patchy" type of segregation which is probably part of the Λ-segregate appearing in the former as masses of dendritic crystals. These crystals have the appearance

* Sulphur prints of the ingots were also made and were considered in preparing these comments, but are not reproduced.

of having been broken off or partially separated from the main crystallisation which succeeds the columnar crystals, owing to the erosive action either of the stream or, more likely, of motion of the liquid after the completion of casting. Thus instead of the Λ-segregate being in the form of two very diffuse lines as in the majority of chill ingots (Fig. 3 (*a*)) or as two sharper lines as in the ingot cast in a sand mould (Fig. 3 (*b*)), the segregate in the present instance is in the form represented in Fig. 3 (*c*), which is a variation or broken-up form of Fig. 3 (*a*). The alternative explanation which had originally to be considered is that the patches represent crystals which have been broken off from the original main side growth and have been transported, possibly over appreciable distances, by the turbulent motion of the liquid. Although it is easy to picture such an occurrence, the first explanation is thought to be the more likely since the patches represent areas of *less* pure and therefore lower-melting-point material, transportation of which would be likely to lead to its re-solution in the liquid. Furthermore, it will

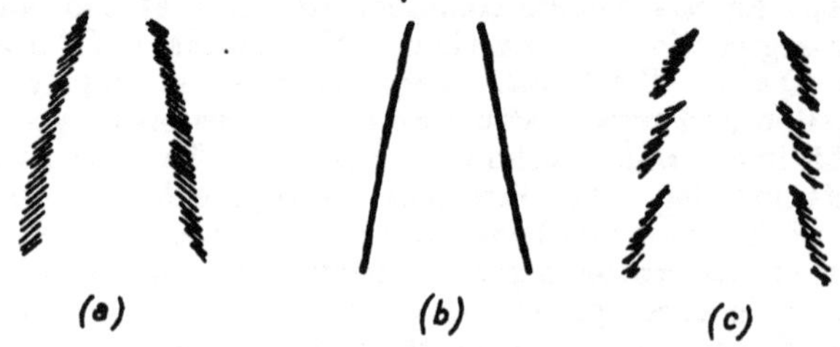

Fig. 3.—Types of Λ-Segregates.

be shown that when a solid stirring rod is brought into contact with the growing crystals, the tips of these are bent over rather than broken off, so that it is unlikely that large masses would be removed by a liquid stream. Another effect of the eccentric pouring has been to move the axis of the V-segregate about 0·5 in. toward the side nearest which the stream was poured.

Ingot QXY, *Stirred.*—The structure of this ingot presents an uncommon appearance typical of that of some centrifugal castings, crystallisation and segregation occurring in bands parallel to the mould wall. Portions of the stirring rods which have broken off and have been entrapped add to the odd appearance of Fig. 26. It will be noted that the majority of these pieces occur near the bottom of the ingot, some even appearing to rest on the layer of columnar crystals at the base. Although it is to be expected that the bottom part of the stirring rod would be the one to be broken off, the partial segregation of these pieces in their present position may be thought to support the hypothesis that falling crystallites of purer steel collect at the bottom of the ingot and are thus responsible for its greater purity. At the same time it should be borne in mind that the bar used for stirring was not only purer in iron than the steel of the ingot but was also colder and would therefore be of higher density than crystallites of the steel. An explanation of the banded structure is offered in a later section of the paper. Although the distance between any one band and the ingot face increased slightly towards the bottom, the close parallelism between the bands and the ingot surface confirms present views on the early stages of solidification of a steel ingot, namely, as an almost parallel-walled shell with steadily increasing wall thickness.

Ingot QXZ, *Tilted Mould.*—This ingot, reproduced in Fig. 27, is outstanding for the comparative absence of columnar structure on its lower side, down which the stream was poured, and the large size of the columnar crystals growing from the opposite face. The difference in structure in the two sides of the ingot is greatest at the top and becomes gradually less prominent as the bottom of the ingot is reached. In spite of the large differences in crystal structure there seems to be rather less segregation of the Λ- or V-type than in most of the previously described ingots.

Columnar Crystal Size and Orientation.

Two features in the macrostructure which were affected by the casting conditions were the size and the orientation of the columnar crystals. Measurements of the lengths of these crystals on both sides of the axial face were obtained at 6-in. intervals from the base of the ingot and are given in Table II.

Attention is drawn to the values for crystal length at the two sides of the tilted-mould ingot, *QXZ*. Of the other ingots the shortest crystals are found in the poker-stirred one, *QXY*, and the longest in the multi-hole-tundish ingot, *QXU*, and, less uniformly, in the bottom-cast ingot, *QXS*. The columnar crystals of *QXX* are also short, as might be expected, but almost as short are those in the sand-mould ingot, *QXW*. The differences in columnar crystal length in the different chill-cast ingots are clearly related to the degree of turbulence, but in the case of the sand-cast ingot it is evident that it is the lower temperature gradient which has enabled crystallisation to occur on the inside soon after it had commenced on the outside of the ingot; the influence of the lower temperature gradient will be discussed later.

Measurements of the orientation of the columnar crystals relative to the normal to the side of the ingot were obtained using a bevel protractor. Although slight differences in orientation occurred between adjacent crystals, the values given in Table III. represent average values and indicate the magnitude of the differences between the ingots.

These figures will be discussed in Section III., but attention may be drawn at this stage to the values for ingots *QXX* and *QXZ* in which it would appear that a downward stream of metal adjacent

TABLE II.—*Length of Columnar Crystals, in Inches.*

Position in Ingot.	*QXS.* Bottom-Cast.*		*QXT.* Single Stream down Centre.		*QXU.* Multi-Hole Tundish.		*QXW.* Sand Mould.		*QXX.* Stream down One Side.		*QXY.* Poker-Stirred.		*QXZ.* Tilted Mould.	
	Left.	Right.	Left.	Right.	Left.	Right.	Left.	Right.	Left.	Right.	Left.	Right.	Left.	Right.
Top of ingot.	0·6	1·0	1·0	1·3	0·8	1·1	1·0	0·8	0·6	0·4	0·9	0·7	0·4	5·8
From base—														
3 ft. 6 in.	0·7 (0·7–1·8)	1·0 (1·0–1·7)	1·2	1·4	1·2	1·4	1·0	0·9	0·8	0·6	0·8	0·7	0·5	5·7
3 ft.	0·8 (0·7–1·7)	1·1 (1·0–1·9)	1·4	1·5	1·3	1·3	0·9	0·9	0·9	0·7	0·8	0·8	0·5	4·7
2 ft. 6 in.	1·8 (0·9–2·0)	2·0 (1·4–2·2)	1·4	1·5	1·5	1·6	1·0	1·0	0·9	0·7	0·9	0·9	0·6	3·5
2 ft.	1·9 (0·9–2·3)	1·9 (1·3–2·3)	1·5	1·5	1·6	1·7	1·0	1·0	1·1	0·8	0·9	0·9	0·9	2·4
1 ft. 6 in.	1·9 (1·6–2·0)	1·9 (1·5–2·1)	1·6	1·6	1·6	1·8	1·0	1·0	1·1	1·0	1·0 (1·3)	1·0 (1·3)	1·1	2·1
1 ft.	1·8	1·9	1·5	1·7	1·8	2·0	1·1	1·2	1·2	1·2	1·0	1·1	1·2	2·0
6 in.	1·7	1·7	1·7	1·8	2·0	2·0	0·9	1·1	1·3	1·3	1·1	1·1	1·3	1·6
Base of ingot.	1·1		1·6		2·0		0·6		1·6		1·3		1·6	

* The demarcation columnar/equi-axial crystals in the bottom-cast ingot was less pronounced than in the other ingots; the figures in brackets show the range of the columnar crystal length at the different positions.

to the mould wall results in an upward inclination of the crystals at that face, and conversely an upward current results in a downward inclination of the crystals.

Table III.—*Inclination of Columnar Crystals from the Normal to the Ingot Face.*

+ indicates upward, − downward inclination.

Ingot Mark and Type.	Left Side.	Right Side.
QXS. Bottom-cast.	−5° at top 1 ft.; 0° in middle; +2° in lower third to +1° at bottom 6 in.	+10° at top 6 in.; then to +7° in lower half to 0° at bottom 6 in.
QXT. Single stream down centre.	+8°	+8°
QXU. Multi-hole tundish.	+10° in top half to +15° in lower half.	+7° in top half to +13° in lower half.
QXW. Sand mould.	+10°	+10°
QXX. Stream down one side.	0° at top to −3° in middle to −7° at bottom.	+2° at top to +7° in middle and +13° at bottom.
QXY. Poker-stirred.	−5° in top 1 ft.; 0° in middle to +4° in bottom 1 ft.	+1° in upper half to +3° in bottom half.
QXZ. Tilted mould.	0° at top 6 in.; +8° for 6 in. where stream touched; +6° to +8° in middle, to +2° in lower third and 0° at bottom.	+7° in top third; −3° in middle third and 0° in bottom third.

Examination of 1-*in. Bars.*

A further study of the macrostructure was carried out on the 1-in. bars cut from the axial slices, across the whole width of the ingot. To facilitate handling, the bars were cut in two along a line corresponding to the axis of the ingot; each bar was stamped 1, 2 or 3 to indicate whether it was from the upper, middle or lower part of the ingot, and also *A* or *B* to indicate the left- or right-hand half of the axial face. The axial face of each bar was then re-polished and etched. Photographs of a few selected structures are reproduced in Figs. 28 and 29. Attention may be drawn to the following characteristics: *QXS.2B* shows the irregular boundary of the columnar crystals in the bottom-cast ingot, *QXW.2A* the open texture of the primary structure of the sand-cast ingot, *QXX.1B* a patch of dendritic structure halfway along the bar and a change in direction of the short columnar crystals, *QXY.2B* bands parallel to the ingot face, *QXZ.0A* small equi-axial crystals, and *QXZ.0B* completely columnar crystals; these last two bars, designated *QXZ*.0, were cut from the top of the ingot below the feeder head, about 2 in. above the No. 1 position. The enlarged photograph of *QXY.1B* (Fig. 29 (*c*)) gives a good impression of the banding near the top of the ingot and also shows the distortion of the primary crystals in the upper part of the section near the first band; *QXX.3A* (Fig. 29 (*d*)) shows a periodic structure near the edge of the ingot and a change in the inclination of the columnar crystals at a faint periodic band at about two-thirds of their length measured from the ingot face. An explanation for such directional changes is offered in Section III.

Microstructure.

Micro-examination showed fairly close correlation between the ferrite-pearlite structure and the primary structure in all the ingots; only a few representative structures will be described or illustrated.

A study of the Λ-segregates showed them to consist for the most part of sharply defined dendrites, although the structure of the purer mass on either side of the segregate was of the nodular or granular type with less contrast in the degree of etching between the cores and inter-core material. The axes of the dendrites in the Λ-segregate appeared very dark against a light ferritic background, the ferrite-pearlite structure, both in the segregated and non-segregated areas, conforming largely with the lay-out of the primary structure. Typical structures are illustrated in Figs. 30 and 31, which show two areas less than 1 in. apart in bar *2B* from the bottom-cast ingot *QXS*. No previous account of this observation has been found in the literature, although several samples have been noticed, and further reference to it will be made later. Other samples of these dendritic patches of Λ-segregate may be seen in Figs. 22, 24, 28, &c.

The sand-cast ingot (Fig. 32) showed least correlation between the primary and tertiary structures; evidence of metal having run between the original ingot shell and the mould, as suggested by the above study of ingot surface, was also given by the appearance of the bar near the ingot face, *see* Fig. 33.

Some difficulty was experienced in the microscopical examination of the bands in the stirred ingot *QXY* owing to the fact that the primary structure mainly responsible for the banded appearance became less pronounced as the magnification was increased. The distortion of the columnar crystals observed in some of the bars from this ingot and illustrated in Fig. 29 was found to have its effect on the ferrite-pearlite structure, the ferrite boundaries curving round between the axes of the curved columnar crystals, *see* Fig. 34. (The structure in the micrograph is reversed.)

It is suggested that the bending over of the ends of the columnar crystal axes is due to mechanical deformation by a blow from the bar used in stirring, since only some of the crystals in this ingot were distorted so and none of the crystals in the other ingots were. It is worth while pointing out that there appeared to be no evidence of any part of the columnar crystals having been broken off, so that it is unlikely that the dendrites in the Λ-segregate of ingot *QXX* are the result of portions of columnar dendritic crystals removed by the liquid stream and carried into the interior of the ingot.

The only remaining microstructures to be considered here are those from the small equi-axial and the wholly columnar regions near the top of the tilted ingot, *QXZ*. Figs. 35 and 36 show representative areas of the bars *QXZ*.0*A* and *QXZ*.0*B* illustrated in Fig. 29, and the dependence of the ferrite-pearlite structures on the primary structures will be observed.

Grain Size.

Determinations of inherent grain size as shown by the McQuaid-Ehn test were made upon samples cut from the outermost and innermost portions of the bar 2*A* from each ingot. The structure of all samples corresponded to a grain size number of 5, although the crystals from the middle of the ingots were less uniform in size than those only ½ in. from the ingot face, the number of grains per sq. in. divided by 100 being, respectively, 15–20 and 20–24.

Tests were also made upon two samples from each of the bars *QXZ*.1*A* and *QXZ*.1*B*, the primary structures of which were, respectively, small equi-axial and wholly columnar. The surfaces examined after the carburising treatment were originally 1½ and 4½ in. distant from the face of the ingot. Here, again, no noticeable difference was observed in the grain size of the four different samples, the grain size number of each being 5.

The results clearly show that in these ingots, which were cast from the same melt, the inherent grain size as determined by the McQuaid-Ehn test was not influenced at all by the primary crystal structure, nor by the method of casting.

Mechanical Properties.

Six tensile test-pieces (3½ in. × ⅝ in. in dia. overall, acting length 1·4 in. × 0·357 in. in dia.) were obtained from the axial slice near the *A* and *F* analysis positions as shown in Fig. 4, in which the dotted line indicates the approximate limit of columnar growth in most of the ingots. An additional set of six test-pieces was cut from the other side of the unsymmetrical tilted ingot *QXZ*. The results, given in Table IV., indicate that (1) the optimum properties in both tensile strength and ductility were shown by the samples composed of columnar crystals of the primary structure; (2) the lowest values were obtained with specimens from position 4 nearest the axis; (3) the ultimate stress values of the sand-cast ingot were 2–5 tons per sq. in. less than those of the other ingots and the ductility was also lower; (4) the properties of the normally cast ingot *QXT* compared favourably with those cast by the other methods and were

TABLE IV.—*Tensile Properties.*

Ingot.	Position of Test-Piece.	0·2% Proof Stress. Tons per sq. in.	Max. Stress. Tons per sq. in.	Elongation. %.	Reduction of Area. %.	Primary Structure.
QXS. Bottom-cast.	1	18·5	40·1	20	19	Columnar
	2	17·6	40·3	18	20	"
	3	17·0	38·4	7	7	Equi-axial
	4	15·5	37·7	7	4	"
	5	17·0	39·9	10	11	"
	6	16·0	39·8	9	7	"
QXT. Single stream down centre.	1	18·0	39·8	19	21	Columnar
	2	17·0	39·8	18	21	Columnar + equi-axial
	3	16·6	38·7	11	14	Equi-axial
	4	16·2	37·3	11	9	"
	5	17·0	39·2	10	9	"
	6	16·0	38·6	10	12	"
QXU. Multi-hole tundish.	1	...	...	...	...	Flaw
	2	17·0	39·5	15	14	Columnar
	3	16·8	38·4	8	7	Equi-axial
	4	15·5	36·6	9	7	"
	5	16·9	39·3	11	11	"
	6	16·5	39·3	11	20	"
QXW. Sand-cast.	1	16·6	38·4	13	14	Columnar
	2	16·5	37·8	9	8	Equi-axial
	3	17·0	30·1	3	3	"
	4	16·3	35·4	8	8	"
	5	16·5	35·6	7	7	"
	6	16·7	33·8	5	4	"
QXX. Stream down one side	1	18·0	39·5	19	22	Columnar
	2	17·5	40·5	16	19	"
	3	17·0	38·0	6	6	Equi-axial
	4	15·5	35·9	9	9	"
	5	16·0	37·8	7	6	"
	6	17·0	37·6	5	7	"
QXY. Poker-stirred.	1	17·5	39·5	17	19	Columnar
	2	16·2	38·5	7	8	Semi-columnar
	3	15·0	35·7	6	7	Equi-axial
	4	14·6	36·3	9	4	"
	5	17·7	39·0	7	6	"
	6	17·5	35·1	4	4	"
QXZ. Tilted mould.	1*A*	18·0	39·2	18	22	Columnar
	2*A*	17·0	40·2	18	23	Semi-columnar
	3*A*	15·5	38·5	10	9	Equi-axial
	4*A*	15·5	36·9	10	11	"
	5*A*	16·2	39·2	9	11	"
	6*A*	17·0	39·3	8	11	"
	1*B*	17·5	39·4	13	13	Columnar
	2*B*	17·0	40·0	15	14	"
	3*B*	16·5	39·1	9	12	Equi-axial
	4*B*	15·5	36·6	8	9	"
	5*B*	16·7	40·0	12	17	Mostly columnar
	6*B*	16·5	39·4	11	12	" "

only fractionally below those of the bottom-cast ingot which was the best of all; and (5) the tensile properties of the steel in the as-cast condition were very much less affected by the methods of casting than by the position in the ingot from which the test-piece was obtained.

In view of the discussion which has taken place from time to time as to the relative properties of columnar and equi-axial crystal aggregates in steel ingots, the upper half of the tilted ingot, *QXZ*, which was composed of columnar crystals on one side and equi-axial on the opposite side, was submitted to a detailed examination, the results of which are listed in Table V. The slice examined was actually cut from that half of the ingot not containing the axial face. The face of this slice adjacent to the axial face was polished and macro-etched and then marked out to be cut into a large number of samples for tensile, hardness and notched-bar impact tests. The positions of the various test-pieces and the division

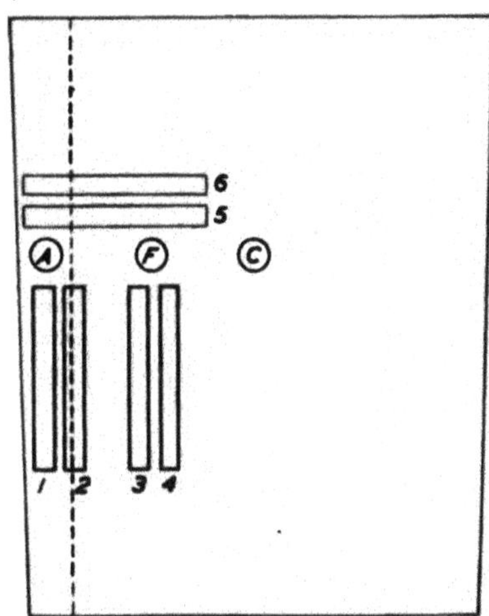

Fig. 4.—Positions of Tensile Test-Pieces taken from near the standard analysis positions *A* and *F*.

between the columnar and equi-axial types of primary crystals are indicated in Fig. 5. Samples which were annealed before testing are indicated by the letter *A*. Photographs of the macrostructures are not reproduced, but reference to Figs. 27 and 37 will show the type of structure examined, bearing in mind that the structure on the slice was reversed from that shown in the photograph, *i.e.*, the columnar portion was on the left of the slice.

There is, in general, a greater uniformity in the properties of the columnar-crystal samples than of the equi-axial ones in both the as-cast and annealed state, the properties of the equi-axial crystal samples being slightly superior near the edge of the ingot, but falling off more rapidly as the centre is approached.

To make the comparison still more complete, a large number of tensometer test-pieces were cut from a section right across the top of the axial slice of this ingot. Test-pieces were obtained parallel and perpendicular to the direction of growth of the columnar crystals. The positions of the samples and types of structure are shown diagrammatically in Fig. 6. Four samples were cut from each position; two were tested in the as-cast condition and two after

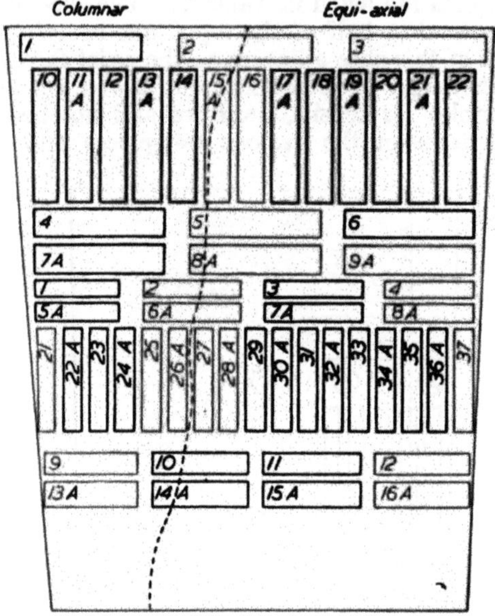

Fig. 5.—Positions of Samples for Tensile (large) and Notched-Bar (small) Tests from Ingot QXZ.

Fig. 6.—Tensometer Test-Pieces from Ingot QXZ.

annealing at 900° for ½ hr. and slowly cooling in the furnace. The results of these tests are given in Table VI.

According to these results it would appear that the advantage, if any, lies with the samples composed of columnar crystals, since the properties across the ingot are more uniform here than in the equi-axial region. The results need to be interpreted with some care,

Plate II.

U, cast with multi-hole tundish; macrostructure. $\times \frac{1}{3}$.

Fig. 23.—Ingot *QXW*, sand-cast; macrostructure. $\times \frac{1}{3}$.

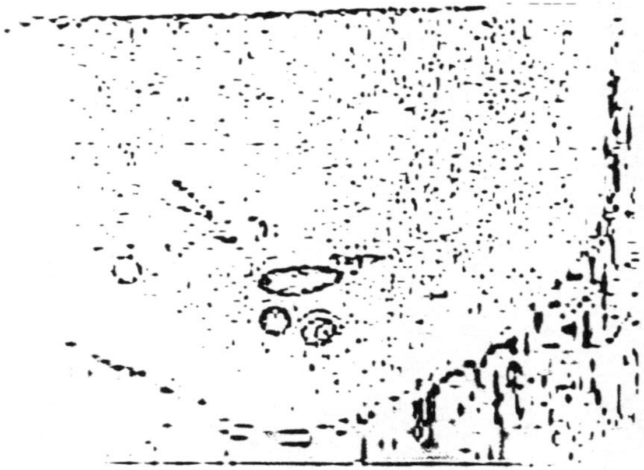

PLATE IV.

FIG. 27.—Ingot *QXZ*, cast in sloping mould; macrostructure. × ½.

FIG. 26.—Ingot *QXY*, stirred with poker; macrostructure. × ½.

PLATE V.

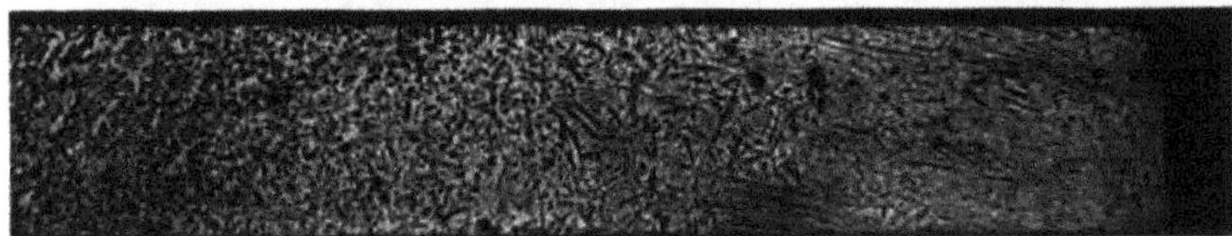

(a) Bar *QXS.2B*, bottom-cast.

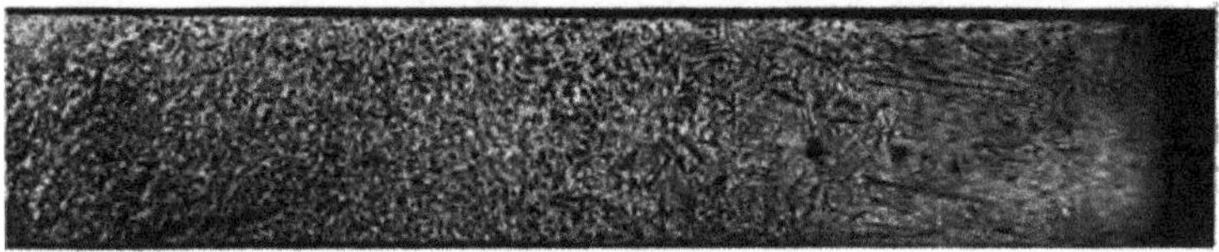

(b) Bar *QXT.2B*, top-cast.

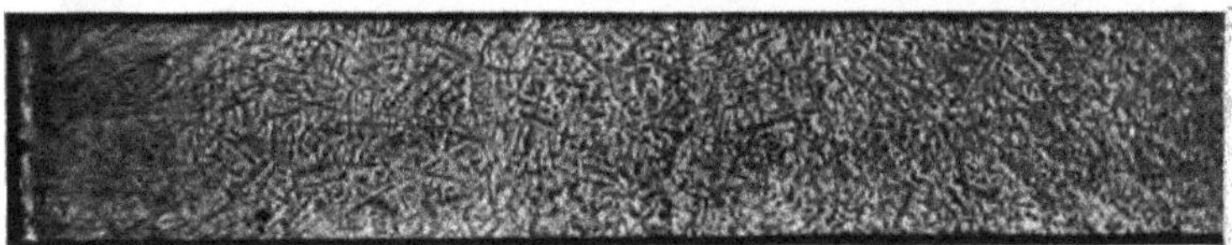

(c) Bar *QXW.2A*, sand-cast.

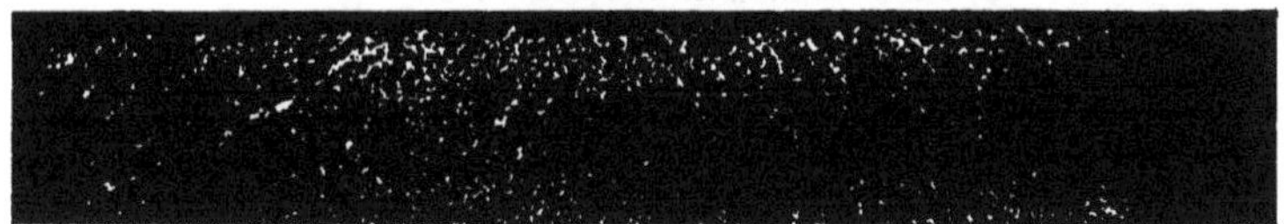

(d) Bar *QXX.1B*, cast with stream near one side.

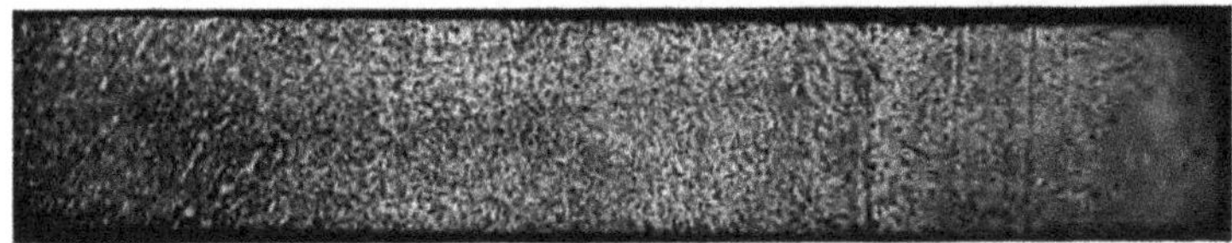

(e) Bar *QXY.2B*, stirred with poker.

FIG. 28.—Half Transverse Slices from Axial Faces of Steel Ingots. × 1.

(Illustrations reduced to two-thirds linear in reproduction.)

[*Northcott.*

PLATE VI.

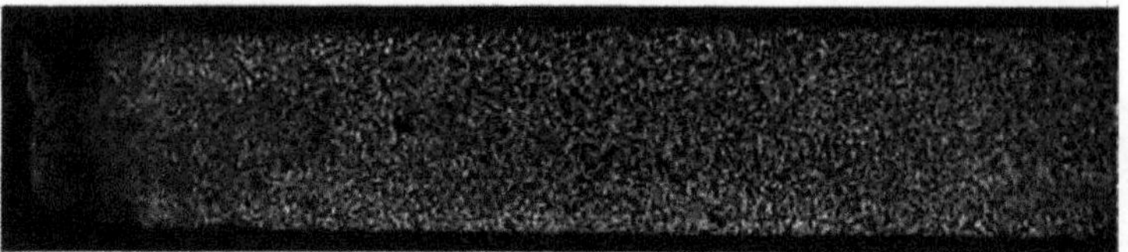

(*a*) Bar *QXZ.0A*, cast in sloping mould. × 1.

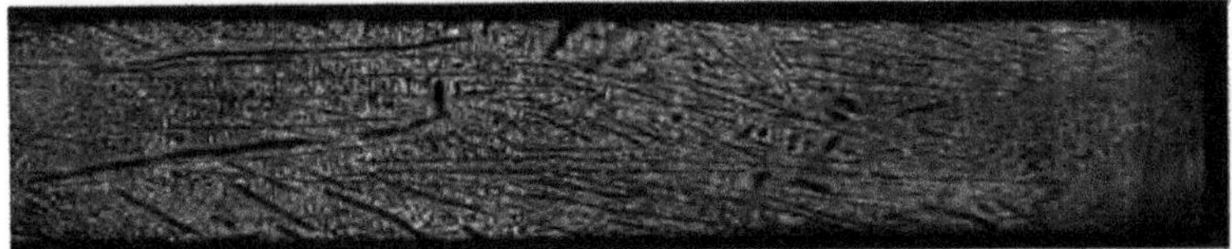

(*b*) Bar *QXZ.0B*, cast in sloping mould. × 1.

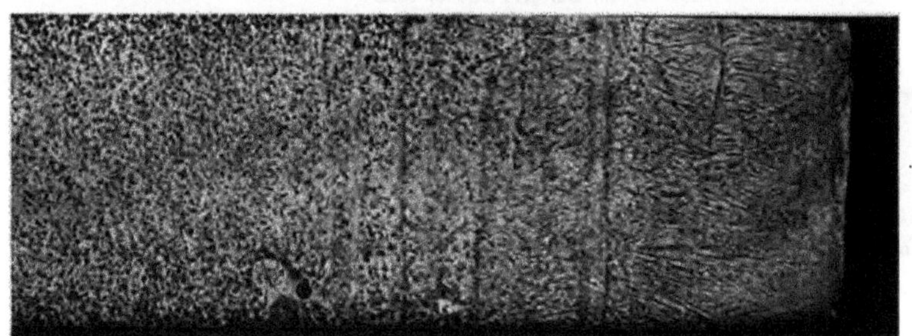

(*c*) Bar *QXY.1B*, stirred with poker. × 2.

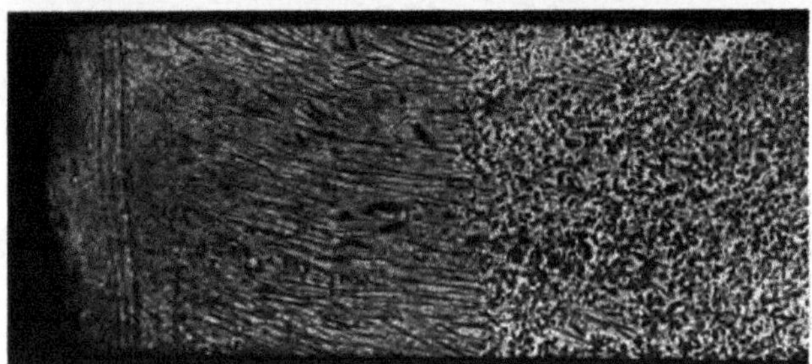

(*d*) Bar *QXX.3A*, cast with stream near one side; entered metal beyond the right-hand margin of the illustration. × 2.

FIG. 29.—Portions of Transverse Slices from Axial Faces of Steel Ingots.

(Illustrations reduced to two-thirds linear in reproduction.)

[*Northcott.*

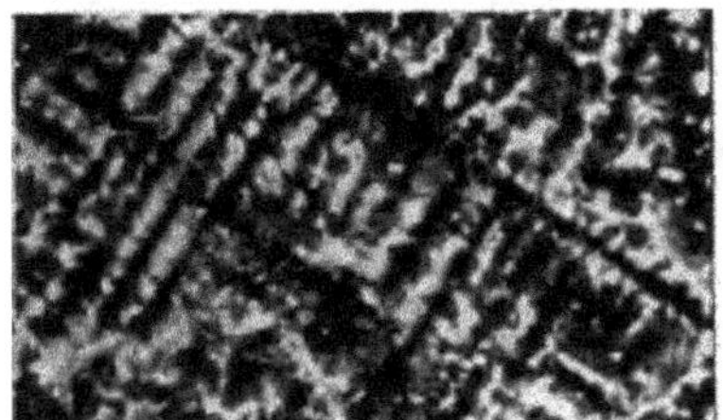

FIG. 30.—Ingot *QXS*; dendritic structure in Λ-segregate. × 15.

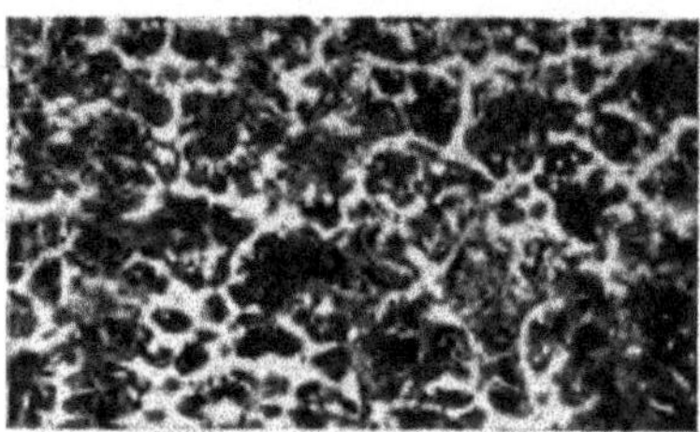

FIG. 31.—Ingot *QXS*; nodular structure away from Λ-segregate. × 15.

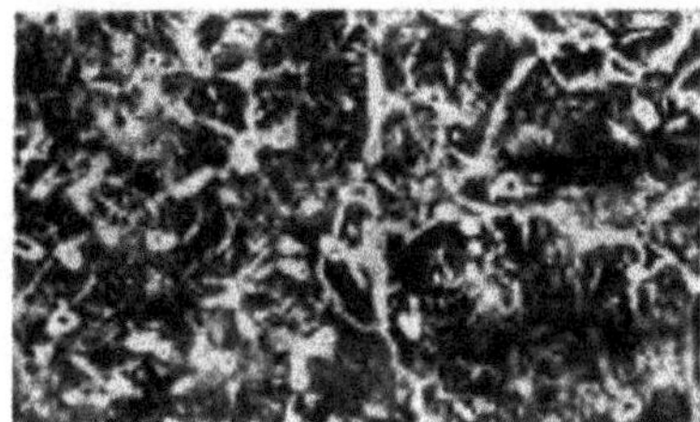

FIG. 32.—Ingot *QXW*, sand-cast. × 15.

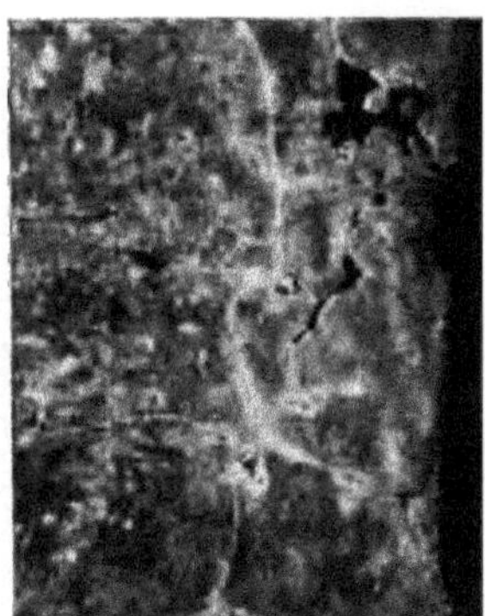

FIG. 33.—Ingot *QXW*, sand-cast; edge. × 15.

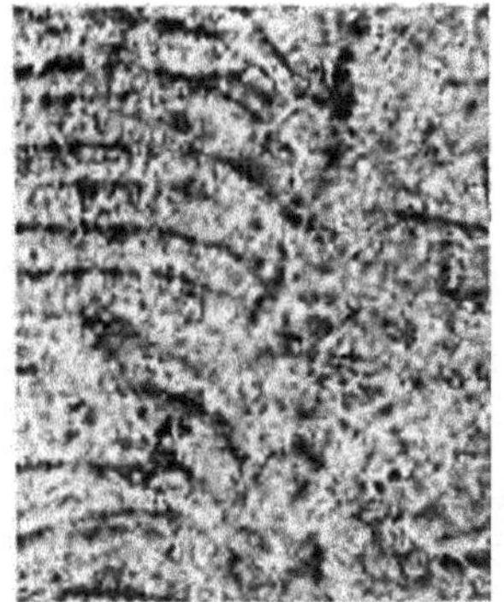

FIG. 34.—Ingot *QXY*, poker-stirred; distortion of dendrites. × 15.

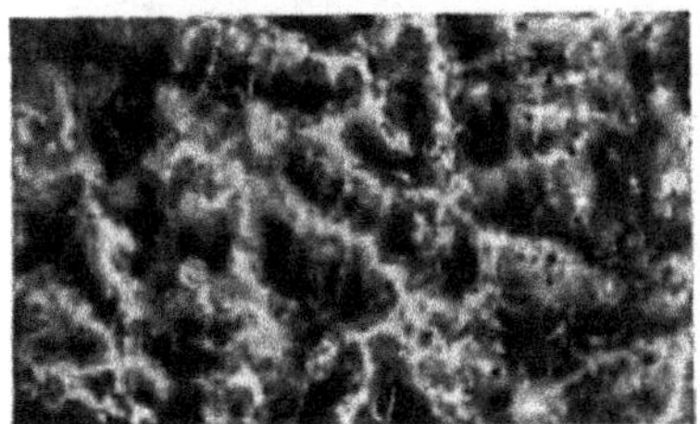

FIG. 35.—Bar *QXZ.0A* from sloping mould ingot; cored granular structure in equi-axial areas. × 15.

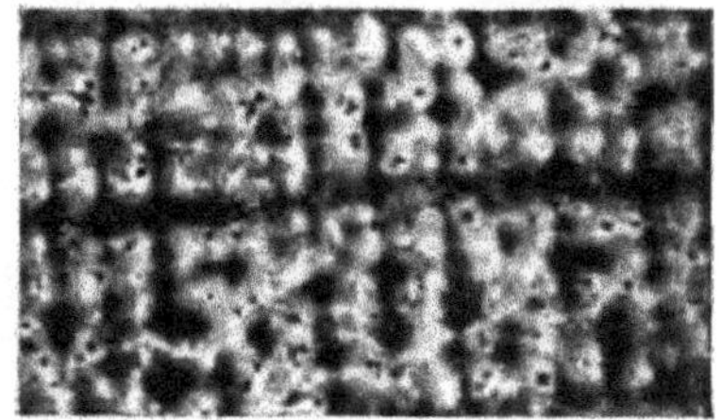

FIG. 36.—Bar *QXZ.0B* from sloping mould ingot; dendritic structure in columnar-crystal areas. × 15.

(Illustrations reduced to two-thirds linear in reproduction.)

[Northcott.

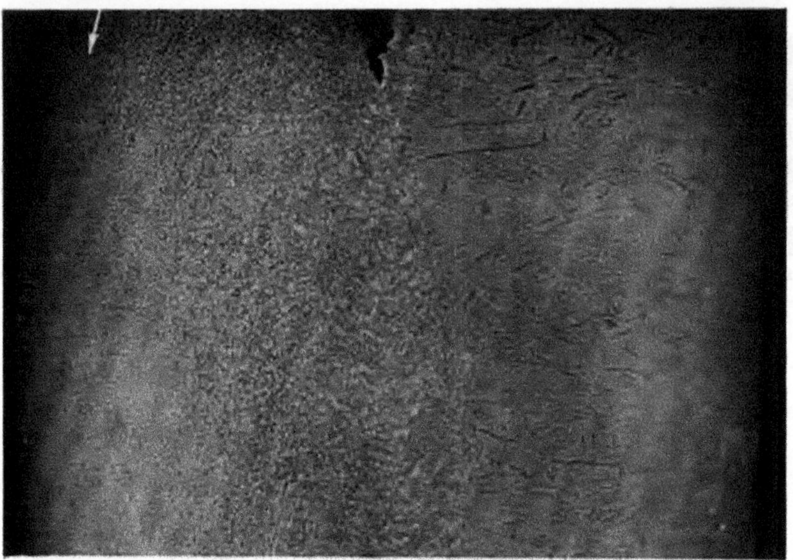

FIG. 37.—Ingot *QXZ*, cast in sloping mould; upper section of axial face. × ½.

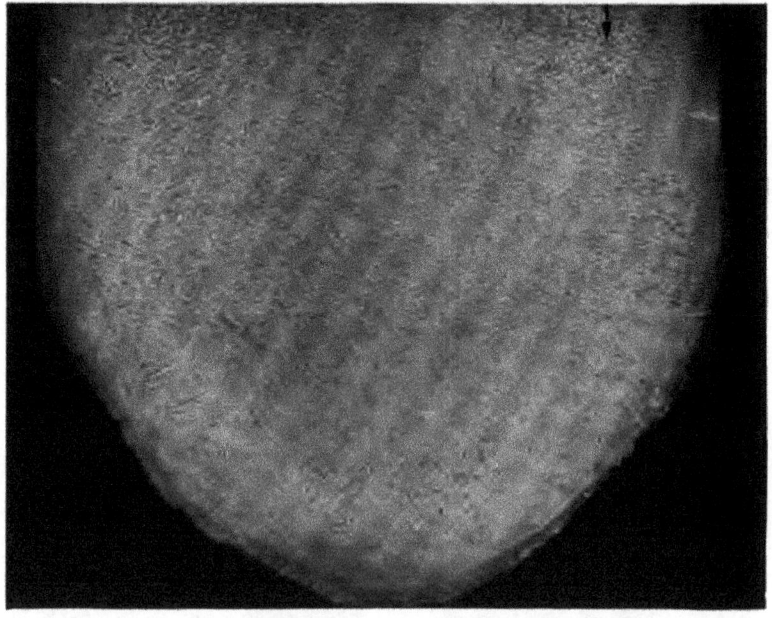

FIG. 38.—Ingot *QXX*, cast with stream near one side; lower section of axial face. × ½.
(Illustrations reduced to four-fifths linear in reproduction.)

[*Northcott.*

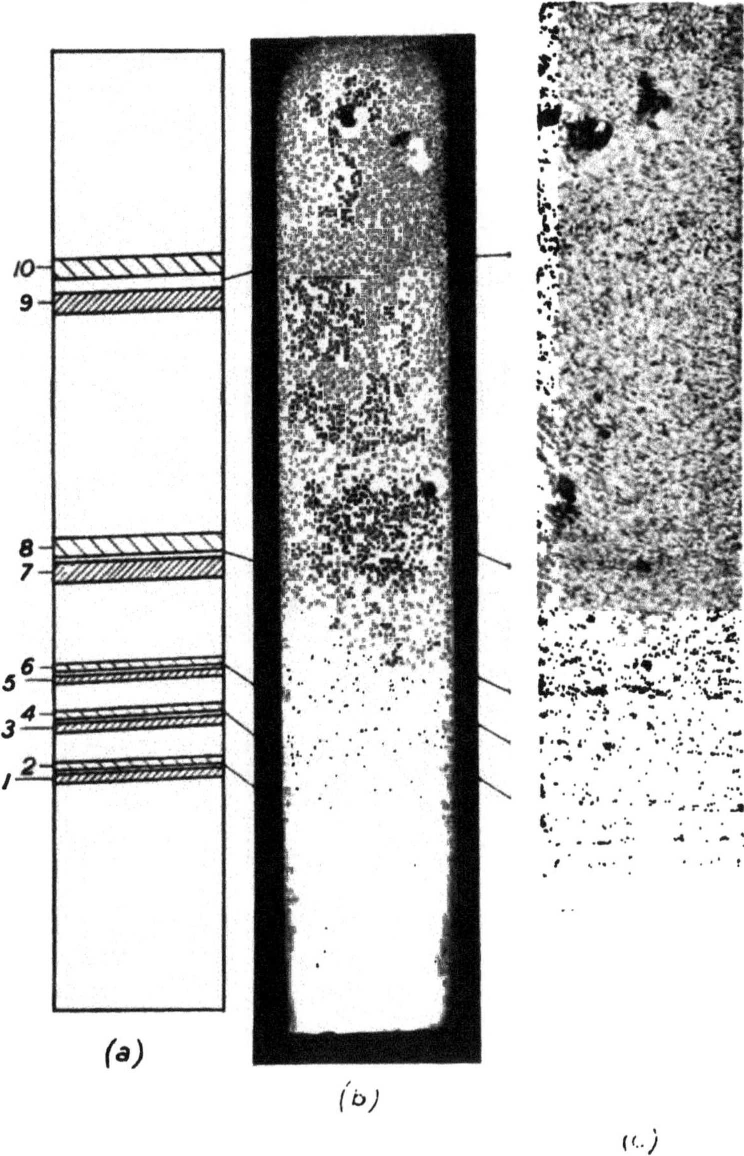

FIG. 39. [illegible]

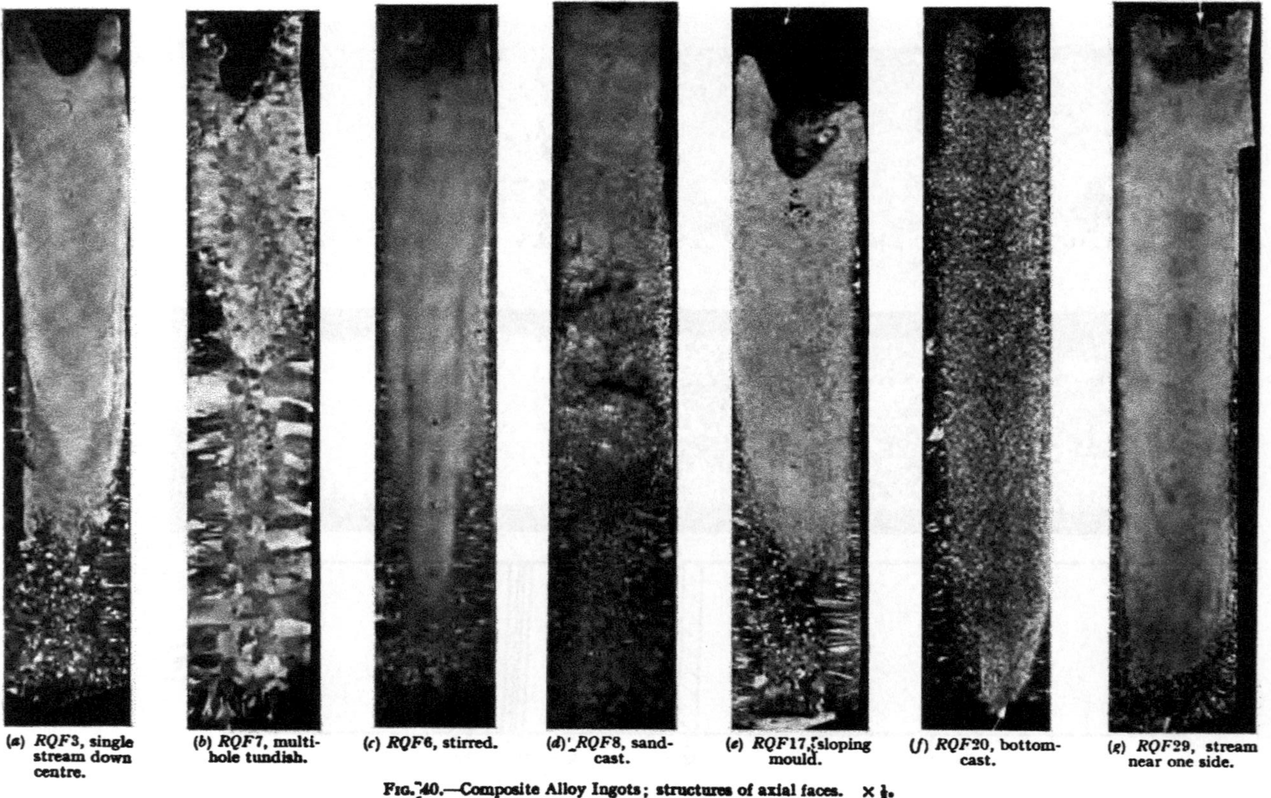

(a) *RQF*3, single stream down centre.

(b) *RQF*7, multi-hole tundish.

(c) *RQF*6, stirred.

(d) *RQF*8, sand-cast.

(e) *RQF*17, sloping mould.

(f) *RQF*20, bottom-cast.

(g) *RQF*29, stream near one side.

FIG. 40.—Composite Alloy Ingots; structures of axial faces. × ⅓.

(Illustrations reduced to two-thirds linear in reproduction.)

[*Northcott.*

. 41.—Water Pouring into a Vessel of Water.

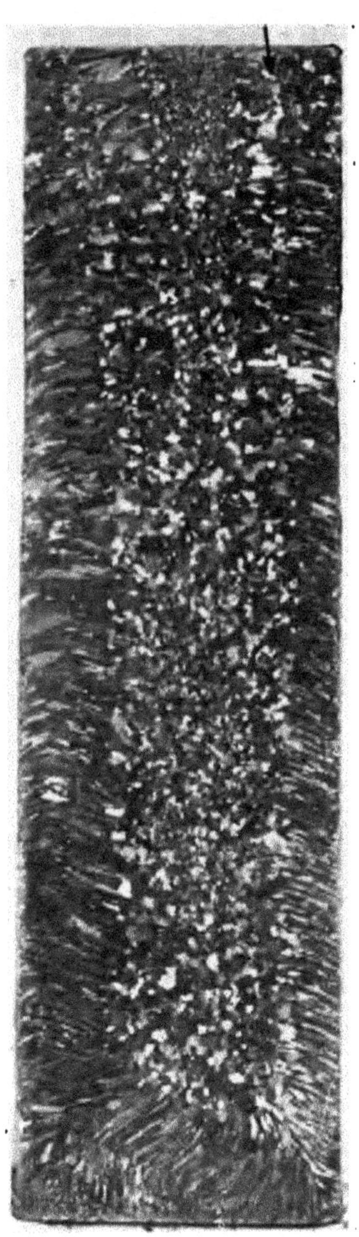

FIG. 42 —Lead Ingot, cast with stream down one side, macrostructure of axial face. × ⅗.

[Northcott.

however, since there are two other factors which have to be borne in mind. In the first place the relative soundness or freedom from porosity of samples of the two types of structures should be considered; the values for ductility indicate a more rapid falling off in soundness in the equi-axial sample as the centre of the ingot is

TABLE V.—*Comparison of Mechanical Properties of Test-Pieces Composed of Columnar or Equi-Axial Primary Crystals.*

Position.	0·2% Proof Stress. Tons per sq. in.	Maximum Stress. Tons per sq. in.	Elongation. %.	Reduction of Area. %.	Position.	0·2% Proof Stress. Tons per sq. in.	Maximum Stress. Tons per sq. in.	Elongation. %.	Reduction of Area. %.
				As cast.					
1	17·0	38·6	7	7	10	18·5	38·4	11	9
2	17·0	33·5	5	2	12	17·2	38·5	8	7
3	17·0	37·9	8	7	14	17·5	39·1	7	5
4	16·7	39·3	8	8	16	16·0	31·3	4	4
5	16·0	34·3	5	5	18	16·7	36·6	8	6
6	17·5	39·9	13	12	20	16·7	38·7	10	8
					22	18·5	39·9	20	25
				Annealed.					
7	22·0	40·9	22	32	15	23·0	41·6	15	16
8	22·0	35·9	9	11	17	22·0	35·0	7	9
9	23·0	41·3	25	36	19	22·0	40·9	20	12
11	23·0	40·7	23	29	21	23·0	41·0	19	32
13	23·0	42·1	24	26					

Position.	Izod Value. Ft. per lb.	Position.	Izod Value. Ft. per lb.	Position.	Izod Value. Ft. per lb.	Position.	Izod Value. Ft. per lb.
	As cast.				*Annealed.*		
1	4	21	3	5	10¾	22	8½
2	2	23	2	6	3¾	24	6
3	2½	25	2	7	9	26	7
4	4½	27	1¾	8	12	28	9
9	6¼	29	2	13	10½	30	7
10	2½	31	2	14	8¾	32	6¾
11	2½	33	2	15	9¼	34	6½
12	4¼	35	4	16	11½	36	7½
		37	2¾				

approached and this is also suggested from the results of tensile tests on samples taken from the *A* and *F* positions, as described above. Micro-examination of representative areas showed an increase in number of inclusions and intercrystalline fissures in the middle of the ingot as compared with the edge, and there was possibly more evidence of unsoundness in the equi-axial than in the

TABLE VI.—*Tensometer Tests on Columnar and Equi-Axial Primary Crystal Samples.*

Position.	Yield Point. Tons per sq. in.	Max. Stress. Tons per sq. in.	Elongation. %.	Reduction of Area. %.	Position.	Yield Point. Tons per sq. in.	Max. Stress. Tons per sq. in.	Elongation. %.	Reduction of Area. %.
As cast.									
Columnar.					*Equi-axial.*				
1	25·0 25·5	44·2 41·2	16 9	20 8	15	22·5 20·5	37·0 42·5	10 14	13 16
2	24·5 23·5	44·9 45·1	20 21	25 25	14	23·1 24·0	46·0 46·4	18 18	18 20
3	26·5 26·7	45·2 44·6	15 15	12 10	13	25·0 23·6	44·8 44·5	14 15	12 12
4	27·8 25·6	46·2 45·9	15 13	10 10	12	25·2 26·0	43·9 40·6	11 8	12 3
5	26·0 26·0	45·4 45·2	14 11	10 8	11	26·8 24·0	42·2 43·0	9 10	5 7
6	20·0 22·5	28·8 32·3	6 5	2 2	10	24·0 25·0	39·4 34·2	6 6	4 3
7	23·7 26·2	43·0 46·3	10 9	5 5	9	23·7 25·0	44·8 44·5	12 10	8 7

Position.	Yield Point. Tons per sq. in.	Max. Stress. Tons per sq. in.	Elongation. %.	Reduction of Area. %.	Position.	Yield Point. Tons per sq. in.	Max. Stress. Tons per sq. in.	Elongation. %.	Reduction of Area. %.
Annealed.									
Columnar.					*Equi-axial.*				
1	26·9 26·0	46·3 45·6	17 25	17 25	15	26·8 25·2	47·3 42·0	23 12	28 10
2	25·5 25·8	46·0 47·0	22 29	32 33	14	27·5 27·0	47·1 46·9	25 27	37 32
3	26·2 26·3	46·8 47·1	20 22	23 25	13	26·3 26·0	46·6 46·7	22 24	24 24
4	25·6 26·5	47·0 46·9	23 21	25 22	12	27·0 26·5	47·0 46·0	21 17	25 17
5	26·8 26·5	47·8 47·5	21 19	22 23	11	26·8 26·2	47·2 46·0	16 15	15 15
6	27·0 27·6	47·9 48·0	15 17	15 13	10	26·4 27·2	43·0 41·5	10 10	12 10
7	27·3 27·7	47·5 40·5	11 9	10 8	9	26·5 26·0	40·3 42·3	10 10	5 7

columnar bar though the difference was not great; also there were rather more small inclusions in the equi-axial samples.

The second point concerns the crystal structure of the material as tested. The columnar or equi-axial nature of the steel only refers to its form as it originally existed at temperatures above the allotropic change points. At room temperature the steel is composed of a mixture of ferrite and pearlite crystals so that, if we assume the metal to be uniformly sound, any difference in mechanical properties will be due to the influence of the primary structure upon either (*a*) the form of the final ferrite-pearlite crystals or (*b*) the disposition of the non-metallic impurities or cavities. The actual grain size of the steel in the cast condition was observed to increase towards the middle of the ingot, but although there was no difference in size between representative areas from the equi-axial and columnar bars there was a greater precipitation of ferrite in the equi-axial samples. In spite of this the average ductility of the equi-axial test-samples is lower, so that they are presumably slightly less sound than the corresponding columnar ones.

It was thought advisable to include these observations, since the results of work being carried out on several non-ferrous engineering alloys which do not undergo polymorphic changes have shown that the mechanical properties of test-pieces composed of a large number of very small crystals are appreciably superior to those of pieces composed of large crystals, either columnar or equi-axial.

Hardness determinations on *QXZ* bars composed of columnar and equi-axial primary crystals in the as-cast condition and after several heat treatments showed no significant differences in hardness between the two structures.

Chemical Segregation.

Examination for heterogeneity was made by the chemical analysis of samples taken from the seven standard positions in each of the seven ingots. The results are given in Table VII.

The different elements are not distributed identically in the different ingots, but the carbon figures show the general trend of the segregation, and as the greatest differences are found with this element, it will suffice to consider these figures only. The purer cone usually found in the lower central region of steel ingots has been displaced in ingots *QXX* and *QXZ* in which the stream was not an axial one. In the tilted-mould ingot *QXZ*, there is evidence of a low-carbon zone along a line joining the *F* and *D* positions and of an impure zone corresponding with the Λ-segregate from the *A* to *G* positions. Ingot *QXT* shows the most uniform distribution of carbon, but it is considered doubtful whether this would be confirmed if a more detailed survey had been undertaken; for example, sample *F* is evidently not representative of the Λ-segregate in this ingot. Actually the bottom-cast ingot *QXS* is the only one in which there is a high carbon content at position *F*, and it is probable that more representative results would have been obtained had positions *F* and *G* been selected for each ingot. High and low values for the different elements are being checked. It is clear, however, that the method of casting has influenced the distribution of the various elements, and viewing the results as a whole it would appear that the multi-hole-tundish ingot shows the least all-round segregation.

TABLE VII.—*Chemical Analyses.*

Ingot.		Element.	*A.*	*B.*	*C.*	*D.*	*F.*	*G.*	*H.*
QXS.	Bottom-cast.	C. %	0·425	0·394	0·420	0·442	0·457	0·417	0·442
		Si. %	0·089	0·082	0·086	0·098	0·074	0·079	0·079
		Mn. %	0·777	0·737	0·782	0·762	0·762	0·787	0·767
		S. %	0·033	0·028	0·027	0·026	0·022	0·020	0·027
		P. %	0·059	0·057	0·052	0·055	0·061	0·065	0·061
QXT.	Single stream down centre.	C. %	0·399	0·401	0·414	0·391	0·397	0·401	0·428
		Si. %	0·072	0·074	0·077	0·076	0·081	0·080	0·077
		Mn. %	0·802	0·762	0·767	0·762	0·772	0·772	0·777
		S. %	0·041	0·032	0·032	0·033	0·038	0·032	0·035
		P. %	0·063	0·052	0·052	0·051	0·053	0·055	0·057
QXU.	Multi-hole tundish.	C. %	0·440	0·391	0·391	0·392	0·398	0·418	0·499
		Si. %	0·078	0·077	0·076	0·074	0·076	0·083	0·078
		Mn. %	0·772	0·757	0·752	0·757	0·792	0·772	0·772
		S. %	0·039	0·032	0·030	0·031	0·034	0·031	0·028
		P. %	0·062	0·052	0·051	0·051	0·059	0·055	0·061
QXW.	Sand mould.	C. %	0·428	0·384	0·388	0·427	0·413	0·413	0·488
		Si. %	0·080	0·079	0·073	0·449*	0·083	0·082	0·083
		Mn. %	0·782	0·757	0·767	0·851*	0·720	0·780	0·770
		S. %	0·021	0·038	0·024	0·036	0·024	0·033	0·040
		P. %	0·051	0·051	0·050	0·055	0·052	0·063	0·067
QXX.	Stream down one side.	C. %	0·439	0·433	0·402	0·360	0·394	0·438	0·522
		Si. %	0·082	0·083	0·082	0·083	0·084	0·082	0·098
		Mn. %	0·790	0·780	0·790	0·800	0·780	0·770	0·735
		S. %	0·030	0·020	0·026	0·036	0·034	0·032	0·048
		P. %	0·063	0·049	0·053	0·054	0·059	0·059	0·065
QXY.	Poker-stirred.	C. %	0·417	0·359	0·405	0·410	0·416	0·369	0·427
		Si. %	0·083	0·084	0·085	0·089	0·084	0·072	0·080
		Mn. %	0·795	0·760	0·750	0·760	0·780	0·670	0·720
		S. %	0·038	0·026	0·029	0·034	0·037	0·036	0·033
		P. %	0·061	0·053	0·055	0·060	0·059	0·049	0·058
QXZ.	Tilted mould.	C. %	0·454	0·424	0·408	0·389	0·376	0·442	0·438
		Si. %	0·087	0·086	0·087	0·085	0·087	0·085	0·086
		Mn. %	0·790	0·780	0·780	0·760	0·770	0·790	0·770
		S. %	0·029	0·033	0·028	0·027	0·032	0·036	0·038
		P. %	0·062	0·055	0·055	0·051	0·059	0·062	0·064

* Sample containing slag.

A further study was made of the stirred ingot, *QXY*, and the tilted-mould ingot, *QXZ*. The 1-in. bar *QXY.2B* was sliced longitudinally into three pieces and *all* the surfaces of these were polished and etched to show the primary structure in order to detect whether the position of the bands was geometrically regular on all faces. This was observed not to be so, but the slice (now less than 0·25 in. thick) showing the most geometric pattern was selected for further tests. Using a fine-pointed shaping tool, small samples for analysis were machined from the slice at the ten positions shown in Fig. 39 (*a*), the depth of cut in all cases being not more than 0·1 in., and the width approximately 0·05 in. for samples 1 to 6 and 0·1 in. for the remainder. The structure of the slice before being machined is illustrated in Fig. 39 (*b*) and the sulphur print in Fig. 39 (*c*).

The results are given in Table VIII.

TABLE VIII.—*Composition of Bands in Stirred Ingot.*

Position.	Carbon. %.	Phosphorus. %.
10	0·43	0·066
9	0·41	0·052
8	0·47	0·063
7	0·45	0·065
6	0·47	0·073
5	0·45	0·058
4	0·43	0·067
3	0·41	0·066
2	0·45	0·065
1	...	0·053

In view of the difficulties in sampling there can be little doubt that the true composition differences are actually greater than those indicated in the Table, but the figures do suffice to show the direction of the changes in chemical composition.

It is evident from Fig. 39 and from the chemical analyses that the bands are less pure, and the metal adjacent to them on the ingot-mould side more pure, than the average composition of that portion of the ingot.

The tilted-mould ingot, *QXZ*, was then examined for variation in composition across the top where the structure was wholly columnar on one side of the axis and equi-axial on the other. The samples were obtained with a $\frac{3}{16}$-in. drill from a 1-in. square bar at intervals of 1 in. along the bar, a margin of $\frac{1}{8}$ in. being left at the ends (1 and 11). The results of these chemical analyses are given in Table IX.

TABLE IX.—*Composition across the Top of the Tilted-Mould Ingot* QXZ.

	Equi-axial.					Columnar.					
	1.	2.	3.	4.	5.	6.	7.	8.	9.	10.	11.
Carbon. % .	0·387	0·447	0·428	0·418	0·432	0·454	0·444	0·461	0·397	0·439	0·427
Silicon. % .	0·088	0·086	0·084	0·084	0·086	0·085	0·087	0·082	0·085	0·084	0·091
Manganese. % .	0·830	0·800	0·780	0·830	0·815	0·805	0·810	0·800	0·750	0·830	0·790
Sulphur. % .	0·034	0·037	0·033	0·035	0·032	0·037	0·036	0·033	0·033	0·037	0·038
Phosphorus. %	0·063	0·066	0·062	0·064	0·065	0·067	0·066	0·064	0·063	0·062	0·060

These results show that there is no significant difference in the range of chemical composition between the columnar and equi-axial samples, and that the equi-axial samples are at least as uniform in composition as the columnar ones.

II.—Composite Alloy Ingots.

One of the factors to be considered in explaining ingot solidification phenomena concerns the motion of the liquid inside the mould in dependence on the method of casting. The effect of the casting conditions on the turbulence of the liquid in the mould may be determined by examining a series of ingots cast by the different methods, as has been done in the present work, and by drawing conclusions from that examination as to the probable conditions existing in the ingot prior to and during solidification. It has also been suggested that the distribution of the liquid metal may be determined by making additions of a radioactive substance to the stream entering the mould, then sectioning the ingot and examining for the distribution of the radioactive material. Hadfield[4] added molten copper to the ingot a short time after casting in order to reveal the segregation, but, as pointed out by Stead, misleading results may be obtained by this method owing to the higher density of the copper.

Another method, now to be described, consists in preparing a series of ingots composed of two alloys of similar freezing point and density but of different colour, cast into the same mould one after the other. A red and a white alloy may be used, the red one being a copper-zinc alloy containing 90% of copper and the white one a nickel silver containing 50% of copper, 30% of zinc and 20% of nickel. Each alloy is melted in a separate crucible and then cast through a tundish held over the mould, one alloy being poured first and the other following before the last of the first alloy leaves the tundish. In this way the mould is filled by an unbroken uniform stream of metal. From an examination of such composite alloy ingots, positive evidence is obtainable as to the final distribution of the first-poured and last-poured metal in the ingot. Varying the proportions of the two alloys under a standard set of casting conditions provides further information as to the distribution of the various portions of the stream. This composite alloy type of casting has been adopted with success in the examination of slab ingots as used for brass strip manufacture.[5]

Casting Conditions.

In the present experiments, small ingots 9 in. long plus a 2½-in. feeder head and 2 in. square were cast under conditions simulating those used for the steel ingots described in the previous section. A tundish having a single hole ¼ in. in dia. was used except in the case of the multi-hole tundish, which had 9 smaller holes, and of the bottom-cast ingot. It is realised that the scale effect and the use of two non-ferrous alloys in place of steel prohibit the reproduction of solidification phenomena identical with those in the much larger steel ingots, and such reproduction was not intended, the purpose of these composite ingot experiments being merely to provide information upon the effect of the different casting conditions on the distribution of the metal stream in the mould. It is freely admitted that the value of the results would have been increased by the use of moulds of the same size as those adopted for the steel ingots, but the facilities were not available.

Ingots were prepared from alloys poured at 1200° C., first in the proportion of 75% red and 25% white, and then 50% red and 50% white, in both instances the red being poured first. The ingots were sectioned down the middle, and the axial faces prepared by polishing and etching to reveal the structure and distribution of the two alloys. The majority of the ingots were actually found to be composed of *three* differently coloured alloys—red, white and pink—this last appearing between the first two and showing the extent to which mixing and diffusion had occurred. In some instances the boundaries between the differently coloured alloys were not well defined.

Examination of Ingots.

Photographs of a few of the ingots are reproduced in Fig. 40. Ingot *RQF*3, single stream down centre, is of interest in showing the penetration of the final portions of the stream into the body of the ingot, the displacement of some of the earlier-poured alloy up the sides of the mould as far as the feeder head, and the fairly large proportion of pink alloy.

In striking contrast is the structure of the multi-hole-tundish ingot *RQF*7; this is characterised by an absence of the pink alloy, a clear demarcation between the white and the red alloys, and large crystal size. Each of these features may be explained as resulting from the considerably reduced turbulence experienced with this method of casting. The structure of the similarly prepared 50-red/50-white alloy was almost identical, differing only in the proportions of the two alloys.

The photograph of the macrostructure of the stirred ingot *RQF*6 reveals the existence of a considerable proportion of the pink alloy, and gives evidence of the extra mixing effect of the increased turbulence; there was also some general streakiness and a certain amount of faint banding along the sides. The sand-cast ingot *RQF*8 was the only one not showing any of the white alloy; the lower rate of solidification obviously permitted ample mixing of the white with the previously poured red material, such that more than half of the ingot was composed of the pink alloy. The ingot cast in the tilted mould, *RQF*17, shows an unsymmetrical structure in the red alloy in the lower half of the ingot with longer crystals growing from the "upper" face of the mould than from the lower face, corresponding to the differences in crystal length in the steel ingot *QXZ* cast in the tilted mould. The white alloy did not show this effect, presumably owing to the pronounced influence which nickel is known to have in reducing grain size.

Several bottom-cast ingots were prepared. In the case of the one illustrated, *RQF*20, for the reason given later, the bottom nozzle was intentionally set at a slight inclination to the vertical so that its axis cut the mould face about one-third to one-half way up from the base. This resulted in a slightly unsymmetrical disposition of the red and white alloys, as well as a difference in the orientation of the short columnar crystals on the two opposite faces of the ingot. The crystals on the right-hand side, where one may assume there was an upward flow of metal, are inclined slightly downwards, whereas on the opposite side the crystals have an upward inclination.

The ingot cast with the stream down one side, *RQF*29, is the only one illustrated which was composed of the two alloys in the 50/50 proportion, and was selected for reproduction in preference to the 75/25 ingot as showing more clearly the directional crystal orientation and the large proportion of the pink mixed alloy. The 75/25 ingot naturally consisted of more of the red alloy at the base, but also showed a larger proportion of red alloy along the side away from the stream than along the one nearest to it.

The significance of the work on composite ingots will be briefly considered in the next Section.

III.—Discussion of Experimental Results.

Assessment of Degree of Turbulence.

It is evident from the examination both of the 15-cwt. steel ingots and of the 13-lb. non-ferrous alloy ingots that the different casting conditions have exerted a pronounced influence on the ingot structure, and, excluding for the time being the sand-cast ingot, this effect is due solely to the alteration in the conditions of turbulence in the mould brought about by varying the method of pouring.

From the data on the lengths of columnar crystals given in Table II., there is reason to believe that the processes involving the least turbulence are those requiring the use of the multi-hole tundish or bottom-casting, since long columnar crystals are found in ingots prepared under either of these conditions. Much shorter crystals are found in the last three ingots, *QXX*, *QXY*, *QXZ*, respectively, stream down one side, poker-stirred and tilted mould, and it is to be expected that, with the exception of the last ingot, *QXZ*, turbulence would be most pronounced in these ingots. Ingot *QXT*, single stream down centre, cast under conditions more representative of standard practice, is intermediate in columnar crystal size between the two main groups. In ingot *QXZ*, cast in the tilted mould, the continuous flow of metal down the lower face constitutes the most pronounced form of turbulence at or near a mould face and has led to the smallest columnar crystal size. The structure of the metal grown from the opposite face indicates quiescent conditions of growth, and these may be likened to those existing in a bottom-casting using a large slowly running stream. The action of stirring after casting (*QXY*), by maintaining turbulent conditions, has reduced the size of the columnar crystals appreciably as compared with those observed in a similar ingot not stirred, *QXT*. That there is such a large difference in structure between ingot *QXX*, stream down one side, and *QXZ*, tilted ingot, which might be thought to have experienced similar turbulent conditions is presumably due to the partial dissipation of the force of the stream in the last ingot, *QXZ*, owing to the stream striking one of the mould faces. The subsequent wider spread of metal of lower energy is then more readily cushioned by the liquid already in the mould. The action may be compared to the method of filling a glass of beer by pouring from a bottle down the lower side of an inclined tumbler as against that in which it is poured into an upright tumbler; less turbulence and hence less frothing is experienced in the former method. The much reduced turbulence near the upper face of the steel ingot *QXZ* has resulted in the growth of long crystals from this face which was farthest from the stream.

The experiments made with the composite alloy ingots, in general, confirm these findings, but a few additional conclusions may be drawn as to the effects of the different casting conditions. For example, it is not possible from an examination of the steel ingots to determine the degree of penetration of the stream into the liquid already in the mould, but this is clearly shown by the composite-ingot method. Its use thus is valuable in demonstrating the comparative absence of major turbulence in the multi-hole-tundish process of casting, and the pronounced penetration of the stream and consequent displacement of the previously poured liquid in the case of ingots cast with a single stream. Furthermore, the reduction of general turbulence by the use of a tilted mould with the single stream is confirmed by the comparatively small proportion of the pink mixed alloy, both in the 75/25 and the 50/50 ingots.

There is one other observation to be made arising from the structure of the tilted mould ingot. The increase in crystal length on the " upper " face as the top of the ingot is approached is clearly due to the reduced turbulence of the liquid in this half of the ingot, together with the smaller temperature increase of the mould face here, since it was in contact for a shorter time with the flow of liquid steel. As the columnar crystals actually reach the *middle* of the ingot at the top, it is evident that under perfectly quiescent or non-turbulent conditions the primary structure of the whole of each ingot would be completely columnar. The difference in turbulence brought about by the different casting conditions has led to the columnar crystals having different lengths, but none of the ingots shows a completely columnar structure. It would appear that the multi-hole-tundish method of casting resulted in the least turbulence of the liquid in the mould, and, although the composite alloy ingot cast by this method shows very large columnar crystals, the steel ingot is by no means completely columnar, and this must be explained as being due to interference from the downward movement of the liquid along the axis of the ingot arising from the feeding of the ingot. The tendency towards completely columnar growth in these particular ingots is therefore insufficient to overcome the disturbing action of this mild flow during feeding.

Inclination of the Columnar Crystals.

In a brief comment on Table III., in which are given the values for the inclination of the columnar crystals, attention was directed to a conclusion drawn from ingots *QXX* and *QXZ* that a downward stream of metal adjacent to the mould wall appeared to result in an upward inclination of the crystals at that face and *vice-versa*, that is to say the inclination of the columnar crystals is in the reverse direction to that of the stream.* Phragmén arrived at this conclusion some years ago and Carlsson and Hultgren [6] have subsequently described a similar effect in small stearine ingots. A reference to the upward inclination of the columnar crystals has also been made by the Committee on the Heterogeneity of Steel Ingots.[7] A good example of the Phragmén effect in the present work is shown by the structure at the bottom of ingot *QXX*, stream down one side, in which the unsymmetrical nature of the crystal orientation on either side of the axis may be observed, *see* Fig. 38.

The structures obtained with two methods of casting, however, did not at first sight appear to fall in line with this principle. These occurred in the bottom-cast ingot *QXS* in which the angles of inclination (*see* Table III.) seemed to be irregular, and the ingots top-cast with single stream down centre, *QXT* and *QXW*, in which the motion of the liquid was expected to be in the form shown in Fig. 2 for *QXT*. According to this the columnar crystals should clearly be inclined downwards, instead of upwards as was actually observed. The inference is that Fig. 2, showing for *QXT* an upward motion of the liquid adjacent to the mould wall, does not represent a true picture of the turbulent conditions. Top-pouring trials were therefore carried out, using a stream of water instead of liquid steel and a large square glass vessel in place of the cast-iron mould, so that the motion of the liquid could be observed through the glass sides of the "mould." Tea leaves suspended in the water helped to depict the liquid motion. Turbulent conditions which had been assumed to operate in ingots *QXX* and *QXZ*, respectively stream

* In common with most steel ingots those examined here showed at the surface a thin chill skin of undefined structure, discussed in brief later. The crystals the inclination of which is now being considered are the columnar crystals on the inside of the chill skin.

down one side and tilted mould, were reproduced in the water experiments, but when the stream was poured down the centre the conditions were different from those assumed in Fig. 2, but were as shown in Fig. 7. The upward current surrounding the central stream was due to the upward rush of air bubbles injected with the stream, as shown in Fig. 41, the air bubbles keeping to paths close to the stream, and the liquid subsequently taking a course downwards on the outside near the mould wall. If the assumption be made that the motion of the steel is in any way similar to that of the water shown in Fig. 7, then the crystal inclination is in accordance with the principle mentioned above.

If the same principle of crystal orientation is held to be operative for the bottom-cast ingot *QXS*, the structure of this ingot may be readily explained by the turbulent motion being of the type shown

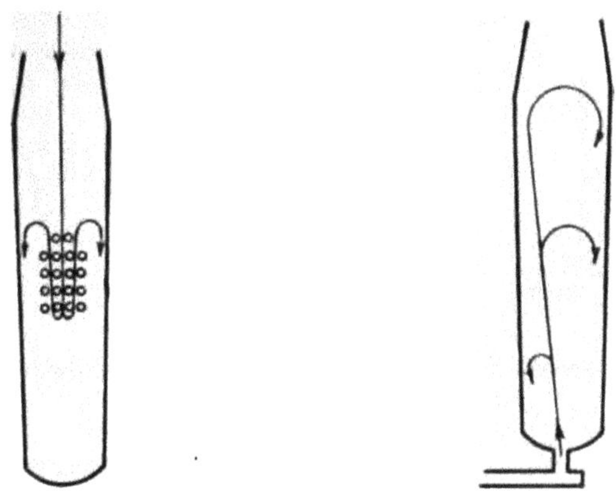

Fig. 7.—Ingot *QXT*, top-cast. Fig. 8.—Ingot *QXS*, bottom-cast.

in Fig. 8. Here the stream is not a vertical one but, owing to slight imperfection either in the bottom central runner or its setting, the final stream is slightly inclined so that it approaches the top of one of the mould faces. Confirmation of this view is found in the bottom-cast composite alloy ingot *RQF*20, Fig. 40 (*f*), in which the stream was intentionally deflected to one side; the columnar crystal inclination follows the general rule described above.

The evidence thus points to the conclusion that the columnar crystals are inclined towards the direction from which the metal flows. The following argument is offered in explanation :

Assume that in an ideal case we have a mould nearly full of quiescent liquid steel, the original temperature of which is the same throughout. As the temperature of the mould is much lower than that of the steel, a temperature gradient will be set up in the liquid at right angles to the mould wall, the isothermals being parallel to the mould, as in Fig. 9 (*a*), with the result that columnar crystals of any appreciable length will tend to grow in a direction normal to the mould face. If into this, now cooling, liquid a stream of metal is poured at the temperature of the original liquid, the stream, irrespective of the direction in which it flows, will lose heat to the liquid already present, and the farther the stream travels the cooler will it become. Reference to the conditions of turbulence in the different ingots shows that in most of them there is a downward current of metal near the mould wall. Since the stream loses heat as it travels downwards, the hotter metal is at the top, so that the isothermals are no longer parallel with the mould wall but are inclined to it in the direction shown, exaggerated, in Fig. 9 (*b*). The isothermals will not be inclined to the mould wall at the same angle throughout their length, because the farther the hot stream

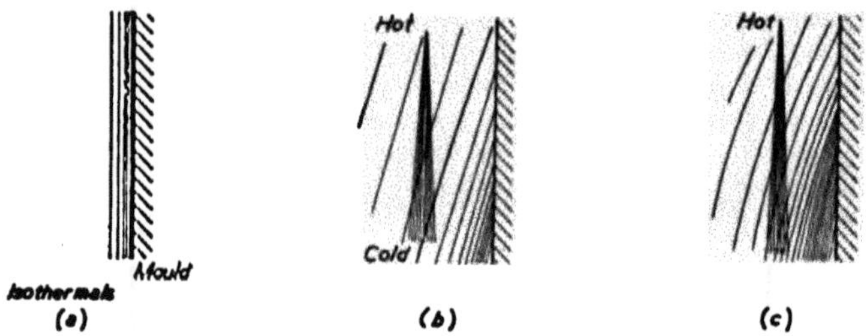

FIG. 9.—Influence of a Stream of Metal on the Isothermals in the Steel into which it is Poured.

travels down into the liquid the greater will be its spread or diffusion and the smaller will be the temperature difference between the stream and the liquid already present. Furthermore, since pouring is completed in about one minute, very considerably less than the time of solidification, turbulence in the liquid brought about by the casting stream will become gradually less, and this, also, will permit the isothermals to return more to the vertical as they approach the middle of the ingot.

Fig. 9 (*c*), a combination of Figs. 9 (*a*) and (*b*), shows the probable form of the isothermals at a particular instant, solidification starting at positions corresponding with the upper part of the diagram where the isothermals are well inclined to the mould wall and columnar growth continuing there in a direction at right angles to the isothermals, that is to say the crystals should be inclined slightly upwards. As in actual practice the crystals are found to be inclined 8–10° upwards, it follows that the isothermals corresponding with the early stages of columnar growth are inclined at about 8–10° to the mould wall. Lower down the ingot crystal growth will continue, though at a less rapid rate, but, according to the scheme outlined above, at any instant the thickness of the solidified shell will be slightly greater in the lower portion of the ingot than it is above. This also is what is found in practice, as, for example, when the liquid is withdrawn from a partially solidified ingot,[8,9] and it is shown by the banding effect in the stirred ingot QXY; it must be added, however, that part of this increase in thickness towards the bottom must be due to casting there first and solidification starting at the bottom whilst the mould is being filled.

The above argument also holds in explaining the *downward* inclination of the crystals in parts of the ingot QXS, bottom-cast, QXX, stream down one side, and QXZ, tilted mould, bearing in mind that in these instances there is locally an *upward* flowing stream of metal, the lower portions of it being hottest.* It would therefore appear that the columnar crystals are likely to be inclined to the normal to the mould wall when the ingot is cast under conditions leading to turbulence of the liquid metal in the mould.

A striking example of the inclination of the crystals towards the direction from which the stream flows is shown by the macrostructure of the axial face of a lead ingot illustrated in Fig. 42. The metal was poured down the right-hand side, and the unsymmetrical crystal orientation which will be observed is similar to that found in the steel ingot QXX, stream down one side, *see* Fig. 38.

According to these views, and as mentioned above, the isothermals tend to approach the vertical as they recede from the ingot-mould face. It should follow that there will be an increasing tendency for the columnar crystals to grow horizontally. It must be remembered, however, that in the crystallisation of a substance such as a metal, where the atomic forces are comparatively powerful even at temperatures just below the melting point, once a dendritic axis has started to grow in one direction, it normally continues growing in the same direction, provided that it meets with no obstruction and that there is no disturbance during solidification. Growth in a slightly different direction usually occurs, not by the bending over of the main axis but by the growth of new axes at right angles to the first followed by systems of other growths of the same type, the *average* direction of these being always normal to the solidus isothermals. With non-metallic materials having much smaller atomic forces the bending over of the main axes is sometimes observed, and examples of this in the case of wax are illustrated in the paper by Carlsson and Hultgren.[6]

Even with metals, conditions may arise which permit a change in the direction of growth of the main axes of the crystal. A study of the columnar crystal structures of the steel ingots described in

* That there may be a downward inclination renders untenable the suggestion that the columnar crystal inclination is due to the direction of the isothermals being modified by the bottom of the mould warming up before the top as the first metal is poured into it.

this paper showed innumerable instances of the dendrites becoming more horizontal with increasing distance from the mould. Lest it be thought that such a change might be due to the bulk movement or sinking of the metal during feeding it should be stated that examples were observed in the lower half as well as in the upper half of the ingot, and also in those crystals originally inclined *downwards*. The change in direction towards the horizontal is then an *upward* one. An example of this is illustrated in Fig. 29, from the ingot *QXX.3A*, in which the stream was poured down one side. Three possible conditions, excluding mechanical movement, may be suggested as enabling the axes to change direction: (1) A change in the direction of the isothermals. This is obviously essential but is not thought to be effective by itself. (2) As (1), together with mechanical shock which interferes momentarily with the course of solidification. (3) As (1), together with variations in the concentration gradient which interfere with the course of solidification. A good example of such a variation is found in the intermittent type of crystallisation characterising periodic structures. These structures were present in all the chill-cast steel ingots examined, and it is clear that an opportunity does arise for an interruption in the directional properties of the forces of crystallisation. It is suggested that this third condition is the prevailing one in the examples observed in the present work, since the change in direction does sometimes, though not always, occur at one of the bands of the periodic structure.

Λ-*Segregates.*

It is not proposed to offer any general explanation for the formation of the Λ-segregates, since the views expressed by the Ingots Committee [7] (*loc. cit.*, p. 11), namely that the segregates are the resultant of two effects, a forward movement of the solid/liquid boundary and an upward movement of the impure liquid arising from differential solidification, are considered to be correct.

A study of the ingots described in this paper has shown that the Λ-segregates are dendritic. A particular example has been described under "Microstructure" in Section I. and illustrated in Fig. 30, but other examples will be observed from the macrographs. Judging from the micro-examination of etched specimens there appears to be a much wider range in composition in the dendritic areas than in adjacent areas on either side of the Λ-segregate. It would therefore appear that the reason for the dendritic structure is that solidification of the segregate material must have occurred by crystallisation from a few centres in a low-melting-point liquid having a wide freezing range—conditions likely to be prevalent in impure strata in the liquid.

Turning now to the patchy appearance in ingot *QXX*, stream down one side, it is clear from the dendritic structure of the patches and their higher content of impurity, judged from the sulphur print and macro-etching, that the patches consist of a slightly broken up form of the Λ-segregate, as pictured in Fig. 3 (c), and do *not* represent portions of columnar crystallites broken off by the stream and subsequently grown. As to the reason why the Λ-segregate has taken this form it can only be conjectured that, even after solidification had proceeded for a little while, there remained a trace of the original pronounced turbulence caused by the method of casting—a trace sufficient to cause slight displacement of the stratified liquid inside the growing shell.

Rate of Crystallisation.

An attempt was made to obtain information on the change in crystallisation velocity during the process of solidification from a study

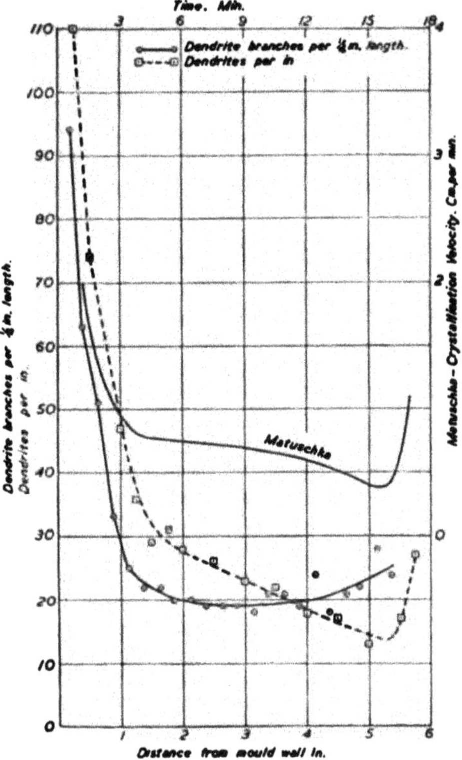

Fig. 10.—Counts of Dendrites and Dendritic Branches.

of the primary structure of the 1-in. bar $QXZ.0B$ illustrated in Fig. 29. The method adopted relies on the assumption that the crystallisation velocity is proportional to the number of dendrites per unit length measured parallel to the mould wall and to the number of dendritic branches per unit length along the dendritic axes. This is an arbitrary assumption, since both these numbers are more truly indicative of the degree of undercooling of the liquid steel at the moment of solidification, so that the assumption really concerns the dependence of rate of nuclei formation and of crystallisation velocity, respectively, upon the degree of undercooling.

Counts of the dendrites and dendritic branches were made at ¼-in. intervals along the bar, and the results plotted in the form of curves (Fig. 10). There is also included in this diagram a curve showing the results obtained by Matuschka [(8)] in his experimental determination of the solidification velocity in 9-in. 7-cwt. ingots of carbon steel, the upper curve being taken from Fig. 3 of Matuschka's paper. The similarity in the form of the curves suggests that there is a linear relationship between the degree of undercooling experienced by these steels and their crystallisation velocity.

A rough estimate of the crystallisation velocity in the 11-in. ingots may be obtained from Fig. 10, if it is assumed that the time of solidification was of the same order as that of Matuschka's 9-in. ingots—and experience with the stirred ingot *QXY* shows this to be so—and the necessary allowance is made for the difference in size of the ingots.

Mechanism of the Solidification of Steel Ingots.

Before attempting an explanation of the banded structure in the stirred ingot *QXY*, it will be necessary to consider the conditions existing during the solidification of an ingot allowed to cool normally. One factor which it is not proposed to deal with here is the composition of the alloy. The influence of different elements upon the crystallisation of iron is being investigated; judging from the results of similar work on copper,[(10)] it is to be expected that different elements will have very different effects upon the crystal structure. Since in the present work all the ingots were cast from the same melt, the assumption will be made that their average composition was the same. Apart from the question of turbulence it is considered that the variation in the crystal structures of the ingots described in this paper may be adequately explained simply by a consideration of concentration gradients and the distribution of temperature in the liquid.

The distribution of temperature in the liquid is certainly one of the principal factors to be taken into account. Owing to the experimental difficulties involved in determining temperatures inside a mass of solidifying steel no direct measurements have been made, but information sufficiently accurate for our purpose is available from the mathematical investigations of Lightfoot [(11)] and from the experimental work of Matuschka [(8)] and of Chipman and Fon Dersmith,[(9)] who studied the rate of solidification of steel ingots. Lightfoot calculated the temperature gradient in an ingot and mould and its variation with time, as well as the position of the solid-liquid boundary at different intervals of time and with different initial mould temperatures. Although the conditions for which Lightfoot's calculations apply are somewhat different from those existing in Chipman's work, the experimental results of the latter plotted on Lightfoot's diagram give a similar type of curve to that computed mathematically. The actual values of thickness for given times are rather less than those proposed by Lightfoot, except in the case of a mould having an initial temperature in the neighbourhood of 650° C. The theoretical and experimental results are reproduced on the same diagram in Fig. 11. Bearing in mind the difference in the two sets of conditions, the computed values appear to be of the correct order and form, and for the purposes of the present discussion the inclusion of Lightfoot's results is considered

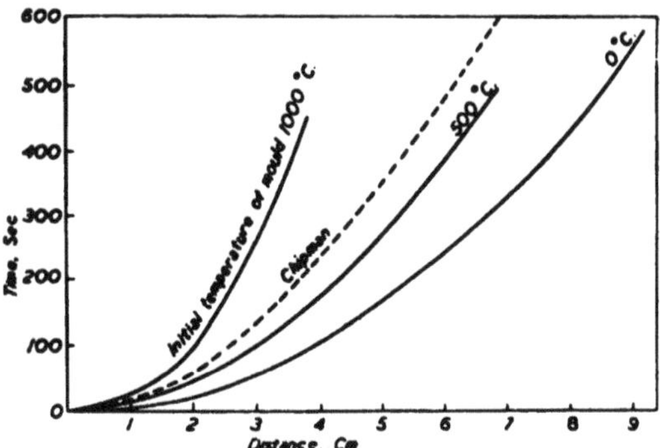

Fig. 11.—Temperature Gradients, as determined theoretically and experimentally.

to be justified, since the argument is not concerned with absolute values.

A portion of the equilibrium diagram of a typical alloy is shown in Fig. 12. A direct interpretation of this diagram at an early stage of solidification permits another of the type shown in Fig. 13 to be constructed in which composition is plotted against distance from the mould face, it being understood that perfect chemical equilibrium is not established. The corresponding temperature-distance curve giving the initial temperature of solidification is shown in Fig. 14. In accordance with the constitutional diagram the layer of liquid metal adjacent to the growing crystal is richer in the lower-melting-point constituent even under equilibrium conditions. In the absence of equilibrium this liquid layer is likely to have a composition to the right of point *C* in Fig. 12, so that it is quite possible for the dip in the composition line for the liquid in Fig. 13 to drop below the composition line *C*, and this is commonly observed in practice. The temperature distribution is shown in

Fig. 15 from Lightfoot's values. The important feature shown by this diagram is that the temperature gradient in front of the growing crystals flattens out with increasing distance from the mould wall and with increase in time from the commencement of solidification.

The effect of this on the manner of solidification may be determined by superimposing Fig. 15 on Fig. 14 and seeing what effect

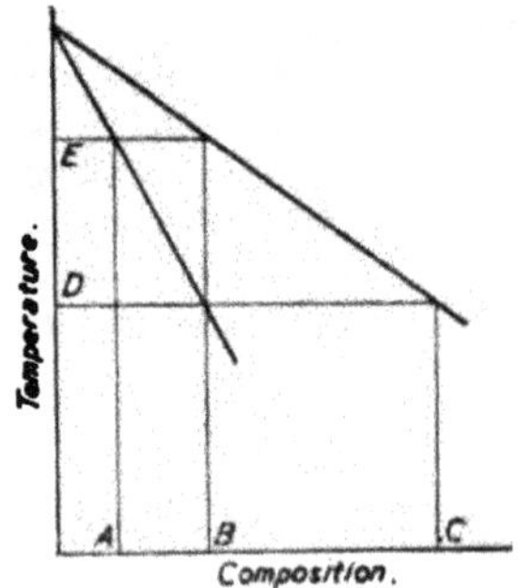

Fig. 12.—Part of Equilibrium Diagram of a Typical Alloy.

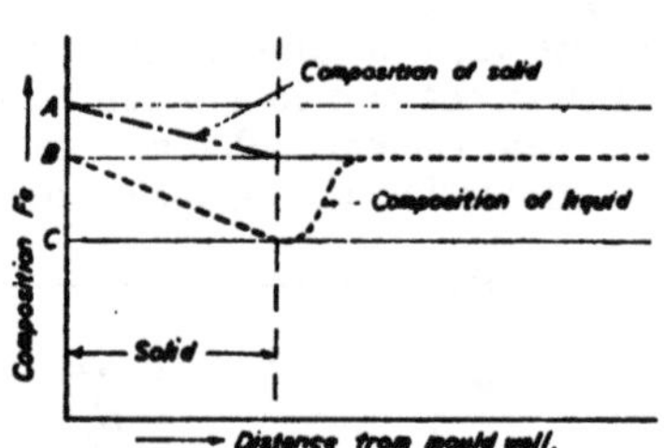

Fig. 13.—Variation of Composition of Solidifying Steel with Distance from the Mould Wall.

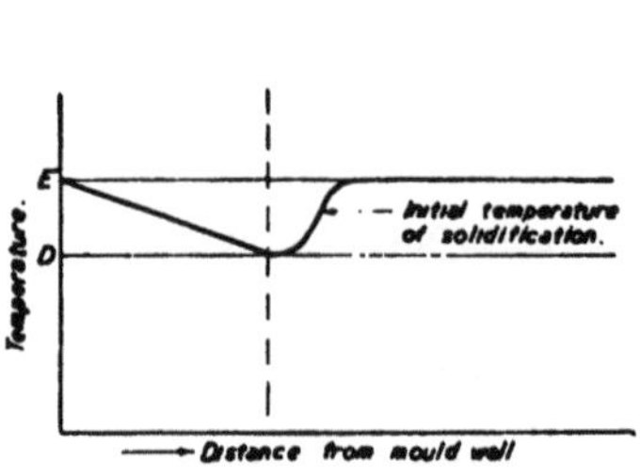

Fig. 14.—Variation of Temperature of Solidifying Steel with Distance from the Mould Wall.

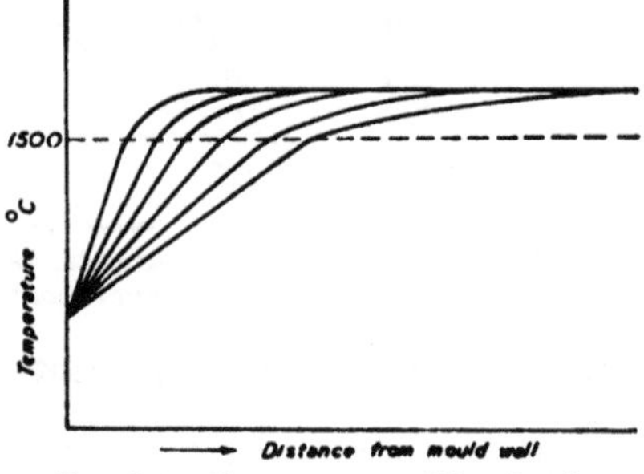

Fig. 15.—Temperature Distribution in Solid and Liquid Steel.

this will have upon the process illustrated in Fig. 13. Explanations for certain of the structures observed in steel ingots are readily obtained when the different values of temperature and composition gradients are taken into account. Typical examples may be considered as follows :

Case I.—When the conditions, *e.g.*, a high casting temperature and massive cold mould, are such as to ensure the temperature gradient remaining fairly steep throughout the course of solidification. In this instance superimposing the appropriate curve from Fig. 15 on to Figs. 14 and 13 gives a diagram of the type shown in Fig. 16 (*a*), with the result that the crystal structure is wholly columnar as represented in Fig. 16 (*b*).

Case II.—When the conditions are such that the temperature gradient is steep at first but flattens out considerably as solidification proceeds; these are the conditions which normally apply to the freezing of the majority of steel ingots and ingots of other alloys. Here it is necessary to consider two diagrams, the first representing the early period when the temperature

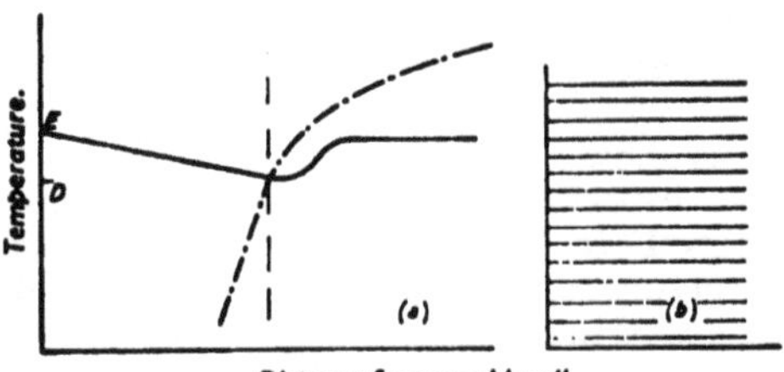

Fig. 16.—Effect on Ingot Structure of Steep Temperature Gradient during Solidification.

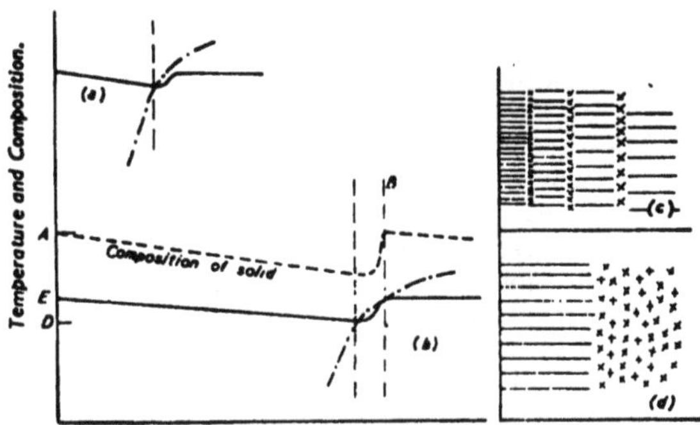

Fig. 17.—Effect on Ingot Structure of Temperature Gradient originally steep but later flattening during Solidification.

gradient is steep—this will be similar to that shown in Fig. 16—and the second representing the later period when the flattened temperature gradient crosses the initial solidification temperature curve at *two* positions simultaneously, which means that solidification can occur on two fronts separated momentarily by a layer, possibly discontinuous, of impure liquid. The two stages are shown in Fig 17 (*a*) and (*b*). Crystallisation occurring independently at position *B* in Fig. 17 (*b*) interrupts the columnar type of growth as at (*a*). In certain circumstances, such as the combination of a very steep temperature gradient and one or more constituents of low diffusing power, this action may be repeated a number of times and results in what have been termed periodic structures;[12] in such cases the interruptions usually occur at increasing intervals and tend to become more and more diffuse, both changes being due to the falling off in the temperature gradient. This type of crystallisation is illustrated diagrammatically in Fig. 17 (*c*). Normally, however, and with a less steep temperature gradient, the structure sooner or later ceases to be columnar, and solidification of the remaining liquid occurs in the equi-axial or nodular manner as explained in case III. below. The final structure then consists partly of columnar crystals and partly of equi-axial crystals, as shown diagrammatically in Fig. 17 (*d*). It is not uncommon, especially in steel ingots, for the structures illustrated in Figs. 17 (*c*) and (*d*) to be combined; in this instance the periodic structure represented in Fig. 17 (*c*) usually develops in the early stages of the columnar growth shown in Fig. 17 (*d*). Examples will be observed in Figs. 28 and 29.

Case III.—When the conditions are such that the temperature gradient remains fairly flat throughout the solidification period, as, for example, when the mould temperature approaches the solidification temperature of the liquid or when a refractory non-conducting mould material is employed; the sand-cast ingot *QXW* comes into this category. In quiet liquid steel, a fairly flat temperature gradient and a non-uniform composition of the liquid in front of the growing crystals lead to independent solidification occurring repeatedly in the liquid beyond that adjacent to the already solidified material. The state of affairs is illustrated diagrammatically by Fig. 18, in which the temperature conditions are represented at (*a*), the composition at (*b*) and the resulting structure, mainly equi-axial, at (*c*). A further point to be mentioned is that with an almost flat temperature gradient in a direction perpendicular to the mould wall there is little difference between this gradient and the temperature gradient parallel to the mould wall, so that there is ample opportunity for side growth of the crystals to occur, as is shown in the sand-cast ingot *QXW*.

According to this scheme it will be seen that the process of equi-axial solidification is not one of a perfectly steady and continuous growth away from the mould wall, but, on the contrary, is rather a discontinuous process. This type of crystallisation applies also to the equi-axial part of the structure of ordinary ingots which consist partly of columnar and partly of equi-axial crystals as considered in case II. above and shown in Fig. 17 (*d*).

It should be mentioned that a *fine* equi-axial structure may result by "shower" crystallisation from the labile state, but, possibly apart from the thin layer of "chill" crystals forming the outer skin of most chill-cast steel ingots, it is not considered that this

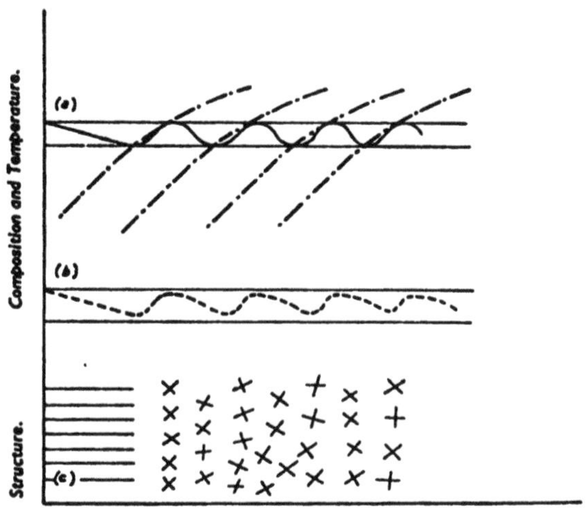

Fig. 18.—Effect on Ingot Structure of Flat Temperature Gradient during Solidification.

type of crystallisation plays any important part in the structure of ordinary steel ingots.

Banding in the Stirred Ingot QXY.

Having considered the process of solidification in general terms, it is now possible to turn to the particular structure of the poker-stirred ingot *QXY* which is characterised by a number of concentric bands approximately parallel to the mould wall. Owing to the rapid heating and softening effect of the liquid steel upon the stirrer, it was not practicable to carry out the stirring operation with one and the same poker and, as a matter of fact, six pokers were used. The dimensions were approximately 7 ft. in length and ½ in. in dia. Each was taken out and discarded when it was devoid of mechanical strength for the stirring action, by which time small portions had broken off from the bottom owing to incipient melting.

Consider now the effect of the stirrers on the distribution of temperature in the liquid steel. Each time a cold stirrer was inserted the temperature of the liquid must have been lowered. The drop in temperature would not be uniform throughout but would reach a maximum in the liquid first in contact with the stirrer. Here the temperature drop was sufficient to bring a small mass of metal below the freezing point so that some of the steel froze on to the stirrer; farther away the reduction in temperature would be much less.

When the general effect on the temperature distribution curve is considered it will be seen that two possibilities may occur, depending upon the relative position of this curve to that representing the initial temperature of solidification. At fairly large distances from the stirrer the two curves may meet at some position away from the already solidified metal; the effect will then be as shown in Fig. 17 (*b*), where a band of independent *purer* crystals forms in the liquid. Near the stirrer the temperature curve will be still more deflected and the two curves may be in contact over a certain distance from the solid as in Fig. 19. This leads to the sudden solidification of the less pure liquid adjacent to the earlier formed solid, so that

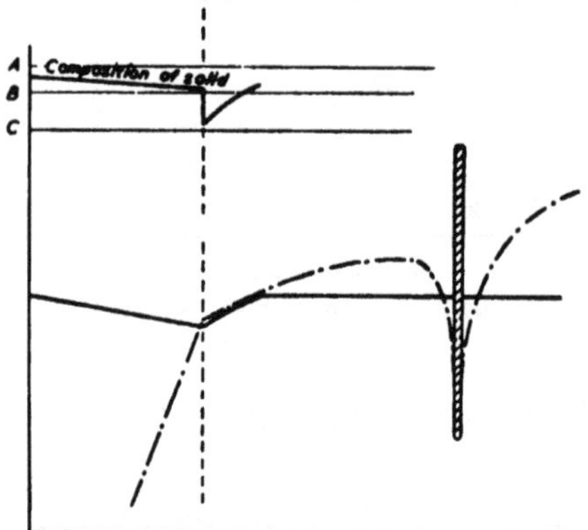

Fig. 19.—Ingot *QXY*, Influence of Stirring on Temperature Distribution.

the structure will be characterised by an *impure* band. Both types are found in the stirred ingot.

It may be thought that where the process is repeated a number of times it would be difficult to decide whether a pure or an impure band had been suddenly solidified. Actually this is not so because in most examples of these banded structures there is a sharp demarcation between pure and impure metal on only *one* side of the bands, and this defines the position at which a sudden change in the mode of crystallisation occurs. Reference to Figs. 26 and 39 shows that in the coarse banding of the stirred ingot the sharp demarcation between pure and impure material occurs with the impure on the inside away from the mould wall, so that the conditions shown in Fig. 19 apply. Evidence of the relative purities of the two portions of several of the bands is given in Table VIII. and Fig. 39 (*b*) and (*c*).

That there are more than six bands at the top of the ingot is to be expected when it is remembered that the poker was moved up and down rather than round and round. The upper part of the poker cooled slightly each time it was removed from the melt and exerted a slight cooling effect when it was re-inserted into the melt. The main chilling action of the lower part of the poker, however, was only exerted when the poker was originally inserted, so that the number of bands in the lower half or so of the ingot corresponds with the number of pokers used. It is assumed that the rather more pronounced nature of the banding on one side of the ingot is due to the fact that this side was nearer the man who inserted the stirrer.

Summary and Conclusions.

Seven 15-cwt. killed steel ingots have been cast from the same melt by different methods to determine the influence of turbulence. The methods employed involved: (1) Bottom casting, (2) top casting, a single stream down the centre of the mould, (3) the use of a multi-hole tundish, (4) a sand mould with a single stream, (5) a single stream near one side of the mould, (6) top casting with a single stream, stirring with a poker after casting, and (7) casting with a sloping mould. Information on the turbulent conditions occurring with these methods was also obtained by casting under similar conditions a series of small composite alloy ingots composed of a red alloy and a white alloy. Axial slices of the steel ingots were examined for chemical segregation, crystal structure and mechanical properties. Some of the results obtained may be summarised thus:

(1) The casting method offering least turbulence was the one using the multi-hole tundish. In order of increasing turbulence this was followed by bottom casting, top-casting with a single stream down the centre, with a tilted mould, with a single stream near one side of the mould and the poker-stirred methods.

(2) The structure of the tilted-mould ingot was very unsymmetrical; the lower side down which the stream was poured was composed of small equi-axial crystals whilst large columnar crystals grew from the opposite face.

(3) The poker-stirred ingot showed a pronounced banded structure, the bands being approximately parallel to the mould wall and extending almost to the middle of the ingot.

(4) The sand-mould ingot was characterised by excessive piping, owing partly to the penetration of the first-formed skin by the liquid on the inside, a coarse primary crystal structure and pronounced Λ-segregation.

(5) The ingot cast with a single stream near one side of the mould showed a patchy form of Λ-segregate.

(6) In general the Λ-segregate showed a strongly dendritic

structure as compared with the nodular structure of the purer metal on either side of the segregate.

(7) Measurements were made of the columnar crystal length and orientation in the different ingots. Most of the columnar crystals were inclined at a small angle to the normal to the mould face. The inclination direction, which may be either downwards or upwards, was towards that *from* which the liquid steel flowed along the mould face. The subject is discussed from the standpoint of the effect of metal flow upon the isothermals near the mould. The length of the columnar crystals was less the greater the turbulence during casting.

(8) The inherent grain size as determined by the McQuaid-Ehn test did not appear to be affected by the method of casting or by the primary structure.

(9) The mechanical properties of samples cut from the chill-cast ingots were more affected by the position in the ingot from which the test-pieces were cut than by the particular ingot. The properties of the sand-cast ingot were somewhat inferior.

(10) There is reason to believe that under completely non-turbulent conditions of casting, the primary structure of the ingots would have been almost wholly columnar.

(11) The distribution of the carbon and other elements is influenced by the method of casting, in particular by those processes making use of a non-axial stream.

Explanations have been put forward for certain of the ingot structures observed based upon the influence of turbulence, concentration gradients and distribution of temperature, and the mode of steel ingot solidification has been discussed.

Acknowledgments.

The ingots described in this paper were cast by Messrs. Hadfields, Ltd., and it is only through the kind co-operation of Mr. W. J. Dawson and his colleagues, Dr. R. J. Sarjant and the late Mr. E. J. Barnes, that this investigation has been possible. Thanks are also due to Dr. R. H. Greaves for his interest and encouragement during the course of the work.

REFERENCES.

(1) L. Northcott: *Journal of the Iron and Steel Institute*, 1939, No. I., p. 297.
(2) R. G. Heggie: *Transactions of the Faraday Society*, 1933, vol. 29, p. 707.
(3) L. Northcott: *Journal of the Institute of Metals*, 1937, vol. 60, p. 229.
(4) R. A. Hadfield: *Journal of the Iron and Steel Institute*, 1912, No. II., p. 40.
(5) R. Genders: *Journal of the Institute of Metals*, 1926, vol. 35, p. 259.
(6) C. G. Carlsson and A. Hultgren: *Jernkontorets Annaler*, 1936, vol. 120, p. 577.
(7) Seventh Report on the Heterogeneity of Steel Ingots, *Iron and Steel Institute*, 1937, *Special Report No.* 16, p. 7.

(8) B. MATUSCHKA : *Journal of the Iron and Steel Institute*, 1931, No. II., p. 361.

(9) J. CHIPMAN and C. R. FON DERSMITH : *American Institute of Mining and Metallurgical Engineers, Technical Publication No.* 812 : *Metals Technology*, Oct., 1937.

(10) L. NORTHCOTT : *Journal of the Institute of Metals*, 1938, vol. 62, p. 101; 1939, vol. 65, p. 173.

(11) N. M. H. LIGHTFOOT : Third Report on the Heterogeneity of Steel Ingots, *Journal of the Iron and Steel Institute*, 1929, No. I., p. 364. Fourth Report, *Iron and Steel Institute*, 1932, *Special Report No.* 2, p. 162.

(12) L. NORTHCOTT : *Journal of the Iron and Steel Institute*, 1934, No. I., p. 171.

CORRESPONDENCE.

Dr. C. H. Desch, F.R.S. (Vice-President; Iron and Steel Industrial Research Council, London), wrote that the paper gave interesting information as to the effects of turbulence in determining ingot structure. It was highly probable that, as found by the writer and Dr. Heggie in experiments with small ingots, if turbulence were entirely excluded the structure would be wholly columnar. On p. 60 P attention was called to the sharp change from the dendritic structure of the Λ-segregates to the nodular structure immediately adjoining. In the "Atlas Metallographicus" of Hanemann and Schrader, Fig. 184 appeared to show a similar sharp transition, although those authors had not attempted to connect the change with segregation.

The account of the convection currents and of their influence on the direction of the dendrites was ingenious, and appeared to fit the facts. It should be emphasised that the bending of dendrites was almost always only an apparent one, and that the deviation from a single direction was due to unsymmetrical additions, the crystal orientation remaining constant. The conclusion that particles broken off from the dendrites were the origin of the nuclei from which the equi-axed crystals grew was shown to be unlikely. In view of the plasticity of steel near its melting point and the support given to the crystals by the surrounding liquid, it was improbable that brittle fracture would occur. The factors which determined equi-axial crystallisation had never been satisfactorily ascertained. If substantial undercooling could be assumed, it might be supposed that the labile range was reached locally, but this seemed improbable. Undercooling of molten metals had been first measured by Roberts-Austen,[1] and could be easily produced in small laboratory melts, but the writer had failed to detect it in large masses of lead, and Bardenheuer [2] had found for steel that a small mass, separated from the walls of the crucible by molten slag, could be undercooled 200° C. or more, but that this did not occur if the metal came into contact with the rough wall, and no measurable undercooling was obtained with large masses. Theory required that there should be some undercooling before crystallisation could begin, but this might be only of the order of a fraction of a degree. The author's conclusion that "shower" crystallisation in the labile range was only likely in the chill-crystal layer, and was improbable in the interior, was almost certainly correct. It would need further direct evidence of temperature gradients

[1] W. C. Roberts-Austen, *Proceedings of the Royal Society*, 1898, vol. 63, p. 447.

[2] P. Bardenheuer and R. Bleckmann, *Stahl und Eisen*, 1941, vol. 61, p. 49. Earlier experiments by the same authors were given in *Mitteilungen aus dem Kaiser-Wilhelm-Institut für Eisenforschung*, 1939, vol. 21, p. 201.

before Figs. 13 to 18 could be accepted as a true picture of the processes of solidification, but they at least presented a basis for discussion, and the paper was a substantial contribution to the knowledge of ingot structure.

AUTHOR'S REPLY.

Dr. NORTHCOTT, in reply, wrote that Dr. Desch's contribution was read with much interest and it was a pleasure to know that he concurred in the general conclusions put forward. It was admitted that further work on the extent of undercooling was required and the author hoped to study this problem when conditions allowed. Experiments carried out since the paper was written would appear to suggest that the degree of undercooling possible with large masses might be greater than that suggested by Bardenheuer and Bleckmann, but the direct measurement of local undercooling was extremely difficult.

[illegible] Figs. 15 to 18 could be accepted as a [illegible] picture of the [illegible]

[illegible]

INTERCRYSTALLINE CRACKING IN BOILER PLATES.[1]

A REPORT FROM THE NATIONAL PHYSICAL LABORATORY (TEDDINGTON).

(Figs. 6 to 20 – Plates XII. to XIV.)
(Figs. 26 to 50 – Plates XV. to XVIII.)
(Figs. 51 to 55 – Plates XIX. and XX.)
(Figs. 62 to 74 – Plates XXI. to XXVII.)

SUMMARY.

Previous work having shown that in the absence of corrosive attack no cracks are formed in boiler-plate steel specimens kept under tension for 5 years, even when concentrations of stress are present, various combinations of stress and exposure to caustic solutions have been investigated. Part I. is an introductory survey of the subject. Part II. describes experiments in which specimens, with or without notches or drilled holes, were kept under tension in concentrated sodium hydroxide solution at 225° C. An electrically heated pressure vessel was used, with special devices for maintaining constant temperature and stress, and for indicating the onset of cracking. The typical form of intercrystalline crack was not obtained, but in regions of concentrated stress breakdown was caused by the growth of non-metallic inclusions. Heavily cold-worked steel resisted better than annealed material. The black magnetic oxide formed had the composition Fe_3O_4.

In Part III. experiments under pressure at temperatures up to 470° C. are described. Using small pressure bombs, intercrystalline cracking was found in boiler-plate steel immersed in a solution of pure sodium hydroxide at 310° C. Decarburisation occurred, the carbon being removed as methane. In other experiments the steel specimens and the solution were enclosed in a steel pressure vessel lined with silver. Under these conditions, no intercrystalline cracks were formed in annealed material at 410° C., but if the steel was cold-worked, cracks similar to those produced by hydrogen at high temperatures were formed. With highly concentrated caustic solutions intercrystalline cracks penetrating from the surface became filled with oxide. Experiments with powdered-silver filters showed that masses of oxide could be precipitated at a distance from the specimen, and this may contribute to the cracking of boilers, by sealing cavities and allowing a pressure of hydrogen to be built up. Pure iron developed cracks of the oxide type. Sodium sulphate in solution did not inhibit cracking. Some alloys of iron were also examined.

In Part IV. observations of strain-etching in acid open-hearth boiler-plate steel are described. Such markings are usually found only in basic steel. They were not produced in plates which had been deformed cold, but were found in material which had been bent at 100° C., and also in a rolled plate which had presumably been finished at a low temperature. They coincided with the stress lines found by magnetic testing, and with the directions of cracking in a marine boiler plate which had developed corrosion cracks in service.

[1] Received February 14, 1941.

Part V. describes the behaviour of specimens of boiler-plate steel and of both riveted and welded joints in the same material, when subjected to slow cycles of alternating bending stress while immersed in a boiling solution of sodium hydroxide. A machine taking plates 2 ft. 3 in. long, ¾ in. thick, and up to 12 in. wide (for riveted joints) was constructed. When failure occurred, it was due to corrosion-fatigue and the cracks were transcrystalline. The typical caustic cracking was thus not obtained, but the cracks observed closely resembled certain defects found in boilers as the result of service.

Part I.—Introduction.

By C. H. DESCH,[1] D.Sc., F.R.S. (London).

The type of failure in boilers which is commonly known as "caustic cracking" is characterised by the intercrystalline path of the cracks. The term "caustic embrittlement" which has often been used is incorrect, as the steel between the cracks may remain quite ductile. It is most common at riveted joints, especially at rivet holes under butt straps, and often extends along a joint through a number of rivets. It is not found in seamless or welded drums, except occasionally around fittings. The cracks are as a rule easily distinguishable from fatigue cracks and from grooving caused by acid corrosion.

The work of Parr and Straub at the University of Illinois, the first report on which was issued in 1917, established the main features. The attention of those authors had been drawn to the subject by the failure of boilers at the University, after the working pressure had been somewhat increased and the boilers worked more continuously. The water of Urbana, where the University is situated, contained 60–70 parts per million of sodium carbonate, sulphates being practically absent. As the water contained calcium salts, it was clear that the salts must be present as bicarbonates, and enquiry showed that numerous other boilers in America of which failure had been recorded had been fed with water of this unusual type. Such a supply does not cause scaling, as carbon dioxide is lost during heating, and the sodium carbonate which remains precipitates calcium carbonate in a flocculent form, so that the water is described as "self-purging."

It was known that sodium hydroxide could act on iron, liberating hydrogen, and a possible relation to "pickling brittleness" was suspected. Appreciable damage to steel by sodium hydroxide was, however, only found when the concentration of the alkali became high, especially when it was 30% or more, a figure far in excess of anything possible in the interior of a boiler. An ingenious explanation was put forward by Parr and Straub to account for this. The usual site of the cracks is where butt straps are attached

[1] Technical Advisor to the Iron and Steel Industrial Research Council; formerly Superintendent, Metallurgy Department, National Physical Laboratory.

to the shell by rivets. It was supposed that slight leakage could occur between the plates and the strap, and also around the rivets, and that, owing to evaporation of water at the surface, a high concentration of dissolved salts, in this case largely alkali, could occur. Laboratory experiments showed that such concentrations could actually be produced by evaporation in capillary spaces. (Subsequent experiments which were claimed to disprove this possibility really failed to reproduce the conditions.) It is possible that concentration of dissolved salts may occur in the crevices of a boiler even in the absence of leakage. The fact must be regarded as established, and when a cracked boiler is dismantled, strongly alkaline incrustations can usually be found between the plates and around the rivets.

It also appeared that the steel must be in a stressed condition for damage to occur, and laboratory experiments indicated that the stress must have exceeded the yield point. This appeared to be confirmed by practical experience. Bad workmanship, represented by bad alignment of rivet holes, so that force had to be used to bring them opposite to one another, or by lack of parallelism between the edges of the plates, necessitating heavy caulking, as well as excessive riveting pressure, have been found to increase the probability of cracking, although there is no evidence that they will cause failure in the absence of the chemical conditions. Of course, cracking may be produced by mechanically defective joints in the absence of alkali, but the cracks are then of a different type. With intermittent working, failure may occur even at well made joints.

The reports of Parr and Straub have been the starting point of much other research work, and their main conclusions have been confirmed. German work, which was undertaken in consequence of a disastrous boiler explosion in 1920, leading to the formation of the Vereinigung der Grosskesselbesitzer, at first followed a different path. The quality of the steel was held to be responsible, but on replacing by plates of known good quality cracking again occurred. The tendency of much of the German work, which for some time ignored the work of Parr and Straub, was to lay stress on processes of ageing in the steel when kept at temperatures of 200° C. or upwards. The Izett non-ageing steel was therefore expected to be immune; but when tested under the American conditions it failed in the same way as ordinary mild steel. Later investigations have come more into line with those of other countries.

Certain other forms of intercrystalline cracking familiar to the chemical industry are evidently related to this type of boiler cracking. Steel vessels in which solutions of caustic soda are concentrated under atmospheric pressure frequently crack, especially at riveted joints, whilst the cracking of pans containing solutions of nitrates, especially ammonium nitrate, is well known. In both cases the cracks pass between the crystal grains.

The important work of Stromeyer on boiler steels must be mentioned. In a series of reports to the Manchester Steam Users' Association, he described many observations on boilers in service and also attempted to reproduce the conditions causing cracking in the laboratory. In one series of experiments, he trepanned rings out of boiler plate, and made them slightly taper, so that one could enclose another with a driving fit, the outer ring then being in tension and the inner in compression. These were then heated for three months in sodium hydroxide solution, of much lower concentration than that described above, and at atmospheric pressure. On sawing specimens from the rings and attempting to bend them, strips from the rings in tension were found to be highly brittle, and they retained this brittleness years later, while those from the inner rings, or from unstressed steel exposed to the same solution, could be bent double. This result pointed to the importance of tensile stress in producing brittleness, although it was not shown that this was identical with caustic cracking.

Stromeyer attached great importance to the quality of the steel and associated the liability to crack more particularly with the presence of nitrogen, having observed that basic Bessemer steel, which contains more nitrogen than other varieties, offered the least resistance. Many failures have, however, been recorded where the material was of acid open-hearth quality.

The conclusion of Parr and Straub, that the alkalinity of the boiler water is an essential factor, has been amply confirmed. Besides naturally alkaline waters, many water supplies are rendered alkaline in the course of softening treatments.

In recent years serious failures in power stations have been reported, in some instances resulting in disastrous explosions. In all these cases the water had received alkaline treatment. The likelihood of failure increases with the working pressure and with the age of the boiler, but some very remarkable failures have occurred very early in the life of boilers, even after a few months. Most of the failures have occurred in power stations, but marine boilers have also been known to suffer from this cause, although less frequently. Lloyd's Register has reported eight cases, and its officers suggested that any appearance of broken rivets must be looked on with grave suspicion. The rarity of the defect in marine boilers may possibly be connected with the presence of sea water, which increases ordinary corrosion, but may act as an inhibitor in this form of attack. The power-station accidents have often been very serious, because when one out of a number of boilers is found to have cracked, an inspection of its neighbours usually reveals that they are also cracked, although the damage has not become obvious, and renewal of the whole set becomes necessary.

The work of the National Physical Laboratory on this subject dates from 1917, and Rosenhain and Hanson gave an account in 1920 of several instances of intercrystalline cracking in boilers, associating the effect with corrosion of an undetermined type, combined with the action of stress on the amorphous cement existing between the crystal grains. In the same paper, the cracking of a steel tube which had been in use in a bath of fused nitrates at 300° C., without applied stress, was described, and the similarity between the two cases was indicated. Experiments were undertaken at a later date with the object of examining, one by one, the factors which might conceivably bring about intercrystalline cracking. The possibility had been considered that an amorphous layer between the crystals, if present, might flow under prolonged stress, so producing intercrystalline rupture as an effect of stress and time alone. With the object of settling this point, test-pieces of mild steel were loaded to one-third, one-half and two-thirds of their normal breaking load in an oven kept at 300° C. in an atmosphere of dry air. After five years of exposure none of the specimens had developed cracks, and only the most heavily loaded showed any appreciable extension. There was some permanent hardening, and it seemed possible that a different result might be obtained if there were local concentrations of stress. Another series of specimens, some with lateral notches and others with drilled holes, was exposed in the same way. The results were again completely negative, none of the specimens cracking, and it followed that in the absence of chemical attack intercrystalline failure would not occur at 300° C. as the result of stress alone.

As boilers in use are subject to " breathing " and during intermittent working bending stresses are imposed and removed from time to time, experiments were also put in hand to ascertain whether repeated bending in water or in various solutions could bring about cracking. Bars of boiler-plate steel were tested in water, air, salt solution and sodium hydroxide solution, being bent beyond the elastic limit at regular intervals and the number of bends needed to break the bar being noted. The intervals between successive bends were varied in an attempt to simulate conditions in a boiler, but at atmospheric temperature, and although the life of the specimens was found to be shorter than when tested similarly in air, in no case was there any sign of intercrystalline cracking. On repeating the experiments in air up to 200° C. and in the liquid baths at 100° C., the number of bends which the steel would withstand was further reduced, but no numerical difference was found in the several solutions, although the type of chemical attack in sodium hydroxide was clearly different from that in water. Again there was no indication of intercrystalline cracking.

These negative results suggested further lines of investigation. Some boiler failures which had been reported showed the characteristics of corrosion-fatigue, the cracks starting from corrosion pits, and this aspect of the problem had been given prominence by McAdam. The breathing and " letting-down " of a boiler produce cycles of stress which might combine with the chemical action of the caustic solution to produce typical corrosion-fatigue. In order to test this possibility, experiments were devised in the Engineering Department. For the sake of convenience, alternating bending stresses were employed. Plain boiler-steel plates, ⅜ in. and ⅝ in. thick, and also both riveted and butt-welded joints in ⅝-in. plate, were taken for testing, and stresses up to a nominal range of 20 tons per sq. in. were applied. The experiments, being conducted in an open tank, were necessarily confined to atmospheric pressure, the plates being completely immersed. The conditions for the concentration of sodium hydroxide in the riveted seams were therefore not present, and the joints had been carefully made, avoiding distortion in the neighbourhood of the rivets. The conditions were thus less drastic than in an industrial boiler.

Typical corrosion-fatigue failures were produced in this apparatus similar to some which have been found in service, but microscopical examination showed that the cracks were not intercrystalline. Corrosion-fatigue, which had been invoked by some writers as the cause of caustic cracking, may therefore be ruled out in this connection, although it may cause failure under other conditions. It is more likely to occur when the water contains free carbon dioxide, but the present report shows that it is not prevented by the presence of even a concentrated solution of caustic alkali.

In the course of these experiments, steel which had been deformed at 100° C. was found to show strain-etching when sections were polished and treated with a suitable reagent. Such strain-etching lines, which indicate regions of local deformation, are well known in plates of basic Bessemer steel, but are not usually found in acid open-hearth material, such as was used in this investigation. The markings were not found when the same steel was deformed at atmospheric temperature. Polished specimens, magnetised and treated with a magnetic powder in suspension, showed that the magnetic stress lines coincided with those produced by etching. A plate from a marine boiler which had developed a corrosion defect in service, leading to a local concentration of stress, showed strain-etching markings corresponding with the distribution of stress around the defect.

Another series of experiments, designed to test the effect of static stress on steel specimens immersed in concentrated sodium hydroxide solution at 225–250° C., is described in Part II. of the present report. The general plan was similar to that employed by Parr and Straub, but, in order to control the applied stress more accurately, stuffing boxes were avoided, and the loading devices were completely enclosed in the pressure vessel. Electrical attachments were provided to control the constancy of temperature and pressure, and also gave warning when a specimen under stress broke or elongated beyond a certain limit.

The results obtained confirmed in the main those of other

investigators. Specimens stressed beyond the yield point (as determined at atmospheric temperature) usually broke in a few hours, but no cracks were found other than the single one causing rupture, and there was no sign of intercrystalline breakdown in its neighbourhood. On examining results published elsewhere, it was found that convincing evidence of intercrystalline cracking was rarely present. The fact that rupture took place under a comparatively low stress was usually accepted as evidence of "caustic cracking" without a thorough microscopical examination. It was therefore decided to examine the effect of concentrations of stress, and with this object in view strip specimens of steel, some with perforations and others with notched edges, were included in the tests. Failure occurred, as would be expected, under lower average stresses than in plain specimens, and cracks were found close to the roots of the notches, but these were mainly transcrystalline. Attack took place along lines of inclusions in the steel, and the growth of oxides along those lines contributed to the destruction of the metal. Much general corrosion occurred, which would probably be less marked in massive plates than in the thin strips used for the tests.

These experiments, using only moderate pressures similar to those in many boilers, having failed to reproduce cracks typical of those found in service, it was decided to produce higher pressures, in view of the fact that a crevice in a riveted joint may become sealed by products of corrosion, allowing a high pressure to develop. Moreover, the temperature in such a riveted seam, exposed to flame, may well rise far above the working temperature of the boiler. In order to study such conditions the experiments described in Part III. of this report were undertaken. Small cylinders of steel were drilled centrally, the hole was filled with sodium hydroxide solution and closed by washers and clamps, and the specimens were heated to various temperatures. At so high a temperature as 470° C. intercrystalline cracks were readily produced in cold-worked material, whether the sodium hydroxide used was pure or contained the usual commercial impurities. On lowering the temperature to 310° C. intercrystalline cracks were still obtained, but at 225° C. no cracks were found even after a month. It was proved that the attack was due to the penetration of hydrogen, and this gas, reacting with the carbide of the steel, produced methane, which accumulated until a high pressure was produced. Fully annealed steels were not attacked.

These small pressure cylinders were rather difficult to control, and negative results were sometimes obtained. An apparatus was therefore constructed in which the conditions could be maintained constant with greater ease. Small cylinders of the steels to be tested were enclosed in a silver-lined bomb, the reaction space in many instances being limited by packing with silver cylinders, so that the solution was not in contact with any steel other than the specimen under test. Pressures were measured by the deflection of a calibrated silver-faced steel diaphragm. Intercrystalline cracks were readily produced in this apparatus.

At high temperatures hydrogen penetration occurred, and methane, in quantities up to ten times the volume of the steel, was formed. That methane was present in the cracks and was responsible for forcing the grains apart was further proved by heating a cracked specimen to 650° C. for some time. Recarburisation took place, and cementite particles were deposited near the edge of the cracks. At a higher temperature patches of pearlite were re-formed. The whole behaviour was similar to that of steels exposed to gaseous hydrogen at much higher temperatures and pressures, as in gas reaction vessels.

With very high concentrations of sodium hydroxide, a different type of intercrystalline attack was obtained, and this occurred in annealed as well as in cold-worked specimens. These cracks were almost independent of the carbides in the steel, often following grain boundaries which were free from carbide. The cracks were filled to some depth with oxide, which was absent from the penetrating cracks of the hydrogen type. At lower temperatures, such as 150° C., these cracks were often transcrystalline.

Two distinct forms of cracking, which may be described as the hydrogen type and the oxide type, respectively, have therefore been obtained. Both may be concerned with the cracking of boilers in service. The oxide probably plays a part in sealing the crevices and so allowing a high pressure to develop. Well-formed crystals of a black oxide, identified as Fe_3O_4, were sometimes obtained. These must have been deposited from solution in the alkali, a conclusion which was confirmed by inserting a filter of silver powder in some of the tube experiments. Iron oxide was deposited on the further side of the filter.

Several matters of chemical interest have arisen in the course of the work. An American investigation led to the surprising conclusion that at 250° C. solutions of sodium hydroxide, even up to 50 g. in 100 c.c., had no more effect on steel of boiler-plate quality maintained under high stress than water alone under the same conditions. The cracking obtained in the earlier experiments was attributed to the use of commercial sodium hydroxide, containing impurities, and the substance claimed to be responsible was sodium silicate, in quantities of the order of 0·2%. When this addition was made to the sodium hydroxide, failure was produced under a stress less than one-third of that which was required in the pure solution. The specimens which failed showed a large number of fine intercrystalline cracks near to the fracture.

This conclusion was widely accepted, although it was difficult to understand, on chemical grounds, why sodium silicate, which might almost be expected to act as an inhibitor, should be so active. It has not been confirmed by the experiments here described in which the purest obtainable sodium hydroxide was found to be quite active in producing failure under the given conditions. Moreover, failures in practice have occurred in recent years, although the sodium carbonate, which is the usual addition, manufactured and sold in Great Britain contains only very small traces of silicate.

Differences in the nature of the attack in solutions containing silicate and in those free from silicate were observed, but these could probably be accounted for by differences in the texture of the oxide coatings. It seems likely that this texture is an important factor, and that the effect of various substances which are known to act as inhibitors may depend on their influence on the texture of the oxide deposited. It is a significant fact that some of them, such as the tannins, are solvents for iron oxide. The known inhibiting effect of sodium sulphate may depend on the relatively low solubility of this salt causing it to crystallise and seal the crevices in a joint before the sodium hydroxide has had time to reach a dangerous concentration. The American investigations have shown that the action of alkalis may be retarded or accelerated by the presence of other salts, and some of them even acted as accelerators at 100° C. but as retarders under high pressure, so that complex factors are evidently involved.

It has been found necessary to terminate the investigation at this point. It may be said that the explanation of caustic cracking originally offered by Parr and Straub has proved to be in the main correct. The following practical conclusions may be drawn :

(1) Intercrystalline cracking in steam boilers is always associated with a high alkalinity of the water. Where this is unavoidable, owing either to the originally alkaline character of the supply or to the softening process which has been applied to a naturally hard water, a method of protection which has proved of value in practice is the maintenance of a ratio of sulphate to alkali above a certain value, depending on the working pressure. No work has been done on the use of other inhibitors, such as phosphates.

(2) The steel must be in a condition of stress, the elastic limit having been exceeded, either through an external constraint, as in a riveted joint which has been forcibly brought together, or by internal stress, as in a rivet or in a plate where it has been subjected to riveting pressure. Experience with concentrating pans suggests that cracking may occur, however, in unstressed steel in the presence of very strong alkali.

(3) There must be opportunities for the concentration of the solution in capillary spaces. This is the case at riveted joints. A seamless drum presents no such cavities, and caustic cracking has not been observed in such drums, unless at joints or fittings.

(4) A high temperature must be reached. This, however,

depends on the composition of the solution. Cracking may occur at or near 100° C., or it may need a much higher temperature. In this connection it should be pointed out that temperatures much in excess of the working temperature of the boiler may be reached in a riveted butt-strap joint exposed to the flame.

(5) The intercrystalline cracks appear to be of two kinds, the one being clearly due to the penetration of hydrogen and formation of methane by its reaction with the carbide of the steel; the other leaving the carbide unchanged, the cracks being filled with black oxide.

(6) Cracks caused by corrosion-fatigue, being transcrystalline, are of a different type. They may usually be distinguished by inspection, but a microscopical investigation is conclusive.

The work summarised in this report and dealt with in detail in Sections II. to V. was carried out at the National Physical Laboratory as part of the programme of the Metallurgy Research Board. The investigation was begun originally at the request of the Board of Trade, but from 1935 onwards the work was accelerated as a result of collaboration with industry, which provided considerable financial contributions to the cost of the experimental work. This assistance from industry, both financial and technical, is gratefully acknowledged.

The work is published by permission of the Department of Scientific and Industrial Research.

Part II.—Prolonged-Stress Tests on Iron and Steel Specimens Immersed in Hot Sodium Hydroxide Solutions.

By C. H. M. JENKINS, D.Sc., A.R.S.M., and FRANK ADCOCK, M.B.E., D.Sc. (National Physical Laboratory).

(Figs. 6 to 20 – Plates XII. to XIV.)

The behaviour of mild steel in regard to the possibility of cracking under the action of prolonged stress at 300° C. was described by Rosenhain and Hanson [1] and on specimens with concentrated stress by Jenkins.[2] No cracking was obtained in these tests on specimens loaded for five years. The specimens were not exposed to sodium hydroxide solutions, as it was desirable to introduce the various factors one at a time. The experiments about to be described were designed to ascertain whether caustic

[1] Rosenhain and Hanson, *Journal of the Iron and Steel Institute*, 1927, No. II., p. 117.

[2] Jenkins, *Journal of the Iron and Steel Institute*, 1935, No. II., p. 281.

cracking occurs as the result of stress in an accelerating medium and were undertaken during the period 1931–1936.

At the time when the present section of the work was begun, Parr and Straub[1] had already conducted prolonged-stress tests on specimens immersed in hot solutions under pressure. For this purpose they employed a type of sealed vessel fitted with a stuffing box in the lid. The specimen was carried inside the vessel, and its upper end was attached to a rod which passed through the stuffing box. By means of a coiled spring outside the vessel tension was applied to the rod which transmitted the stress to the specimen.

In the present experiments the exact method of Parr and Straub was not adopted, as it was considered that, owing to friction at the gland after corrosion had taken place, the specimen might not be subjected to the full tension applied by the spring, while it would be difficult to avoid slight leakage at the gland, which, although negligible in ordinary practice, would deplete the limited quantity of liquid during a prolonged test.

Description of Apparatus.

Pressure Cylinder and Internal Fittings.

The apparatus, shown in Fig. 6, consisted essentially of a short thick-walled cylindrical vessel, electrically heated, with internal fittings for applying tension to specimens. It was essential to avoid any form of joint in the vessel below the liquid level during the experiments. Accordingly, a hot-drawn mild steel cylinder was employed. This cylinder, which had been tested hydraulically

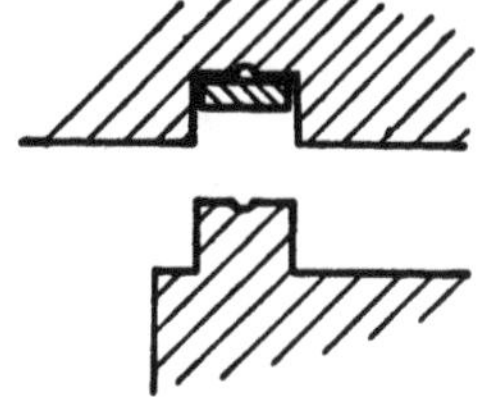

Fig. 1.—Cross-Section of Steam-Tight Joint. Three-quarters natural size.

at 1000 lb. per sq. in., measured 9¾ in. in dia. by 12 in. in height internally, the minimum wall thickness being 1 in. The upper part was threaded externally and carried a heavy flange with sixteen 1-in. dia. bolt-holes. A cover, 17 in. in dia. and 1¾ in. thick, was bolted to the flange by high-tensile steel bolts. A leak-proof joint was made by means of a carefully machined tongue on the top end of the cylinder, which entered a groove in the under-

[1] Parr and Straub, *University of Illinois Engineering Experiment Station*, 1926, *Bulletin No.* 115, p. 34.

side of the cover. When the holding-down bolts were tightened the tongue bedded firmly on a copper washer carried in the groove. The original copper washer had been used for thirty-four experiments, lasting in the aggregate for four hundred and ten days approximately, and a leak-proof joint was readily attained at the end of this period. Fig. 1 shows the cross section of the joint. Two mild-steel brackets were attached to the underside of the

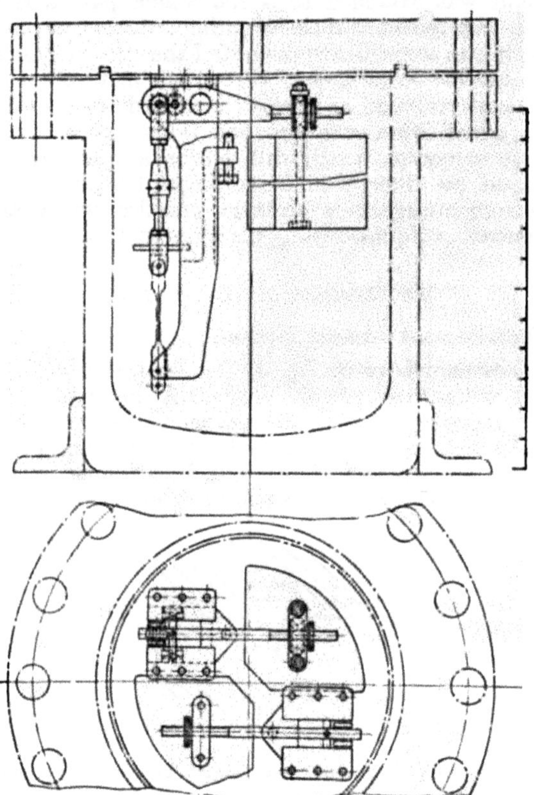

FIG. 2.—Pressure Cylinder with Lid and Internal Fittings. Scale in inches.

cover and reached almost to the bottom of the pressure vessel when the cover was bolted in position. Each bracket carried a lever or beam, normally horizontal and mounted close under the lid. The short end of each lever was connected to the top end of the specimen under test by a suitable link, while the long end carried a weight, the bottom of each specimen being attached to the slight projection at the foot of the bracket (Figs. 2 and 7).

Cast-iron weights were employed and shaped so as to utilise to the best advantage the limited space available. Final adjustment of the load was made by slightly altering the distance from the fulcrum of the grooved support in which rested the knife-edge fastened to the weights. The tension applied to the specimen was about eight times the direct load of the weight on the long arm of the lever. Loads of up to 100 lb. could be conveniently applied in this way, and the degree of movement of the lever permitted of a maximum extension of the specimen of $\frac{3}{16}$ in. before the beam rested on the stop and the stress on the specimen was relieved. The tension applied was measured by placing the cover with the attached brackets, beams, weights, &c., on a frame with a clear space beneath. The upper end of a dummy specimen was fixed to the upper shackle

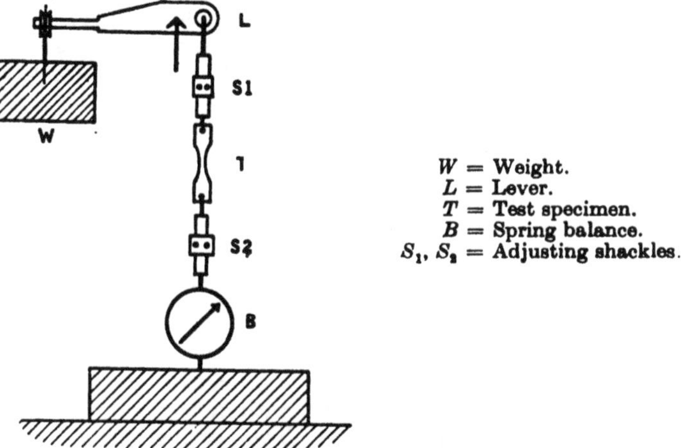

Fig. 3.—Method of Estimating the Load on the Specimen.

and its lower end was attached to a spring balance anchored to a heavy weight resting on the floor. Fig. 3 shows the arrangement diagrammatically. By adjusting the strainers S_1 and S_2 the tension on the specimen could be ascertained from the dial reading, provided that a small allowance was made for the weight of the upper part of the balance, &c. Once the specimens were in position the weights were removed and only replaced when the cover was being finally lowered, as smoothly as possible. Small ball-bearings of stainless steel were used to support the lever and to attach the upper shackle to its short end. New ball-bearings were found to bind during a prolonged experiment, owing to loose black oxide being washed from the cover into the ball races. This difficulty was overcome by acid-pickling the ball-bearings before use so as make them slightly "slack."

These ball-bearings proved quite suitable and no variation in load exceeding 2% of the total could be detected in any working position of the beam. An additional advantage was that the components of the lever system could not be accidentally separated.

External Fittings to the Pressure Vessels.

The cover was fitted with a steam pressure gauge and a dial thermometer, the steel stem of which dipped into the liquid surrounding the test specimen. A pipe connection, with a stop-valve attached, was also provided either for releasing the steam and gases or for injecting water into the cylinder by a hydraulic pump. An emergency pressure diaphragm release was fixed centrally in the cover and connected to a separate pipe (not shown in Figs. 6 and 7) discharging outside the building.

Five insulated steel electric leads were passed through the cover, each consisting of a mushroom head with a threaded stem attached, which was centred and prevented from making metallic contact by two mica washers carried in carefully surfaced recesses in the cover.

Heating Arrangements and Associated Control Gear.

The base of the pressure cylinder was firmly bedded on a hot-plate of heat-resisting iron, which was screwed (with Monel-metal screws) to a heavy steel rim attached to the cylinder. A heating element of nickel-chromium alloy tape was carried in grooves in the hot-plate casting and insulated from it by strips of mica. The hot-plate was supplied with alternating current, about 3 kW. being used to heat up the cylinder from the cold. Once the working temperature was attained, an input of 1¼ kW. was sufficient to maintain it. Thermal lagging was provided by placing the apparatus in a metal container lined with refractory bricks and strips of asbestos cloth on the cover. The heating current was in the first instance regulated by means of an adjustable resistance. Automatic thermostatic control was effected by short-circuiting at intervals another resistance in the heater circuit by the heavy current relay, the pilot coil of which was electrically connected to light contacts incorporated in the dial thermometer fixed in the cover.

Another relay was arranged to interrupt the heating current completely should the normal working pressure in the cylinder be greatly exceeded. The pilot coil of this relay was operated by a pair of light contacts attached to the steam pressure gauge. A further arrangement was made whereby the heating current was interrupted should the temperature of the cylinder rise too high, irrespective of pressure conditions.

Equipment for Indicating Conditions Inside the Pressure Vessel.

To determine the level of the liquid inside the cylinder, three vertical mild-steel rods of different lengths were each attached to the underside of an insulated lead passing through the cover. A small potential applied to these rods caused a current to flow through those rods only which made contact with the liquid. Electrical polarisation effects were avoided by using alternating current.

A knowledge of the condition of the specimens under test was of primary importance, as this was the chief factor in determining the duration of a particular experiment. This was accomplished electrically as follows: A piece of silver strip was attached to the underside of each of the remaining insulated leads passing through the cover and bent so that the long arm of each beam made contact with its corresponding strip when near the end of its normal travel downwards. Thus, the breaking of a specimen would allow the end of the beam, to which the weight was attached, to fall and press tightly on the silver strip. A silver contact was fixed on the beam so as to ensure good electrical connection with the strip when this occurred. A small alternating potential was continuously applied to the strips, and the time of failure of a specimen was indicated by the passage of a current.

Types of Test-pieces and Material used for Tests.

The test-pieces used were in the form of plain, perforated or notched strips, about 4·5 in. in length. Except for two round specimens, the tensile tests in air at room temperature were conducted with plain strip specimens. The shape and dimensions of these test-pieces are indi-

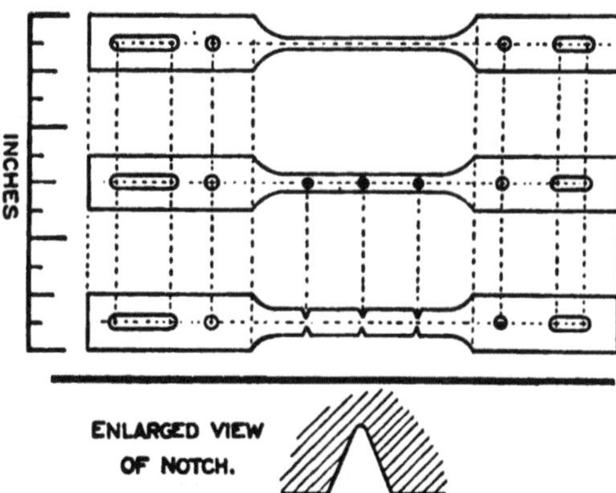

Fig. 4.—Types of Strip Test-Pieces Employed. Notch at four times natural size.

cated in Fig. 4. Although in the form of flat strip, the specimens of boiler plate were not made by rolling a thicker plate. Blanks were first cut from the plate and ground to a thickness of 0·02 in. A number of the ground strips were then clamped together and cut to shape. Except in tests on cold-worked material, all the finished specimens were heat-treated *in vacuo* in a sealed silica tube before testing.

The test material consisted of a portion of a boiler plate which had developed general intercrystalline breakdown from caustic cracking during service. Analyses of the material used for making most of the tests gave:

Carbon	0·14%	Manganese	0·73%
Silicon	0·02%	Copper	0·067%
Sulphur	0·038%	Molybdenum	Trace%
Phosphorus	0·040%		

The microstructure after heat treatment is shown in Figs. 8 and 9. Examination of sections in the as-received state indicated that the plate had been cross-rolled. It was consequently not considered necessary to cut test specimens from the plate in different directions.

Preliminary tensile tests on material cut from the boiler plate were conducted at room temperature in air. The ultimate strength showed some variation, which is thought to be mainly due to the small size of the strip test-pieces, which had a cross-section of only 0·002 sq. in. Thus, a large non-metallic inclusion or ferrite band in the test length might appreciably affect the results. Tests on two larger test-pieces of circular section were in good agreement.

Prolonged-stress tests were also made in air at 225° C. The specimens were loaded by weights and levers, contained in a large oven at the required temperature. A test-piece of boiler plate (carbon 0·14%) stressed at 16 tons per sq. in.—close to, but just below, the yield stress for this material at room temperature—gave no measurable extension in 6 months. Another specimen, stressed at 20 tons per sq. in. for 17 months—somewhat above the yield stress at room temperature—showed some extension at the end of the first month, but no further extension could be detected during the remaining 16 months. Comparative tests of 8 months each were made at 225° C. on steels containing respectively 0·11%, 0·14% and 0·23% of carbon. In these tests care was taken not to apply the full stress until the specimen had reached a steady temperature of 225° C. The yield and ultimate stresses for the three steels at room temperature were known, and the stresses imposed for the prolonged tests at 225° C. were calculated from the arbitrary formula which allowed for varying strength in the material:

$$\text{Stress} = Y + \tfrac{1}{3}(U - Y),$$

where Y = yield stress in tons per sq. in. at room temperature
U = ultimate stress " " " "

None of these specimens showed any measurable extension during the last 6 months, indicating that the specimens would withstand indefinitely the imposition of stresses in accordance with the above formula. For the boiler plate (carbon 0·14%) which was used for many of the experiments with the pressure vessel, this stress was 21·9 tons per sq. in.

Experimental Procedure and Results of Tests in Pressure Cylinder.

Most of the tests were carried out with sodium hydroxide solution of definite strength and maintained at a single fixed temperature. With a few exceptions, the test specimens were made from one piece of used boiler plate, which was known to be susceptible to intercrystalline cracking under service conditions. Some of the fifty tests were terminated in a few hours, while others extended over periods of up to 65 days. The duplication of the loading arrangements inside the cylinder permitted of certain comparative tests, but for general use any advantage was largely negatived by the necessity of opening up the cylinder as soon as the first specimen failed. Ten tests were conducted with normalised plain strip specimens subjected to steady tension while immersed in commercial sodium hydroxide solution of density 1·29 (measured in the cold) maintained at 225° C. A steam pressure of 300 lb. per sq. in. approximately existed over the solution in these circumstances.

Of three specimens stressed at 19 tons per sq. in., two broke in a few hours, while the third remained unbroken after 15 days. A specimen stressed at 22½ tons per sq. in. broke in 1½ hr. No breakages occurred in specimens stressed at 17½ tons per sq. in. or at lower loads.

No small cracks were found in either the broken or the unbroken plain strip specimens after these tests. The results suggest that the margin between a stress which causes rapid failure and one which can be resisted indefinitely is very narrow. Under such conditions it is probable that once a small crack has formed the resulting concentration of stress causes its rapid propagation and fracture of the specimen. The uniform nature of the oxide coating formed on the steel during the test is shown in Fig. 10, which represents a polished and etched section of the specimen that had resisted a stress of 19 tons per sq. in. for 15 days.

At a later stage in this series of tests the question of the influence of sodium silicate in the sodium hydroxide solutions was raised by American workers.[1] In view of their publication, the solid commercial sodium hydroxide used in making up the later solutions was analysed and was found to contain 0·158% of SiO_2. This corresponded with approximately 0·054 g. of SiO_2 per 100 c.c. of the solution used. Tests were also made at 225° C. in a solution, of density 1·29, of a purified brand of sodium hydroxide, the specified maximum quantity of SiO_2 in the solid material being 0·01% (corresponding with, say, 0·003 g. of SiO_2 per 100 c.c. of solution). Two specimens of annealed boiler plate stressed at 22 tons per sq.

[1] Schroeder, Berk and Partridge, *Proceedings of the American Society for Testing Materials*, 1936, vol. 36, Part 2, p. 721.

in. remained unbroken after tests lasting respectively 12 and 19 days. These specimens would undoubtedly have failed had they been tested under similar conditions in a solution of the commercial sodium hydroxide.

Further tests were then made in which the specimens were immersed in a high-purity sodium hydroxide solution to which definite quantities of sodium silicate were added. In tests on normalised specimens at 225° C. in a solution of density 1·29 which contained approximately 0·027 g. of SiO_2 per 100 c.c., one specimen loaded to 22 tons per sq. in. broke after 4 days, while the specimen stressed at 15 tons per sq. in. remained unbroken. A solution of high-purity sodium hydroxide to which had been added 0·2 g. of SiO_2 per 100 c.c. was used for four additional tests. All these specimens, which were stressed at 22 tons per sq. in., fractured in from 17 to 50 hr.

These tests, all made on plain strip specimens of normalised boiler plate, indicate that for the particular conditions of these experiments the presence of a small amount of sodium silicate in the sodium hydroxide solution definitely lowers their resistance to prolonged tensile stress. Thus, a specimen stressed at 22 tons per sq. in. in the high-purity sodium hydroxide solution remained unbroken after 19 days; on the other hand, the average life of four specimens also stressed at 22 tons per sq. in. in sodium hydroxide solution containing 0·2 g. of SiO_2 per 100 c.c. was 30 hr.

Most of the specimens were examined microscopically after test. A small portion of metal was cut (usually from the test length), and after mounting in synthetic resin was polished and etched. In spite of the relatively short exposure, the specimens immersed in the solution containing sodium silicate had suffered considerable general surface attack and were coated heavily with black oxide. This general surface attack was less marked on specimens tested in solutions low in sodium silicate. The test-pieces were only 0·02 in. thick, and any surface attack was an important factor in causing fracture, since the stress acting on the remaining metal was necessarily increased. The conditions for thick material under stress and exposed to hot sodium hydroxide solution containing sodium silicate are somewhat different, however, since any general surface corrosion would need to proceed to a greater depth before it gave rise to an appreciable increase in stress in the remaining metal.

During experiments inside the pressure vessel it was found that frequently a more intense general surface corrosion took place on steel parts which were not in good metallic contact with the adjacent metal masses. Thus, it was found essential to connect the opposite ends of specimens under test with a loop of steel wire. Unless this had been done, a specimen, once fractured, suffered rapid corrosion, particularly near the fractured ends, even in the few hours required for the cylinder to cool down. Specimens electrically insulated (apart from the conductive path through the liquid) from the rest of the cylinder and fittings suffered intense general corrosion in spite of the fact that the very small negative potential which was incidentally applied to them during the experiments would tend to restrain the corrosive action of the liquid. Accelerated surface corrosion is not only caused by the presence of sodium silicate in the caustic solution, but may occur merely as a result of the mechanical arrangement of the steel parts in contact with solution which contains practically no sodium silicate. Apart from the general surface corrosion, the presence of sodium silicate in the solution is responsible for the formation of fine fissures in the steel. These fissures are apparently filled with black oxide, and, as will be seen from Fig. 11, are both inter- and trans-crystalline in character. Another specimen tested under similar conditions showed a large number of short spurs of oxide which projected into the metal. These spurs, which displayed a tendency to penetrate a very short distance along the grain boundaries, were detected not only in the highly stressed metal of the test length but also at the ends which were not appreciably stressed.

Although the addition of sodium silicate to the sodium hydroxide solution accelerated the general surface corrosion of steel specimens under pressure-cylinder conditions, no such action occurred with high-purity iron. Fig. 12 shows, after polishing and etching, a highly-stressed region of a high-purity iron specimen in the form of notched strip. In spite of five days' exposure, the oxide coat was thin and there were no fissures either at the grain boundaries or elsewhere.

In general, the results for steel specimens tend to confirm Schroeder, Berk and Partridge (*loc. cit.*), who found, when working under somewhat different conditions, that exposure to pure sodium hydroxide solution did not reduce the load-carrying capacity of uniformly-stressed specimens. Two of those authors [1] established that sodium silicate additions to the sodium hydroxide solution caused failure of uniformly-loaded steel specimens which would not have failed in pure sodium hydroxide solution. Their specimens were all very highly stressed, however; it should be noted that the minimum stress adopted was just below the ultimate stress for the material at room temperature. The results of the present authors show that additions of sodium silicate may cause failure of uniformly-loaded steel specimens subjected to much lower stresses—stresses in the neighbourhood of the room-temperature yield stress of the material—but that this is due to an increase of general corrosion and not to intercrystalline cracking. All the tests of the present authors, as well as those of Schroeder, Berk and Partridge, were made on relatively thin specimens, and their significance regarding boiler practice is a matter for conjecture.

[1] Schroeder and Berk, *American Institute of Mining and Metallurgical Engineers, Technical Publication No.* 691; *Metals Technology*, 1936, vol. 3, Jan.

Periodic Relieving of Stress.

It was thought possible that the elastic dimensional changes which occurred in the metal when the stress was relieved or re-imposed would break up any protective oxide coat, and so expose more metal to the attack of the sodium hydroxide solution. Accordingly the arrangement shown in Fig. 5 was made. By rotating a screwed rod *B*, which passed through the cylinder cover, pressure was brought to bear on the small projection *F* attached to the beam *G*. By this means the end of the beam attached to the specimen was forced down and the stress was relieved. A gland

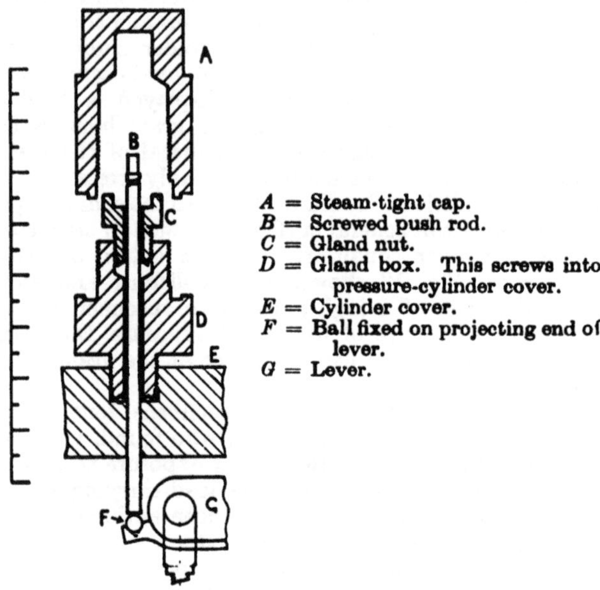

Fig. 5.—Apparatus for Relieving Stress on Specimen inside Pressure Vessel.

prevented serious leakage of steam when the steam-tight cap *A* was removed to rotate the screwed rod.

Two plain strip specimens of normalised boiler plate were stressed at 17½ tons per sq. in. at the same time in the cylinder, this being approximately the highest stress which could be resisted indefinitely in the commercial sodium hydroxide solution. The stress on one specimen was relieved for 5 min. daily while that on the other was steadily applied. Neither specimen broke in 65 days.

Cold-Worked Strip Specimens.

Two plain strips of normalised boiler plate were reduced 20% in thickness by cold-rolling and then stressed at 17½ tons and

14 tons per sq. in., respectively, in commercial sodium hydroxide solution. The former specimen failed in 3 days, while the latter remained unbroken after 48 days. This suggested that the cold-rolling operation lessened the resistance, since normalised specimens were known to resist a stress of 17½ tons per sq. in. for long periods. Microscopical examination after fracture showed several penetrating oxide masses. Some were of the "blunt" type shown in Fig. 13, while others occupied narrow cracks, mainly transcrystalline, as in Fig. 14. Further tests were made on boiler plate which had been normalised and then reduced 65% in thickness by cold-rolling. The highest stress applied to a specimen of this series was 24½ tons per sq. in. This was resisted successfully for 12 days. The extension on the test length was certainly less than 0·05 cm.

Although the preliminary tests with the 20%-reduced specimens suggested that cold-working reduced the resistance under pressure-cylinder conditions, the heavily cold-worked material offered a greater resistance than the normalised specimens. The latter results are in agreement with those of Parr and Straub (*loc. cit.*). The different behaviour under stress of the 20%-reduced specimens may perhaps be attributed to the fact that they were cold-rolled after being made into specimens, and the metal in different zones may have undergone very different amounts of cold-working.

Application of Negative Potential during Test.

A plain strip of normalised boiler plate was stressed at 17½ tons per sq. in. and an insulated steel rod, metallically connected to one of the insulated electric leads passing through the cylinder cover and made electro-positive in respect to the specimen under tension (as well as the rest of the cylinder and fittings), was placed close to it in the solution. A current of 0·1 amp. was passed through this electrode and it was expected that a considerable proportion of this current would pass through the solution to the specimen, which was provided with a special electrical connection to the main part of the cylinder. The specimen fractured after 5½ days, and, although no microscopical cracks (apart from the main fracture) were detected, the oxide coating separated readily from the underlying metal, which appeared quite bright.

Tests on Perforated Specimens.

The effect of localised stress was investigated by means of the perforated type of specimen shown in Fig. 4. Perforated strips of normalised plate were stressed at 15 tons and 17 tons per sq. in. average stress, calculated for the minimum cross-section. The specimen stressed at 17 tons per sq. in. remained unbroken after 35 days. Microscopical examination of sections near the holes revealed numerous non-metallic inclusions. The preferential corrosive attack on these inclusions near a hole is indicated in Fig. 15. Under higher powers, intrusions of oxide were detected at the edge of a hole, and near the tip of one of these oxide intrusions fine intercrystalline cracks were detected (Fig. 16).

The work with specimens subjected to moderate concentrations of stress suggests that failure would ultimately occur owing to the slow growth of oxide masses in the regions of small original inclusions in the steel. Signs of intercrystalline attack of the metal were rarely observed.

Tests on Notched Specimens.

Experiments were conducted on steel specimens of the notched-strip type, giving a high concentration of stress, as shown in Fig. 4. An average stress of 12 tons per sq. in. on the minimum cross-section was resisted successfully for 48 days, although appreciable extensions had occurred at the notches. A stress of 14 tons per sq. in. brought about failure in 50 hr., and with higher stresses the specimens fractured in shorter times. These results indicate that severe local concentration of stress is an important factor in failure.

A microscopical examination of the broken notch of the specimen which failed under test (Fig. 17) showed intrusions, some of which were transcrystalline, while others tended to follow the crystal boundaries. A further example of a thin penetrating oxide intrusion was detected at an unbroken notch of another specimen after test (Fig. 18). Although the line of this intrusion appears to be slightly influenced by crystal boundaries and the presence of pearlite, the straightness of its course indicates its transcrystalline nature.

Tests on Notched Specimens of Steel Prepared in the Laboratory.

In a laboratory investigation of a boiler plate which had failed by caustic cracking in service, it was noticed that a strong smell characteristic of hydrogen phosphide was evolved when pieces were being ground or polished. It seemed possible that the phosphorus in the steel might be a factor conducing to caustic embrittlement. A melt was accordingly prepared from ingot iron with additions of carbon and calcium phosphide in a high-frequency induction furnace. After forging and rolling the small ingot thus produced, several notched strip test-pieces were prepared and normalised. No analysis was made, but the typical smell of a phosphide was present during grinding and polishing. Microscopical examination showed that the material was almost devoid of pearlite and that an unusual amount of cementite was present at the ferrite grain boundaries. A notched strip specimen stressed at 15 tons per sq. in. (average stress calculated on the minimum area of cross-section at the notch) failed in 36 hr. under pressure-cylinder conditions. After this test the specimen was ground and polished near one of the notches and examined microscopically after etching. Some rather broad oxide intrusions were detected. From Fig. 19, which shows a typical oxide intrusion, it will be seen that there is little indication of *any* intercrystalline penetration.

Tests on Notched Specimens of High-Purity Iron.

Specimens of high-purity iron stressed at 10 tons and 7 tons per sq. in., respectively (calculated on the area of minimum cross-section), and immersed in commercial sodium hydroxide solution fractured in less than one day. Apart from the main fractures, small cracks had formed near the roots of most of the notches. Fig. 20 shows an area near the root of a notch after polishing and etching. Of the three cracks visible, only the centre one appears to follow in any way the crystal boundaries. Another type of breakdown characterised by several closely associated fine almost parallel cracks was also observed in the iron.

Composition of the Black Oxide Produced by Corrosion.

Two samples of magnetic black oxide of iron were chemically examined. The first was obtained by exposure of a boiler-plate specimen to sodium hydroxide (sp. gr. 1·29, cold) at 100° C.; the second was removed from the wall of the high-pressure vessel after the action of sodium hydroxide (sp. gr. 1·40, cold) at 225° C. The first contained 69·2% and the second 69·1% of metallic iron. These figures agree closely with the iron content of Fe_2O_3 (69·4%) and not with that of Fe_3O_4 (72·3%). On ignition at 800° C., the first sample of oxide became red in colour and contained 69·8% of metallic iron. Thus, it appears that the black oxide of iron produced in these particular circumstances is largely Fe_2O_3. Under different experimental conditions, involving higher temperatures, an oxide corresponding with Fe_3O_4 has been obtained (*see* Part III.).

General Conclusions.

(1) The present work on specimens of normalised boiler plate involved the simultaneous action of prolonged stress and the corrosive influence of hot strong solutions of sodium hydroxide. In some experiments the stress was concentrated by notches and by drilled holes. No important attack on the crystal boundaries occurred, a fact which disagrees in some measure with the conclusions of Parr and Straub. The authors consider that some other factor, which is necessary to give rise to intercrystalline cracking, was not introduced into the experiments. For example, such a factor might be one which would permit the ready entrance of hydrogen into the steel. Attempts to introduce hydrogen into the steel by electrolytic means have not so far produced general intercrystalline breakdown. The results of these experiments have been difficult to interpret on account of the severity of the general corrosion.

(2) When specimens of normalised boiler plate were uniformly stressed at 225° C. in a solution of high-purity sodium hydroxide to which a small amount of sodium silicate had been added, a

very slight amount of intercrystalline cracking took place. This attack was of a shallow type and occurred both in the highly-stressed metal of the test length and in practically unstressed metal near the end of the test-pieces. Specimens stressed appreciably beyond the room-temperature yield stress fractured in a few hours, whilst specimens stressed under similar conditions in a solution of high-purity sodium hydroxide resisted the attack for many days. Owing to the excessive influence of general surface corrosion in these strip specimens, it is doubtful to what degree these results apply to material of plate thickness.

(3) In order to study the effect of stress concentration, both perforated and notched specimens were stressed at 225° C. in a solution of commercial sodium hydroxide. In the perforated specimens where regions of moderate stress concentration existed, deterioration of the metal was caused by the steady growth of the original non-metallic inclusions in the steel. A different effect, however, occurred in the regions of highly-localised stress in notched specimens; here the material failed owing to cracking, mainly transcrystalline, but occasionally influenced in direction by the presence of a crystal grain boundary.

(4) Heavily cold-worked boiler-plate material, uniformly stressed at 225° C., and exposed to commercial sodium hydroxide solution resisted for long periods a stress which would have caused fracture of a normalised specimen of the same material in a few hours when tested in a similar manner.

(5) The black magnetic oxide formed on specimens of mild steel when exposed to sodium hydroxide solution at 225° C. was found by analysis to correspond closely in composition with Fe_2O_3.

The authors desire to acknowledge the assistance of Messrs. C. A. Harvey and A. J. Cook in the constructional and observational work. The analyses of specimens, solutions, gases and oxides of iron were undertaken by the Chemistry Division of the Metallurgy Department, while the machining of parts and specimens was undertaken at various times by the workshops attached to the Engineering, Metallurgy and Aerodynamics Departments. The authors desire to acknowledge the suggestions of Mr. A. Bailey, M.Sc., A.M.Inst.C.E., concerning the diaphragm safety pressure release.

Part III.—Exposure of Iron and Steel Specimens to Sodium Hydroxide Solutions at High Temperature and Pressure.

BY FRANK ADCOCK, M.B.E., D.Sc. (ASSISTED BY A. J. COOK)
(NATIONAL PHYSICAL LABORATORY).

(Figs. 26 to 50 – Plates XV. to XVIII.)

Experiments described in the previous section (Part II.) showed that when boiler-plate material was immersed in sodium hydroxide solution of density 1·29 at temperatures up to 250° C. and maintained under a stress sufficient to cause definite yielding, rupture occurred in a few hours. This confirmed the work of Parr and Straub.[1] On the other hand, inspection of many test-pieces subjected to combined stress and caustic attack failed to reveal intercrystalline cracks. In this respect the results of the authors differed from those of the earlier workers. The position has been rendered more confused by the recent publication of work which has raised doubts concerning several factors that at one time were accepted as proven. It was clear that additional experimental evidence would be required before the mechanism of the caustic cracking of boiler plates could be elucidated. It seemed that the problem was a complex one and that a simple solution was unlikely.

It was decided, in the first instance, to study the reaction between iron and steel and sodium hydroxide solution at high temperatures in small sealed cylinders. Further work was to depend chiefly on the preliminary results obtained. Actually, some interesting but erratic results were obtained from the sealed cylinders, and the research was continued by making use of a silver-lined steel tube as a pressure vessel. A certain amount of attention was also devoted to subsidiary problems which arose from time to time.

TESTS WITH SMALL SEALED CYLINDERS.

The cylinders were $\frac{1}{2}$ in. in dia. by $\frac{3}{4}$ in. long, with a central hole $\frac{1}{8}$ in. in dia. by $\frac{9}{16}$ in. long. Sodium hydroxide solution or some other substance was placed in the cavity, which was then sealed by a cupro-nickel washer and an external clamp. Details of the cylinder and clamps are given in Fig. 21. In order to secure a tight joint it was found necessary to machine carefully the end of the cylinder and to coat it electrolytically with copper. Once the cylinder was sealed, the clamp and the cylinder within it were placed inside a thermostatically-controlled furnace. Observations of temperature were made by means of a thermocouple contained in a silica sheath, which passed through the central hole in the clamp and rested on the end of the cylinder. That portion of the

[1] Parr and Straub, *University of Illinois Engineering Experiment Station*, 1928, *Bulletin No.* 177.

clamp which compressed the cupro-nickel washer was made of high-tensile steel, the other parts of the clamp being of mild steel.

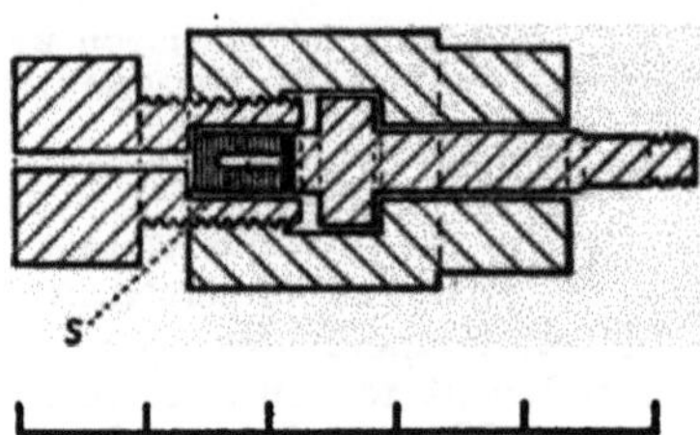

Fig. 21.—Cylinder under Test (marked *S*) and Clamp (diagrammatic). Scale in inches.

Although some tests were made at temperatures as low as 225° C., most of the early experiments were conducted at 470° C.

Tests on Iron and Steel at 470° *C.*

In general, the specimens were attacked rapidly at 470° C. and a large number of comparative tests could be quickly made. It was concluded from these experiments that ingot iron and straight steels of various carbon contents all rapidly develop intercrystalline cracking when subjected to the action of either pure or commercial sodium hydroxide solution. The depth of cracking around the central hole in the hardened and tempered steel (carbon 1·2%) will be seen from Fig. 26. In this case the magnification is too low to show the nature of the cracking.

Tests on Boiler-Plate Material at Temperatures Ranging from 225° *to* 470° *C.*

As it was of considerable interest to ascertain the lowest temperature at which this form of intercrystalline attack occurred, further tests were made on cylinders of boiler-plate material over a temperature range of 225° to 470° C. No intercrystalline penetration was detected in a specimen held for one month at 225° C. On the other hand, a cylinder held at 310° C. for 3 weeks showed considerable intercrystalline penetration. Fig. 27 depicts the appearance of the polished and etched metal close to the central hole. The oxide is present both in the massive form and as relatively fine intercrystalline films. At 360° C. considerable intercrystalline cracking occurred in 6 days (Fig. 28). When the temperature was raised to 400° C., apparently isolated intercrystalline cracks appeared at a distance from the central hole, often at the edge of the pearlite, the structure of which was somewhat modified close to the crack. A typical example is given in Fig. 29. Tests at 470° C. resulted both in intercrystalline cracking and in decarburisation as indicated by the disappearance of the pearlite.

As will be discussed later, some of the carbon is removed in the form of hydrocarbons. The depth from which the pearlite was removed is shown in Fig. 30, but the magnification is not high enough to show clearly the intercrystalline cracks which extend well beyond this decarburised zone. Appreciable decarburisation as shown by the removal of the pearlite only occurred in tests lasting up to 48 hr. at temperatures of 400° C. or above.

From this series of experiments it appears that destructive intercrystalline cracking of boiler-plate material may occur in the presence of sodium hydroxide of high purity at temperatures as low as 310° C.

Tests at 420° to 470° C. on Boiler-Plate Material with Various Substances in the Cavity.

In order to ascertain to what degree sodium hydroxide solution was peculiar in causing intercrystalline cracking, tests were made at 420–470° C. on cylinders of boiler-plate material with the following substances in the cavities: Distilled water; solid sodium hydroxide; metallic sodium; distilled water and mercury; distilled water and tin; sodium nitrate solution (25 g. + 100 c.c. of water); strong sodium silicate solution; and potassium hydroxide solution equivalent in strength to sodium hydroxide solution of density 1·29. It was concluded that intercrystalline attack only occurs in short-time tests when both caustic and water (vapour) are present in the sealed cavity. Further, the action of potassium hydroxide is similar to that of sodium hydroxide.

Tests on Various Ferrous Materials at 420° to 470° C. with Commercial or High-Purity Sodium Hydroxide Solution in the Cavity.

Both cold-rolled and nitrided boiler-plate material suffered intercrystalline attack at 420° C. in the presence of commercial sodium hydroxide solution. A 5% nickel steel was next tested at 470° C., and Figs. 31 and 32, at low and high magnifications, respectively, show the polished and etched specimen after test. The general zone of attack is indicated in Fig. 31, while Fig. 32 shows that the cracking is intercrystalline. Two copper-bearing steels containing respectively 0·28% and 0·98% of copper, also developed serious intercrystalline cracking during the tests. An austenitic steel (chromium 18%, nickel 8%, with additional elements) developed several large cracks under test, which appeared to be intercrystalline. Fig. 33 shows a polished and etched section of this material containing a typical crack.

Nickel washers were used to seal the cylinder cavities in some of the earlier tests, but this material did not prove satisfactory. The reason will be apparent on inspection of Fig. 34, which shows the polished and etched section of a nickel washer after test. The material, which was subjected to considerable mechanical stress, had developed general intercrystalline weakness.

General Observations Based on the Experiments with Small Pressure Cylinders.

At the conclusion of most of the experiments at high temperature, gas under high pressure was present in the cavity. By means of the apparatus shown in Fig. 22, a small hole was drilled through the cupro-nickel sealing washer and the escaping gas collected over water. A cylinder of boiler-plate material contained a mixture of 23% of methane and 77% of hydrogen under an estimated pressure of 70 atm. (when cold). The gas remaining in the cavity of a cylinder of drill-temper steel (carbon 1·2%) appeared to be entirely methane. These results must be regarded as approximations, since methane is rather soluble in water.

All the materials tested in the form of sealed cylinders were shown to be susceptible to intercrystalline cracking, but it was

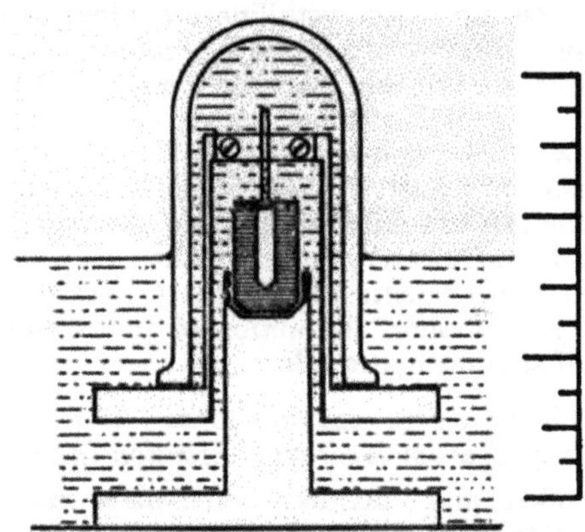

FIG. 22.—Arrangement for Drilling Cupro-Nickel Washer of Cylinder and Collecting Gas (diagrammatic). Scale in inches divided into quarters.

impossible to ignore the fact that many experiments did not give rise to any attack beyond superficial oxidation. At first these failures were attributed (with some justification) to leakage due to faulty sealing, but this explanation was not valid for later experiments in which liquid still remained in the cavity after an unsuccessful run. When pairs of cylinders of the same material were subjected to identical temperature and chemical conditions, one cylinder would frequently show heavy intercrystalline penetration while the other would remain practically unattacked.

An attempt was made to simulate boiler conditions more closely by means of long narrow and rather thin-walled cylinders subjected to a tensile stress of 6 tons per sq. in. during the action of sodium hydroxide solution (in the cavity) at high temperature. None of these cylinders, however, suffered appreciable intercrystalline attack. In view of the erratic results and the complete lack of knowledge of conditions existing in the cavities at high temperature, it was decided to make further experiments in which the specimen under test was placed inside a container resistant to the action of sodium hydroxide solution.

Tests made in Connection with a Silver-Lined Pressure Vessel.

The pressure vessel (Fig. 23) consisted essentially of an alloy-steel tube 18 in. long, $\frac{3}{4}$ in. in external dia. and with a bore of 0·312 in. after slight machining. It was intended that the reactions

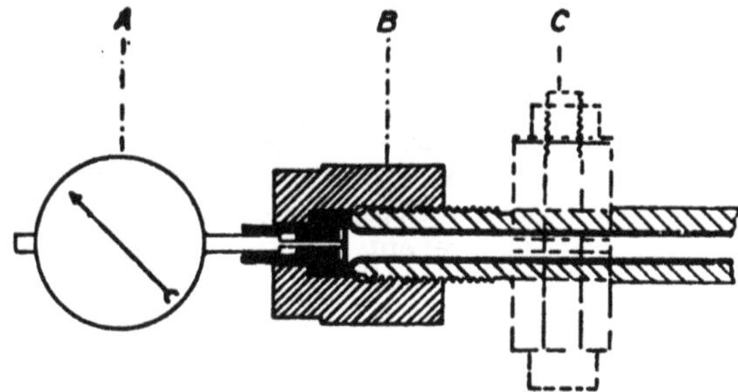

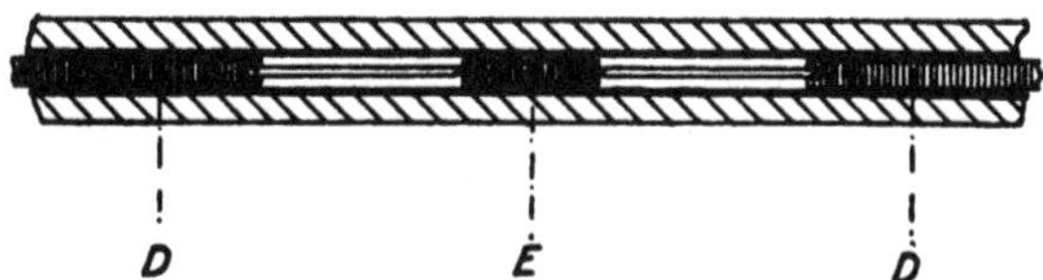

A = Dial gauge to indicate deflection of diaphragm.
B = Sealing cap.
C = Water-cooled clamp for introducing heating current into tube.
D = Silver plugs.
E = Specimen under test held in central position by silver wires extending from silver plugs.

Fig. 23.—Silver-Lined Pressure Tube. Sections of one end and central portion of tube.

to be studied should take place inside the tube, and to prevent contact between the steel and the sodium hydroxide solution a thin-walled silver lining was fitted inside the tube. At both ends of the steel tube the bore was chamfered internally and the silver liner expanded or flanged over the ends in such a way that seals could be readily made by diaphragms of spring steel. These diaphragms were faced on the solution side with a thin disc of silver and held in position by the screwed end-caps. For many experiments the reaction space was limited to a length of 4 in. in the centre of the tube by inserting closely-fitting rods of silver 7 in. in length at each end. The solution was thus in contact with silver except for the specimen under test. One end-cap carried a dial gauge (of a type used in measuring the thickness of material), and this was arranged to measure the deflection of the diaphragm. As diaphragms of the same material and thickness had been already calibrated with the gauge by means of a pressure testing machine, it was possible to estimate the pressure in the tube at any stage of the experiment.

The diaphragm at the other end of the tube was somewhat thinner, and acted as a safety release by bursting should the pressure attain 5 tons per sq. in. As suitable electrical equipment was already available, it was found convenient to heat the tube by passing along its length an alternating current of about 1000 amp. at less than one volt. The current was carried to the tube by water-cooled steel clamps fixed near the ends, the tube being slightly lagged with asbestos string between the clamps. These clamps were also used to prevent the tube from rotating when screwing or unscrewing the sealing caps. Temperatures were closely maintained by means of a thermocouple bound closely to the tube and operating in conjunction with a furnace temperature-controller. During experiments which lasted overnight, or longer, evidence that the temperature had been properly regulated was obtained by means of an independent thermocouple connected with a temperature recorder. It was assumed that the temperature of the reaction space was very close to that of the outside of the tube, and an exploration made by thermocouples tightly bound on to the outside of the tube revealed that the variation of temperature over the central 4-in. length (the reaction space) was less than 10° C.

The intercrystalline cracking of boiler plates in service may occur gradually and only become appreciable after a considerable time. It was impossible to conduct many long-period experiments imitating such conditions with the single piece of apparatus available, but one test lasting 17 days was made at 300° C. In this experiment a piece of boiler plate, previously cold-worked, was subjected to the action of sodium hydroxide solution of density 1·29. On micro-examination, after test, the polished and etched specimen was found to contain intercrystalline cracks near areas of pearlite which had in places suffered slight decarburisation. A number of experiments lasting 17 hr. approximately and at a temperature of 410° C. demonstrated that the same cold-worked material when in contact with this solution developed intercrystalline cracks similar to those produced at 300° C. Although the short-time tests at the higher temperature caused somewhat more cracking, it was thought that such tests would give a trustworthy indication of the results likely to be obtained with longer tests at lower temperatures. Accordingly many short experiments were made at 410° C.

Tests on Cold-Worked Steel at 410° *C.*

The following specimens, all of which had been previously cold-worked, were subjected to the action of 1·29 density sodium hydroxide solution at 410° C. for 17 hr.:

> Samples of boiler plate containing respectively 0·14% and 0·20% of carbon.
>
> Steels containing respectively 0·09%, 0·19% and 1·1% of carbon.
>
> Ingot iron.
>
> Four iron-carbon alloys of high purity containing respectively 0·031%, 0·15%, 0·16% and 1·01% of carbon.
>
> Mild-steel tubing.
>
> Two samples of mild-steel tubing, not deliberately cold-worked but probably not fully annealed after drawing.

All these samples developed intercrystalline cracks. Fig. 35 shows the cracks in a piece of steel tubing after treatment, the boundary cementite and pearlitic areas being almost entirely destroyed.

Tests on Annealed Material at 410° *C.*

The boiler-plate materials and mild steels behaved quite differently when fully annealed, and were found to be free from cracks after 17 hr. at 410° C. in sodium hydroxide solution of 1·29 density. On the other hand, the 1·1% carbon steel and the four high-purity alloys developed intercrystalline cracks, but less than the corresponding cold-worked samples. The fact that the commercial mild steels and boiler plate behaved similarly to each other but differently from the high-purity iron-carbon alloys will be discussed later.

Tests on Specially Treated Steels.

As the intercrystalline cracks produced in the preceding experiments were associated with the carbon-bearing regions, it was thought advisable to study the effect of treating the steels in such a way as to alter the size of the carbide particles before test.

A cold-worked specimen of heavily spheroidised steel containing 0·13% of carbon was treated with sodium hydroxide solution, density 1·29, for 17 hr. at 410° C. The cracks formed were intercrystalline and not associated in any way with carbide particles remote from crystal boundaries. Another sample containing 0·19% of carbon was held for 15 min. at 950° C., water-quenched and tempered at 410° C. for 24 hr. One end was cold-worked, and the whole specimen was then subjected to the caustic soda attack for 17 hr. at 410° C. After test the cold-worked end was found to contain many cracks (Fig. 36).

As far as could be judged in a martensitic material, these cracks were intercrystalline and were confined to the cold-worked end. A certain amount of the martensitic structure was destroyed near the cracks. It was concluded from these two experiments that the state of division of the carbide had little influence on the degree of cracking.

Investigation of the Mechanism of Cracking of Specimens Subjected to Caustic Attack in a Pressure Tube.

A specimen of boiler plate which had developed cracks as the result of sodium hydroxide attack in the pressure tube was drilled

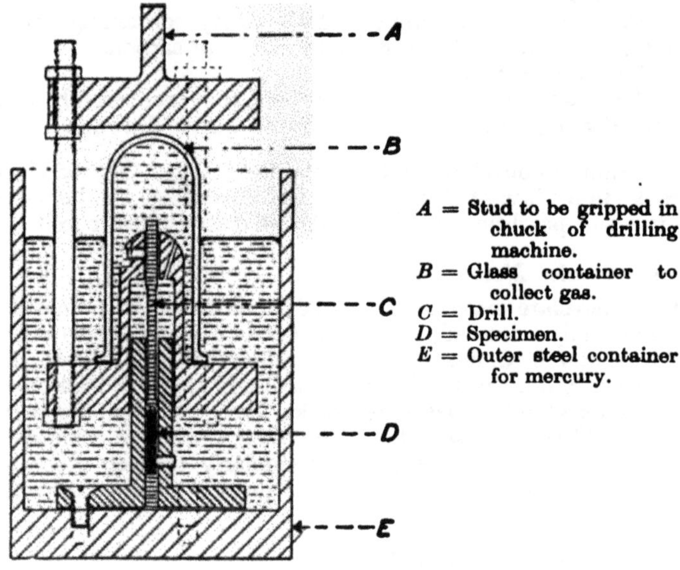

Fig. 24.—Apparatus for Drilling Specimens under Mercury so as to collect the gas evolved.

to pieces in the apparatus shown in Fig. 24. The gas evolved was collected over mercury and gave the following analysis :

Methane.	Hydrogen.	Nitrogen.	Carbon Monoxide.
72·9%	7·3%	13·8%	5·9%

Altogether the gas collected amounted to about ten times the volume of the specimen and proved to be mainly methane. A similar test on the original steel, not cracked by the caustic action, gave an insignificant quantity of gas.

In general, the chemical conditions existing inside the specimen during the caustic attack appeared to be quite different from those at the surface. Although the surfaces after test were coated with oxide, yet no oxide could be detected in the cracks which had formed in the interior. The mechanism of cracking appeared to be as follows: Hydrogen was produced near the surface and diffused into the metal, where, under some circumstances, it attacked the carbide particles. This reaction produced methane, which was unable to diffuse out of the metal and formed the intercrystalline cavities, which contained the gas under high pressure. The oxidising attack was confined to the surface, and in this region the carbide particles were more resistant than the ferrite, as was most clearly shown in a high-carbon steel (Fig. 37). The oxide coat in its gradual advance into the metal reached some of the cavities, formed earlier by the hydrogen attack, and filled them with oxide. That the cracks were formed by the accumulation of methane was confirmed by further experiments. A piece of boiler plate which had been cracked by caustic treatment was heated to 650° C. for 45 min. Microscopic examination after polishing and etching showed that the cracks had become somewhat wider and that minute particles of cementite had formed near their edges (Fig. 38). This suggests that the methane had first widened the cracks by increased pressure on heating and later recarburised the surrounding iron. Another sample (after caustic attack) was heated momentarily to about 900° C., and, as will be seen from Fig. 39, small, newly-formed pearlite masses were present at the edges of the cracks. When the sample was heated for an appreciable time over 900° C. the pearlite areas were of normal size and no longer associated with the cracks. As mentioned earlier, the cold-worked samples of commercial mild steels developed intercrystalline cracks under caustic attack, while the corresponding annealed samples did not suffer. Both cold-worked and annealed samples were coated with oxide.

High-purity iron-carbon alloys which had been allowed to solidify in the crucible and subsequently annealed cracked under caustic attack, but not so rapidly as the same materials when cold-worked. This suggests that the hydrogen diffusing through the ferrite can only attack the carbide when minute (sub-microscopic) cavities are available for the initial quantities of methane produced. Such cavities may exist in cast alloys and in cold-worked commercial mild steels, but not in steels which have undergone considerable hot-working and subsequent annealing. This view is supported by the behaviour of a high-purity iron-carbon alloy (carbon 0·13%). This material when hot-forged and later annealed cracked much less readily than did the metal in the as-cast and subsequently annealed state.

At this stage it appeared that the caustic attack in the pressure tube could give rise to intercrystalline cracking accompanied by decarburisation of a type closely resembling that produced at high temperatures by hydrogen under pressure. To test this, a small-diameter mild-steel tube was fixed inside the silver-lined pressure vessel, as shown diagrammatically in Fig. 25 (a hydrogen inlet was provided in one end-cap, but is not shown in the figure). When a pressure of 100 atm. of hydrogen was maintained in the annular space between the tubes and the temperature was gradually raised, an increasing amount of hydrogen passed through the walls of the small tube. At 400° C. this rate was found to be approximately 0·05 c.c. of gas per min. (at atmospheric pressure) per sq. cm. of tube surface, the metal being about 1 mm. thick. There was no noticeable difference in the rate of flow through a tube in the as-received and the annealed state. On the other hand, typical intercrystalline cracks accompanied by decarburisation occurred

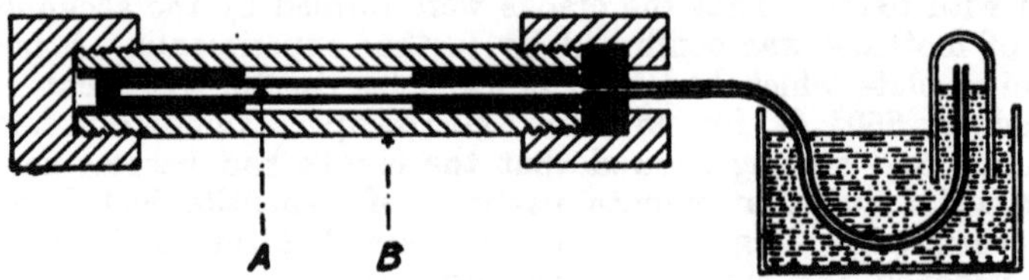

A = Thin-walled steel tube through walls of which hydrogen passes.
B = Silver-lined steel tube sealed by end-caps.

Fig. 25.—Passage of Hydrogen through Steel undergoing Caustic Attack (diagrammatic).

in the as-received tube, but the annealed tube appeared to be quite unchanged.

These experiments confirmed the belief that the cracking produced in the pressure tube by the action of sodium hydroxide was almost identical with that produced by gaseous hydrogen under high pressure. The exhaustive article by Neumann [1] describes the latter form of attack. A further test was made in a pressure tube in which a specimen of boiler plate was placed in contact with distilled water and heated to 410° C. for 5 days. Intercrystalline attack and decarburisation were produced. The sodium hydroxide solution of density 1·29 thus acted in the same manner as steam but more rapidly. This was probably connected with the solubility of iron oxide in caustic soda solution at high temperatures.

Investigation of Oxide Formed in the Pressure Tube.

On opening the pressure tube after most experiments with caustic solution some well-formed glistening octahedral black crystals of iron oxide were found adhering to the ends of the silver

[1] Neumann, *Stahl und Eisen*, 1937, vol. 57, Aug. 12, p. 889.

plugs (Fig. 40). These crystals were about 1 mm. in dia. and had been apparently deposited from solution. The crystals were magnetic, and X-ray examination showed them to be single crystals of cubic structure. Heating for 2 hr. in air at 600° C. failed to alter the X-ray structure or to destroy the magnetic properties. A microchemical investigation showed the presence of ferrous iron. The absence of hydrogen generation when the oxide was treated with acid indicated the absence of metallic iron. Microscopic examination of a crystal showed it to be homogeneous and free from cavities. It was concluded that the material was cubic Fe_3O_4 deposited directly from the liquid. In view of the possible importance of the solubility of iron oxide in sodium hydroxide solutions under pressure tube conditions—in connection with the caustic cracking of actual boiler plates—the following experiment was undertaken. A small hollow silver cylinder, fitted with a powdered-silver filter at each end, was placed in a slightly cooler part of the pressure tube while a steel specimen was subjected to caustic attack. After the experiment octahedral crystals of iron oxide of appreciable size were found inside the cylinder, thus establishing that the iron oxide had been transferred in some extremely divided form (or solution) from one place to another. (The filters were found still effective after test.) The deposition of iron oxide at a distance from the seat of formation might conceivably result in the complete sealing of a crevice in a boiler and the subsequent development of excessive pressures in the enclosed space.

Tests made with Concentrated Solutions of Sodium Hydroxide.

During some of the earlier experiments in which specimens were exposed to sodium hydroxide solution of density 1·29, two mild-steel tubes, one fitting closely but not tightly inside the other, had been placed in the pressure tube. At the finish of the experiment intercrystalline attack, as shown in Fig. 41, was found near the outside surface of the inner tube. This attack was very different in type from that produced by the caustic of density 1·29 and was believed to be due to the gradual concentration of the sodium hydroxide solution in the crevice between the tubes. Accordingly, a few experiments were carried out with solid sodium hydroxide and with very strong solutions. First, a sample of cold-worked boiler plate was treated with a caustic solution (3 parts of NaOH + 1 part of H_2O by weight, approx.) at 410° C. for 46 hr. When removed from the tube after the experiment the specimen was dull white and almost free from adhering oxide. Under the microscope the interior showed the usual form of short intercrystalline attack associated with carbon-bearing regions, but near the edges, as will be seen in Fig. 42, the intercrystalline cracks were more continuous and there was no preferential attack in the carbon-bearing zones. In further tests lasting 41 hr. and conducted at

410° C. a study was made of the action of strong sodium hydroxide solutions (3 parts of NaOH + 1 part of H_2O by weight, approx.) on the following materials:

Boiler plates containing 0·14% and 0·20% of carbon, respectively.

Steels containing 0·09%, 0·19% and 1·11% of carbon, respectively.

Ingot iron.

High-purity iron-carbon alloy, carbon 1·01%.

High-purity iron-chromium alloy, chromium 11·9%.

High-purity iron-chromium-nitrogen alloy, chromium 11·8%, nitrogen 0·12%.

Boiler plate, carbon 0·14%, previously worked at 250° C.

Except for the last-mentioned specimen, all the samples were fully annealed and then one end was cold-worked prior to testing. The commercial steels all behaved similarly, showing the hydrogen type of cracking in their interiors where the metal had been cold-worked and the more continuous type of cracking at the edges. This latter type of intercrystalline cracking associated with the attack by strong caustic solutions occurred without distinction in the annealed and worked metal. At certain stages of the attack, the cracks often stopped short at the pearlite areas, as shown in Fig. 43, which depicts at a high magnification a polished and etched section of a mild steel tube after strong caustic attack. A form of intercrystalline cracking which often avoided pearlitic areas was found in a specimen of boiler plate after attack at 410° C. by strong sodium hydroxide solution. The cracks (Fig. 44) were characterised by many spurs or pits which projected into the crystal grains. Occasionally the pits were well defined and their orientation was the same throughout individual crystal grains.

Steel specimens which had been previously cold-worked and then exposed to strong sodium hydroxide solution frequently revealed zones where both forms of intercrystalline attack were present. Thus, Fig. 45 shows that cracking not only occurs alongside pearlitic areas (hydrogen type of cracking) but also at crystal boundaries where carbide had apparently been absent (oxidising type of cracking). This dual form of intercrystalline attack would necessarily reduce the mechanical strength to negligible proportions.

Under the action of the alkali the high-purity 1·01% carbon iron alloy was disintegrated into what appeared to be large crystal grains with facets. On the other hand, the iron-chromium and the iron-chromium-nitrogen alloys suffered no intercrystalline attack under similar conditions. The sample of boiler plate which had been worked at 250° C. developed intercrystalline attack similar to that undergone by the cold-worked metal.

Fig. 7.—Fittings on Lid of Pressure Cylinder. Weight removed from lever in foreground.

Fig. 6.—General View of Pressure Cylinder. Heater and lagging removed.

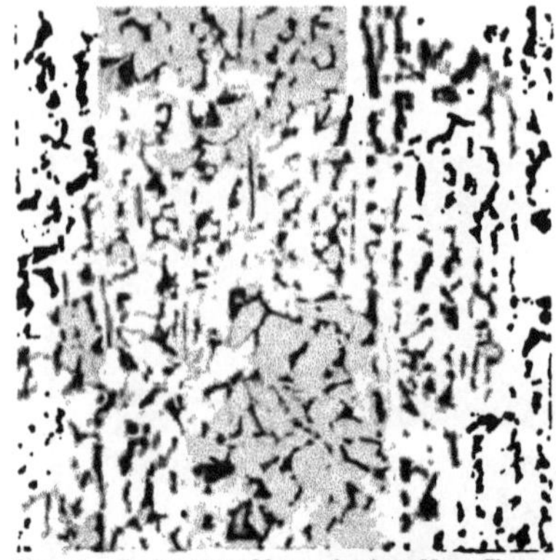

Fig. 8.—Boiler-Plate Material after Heat Treatment. × 150.

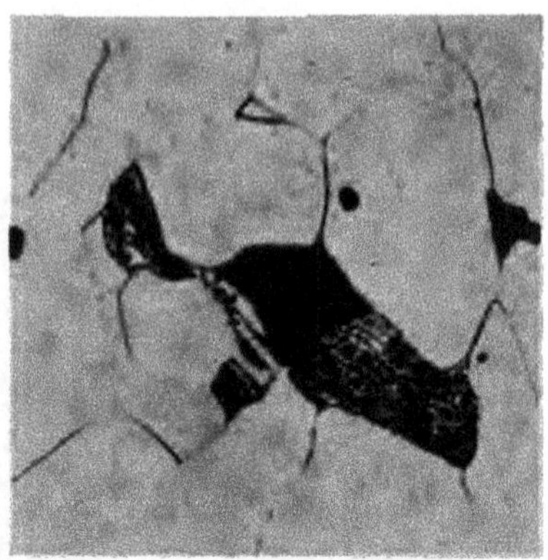

Fig. 9.—Boiler-Plate Material after Heat Treatment. × 1500.

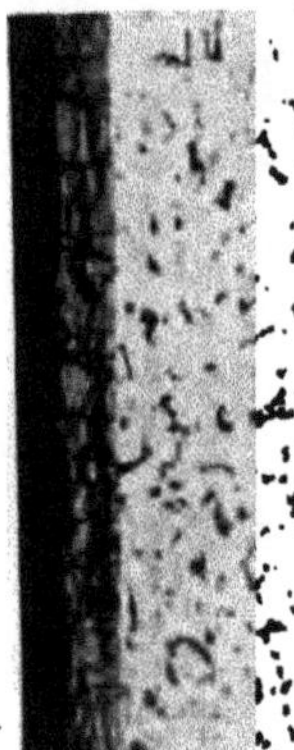

Fig. 10.—Oxide Coat on Stressed Specimen of Boiler Plate after 15 days in hot sodium hydroxide solution. × 150.

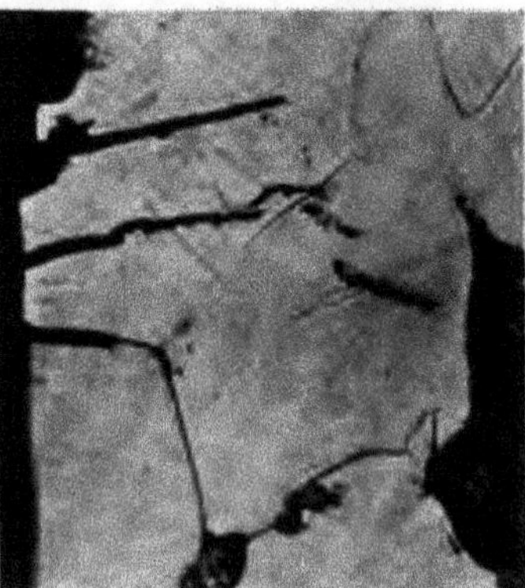

Fig. 11.—Fine Fissures in Stressed Specimen of Boiler Plate, caused by sodium hydroxide solution containing sodium silicate. × 1500.

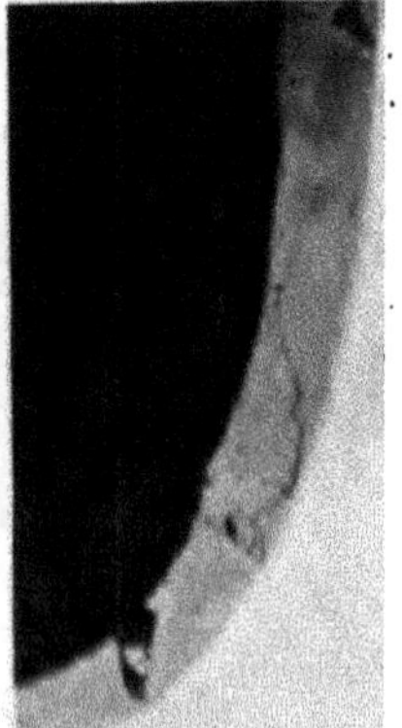

Fig. 12.—Stressed Specimen of High-Purity Iron afte exposure to sodium hydr oxide solution containin sodium silicate. (Note uni form oxide coat and absence of fissures.) × 500.

Fig. 13.—Cold-Worked Specimen of Boiler Plate. "Blunt" oxide penetrations due to action of hot sodium hydroxide solution. × 1000.

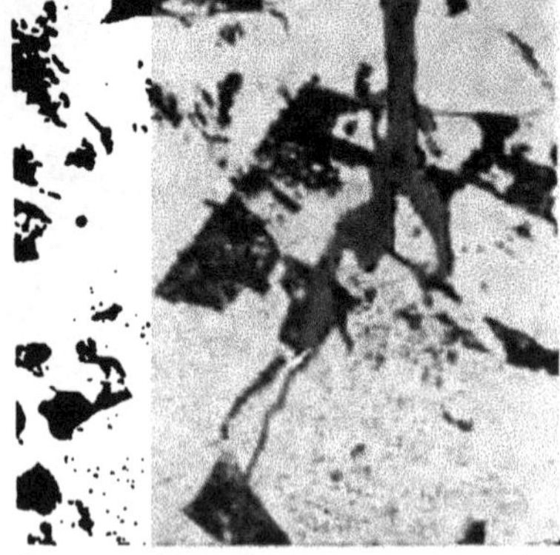

Fig. 14.—Cold-Worked Specimen of Boiler Plate. Crack formed by action of hot sodium hydroxide solution. × 1000.

(Micrographs reduced to two-thirds linear in reproduction.) [*N.P.L. Report, Part II.*

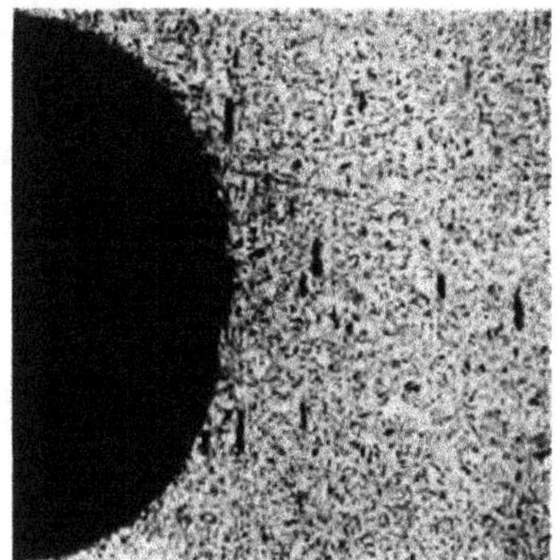

FIG. 15.—Oxide Inclusions formed near drilled hole in specimen of boiler plate. × 45.

FIG. 16.—Fine Intercrystalline Cracks near drilled hole in specimen of boiler plate. × 2000.

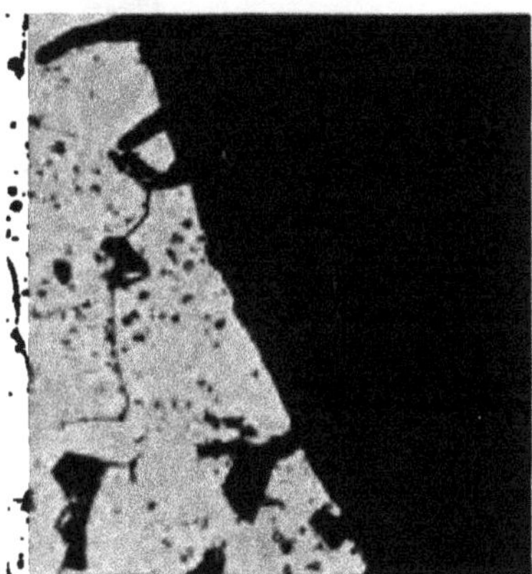

FIG. 17.—Oxide Penetrations near machined notch in specimen of boiler plate. × 1000.

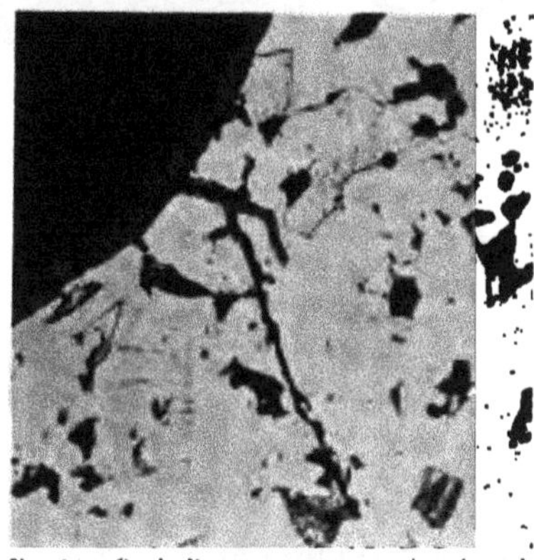

FIG. 18.—Oxide Penetration near machined notch in specimen of boiler plate. × 500.

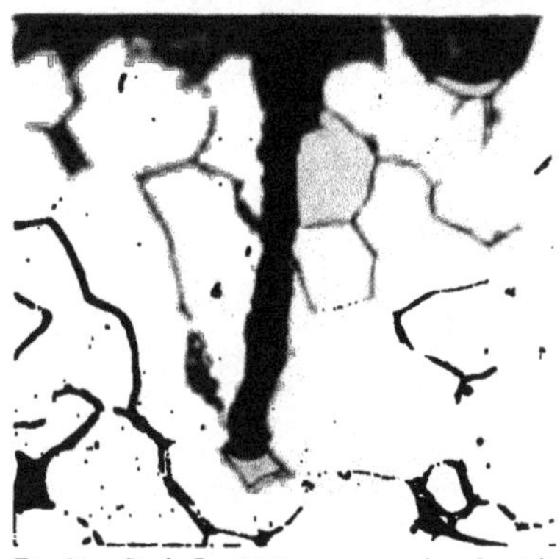

FIG. 19.—Oxide Penetration near machined notch in steel specimen containing phosphorus. × 1000.

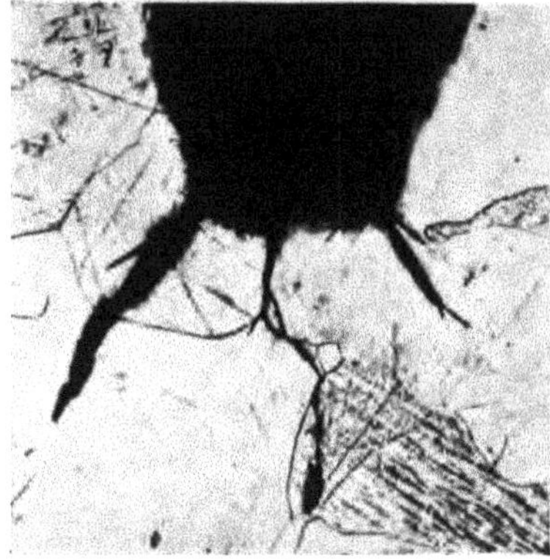

FIG. 20.—Oxide Penetrations near machined notch in specimen of high-purity iron. × 100.

(Micrographs reduced to two-thirds linear in reproduction.)

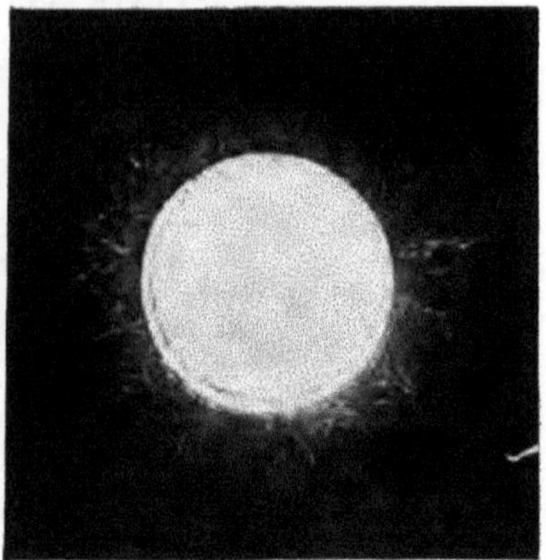

Fig. 26.—Drill Temper Steel (Carbon 1·2%). Depth of cracking produced in 12 hr. at 470° C. × 10.

Fig. 27.—Boiler Plate. Oxide penetrations produced in 3 weeks at 310° C. × 1000.

Fig. 28.—Boiler Plate. Oxide penetrations produced in 6 days at 360° C. × 1000.

Fig. 29.—Boiler Plate. Oxide films and modified pearlite produced in 48 hr. at 400° C. × 2000.

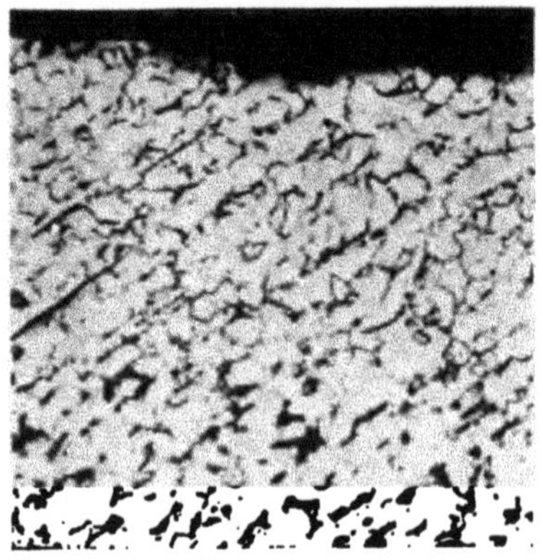

Fig. 30.—Boiler Plate. Depth of decarburisation produced in 48 hr. at 410° C. × 150.

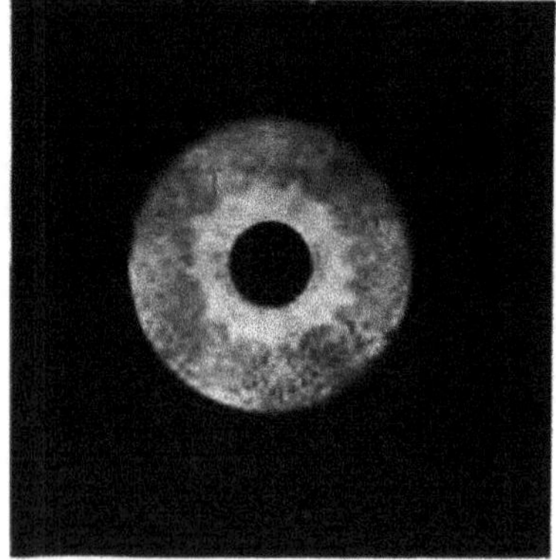

Fig. 31.—5% Nickel Steel. Depth of attack produced in 20 hr. at 470° C. × 3.

(Micrographs reduced to two-thirds linear in reproduction.)

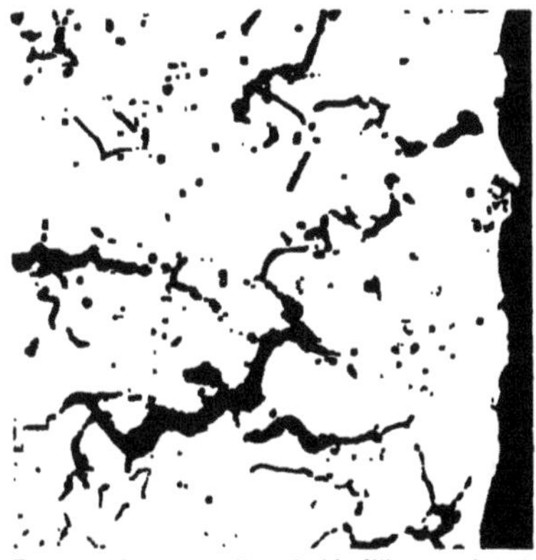

Fig. 32.—Intercrystalline Oxide Films produced in 20 hr. at 470° C. 5% nickel steel. × 1000.

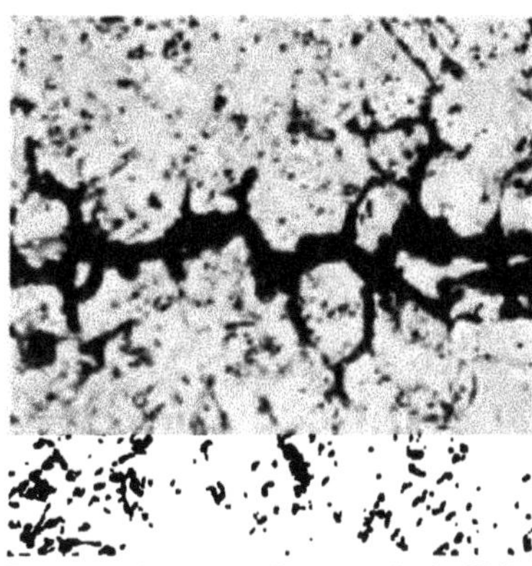

Fig. 33.—Austenitic Stainless Steel (18/8). Crack, apparently intercrystalline, produced in 20 hr. at 470° C. × 1000.

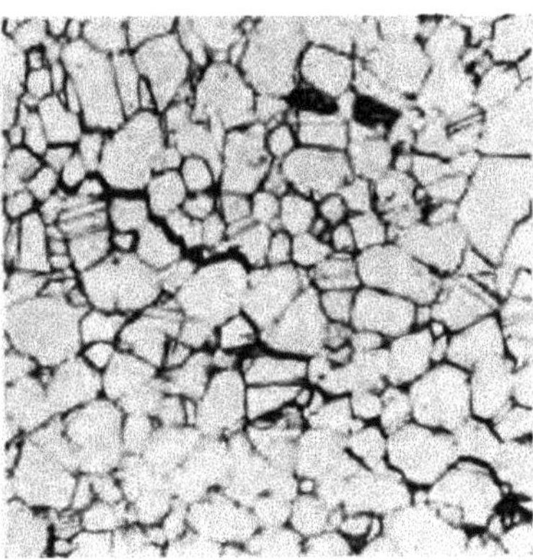

Fig. 34.—Nickel. Intercrystalline cracks in washer used to seal cavity in cylinder (12 hr. at 420° C.). × 150.

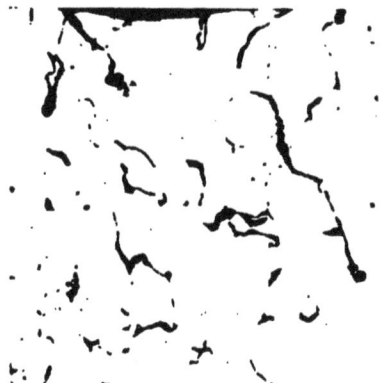

Fig. 35.—Intercrystalline Cracks in Mild-Steel Tubing produced in 17 hr. at 410° C. × 500.

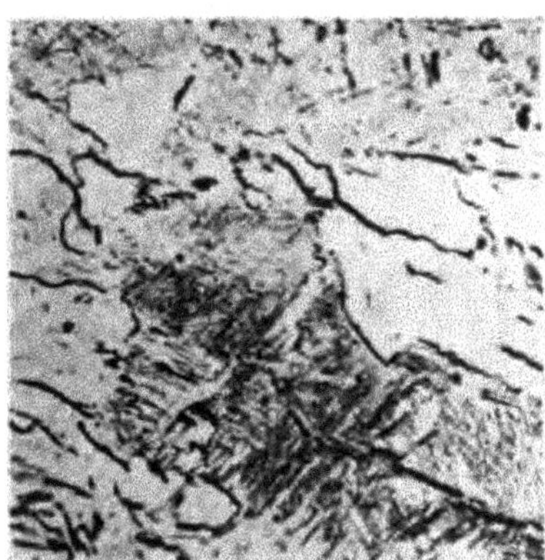

Fig. 36.—Cracks in Quenched and Tempered Steel produced in 17 hr. at 410° C. × 500.

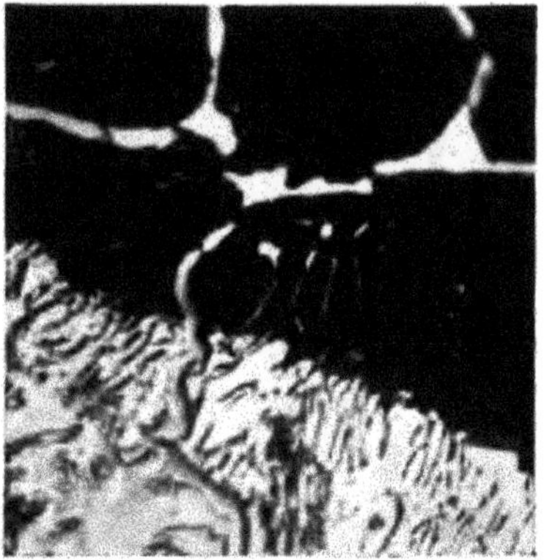

Fig. 37.—Preferential Attack on Ferrite by Caustic. Carbide plates remaining unattacked in oxide layer. × 1500.

(Micrographs *reduced to* two-thirds linear in reproduction.)

Fig. 38.—Cracked Specimen after Annealing 45 min. at 650° C. Small carbide particles formed by reaction of methane inside cavities. × 1500.

Fig. 39.—Cracked Specimen heated momentarily to 900° C. approx. Pearlite masses associated with cavities which contained methane. × 2000.

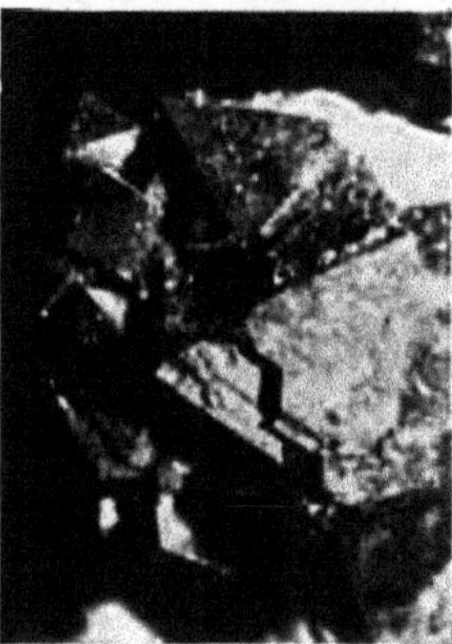

Fig. 40.—Crystals of Iron Oxide formed on silver rod. × 15.

Fig. 41.—Intercrystalline Attack on Mild-Steel Tube. × 500.

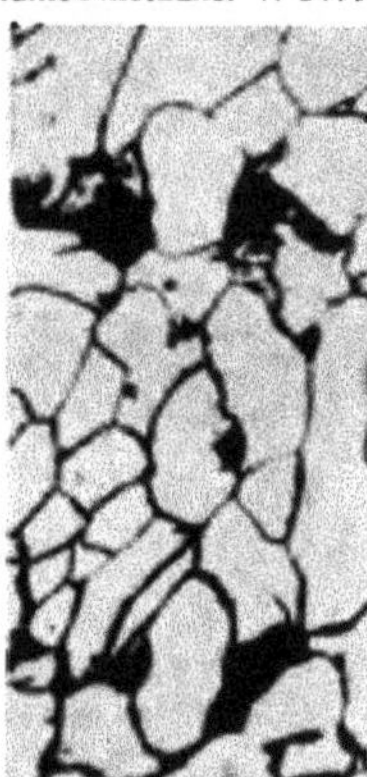

Fig. 42.—Intercrystalline Attack on Cold-Worked Boiler-Plate Material, produced by strong caustic in 46 hr. at 410° C. × 500.

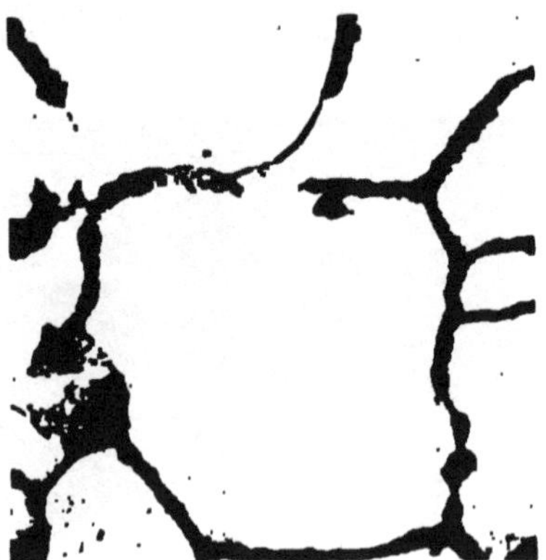

Fig. 43.

Fig. 44.

Figs. 43 and 44. Intercrystalline Attack in Mild-Steel Tube, produced by strong sodium hydroxide solution in 41 hr. at 410° C. Pearlitic areas often avoided. × 1500.

(Micrographs reduced to two-thirds linear in reproduction.) [*N.P.L. Report, Part III.*

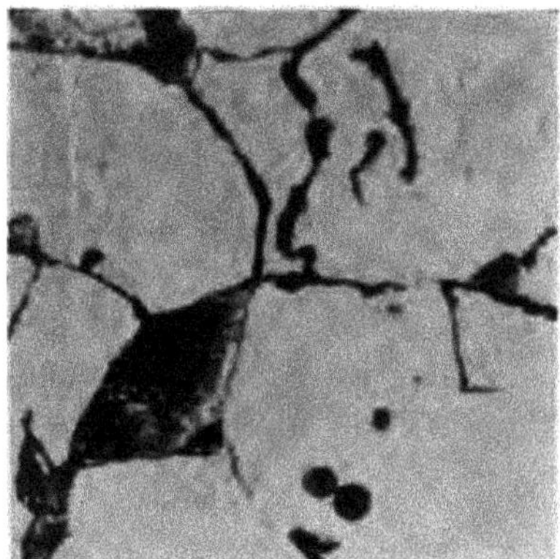

FIG. 45.—Intercrystalline Attack on Cold-Worked Steel Specimen (Carbon 0·19%), produced by strong sodium hydroxide solution in 41 hr. at 410° C. × 1500.

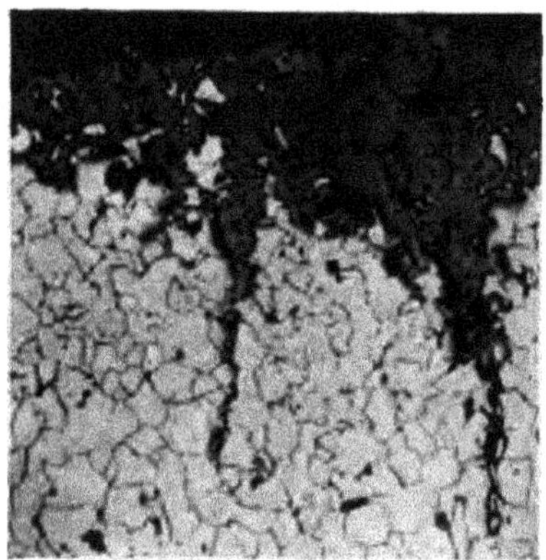

FIG. 46.—Caustic Attack on Mild Steel, 39 days at 275° C. × 200.

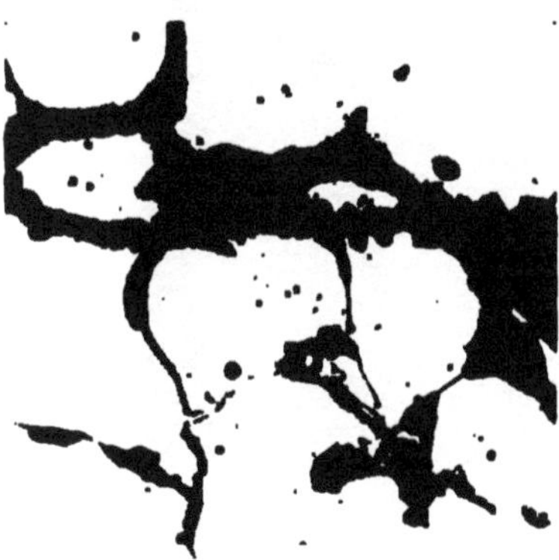

FIG. 47.—Caustic Attack on Mild Steel; 39 days at 275° C. × 1500.

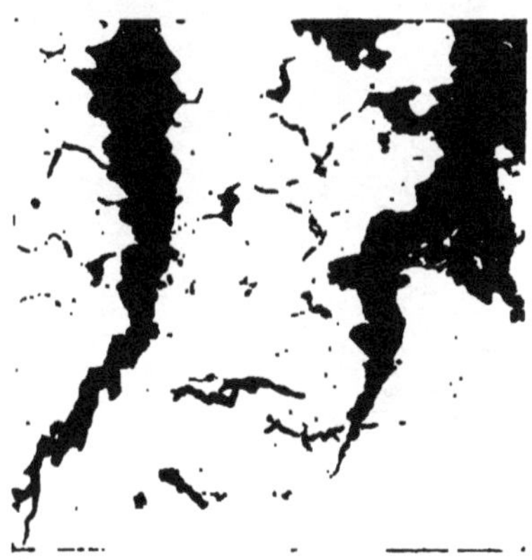

FIG. 48.—Caustic Attack on Mild Steel, 60 days at 150° C. × 500.

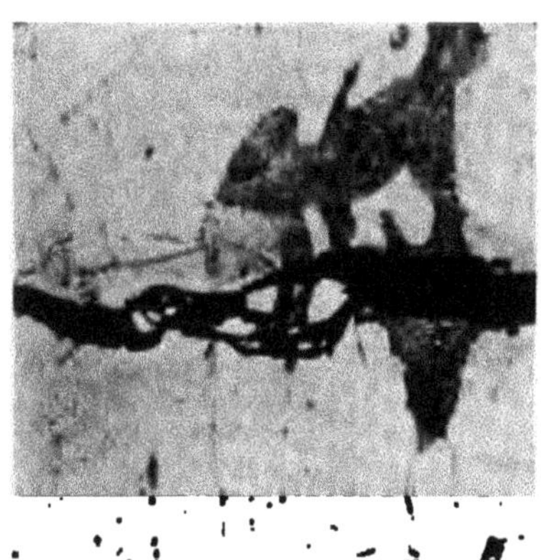

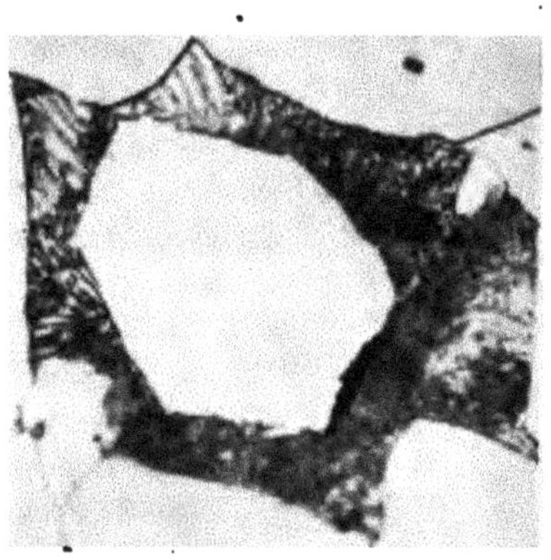

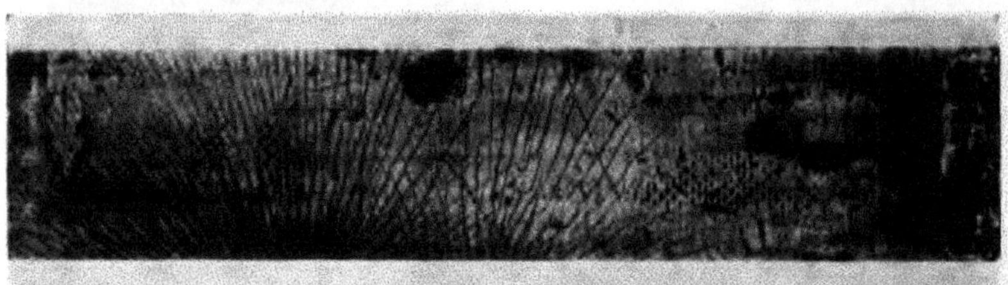

FIG. 51.—Steel Bar Etched with Fry's Micro-reagent after repeated bending at 100° C. × ½ approx.

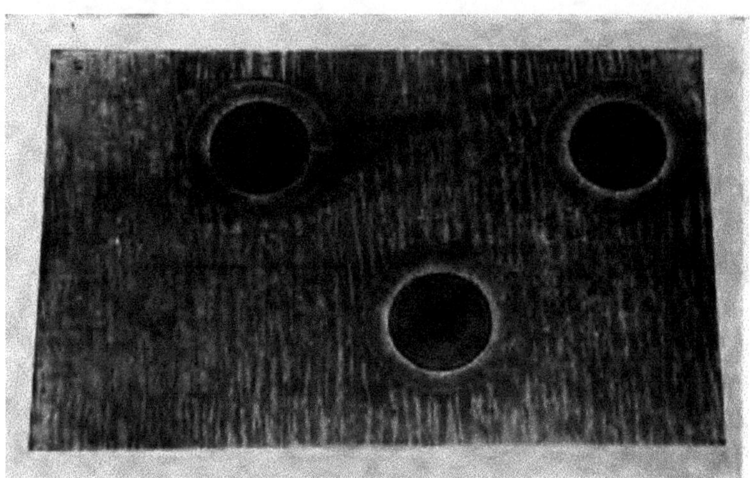

FIG. 52.—Portion of Butt Strap of unused riveted joint etched with Fry's macro-reagent. (Less than full size.)

FIG. 53.—Same Steel Bar as in Fig. 51, repolished and tested by magnetic powder method. × ½ approx.

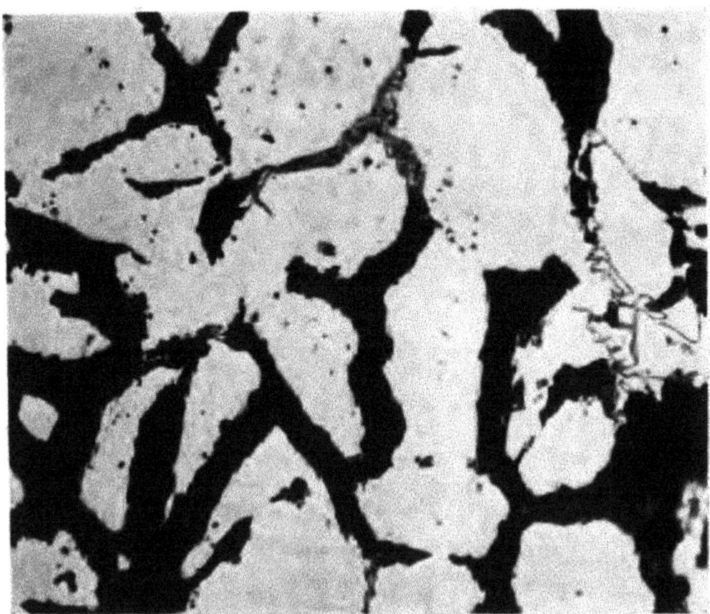

FIG. 54.—Steel Lightly Etched with Fry's Macro-reagent, then etched with 1% solution of nitric acid in alcohol. × 1000.

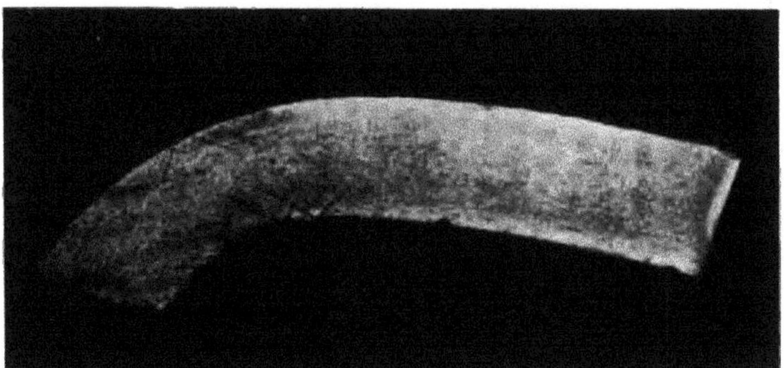

FIG. 55.—Section of Plate from Marine Boiler. Etched with Fry's macro-reagent. (Slightly under full size.)

FIG. 62.—General View of Testing Machine.

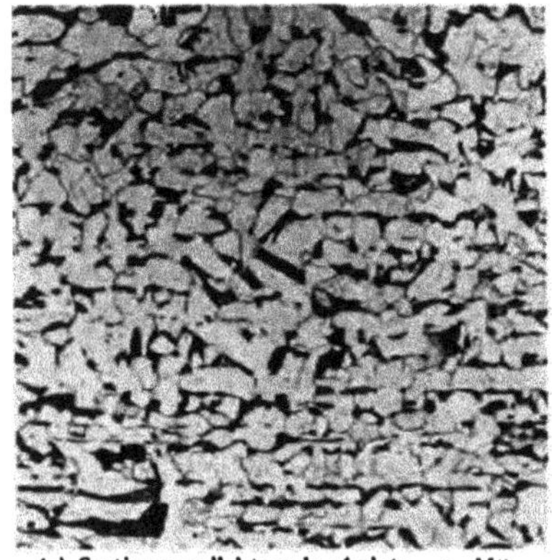

(a) Section parallel to side of plate. × 150.

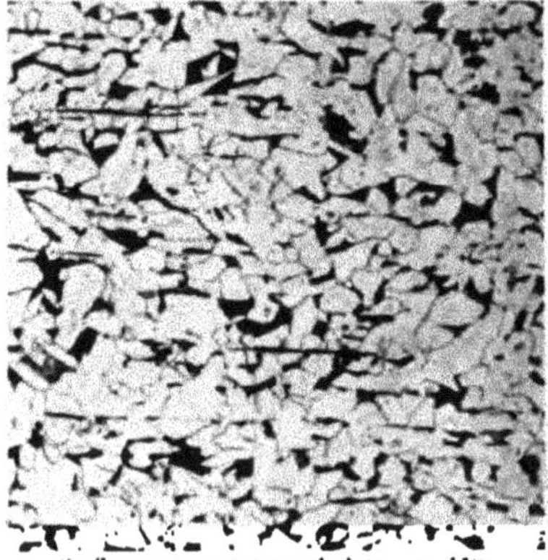

(b) Transverse section of plate. × 150.

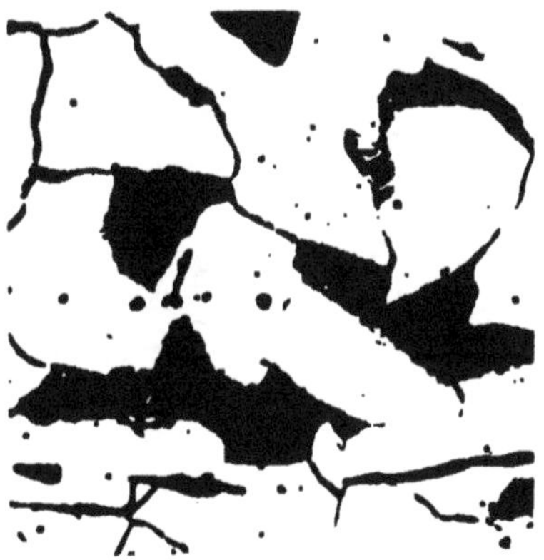

(c) Section parallel to side of plate. × 1000.

FIG. 64. Specimen *HH*.130, ⅝-in. plate.

(a) Sulphur print of section parallel to side of plate.

(b) Section parallel to side of plate, deeply etched to show phosphorus distribution.

FIG. 65.—Specimen *HHA*30, ⅝-in. plate.

(Illustrations reduced to two-thirds linear in reproduction.)

(*a*) Longitudinal section parallel to edge of plate. Length of plate horizontal. × 150.

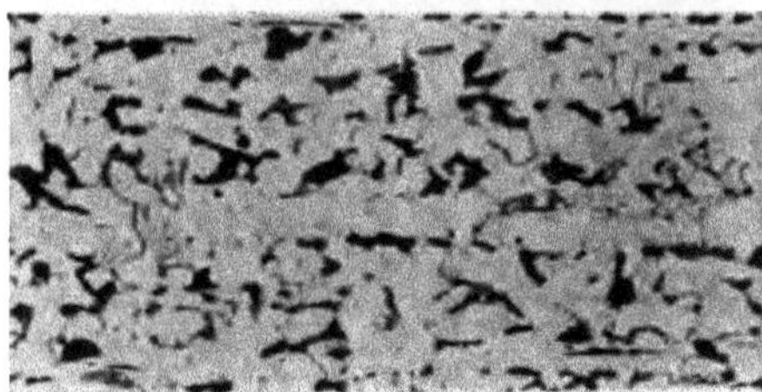

(*b*) Transverse section of plate. Width of plate horizontal. × 150.

FIG. 66.—Specimen *HHB*4, ⅜-in. plate.

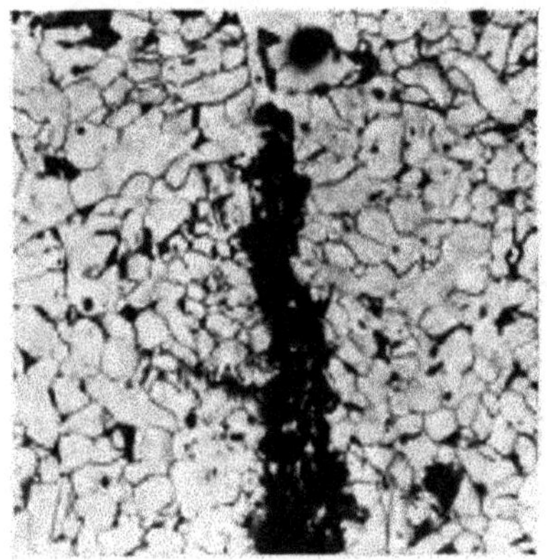

(*a*) Tension surface of plate. × 150.

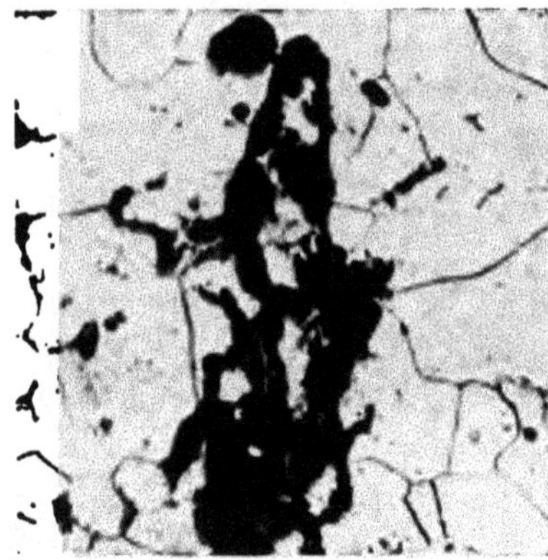

(*b*) Tension surface of plate. × 500.

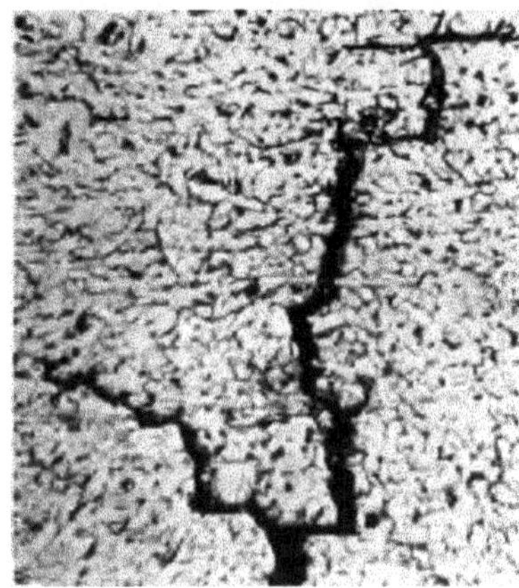

(*c*) Side of plate, rolling direction horizontal. × 75.

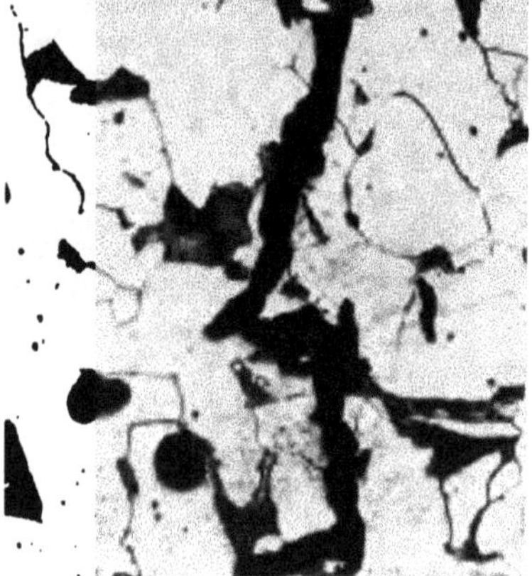

(*d*) Side of plate: rolling direction vertical. × 500.

FIG. 67.—Specimen *HH*.11, ⅜-in. plate.

(Illustrations reduced to two-thirds linear in reproduction.)

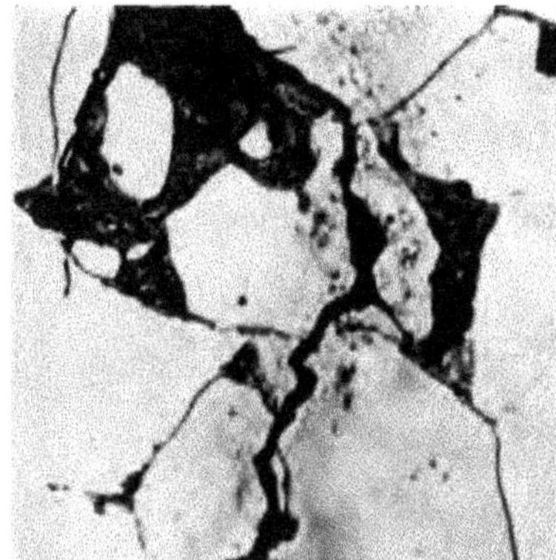

(a) Main crack near tip. Longitudinal section cut parallel and close to the tension side of the plate. × 1000.

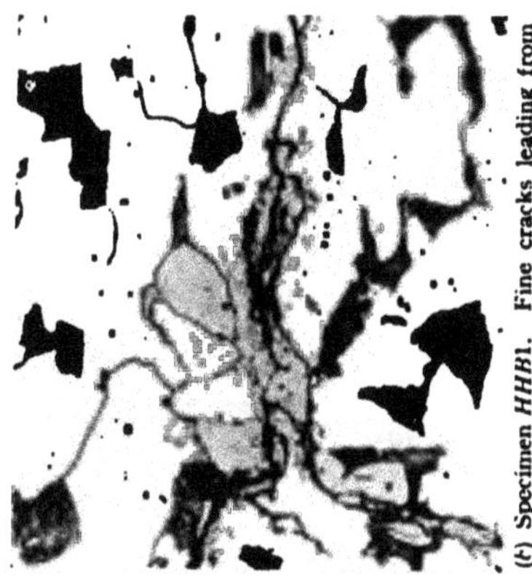

(e) Specimen *HHB1*. Fine cracks leading from main crack. × 500.

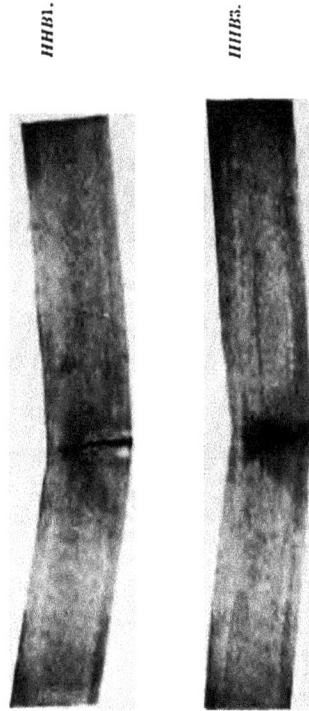

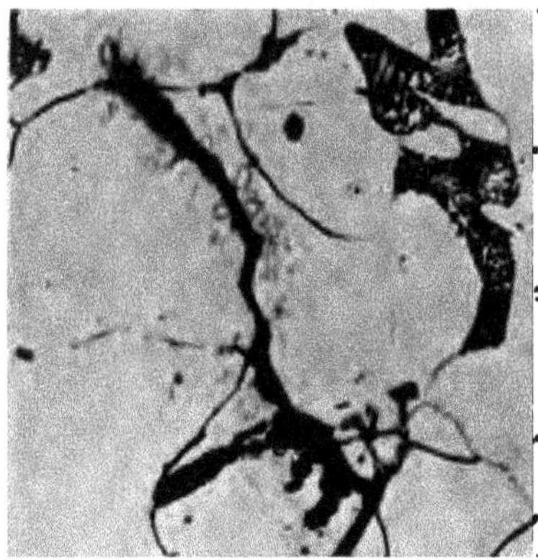

(b) Main crack near tip. Longitudinal section cut parallel and close to the tension side of the plate. × 1000.

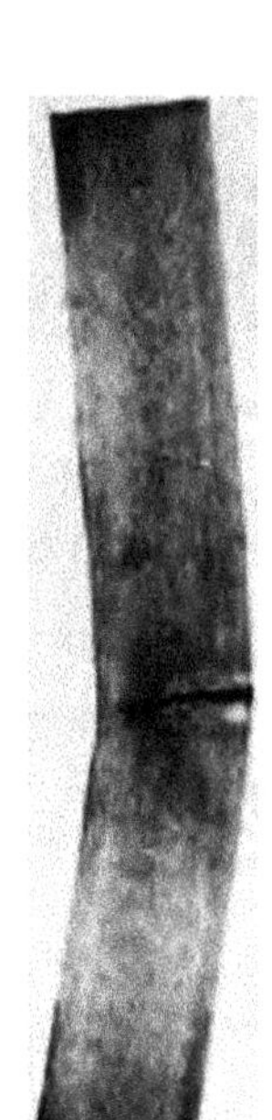

(d) Longitudinal section.

FIG. 69.—Plate Specimens, ¾-in. thick.

(c) Main crack near tip. Longitudinal section cut at right angles to the tension side of the plate. × 1000.

(*a*) Specimen *HMH1*.

(*b*) Specimen *HMH2*.

(*c*) Specimen *HMH3*.

(*d*) Specimen *HMH4*.

FIG. 70.—Riveted Joints, ⅜-in. plate.

(*a*) Specimen *HMH2*. Main crack near tip. × 1000.

(*b*) Specimen *HMH4*. Tip of main crack. × 1000.

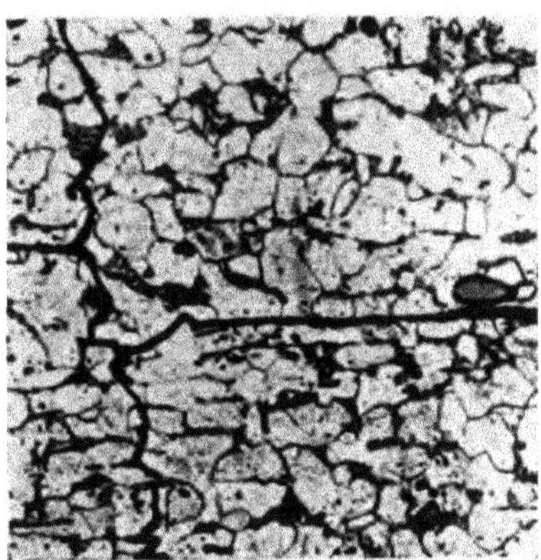

(*c*) Specimen *HMH4*. Branch crack. × 250.

Fig. 71.—Riveted Joints.

(*a*) Section showing major crack and crack near rivet hole. × ⅔.

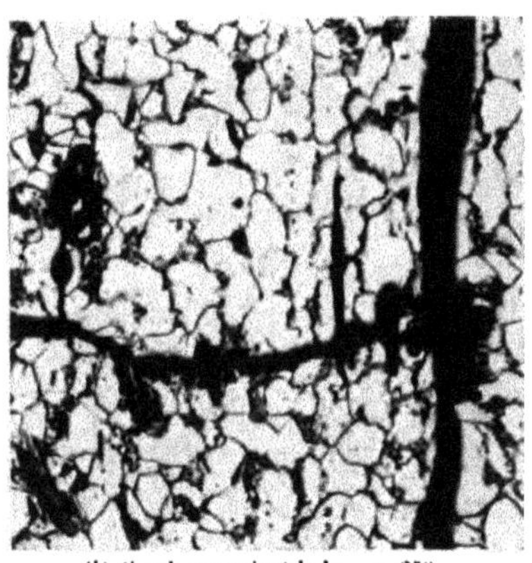

(*b*) Crack near rivet hole. × 250.

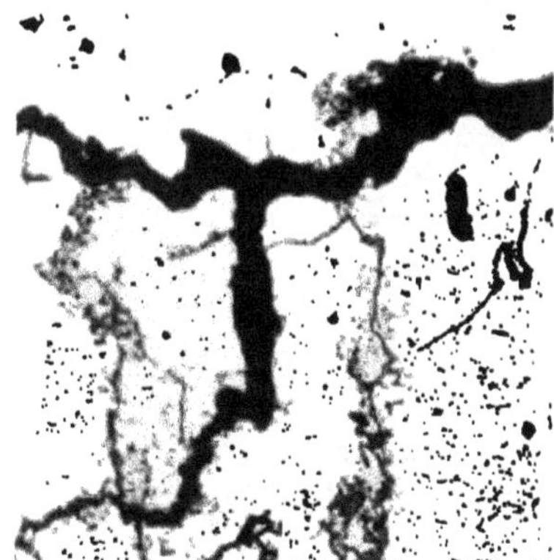

(*c*) Crack near rivet hole. × 1000.

Fig. 72.—Specimen *HMH3*, riveted joint.

(Illustrations reduced to two-thirds linear in reproduction.)

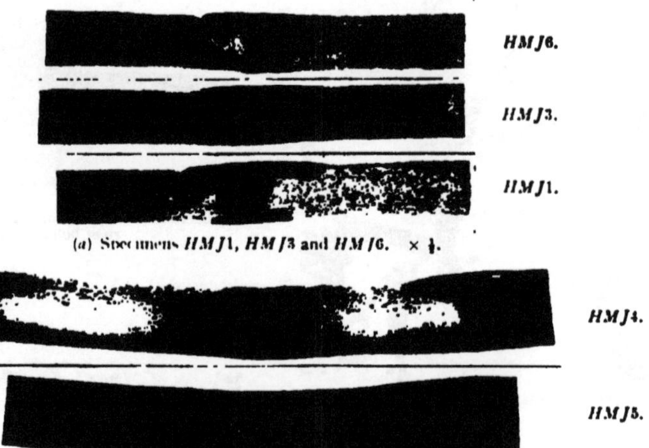

(a) Specimens *HMJ*1, *HMJ*3 and *HMJ*6. × ½.

(b) Specimens *HMJ*4 and *HMJ*5. × ⅔.

FIG. 73.—Welded Joints, ¾-in. plate.

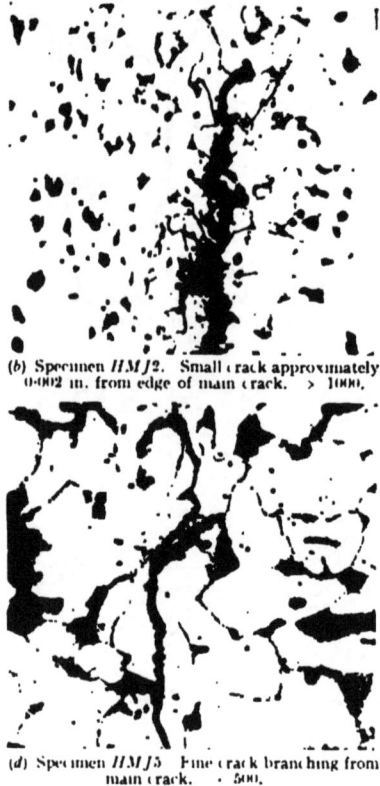

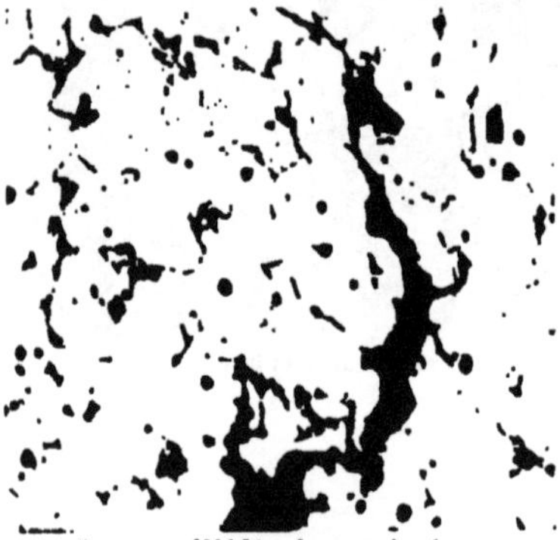

(a) Specimen *HMJ*6. Longitudinal section parallel to edge of plate. × 1000.

(b) Specimen *HMJ*2. Small crack approximately 0·002 in. from edge of main crack. × 1000.

(c) Specimen *HMJ*4. Section near tip of main crack. × 250.

(d) Specimen *HMJ*5. Fine crack branching from main crack. × 500.

FIG. 74.—Welded Joints, ¾-in. plate.

(Illustrations reproduced to two-thirds linear in reduction.) [*N.P.L. Report, Part V.*

[*To face p.* 129 P.

Miscellaneous Tests in the Silver-Lined Pressure Tube.

In practice, sodium sulphate is sometimes added to boiler waters to avoid the risk of caustic cracking of the boiler plates. It has been suggested that the protecting action of sodium sulphate might in some way be connected with the production of iron oxide of a different form from that obtained by the action of pure caustic solution. Accordingly, a mild-steel specimen was acted on at 410° C. for 17 hr. by a solution of the following composition: H_2O, 25 c.c.; NaOH, 9·2 g.; $Na_2SO_4.10H_2O$, 16·3 g. Much intercrystalline cracking occurred, and the iron oxide crystals deposited in the cooled parts of the tube were (as far as could be judged by using a lens) identical with those deposited from the pure sodium hydroxide solutions.

An experiment was also made in which the pure sodium hydroxide solution was replaced by one containing: H_2O, 25 c.c.; NaOH, 9·2 g.; Na_2CO_3, 4·6 g. The usual intercrystalline cracking occurred, but it was accompanied by an intense corrosion at the surface of the metal. Many oxide spurs of the blunt type, often found in corroded boiler plates, were observed in this specimen.

Long-Period Tests Conducted in Sealed Steel Tubes.

Owing to lack of time it was impossible to conduct a number of long-period tests in the silver-lined pressure vessel. Three tests, one lasting 60 days, were, therefore, made with mild-steel specimens in contact with sodium hydroxide solution of density 1·29 contained in long thick-walled steel tubes. Each tube was plugged with steel rods, so as to leave only a short length of the bore unobstructed near the centre of the tube, this space containing the specimen and the reacting solution. Finally, the tubes were sealed by screwed caps silver-soldered in position. Thermocouples for indicating temperatures were tightly bound on the tubes and the temperatures maintained within $\pm$ 10° C. by electric heating. The central parts of the tubes were heated, but the ends remained relatively cold. The first test conducted at 310° C. for 20 days resulted in decarburisation and cracking of the specimen very similar to that obtained in the short-time test with sodium hydroxide solution, density 1·29 (18 hr. at 410° C.). On the other hand, the specimens treated at 275° C. for 39 days and at 150° C. for 60 days showed no appreciable decarburisation or cracking of the hydrogen type. In both these samples the heavy surface corrosion had extended into the body of the metal in the form of oxide spurs which were partly intercrystalline and partly transcrystalline. Figs. 46 and 47 show polished and etched areas of the specimen after the test at 275° C. The treatment at 150° C. resulted in cracks mainly, but not entirely, transcrystalline (Fig. 48).

These long-period tests revealed that destructive cracking might occur during the caustic attack on steel at quite low temperatures, but a considerable number of experiments would be desirable before detailed conclusions could be drawn. One complication is the progressive concentration of the sodium hydroxide solution which must occur as a result of the gradual diffusion of hydrogen through the steel tube and the formation of iron oxide.

The Action of Concentrated Sodium Hydroxide at Low Steam and Hydrogen Pressures.

A specimen of boiler-plate material and some solid sodium hydroxide were placed in a vertical silver-lined steel tube, closed at the base. The tube was heated to 410° C. and was of sufficient length to impart considerable superheat to a small supply of steam which was passed down a narrow steel tube and blown on to the specimen. After a test lasting 67 hr. the boiler-plate sample was coated with reddish oxide and showed considerable surface corrosion. No intercrystalline penetration of any kind could be detected. In this single experiment, to the result of which too much weight should not be given, it is shown that cracking does not take place in the absence of hydrogen and steam pressures.

An attempt was made to differentiate between the influence of steam and of hydrogen pressures in an experiment in the silver-lined pressure tube, where a specimen of boiler material was subjected to the action of strong caustic solution at 410° C. During an experiment lasting 37½ hr. the tube was cooled temporarily every 4 hr. and the hydrogen released by means of a special valve. Examination of the specimen at the conclusion of the test showed that intercrystalline penetration had occurred to a depth of only one or two crystal grains. This suggests that some pressure of hydrogen is essential for the development of intercrystalline cracking during caustic attack.

Tests on Steel Wire at Room Temperature.

Steel wire containing 0·13% of carbon was vacuum-annealed, and then had a tensile strength of 28·3 tons per sq. in. When an annealed wire was loaded to 19·3 tons per sq. in. and made the cathode in a 10% sulphuric acid bath (c.d. 1 amp. per sq. in.) fracture took place in 25 min. A slight initial extension of the wire occurred on loading, the diameter being reduced from 0·036 to 0·035 in., but the fracture was not accompanied by any further general extension or local "necking." Fig. 49, which represents a region quite close to the main fracture, demonstrates that the rupture was transcrystalline.

Conclusions.

The experiments in the silver-lined pressure tube show that when fabricated steel plate was acted on, at 410° C., by sodium hydroxide solution of medium strength (35 parts of NaOH + 100 parts of H_2O, by weight) no intercrystalline cracking occurred, provided that the material was in the annealed condition. On the other hand, the corresponding cold-worked material developed, throughout most of the sample, intercrystalline cracking of a somewhat discontinuous type associated with the carbon-bearing areas and probably identical with that produced by high-pressure hydrogen at high temperatures. This type of cracking was also produced in 17 days at 300° C. Although the cracking at 410° C., and to a lesser extent at 300° C., was associated with the decarburisation of the pearlitic areas, this effect was often absent in the early stages of the attack, as indicated in Fig. 50. The combined pressure of hydrogen and steam often rose to 3 tons per sq. in. in the silver-lined tube towards the end of an experiment at 410° C., and it is possible that the difference in behaviour between annealed and cold-worked material may hold strictly only for these particular experimental conditions. It is safe to assume, however, that cold-working the steel increases its susceptibility to the hydrogen type of intercrystalline cracking. As mentioned previously, this type of intercrystalline cracking is apparently caused by hydrogen, generated at the metal surface, diffusing into the metal and reacting with iron carbide at or near crystal boundaries. Methane is formed and, being unable to escape by diffusion, collects under great pressure in the intercrystalline crevices, which are thereby extended. The initial stages of this process are favoured by the presence of sub-microscopic cavities believed to be present at the crystal boundaries of as-cast and cold-worked steels. Unless the cracks were in communication with the outer oxide layer on the metal surface they did not contain oxide.

With caustic solutions of much higher concentration, intercrystalline attack of another type was produced in steel specimens treated at 410° C. These cracks tended to form a more continuous network penetrating the metal from the outer oxide coat. In the early stages, however, the cracks often stopped short at the carbon-bearing regions (*see* Fig. 43). These cracks were usually filled with oxide (termed for convenience cracks of the oxidising type), and differed from the hydrogen type of cracking, as they penetrated annealed or cold-worked steel, as far as could be judged, with equal facility. Again, cracks of the hydrogen type were often found throughout the whole mass of a cold-worked specimen, while those of the oxidising type existed in patches always in communication with the outer oxide coat. Both types of cracking may be present in a steel sample, as is demonstrated in Fig. 45, where cracks will be seen not only alongside pearlitic areas but also in regions where the carbon-bearing constituent is absent. The effect on the mechanical properties of such a dual attack would be very serious.

It was not possible to make long-period experiments with strong caustic solutions in the silver-lined pressure vessel at temperatures below 410° C., but the experiments made at 350° C., 275° C. and 150° C. in the sealed steel tubes showed how destructive cracking might occur at these temperatures. Although these tubes originally contained the weaker caustic solution (density 1·29), a gradual concentration occurred, owing to the formation of iron oxide and the escape by diffusion of hydrogen.

Since the experiments with superheated steam at atmospheric pressure in the presence of caustic and also in the silver-lined pressure tube with a periodic release of hydrogen did not give rise to serious cracking, it seems probable that a degree of hydrogen pressure is required at some stage in the attack to bring about intercrystalline cracking even of the oxidising type. That the penetration of hydrogen may rapidly weaken steel even at room temperature was shown by the experiment in which a stressed steel wire was made the cathode in an electrolytic bath. Experiments with powdered-silver filters demonstrated that dense masses of iron oxide could be deposited at a distance from the place of original formation of this material. This is regarded as important, since the crevices in a boiler may become sealed in this manner and high hydrogen or steam pressures may develop within them. The oxide is apparently cubic Fe_3O_4.

At lower temperatures the oxide penetration in the steel is often partly transcrystalline and partly intercrystalline. The transcrystalline nature of the cracks was especially marked in the specimen treated for 60 days at 150° C. In this connection the author ventures to express the opinion that destructive caustic cracking in boilers is not always entirely intercrystalline—the structures are often difficult to interpret and there has been, perhaps, a tendency for writers to select for their micrographs fields emphasising the intercrystalline character of the attack.

In view of the part played by the carbon-bearing regions of steel in giving rise to cracks during caustic attack, the behaviour of carbon-free high-purity iron is of special interest. Unfortunately, the samples of high-purity iron available revealed intercrystalline weakness in cold-working before exposure to caustic solution. In spite of this, no certain evidence could be obtained of the occurrence of the hydrogen type of intercrystalline attack. On the other hand, a sample of high-purity iron readily developed a network of intercrystalline cracks of the oxidising type on exposure to strong sodium hydroxide solution.

Finally, this investigation, although incomplete, has shown that the reaction between steel and sodium hydroxide solution at elevated temperatures is complex. At least two distinct modes of intercrystalline cracking were detected under laboratory tests, and it is probable that both forms contribute to the caustic cracking of boiler plates under service conditions.

Acknowledgment is due to Dr. C. H. M. Jenkins, A.R.S.M., for his advice during the conduct of the investigation and also to other colleagues in the Metallurgy and Engineering Departments. The author also desires to acknowledge the advice of Mr. A. Bailey, M.Sc., A.M.Inst.C.E., in the design of suitable sealing caps for the silver-lined pressure tube.

Part IV.—Strain-Etch Markings in Boiler-Plate Material of Acid Open-Hearth Manufacture.

By FRANK ADCOCK, M.B.E., D.Sc., and C. H. M. JENKINS, D.Sc., A.R.S.M. (National Physical Laboratory).

(Figs. 51 to 55 – Plates XIX. and XX.)

Although it is well known that strain-etch markings are readily produced in strained basic Bessemer steel, their development, by a short-time etching process such as was used by the authors, appears to be unusual in the case of steel made by the open-hearth process. Jevons [1] tested more than a hundred steels which were presumably of acid open-hearth origin without obtaining strain-etch markings, but was able to etch a German boiler-plate steel of basic origin in less than one minute and produce the characteristic markings.

The observations in this short paper are selected from the results of a larger investigation; it was thought desirable to publish them in their present form in view of the general interest in this subject. Except for one specimen cut from a corroded marine boiler, all the work was conducted on bars and riveted joints made of boiler plate of acid open-hearth manufacture. The analysis made in the Laboratory was as follows:

Carbon.	Silicon.	Sulphur.	Phosphorus.	Manganese.	Nitrogen.
0·20%	0·03%	0·028%	0·036%	0·50%	0·004%

The bars and joints were from the same cast, but had been subjected to different rolling processes during manufacture, the bars being ¾ in. thick, while the experimental joints were fabricated from ⅝-in. plate. A number of bars were first subjected to various mechanical tests, then the surfaces were ground, polished and etched with Fry's macro-reagent [2] for 5–10 min. at room temperature. Table I. gives particulars of these tests, together with the results of the etching treatment.

The results in Table I. suggest that well-defined strain-etch markings are produced only in material which has undergone

[1] Jevons, *Journal of the Iron and Steel Institute*, 1925, No. I., p. 191.

[2] Fry's macro-reagent: $CuCl_2.2H_2O$ . . 90 g.
H_2O . . . 100 c.c.
HCl . . . . 120 c.c.

deformation at elevated temperatures. Repeated bending is unnecessary, as item 4 in the Table indicates that a single bend will enable strain-etch markings to be produced.

TABLE I.—*Effects of Mechanical Treatment, with or without the Simultaneous Action of Sodium Hydroxide Solution.*

Mechanical and Thermal Treatment of Bars of Boiler-Plate Material.	Result of Subsequent Etching with Fry's Macro-reagent.
1.—As received from manufacturer.	No strain-etch markings.
2.—Repeated bending in sodium hydroxide solution at 100° C. (10 million cycles). (Stress range, 2·3 to 18·3 tons per sq. in.)	Well developed strain-etch markings (see Fig. 51).
3.—Repeated bending in sodium hydroxide solution at 15–20° C. (559,800 cycles). (Stress range, 2·3 to 22·3 tons per sq. in.)	No definite strain-etch markings. Etching of bar after an interval of 3 months showed feeble strain-etch markings.
4.—Bar maintained under steady stress of 25 tons per sq. in. in sodium hydroxide solution at 100° C., for 12 days.	Well defined strain-etch markings on large part of specimen.
5.—Bar bent once at room temperature by plunger of small radius.	No strain-etch markings.

It was possible, with some difficulty, to develop the strain-etch markings on the narrow sides of the bars and it was concluded that the markings were really traces, on the surface, of planes of disturbance passing through the metal and probably related to the directions of maximum shear stress. Hartmann,[1] as far back as 1896, made use of acid attack on steel to show that the disturbed zones of a steel specimen are not confined to the surface. Fig. 51 shows, after etching with Fry's reagent, a specimen which had been subjected to repeated bending in caustic soda solution at 100° C.

A specially fabricated joint of boiler-plate material was next examined to ascertain the conditions existing in riveted construction. The joint had not been in service nor subjected to any mechanical tests, and Fig. 52 shows the appearance of a portion of the butt strap, etched with Fry's reagent, after machining away the rivet heads and a small depth of metal. Besides the dark and light rings round the rivets and the fine curved lines resulting

[1] Hartmann, "Distribution des déformations dans les métaux soumis à des efforts," p. 81. Paris, 1896 : Berger-Levrault & Cie.

from the riveting of the plates, broad etch markings, all lying roughly in the same direction, cover most of the plate. These are believed to be due to a high finishing temperature during the rolling of the plate. Fell's [1] results suggest that this temperature is below 500° C.

It has been previously shown [2] that when steel, presumably of basic manufacture, has been mechanically treated so that strain-etch markings would develop on etching, the positions of the more important markings are indicated when the specimen is tested by the magnetic-powder method. For this purpose the specimen is maintained in a magnetised state, and a liquid containing magnetically susceptible powder of reduced iron oxide is poured over it. The authors have found that in acid open-hearth steel a similar relationship exists between the magnetic-powder patterns and the strain-etch markings in the bars which had undergone mechanical deformation at an elevated temperature. This is apparent from a comparison of Figs. 51 and 53. As is well known, when cracks are present the magnetic-powder method gives sharply defined markings. These are different in character from those referred to in the present paper.

Strain-etch markings of the type developed by Fry's macro-reagent are not produced by the majority of etching reagents. The mechanism of the attack by Fry's reagent is obscured by the drastic action. of the etching solution. By adopting a special technique, however, it has been possible to observe at a high magnification the mode of attack on steel. The specimen was first etched for a short time in Fry's reagent (the strain-etch markings being developed faintly by this process), and then in a 4% solution of nitric acid in alcohol. In the immediate region of a dark strain-etch marking, examination under high power reveals that the attack bears some relationship to the crystal boundaries. This is demonstrated in Fig. 54. In the regions where the strain-etch markings were not developed, the steel had not been attacked in this manner, although general etching had occurred.

The condition of the steel responsible for the production of strain-etch markings when treated with Fry's reagent is associated with the development of corrosion defects of boiler plates during service. This is strongly suggested by Fig. 55, which represents a polished and etched section of plate cut from a corroded marine boiler. It is probable that the plate had either been deformed or strained at some elevated temperature (but below 500° C.) or had been cold-worked during the construction of the boiler and then exposed to a similar range of temperature in service. The method of manufacture of this boiler plate could not be definitely ascer-

[1] Fell, *Iron and Steel Institute, Carnegie Scholarship Memoirs*, 1927, vol. 16, p. 101.

[2] Wever and Otto, *Mitteilungen aus dem Kaiser-Wilhelm-Institut für Eisenforschung*, 1930, vol. 12, p. 373.

tained, although it was believed to have been made by the acid open-hearth process.

The authors desire to acknowledge the assistance of Mr. C. A. Harvey, of the Metallurgy Department, in the experimental work. The mechanical treatment of most of the specimens was undertaken in the Engineering Department.

Part V.—Some Experiments on the Behaviour of Specimens of Boiler Plate and Boiler Joints Subjected to Slow Cycles of Repeated Bending Stresses while Immersed in a Boiling Aqueous Solution.

By H. J. GOUGH,[1] M.B.E., D.Sc., F.R.S., M.I.Mech.E. (London), and H. V. POLLARD, A.M.I.Mech.E. (National Physical Laboratory).

(Figs. 62 to 74 – Plates XXI. to XXVII.)

I.—Statement of General Objects of the Research and Summary of Principal Results.

The causes of failure of boiler plates have received much experimental attention. Of the various types of failure encountered in service, probably the most important is that in which the fracture is essentially intercrystalline in character, the cracking showing a marked preferential—in some cases, an exclusive—tendency to follow the grain boundaries of the steel.

It is well known that boiler failure is *not* always characterised by failure of an intercrystalline type. In many cases marked grooving and fissuring have been encountered in various parts of boilers, which features have been responsible for complete failure or withdrawal from service. These defects are usually found at sharply curved portions or other places where local stress concentrations exist; cases of this kind which have been examined at the National Physical Laboratory exhibit no evidence of preferential attack along grain boundaries of the type which suggests a general breakdown between the crystal grains. Owing to " breathing " or " letting-down " of the boiler, such regions will be submitted to cyclic stressing in the presence of a corrosive agent, which supplies the conditions requisite for the type of accelerated failure generally described as corrosion-fatigue,[2] which, usually much more destructive in its action than the ordinary form of fatigue, generally produces a fracture of the same essentially trans-

[1] Director of Scientific Research, Ministry of Supply (formerly War Office); formerly Superintendent, Engineering Department, National Physical Laboratory.
[2] H. J. Gough, *Journal of the Institute of Metals*, 1932, vol. 49, p. 17.

crystalline type, often accompanied, however, by such fissures or grooves at the surface of the fracture or crack.

It is, therefore, *possible* that corrosion-fatigue failure may be a factor in some cases of boiler failure, and McAdam [1, 2, 3] has attached considerable importance to this possibility. He describes [1] an interesting case of a boiler which had been used intermittently—about 1100 times, each for a period of one day—during its eleven years of service. The boiler exploded, failure occurring at the region of most abrupt curvature in the head of a steam drum. Many accessory cracks, all located entirely below the water-level, were of the type associated with corrosion-fatigue and consisted of a deep corrosion pit, lined with corrosion products, terminating in a sharp fatigue crack. The mechanical conditions of the service of this boiler were equivalent to the application of 1100 stress cycles ranging between zero and a maximum at an average frequency of about 2 cycles per week; the water contained calcium carbonate. McAdam also described serious pitting damage of this type encountered in water drums of Yarrow boilers in service for about 10 years. In discussing these failures in relation to his experiments on corrosion-fatigue at low cyclic frequency and to the cases of failure by caustic embrittlement, McAdam suggests that "it seems probable that intercrystalline failure at seams and rivet holes are special examples of failure under combined stress and corrosion." With regard to this suggestion it may be mentioned that, within the knowledge of the present authors, McAdam has not yet published any photomicrographs or other evidence that the fractures which he records were, in fact, predominantly intercrystalline in character as is usually associated with the term "brittle failure by caustic embrittlement" as used in Great Britain. Nevertheless, McAdam, in presenting his experiences and experimental data, has brought to notice the possible influence of corrosion-fatigue as a factor in the general problem of boiler failure, which suggestion ought not to be ignored, even though that factor may have little to do with the peculiar type of failure which has hitherto attracted major attention. His work and suggestions were very carefully considered, and led to the initiation of the experiments recorded in the present paper.

It was considered rather unlikely, in view of previous experience, that corrosion-fatigue conditions would result in a definitely intercrystalline type of failure; more probably, studies of the behaviour of specimens exposed to simultaneous *static* stress and corrosion would achieve this result, and this aspect

[1] D. J. McAdam, *Proceedings of the International Congress on Applied Mechanics, Stockholm*, 1930, vol. II., p. 269.

[2] D. J. McAdam, *Proceedings of the Zurich Congress of the International Association for Testing Materials*, 1931, p. 228.

[3] D. J. McAdam and R. W. Clyne, *Journal of Research of the Bureau of Standards*, 1934, vol. 13, No. 4, Oct., p. 527.

of the problem has received close study in the Metallurgy Department of the National Physical Laboratory. As a correlative investigation, it seemed well worth while to carry out a few tests aimed at obtaining some information as to the extent to which corrosion-fatigue conditions may be of importance in boiler failure, in view of the many points of resemblance which have been observed or suspected. The essential difference between corrosion-fatigue conditions and a combination of static stress and corrosion is, of course, that in the former a *range* of cyclic stress or strain is involved, which is absent from the latter; in many cases, a range of cyclic stressing produces much greater destruction and quicker failure than a much higher value, numerically, of static stress. It is extremely difficult to determine—and very easy to overestimate—the extent by which boilers are, in fact, submitted to cyclic stressing; it may, however, be considered as entering into the boiler problem under at least three conditions: (1) In which the boiler shell is considered as being subject mainly to a steady stress, but also to superimposed stress variations due to pressure differences; (2) in which the boiler shell is considered as normally under a steady stress, which is released at intervals when the boiler is "let down"; and (3) a combination of conditions (1) and (2).

The first condition corresponds to the application of a fairly small range of pulsating stress, but applied very much more frequently, in general, than the much greater range of repeated stress arising under the second condition. As repeated applications of a large range of stress would be expected to produce accelerated damage, this type of stressing was selected for the experiments. Careful consideration was also given to the most suitable cyclic frequency to be employed. Given a certain convenient total duration of test, increased cycles can be obtained by raising the testing frequency. It has been established that, at air temperature, what may be termed a "speed effect" exists in corrosion-fatigue phenomena; more destructive effects are obtained, in an equal number of cycles, as the cyclic frequency decreases. This is probably merely stating indirectly that, under corrosion conditions, the "time" factor is of great importance and cannot be accelerated. Further, cyclic stressing at high frequency does not appear to be a condition operating in boiler plant. Weighing up these various considerations, it was concluded that a study of the effect of cycles of *repeated* stress employing a low cyclic frequency would be the most practicable form to use in an investigation designed to examine the degree of correlation existing between boiler failure and corrosion-fatigue.

Concerning the type of stressing to be employed, repeated tensile straining may appear to be the most informative. But, to induce the same value of stress, larger loads must be applied when using direct stressing than when bending moments are employed, while experience has shown that essentially the same type of damage is produced, under corrosion-fatigue conditions, when either stress system is used (this would be expected, as corrosion, in its early stages at least, is a surface effect); hence, cycles of repeated bending stresses were chosen.

There remained for consideration the shape and size of the test-pieces to be used. It appeared desirable to use riveted and welded joints in preference to simpler forms of test-piece; further, in view of the difficulty of obtaining scale models exactly similar, metallurgically, to full-sized joints, it was decided to use the latter made under exactly the same conditions as service joints.

Accordingly, a suitable machine was specially designed and constructed, capable of applying repeated loading, varying from nearly zero to a maximum of 10 tons, at an operating frequency of 35 cycles per min. The specimens, which may consist of one, two or three samples, depending on their size, are stressed as simply supported beams of which the middle half is subjected to cycles of a uniform bending moment. The specimen under test is so arranged that it can be tested either in air, or totally immersed in a boiling aqueous solution to produce corrosion-fatigue conditions; all the tests reported have been made under the latter conditions.

All the tests described were made on material taken from one cast of Siemens-Martin open-hearth acid mild steel, conforming to a specification for boiler material of good quality. The specimens fall into three classes: (*a*) 3-in. wide strips of plate, ⅝ in. and also ¾ in. in thickness; (*b*) 12-in. wide riveted joints made of ⅝-in. plate and fitted with double-riveted double cover straps; and (*c*) 3-in. wide strips of electrically welded butt joint. With regard to the riveted joints, the rivet heads were caulked on both sides of the joint, while the edges of both butt straps were lightly caulked; ⅝-in. plate was used for the riveted joints as representative of common practice, but ¾-in. plate was used for the welded joint, as it was considered that this larger size would give a joint more representative, metallurgically, of practice than the smaller size.

The tests were made on specimens of the three types mentioned above while immersed in a boiling aqueous solution of sodium hydroxide in distilled water. The initial strength of the solution was 6000 grains per gal.; this high degree of concentration was chosen as probably comparable with that occurring locally in service after evaporation has taken place, permitting a high concentration within joints, &c. Considerable difficulty was encountered, owing to various causes, due to the changing conditions of alkalinity during any individual test; these changes have been reported in detail and must be considered in assessing the results of the tests. Each specimen was, after test, submitted to metallurgical examination, particular attention being devoted to the *type* of fracture, careful search being made for evidences of intercrystalline failure.

The repeated-loading tests were carried on, with each type of specimen, to endurances extending up to from 2 to 10 million cycles, corresponding to total corrosion times not exceeding about 200 days; if failure had not occurred in a period of this duration, the test was discontinued and the specimen removed and examined.

The performance of the specimens may be briefly summarised as shown in Table II.

TABLE II.—*Summary of Performance of Specimens.*

Type of Specimen.	Maximum Nominal * Range of Repeated Bending Stress under which Samples remained Unbroken (reckoned on Plate). Tons per sq. in.	Corresponding Number of Stress Cycles. Millions.	Corresponding Total Corrosion Time. Days.
(1) $\frac{5}{8}$-in. plain boiler plate .	19	8–10	178–205
(2) $\frac{3}{4}$-in. plain boiler plate .	21	$5\frac{1}{2}$	117
(3) Riveted joints of $\frac{5}{8}$-in. plate . . .	17	5	102
(4) Electrically welded butt joints of $\frac{3}{4}$-in. plate .	$19\frac{3}{4}$	$7\frac{1}{2}$	163

* No account has been taken of any local stress-concentration effect at the points of application of the load or at the sudden change of section in the riveted joints at the edge of the cover straps.

Visual examination of the surfaces of the plates and joints after test showed very little evidence of any general corrosive attack or of serious pitting, while the intersections of the fractures with the surface were of the irregular and stepped type which is typical of failure under corrosion-fatigue. The metallurgical examination of the fractured specimens established quite clearly that these are essentially of the type associated with failure under corrosion-fatigue; no evidences were obtained of any tendency for cracks either to form at or to follow intercrystalline boundaries. Hence, the test conditions employed did *not* produce the peculiar type of intercrystalline failure usually associated with the term "caustic embrittlement."

Considering the summary of numerical results (Table II.) merely as corrosion-fatigue data, the values as recorded and relating to a mild steel of about $26\frac{1}{2}$ tons per sq. in. tenacity, do not give cause for serious alarm. The performance of the riveted and welded joints, in relation to that of the plates, appears to be satisfactory. It must, of course, be realised that endurances of the order of from 10^6 to 10^7 cycles, with maximum corrosion times not exceeding 200 days, undoubtedly underestimate the probable effects of much longer periods of immersion. In fact, the results convey a distinct warning that, as far as possible, precautions must be taken to avoid those conditions suitable for corrosion-fatigue, and suggest that, if those conditions are not avoided, then considerable damage may result. The type of fracture obtained in these tests is in marked agreement with damage which has often been observed in boilers in the form of pitting and grooving. On the other hand, the numerical stress values given in the above Table are much in excess of stresses operating in boilers in good practice. Under service conditions, therefore, fairly long times would be necessary to cause serious damage, and careful periodic inspection should go far to reduce the danger of actual fracture in service.

Nevertheless, some caution should be exercised in applying the results of the present experiments to actual boilers, and the differences between the test conditions employed and operating conditions must be clearly recognised. In these laboratory tests, a temperature of just under 100° C. only was employed, while pressure effects—except the induced stresses normally caused by boiler pressure—were absent. Again, the joints were totally immersed; this may have an influence on the effect of "seepage" through a boiler joint. There is also no doubt that the riveted and welded joints used were very carefully made and had not been subjected to the abuse to which some joints are subjected in ordinary construction. Then, turning to the solution used, this cannot be directly compared with water present in service boilers; every effort was made to overcome, as far as possible, the difficulties encountered during the progress of the tests, but, to avoid any misunderstanding, the chemical history of the solution has been presented in detail.

It may still be true that, under certain critical conditions of concentration, the presence of cyclic stress may produce a dangerous acceleration of damage; on this point, the present experiments offer no evidence.

It has been considered desirable to give, at the commencement of this paper, this statement of the general objects of the experiments, the summary of results and a criticism of the results obtained. A detailed description of the investigation now follows.

II.—Description of the Testing Machine and Aqueous Solution.

Testing Machine.

The machine is designed to apply repeated bending strains (zero to a maximum) to a number of specimens (one, two or three) tested in air, or immersed in boiling water or boiling solutions of suitable salts such as sodium hydroxide, &c.; the ambient fluid is not under pressure. Each specimen is tested as a beam, the loading being applied at two points, so that the middle half of the specimen is subjected to a uniform bending moment. Strips of boiler plate, 27 in. long, 3 in. wide and up to ¾ in. thick, can be accommodated in each of the three units of the machine. By combining two or three units, riveted joints—double-riveted butt joints with double cover straps or lap joints—of $\frac{3}{4}$-in. plate and up to 12 in. in width can be tested.

The designed maximum load capacity of each unit is 7500 lb., giving a total of 22,500 lb. for the combination. The machine operates at 35 loadings per min. and is illustrated in Figs. 62 and 63.

The specimen, S_p, under test is supported on rollers retained in cylindrical seatings. The seatings are fixed in a steel tank, T, which contains the fluid. The tank and fluid are heated by three gas burners, the temperature being maintained constant by means of a thermostatic control. The fluid circulates through the tank by convection, three inclined pipes being fitted to the underside of the tank immediately above the gas burners. In the tests to be described, the temperature of the solution was maintained between 96° and 100° C.

The load is applied as follows: A motor-driven rotating cam shaft and cams, C, operate the rocking levers L_1 pivoted in ball bearings at P_1. The motion of L_1 is transmitted to the levers L_2 through vertical rods and calibrated springs, S_1, thus applying predetermined maximum forces to each of the levers L_2, which are pivoted on knife-edges attached to the cross member M. From the levers L_2 the load is applied to the specimen S_p through the struts B and the two-point loading brackets D. The cams are so designed that the load on the springs is completely released during each revolution. In order to keep the knife-edges always in contact with their seatings, the weight of the levers L_2, struts B and brackets D is never removed from the specimen; this weight accounts for the small positive value of the minimum stress employed in the tests. The tank is fitted with a cover and two condenser tubes (not shown) for the purpose of reducing evaporation.

Ambient Solution.

The solution used in the tank was distilled water (condensed boiler water) to which had been added 6000 grains of sodium hydroxide per gallon of water, the level in the tank being kept constant by means of a ball-valve. This figure was chosen as being considered comparable with that which might prevail in practice when evaporation takes place, permitting a high concentration within the joints and cracks. In the testing machine the specimen is completely immersed, and evaporation at the joints or cracks cannot occur. It was at first intended to maintain a constant alkali content, but as it was found that this would mean no sodium hydroxide being present during tests on later specimens it was then decided to renew the solution once a week, the sodium hydroxide content falling from 6000 to 4000 grains per gal. in that period.

Although a cover and condenser tubes had been fitted to the tank, the evaporation encountered was considered to be rather

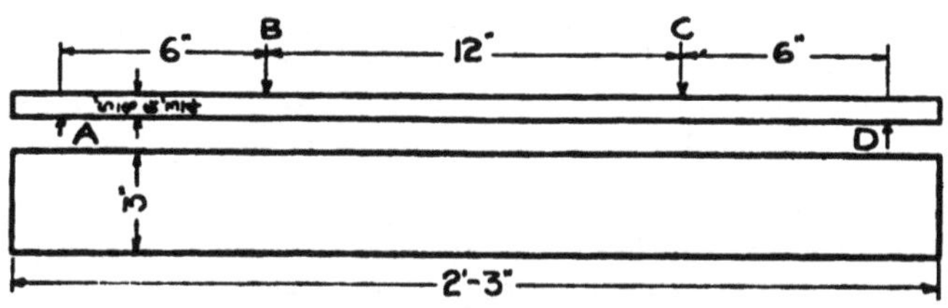

SPECIMEN - TYPE "A" PLAIN STRIP

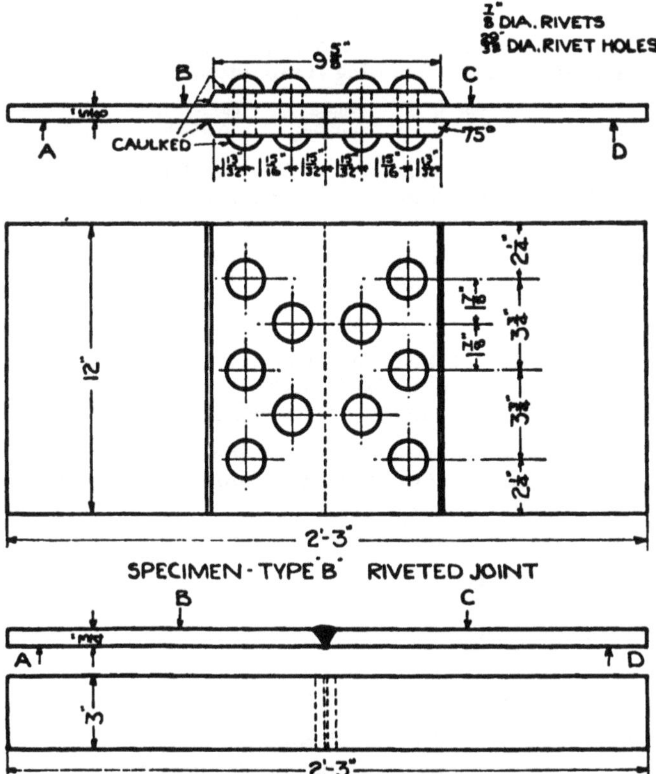

SPECIMEN - TYPE "B" RIVETED JOINT

B C A D 3" 2'-3"

SPECIMEN - TYPE "C" WELDED JOINT

Note - Loads applied at A B C and D

Fig. 56.—Types of Specimens.

excessive, amounting to about 4½ gal. per day. It was then decided to reduce this and to maintain a higher concentration over longer periods by pouring a film of oil on the top of the solution. A ¼-in. thick film of B.P. paraffin was used for this purpose. Under these conditions evaporation was reduced to the order of ¼–½ gal. per day; the procedure was also changed so that the solution was renewed when the concentration of sodium hydroxide had fallen to 3000 grains per gal. Whenever a specimen fractured, or a test was discontinued, the bath was cleaned out and a new supply of the original strength (6000 grains per gal.) was used.

III.—Range of Investigation.

The material used in the investigation was steel boiler plate made by the Siemens-Martin acid open-hearth process, stated by the makers to be of the following composition:

Carbon.	Silicon.	Sulphur.	Phosphorus.	Manganese.
0·21%	0·05%	0·039%	0·036%	0·51%

It was supplied in the form of plate and riveted and welded joints, to the form and dimensions given in Fig. 56, in the following quantities:

30 specimens of type *A* . .	⅝ in. thick		
10 " " *A* . .	¾ " "		
30 " " *B* . .	⅝ " "	plate	
30 " " *C* . .	¾ " "	"	

In addition to the main series of corrosion-fatigue tests in a boiling aqueous solution of NaOH made on specimens of each type, the following additional tests were carried out:

(1) Chemical analysis of the material.
(2) Metallurgical examination of untested material.
(3) Static tensile tests on:
 (*a*) Plain specimen ⅝ in. thick
 (*b*) " " ¾ in. "
 (*c*) Welded joint.
(4) Metallurgical examination of tested specimens.
(5) Analysis of the solution.

IV. Results of Investigation.

(1) *Chemical Analysis of the Material.*

Particulars of the material and its chemical analysis as determined at the National Physical Laboratory are given in Table III.

(2) *Metallurgical Examination of Untested Material.*

A plain specimen (*HHA*30) of type *A*, ⅝ in. thick, was submitted to micro-examination. Figs. 64(*a*) and 64(*b*) represent the

TABLE III.—*Particulars and Chemical Analysis of the Material.*

Reference Mark and Type of Specimens.	Description of Material Used.	Chemical Composition.					
		C. %.	Si. %.	S. %.	P. %.	Mn. %.	N. %.
HHA. Type *A*. Plate, ⅝ in. thick. *HHB*. Type *A*. Plate, ¾ in. thick. *HMH*. Type *B*. Riveted joint of ⅝ in. plate. *HMJ*. Type *C*. Welded joint of ¾ in. plate.	Siemens-Martin acid open-hearth boiler steel.	0·20	0·03	0·028	0·036	0·50	0·004

polished and etched "side" and "end" sections, respectively. The side section is also shown in Fig. 64(*c*). The structure of this material is that of typical boiler plate.

Fig. 65(*a*) represents a sulphur print prepared in the usual manner from a section of the material parallel to the side of the same plate and indicates the distribution of the sulphur. A macrograph, Fig. 65(*b*), was prepared from a section of the material deeply etched in acid copper chloride solution, in order to reveal the distribution of phosphorus and possibly oxygen. Both the sulphur and phosphorus are evenly distributed.

A plate, *HHB*4, of type *A*, ¾ in. thick, was also submitted to microscopic examination. Figs. 66(*a*) and 66(*b*) represent corresponding sections to those shown in Figs. 64(*a*) and 64(*b*) for the ⅝-in. thick bar. The banded or laminated structure of the material is clearly indicated in Figs. 64(*a*) and 66(*a*), and appears somewhat more pronounced in the ⅝-in. thick material. This agrees with the fact that *HHA*30 (⅝-in.) and *HHB*4 (¾-in.) both originated from the same cast and that the former received the greater reduction in thickness in rolling during manufacture.

(3) *Static Tensile Tests.*

These were carried out on strips cut from the plain plates, also on a strip cut from a welded plate. The specimens were of the standard British Standards Institution form for tests on plate. The proportional limits were determined, using a Marten's extensometer. The results of these tests are as stated in Table IV.

(4) *Corrosion-Fatigue Tests.*

The results of the tests carried out in the specially designed machine, in which the specimens were subjected to repeated applications of stress while totally immersed in a boiling sodium hydroxide solution, are given in Table V. (tests on plate specimens, type *A*), Table VI. (tests on riveted joints, type *B*) and Table VII. (tests on welded joints, type *C*). In these Tables the given estimated values of nominal stress relate to the *main plate* only.

TABLE IV.—*Results of Static Tensile Tests.*

Type of Specimen : Specimen Reference Mark :	Plain ⅝-in. Material. *HHA11A.*	Plain ¾-in. Material. *HHB4A.*	Electrically-Welded Joint.* *HMJ7A.*
Cross-section of specimen. In.	0·645 × 2·002	0·764 × 2·002	0·765 × 2·001$_5$ (plate) 0·862 over weld
Gauge length. In.	8	8	8
Limit of proportionality. Tons per sq. in.	13·2	12·4	8·2
Stress at yield point. Tons per sq. in.	15·3	15·7	15·2
Ultimate tensile strength. Tons per sq. in.	26·4	26·6	26·5
Elongation on 8 in. %	31	31	24
Reduction of area. %	49½	48	47
Young's Modulus, *E*. Lb. per sq. in. × 10^{-6}.	30·4	30·1	29·9
Remarks on fracture	Silky fibrous.	Silky fibrous.	Silky fibrous; fractured in *plate* at point 4 in. from centre of weld at a slight internal flaw.

* The welded section was situated at the mid-section of the test length.

TABLE V.—*Results of Tests on Specimens of Boiler Plate, Type* A, *in Solution initially containing* 6000 *Grains of Sodium Hydroxide per Gal. of Distilled Water at Temperatures of* 96–100° *C.*

Specimen.	Thickness. In.	Nominal Stress Cycle. Tons per sq. in. Min.	Nominal Stress Cycle. Tons per sq. in. Max.	Range of Stress. Tons per sq. in.	Number of Applied Stress Cycles. Millions.	Duration of Test.* Days.	Remarks on Specimen.
*HHA*10	0·64	2·4	24·4	22·0	0·5376	10¾ B	Broken.
9	0·65	2·3	23·3	21·0	0·4645	9¾ B	Broken.
8	0·65	2·3	22·3	20·0	0·8427	19 B	Broken.
6	0·65	2·3	22·3	20·0	4·5608	90 B	Unbroken.
1	0·65	2·3	21·6	19·3	1·2565	27 A	Broken.
7	0·65	2·3	21·6	19·3	3·0470	58 B	Broken.
5	0·65	2·3	21·6	19·3	3·2281	63 B	Broken.
4	0·65	2·3	20·3	18·0	8·8642	178 {83 A, 95 B}	Unbroken.
2	0·65	2·3	18·3	16·0	10·1207	205 {110 A, 95 B}	Unbroken.
3	0·65	2·3	14·3	12·0	10·1207	205 {110 A, 95 B}	Unbroken.
*HHB*3	0·77	1·6	24·6	23·0	1·1404	23 B	Broken.
1	0·77	1·6	23·6	22·0	1·9062	41 B	Broken.
2	0·77	1·6	22·6	21·0	5·5823	117 B	Unbroken.

* A = without, B = with oil film on solution.

TABLE VI.—*Results of Tests on Specimens of Riveted Joints, Type* B, *in Solution initially containing* 6000 *Grains of Sodium Hydroxide per Gal. of Distilled Water at Temperatures of* 96–100° *C.*

Solution covered with a film of oil.

Specimen.	Plate Size. In.	Nominal Stress Cycles. Tons per sq. in.		Range of Stress (on Plate). Tons per sq. in.	Number of Applied Stress Cycles. Millions.	Duration of Test. Days.	Remarks on Specimens.
		Min.	Max.				
*HMH*1	12 × 0·66	1·7	25·6	23·9	0·2295	5	Broken.
2	11·9 × 0·66	1·7	23·7	22·0	0·6700	20 *	Broken.
3	12 × 0·65	1·7	22·7	21·0	0·6532	14	Broken.
4	11·9 × 0·66	1·7	19·7	18·0	1·2477	25	Broken.
5	12·2 × 0·65	1·7	18·7	17·0	5·1468	102	Unbroken.

* Machine idle for 6 days; corrosion-fatigue conditions operating for 14 days only.

TABLE VII.—*Results of Tests on Specimens of Electrically Welded Butt Joints, Type* C, *in Solution initially containing* 6000 *Grains of Sodium Hydroxide per Gal. of Distilled Water at Temperatures of* 96–100° *C.*

Solution covered with a film of oil.

Specimen.	Plate Thickness. In.	Nominal Stress Cycle. Tons per sq. in.		Range of Stress (estimated on Plate). Tons per sq. in.	Number of Applied Stress Cycles. Millions.	Duration of Test. Days.	Remarks on Specimens.
		Min.	Max.				
*HMJ*6	0·76	1·7	23·7	22·0	0·7213	14	Broken at weld.
5	0·76	1·7	22·7	21·0	2·0829	50	Broken 1⅞ in. from weld.
4	0·76	1·7	21·7	20·0	4·2804	94	Broken 2⅛ in. from weld.
1	0·76	1·7	21·7	20·0	7·4132	163	Unbroken.
2	0·76	1·7	20·2	18·5	1·5480	32	Broken at weld.
3	0·76	1·7	18·7	17·0	7·4132	163	Unbroken.

Previous mention has been made of the changes occurring in the strength of the aqueous solution; the recorded concentrations and evaporation of the solution while the specimens were under test are shown in Figs. 57A and 57B (plain specimens, ⅝ in. thick), Fig. 58 (plain specimens, ¾ in. thick), Fig. 59 (riveted joints), and Fig. 60 (welded joints). These diagrams probably require the explanation that the curves relate to the bath as a whole, which may have contained one, two or three specimens at any particular time in the period covered by the diagram; the portion of this period covered by particular specimens is therefore indicated at the bottom of each diagram. An analysis of the solution was made on a sample taken from the bath at each time indicated by a black circle on these diagrams.

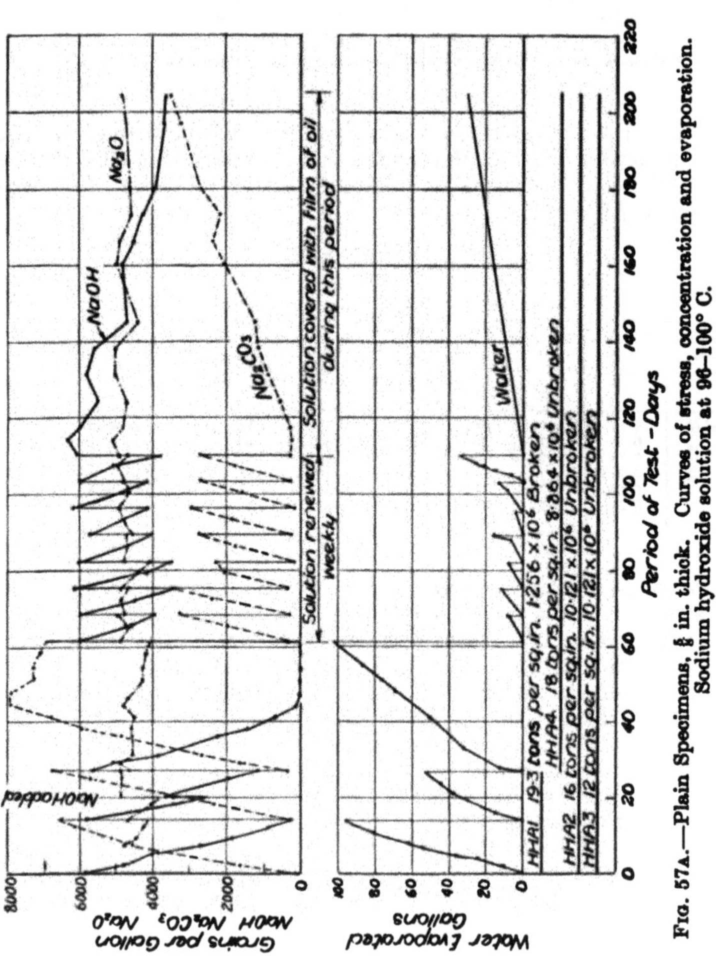

FIG. 57A.—Plain Specimens, $\frac{5}{8}$ in. thick. Curves of stress, concentration and evaporation. Sodium hydroxide solution at 96–100° C.

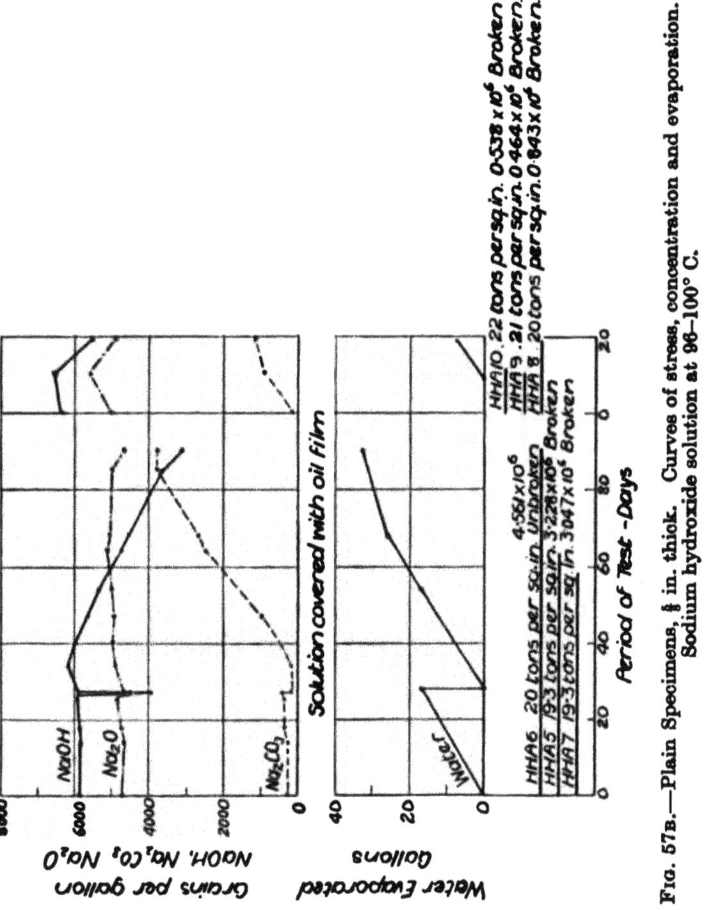

FIG. 57B.—Plain Specimens, ⅝ in. thick. Curves of stress, concentration and evaporation. Sodium hydroxide solution at 96–100° C.

Stress-Endurance Curves of the specimens tested under these conditions are plotted in Fig. 61. In these curves, a sharp knee has been drawn, as it appears to represent the trend of the results, but this should not be taken to indicate the existence, or belief in the existence, of a " corrosion-fatigue limit."

(5) *Metallurgical Examination of Tested Specimens.*

Photographs and micrographs of tested specimens and fractures are given in Figs. 67 and 68 (plain specimens, ⅝ in. thick), Fig. 69

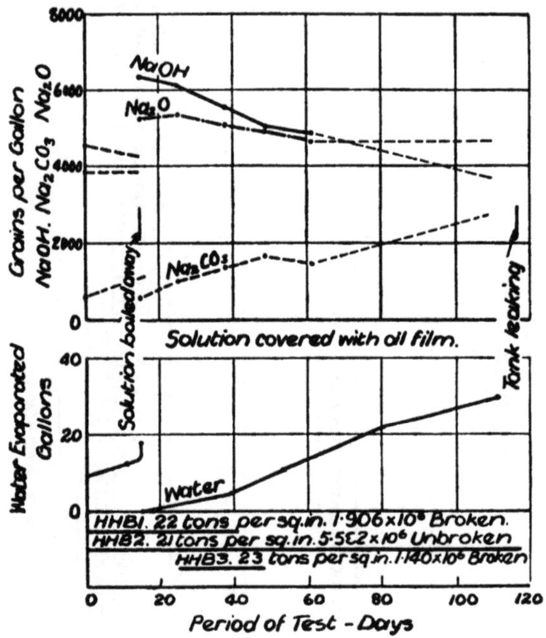

FIG. 58.—Plain Specimens, ¾ in. thick. Curves of stress, concentration and evaporation. Sodium hydroxide solution at 96–100° C.

(plain specimens, ¾ in. thick), Figs. 70 to 72 (riveted joints) and Figs. 73 and 74 (welded joints).

(6) *Analysis of Solution.*

Figs. 57 to 60 show the concentrations of the solution, expressed as grains per gallon of distilled (condenser) water, of sodium hydroxide, sodium carbonate and the total alkalinity, also the amount of water evaporated during the period of test, while the lower parts of the diagrams indicate the chronological disposition of the specimens. From these figures the condition of the solution during the test on each specimen can be readily seen.

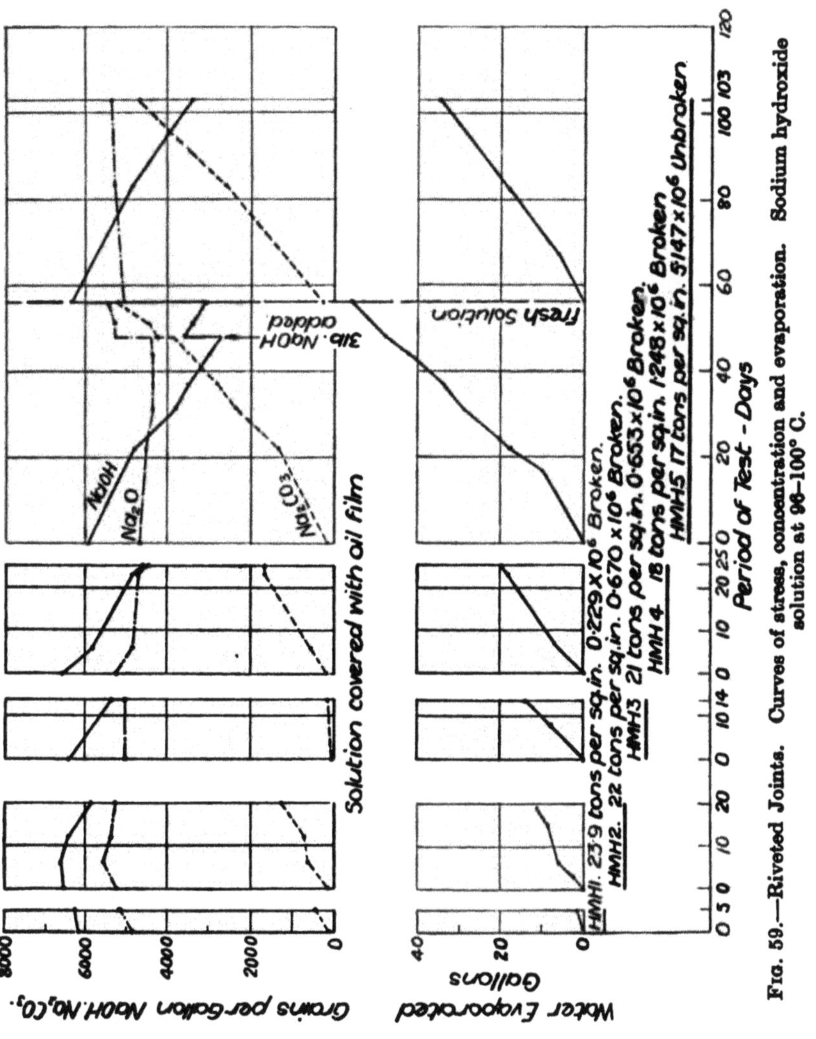

Fig. 59.—Riveted Joints. Curves of stress, concentration and evaporation. Sodium hydroxide solution at 96–100° C.

At first the solution was renewed when the sodium hydroxide content fell nearly to zero, then for a period of 40 days an attempt was made to keep the total alkali constant, but the evaporation of the solution was so great that after about 20 days the sodium

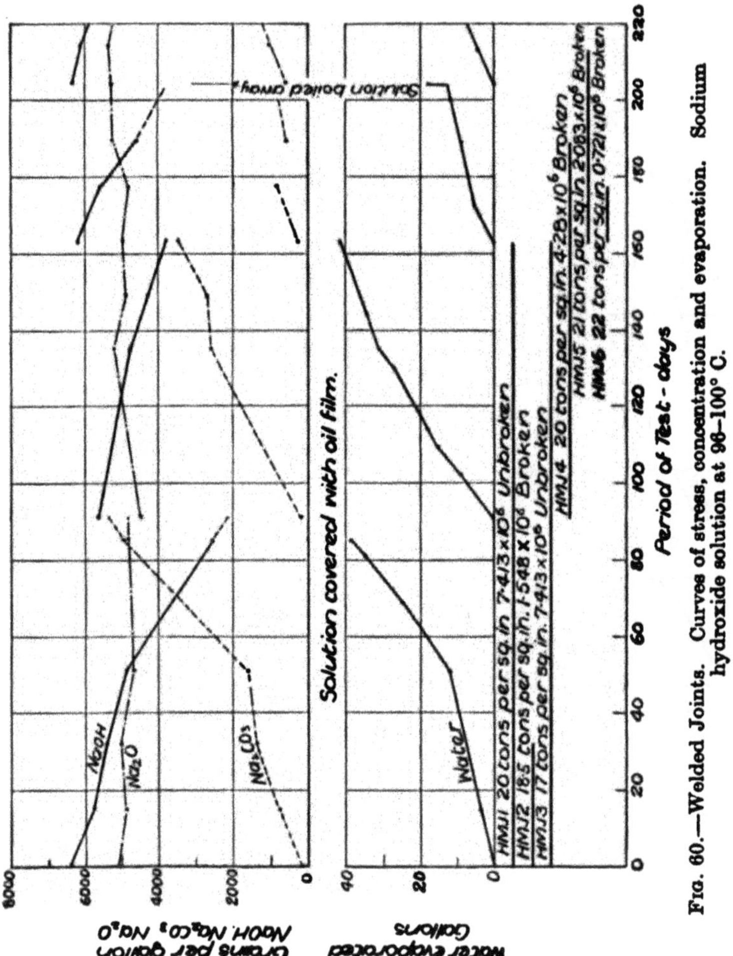

Fig. 60.—Welded Joints. Curves of stress, concentration and evaporation. Sodium hydroxide solution at 96–100° C.

hydroxide content was reduced to zero while sodium carbonate was found to be present to the extent of 8000 grains per gal. This condition was considered unsatisfactory, because, had a new specimen been inserted and broken, no sodium hydroxide would have been present during the test. At this stage the procedure was altered, the solution being renewed weekly for a period of 7 weeks. It was then considered possible that, owing to the high rate of evaporation (which loss was automatically replaced by

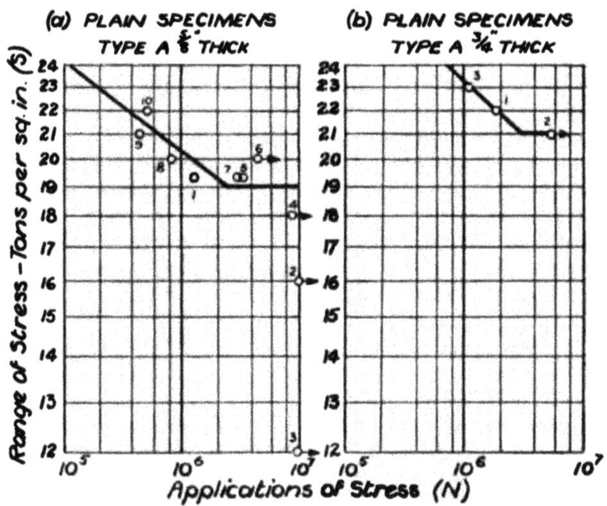

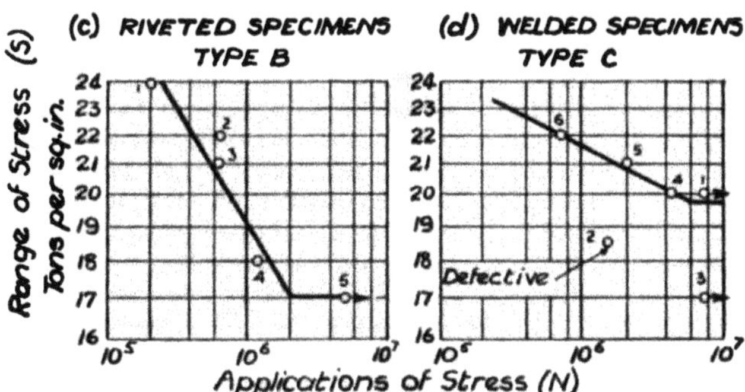

FIG. 61.—Stress-Endurance Curves; ambient solution of 6000 grains of sodium hydroxide per gal. of distilled water at 96–100° C.

Note.—Numbers near the experimental points refer to the reference numbers of the specimens. They should be prefixed by the reference letters as follows: (*a*) *HHA*; (*b*) *HHB*; (*c*) *HMH*; (*d*) *HMJ*.

addition of distilled water), the access of oxygen to the solution might result in preventing any intergranular attack by hydrogen. From this date onwards, therefore, the solution was covered with a layer of inert oil, a ½-in. thick layer of B.P. paraffin being used.

(At one time a thick gear-oil was tried, but owing to the difficulty of removing all traces of the oil from fittings in the tank, the use of paraffin was preferred.) The effect of using the layer of oil was to reduce considerably the evaporation of water and maintain the sodium hydroxide strength over much longer periods. (Slight increases sometimes observed in the sodium hydroxide content were due to the ball-valve controlling the water level not functioning properly, and resulting in a decrease in the quantity of water in the tank.)

A procedure was therefore adopted in which the solution was renewed when the sodium hydroxide content had fallen to about 3000 grains per gal., or, alternatively, when a specimen was removed from the tank and a new specimen inserted. It may be mentioned that a complete analysis showed that the solution also contained the following salts: sodium chloride, 0·71 g. per litre; sodium silicate, 0·16 g. per litre; sodium sulphate, 0·50 g. per litre.

V.—Discussion of Results and Conclusions.

(*a*) *Plain Specimens.*

In specimen *HHA*1, a crack had developed near the centre of the plate. This crack lay chiefly in a plane at right angles to the axis of the plate and originated in the side subjected to tensile stressing; the crack did not penetrate the full thickness of the material nor extend to the full width of the under surface.

The terminal portion of the crack was examined after removing the scale and polishing. Figs. 67(*a*) and 67(*b*) show that the crack is of considerable width even quite near the tip and that there is no marked tendency for the "spurs" leading out of the main crack to penetrate along the grain boundaries. Fig. 67(*c*) shows a polished and etched portion of the side of the plate in the vicinity of the crack, which is "branched," and of smaller width than that shown in Figs. 67(*a*) and 67(*b*). The material exhibits a banded structure, presumably associated with rolling during manufacture; the cracks are influenced in direction by this banded structure for short distances. None of the photographs of Fig. 67 reveal any tendency for the cracks to follow grain boundaries.

Specimens *HHA*2, 4 and 5 were also examined. After removal of scale, specimens *HHA*2 and *HHA*4—which were unbroken under test—were examined with a magnetic crack detector. No cracks were found in these specimens nor in specimen *HHA*5 at positions away from the main crack. The fracture of the latter specimen did not extend completely through the plate, and suitable longitudinal sections, parallel to either the faces or the sides of the plate, were cut in order to examine the crack near its root. Fig. 68 illustrates the general nature of the cracking, which is essentially of the corrosion-fatigue type with no evidence of a general intercrystalline breakdown.

Specimens *HHA*8, 9 and 10 were also submitted to microscopical examination. The essential features of the type of cracking observed were exactly similar to those encountered in the specimens previously described; it is therefore unnecessary to reproduce any micrographs relating to these specimens.

Turning to the specimens of ¾-in. plate (*HHB*), the microstructures of these specimens were examined in detail at magnifications of 500 and 2000 diameters, with results of the same general nature as those encountered in the specimens of ⅜-in. plate, the chief characteristics revealed being main and subsidiary branching cracks of a transcrystalline nature. Fig. 69(*a*) is included as representative of the transverse appearance of main cracks after polishing and etching the sectional plates with Fry's reagent. Fig. 69(*b*) is also reproduced, as it exhibits so faithfully the intricate system of cracking entirely characteristic of failure by corrosion-fatigue.

(*b*) *Riveted Joints.*

The position of cracking in the riveted-joint specimens can be seen in Fig. 70. In one specimen, cracking occurred very close to the edge of the cover strap; in the other three cases failure took place at the section corresponding to one of the points of application of the external load, situated about 1 in. from the edge of the cover strap. Failure at either of these positions was, therefore, influenced, to some extent, by stress-concentration effects. Each of the four joints was sectioned and submitted to microscopical examination. Features common to all may be illustrated by a few typical micrographs. In each case, the main crack exhibited branching characteristics, and terminated finally in a fine crack, as illustrated at high magnification in Fig. 71(*a*), which relates to specimen *HMH*2. This field also shows two other features typical of this batch of specimens, the spheroidisation of the pearlite and the "faulting" of the material caused by the deformation at fracture. Fig. 71(*b*), showing the end of the main crack causing failure in specimen *HMH*4, is an exceptionally clear example of the predominantly transcrystalline nature of the cracking encountered. It was also observed, in many cases, that the cracking tended to follow lines of non-metallic inclusions; Fig. 71(*c*), also from specimen *HMH*4, is a typical example.

When specimen *HMH*3 was sectioned, an extremely interesting feature was disclosed. In addition to the main fracture—which occurred in the plate outside the joint—a crack was detected *within* the joint and near to one of the rivet holes; this internal crack, and also the main crack, are clearly visible in Fig. 72(*a*). Two micrographs, Figs. 72(*b*) and 72(*c*), of this internal crack are reproduced. Both indicate that corrosion-fatigue is the cause, the crack in general following a transcrystalline path, with the usual branching characteristics and a tendency to follow lines of non-metallic inclusions. No evidence of a faulty joint could be detected, while the corrosion-fatigue history of this specimen was not unduly great ($0 \cdot 6 \times 10^6$ stress cycles and 14 days' immersion). Failure at this position in a joint has, on several occasions, been observed in cases of boiler failure in service, and the present example is of great interest. It will be noted that it occurred on the tension side of the main plate and the corrosive solution could have found access between the main and cover plate, had relative movement occurred between these. There will also be some stress-concentration effect due to the presence of the rivet hole. Specimen *HMH*5 had remained unbroken after a much *longer* corrosion-fatigue history (5×10^6 stress cycles and 102 days' immersion), although the applied range of stress (17 tons per sq. in.) was less than that applied to specimen *HMH*3 (21 tons per sq. in.); specimen *HMH*5 was submitted to very careful examination to discover if any cracking were present. The tension face of the joint was machined so as to remove three rivet heads and a thin layer of plate material. A complete transverse section of the joint was also made along the line of the three rivets near one of the outer edges of the butt-straps. Lastly, the main plate outside and near the edge of the butt-strap was sectioned and examined. No evidence whatever of cracking, of any sort, was found in any portion of the specimen.

Although the cracking of the joint HMH3 *was, apparently, not repeated in similar specimens, it is of some significance that the conditions of these tests have, even in one specimen, produced cracking within a joint of a type that had previously been observed in some service failures.*

(c) *Welded Joints.*

As recorded in Table VII., of the six specimens tested, two (*HMJ*2 and 6) fractured through the weld, two others (*HMJ*4 and 5) fractured away from the weld, while the remaining two (*HMJ*1 and 3) remained unbroken. The appearance of longitudinal sections, after polishing and etching, of five of these specimens is shown in Fig. 73. Reference to Fig. 61(*d*) indicates that the resistance of specimen *HMJ*2 was very low compared with that of the remaining specimens. A visual examination of the fracture of this specimen showed the presence of a layer occupying, approximately, the middle third or middle half of the cross-section, which gave the appearance of lack of complete fusion of the weld metal; the fracture was distinctly " stepped " at the boundaries of this central zone. This specimen has, therefore, been regarded as defective and ignored in assessing the resistance of this batch of joints considered as a whole, but such a defect could, presumably, occur in commercial production. The type of cracking observed in specimens *HMJ*2 and 6 was, however, essentially similar, consisting of branching cracks following, mainly, transcrystalline paths. Thus Fig. 74(*a*) shows a typical portion of the main crack in *HMJ*6, while Fig. 74(*b*) illustrates a small crack in *HMJ*2.

Turning to specimens *HMJ*4 and 5, both of which fractured away from the weld (*see* Fig. 73), examination of various polished and etched sections revealed characteristics of the usual corrosion-fatigue type previously discussed. Fig. 74(*c*), showing one of the small branching cracks in the fracture of *HMJ*4, is of some interest, as the crack consists of almost straight portions sharply inclined to each other; this is an example of this type rarely met with in the present experiments, although it has been encountered elsewhere in failures caused by corrosion-fatigue. It is sufficient to record Fig. 74(*d*) as a typical portion of the cracking in specimen *HMJ*5; no new features are revealed.

Specimens *HMJ*1 and 3, which remained unbroken after test, were also sectioned and examined; no signs of cracking were observed.

(*d*) *General Conclusion.*

Summarising broadly the results of the investigation, the principal fact that emerges is that, *under the test conditions employed*, samples of boiler plate, riveted joints and welded joints failed by a process which is essentially that associated with corrosion-fatigue; the predominantly intercrystalline type of cracking ascribed to failure by caustic embrittlement was *not* produced. Considerable evidence is available, from certain service boilers, that corrosion-fatigue is a possible cause of danger, and similar fissuring and cracking of this type was reproduced in the present tests. In these tests, the applied ranges of stress which were found necessary to produce fracture were relatively so high—in comparison with nominal working stress adopted in practice—that little cause for alarm would arise from the actual results, although, in this connection, it should be remembered that the test conditions of temperature were much lower than those prevailing in modern practice; also, the maximum period occupied by any one test was small in comparison with normal boiler life. It is therefore possible that under service conditions much more rapid attack might result, and, hence, every effort should be made in operation to exclude, as far as possible, those conditions which would be conducive to such attack. It is of interest to find that, even with the somewhat artificial conditions used in the tests, one example was encountered of the occurrence of cracking in a riveted joint at a position in the interior of the joint and near to a rivet; such cracks have been found in service failures, the crack being of a transcrystalline character as in the present tests.

Acknowledgments.

The whole of the material—plates, riveted joints and welded joints—used in the investigation were specially prepared and presented by Messrs. Babcock & Wilcox, Ltd., through the courtesy of Mr. C. H. Davy, Chief Research Engineer of that firm, who also showed great personal interest in the investigation. The very extensive work involved in the chemical analyses and metallurgical examination of material and fractured samples was carried out in the Metallurgy Department of the National Physical Laboratory by Dr. F. Adcock, Mr. A. J. Cook and Mr. C. A. Harvey, under the general direction of Dr. C. H. Desch, F.R.S.

CORRESPONDENCE.

Dr. U. R. Evans (Cambridge University) wrote that, in his interesting Introduction, Dr. Desch had suggested that the cracking produced by hot ammonium nitrate solution might be related to the cracking of boiler plates by strongly alkaline solutions. It might be of interest to refer to some unpublished work by Dr. Wooster and Dr. Nockolds in the Department of Mineralogy and Petrology of Cambridge University. It was previously known that stressed steel was cracked by boiling ammonium nitrate solution but that the effect was not produced at lower temperatures. Wooster and Nockolds had observed that the boiling solution produced a black, compact layer, probably of oxide, which cracked under tension, whereas the colder solutions gave a greenish, more slimy mass, probably hydrated magnetite. Presumably the high-temperature oxide would be sufficiently a conductor of electricity to act as cathode towards the iron exposed at the network of cracks. This combination of large cathode (the oxide-covered surface) and small anode (the iron exposed at the cracks) was one which was always apt to lead to intensified corrosion—in accordance with well-established electrochemical principles. The greenish body formed at lower temperatures was physically unsuited to act in this way, thus explaining why colder solutions produced no cracking.

When anodic attack had once started at the breaks in the oxide film, it would tend to continue downwards in a direction normal to the stress direction, since the stresses would keep cracking any freshly formed film at the bottom of the fissures produced. Other things being equal, the anodic attack would follow the grain boundaries, since the energy change involved in the removal of an atom situated at the junction of two lattice systems would be different from that involved in the removal of an atom situated within an undisturbed lattice. It might be thought that this would imply that all ordinary corrosion should be intergranular in character, but in attack by ordinary waters or salt solutions, where the anodic and cathodic products interacted to yield a sparingly soluble substance by precipitation, such fissure corrosion would soon stifle itself, even if considerable tensions were applied to the metal. Thus, intergranular attack was favoured by spatial separation of anodes and cathodes, as in the case of lead cable-sheathing, where stray currents gave rise to intergranular attack, whilst "natural" soil corrosion gave a more general type. Even in the absence of stifling, the accumulation of metal ions in intergranular cavities would normally shift the potential sufficiently to divert the attack elsewhere. Exceptions would occur—where the intergranular material was, owing to chemical differences from the interior of the grains, naturally anodic, as in the case of austenitic chromium-nickel steel which had been heat-treated in such a way as to cause intergranular impoverishment in chromium. Here intergranular attack might arise, as was well known. Another exception would occur where bodies were present which removed the cations as complexes; thus, ammonia promoted the intergranular corrosion of copper under stress, and nitrites that of lead.

A particularly favourable case for intergranular attack under stress would occur where the complex-forming body was one which came to be concentrated in crevices. Sodium hydroxide was a chemical of this character; its creeping power into crevices was well known to those who had studied the corrosive actions of salts which form sodium hydroxide as a cathodic product. Within his (the writer's) experience no other inorganic substance had the same power of creeping into capillary crevices. Now, this tendency of sodium hydroxide solution to penetrate into crevices was itself a sign that the concentration of sodium hydroxide was greater in crevices than in the body of the liquid. Thus, when once cracking had started in iron immersed in hot alkaline water, one might expect the cell:

Iron|concentrated sodium hydroxide|dilute sodium hydroxide|iron.[1]

In this cell the iron in the crevice would be the anode and the iron outside the crevice the cathode, since dilute alkali would depress the solubility of iron oxides and thus tend to keep the iron passive, whereas the concentrated alkali would at high temperature dissolve them, forming such compounds as ferroates, and making the iron within the crevice active and anodic. The ferroates could only exist in the presence of a great excess of alkali, and when the ferroate formed at the tip of the crack began to diffuse outwards into the parts where the concentration of alkali was lower, iron oxide was likely to separate out, thus explaining the precipitation at a distance noted by Dr. Desch.

Dr. Desch's statement that there were two distinct forms of cracking was interesting. He denoted them as the hydrogen and the oxide types. According to the theory developed tentatively above, they could be referred to as the cathodic type and the anodic type, respectively. The hydrogen or cathodic type of cracking was almost exactly analogous to pickling embrittlement, whereas the oxide or anodic type would have much in common with the intergranular corrosion of lead cable-sheaths produced by stray electric currents.

The suggestions made above were necessarily speculative, but might serve to indicate the general lines which might be expected to lead to the correct explanation.

Mr. William Barr (Colvilles, Ltd., Glasgow) wrote that, in his experience, caustic cracking in boiler plates was invariably intercrystalline in part, and indeed it was this feature that distinguished

[1] The analogous cell Al|N NaOH|$N/100$ NaOH|Al certainly generated a current, the aluminium in the more concentrated solution being the anode.

caustic cracking from corrosion-fatigue. Intercrystalline cracks were usually associated with fissures and wide cracks of the trans-crystalline type. In cracked and broken rivets which frequently accompanied the cracking of plates it was impossible to define the nature of the cracks, owing to the severely distorted structure of the rivets.

Dr. Adcock had suggested that there had been a tendency to overemphasise the intercrystalline character of caustic attack. A more unfortunate tendency had, however, crept in and that was to assume that hydrogen was invariably the cause of intercrystalline failure. This was exemplified in Part II. of the paper, in which was described the application of a *negative* potential during test. Other assumptions that had been made and which had led to the construction of elaborate research apparatus were that high stresses and temperatures were necessary for the production of inter-crystalline cracking. That none of these assumptions was correct was borne out by work which the writer had carried out. He would, perhaps, have an opportunity of publishing this work when completed, but would meantime summarise briefly some of the results already obtained.

Chemical intercrystalline cracks of the "oxygen" type had been produced in 60 hr. in a machined bar of mild steel in the as-rolled condition, which was made the anode in a solution of potassium nitrate (20 g. per 100 c.c.). The current density was approximately 0·03 amp. per sq. cm. and the temperature was maintained at approximately 90° C. Similar results had been obtained with solutions of different concentrations of potassium nitrate, but the quantitative relationship between time to produce visible cracking and concentration of solution had not been determined.

No cracks had been observed when the sample (whether stressed or unstressed) was made the cathode. The presence of some internal stress, as in a test-piece strained by compression, did not accelerate the cracking; on the other hand, strain-age-hardening at 200° C. appeared to do so. Over-ageing had no accelerating effect.

In the absence of an externally applied e.m.f., cracks had only been obtained with strained specimens.

Concentrated solutions of sodium hydroxide acted similarly to solutions of potassium nitrate, but the time required for production of cracks was considerably longer, and general corrosion and fissuring were more marked.

From the above considerations, the writer would venture to suggest that chemical intercrystalline attack of the "caustic" type could only take place when the material was electro-positive in an electrolyte which could maintain the passage of a current. Caustic cracking in service occurred because these conditions were encountered. As was well known, strained metal resulted in potential differences between various parts. The chemical constitution and the concentration of the electrolyte, its temperature and its motion, all contributed to determine whether the passage of a current could be maintained against polarisation and insulation effects. Inhibitors, for example, acted by reducing the ionisation of the solution or/and, by precipitation, as insulators.

It would readily be admitted that the reactions were very complex, hence the frequent simultaneous occurrence of general corrosion, fissuring and intercrystalline cracking.

As suggested by Dr. Adcock, the hydrogen type of intercrystalline cracking produced at high temperatures and pressures had a different mode of formation, in which electrolysis was probably not the dominating factor.

AUTHORS' REPLIES.

Dr. C. H. Desch, F.R.S., in reply to Dr. Evans, wrote that his own observations were in agreement with those of Wooster and Nockolds, and there was no doubt that the texture of the iron oxide produced, as well as its composition, had a great effect on the nature of the corrosive attack. This was probably the way in which some of the organic inhibitors exerted their effect. On the other hand, he did not think that the suggested explanation of the localisation of attack in the grain boundaries was a complete one. If it were, corrosion should be much more frequently intercrystalline than it was. The most marked characteristic of such corrosion was its highly specific character, only a very few reagents being active. Many brasses, even under stress, could be corroded by ammonia without any preferential attack on the boundaries, and brasses susceptible to such attack by ammonia or mercury would corrode quite evenly in other reagents.

Mr. Barr's experiments were very interesting. There was no doubt that both the oxygen and the hydrogen type of attack could occur, and the conditions determining them would well repay investigation. Why was it, also, that sometimes a susceptible steel under stress could be kept in a boiling solution of ammonium nitrate until a large part of it had dissolved, without any indication of intercrystalline attack?

Dr. Frank Adcock wrote that he did not feel that he could make any useful reply to the correspondence on the papers, but wished to thank Mr. Barr for his most interesting observations.

THE THERMAL RELATIONS BETWEEN INGOT AND MOULD.*

By T. F. RUSSELL (Sheffield).

Paper No. 4/1940 *of the Committee on the Heterogeneity of Steel Ingots (submitted by the Stresses in Moulds Panel of the Ingot Moulds Sub-Committee).*

Synopsis.

Saitô's formulæ for the distribution of heat between ingot and mould are examined quantitatively. The practical variations from the ideal case are discussed, and it is thought that the possibility of deriving more exact formulæ than Saitô's is very remote.

Examples are worked out for circular and square ingots of the same cross-sectional area cast into moulds of four different thicknesses. Curves are drawn showing the temperature at different points in the ingot and moulds, the temperature distribution across a diameter and the total quantities of heat in the ingot and mould at different times. Curves are also drawn to show the effect of "mould ratio" on the temperature cycle occurring in the ingot near to the mould wall and on the time taken for the temperature at the centre of an ingot to fall certain amounts representing solidification. The latter show that solidification is accelerated by increasing the mould thickness until the mould ratio is about 0·8–1·0, but that further increase in mould thickness has an inappreciable effect.

Four sets of experimental results on the measurement of mould temperature are examined, and they show that the greatest difference between theory and practice is found at positions in the mould near to the inner face. This is attributed to the effect of the air-gap forming between the ingot and the mould.

Introduction.

One of the terms of reference of the Stresses in Moulds Panel of the Ingot Moulds Sub-Committee (a Sub-Committee of the Heterogeneity of Steel Ingots and Open-Hearth Committees) reads as follows: "To determine, if possible, the magnitude of the stresses developed in the wall of an ingot mould by a mathematical analysis of the temperature gradients and other conditions set up during and after casting the ingot."

Obviously, the first step must be to examine thoroughly known formulæ for temperature distribution in the mould. As a fairly large number of examples had already been worked out in the metallurgical laboratories of the English Steel Corporation, the author obtained permission to prepare a paper, which is now presented.

The determination of the exact mathematical relationship

* Received October 4, 1940.

between the temperature distribution in an ingot and mould and the time which has elapsed since cooling commenced presents a problem of such complexity that it is safe to prophesy that many years must pass before any real advance is made on the present position of knowledge, established by the outstanding works of Saitô [2] and of Lightfoot.[4–7]

Perhaps the chief obstacle to further progress is the variation in the thermal diffusivity of a steel with temperature. Even if the exact functional relationship could be determined by such researches as those of Sub-Committee *A*, Thermal Treatment, of the Alloy Steels Research Committee on the thermal properties of steels, it would be extremely difficult, if not impossible, to deal with such functions in a mathematical treatment; up to the present time all theoretical investigations have presupposed a constant diffusivity.

There are many other factors, most of which are enumerated below, which disturb the purely theoretical considerations dealt with in this paper and increase enormously the difficulty of finding more exact solutions than those developed by Saitô nearly twenty years ago. In fact, it can be confidently stated that, as long as casting procedure and pit practice are what they are to-day, a rigid mathematical solution will *never* be found, as *the metallurgist is totally unable to state his problem to the mathematician.*

Nevertheless, even such an apparently formidable array of disturbing influences should not deter us from examining thoroughly those formulæ which are based on theoretical assumptions at least approximating to practical conditions. It should then be possible to deduce some fundamental principles which can be compared with works experience, or to compare calculated time-temperature curves for certain points in a mould with experimental results. These comparisons may show that there is no satisfactory agreement between theory and practice, but at the same time they may show in which direction the disturbing influences work, or even that the total effect of them is neglectable in certain cases. The reader is left to judge for himself.

The reason why known formulæ have not hitherto been thoroughly examined is undoubtedly that graphical determination of the roots of the equations given below is not a satisfactory process and that an enormous amount of laborious work is required to interpret the formulæ arithmetically. The calculations were, however, facilitated by the use of a Monroe eight-figure calculating machine and the Mathematical Tables published by the British Association for the Advancement of Science, volume 1 containing circular functions and volume 6, part 1, Bessel functions of the order zero and unity.

Symbols Used.

T_1 = initial temperature of the steel.
T_0 = initial temperature of the mould and surroundings.
T = variable temperature at any point in the ingot or mould.

$U \quad = \dfrac{T - T_0}{T_1 - T_0}$.
$t \quad$ = time, measured from the commencement of cooling.
$Q \quad$ = quantity of heat.
$c \quad$ = constant of proportionality in Newton's law of cooling.
$k \quad$ = thermal conductivity.
$h \quad = \dfrac{c}{k}$.
$2b$ or $2w$ = inner dimensions of round or rectangular moulds.
$2B$ or $2W$ = outer dimensions of round or rectangular moulds.
$H \quad = hB$.
$r \quad$ = distance of any point in a circular ingot or mould from the centre.
x, y = co-ordinates of any point in a rectangular ingot or mould—origin at the centre of the ingot and axes parallel to the sides.
$\sigma \quad$ = specific heat of the ingot and mould materials.
$\rho \quad$ = density of the ingot and mould materials.
$a^2 \quad = \dfrac{k}{\sigma\rho}$ = thermal diffusivity.
$J_0 \quad$ = the Bessel function of the first kind—order zero.
$J_1 \quad$ = the Bessel function of the first kind—first order.

Basic Assumptions in Thermal Formulæ.

It is not considered necessary to deal in this place with those formulæ which, whilst being of mathematical interest, deal with cases far removed from those met with in practice. Examples of such cases are the cooling of an ingot in a mould the outer face of which is adiabatic, the cooling of a finite ingot in an infinite mould, or the cooling of the face of a semi-infinite ingot in contact with a finite slab. They represent the extreme or limiting cases of the one considered here, namely, the more practical case of a finite ingot and a finite mould which loses heat from its outside surface.

The formulæ given below are based on the following assumptions :

(1) *That there is no latent heat of solidification, or latent heat of phase changes in the solid state.*

Actually, both these latent heats are present. Saitô studied the problem of the latent heat of phase changes, but only for the cooling of infinite slabs and cylinders the outsides of which were kept at constant temperature, 0° C., so that his formulæ are not applicable to ingots and moulds. On the other hand, Lightfoot dealt with the latent heat of solidification, but chiefly from the point of view of its effect on the rate of solidification of molten steel. Even this problem proved to be so difficult that complete solutions could be found only for the purely hypothetical cases of a semi-infinite mass of molten metal, originally at constant temperature, the boundary $x = 0$ being either kept at 0° C. or placed in contact with a semi-infinite mass of steel originally at 0° C.

The difficulties met with when dealing with the latent heat of solidification must be very much increased if the latent heat of phase changes in the solid state is to be considered, for both the quantity of this latent heat and the range of temperature over which it is given up depend upon the rate of cooling. Moreover, in the same ingot, particularly if the ingot is of alloy steel, it may happen that different parts of the ingot will give up different amounts of latent heat in either the Ar′ or the Ar″ or both ranges of temperature. How can one ever hope to deal mathematically with this aspect of the case?

(2) *That the initial dimensions do not alter, and the ingot and mould are in contact the whole time during the cooling of the ingot.*

It is known, however, that these dimensions are varying during the whole of this time, so much so that shortly after the solidification of the parts of the molten metal in contact with the mould there is an air-gap, which itself changes in thickness, between the ingot and the mould. Both Lightfoot and McCance [10] have considered the effect of the air-gap and there seems to be no doubt that its presence will slow up both the cooling of the ingot and the heating of the mould.

(3) *That the liquid steel in any horizontal layer is tranquil.*

This assumption cannot be correct, for there are convection currents in the layer as long as any portion remains liquid. These currents are caused by the pouring process, the evolution of gases when present, the movement of liquates and segregates and the downward movements of crystallites during the differential freezing of the metal.

(4) *That heat transfer takes place parallel to one plane only,* viz., *a plane at right angles to the axis of the ingot, and there is no flow of heat across the axis.*

As stated in paragraph (3) above, there is an exchange of heat between different layers due to convection currents, but the total effect on the rate of cooling may be neglected. There is also a flow of heat downwards due to the effect of the bottom plate. A quantitative allowance may be made for this, but if the considered layer be sufficiently far removed from the bottom plate, the effect of bottom-plate cooling is very small. The loss of heat from the layer due to the air is of importance only inasmuch as it alters the value of T_1 (*see* next paragraph), or if the layer is very near to the top of the ingot.

(5) *That the molten steel is originally at a uniform temperature and the whole of the mould is at the temperature of the surroundings.*

As time is required to fill a mould, and as during this time the molten steel is losing heat by cooling in the ladle and by air-cooling of the stream from the ladle to the ingot, the value of the steel temperature will tend to decrease, as the steel rises, from the bottom to the top of the ingot; similarly the mould temperature tends to increase. T_1 is the temperature of the steel when it comes in contact with the mould. T_0 is the original mould temperature.

(6) *That the thermal diffusivities of molten steel, solid steel and the cast-iron mould material are equal to one another and independent of temperature.*

As already stated, this assumption cannot be justified. It is probable, however, that there is not much difference at any one temperature between the diffusivities of the pearlitic steels and of cast iron. An average value over a range of temperature has been used.

(7) *That the temperature gradient at the surface of the mould is proportional to the difference in temperature between the mould surface and the surroundings.*

This is a form of Newton's law of cooling, which states that the rate of loss of heat from a surface is proportional to the difference in temperature between that surface and the surroundings; *i.e.*, if T is the temperature of the mould at any time:

$$\frac{dQ}{dt} = k\left(\frac{dT}{dr}\right)_s = -c(T_s - T_0),$$

where suffix s denotes quantities taken at the surface of the mould. Put

$$U_s = \frac{T_s - T_0}{T_1 - T_0},$$

then

$$\left(\frac{dU}{dr}\right)_s = \frac{dT_s}{dr} \cdot \frac{1}{T_1 - T_0}$$

$$= -\frac{c}{k} \cdot \frac{T_s - T_0}{T_1 - T_0} = -hU_s,$$

where

$$h = \frac{c}{k}.$$

Now, it is known that Newton's law holds only for small temperature differences and also that the conductivity k varies with temperature, so that again this theoretical assumption cannot be fully justified. Even if this seventh assumption were reasonably correct, it is inconceivable that the value of h would not be modified to some extent by external influences, such as weather conditions, the type of casting pit, the presence of other ingot moulds, &c. Saitô derived his value for h (h/k in Saitô's notation) theoretically from an approximation to Stefan's fourth-power radiation law, using the values $4{\cdot}3 \times 10^{-5}$ erg per cm.^{2} deg.^{4} for the radiation constant and 0·09 cal. cm.$^{-2}$ sec.$^{-1}$ (° C. cm.$^{-1}$)$^{-1}$ for the thermal conductivity. This gives $h = 0{\cdot}0102$ for an average temperature of 610° K. (337° C.). From experiments in the laboratories of the English Steel Corporation, Ltd., on the air-cooling of bars of different steels varying from $\frac{1}{2}$ in. to 6 in. in dia., it has been found that a good average value of h over a range of temperature from 700° C. to atmospheric temperature is 0·015 c.g.s. unit. This value has been used throughout the present work, but it is a simple matter to interpret quantitatively all the curves given for any other value of h, as will be explained later. One example only is given of the effect on mould-temperature curves of altering h from 0·015 to 0·010, *see* Figs. 29 and 30.

Formulæ Used for the Calculations.

The cases dealt with are those of ingot and mould, either both circular or both rectangular. No attempt has been made to deal with other sections; the simple ones are troublesome enough. Rounded corners and concave or convex sides have their effect localised near the surface, and their effect becomes less as the distance from the face of the ingot increases. Slab ingots may therefore be treated as their equivalent rectangular ingots and multi-sided ingots

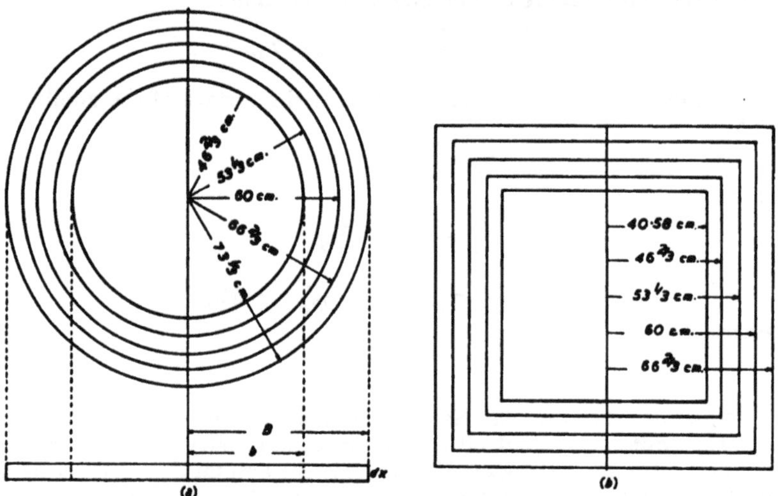

Fig. 1.—Dimensions of the Ingots and Moulds Considered.

as circular ones, in both cases the mould ratio being kept the same as in the one considered, where "mould ratio" is defined as the cross-sectional area of the mould per unit area of the ingot, *e.g.*, $(B^2 - b^2)/b^2$ in Fig. 1 (*a*).

The concept of mould ratio for describing the relative sizes of the mould and ingot is far from satisfactory and should only be used for comparing among themselves ingots which have moulds the outside dimensions of which are geometrically similar to those of the ingots. This arises from the fact that the cooling of an ingot depends on, amongst other things, the area of the ingot in contact with the mould, the area of the mould exposed to the atmosphere and the thickness of the mould wall. If a square ingot and a circular ingot have the same cross-sectional area and are surrounded by moulds which give the same mould ratio, then the square mould has $2/\sqrt{\pi}$ times more area in contact with the ingot, and exposes $2/\sqrt{\pi} = 1{\cdot}128$ times more surface to the atmosphere, but the thickness of the wall is only $\sqrt{\pi}/2 = 0{\cdot}886$ of that of the circular mould. A similar comparison with a rectangular ingot in a geometrically similar mould of which the long side is n times the short side shows that the rectangular mould exposes $(1 + n)/\sqrt{\pi n}$ times more area than the circular mould, and has $(1 + n)/\sqrt{\pi n}$ times more area in contact with the ingot; the thickness of the long face is $\frac{1}{2}\sqrt{\pi/n}$, and of the short face $\frac{1}{2}\sqrt{n\pi}$ times the thickness of the circular mould.

The formulæ used are essentially those developed by Saitô. Those for rectangular ingots and moulds have been modified by moving the origin of the axes of the co-ordinates from the corner of the mould to the centre of the ingot; this simplifies the arithmetical work and makes the formulæ more flexible.

Formulæ for Circular Ingots and Moulds.

The temperature at any point r *in the ingot or mould* is given by:

$$U = \Sigma_{n=1}^{n=\infty} A_n' e^{-\frac{\mu_n^2 a^2 t}{B^2}} J_0\left(\mu_n \frac{r}{B}\right),$$

where μ_n is given by

$$H = \frac{\mu_n J_1(\mu_n)}{J_0(\mu_n)}$$

and

$$A'_n = 2\frac{b}{B}\mu_n \frac{J_1\left(\mu_n \frac{b}{B}\right)}{J_0^2(\mu_n)\{\mu_n^2 + H^2\}}.$$

The fraction of the initial total heat in the mould at time t is:

$$\Sigma_{n=1}^{n=\infty} A_n' e^{-\frac{\mu_n^2 a^2 t}{B^2}} 2\left(\frac{B}{b}\right)^2 \frac{1}{\mu_n}\left\{J_1(\mu_n) - \frac{b}{B} J_1\left(\mu_n \frac{b}{B}\right)\right\}.$$

The fraction of the initial total heat in the ingot at time t is:

$$\Sigma_{n=1}^{n=\infty} A_n' e^{-\frac{\mu_n^2 a^2 t}{B^2}} 2\frac{B}{b} \cdot \frac{1}{\mu_n} J_1\left(\mu_n \frac{b}{B}\right).$$

In the above formulæ b and B indicate the inner and outer radii of the mould and a^2 is the thermal diffusivity.

Formulæ for Rectangular Ingots and Moulds.

The temperature at any point x, y *in the ingot or mould* is given by:

$$U = \Sigma_{n=1}^{n=\infty} \Sigma_{m=1}^{m=\infty} A_n' A_m'' e^{-\frac{a^2\mu_n^2 t}{B^2} - \frac{a^2 c_m^2 t}{W^2}} \cos\left(\frac{x}{B}\mu_n\right) \cos\left(\frac{y c_m}{W}\right),$$

where μ_n is given by $\mu_n \tan \mu_n = hB = H'$
and c_m ,, $c_m \tan c_m = hW = H''$,

and

$$A_n' = \frac{2H' \sin \frac{\mu_n b}{B}}{\mu_n(H' + \sin^2 \mu_n)}$$

$$A_m'' = \frac{2H'' \sin \frac{c_m w}{W}}{c_m(H'' + \sin^2 c_m)}.$$

The fraction of the initial total heat in the ingot and mould at any time t is:

$$\Sigma_{n=1}^{n=\infty} \Sigma_{m=1}^{m=\infty} A_n' A_m'' e^{-\frac{a^2\mu_n^2 t}{B^2} - \frac{a^2 c_m^2 t}{W^2}} \frac{BW}{bw\mu_n c_m} \sin \mu_n \sin c_m.$$

The fraction of the initial total heat in the ingot at any time t is:

$$\Sigma_{n=1}^{n=\infty} \Sigma_{m=1}^{m=\infty} A_n' A_m'' e^{-\frac{a^2\mu_n^2 t}{B^2} - \frac{a^2 c_m^2 t}{W^2}} \frac{BW}{bw\mu_n c_m} \sin \frac{\mu_n b}{B} \sin \frac{c_m w}{W}.$$

In the above formulæ $2B$ and $2W$ are the external dimensions of the mould and $2b$ and $2w$ the inner side lengths.

All the terms used have been defined except the "fractional quantity of heat." If a circular mould and the surrounding atmosphere are at temperature T_0 and the mould is filled with molten metal at temperature T_1 then one can say that the heat added per unit length of the mould is:

$$\pi b^2 \sigma \rho (T_1 - T_0),$$

where σ is the specific heat and ρ the density. This quantity of heat passes through the mould into the air, and it is the fraction of this quantity present in the ingot and mould at any time which has been calculated.

It is obvious that this fractional quantity of heat when multiplied by $(T_1 - T_0)$ gives the average temperature of the ingot and mould.

In work of this kind the numerical values apply only to particular cases and are of no use for others, so for this reason the actual values have not been recorded in this paper, but they are available and may be consulted if desired.

The Effect of Mould Thickness on the Rate of Solidification of the Ingot.

(*a*) *Circular Ingots and Moulds.*

To study the theoretical aspect of the effect of mould thickness on the rate of solidification and the subsequent cooling of an ingot, four cases have been considered; full particulars are given in Table I.

Table I.—*Dimensions of Circular Ingots and Moulds.*

$H(= h \times B)$:	0·8	0·9	1·0	1·1
b, radius of ingot. Cm.	$46\frac{2}{3}$	$46\frac{2}{3}$	$46\frac{2}{3}$	$46\frac{2}{3}$
B, external radius of mould. Cm.	$53\frac{1}{3}$	60	$66\frac{2}{3}$	$73\frac{1}{3}$
b/B	$\frac{7}{8}$	$\frac{7}{9}$	$\frac{7}{10}$	$\frac{7}{11}$
Mould ratio	0·306	0·653	1·041	1·469
Dia. of ingot. In.	36·75	36·75	36·75	36·75
Thickness of mould. In.	2·625	5·249	7·874	10·499

All the ingots are of the same diameter and are supposed to be cast into moulds of different thicknesses. The necessary calculations for drawing the curves were then made. The value of h has been taken as 0·015 throughout, but the curves are equally applicable for any assumed value of h by a suitable modification of the ingot and mould size. Supposing, for example, that a more accurate determination of the best average value of h is found to be $h = 0{\cdot}010$, then the new diameter of the mould B_1 for the first ingot, where $H = 0{\cdot}8$, is $B_1 = 0{\cdot}80/0{\cdot}01 = 80$ cm. As the mould ratio must be kept constant, the new value of the ingot radius $b_1 = 70$ cm., and the new value of $(a^2t)_1$ is $a^2t \times B_1{}^2/B^2$. This must *not* be misinterpreted into a general statement that, for geometrically similar ingots and moulds, the time required, say, for any part to fall through a given range of temperature is proportional to the square of a linear dimension; that statement would be true only if H, *i.e.*, the product of h and B, could be kept constant.

On the so-called time-temperature curves the co-ordinates are U (non-dimensional) and a^2t in c.g.s. units, the latter being used because the best average value of a^2 is not known.

The curves may now be examined in more detail. A question which is often discussed is: What is the effect of the mould thickness on the rate of solidification of a steel ingot? A study of the published literature shows that there are, or were, two distinct schools of thought. Experimental evidence has been put forward which proved that an increase of mould thickness increases the rate of solidification; other experimental evidence has shown equally convincingly that an increase of mould thickness has no effect on the rate of solidification, and at least one paper has been published which purported to show that heat is extracted more quickly by a thin mould than by a thick mould.[11] The latter view does not seem to follow logically from the experimental results and will be reserved for discussion later, but the other two conflicting opinions can now be reconciled. The problem may be approached from either one of two directions: The effect of mould thickness on the time required

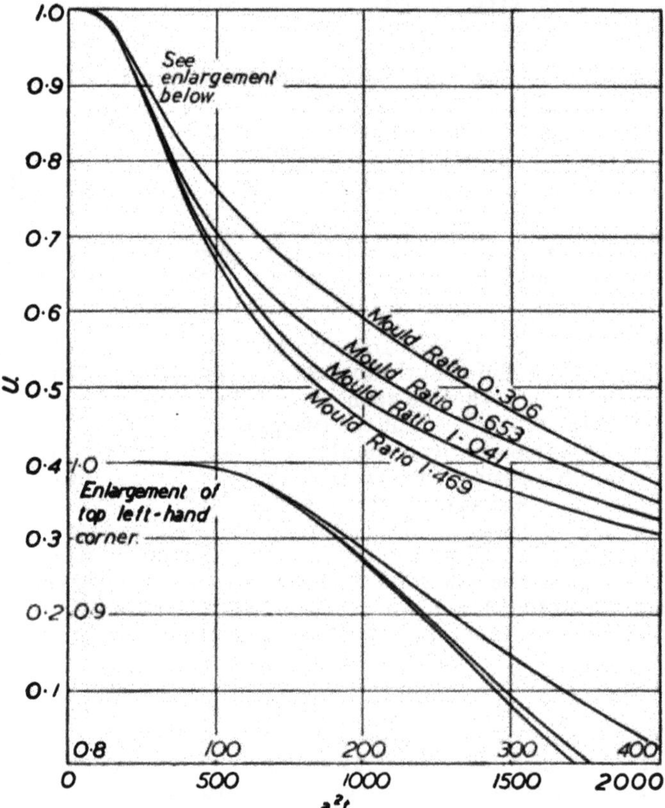

Fig. 2.—Temperature Fall at Centre of Circular Ingot.

for the centre of the ingot to solidify, or its effect on the total quantity of heat extracted from the ingot in a given time. An allowance must be made for the latent heat of solidification and for some degree of superheat. It has been suggested that the latent heat may be considered as an increase in initial temperature, and this suggestion is used here. Now Fig. 2, which gives the temperature fall at the centre of the circular ingot, shows that the time required for U to fall from 1 to as low as 0·8 is appreciably less when the mould ratio is 0·65 than when it is 0·3; the time is slightly less still when the mould ratio is 1·0, but this time decreases by only a very small amount when the mould ratio is increased to 1·47. If it be assumed that the ingot has solidified to the centre when $U = 0{\cdot}8$, this would correspond to a solidus temperature of, say, 1450° C., and if the temperature of the mould be taken as 25° C., then the combined effect of latent heat and superheat in this case would be equivalent to an initial temperature 356° C. above the solidus temperature, as:

$$\frac{1450 - 25}{(1450 + 356) - 25} = 0{\cdot}8.$$

The effect of the mould ratio on the time required for the centre

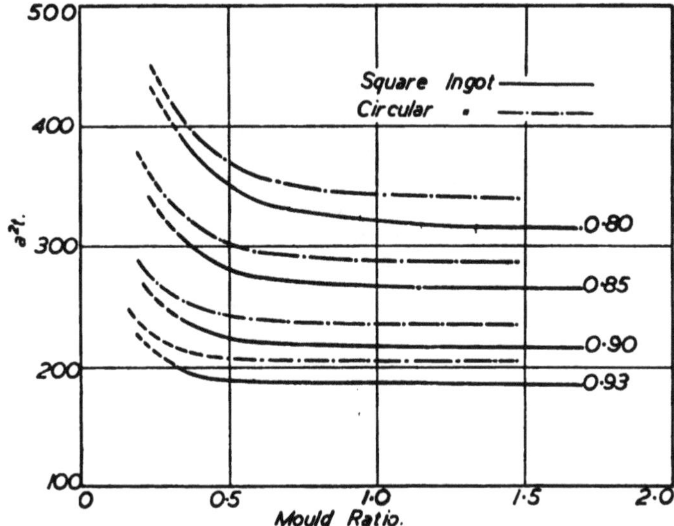

FIG. 3.—Effect of the Mould Ratio on the Time required for an Ingot to Solidify (or for U to fall to 0·93, 0·90, 0·85, 0·80).

of an ingot to solidify is shown more clearly in Fig. 3, where the times —in terms of a^2t—required for the centre of the ingot to reach $U = 0{\cdot}93$, 0·90, 0·85 and 0·80 are plotted against the mould ratio.

This theoretical treatment shows that there can be no advantage, from the point of view of promoting rapid solidification of an ingot, in increasing the mould ratio above 1·0, even for casting temperatures far in excess of those ever used, and above, say, 0·8 for normal practice.

Fig. 3 appears to throw some light on the disputed question of whether thick or thin moulds promote solidification more rapidly.

If an investigator had examined two ingots of the same size cast into two moulds, one having a mould ratio of, say, less than 0·5 and the other having a mould ratio of 1·0 or more, he would probably have reported in favour of the heavy mould. On the other hand, if both moulds, whilst differing appreciably in thickness, had mould ratios in excess of, say, 0·8, the conclusion might have been that there was no advantage in the thicker mould. A similar reasoning also applies to a more or less parallel-sided ingot cast into a tapered mould with its thick walls at the bottom to promote solidification from the bottom upwards. Such a mould would probably be more effective for its purpose if designed on data such as those given in Fig. 3.

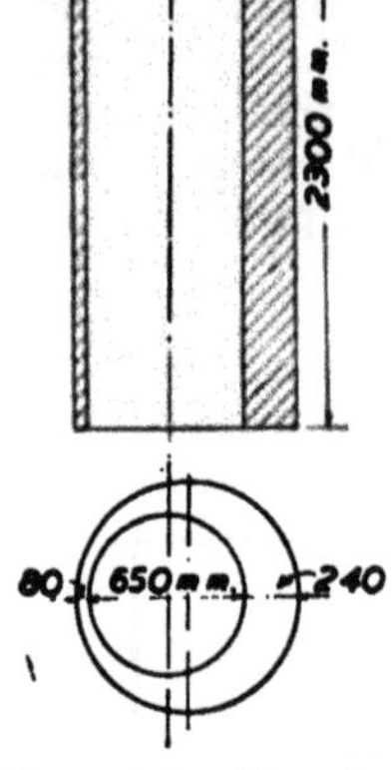

Fig. 4.—Experimental Mould Used by Granat and Bezdenezhnykh.

The only experimental work known to the author which purports to show that ingots cool more quickly in thin-sided moulds is that of Granat and Bezdenezhnykh.[11] These workers used a specially designed mould, the dimensions of which are shown in Fig. 4, and they inserted thermocouples through the thinnest and thickest parts of the mould so that the hot junctions extended 30 mm. into the liquid steel. Time-temperature curves

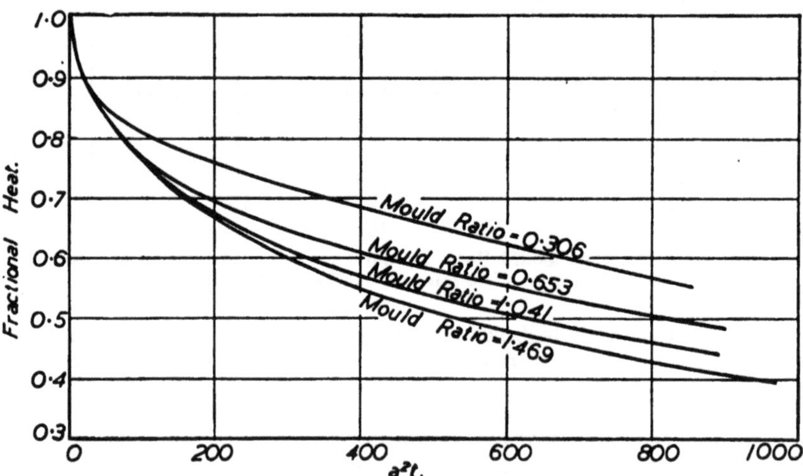

Fig. 5.—Fractional Total Heat in Circular Ingot Plotted against Time.

taken at these points undoubtedly show that *in this ingot mould* the metal near to the thin face cooled more quickly than that near the thick face, but it seems illogical to conclude that the same effect will be found in moulds of constant mould thickness. In the latter, heat is lost by the mould only by radiation to the air, whereas in the former quite an appreciable quantity of heat must be lost by the thin side, owing to conduction to the thick side. The effect then is that of artificially cooling the thin wall and heating the thick wall while the cooling of the steel is taking place.

Fig. 5 shows the total heat in the ingot—as a fraction of the initial heat—plotted against time. If solidification is complete at $U = 0{\cdot}93$ then it is seen from Fig. 2 that the ingots with mould ratios of 1·04 and 1·47 have solidified at a^2t equal to about 210, and Fig. 5 shows that by increasing the mould ratio from 1·04 to 1·47

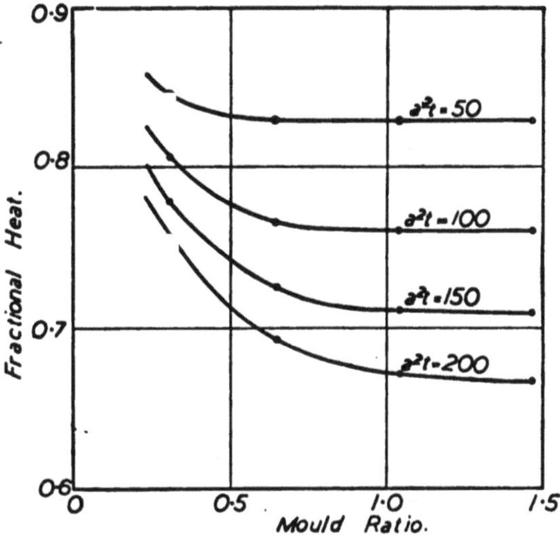

Fig. 6.—Effect of the Mould Ratio on the Fractional Total Heat n a Circular Ingot.

the total heat has fallen in this time only from 0·665 to 0·660; even in what must be considered the most extreme case, *i.e.*, when $U = 0{\cdot}80$ and a^2t is about 370, the total heat in the ingot has fallen only from 0·58 to 0·572.

Fig. 6 shows these general features in a different way; the total heat in the ingot at constant times is plotted against the mould ratio, whilst Fig. 7 shows similar curves for the total heat in the mould.

Time-temperature cooling curves for points on the diameter of the ingot and mould are given in Figs. 8–11 and the temperature distribution across the ingot and mould is shown in Figs. 12–15. The temperature-distribution curves are for the same time periods

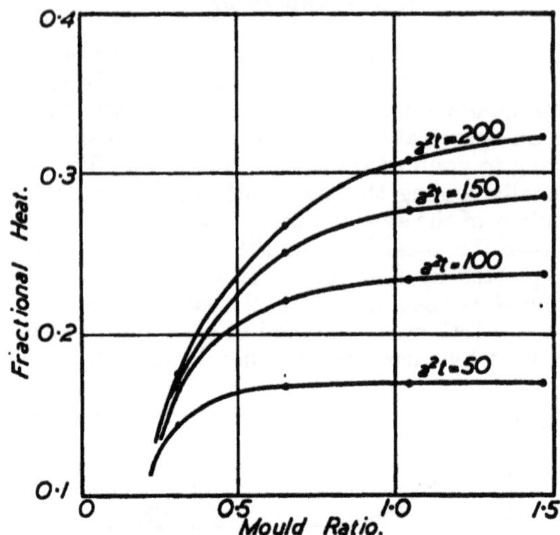

Fig. 7.—Effect of the Mould Ratio on the Fractional Total Heat in the Mould of a Circular Ingot.

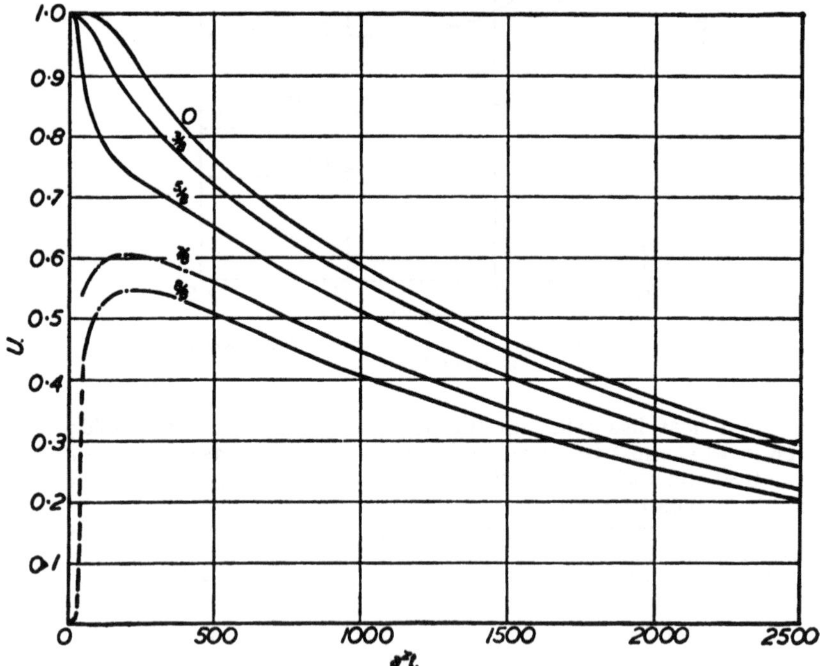

Fig. 8.—Time-Temperature Cooling Curves for Points on the Diameter of a Circular Ingot with a mould ratio of 0·306.

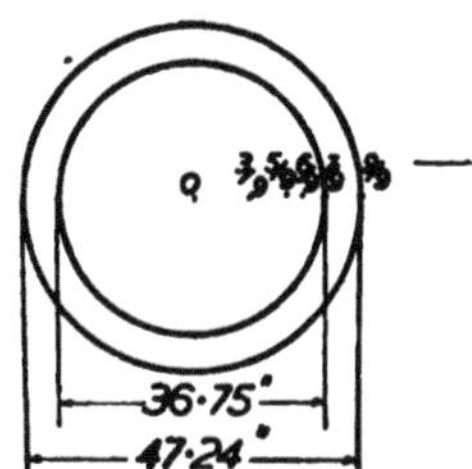

0 500 1000 1500 2000 2500

a^2t.

Fig. 9.—Time-Temperature Cooling Curves for Points on the Diameter of a Circular Ingot with a mould ratio of 0·653.

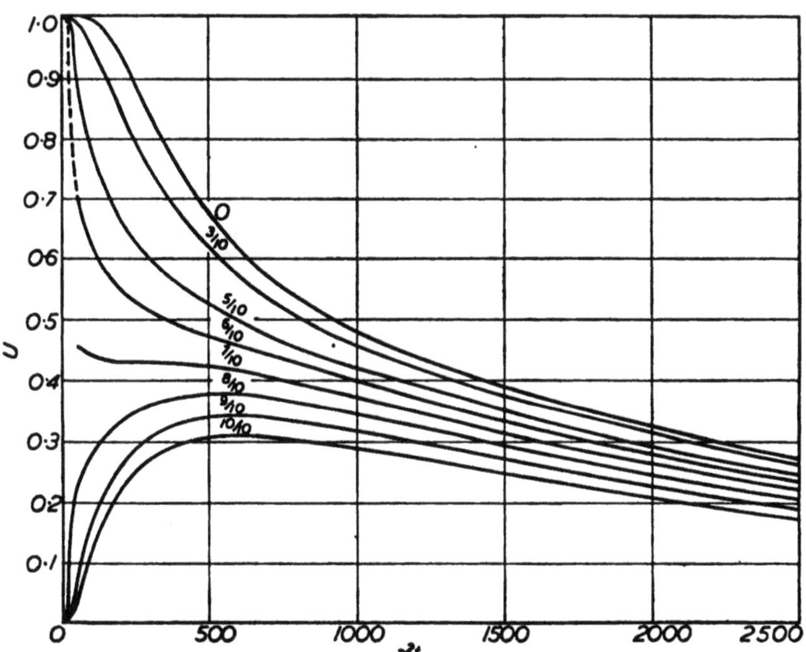

Fig. 10.—Time-Temperature Cooling Curves for Points on the Diameter of a Circular Ingot with a mould ratio of 1·041.

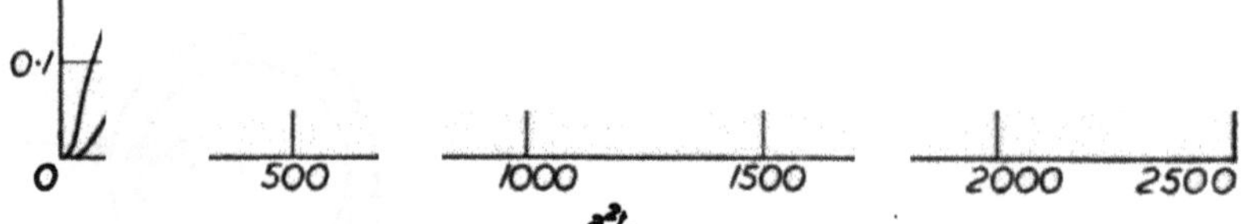

FIG. 11.—Time-Temperature Cooling Curves for Points on the Diameter of a Circular Ingot with a mould ratio of 1·469.

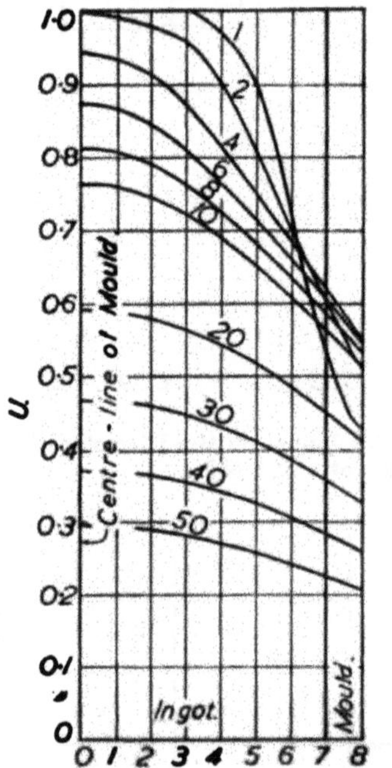

FIG. 12.—Temperature Distribution across Ingot and Mould; mould ratio, 0·306.

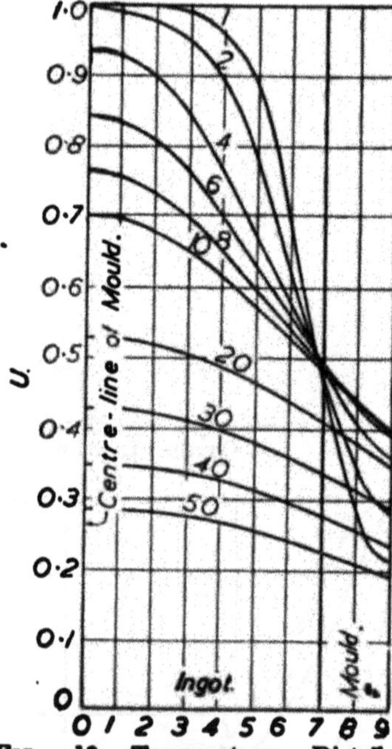

FIG. 13.—Temperature Distribution across Ingot and Mould; mould ratio, 0·653

for each ingot, so that the effect of mould thickness is readily seen. The unit interval of time is $a^2t = 50$, so that a curve marked 20 is for a time given by $a^2t = 1000$ after cooling has commenced. One of the most interesting points brought out by these curves is that for mould ratios of 0·306 and 0·653—Figs. 12 and 13—there are, near to the mould wall, some portions of the ingot which fall in temperature and then rise again before they finally cool off. With mould ratios of 1·041 and 1·469 all parts of the ingot cool continuously, although parts near to the mould wall fall very slowly

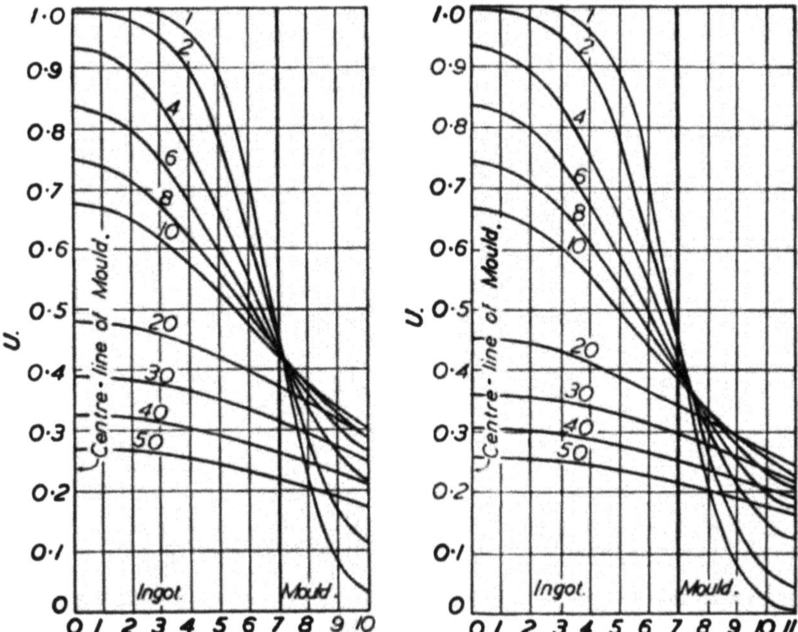

Fig. 14.—Temperature Distribution across Ingot and Mould; mould ratio, 1·041.

Fig. 15.—Temperature Distribution across Ingot and Mould; mould ratio, 1·469.

over the time interval $a^2t = 50$ to $a^2t = 500$. Saitô, by calculation, also found that in a circular ingot of 80 cm. radius (63 in. in dia.) surrounded by a mould 20 cm. (7·87 in.) thick and therefore having a mould ratio of 0·5625, there are, near to the mould wall, parts of the ingot which go through this temperature cycle. The influence of mould thickness on the cooling at a point in the ingot 2 cm. from the mould wall is shown in Fig. 16. The rise in temperature even at this distance from the mould face is still evident with a mould ratio of 0·653.

(*b*) *Square Ingots and Moulds.*

The square ingot considered had a cross-sectional area approximately equal to that of the circular ingot already described. At first it was intended to surround this ingot with four moulds which would have mould ratios equal to those of the circular ingot, but this led to such "awkward" values of *H* that it was ultimately

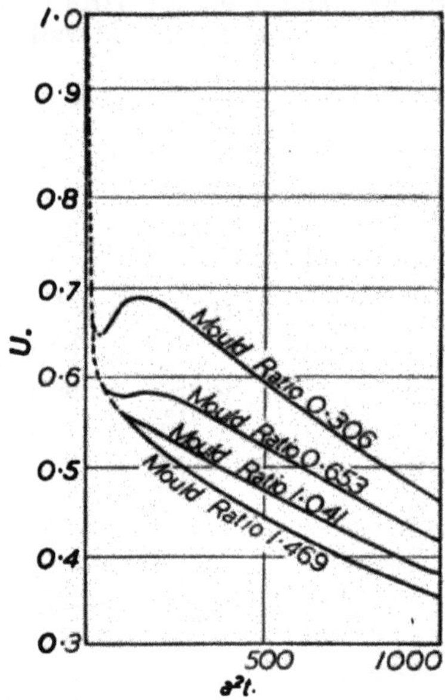

FIG. 16.—Effect of the Mould Ratio on the Rate of Cooling at a Point in a Circular Ingot 2 cm. from the Mould Wall.

decided to use simple values of *H* (0·7, 0·8, 0·9 and 1·0), although this meant different mould ratios. A direct comparison between round and square ingots could then be obtained by plotting principal values against mould ratios. Particulars of the ingots and moulds —*see* Fig. 1 (*b*)—are given in Table II.

TABLE II.—*Dimensions of Square Ingots and Moulds.*

$H (= h \times B)$:	0·7	0·8	0·9	1·0
b, half side of ingot. Cm.	40·58	40·58	40·58	40·58
B, half side, external, of mould. Cm.	$46\frac{2}{3}$	$53\frac{1}{3}$	60	$66\frac{2}{3}$
Mould ratio	0·323	0·727	1·186	1·699
Side of ingot. In.	31·95	31·95	31·95	31·95
Thickness of mould. In.	2·396	5·021	7·646	10·270

Fig. 17 shows the time-temperature cooling curves for the centre of each ingot, and Fig. 18 the fraction of the initial heat remaining in the ingot plotted against time.

Fig. 19 gives the amount of heat remaining in the *ingot* and Fig. 20 the amount of heat in the *mould* at certain times plotted against the mould ratio.

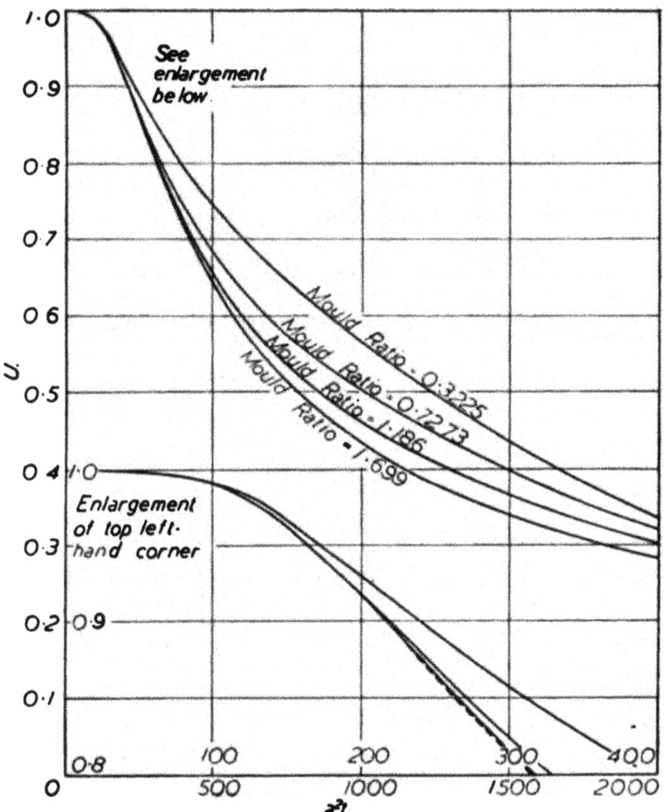

Fig. 17.—Temperature Fall at the Centre of a Square Ingot.

Fig. 3 indicates the times required for the centre of the ingot to reach temperatures corresponding to $U = 0{\cdot}93$, $0{\cdot}90$, $0{\cdot}85$ and $0{\cdot}80$ plotted against the mould ratio.

These curves show that for moulds of the same cross-sectional area and the same mould ratio, solidification at the centre takes place more quickly for a square ingot than for a round ingot; *e.g.*, from Fig. 3, if the mould ratio is 1, then the time for the centre to fall to $U = 0{\cdot}93$ is $a^2t = 186$ for a square ingot and $a^2t = 205$ for a round ingot. This is in spite of the fact that the thickness

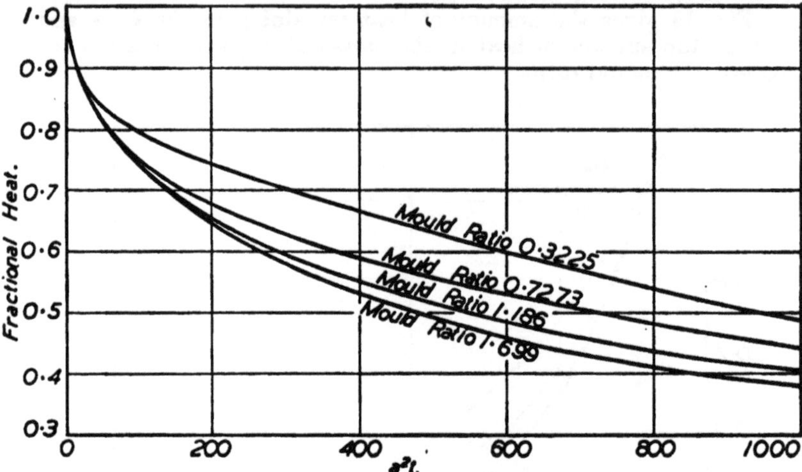

FIG. 18.—Fractional Total Heat in a Square Ingot Plotted against Time.

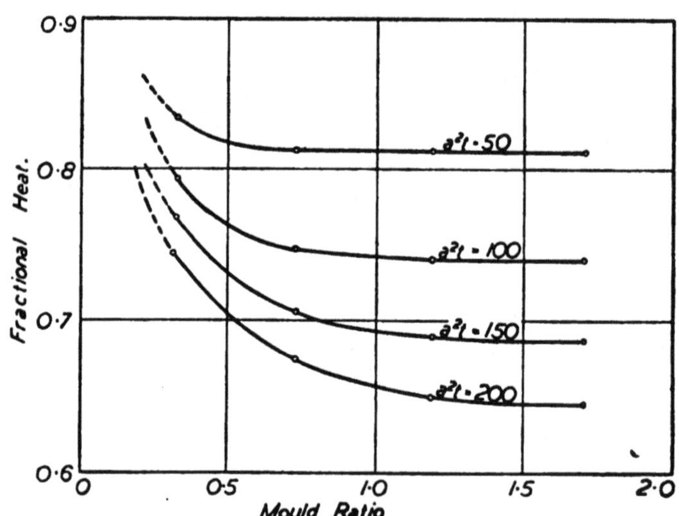

FIG. 19.—Effect of the Mould Ratio on the Fractional Total Heat in a Square Ingot.

of the square mould is only 0·886 of that of the round mould; the more rapid solidification must be due to the greater surface of contact between the ingot and the mould, and, perhaps to a less extent, to the greater surface exposed to the air. As distinct from circular moulds, where parts of the ingot near to the mould wall will fall in temperature, then rise and fall again only if the mould ratio is

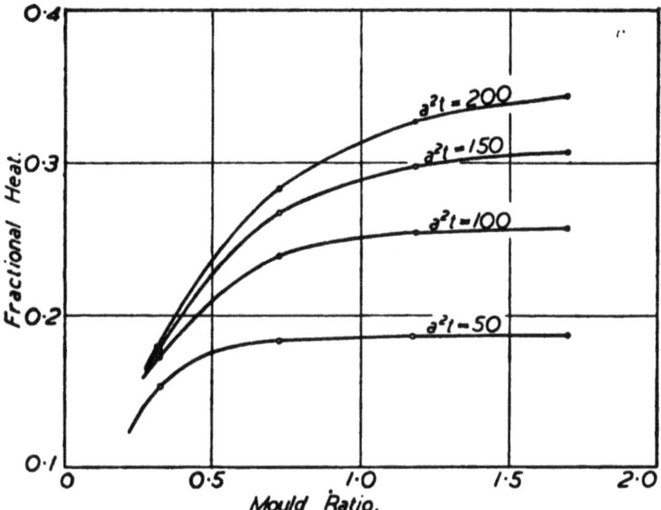

Fig. 20.—Effect of the Mould Ratio on the Fractional Total Heat in the Mould of a Square Ingot.

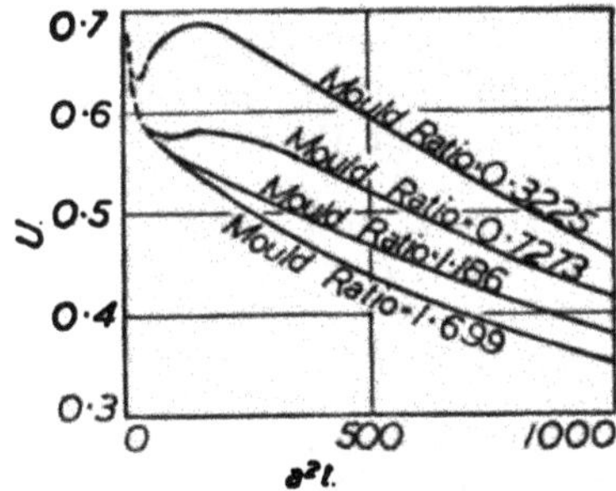

Fig. 21.—Effect of the Mould Ratio on the Rate of Cooling at a Point in a Square Ingot 2 cm. from the Middle of a Face.

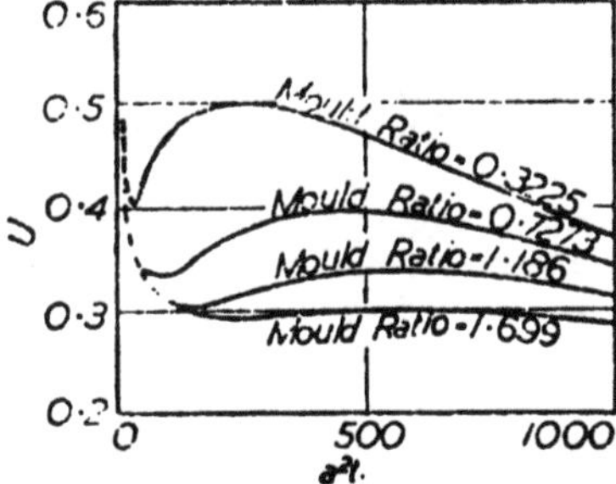

Fig. 22.—Effect of the Mould Ratio on the Rate of Cooling at a Point in the Corner of a Square Ingot 2 cm. from each Face.

small, there is always some part of a square ingot, particularly near the corners, which will undergo this cycle of temperature changes. Fig. 21 shows the effect of mould ratio on the time-temperature cooling curves for a point in the ingot 2 cm. from the middle of a face, whilst Fig. 22 contains similar curves for a point in the corner 2 cm. from each face.

Comparison of Calculated with Practical Results.

The value of any theoretical work cannot be appraised until it is put to a practical test. Unfortunately, the author knows of only two published works where the investigators have compared theoretical and practical results, but neither of these is wholly satisfactory. Saitô calculated the temperature rise at a point at mid-height on the outside of a mould having radii of 27 and 40 cm., using an initial temperature of 1500° C. and diffusivity value of 0·0974, and compared his curve with the practical results obtained by T. Kikuta (reported by Saitô [2]). The agreement is promising, but the curves would have agreed better still if a lower value for the diffusivity had been used. Matuschka [9] measured the temperature rise at four points in an ingot mould and, without giving any details of the calculation or reproducing the calculated curves, says that the practical results show good agreement with the curves calculated by the methods of Saitô and Lightfoot. Some of Matuschka's curves have been re-examined and are referred to later.

There are, however, several sets of results of experimentally determined time-temperature curves for points in a mould, and at least one attempt has been made to determine the cooling curves for a point inside the ingot and near to the mould wall. Some of these have been examined theoretically.

In 1931 two experiments on mould temperatures were made at the Vickers Works of the English Steel Corporation, Ltd. One of these was made on an octagon ingot 36 in. across the flats, and temperature measurements were taken at a point half-way in the ingot mould. Planimetric measurements showed that the mould ratio was 1·211. The size of a circular ingot having the same area as the octagon ingot was calculated, and also the size of a mould giving the same mould ratio. This gave a value of $H = 1{\cdot}047$ when h was assumed to be 0·015. Time-temperature curves were calculated for a point half-way in the mould, and plotted in the usual way, U against a^2t. After several trials, *it was found that with a mould temperature of* 50° C. *the best values of initial temperature and thermal diffusivity which fitted the practical curve were* 1650° C. *and* 0·06, *respectively*. These curves are shown in Fig. 23, plotted with two different time scales. The curves agree very well up to some time after the maximum temperature is reached, when they begin to diverge. This indicates that the value of 0·015 for h may be too high.

The second experiment was made on a large ingot having about five times the area of the smaller one. It was an octagonal ingot, 80 in./77 in., and the mould ratio was 1·07. This was treated in

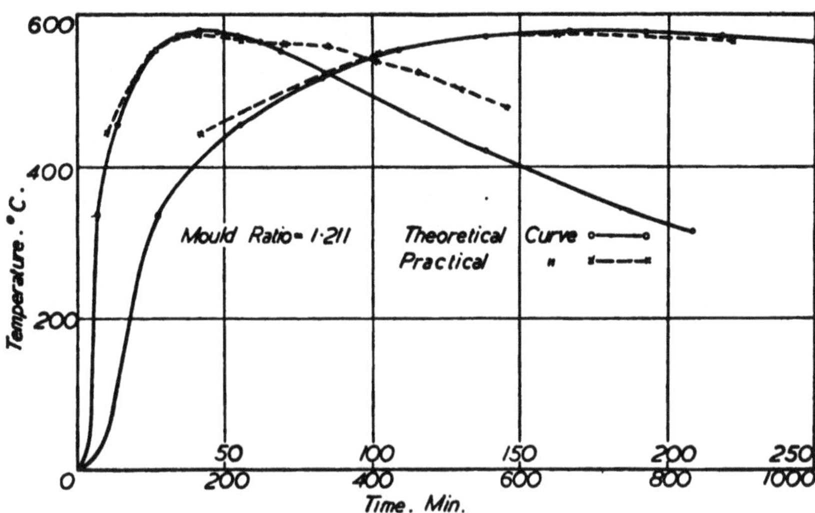

Fig. 23.—Time-Temperature Curves for a Mould for an Octagonal Ingot 36 in. across the Flats.

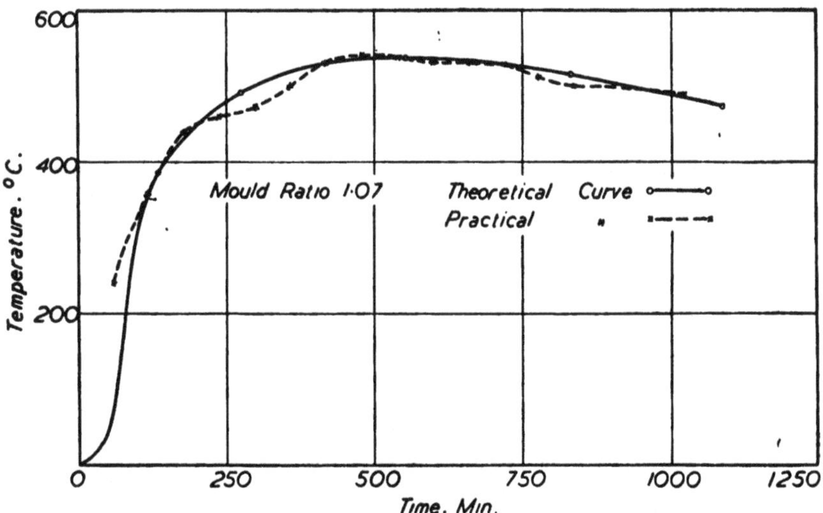

Fig. 24.—Time-Temperature Curves for a Mould for an Octagonal Ingot 78½ in. across the Flats.

exactly the same way as the first ingot and, with $h = 0{\cdot}015$, gave $H = 2{\cdot}183$. The calculated curve—taking the initial temperature as 1650° C., the mould temperature as 50° C. and the diffusivity as 0·06—and the practical curves are given in Fig. 24. Temperature readings were discontinued after about 17 hr.

Figs. 23 and 24 raised the hope that it might be possible, with a suitable choice of values of h, maximum temperature and diffusivity, to estimate approximately the temperature distribution in a mould, and to test this the experimental data given by Matuschka were examined.

Compared with the moulds already considered, Matuschka's circular mould was very small, being only 250 mm. (9·84 in.) in

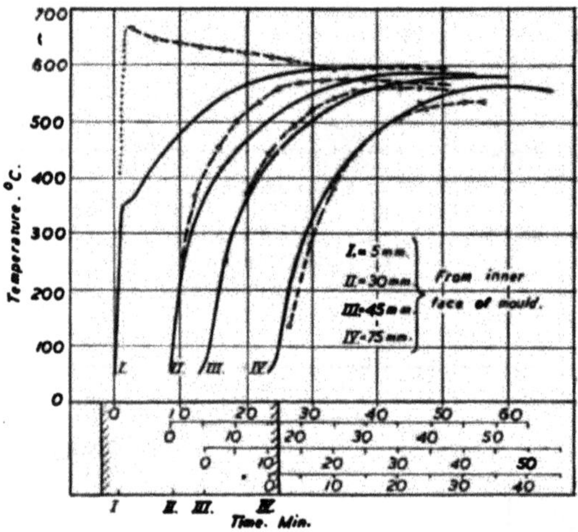

FIG. 25.—Time-Temperature Curves. Full lines, Matuschka's experimental results; dotted lines, calculated results. Ingot, 250 mm. in dia.; mould thickness, 80 mm.; casting temperature (uncorrected), 1360° C.; mould temperature, 40° C.

dia. and the mould thickness 80 mm. (3·15 in.). The mould ratio was therefore 1·69. Temperatures were taken at points 5 mm., 30 mm., 45 mm. and 75 mm. from the inner face of the mould. Matuschka's curves are redrawn in Fig. 25, which also shows the calculated curves—dotted lines—with $h = 0{\cdot}015$, maximum temperature 1650° C., mould temperature 40° C. and diffusivity 0·05 c.g.s. unit. The curves for points II., III. and IV. are not too bad, but for point I., which is only 0·2 in. from the inner surface, the curves have not even a similarity of form. The effect of the air gap between the ingot and mould can be traced from these curves. Up to the time when any part of the mould reaches its maximum temperature, the difference between the calculated and actual temperatures is very appreciable at points near the inner wall, but the difference decreases as the distance from the inner face increases.

The experimental results obtained on a slab mould by the Ingot Moulds Sub-Committee in co-operation with the Appleby-Frodingham Steel Co., Ltd.,[12] were also examined. Fig. 26 gives the dimensions at mid-height of the slab mould used and the dimensions of the equivalent rectangular mould. The points marked *A* to *F*

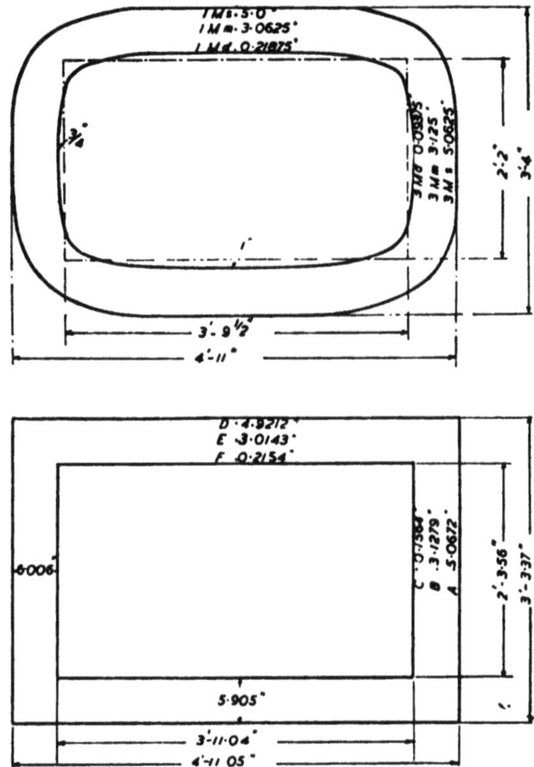

Fig. 26.—Dimensions of a Slab Mould Reported on by the Ingot Moulds Sub-Committee.

are the positions of the thermocouples or the calculated points, and the distance from the inner mould face is given. The points at mid-height were chosen, as the end effects could then be ignored.

Figs. 27 and 28 show the practical curves (first test) and calculated curves for the narrow face and the broad face, the zero in both cases being 6 min. before teeming commenced. The calculations were based on the assumption that $h = 0{\cdot}015$, initial temperature =

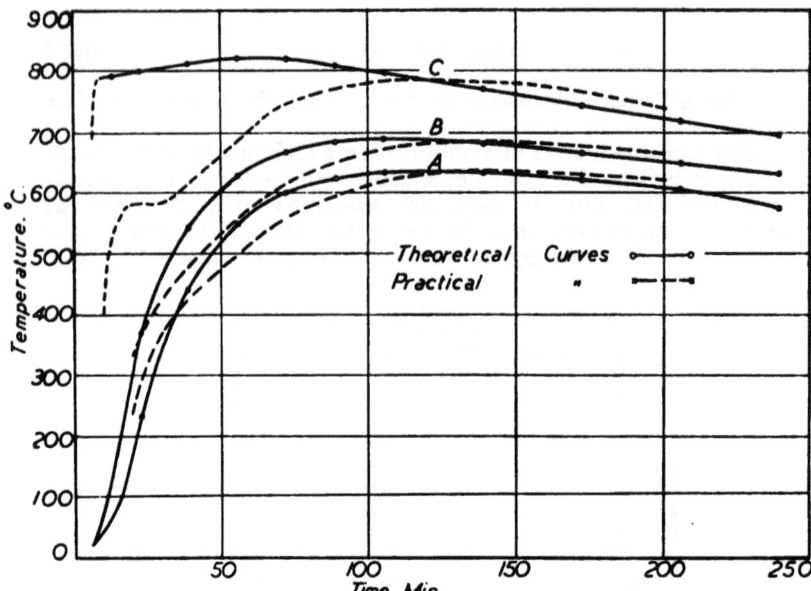

Fig. 27.—Slab Mould. Calculated and practically determined curves for the middle holes in the narrow face.

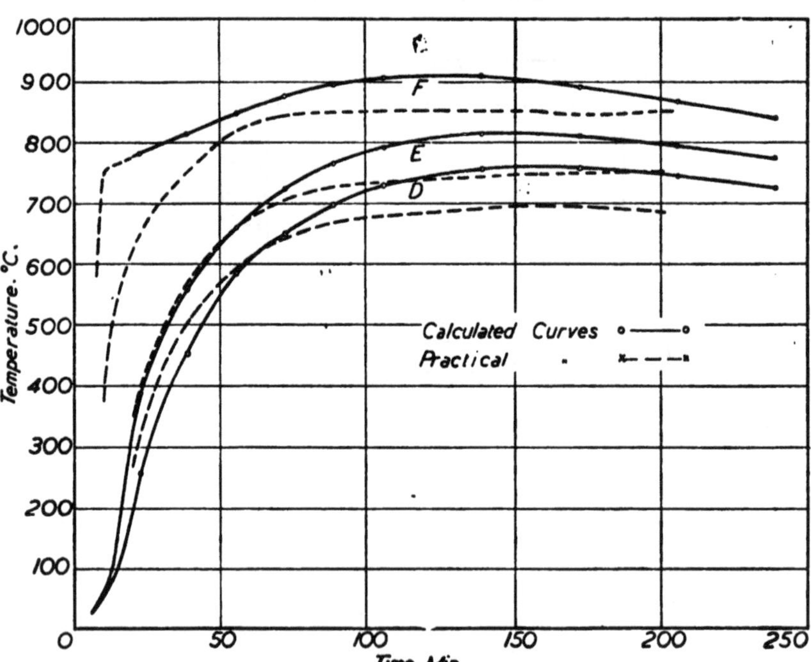

Fig. 28.—Slab Mould. Calculated and practically determined curves for the middle holes in the broad face.

1600° C., mould temperature = 25° C. and thermal diffusivity = 0·05 c.g.s. unit. Again it is seen that the greatest discrepancy between theory and practice is found in the early stages of curves C and F, which are nearest to the inner walls of the mould on the narrow face and broad face, respectively. Taken generally, the

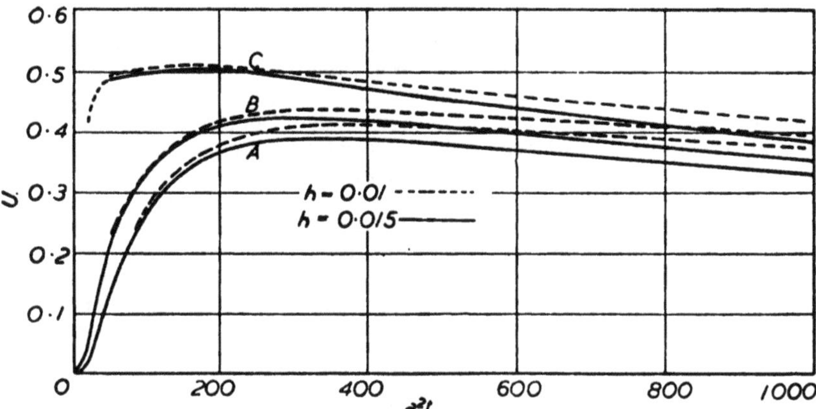

Fig. 29.—Slab Mould, middle holes in the narrow face; effect of variation of h.

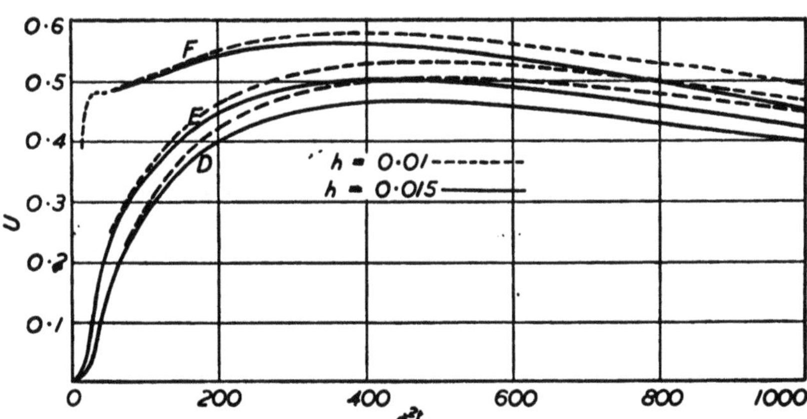

Fig. 30.—Slab Mould, middle holes in the broad face; effect of variation of h.

curves for the narrow face *fit* better than those for the broad face, particularly for holes A and B as compared with D and E. To show the effect of a variation in the value of h, all the curves for points A to F were re-calculated with $h = 0·010$, and are given in Figs. 29 and 30.

The sudden drop in the rate of heating at the points C and F is noteworthy. If actual temperature measurements had given curves of this type, the sudden drop in the rate of heating would have been attributed to the moving away of the ingot from the mould wall, whereas the theoretical curves assume continuous contact of ingot and mould. In this connection it is worth while to turn back to Fig. 9. If the curve for $r/B = 7/9$—which is the surface of contact between the ingot and the mould—be continued backwards so that $U = 0{\cdot}5$ at zero time, then it is evident that for a point in the mould near to its inner face the temperature will first rise to a maximum and then fall rather suddenly before rising to its second maximum. Unfortunately, it is not an easy matter to give a calculated example, as not more than eight roots—or values of μ_n—can be obtained accurately enough from the British Association's table of Bessel functions, and eight terms are not sufficient to prove this point. It is, however, suggested that to determine the time of separation of an ingot from the mould by noting the time of slowing up of the rise in temperature—or even a fall of temperature—at a point in the mould near to the inner surface is not a satisfactory procedure.

Acknowledgments.

The author wishes to place on record his indebtedness to one of his assistants, Miss K. E. Bolton, who has patiently worked out most of the arithmetical examples, and to express his thanks to Mr. H. H. Burton, Chief Metallurgist, and to the Directors of the English Steel Corporation, Ltd., for permission to issue this paper.

SELECTED BIBLIOGRAPHY.

(1) J. E. Fletcher: "On the Cooling of Steel in Ingot and other Forms," *Journal of the Iron and Steel Institute*, 1918, No. II., p. 231.

(2) S. Saitô: "On the Distribution of Temperature in Steel Ingots during Cooling," *Science Reports of the Tôhoku Imperial University*, 1921, vol. 10, p. 305.

(3) A. L. Feild: "Solidification of Steel in the Ingot Mould," *Transactions of the American Society for Steel Treating*, 1927, vol. 11, p. 264.

(4) N. M. H. Lightfoot: "The Effect of Latent Heat on the Solidification of Steel Ingots," *Journal of the Iron and Steel Institute*, 1929, No. I., p. 364.

(5) N. M. H. Lightfoot: "The Solidification of Molten Steel," *Proceedings of the London Mathematical Society*, 1930, II., vol. 31, part 2, p. 97.

(6) N. M. H. Lightfoot: "Some Further Mathematical Considerations concerning the Cooling and Freezing of Steel Ingots," Fourth Report on the Heterogeneity of Steel Ingots, *Iron and Steel Institute*, 1932, *Special Report No.* 2, p. 162.

(7) N. M. H. Lightfoot: "Estimation of the Time of Separation of the Ingot from the Mould," Fifth Report on the Heterogeneity of Steel Ingots, *Iron and Steel Institute*, 1933, *Special Report No.* 4, p. 64.

(8) B. Matuschka: "Heat Equilibrium between Ingot and Ingot-Mould Wall," *Archiv für das Eisenhüttenwesen*, 1929, vol. 2, p. 405.

(9) B. MATUSCHKA : "The Solidification and Crystallisation of Steel Ingots : The Influence of Casting Temperature and Undercooling Capacity on the Steel," *Archiv für das Eisenhüttenwesen*, 1932, vol. 6, p. 1.

(10) A. McCANCE : "Ingots and Ingot Moulds," *Journal of the West of Scotland Iron and Steel Institute*, 1930, vol. 37, p. 101.

(11) I. YA. GRANAT and A. A. BEZDENEZHNYKH : "Effect of Mould Wall Thickness on the Process of Cooling and on the Quality of Steel Ingots," *Metallurg*, 1938, No. 10, p. 19 : *Foundry Trade Journal*, 1939, vol. 61, Nov. 16, p. 335.

(12) "Second Report of the Ingot Moulds Sub-Committee," Eighth Report on the Heterogeneity of Steel Ingots, *Iron and Steel Institute*, 1939, *Special Report No.* 25, p. 265.

CORRESPONDENCE.

Mr. A. L. Feild (Rustless Iron and Steel Corp., Baltimore, Md., U.S.A.) wrote that his own study of the author's paper had convinced him that assumptions (1), (3) and (6) differed too widely from those conditions which actually prevailed to permit of a valid application of theory to practice.

Assumptions (1), (3) and (6) considered together were equivalent to the broad assumption that neither molten steel nor the solidification process was involved in the problem under consideration. Instead, the author assumed, in effect, that the mould was initially filled with a homogeneous solid body having the properties of steel in the solid state. If the molten steel had no latent heat of solidification (assumption (1)), if it was not subject to convection currents and was thus unable to maintain a relatively uniform temperature throughout (assumption (3)), and if it had the same density, thermal conductivity and specific heat as solid steel (assumption (6)), what properties were left to it which distinguished it from solid steel? The answer was none, so far as the author's treatment was concerned.

The following values represented the latent heats of pure iron as commonly accepted : [1]

Fusion	65 g. cal. per g.
A_4 point	1·7 ,, ,,
A_3 point	3·86 ,, ,,
A_2 point	None.

In view of the relatively minor heat effects at the critical points as compared with the latent heat of fusion, it was difficult to maintain the position that all latent heat effects could be ignored, because it was either impracticable or impossible mathematically to take all of these effects into consideration.

The heat content of pure iron in the solid state at the melting temperature (1535° C.) was about 280 cal. per g. [2] At 1175° C. its heat content was 215 cal. per g. Hence, the heat set free by the solidification of one gramme of pure iron (65 cal.) was equal to that lost by one gramme of solid iron in cooling from 1535° to 1175° C. —a temperature interval of 360° C.

The author's belief that his formulæ were based on theoretical consideration at least approximating to practical conditions would appear to require more careful analysis. The writer believed with Lightfoot that the latent heat of solidification was important in its effect on the rate of solidification. Further, the rate of solidification determined in a fundamental manner the rate at which heat was absorbed by the mould. Obviously, there could be no exchange of heat between liquid steel and the innermost layer of the con-

[1] O. C. Ralston, *United States Bureau of Mines*, 1929, *Bulletin No.* 296. H. Klinkhardt, *Annalen der Physik*, 1927, Series 4, vol. 84, pp. 167–200.

[2] J. B. Austin, *Journal of Industrial and Engineering Chemistry*, 1932, vol. 24, pp. 1225–1235.

fining solid shell except by means of the heat effect which was represented by the latent heat of fusion. At the interface solid and liquid were both at the melting temperature. If one accepted the author's assumptions, it was true that superheat might be considered as an increase in initial temperature; latent heat of fusion, however, fell in an entirely different category.

The writer regretted that he had not familiarised himself in detail with the published work of Lightfoot. Nevertheless, he would remind the author that Lightfoot's general procedure of assuming boundary conditions which were hypothetical did not necessarily lead to conclusions which, over the entire time interval of the solidification and cooling process, were to be dismissed as lacking in approximate accuracy. The author, as well as Lightfoot, were dealing with hypothetical assumptions, and the main issue was which set of assumptions within their common field of application yielded conclusions most closely in accord with the phenomena under consideration.

The author's Fig. 2, which related to the temperature fall at the centre of the ingot, was particularly puzzling, because as long as any liquid steel remained in the ingot the temperature at the centre did not fall below the melting temperature.

Possibly the author might find in a recent paper by Chipman and FonDersmith [1] experimental data on the rate of solidification of steel ingots which would be useful in comparing theory with practice. Those investigators determined the rate of solidification of ingots by spilling the liquid contents and measuring the thickness of the solidified wall. The ingot's measured 18 by 39 in. in cross-section. The thickness solidified at various times was :

Time. Min. .	1	2	3	5	10	15	20	25	30
Thickness. In.	0·78	1·15	1·44	1·89	2·73	3·36	3·90	4·38	4·81

Chipman and FonDersmith stated that a straight-line relationship between the thickness of solidified metal and the square root of the time was valid up to 30 min. Such a relationship had been previously determined by the writer from purely theoretical considerations in the paper which was listed in the bibliography at the end of the author's paper. These theoretical considerations included a hypothetical boundary condition and employed as their leading motif the latent heat of fusion.

AUTHOR'S REPLY.

Mr. RUSSELL, in reply, wrote that he felt gratified that his paper had attracted the attention of Mr. Feild, whose own work, published in 1927, had done so much to stimulate interest in the rates of solidification of steel in ingot moulds.

[1] *Transactions of the American Institute of Mining and Metallurgical Engineers*, 1937, vol. 125, pp. 370–377.

It was difficult to see why Mr. Feild had found Fig. 2 particularly puzzling, as he, himself, stated that the calculations were based on the assumption, "in effect, that the mould was filled with a homogeneous solid body having the properties of steel in the solid state." This was correctly stated, but there should also be added "*if one imagined that the initial temperature of the solid body was* 356° *C. above the temperature of the solidus of steel.*" This high initial temperature was to allow for the effect of latent heat and any superheat, and was in reasonable agreement with the range of temperature, 360° C., given by Mr. Feild. The suggestion was then made that the times required for the centres of these hypothetical solid bodies to fall the first 356° C., 1806° C. to 1450° C., or from $U = 1{\cdot}0$ to $U = 0{\cdot}8$, represented the relative times for the centres of steel ingots to solidify. The curves in Fig. 3 were drawn from data obtained from Fig. 2.

He was in almost full agreement with the remainder of Mr. Feild's comments. In the paper he had tried to deal fairly and fully with the limitations of the application of mathematics to the problem. There was no known method of even attempting to develop a formula which would allow for the effects of convection currents or for the changes of thermal constants with temperature.

With regard to the latent heat, he would say that no formula that was capable of arithmetic interpretation had yet been obtained which allowed for the latent-heat effects in either a rectangular or a circular ingot, even for the theoretical case of the latent heat being evolved at a constant temperature. The problem was much more difficult when applied to the cooling of a commercial steel containing, say, 0·45% of carbon. In such a steel the process of solidification was spread over a range of temperature of some 60–70° C. and was further complicated by a peritectic reaction which took place in this range. During the transformation there was an evolution of about 2·8 cal. per g. spread unevenly over the ferrite-precipitation range and a further evolution of about 10 cal. at the eutectoid temperature. He (Mr. Russell) reaffirmed his statement that the possibility of developing rigid formulæ for the cooling of steel ingots was very remote at the present time. However, he had thought it worth while to examine quantitatively those formulæ which had been published, and apparently the Committee on the Heterogeneity of Steel Ingots, who passed the paper for publication, thought so too. Of the value of the mathematical method he had asked readers to judge for themselves, and there was no doubt that Mr. Feild's opinion that "his own study of the author's paper had convinced him that assumptions (1), (3) and (6) differed too widely from those conditions which actually prevailed to permit of a valid application of theory to practice" would carry much weight with people who did not study the paper themselves. He (Mr. Russell) preferred to withhold his final judgment until more experimental results had been examined.

THE POLISHING OF CAST-IRON MICRO-SPECIMENS AND THE METALLOGRAPHY OF GRAPHITE FLAKES.*

BY H. MORROGH (BRITISH CAST IRON RESEARCH ASSOCIATION, BIRMINGHAM).

(Figs. 1 to 36 = Plates XXVIII. to XXXVII.)

SUMMARY.

A polishing technique has been developed for grey cast irons, whereby the graphite flakes can be obtained perfectly preserved and smoothly polished. The polishing medium used is either Diamantine or magnesium oxide. The method owes its success to the correct application of a repeated polishing and etching operation.

With specimens prepared in this manner, it is possible to examine the internal structure of the graphite flakes and temper carbon nodules. Secondary graphite is shown to be deposited either on the existing eutectic graphite flakes or in a Widmanstätten pattern. Graphite flakes show complicated internal structures, which are illustrated micrographically using polarised light. "Inclusions" are also shown intimately associated with graphite flakes.

PART I.—THE POLISHING OF CAST-IRON MICRO-SPECIMENS.

Introduction.—That cast-iron micro-specimens are difficult to polish and that there is a very urgent need for standard methods of preparation of metallographic specimens, is indicated by the many and varied results obtained by different investigators. Most text-books of metallography give a description of a process whereby metal micro-specimens can be obtained flat, polished free of scratches and suitable for observation under the microscope. In general, metallographists have acquired this technique and can now produce, fairly easily, metal specimens so polished. Unfortunately, from the point of view of the preparation of micro-specimens, cast irons have a combination of constituents which among the common metals is unique in so far as a soft and very friable substance, namely, graphite, occurs surrounded by a relatively hard, but metallic, matrix which nevertheless has a measure of plasticity. A perfectly prepared micro-specimen of grey cast iron † would have a metallic matrix polished free from scratches, and in this matrix would be graphite flakes also polished smooth and free from scratches, their surfaces being level with the polished metal surface. When a

* Received November 8, 1940.

† This paper is concerned exclusively with the preparation of specimens of grey cast iron. White irons not only avoid the special difficulties due to graphite, but are generally quite easy to prepare.

"general" metallographic technique (*i.e.*, one recommended for any ferrous metal) is applied to grey cast iron, even though the metallic matrix may be polished correctly, the graphite flakes are seldom, if ever, polished at all. The peculiar friability of graphite flakes results in the substance being torn out wholly or in part from the cavities which are either opened up or burnished over, according to the particular type of polishing procedure. This has been discussed in detail by Norbury and Bolton.[1,2]

As a result of the tendency for graphite to be torn out, metallographists have almost become resigned to the fact that in a cast-iron specimen, one must see not the actual graphite flakes, but cavities where the graphite flakes existed prior to the polishing operation. It is the purpose of this paper to point out that perfectly polished micro-specimens of grey or malleable cast iron can be obtained with the graphite intact, to give a description of the process used, and to indicate what parts of a general technique should be avoided if perfectly prepared specimens are to be obtained.

Use of Loose Abrasives.—Most metallographic specimens are, in the initial stages, either ground or filed flat and then treated in some way, so that the relatively coarse scratches produced by the grinding wheel or file are removed and replaced by finer scratches, which can in turn be easily removed by the polishing powder. The methods by which the specimen is smoothed prior to polishing may be roughly divided into two classes. The first class includes those methods which use a series of loose abrasives, and the second includes those methods which use successively finer grades of emery paper. For our purpose the second method is to be preferred, the objection to the first method being that while satisfactory metal surfaces can be obtained, the graphite flakes are nearly always more badly damaged than in specimens prepared by an emery paper method. This fact has been pointed out previously by Vilella.[3] Fig. 1 shows a micrograph of an ingot mould specimen after the removal of grinding marks by carborundum and coarse alumina; Fig. 2 shows a micrograph of the same specimen after regrinding and rubbing successively through finer grades of emery paper. It is seen that in the latter case the graphite flakes are in a much better state of preservation than in the former, which shows cavities and not graphite. We can therefore say that any methods using loose abrasive in the preliminary stages are to be avoided if perfectly polished cast irons are to be obtained.

Use of Emery Papers.—With regard to the use of emery papers, the following procedure has been found to give good results: Starting from a freshly ground surface, rub the specimen through a series of three emery papers, grades 1*G*, 1*F* and 00 (or their equivalents). The rubbing should be slow and the specimen should be pressed very firmly to the paper so that the operator can feel the surface being cut. The presence of small particles of grit between the paper and the specimen tends to tear out the graphite, and these can be removed by blowing or shaking the paper. It is not necessary to introduce more papers into the series mentioned above, since they introduce more complications and increase the time of preparation per specimen. Papers finer than 00, such as 000 and 0000, are frequently uneven and gritty, and a fine paper can be prepared from a piece of used 00 paper. Methylated spirits is dropped from a spotting bottle on to the paper. This is then rubbed with cotton wool until the required type of paper is obtained. A final paper produced in this way is uniform, free from grit and it will give a finer finish to the specimen than 000 or 0000 papers. Immediately after treatment on the final emery paper, the specimen should be etched in either a 5% solution of nitric acid in alcohol, or a saturated solution of picric acid in alcohol. A specimen rubbed down and etched in this manner is ready for polishing.

Final Polishing.—About the final polishing there is a very general agreement as to how it should be performed. Most polishing machines have a revolving wheel to which is attached a piece of soft cloth impregnated with polishing powder.

Polishing cloths with a high pile tend to dislodge the graphite in a similar manner to loose abrasives and free particles of grit on the emery papers. To preserve the graphite, it is best to polish with a cloth having very little pile. Vilella has suggested the use of the dull side of a heavy pure silk satin, but this type of material does not retain the polishing powder during the rotation of the pad. If, however, Selvyt cloth is used, the pile becomes flattened and the cloth apparently smooth in a short time after the beginning of polishing. A suitable speed for the polishing wheel is about 600 r.p.m. Higher speeds give faster polishing, but the polishing powder is thrown off the wheel more easily.

The possible polishing powders, namely, chromium oxide, rouge, zinc oxide, manganese dioxide, tin oxide, magnesium oxide and Diamantine (a specially prepared form of alumina), vary considerably in their ability to polish cast irons without tearing out the graphite. Diamantine and magnesium oxide tend to leave the graphite undisturbed, polishing it at the same time. The other polishing powders, chromium oxide in particular, produce similar effects to loose abrasive. Fig. 3 shows the same spot as Fig. 2 after polishing on chromium oxide. It is seen that the surface of the graphite has been badly damaged and the width of the graphite lamellæ increased. Of the two polishing powders, Diamantine and magnesium oxide, the former is to be preferred, since it polishes very much more quickly, although where the price of Diamantine prohibits its use, magnesium oxide can be used satisfactorily. The latter requires more polishings and etchings than the former and hence takes more time.

Alternate Polishing and Etching.—Even with the use of Diamantine or magnesium oxide, it is only rarely that perfectly polished specimens are obtained after one polishing operation. To overcome this difficulty, Hanemann and Schrader[4] (*loc. cit.*, Part 1, p. 16) have recommended the use of alternate etching and polishing. This method has been widely used for a long time by metallographists to remove a distorted surface layer from micro-specimens of various metals. In the case of cast iron, the use of this method depends on the removal of layers of the metallic matrix until the polished surface is level with the undamaged regions of the graphite flakes. The method works excellently with grey or malleable cast irons (Fig. 4) and the technique of the process has been worked out by the author and satisfactorily employed on a large number of varied types of irons, typical examples of which are shown in Figs. 5, 6 and 7, in all of which Diamantine has been used. Particulars of this process are given in the following:

The polishing powder, 250-mesh Diamantine, is mixed with water in the proportions 50 g. of Diamantine to 100 c.c. of water, so that a fairly thick paste is formed. Frequently in the literature it is recommended that the polishing powder should be mixed with much water.[5] The thick paste is preferred, however, because the pad is likely to be used several times for each lot of Diamantine applied to it, and when in paste form it tends to be retained by the wheel more than when used in the watery condition. Diamantine paste in an amount equal to about 15 c.c. for a 6-in.-dia. wheel, is put on the pad and rubbed well into the pile with the fingers. By this means the operator is able to detect any particles of grit. A pad prepared in this way will polish out the scratches from the emery paper in times varying from about 20 sec. to 2 min., depending on the type and size of the specimen.

The best procedure is to keep the orientation of the specimen with respect to the operator constant, rotating it in the opposite direction to the pad. A fair amount of pressure should be exerted so that the wheel can be felt pulling the specimen. Fine and medium-sized graphite will generally be polished better than coarse graphite. It is frequently only necessary to polish a specimen containing medium graphite once for a successful exmination of its structure. After the first polishing operation, the pile of the pad should not be disturbed; one made up in the above way will last for twelve additional polishing operations, any further polishing making the pad more nearly ideal for the preservation of the graphite.

If, after the first polishing operation, the graphite is torn out, opened up or burnished over, as is usually the case, the specimen should be re-etched and polished again; the time required to polish off the etching depends on the specimen, but generally 10 sec. is ample. It is not even necessary to wait for the removal of the emery paper scratches before the etching and polishing operations begin; in fact, the disappearance of the scratches is made more rapid by etching before the scratches have disappeared. By protracted etching and polishing the graphite flakes gradually assume a light grey to brown colour, serrated edges tend to disappear and any burnishing or opening is removed.

The choice of etching reagent is of some importance in the repeated polishing and etching operations. It is necessary to use an etching reagent which actually etches the metallic matrix. For pearlitic grey irons or for irons containing more than 50% of pearlite in the matrix, both nitric acid and picric acid produce similar effects with regard to the polishing of the graphite. In the case of ferritic grey irons and ferritic blackheart malleable irons or irons containing less than 50% of pearlite in the matrix, nitric acid is definitely preferable and produces the desired effects more rapidly than picric acid. In some cases, when the specimens have very coarse graphite in a ferritic matrix, it is almost impossible to remove the burnished and flowed layer from the graphite using picric acid; nitric acid will do this quite readily.

When two samples of closely similar graphite size have to be compared, there is frequently some doubt cast on the verdict, since one specimen may be slightly burnished, giving the appearance of a fine graphite size, or it may be "opened," giving the appearance of a coarse type of graphite. If specimens are prepared by alternate etching and polishing, any burnishing or opening of the graphite is removed and accurate comparisons of the length and width of graphite can be made.

Phosphoric irons are more difficult to polish by this method than hematite irons, because the phosphide eutectic tends to stand up in relief; this can be avoided if the times of actual polishing are increased.

Correctly prepared samples bring into photomicrography another element of difficulty. Improperly prepared samples show the graphite as a black constituent and the matrix white, forming an easy combination to photograph with a contrasty plate; on the other hand, the photography of samples prepared by the improved technique requires a less contrasty plate and more accurate control with respect to exposure and development.

Part II.—Some Observations on the Internal Structure of Graphite in Cast Iron Micro-Specimens Prepared by the Improved Technique.

When the graphite flakes in a carefully prepared sample of cast iron are examined at a low magnification (50–200 diameters) they appear to be light grey to brown in colour; any defect on the surface of the graphite gives it a black appearance (see Figs. 8 and 9).

At high magnifications (generally above 500 diameters) the colour of any particular graphite flake depends on its position relative to the glass-slip reflector of the microscope, with the result that some graphite flakes appear light grey, some black and some of intermediate tone. On rotating the specimen through 90°, keeping the glass-slip reflector still, those flakes formerly black become light brown and *vice versa* (Figs. 10, 11, 12 and 13).

When a beam of light is reflected from the surface of a transparent substance, it is partly polarised, the vibrations being parallel to the reflecting surface; in this way the beam of light from the illuminating train of a microscope after reflection from the glass slip is partially polarised. Certain portions of the spectrum are absorbed when a beam of light is reflected from a coloured substance, and the complementary colours are seen. With some anisotropic substances, the wave lengths removed depend upon the plane in which the incident light is vibrating and its relation to the optic axis of the crystal; this latter phenomenon is referred to as reflex pleochroism. Graphite exhibits reflex pleochroism [6] and this accounts for the changes in colour of graphite flakes on rotating the stage when the illuminating system is of the glass-slip reflector type. That the manifestation of this optical property of graphite is due to the glass slip alone is shown in Figs. 14 and 15.

Fig. 14 shows two graphite flakes at right angles; the micrograph is taken with the usual type of glass-slip reflector and it is seen that one flake is dark and the other is light; Fig. 15 shows the same spot taken using a metallic reflector,[7] which fills only a portion of the back lens of the objective with light. In this case, both flakes show the same absorption, since this reflector does not polarise the light. The relation between the plane of polarisation of the incident light and the positions of maximum and minimum brightness of graphite flakes is such that when the plane of polarisation is at right angles to the graphite flakes, they appear dark, and when parallel to the graphite flakes they appear bright. If the glass slip is rotated through the same angle as the specimen, no change in brightness of the graphite is observed. If a single nicol prism is inserted in the illuminating beam so that the plane of polarisation of the light passing through it is parallel to that of the light reflected from the glass slip, then the effect of the pleochroism is more marked, since the light incident on the specimen is almost completely polarised.

When viewed under ordinary illumination, a number of graphite flakes appear to be uniform and similarly oriented throughout their lengths and breadths; that is, in one particular position with respect to the illuminating system, the whole of a graphite lamella is of the same colour (Fig. 18).

When such a graphite flake is examined under polarised light between crossed nicols, on rotation of the specimen through 360° it lights up four times; this is expected from the hexagonal anisotropy of graphite (Figs. 16 and 17). The actual position of lighting up varies slightly according to the orientation of the graphite flake in relation to the surface of the specimen. However, if the cross

FIG. 1.—Ingot Mould Specimen, after preparation on carborundum and alumina. Graphite torn from specimen. × 200.

FIG. 2.—Same Specimen as Fig. 1, after regrinding and rubbing successively on 1G, 1F and 00 emery paper. Graphite relatively undamaged. × 200.

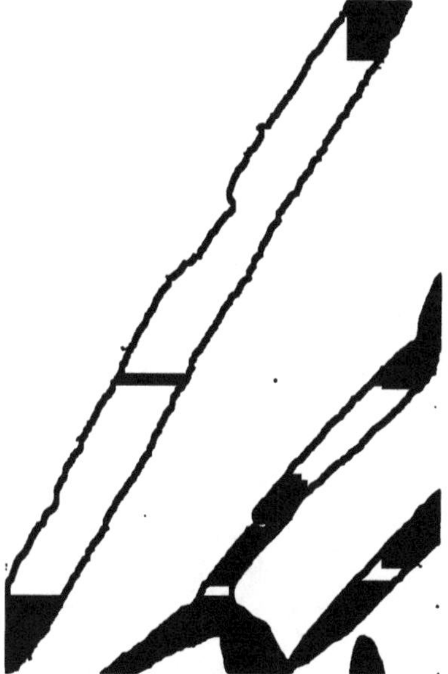

FIG. 3.—Same as Fig. 2, after polishing on chromium oxide. Graphite torn out. × 200.

FIG. 4.—Same as Figs. 2 and 3, after 30 polishings and etchings. Polishing powder, magnesium oxide. Graphite intact. × 200.

(Micrographs reduced to four-fifths linear in reproduction.)

[Morrogh.

[To face p. 200.

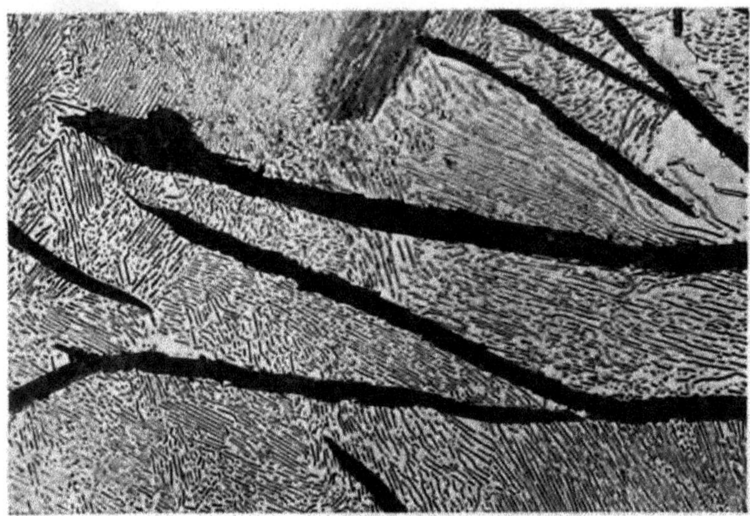

FIG. 5.—Ingot Mould Specimen; brown graphite lamellæ in pearlitic matrix. Etched in picric acid. × 500.

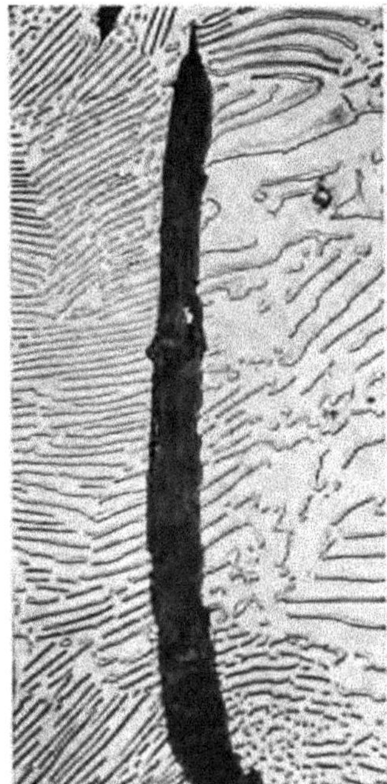

FIG. 6.—Smoothly Polished Graphite Flake in sample of pig iron. Etched in picric acid. × 1000.

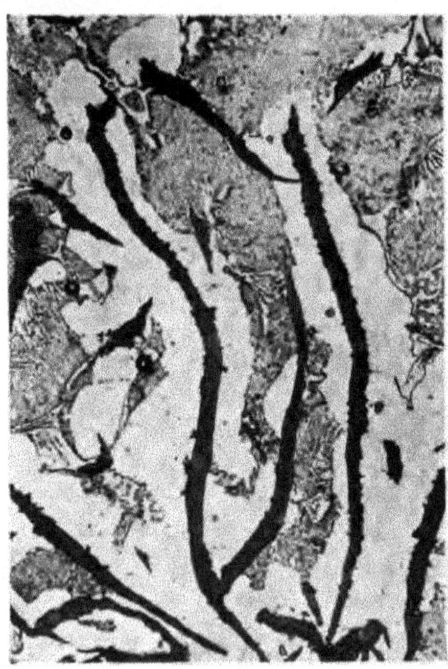

FIG. 7.—Common Iron; grey-brown graphite lamellæ in ferrite with pearlite and phosphide eutectic. Secondary graphite in the form of "fluffy" markings along the edges of the graphite. Etched in picric acid. × 500.

(Micrographs reduced to two-thirds linear in reproduction.)

[*Morrogh.*

Fig. 8.—Graphite Flakes in Ingot Mould Sample. Black spots due to their slightly damaged surfaces. Unetched. × 50.

Fig. 9.—Section of a 1·2-in. dia. Test Bar. Not smooth surfaces of graphite flakes. Unetched × 200.

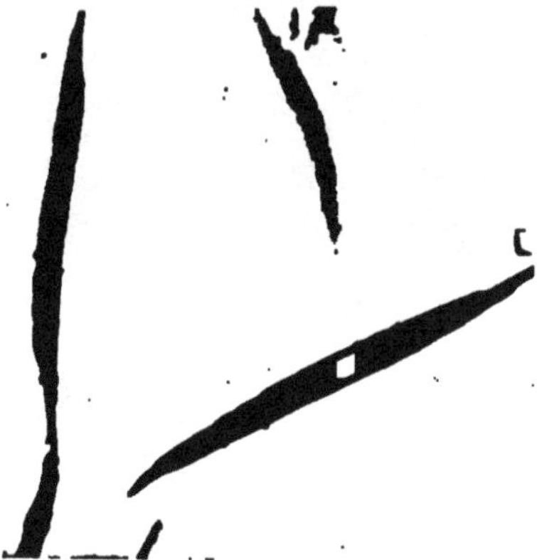

Fig. 10.—Graphite Flakes at a high magnification. Some flakes appear light grey, others black. × 1500.

Fig. 11.—Same as Fig. 10 with specimen rotated through 90°. Graphite flakes formerly black now light grey, and *vice versa*. × 1500.

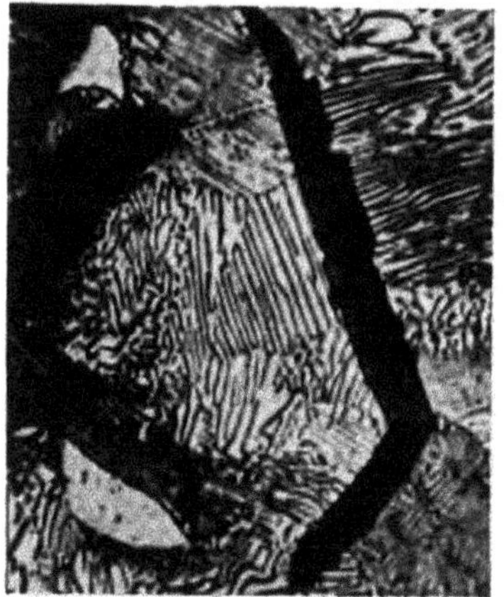

Fig. 12.—Graphite Flake in its Brightest Position. Etched in picric acid. × 2000.

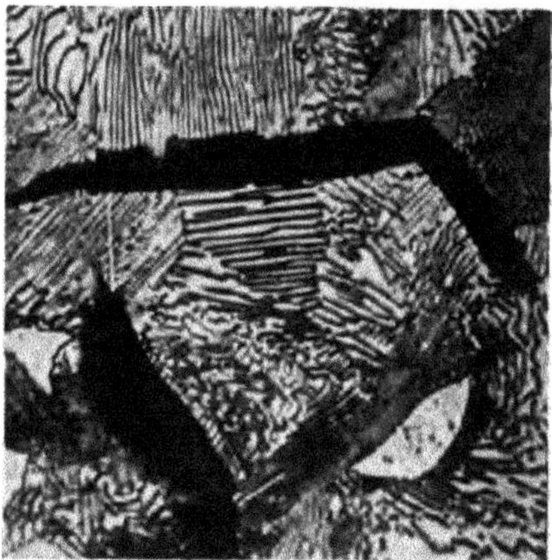

Fig. 13.—Same as Fig. 12 after rotating the specimen through 90°. × 2000.

(Micrographs reduced to four-fifths linear in reproduction.)

[Morr

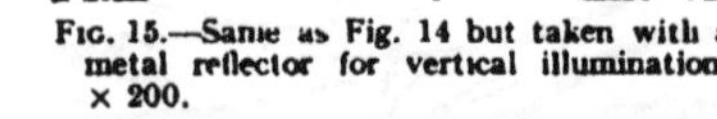

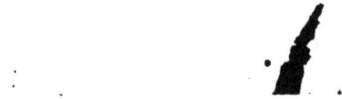

Fig. 14.—Graphite Flakes in Ingot Mould Sample. Taken with a glass slip reflector for vertical illumination. Unetched. × 200.

Fig. 15.—Same as Fig. 14 but taken with a metal reflector for vertical illumination. × 200.

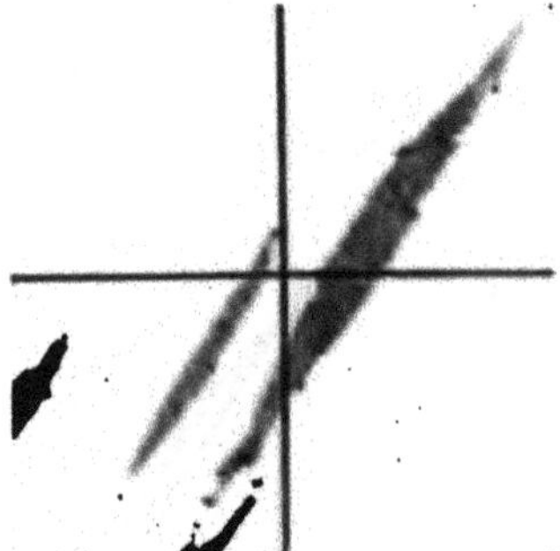

Fig. 16.—Uniformly Coloured Graphite Flake. Unetched. × 2000.

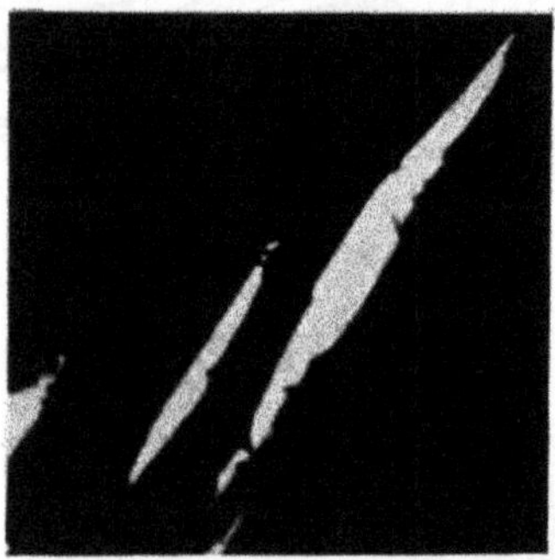

Fig. 17.—Same as Fig. 16 with crossed nicols. Graphite flake in position of maximum brightness. × 2000.

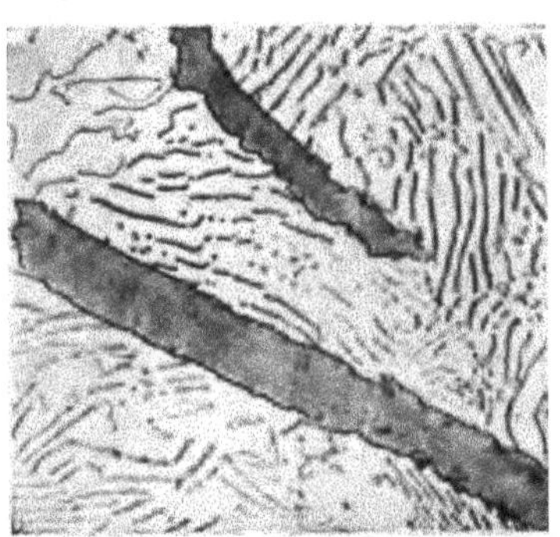

Fig. 18.—Cylinder Liner Sample; uniformly coloured graphite flake in background of pearlite. Etched in picric acid. × 1000.

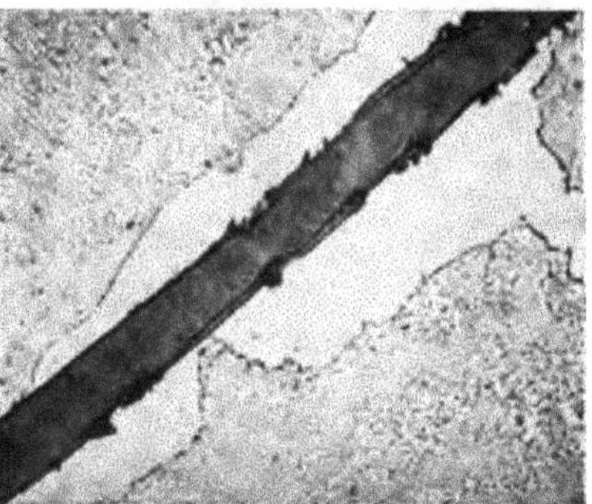

Fig. 19.—Ingot Mould Section; graphite flake surrounded by ferrite and pearlite. × 1000.

(Micrographs reduced to two-thirds linear in reproduction.)

[*Morrogh*.

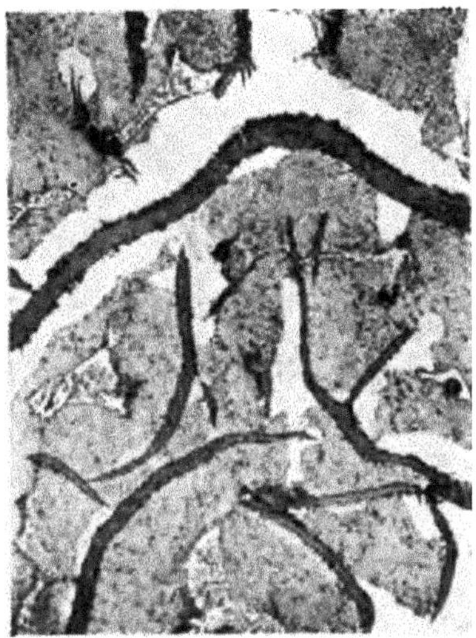

Fig. 20.—Common Iron; "fluffy" appearance of secondary graphite. Etched in picric acid. × 200.

Fig. 21.—Same as Fig. 20 at higher magnification. × 500.

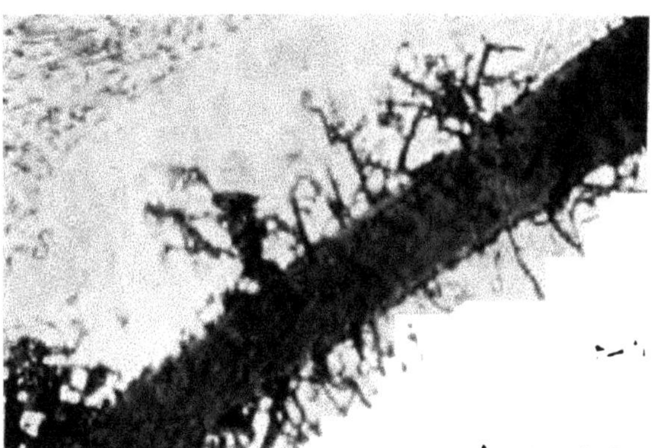

Fig. 22.—Same Sample as Figs. 20 and 21 at higher magnification Secondary graphite, which appeared "fluffy" at lower powers, seen to be arranged in Widmanstatten formation. × 1000.

[Morrogh.

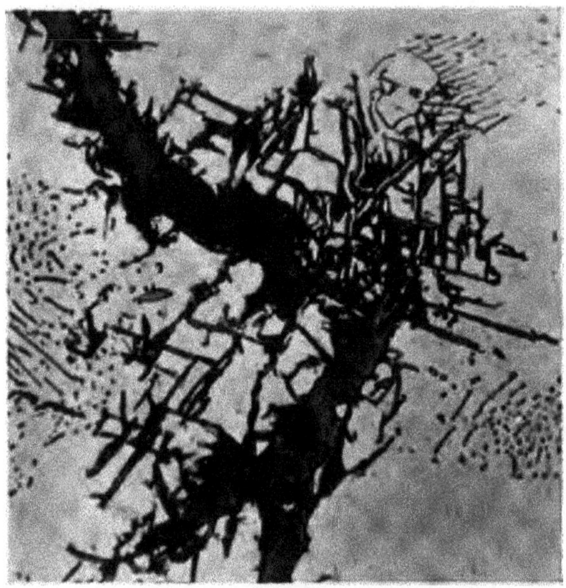

FIG. 23.—Well-Developed Widmanstätten Structure in Hot-Mould Iron; secondary graphite obviously attached to flake graphite. Matrix, ferrite with small amount of pearlite. × 1000.

FIG. 25.—Small Crystal of Secondary Graphite. Ordinary vertical illumination. × 750.

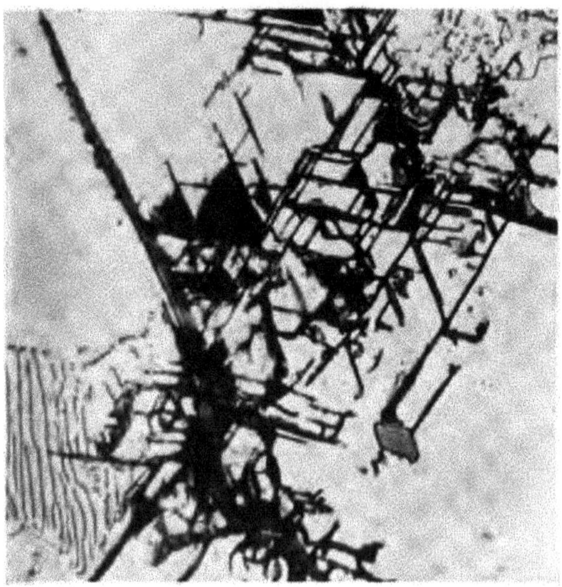

FIG. 24.—Same Specimen as Fig. 23. Graphite flake apparently out of the plane of the surface of the specimen. × 1000.

FIG. 26.—Same as Fig. 25 with crossed nicols. × 750.

[Morrogh.

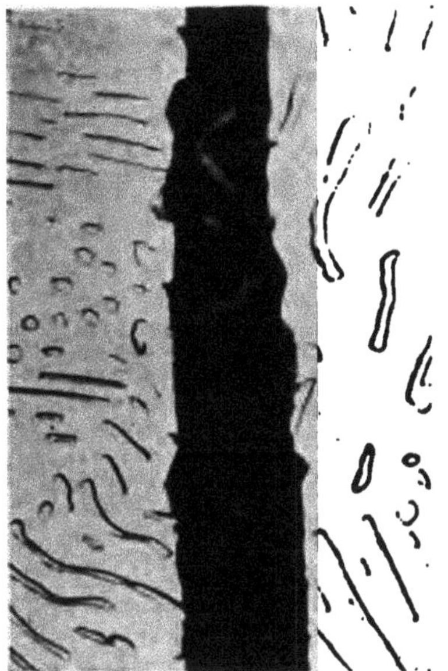

FIG. 27.—Needle Formation in Graphite Flake in ingot mould specimen. Flake in its dark position. Etched in picric acid. × 2000.

FIG. 28.—Same as Fig. 27 between crossed nicols. × 2000.

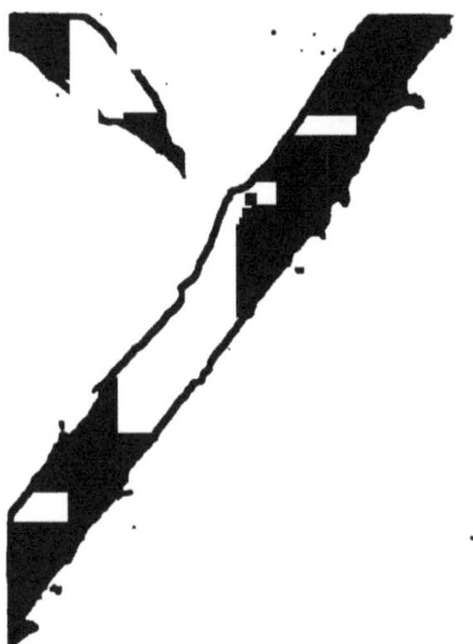

FIG. 29.—Simple Set of Needles in a Graphite Flake. Unetched. × 2000.

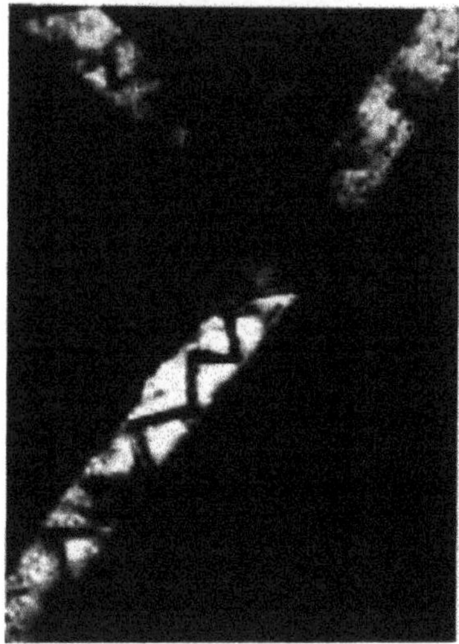

FIG. 30.—Same as Fig. 29 between crossed nicols. × 2000.

(b) Flake in
position; needles

-100-Ton Anvil Block section;
confined to "primary" graphite
n secondary

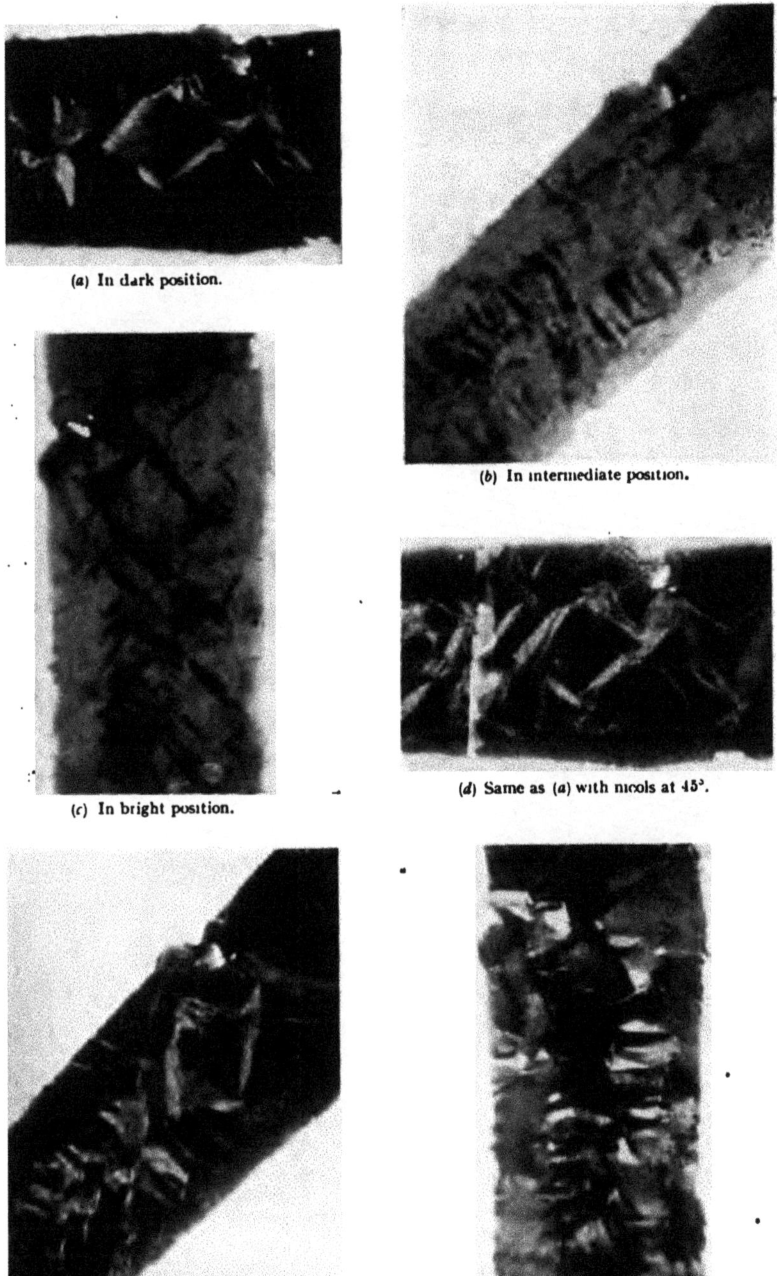

(*a*) In dark position.

(*b*) In intermediate position.

(*c*) In bright position.

(*d*) Same as (*a*) with nicols at 45°.

(*e*) Same as (*b*) with nicols at 45°.

(*f*) Same as (*c*) with nicols at 45°.

Fig. 33.—Flake of Kish Graphite. Needle formation not of the simple type shown in Fig. 32. Unetched. × 2000.

(Micrographs reduced to four-fifths linear in reproduction.) [Morrogh.

PLATE XXXVII.

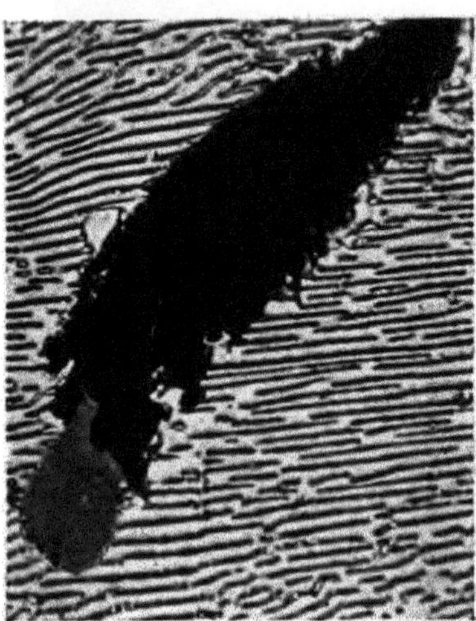

FIG. 34.—Graphite and Manganese Sulphide in Matrix of Pearlite. Etched in picric acid. × 1000.

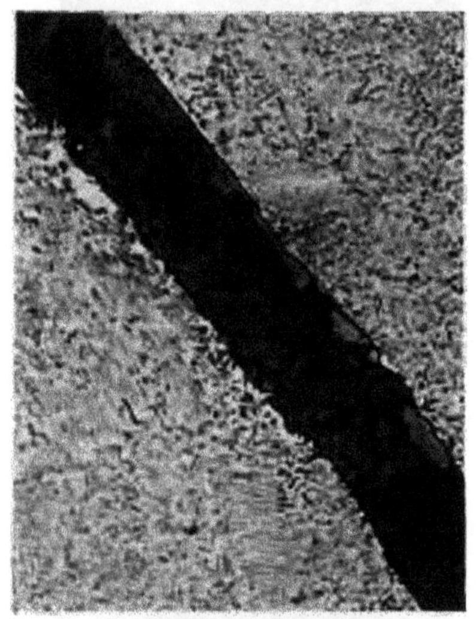

FIG. 35.—Graphite Flake in Ingot Mould; inclusion attached to the side. Etched in picric acid. × 1000.

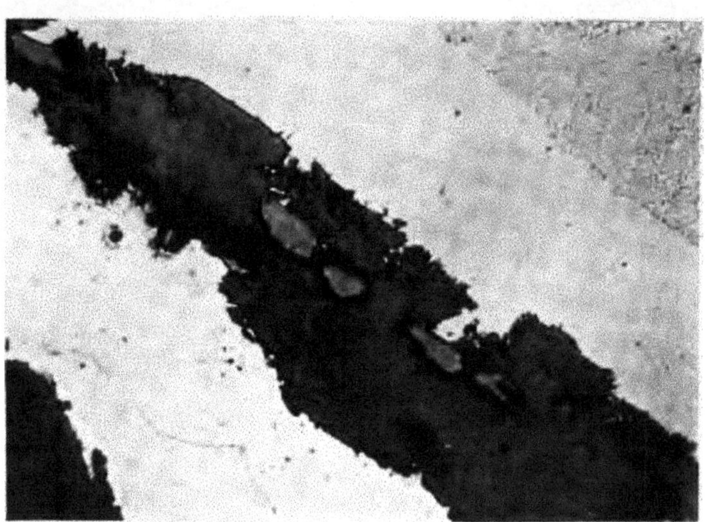

FIG. 36.—Same Specimen as Fig. 35. Inclusions apparently in the centre of a graphite flake. Etched in picric acid. × 1000.

[*Morrogh.*

[*To face p.* 201 P.

lines in Fig. 16 represent the planes of polarisation of the analyser and polariser, then the position of the graphite flake shown is one of the positions of brightness.

The only graphite flakes to show this " ideal " behaviour under polarised light are generally the short, straight ones. Curved flakes show structural complications.

Secondary graphite, that is, hypereutectoid and eutectoid graphite, is generally deposited on the existing graphite flakes in the form of a smooth band which is similarly oriented to the main body and is therefore difficult to show. A band of such secondary graphite is seen in Fig. 19.

Sometimes, the secondary graphite is deposited on the graphite lamellæ in an irregular manner, so that the edges have a " fluffy " appearance (Figs. 20 and 21). At high magnifications, it appears that this fluffy secondary graphite is arranged in a Widmanstätten formation (Fig. 22). In fact, in rare cases, a well developed Widmanstätten structure is obtained. This is shown in Figs. 23 and 24. The reason for the formation of the graphite Widmanstätten structure may seem obscure at first sight, but while in certain respects its formation is unique, the explanation may be very simple. When a solid solution stable at high temperatures precipitates a new phase over a range of temperature on casting, then the new phase may be deposited in one of several ways. For instance, it may be deposited at the grain boundaries of the parent solid solution, or it may be deposited in such a way that its lattice bears a definite crystallographic relation to the lattice of the parent solid solution, that is, it forms a Widmanstätten structure. Further, if the phase which is being precipitated from the solid solution already exists in the structure, then it may be deposited on the already existing areas of this phase. A typical example of the latter mode is the deposition of the hypereutectoid cementite on the already existing eutectic cementite in hypo-eutectic white irons. The Widmanstätten mode of deposition of the new phase is an alternative to both the grain boundary type of deposition and the deposition on the already existing phase. Of course, when there is an already existing phase, then deposition of the phase from the solid solution may occur in three different ways—it may be deposited on the already existing phase, it may be deposited at the grain boundaries of the parent solid solution, or it may be deposited in a Widmanstätten structure in that solid solution. The simultaneous deposition of hypereutectoid cementite in all these three ways has, in fact, been observed in certain low-carbon, hypo-eutectic white irons. When the new phase is being deposited from a solid solution with no other phases in the structure, then a controlled rate of cooling, neither extremely fast nor extremely slow, and a large grain size are the conditions favourable for the development of a well-defined Widmanstätten structure. Now, the graphite Widmanstätten structure shown here consists of hypereutectoid graphite, which has been deposited from the austenite. It may be said that the normal mode of deposition of this hypereutectoid graphite is in the form of a smooth band on the already existing eutectic flake graphite. From these points it is seen that the grain size of the original austenite does not vitally determine whether the graphite will be deposited normally or in the Widmanstätten form, and the conditions favourable for the development of the graphite Widmanstätten pattern are a controlled rate of cooling, neither extremely slow nor extremely fast, and a eutectic graphite structure consisting of a few very coarse flakes. The latter condition is most easly met in low-carbon irons of large section. The actual rate of cooling required to produce the graphite Widmanstätten pattern must be very critical, as this structure is seen in a fully developed form only very rarely.

Hanemann and Schrader[(4)] (*loc. cit.*, Part 2, Table 3) have shown that secondary graphite sometimes forms small needle-like crystals of different orientation from the graphite flake. An example of this is shown in Fig. 25. The arrow points to a small crystal of secondary graphite and it is seen to have a different orientation from the flake, since it is light while the flake is dark. Fig. 26 shows the same spot taken between crossed nicols; the flake is black and the secondary graphite has lit up.

At high magnifications, a series of needles differently coloured and hence differently oriented from the rest of the graphite is apparent. Typical examples of these needle-like formations are shown in Figs. 27, 28, 29 and 30. This effect is most marked near to and at bends in graphite flakes. If the flake is in its dark position, then the needles appear bright, and if the flake is in its bright position, the needles appear dark. Vilella[(3)] observed a microstructure in graphite flakes of carefully prepared samples, and Hanemann and Schrader recognised the needle-like formation, attributing it to mechanical stressing of the graphite flakes during solidification[(4)] (*loc. cit.*, Part 2, p. 22). In the simplest cases the needles are arranged end to end, but at an angle to one another.

By heat-tinting methods, it has been found possible to indicate the presence of both the secondary graphite and the needle formations in the graphite flake, and it is interesting to observe that this formation is confined almost entirely to the "primary" graphite flake and is usually not present in the secondary graphite (Fig. 31).

Frequently, and more particularly in very large graphite flakes, the needle-like structure takes on more complicated forms. In this respect the micrographs in Fig. 32(*a*, *b*, *c*) are interesting. They illustrate a flake of kish graphite in a piece of dead-annealed pig iron. Fig. 32(*a*) shows the main body of the graphite in its bright pleochroic position; in this case the needles are dark. Fig. 32(*b*) depicts the same spot with the flake in the intermediate pleochroic position; the needles are now bright. Fig. 32(*c*) illustrates the same area with the main body of the graphite flake in its position of maximum pleochroic absorption. It is seen in Fig. 32(*a*) that the flake is made up of five bands and that the large system of needles is confined to the centre band, which is presumably the primary graphite crystal. The two bands on either side of the centre appear to have been deposited in the manner of secondary graphite; the extreme outer bands are probably eutectoid graphite produced by the annealing treatment, and the intermediate bands are probably hypereutectoid or even eutectic graphite. The intermediate band in the lower part of the flake in Fig. 32(*c*) is seen to contain a series of needles. If Hanemann and Schrader's assumption is correct, then it seems that some mechanical stressing occurred after the deposition of this band. It is interesting to note that a rotation of 90° introduces a complete reversal of pleochroic contrasts between the main body of the flake and the needle formations.

In some graphite flakes, the simple end-to-end needle formation is not apparent (Fig. 33(*a*, *b*, *c*)), the whole of the graphite flake being a conglomerate of small needle-like crystals of graphite with various orientations, this being particularly obvious when the specimen is viewed under polarised light with the nicols at 45° (Fig. 33(*d*, *e*, *f*)).

Non-Metallic Inclusions.—The effect of non-metallic inclusions on the structure of grey cast iron is a much-debated topic. While there is little direct microscopical evidence of inoculation or nucleation of melts by inclusions, it is significant to note that they do occur attached to and perhaps inside graphite flakes. Both titanium cyano-nitride and manganese sulphide crystals are frequently observed attached or adjacent to the ends of graphite flakes. A typical example of a manganese sulphide crystal attached to the end of a graphite flake is illustrated in Fig. 34.

Dove-grey, apparently allotriomorphic inclusions, which appear to be manganese sulphide, have been found very intimately associated with graphite in the top of a large ingot mould. This is illustrated in Figs. 35 and 36. Fig. 36 shows the sulphide to be actually inside the graphite flake; this may, of course, be only an accident caused by the position of the surface of the specimen in relation to the flake which makes the inclusion have the appearance of being inside the graphite. It is also possible that the occurrence of the sulphide so intimately associated with graphite has something to do with the flotation of the sulphide to the top of the ingot mould during solidification, where it became enmeshed with the growing eutectic flake graphite.

Temper Carbon.—Using the repeated etching and polishing process, it has been possible to observe the internal structure of temper-carbon nodules in malleable cast iron. The structure of the temper-carbon nodules has been dealt with in the author's paper on "The Metallography of Inclusions in Cast Irons and Pig

Irons "[8] from the point of view of the effect of iron and manganese sulphide on the type of temper carbon formed. In malleable irons containing iron sulphide—for instance, whiteheart malleable irons—the temper carbon is in a spherulitic form; a typical example of this is shown in Fig. 22 of the above paper. The basal planes of the graphite crystals composing the temper carbon nodule are at right angles to the radii of the spheroid of which the nodule can be considered to be composed. It has been possible to show crystals of iron sulphide in the centre of some spherulitic temper-carbon nodules as though providing nuclei for the temper-carbon formation (see Fig. 24 of the same paper). In malleable irons which contain all the sulphur in the form of manganese sulphide, the temper carbon appears to be an aggregate of very small graphite "flakes" having no particular orientations with respect to each other. A typical example of this is shown in Fig. 23 of the above paper.

Conclusion.

It has been shown that the graphite flakes in grey cast iron can be retained quite easily during the polishing operation with but few changes in normal recommended procedures. The ideal to aim for is a smoothly polished metallic matrix free from surface distortions and smoothly polished graphite flakes with no burnishing or flowing of the metallic matrix over the graphite flakes. Only with specimens prepared in this way can any accurate idea be obtained of the size, shape and internal structure of graphite flakes. The process which has been described here enables the complete preservation of the graphite flakes and the polishing operation, if correctly adhered to, produces specimens perfectly prepared for photomicrography.

The polishing procedure given here has the advantage that it is a comparatively rapid method of specimen preparation and the stages are quite simple. It is rather difficult to give any accurate times for the preparation of specimens, but as a rough guide it can be said that it should be possible to prepare a specimen of a cross-section from a 0·875-in.-dia. test bar in 15 min., that is including cutting, grinding, rubbing down, and polishing and etching. The actual time required for specimen preparation by this method is to some extent a function of the number of times the specimen is etched and repolished, and this again depends on the structure of the specimen and on the perfection required. For the routine visual examination of cast irons, it is frequently unnecessary to have the graphite flakes perfectly prepared. In these cases only a rough idea of the general structure is needed and the number of polishing and etching operations can be reduced to a minimum. However, no specimen should have less than three polishings and etchings, and in only a very few cases should it be necessary to polish and etch more than ten times to obtain a perfectly prepared specimen. The repeated etching and polishing operation not only improves the appearance of the graphite structure, but also gives a better finish to the metallic matrix and the adoption of this type of technique is advisable in the preparation of specimens taken from nearly every type of ferrous and non-ferrous alloy.

The author thanks the Council of the British Cast Iron Research Association for permission to publish this paper.

REFERENCES.

(1) A. L. Norbury and L. W. Bolton : *Foundry Trade Journal*, 1939, vol. 41, p. 265.
(2) A. L. Norbury and L. W. Bolton : *Bulletin of the British Cast Iron Research Association*, 1928, vol. 2, p. 52.
(3) J. R. Vilella : *Metals and Alloys*, 1932, vol. 3, p. 205.
(4) H. Hanemann and A. Schrader : "Atlas Metallographicus," vol. 2. Berlin, 1937 : Gebrüder Borntrâger.
(5) H. M. Boylston : *National Metals Handbook*, 1933, p. 611.
(6) H. Nipper : *Foundry Trade Journal*, 1934, vol. 51, p. 7.
(7) "Photomicrography with the Vickers Projection Microscope," p. 26. York, 1937: Cooke, Troughton and Simms, Ltd.
(8) H. Morrogh : *Journal of the Iron and Steel Institute*, 1941, No. I., p. 207 P (this volume).

[This paper was discussed jointly with the following one by H. Morrogh on "The Metallography of Inclusions in Cast Irons and Pig Irons."]

THE METALLOGRAPHY OF INCLUSIONS IN CAST IRONS AND PIG IRONS.*

By H. MORROGH (British Cast Iron Research Association, Birmingham).

(Figs. 9 to 74—Plates XXXVIII. to XLVII.)

Summary.

A preliminary scheme of classification for inclusions in cast irons and pig irons has been developed. Using this classification, the various inclusions are dealt with under the appropriate headings. Various experiments have been performed to elucidate the nature and mode of occurrence of these particles.

The effect of pouring temperature on the morphology of manganese sulphide is discussed. Both manganese and iron sulphide were found to behave as nuclei for the formation of temper carbon in malleable iron. Manganese sulphide gives "graphite-flake-aggregate" temper carbon and iron sulphide gives spherulitic temper carbon.

A blue-pink inclusion has been observed in various cast irons containing titanium and insufficient manganese to neutralise all the sulphur as manganese sulphide. This constituent has been prepared in a number of melts and shown to be probably titanium sulphide. Two forms of the titanium sulphide inclusion occur, one allotriomorphic and one idiomorphic. The complicated optical properties of this inclusion, as revealed by the metallurgical polarising microscope, are described in detail.

The effects of test-bar diameter and titanium content on the number of titanium carbide and titanium cyano-nitride crystals have been determined by means of inclusion counts. An attempt was made to determine whether the solubility of titanium carbide in austenite could be detected by the inclusion count method.

The effect of zirconium, in amounts up to about 0·5%, on the inclusions in cast irons was studied. With increasing zirconium contents it was found that the manganese sulphide in the base iron was gradually replaced by an orange-yellow to grey inclusion. When all the manganese sulphide had been removed from the structure, blue-grey cubes of zirconium carbide appeared, which combined with the titanium carbide present to give a complex titanium-zirconium carbide. The optical properties of the orange-yellow to grey inclusion, as revealed by the polarising microscope, are given in detail. In melts carried out in a rocking arc furnace, the yield of zirconium from ferro-silicon-zirconium additions was very poor and most of the zirconium appeared to be fixed as lemon-yellow zirconium nitride. An attempt to introduce this inclusion into crucible-melted cast iron by bubbling nitrogen through the melt failed.

Very little analogy was found between the inclusions in steels and cast irons, the latter being characterised by the almost complete absence of visible oxides or silicates. In conclusion, it is suggested that the small particles referred to in the paper could be termed "minor phases" to great advantage with regard to definition.

* Received October 3, 1940.

INTRODUCTION.

THERE has been considerable discussion on the effect of inclusions in cast iron, and theories have been put forward implying that these inclusions exert effects on the structural characteristics of this alloy. The majority of these theories have been enunciated without any effort to define the precise meaning of the term "inclusions." The terms "non-metallic inclusion" and "slag inclusion" are more specific in themselves, but throughout the literature both have been used very loosely. The word "inclusion" is used here, not because it is a good one, but rather for lack of a better one and because it is the commonly accepted word for these minor structural constituents. An attempt will be made in this paper to show how misleading this nomenclature is.

With regard to theories on the effect of inclusions in cast iron and pig iron, by far the most important are those which postulate their effect on the size and type of graphite formation. The theory put forward by von Keil and his collaborators [(1)] says that the inclusions are sub-microscopic, while others do not definitely say that the inclusions are either sub-microscopic or microscopic. It is obvious that there can be no direct microscopic metallography of the sub-microscopic inclusions, although Mitsche [(2)] has examined the possibility of detecting these by means of fluorescence effects.

The process of cataloguing, recognising and tentatively naming the inclusions in plain carbon steels, alloy steels and ferro-alloys by the use of the microscope has been proceeding very rapidly, with the result that a large amount of metallographic information is available in this field. In the case of some inclusions there is a considerable analogy between steels and cast irons, but in other cases the opposite is the fact. While oxides and silicates are very important inclusions in steels, they are virtually non-existent from the point of view of the metallography of cast iron.

Bolton [(3)] has suggested that a rough classification of inclusions as foreign and inherent might be made. Foreign inclusions would be entrapped furnace slag, refractories and foundry sand, while inherent inclusions would be small particles inherent in the melting stock or deposited therein by the reaction of alloy additions with elements already in the base iron, or by the reaction of atmospheric or furnace gases with the constituents of the alloy. This preliminary classification seems logical and quite satisfactory. The metallography of the most important foreign inclusions in cast iron has been dealt with previously by the author,[(4)] and only inherent inclusions will be discussed here.

The effect of foreign inclusions on the inherent inclusions and on the structure of the metal must not, however, be overlooked. In this respect the results obtained recently by Scott and Joseph [(5)] are interesting. These investigators, using the iodine method for the determination of "non-metallic inclusions," found very much higher silica figures in sand-cast samples than in samples cast in a graphite mould. It has been noted that in the case of cast irons with border-line compositions (that is to say, those which, had they been cast in a slightly larger section, would have solidified grey or mottled), where particles of moulding sand have been included at the edge of the castings, the metal in the immediate vicinity of the inclusion tends to solidify grey. It has also been observed that the presence of such inclusions in irons containing supercooled graphite causes a slight local reversion to the normal flake graphite structure.

A large number of isolated researches have been carried out which are concerned directly and indirectly with inclusions in steels and cast irons, but so far no attempt has been made to draw together the threads of these separate and important researches in a comprehensive manner with respect to cast iron. It is proposed here to consider these researches and to correlate them, wherever possible, with results obtained by the routine microscopic examination of cast irons over a number of years.

A Preliminary Scheme of Classification for Inclusions in Cast Irons and Pig Irons.

From the points of view of their colour, morphology and chemical constitution, the inherent inclusions in cast irons and pig irons may be classified arbitrarily into a few groups which greatly facilitate the study of this problem. Such a classification is given below; this was actually drawn up before any experimental work had been done which subsequently indicated that inclusions in different groups in the present classification might be grouped together solely according to chemical constitution. However, it has been used here because it provided the basis on which the constitution and mode of formation of the inclusions were investigated. It is realised that the classification does not, in itself, solve the ultimate identity of the inclusions concerned.

Group I.—Manganese sulphide, iron sulphide and composite inclusions of these two.

Group II.—Pink Inclusions:
- (*a*) Red-pink and idiomorphic.
- (*b*) Blue-pink and allotriomorphic
- (*c*) Blue-pink and idiomorphic.

Group III.—White or faint grey inclusions:
- (*a*) Idiomorphic
- (*b*) Allotriomorphic.

Group IV.—Inclusions produced by alloying with zirconium.

Group V.—Inclusions produced by alloying with aluminium.

The various groups of inclusions are dealt with in detail in the next section. In the course of the preparation of this paper, experiments were performed to elucidate some of the problems raised by a critical consideration of the literature and by routine observations. The details and results of these experiments are also given.

Metallography of the Individual Inclusion Groups.

Group I.—Manganese Sulphide, Iron Sulphide and Composite Inclusions of these Two.

It is convenient to deal, in this section, with both iron sulphide and manganese sulphide. These inclusions are very well known, and it is generally recognised that the dove-grey idiomorphic inclusions in cast irons are manganese sulphide, and that the khaki-coloured inclusions, present in irons having low manganese contents, are iron sulphide. In commercial materials the effects of manganese and sulphur are intimately connected. Information regarding these inclusions is to be found in three lines of investigation:

(*a*) Descriptive metallography of the mode of occurrence of the inclusions;

(*b*) Study of systems relevant to these inclusions, for instance, the systems Fe–S, FeS–MnS, Fe–FeS–MnS–Mn, Fe–S–C, Fe–S–P, &c.;

(*c*) Consideration of the facts known about the desulphurising action of manganese.

Much purely metallographic work has been done on manganese and iron sulphide inclusions, equilibrium diagrams pertaining to systems connected with them have been extensively investigated, and many investigations have been performed and much theory has been propounded on the desulphurising action of manganese.

In irons which are completely free from manganese, the sulphur is in the form of iron sulphide (FeS). Additions of manganese to such irons result in the disappearance of iron sulphide, owing to the formation of manganese sulphide. The reaction may be represented by the following equation: [6, 7]

$$FeS + Mn = MnS + Fe.$$

In order to neutralise all the sulphur as manganese sulphide, it is necessary to have an excess of manganese over the theoretical percentage indicated by the formula (MnS). This fact is due to the reversibility of the reaction, as was originally suggested by Baykoff,[8] and confirmed experimentally by Herty and True: [9]

$$FeS + Mn \rightleftharpoons MnS + Fe.$$

In irons (or steels) containing insufficient manganese to neutralise the sulphur entirely as manganese sulphide, the dove-grey manganese sulphide occurs associated with the khaki iron sulphide in the form of duplex inclusions.

It has been shown by numerous investigations [6, 10, 11, 12] that additions of manganese to cast irons and steels promote the removal of sulphur, owing to the upward settling of manganese sulphide. The extent of desulphurisation was shown by Heike [13] to be proportional to the manganese content and inversely proportional to the temperature. The process of desulphurisation by manganese can be attributed to two main factors. First, additions of manganese cause the above reaction to proceed to the right and, secondly, owing to the very slight solubility of manganese sulphide in molten cast iron and its low relative density, the solid sulphide so formed rises to the surface of the metal. Correspondingly, the effect of the temperature is manifest in two ways. At higher temperatures the reaction tends to proceed more to the left, with a decrease in the concentration of manganese sulphide and an increase in the concentration of iron sulphide,[14] which is very soluble in molten cast iron. Although manganese sulphide is only slightly soluble in molten iron, its solubility does increase appreciably with increasing temperature,[15] thus limiting the extent of possible desulphurisation.

Herty and Gaines [16, 17] found that in ladles of molten iron the equilibrium between manganese and sulphur can be expressed by the product $Mn\% \times S\% = K$. The product was found to vary within the limits 0·03 at 1100° C. and 0·656 at at 1440° C.

Wohrman [18] prepared synthetic sulphide inclusions in pure iron melts. He gave an extensive and very thorough description of the morphology and mode of occurrence of iron and manganese sulphide in the samples obtained. It is interesting to note that this investigator found large, well-formed dendrites of manganese sulphide in the metal at the tops of the melts containing this inclusion.

Andrew and Binney [19] investigated the solubility of iron and manganese sulphides in solid steels. They concluded, as a result of their experiments, that iron sulphide may be retained in solution in iron by rapidly chilling the metal and that a deposition of iron sulphide occurs on reheating the alloy to a temperature of 900° C. They were unable to show manganese sulphide to be soluble in solid iron.

Andrew, Maddocks and Fowler [20] investigated the partial systems manganous-oxide/manganese-sulphide, manganous-silicate/manganese sulphide and ferrous-silicate/manganese-sulphide. The diagrams given by them for these systems are of the simple eutectiferous type and indicate manganese oxide to be appreciably soluble in solid manganese sulphide at room temperature and manganese and ferrous silicates only slightly so.

The iron-sulphur-oxygen system,[21] the iron-sulphur-carbon system [22, 23, 24] and the iron-sulphur-phosphorus system [25] have been investigated and shown to exhibit miscibility gaps in the liquid state. In the first system, the miscibility gap extends from the iron/iron-oxide side into the ferrous sulphide corner of the triangle. In the two latter systems, the miscibility gaps extend from the ferrous-sulphide/iron-carbide and from the ferrous-sulphide/iron-phosphide sides into the iron corners.

The system Fe–FeS–MnS–Mn has been investigated [22, 26, 27, 28] and equilibrium diagrams have been drawn up for this system, showing a miscibility gap which extends from the manganese/manganese-sulphide side far into the iron corner. Wentrup [29] summarised and closely analysed all the published information pertaining to this system. Additions of manganese to the iron-sulphur-carbon system were shown by Satô [30] to extend the region of liquid immiscibility and so promote desulphurisation. In his own experimental work Wentrup studied the effect of carbon, silicon and phosphorus on the extent of desulphurisation. Previously some indirect evidence on this subject had been obtained by Daeves [31] in a statistical study of silicon, phosphorus, manganese and sulphur analyses of pig iron from three blast-furnaces. He found that with increasing silicon or phosphorus contents the sulphur contents fell. Wentrup observed that an increase in the amounts of carbon, silicon or phosphorus favoured an increase in the extent of manganese desulphurisation. This investigator also pointed out the desulphurisation of pig iron by manganese to be purely a process of segregation by the crystallisation of solid sulphide from the melt.

Joseph and Holbrook [32] studied the effect of temperature on the desulphurising action of manganese in irons with various silicon contents. In presenting the data, the effect of temperature was related to the product Mn% × S%. The results indicated that the product was fairly constant at the same temperature, and, when plotted graphically against the temperature, the points so obtained fell approximately on a smooth curve. The results of Joseph and Holbrook are given in Fig. 1 and Table I.

Shibata [14] investigated the partial system FeS–MnS and produced an equilibrium diagram for this system. Very pure materials were used. It was found that the system gave a simple binary eutectic of two solid solutions at 93·5% iron sulphide and 6·3% manganese sulphide at a temperature of 1164° C. At the eutectic temperature, the iron-sulphide-rich solid solution was found to dissolve 2·0% of manganese sulphide and the manganese-sulphide-rich solid solution to dissolve 75% of iron sulphide. On cooling to room temperature, the mutual solubility of these two compounds decreased, so that the α solid solution would only dissolve 0·5% of manganese sulphide, and the β solid solution 24·0% of iron sulphide. For the melting point of iron sulphide, Shibata obtained the value 1173 ± 2° C., and for pure manganese sulphide the value 1610 ± 3° C.

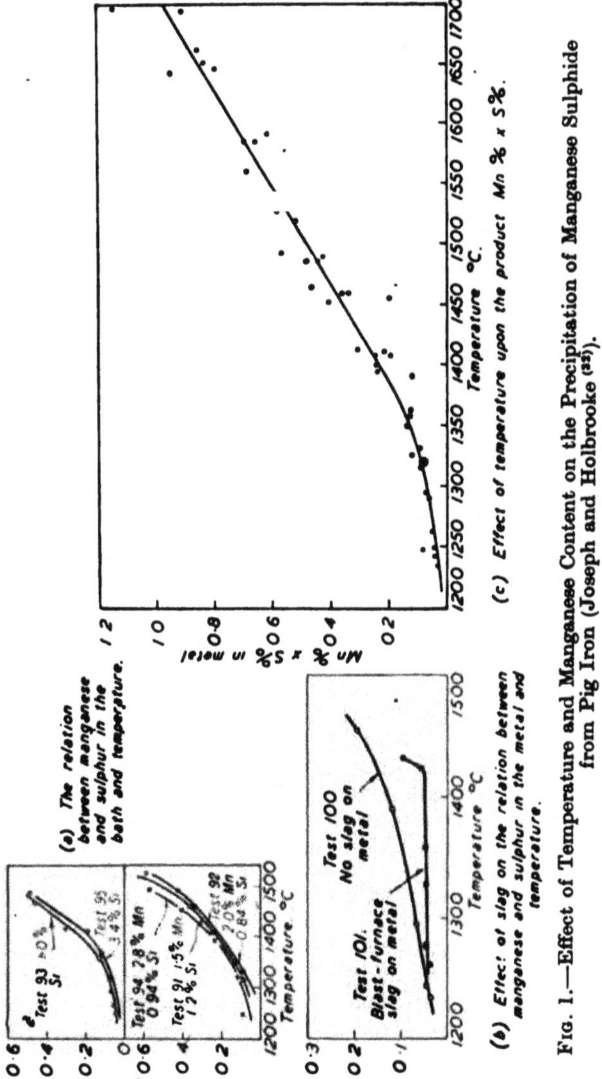

FIG. 1.—Effect of Temperature and Manganese Content on the Precipitation of Manganese Sulphide from Pig Iron (Joseph and Holbrooke [32]).

TABLE I.—*Effect of Temperature and Manganese Content on the Precipitation of Manganese Sulphide from Pig Iron* (*Joseph and Holbrook* [32]).

Sample No.	Mn. %.	S. %.	Temp. °C.	Mn% × S%.
Test 91 (*Silicon* 1·2%).				
1	1·81	0·322	1527	0·583
2	1·77	0·318	1493	0·563
3	1·63	0·216	1460	0·352
4	1·56	0·153	1408	0·238
5	1·48	0·089	1349	0·132
6	1·39	0·064	1315	0·089
Test 92 (*Silicon* 0·84%).				
1	2·13	0·199	1489	0·424
3	2·05	0·116	1399	0·238
4	1·97	0·045	1331	0·089
Test 93 (*Silicon* 1·0%).				
1	1·18	0·41	1485	0·484
2	1·01	0·30	1413	0·303
3	0·80	0·16	1358	0·128
4	0·70	0·11	1320	0·077
5	0·66	0·086	1263	0·057
6	0·66	0·067	1249	0·044
Test 94 (*Silicon* 0·90%).				
1	2·90	0·139	1452	0·403
2	2·84	0·084	1395	0·238
3	2·74	0·042	1325	0·115
4	2·70	0·032	1247	0·086

Sample No.	Mn. %.	S. %.	Temp. °C.	Mn% × S%.
Test 95 (*Silicon* 3·4%).				
1	1·31	0·355	1465	0·465
2	1·05	0·182	1408	0·191
3	0·96	0·125	1363	0·120
4	0·89	0·087	1320	0·077
5	0·86	0·072	1291	0·062
Test 100 (*No Slag on Metal*).*				
1	1·61	0·12	1455	0·193
2	1·64	0·07	1390	0·116
3	1·61	0·04	1294	0·065
4	1·58	0·027	1243	0·043
5	1·57	0·021	1234	0·033
Ingot	1·56	0·032	...	0·050
Test 101 (*Metal covered with Typical Blast-Furnace Slag*).*				
1	1·66	0·058	1432	0·096
2	1·64	0·035	1424	0·057
3	1·59	0·030	1359	0·048
4	1·62	0·027	1327	0·044
5	1·65	0·027	1277	0·045
6	1·62	0·023	1261	0·037
Ingot	1·52	0·024	...	0·036
	1·52	0·023	...	0·035

* Temperature of bath was measured with a noble-metal thermocouple.

Whiteley [33] heat-treated and quenched samples of steel containing 0·19% of manganese and 0·019% of sulphur. The original steel showed both iron and manganese sulphide, but the heat-treatment experiments indicated that at temperatures below 1000° C. the iron sulphide was gradually dissolved in the manganese sulphide until by heat-treating at 1150° C. for 5 min. the iron sulphide disappeared completely. The results obtained by Whiteley were in agreement with the requirements of the equilibrium diagram for the system FeS–MnS proposed by Shibata. This diagram indicated, as pointed out previously, that the solubility of solid iron sulphide in solid manganese sulphide decreases rapidly with fall in temperature. By a further experiment Whiteley was able to show that at temperatures above the solidus the sulphide consisted predominantly of FeS. The appearance of iron sulphide at these temperatures is due to the fact that the reaction $FeS + Mn \rightleftharpoons MnS + Fe$, will tend to proceed towards the left at higher temperatures. By very drastic quenching from the liquid state, this investigator was able also to show that iron sulphide was present in liquid steel containing 0·06% of sulphur and 0·6% of manganese.

When examined with a dry objective, manganese sulphide presents the typical uniform dove-grey colour. When this constituent is examined with an oil-immersion or a mono-bromo-naphthalene-immersion objective, it appears much darker, with a definite greenish coloration, and occasionally exhibits translucency. This latter property is manifested by the occurrence of light greenish-yellow spots and sets of coloured wavy lines, suggestive of interference phenomena, on the surface of the inclusions. A typical example of the greenish-yellow spots is illustrated in Fig. 9, while Fig. 10 illustrates the occurrence of the coloured wavy lines. In cases where the manganese/sulphur ratio is low and the sulphide inclusions consist of manganese sulphide containing dissolved iron sulphide, the colour under oil-immersion objectives is light dove-grey, without any greenish appearance, and the translucency effects are not marked, and in some cases are not even detectable.

Manganese sulphide has a cubic crystal structure,[34] and, as would be expected, inclusions of this compound remain dark in all positions of the stage when they are examined under polarised light between crossed nicols.[35] When, however, a crystal showing light green spots or a set of wavy lines is examined in this manner, a greenish coloration is observed which persists in all positions of the stage. This coloration makes the outline of the inclusion somewhat indefinite. The fact that it appears in all positions of the stage indicates that the brightness is not an effect due to anisotropy. This is in agreement with the crystal structure of manganese sulphide. When the inclusion is spherical or approximately spherical, and also shows these light green spots under normal bright-field vertical illumination, then examination under polarised light between crossed nicols reveals an optical cross and greenish concentric rings. The more nearly spherical the inclusion is, the more definite does this optical formation become. A nearly spherical manganese sulphide inclusion is shown in Fig. 11. This inclusion had a light green coloration in its centre under bright-field vertical illumination, and when examined under polarised light between crossed nicols the optical cross and concentric rings were clearly developed. This latter point is illustrated in Fig. 12. To the left of the spherical inclusion is one more irregular in outline, but also showing a greenish coloration. In the micrograph taken under polarised light, it is seen that the optical-cross and concentric-ring formations are not so clearly defined in this inclusion as in the spherical one.

Dayton[36] showed that spherical, transparent isotropic inclusions manifest the optical-cross and concentric-ring phenomena when examined under polarised light between crossed nicols. The position of the dark cross remains constant on rotating the stage of the microscope, the arms of the cross being parallel to the planes of polarisation of the incident light and the analyser nicol. Further, Dayton showed that these phenomena are due to the production of circularly polarised light by reflection in isotropic media. These facts lead to the conclusion that the above observations indicate manganese sulphide to be transparent to green light. It would thus appear that the greenish spots seen under ordinary vertical illumination are due to the fact that some of the incident light is transmitted by the inclusion, reflected at the inclusion-metal interface, and transmitted back through the inclusion to the microscope objective. The transmission of light by the inclusion results in the absorption of all wavelengths other than the green or yellow-green. The green coloration seen under polarised light between crossed nicols is due to the production of elliptically polarised light by reflection at the base of the inclusion. The transparency of manganese sulphide also explains the wavy lines. These lines exhibit all the colours of the spectrum, and it would seem that they are interference bands produced by the interference of light reflected from the surface of the inclusion with light reflected from the base of the inclusion. It is obvious that the inclusion must be very thin for this to occur and, in general, it must be wedge-shaped. Fig. 13 illustrates this latter point. The manganese sulphide crystal showing the interference lines appears to be a thin wedge which increases in thickness from right to left.

From the review of the literature on the mode of occurrence of manganese sulphide and on the desulphurisation of iron by manganese, it is seen that, in general, the following facts emerge: Manganese sulphide is only slightly soluble in molten cast iron, the solubility increasing with increasing temperature. At any given temperature, any excess manganese sulphide will tend to rise and segregate at the surface of the molten metal. On cooling from such a specified temperature, the dissolved manganese sulphide will be deposited in a dendritic form. These points are entirely confirmed by the observed morphology of manganese sulphide in cast irons.

In cast irons poured at fairly low temperatures (<1350° C.), manganese sulphide occurs in typical compact idiomorphic crystals, whereas in irons poured at high temperatures (>1400° C.) the manganese sulphide tends to occur in recognisable dendritic forms in the case of high-sulphur irons, and as "anchor-shaped" crystals in low-sulphur irons. Portevin and Castro [37] have observed this latter form in cast irons. A typical example of an anchor-shaped manganese sulphide dendrite is shown in Fig. 14, and a well-formed manganese sulphide dendrite in a hot-poured high-sulphur iron is illustrated in Fig. 15.

Norbury [38] has shown that the primary microstructure of cast iron is considerably affected by the pouring temperature. For instance, in the case of hypo-eutectic cast irons, the primary austenite dendrites decrease in length and become more compact with decreasing pouring temperature. It is obvious from the above observations that manganese sulphide behaves in a similar manner (with regard to pouring temperature), with the one difference that the manganese sulphide dendrites formed at high temperatures will tend to float up to the surface of the metal, by virtue of their low relative density. The manifestly dendritic form of manganese sulphide in hot-poured samples is the exact analogue of the long, well-formed, primary austenite dendrites, while the small, typically idiomorphic crystals correspond to the short primary austenite dendrites. The fact that manganese sulphide exhibits the same morphological changes with decreasing pouring temperature as do the primary metallic dendrites, still further confirms that manganese sulphide is soluble in molten cast iron, from which it is deposited in the solid form as cooling proceeds.

Occasionally, small spherical holes filled with metal are observed in the centres of manganese sulphide crystals. This fact has been recorded photomicrographically by Allen.[39] The examination of a large number of samples of cast iron containing well-formed manganese sulphide crystals has suggested quite a simple explanation for this curious morphology. This is illustrated in Figs. 16(*a–d*), which were taken from the same sample as that used for Fig. 15. Manganese sulphide begins to grow in the manner characteristic of dendrites having a cubic crystal structure. The initial dendrite has six outgrowing branches radiating from a common centre. This is shown in Fig. 16(*a*). At first these small manganese sulphide dendrites are rounded, but with a fall in temperature, the tips of the dendrite branches gradually take on an "arrow-head" shape (Fig. 16(*b*)). The next stage of growth results in the tips of the arrow-heads becoming angular, and the arrow-heads on different dendrite branches grow out to meet each other (Fig. 16(*c*)). Finally, the joining up of the dendrite branches is completed at the extreme ends of the dendrites, and a perfect idiomorphic crystal is obtained with small holes containing entrapped metal (Fig. 16(*d*)). The number of such spheroids of entrapped metal which are visible must obviously depend on the particular section which the surface of the micro-specimen cuts through the manganese sulphide crystal.

Very rarely, minute black specks have been observed in manganese sulphide, but they were so small that it was impossible to say definitely whether they were holes or small inclusions, even by the use of the highest available magnifications. However, on one occasion, samples of manganese sulphide containing large, well-developed dark crystals were obtained. The melt in question was carried out as follows: 195 g. of Swedish white iron and 6½ g. of 80% ferromanganese were melted in an alundum crucible, using a high-frequency induction furnace; the metal was superheated as far as possible, then 3 g. of iron sulphide were added and the furnace was shut off, the crucible and its contents being allowed to cool in air. The resultant ingot was sectioned from top to bottom for microscopic examination. The original purpose of the experiment was to obtain some well-formed manganese sulphide dendrites. However, not only were dendrites present, but also large spheroids of this compound, indicating that the composition and temperature of superheating were such that the region of liquid immiscibility had been reached. A large number of idiomorphic and dendritic manganese sulphide crystals either contained, or were associated with, quite large crystals of a dark, grey-green compound, although the spheroids of manganese sulphide did not show this. Generally, these dark crystals were cubic in outline and they appeared to be much harder than manganese sulphide, above which they stood in considerable relief. A typical example of this inclusion is shown in Fig. 17. Occasionally the inclusion occurred in an elongated form, and when the length of the manganese sulphide dendrite happened to be sectioned by the polished surface, it was seen to occur in a particularly complex form along the dendrite. Figs. 18 and 19 illustrate these two latter modes of occurrence.

The chemical analysis of the sample gave 2·41% of manganese and 0·40% of sulphur. A comparison of the morphology of the dark inclusion with that of the manganese sulphide suggested that the cubic appearance in a compact idiomorphic manganese sulphide crystal was due to the fact that, actually, a cross-section of a manganese sulphide dendrite was being examined, and that the actual form of the dark inclusion was acicular, running the length of the manganese sulphide dendrites and having a cubic cross-section.

Careful examination of the inclusion revealed it to have a faint greenish translucency, which was confirmed by its behaviour under polarised light between crossed nicols, when a bright green coloration was obtained, persisting in all positions of the stage. The conclusions to be drawn from the polarised-light examination are that the inclusion is either completely isotropic or cubic, and that it is at any rate partially transparent to green light. The observations made so far would seem to indicate this new inclusion to be probably manganese oxide, containing, perhaps, a small amount of iron oxide, FeO. Portevin and Castro [37] have reported that the addition of manganese (of the order of 1%) to oxidised iron results in the double-oxide inclusions becoming darker, and with high manganese contents the colour changes to green, with a high transparency.

In whiteheart malleable irons, where, as a rule, the manganese/sulphur ratio is low, iron sulphide is frequently found alone or associated with varying amounts of manganese sulphide. The line of demarcation between the two constituents is usually very sharp and straight, and sometimes iron sulphide occurs in a Widmanstätten formation in the manganese sulphide. These facts seem to indicate that either the iron sulphide or the manganese sulphide is formed in the solid state from either a manganese-sulphide-rich solid solution in the first case, or an iron-sulphide-rich solid solution in the second case. Such behaviour is, of course, indicated by the accepted equilibrium diagram for the partial system FeS–MnS, and also by the experiments of Whiteley, which have already been referred to. Further evidence on this point was obtained in the following manner: Pieces of white iron (used in the manufacture of whiteheart malleable iron) containing 0·10% of manganese and 0·187% of sulphur, were heated for 3 hr. at temperatures ranging from 750° to 1000° C. and quenched in water. The as-cast structure had fairly large amounts of iron sulphide with a small amount of manganese sulphide; the samples quenched from 950° and 1000° C. showed no iron sulphide, and those quenched from 900°, 850° and 800° C. contained only traces of iron sulphide, while the sample quenched from 750° C. was very similar to the as-cast sample. It is interesting to remember this point in connection with the malleabilising of white iron. Although an iron may show iron sulphide in the as-cast state, it does not necessarily mean that free iron sulphide will be present in the iron at the annealing temperature.

Iron sulphide has a hexagonal crystal structure [34] and, as would be expected, when examined under polarised light between crossed nicols it lights up four times for a complete revolution of the stage. This method of examination can be used very effectively to distinguish between iron sulphide and manganese sulphide, or between iron sulphide and manganese sulphide rich in iron sulphide. A typical example of a duplex inclusion of iron and manganese sulphide is shown in Figs. 20 and 21. The effect of the annealing on the inclusions in irons for malleablising is to spheroidise them. This is particularly noticeable in irons containing the duplex sulphide inclusions.

An improved technique has been developed for the preparation of cast iron and malleable iron microsections, which enables the actual internal structure of the graphite to be observed.* The application of this technique to malleable iron samples has revealed that at least two different forms of temper-carbon nodules can exist. Generally speaking, the temper carbon occurring in whiteheart malleable iron is in a compact spherulitic form, in which the graphite crystallites have their basal planes orientated at right angles to the sphere of which the temper-carbon nodule can be considered to be composed. A typical example of this type of temper carbon is illustrated in Fig. 22. The "fluffy" graphite around the main body of the nodule is the hyper-eutectoid graphite, the amount and "fluffiness" of which depend on the annealing temperature, the stability of the cementite and the rate of cooling given to the sample. In blackheart malleable irons it was observed that the temper carbon generally existed in a form which, when examined by the improved technique, was revealed to be virtually

* The author's paper describing this method of preparing samples of cast and malleable irons for micro-examination is on p. 195 P of this volume.

an aggregation of small graphite flakes. At first it was thought that the two different types of temper-carbon nodule were the result of the two different types of annealing used in the production of whiteheart and blackheart malleable iron. This was disproved, however, when a piece of whiteheart malleable iron was discovered with the "graphite-flake-aggregate" type of temper carbon (Fig. 23). Examination of the sample revealed that the manganese content of the iron was sufficient to balance the sulphur content and that only manganese sulphide was present. Thus it became apparent that the real causes of the production of the two different forms of temper carbon lay, not in the different annealing treatments which blackheart and whiteheart malleable irons receive, but rather in the fact that, in general, whiteheart malleable irons contain iron sulphide, and blackheart malleable irons contain only manganese sulphide. This became more apparent when the mode of occurrence of the temper carbon was more closely examined. In whiteheart malleable irons containing iron sulphide, it was found that a large number of temper-carbon nodules had nuclei of iron sulphide. A typical example of this is illustrated in Fig. 24. So far, it has not been possible to determine whether every spherulitic temper-carbon nodule contains an iron sulphide nucleus, since it appears that the nucleus will only be visible when the temper-carbon nodule is so sectioned that the polished metal surface cuts through the centre of the nodule. When the white iron used for the manufacture of whiteheart malleable is poured at a fairly high temperature, the resultant malleable iron frequently contains groups of spherulitic temper carbon joined together, forming an almost continuous and straight band of graphite. A typical example of this is illustrated in Fig. 25. The explanation of this phenomenon is fairly obvious when it is remembered that iron sulphide can act as a nucleus for the temper carbon. In hot-poured white irons, the primary dendrites will tend to be long, and in thin sections (such as are frequent in the malleable-iron industry) columnar. As a result of this, the extra-dendritic or eutectic regions will be likewise striated or columnar. Now, iron sulphide segregates to the eutectic regions of white irons and so it will be arranged more or less in straight lines radiating approximately at right angles to the cooling surface. These iron sulphide inclusions will then act as nuclei for the formation of spherulitic temper carbon, which, if the inclusions are sufficiently close together, will give a continuous band of graphite. It is important to observe at this stage that only a few of the available iron sulphide nuclei actually behave as nuclei for the temper-carbon nodules. There are always far more iron sulphide inclusions than temper-carbon nodules. The question becomes further complicated when it is remembered that, although iron sulphide may be present as the free compound in the iron at room temperature, it is also possible that it is dissolved to some extent in the manganese sulphide at the annealing temperature. In this fact may lie the explanation of the observation that only a few of the iron sulphide inclusions behave as nuclei. In the case of irons containing duplex sulphide inclusions, some of the iron sulphide will tend to dissolve in the manganese sulphide, giving an apparently homogeneous type of inclusion. It is probable that with this state of affairs the manganese sulphide inclusions rich in iron sulphide do not act as nuclei for the graphite, but the relatively small amount of iron sulphide remaining undissolved does act in this manner, and so at room temperature, when the iron sulphide is reprecipitated, only few of the visible iron sulphide inclusions appear to have nucleated the graphite.

Schwartz [(40)] has reported and recorded photomicrographically the nucleisation of temper carbon by manganese sulphide. Unfortunately, the micrograph which he reproduced does not reveal the structure of the temper carbon. This investigator claims that graphitisation usually begins at a cementite/solid-solution interface.[(41)] Although no actual analyses are given, these remarks presumably apply to irons in which the sulphur content is balanced by the manganese content. The use of the improved polishing technique has, however, made it possible to observe the process of graphite-flake-aggregate temper carbon forming around manganese sulphide crystals.

A piece of a 1·2-in. bar was annealed at a temperature of 1050° C. for 2 hr. and then air-cooled. The analysis of the bar in the as-cast condition was as follows:

Total carbon . .	1·92%	Sulphur . .	0·060%
Silicon . . .	0·77%	Phosphorus . .	0·048%
Manganese . .	0·38%		

Figs. 26 and 27 illustrate a typical graphite formation around manganese sulphide in this sample. The temper-carbon structure is, in this case, very different from the spherulitic structure. Fig. 27, taken under polarised light between crossed nicols, reveals the outlines of the numerous graphite crystallites, which, under normal vertical illumination, are not differentiated.

An examination of the etched structure of the annealed sample showed that a large number of the manganese sulphide crystals surrounded by graphite were actually situated in a eutectic cementite region. A typical example of this is illustrated in Fig. 28. It was also found that a considerable amount of the hypereutectoid cementite had apparently graphitised *in situ* without the presence of manganese sulphide (Fig. 29). Generally, the hypereutectoid graphite so formed was continuous with its cementite counterpart.

The joining up of several adjacent flake-graphite-aggregate-type temper-carbon nodules has been observed in blackheart malleable irons. Although this phenomenon is apparently analogous to that occurring with the spherulitic temper-carbon nodules, it has not, so far, been possible to demonstrate that it is caused in the same manner. A typical example of a continuous band of such temper carbon is illustrated in Fig. 30.

From the above remarks it is seen that the question of the nucleisation of graphite by manganese sulphide is more complex than that by iron sulphide. As Schwartz [42] has pointed out, the number of nuclei present for the formation of temper carbon is greatly affected by variations in other conditions, and by the introduction of a quenching treatment prior to the annealing. However, as far as is known at present, these facts only apply to irons in which the manganese is balanced by the sulphur, or in which no sulphides other than manganese sulphide exist. From the observations made above, the one certain and significant fact emerges, that malleable irons made from white irons containing iron sulphide, or a mixture of iron and manganese sulphide, have a very different form of temper carbon from irons containing only manganese sulphide. The fundamental explanation of this behaviour is still obscure, the more so since the presence or absence of these inclusions is not the sole factor affecting the number of graphite nuclei formed. Nevertheless, inclusions of this nature obviously affect the graphitisation of white iron to a great extent, and it may be important to try to reconcile the effects of manganese sulphide and iron sulphide with their different crystal structures.

The presence of iron sulphide in grey or mottled irons of fairly large section frequently induces some very characteristic structural features in the pearlitic matrix. It is not necessary for all the sulphur to be in this form—a trace of iron sulphide will suffice. The structural characteristic referred to is the occurrence throughout the section of very coarsely laminated pearlite along with very fine pearlite, which is sometimes in a network formation. A typical example of this is illustrated in Fig. 31. The sample from which this micrograph was taken had the following analysis :

Total carbon . .	3·93%	Sulphur . .	0·145%
Silicon	0·88%	Phosphorus . .	0·03%
Manganese . .	0·30%		

The general structure of the sample consisted of a mixture of medium-sized flake graphite and supercooled graphite in a matrix of pearlite, with a small amount of eutectic cementite. Both manganese and iron sulphides were present. The section of the sample concerned was approximately 2 × 2 in. Examination at a high magnification revealed that fine pearlite usually radiated from a very fine band of hypereutectoid cementite (Fig. 32).

This type of structure also occurs to a different degree in white-heart malleable irons which contain an unusually high sulphur/manganese ratio. In spite of the fact that in these malleable irons the eutectic cementite has broken down, the hypereutectoid cementite, remaining dissolved in the austenite at the annealing temperature, appears to be exceptionally stable, with the result that it is deposited at the grain boundaries in a network formation. Pearlite forming adjacent to this hypereutectoid cementite exists in a very fine form, while that towards the centres of the grains occurs in a very coarse and abnormally divorced form. A typical example of this structure in a whiteheart malleable iron is illustrated in Figs. 33 and 34. Such a structure in malleable iron is always associated with inferior mechanical properties. At the moment it is impossible to tell whether the occurrence of these fine and coarse forms of pearlite is the direct result of the presence of iron sulphide or an indirect result brought about by the stabilising effect of the iron sulphide on the hypo-eutectoid cementite. It is certain, however, that this structural characteristic can be used as a preliminary indication of the presence of iron sulphide in any particular sample and of its tendency to throw a "sulphur chill."

Group II.—Pink Inclusions.

(a) *Red-Pink and Idiomorphic Inclusions.*

This group comprises the inclusions usually referred to in the literature as titanium cyano-nitrides. These occur in nearly all cast irons in the form of small cubes. They are considerably harder than the metallic matrix and tend, usually, to stand up in appreciable relief. Whether viewed under dry or immersion objectives, they present brilliant surfaces. With regard to the occurrence of this constituent, it can be said that in cast irons it shows a marked tendency to occur associated with manganese sulphide in the extra-dendritic regions of the structure. This may be due to the fact that it forms from the melt in a similar manner to manganese sulphide. Titanium cyano-nitride is shown associated with manganese sulphide in Fig. 35. This titanium inclusion has also been observed in association with graphite. An example of this is illustrated in Fig. 36.

The term "titanium cyano-nitride" does not necessarily signify a definite chemical compound, but rather a solid solution of titanium carbide in titanium nitride, and so the ratio of titanium carbide to titanium nitride does not conform to a definite stoichiometrical relationship. The partial system titanium-carbide/titanium-nitride has been studied [43] and shown to yield a continuous series of solid solutions. These facts are in agreement with the observation that the colour of titanium cyano-nitrides in cast iron vary in colour from deep salmon-pink to very faint pink. It is significant to note that the faint-pink cyano-nitrides usually occur with relatively high titanium contents and consequently with relatively large amounts of titanium carbide.

Both titanium carbide and titanium nitride have cubic crystal structures, and, as would be expected, titanium cyano-nitride inclusions appear to be optically isotropic. Actually, when examined under polarised light between crossed nicols, they show a slight red coloration (particularly when an oil-immersion objective is used), which persists in all positions of the stage. Generally, they also show elliptical polarisation effects at their edges, owing to the fact that they tend to stand up in relief from the metallic matrix.

(b) Blue-Pink and Allotriomorphic Inclusions, and (c) Blue-Pink and Idiomorphic Inclusions.

Quite frequently, allotriomorphic pink inclusions, quite different in colour and morphology from titanium cyano-nitride, have been observed in certain pig irons and cast irons. These inclusions, while being predominantly pink, were observed to have a faint bluish tinge, in contrast to the definite red appearance of the titanium cyano-nitrides. Again, examination between crossed nicols under polarised light revealed these inclusions to be anisotropic—titanium cyano-nitride is isotropic. They stand up in relief from the metallic matrix, and on this account it was difficult, at first, to detect any optical anisotropy, owing to the elliptically polarised light produced by reflection at large angles of azimuth at the curved edges of the inclusions. (A full description of their colour and optical properties is given later in this paper when experiments which produced them synthetically are described.) The analyses of these samples were then carefully examined and it was found that they all had the following characteristics: A low manganese/sulphur ratio and small amounts of titanium. In every case the manganese content was insufficient to balance the sulphur content according to the formula: [44]

$$\text{Mn}\ \% = (1{\cdot}7 \times \text{S}\%) + 0{\cdot}3.$$

An example of a group of these inclusions is shown in Fig. 37. This micrograph was taken from a pig iron having the following analysis:

Total carbon . .	3·27%	Phosphorus . .	1·42%
Silicon . . .	2·86%	Titanium . .	0·21%
Manganese . .	0·29%	Vanadium . .	0·04%
Sulphur . .	0·026%		

Inclusions of this type have been classified as those of Group II.(*b*).

Also present in this sample were a number of perfectly idiomorphic inclusions of exactly the same colour and optical properties as the allotriomorphic blue-pink inclusions just described. These inclusions have been classified here as those of Group II.(*c*). An example of the idiomorphic blue-pink inclusions is shown in Figs. 38 and 39.

With regard to the allotriomorphic blue-pink inclusions, it was observed that they occurred in a characteristic formation consisting of groups of parallel lamellæ, with each lamella tending to become bulbous at the extreme ends. This morphology is illustrated

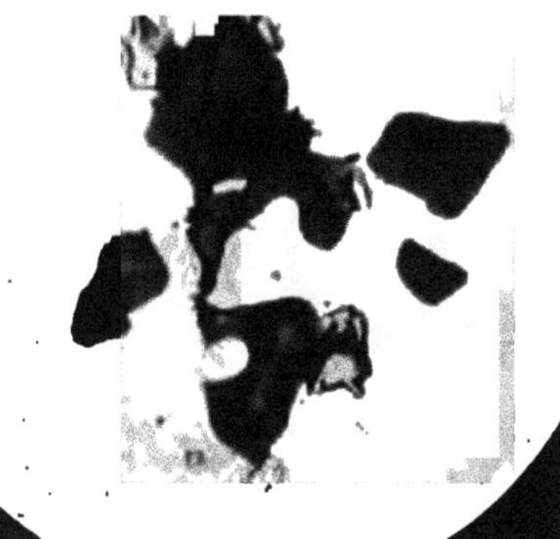

Fig. 9.—Manganese Sulphide (associated with titanium carbide) with light green spots. 2-mm. oil-immersion objective. Unetched. × 1200.

Fig. 10.—Manganese Sulphide, showing wavy interference lines. 2-mm. oil-immersion objective. Unetched. × 1000.

Fig. 11.—Manganese Sulphides showing light green colorations. 2-mm. oil immersion objective. Unetched. × 1200.

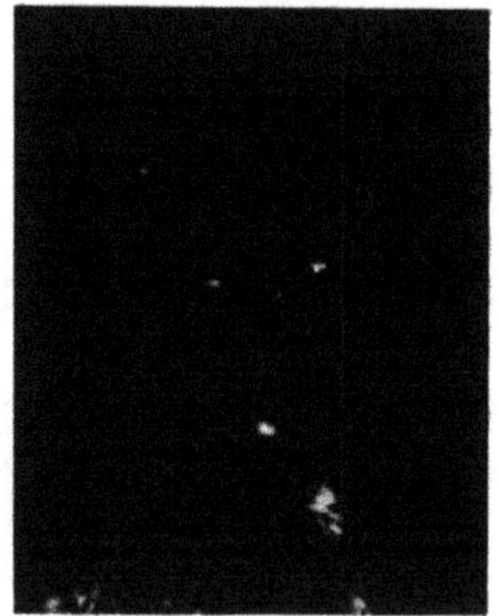

Fig. 12.—Same Spot as Fig. 11 under polarised light between crossed nicols.

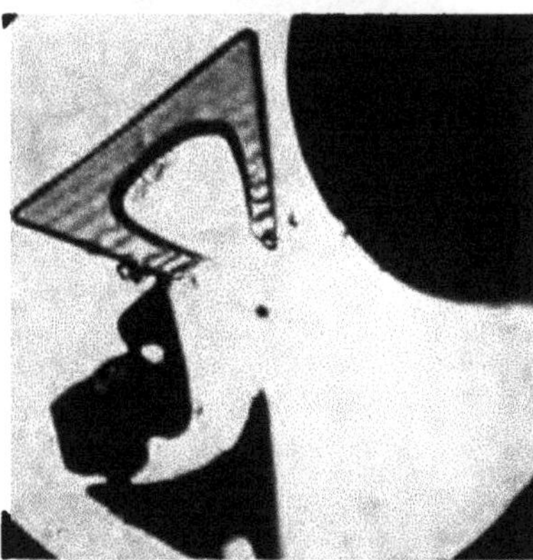

Fig. 13.—Thin Wedge of Manganese Sulphide with interference lines. 2-mm. oil-immersion objective Unetched. × 1200.

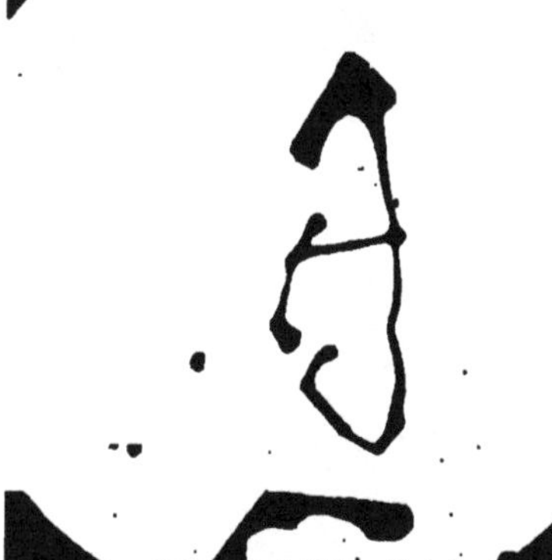

Fig. 14. "Anchor-Shaped" Manganese Sulphide Dendrite in low-sulphur, hot-poured cast iron. 2-mm. oil-immersion objective. Unetched. × 1200.

(Micrographs reduced to four-fifths linear in reproduction.)

[Morrogh.
[To face p. 224 P.

FIG. 15.—Manganese Sulphide Dendrite in sample containing 0·280% S and 0·83% Mn. Pouring temperature approx. 1600° C. 4-mm. objective. Unetched. × 500.

FIG. 17.—Manganese Sulphide containing dark grey-green cubic inclusions. 2-mm. oil-immersion objective. Unetched. × 1200.

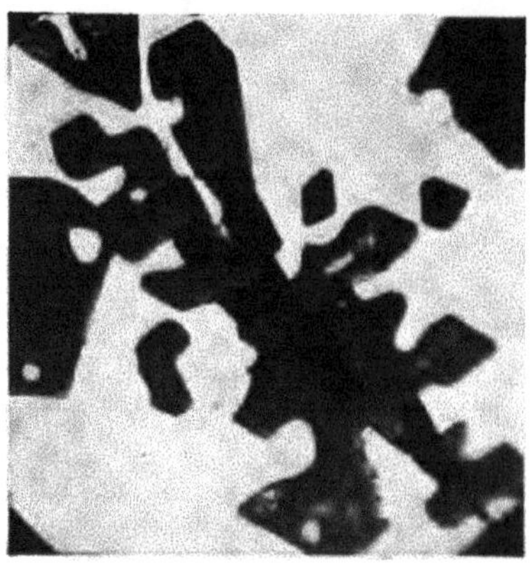

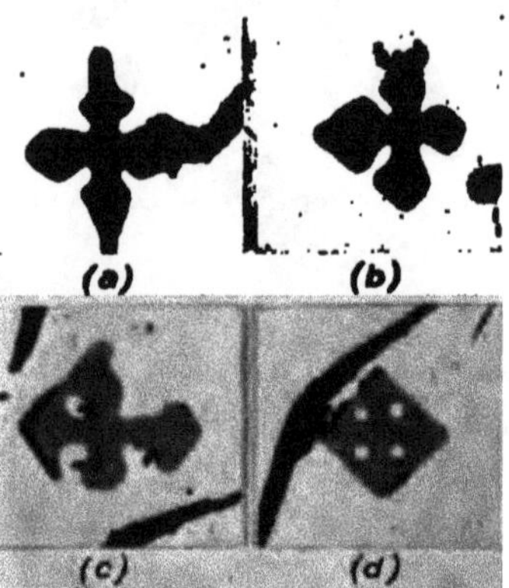

FIG. 16.—Mode of Occurrence and Formation of Small Circular Holes in Manganese Sulphide Crystals. 4-mm. objective. Unetched. × 500.

FIG. 18.—Elongated Dark Grey-Green Inclusion in manganese sulphide. 2-mm. oil-immersion objective. Unetched. × 1200.

FIG. 20.—Duplex Inclusion of Iron and Manganese Sulphide in whiteheart malleable iron. Note sharp line of demarkation. 2-mm. oil-immersion objective. Unetched. × 1200.

FIG. 22.—Spherulitic Temper-Carbon Nodule in whiteheart malleable iron containing free iron sulphide. 2-mm. oil-immersion objective. Etched in picric acid. × 1000.

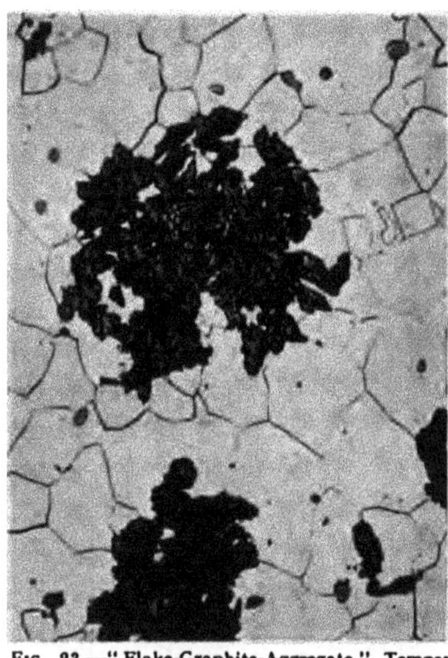

FIG. 23.—" Flake-Graphite-Aggregate " Temper Carbon in whiteheart malleable iron containing sufficient manganese to balance the sulphur content. 8-mm. dry objective, N.A. 83. Etched in 5% alcoholic nitric acid. × 300.

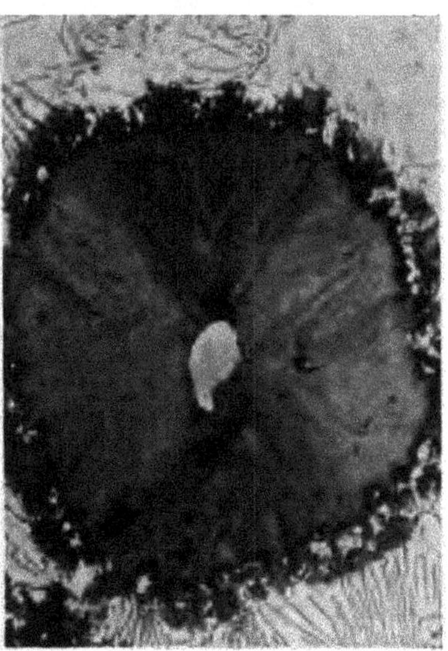

FIG. 24.—Spherulitic Temper Carbon Nodule containing a nucleus of iron sulphide. 2-mm. oil-immersion objective. Etched in picric acid. × 1000.

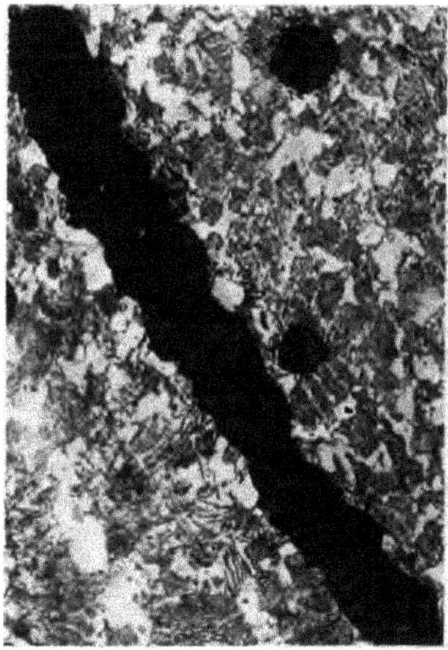

FIG. 25.—Group of Spherulitic Temper-Carbon Nodules joined together owing to close proximity of their iron sulphide nuclei. 16-mm. dry objective. Etched in picric acid. × 150.

(Micrographs reduced to four-fifths linear in reproduction.)

[Morrogh.

Fig. 26.—Temper Carbon forming round a crystal of manganese sulphide. 2-mm. oil-immersion objective. Unetched. × 1200.

Fig. 27.—Same Spot as Fig. 26 under polarised light between crossed nicols.

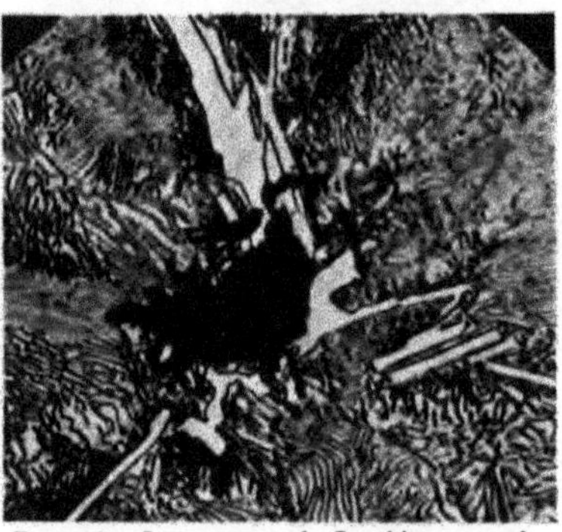

Fig. 28.—Occurrence of Graphite around a Manganese Sulphide Crystal in eutectic cementite area. 2-mm. oil-immersion objective. Etched in picric acid. × 1200.

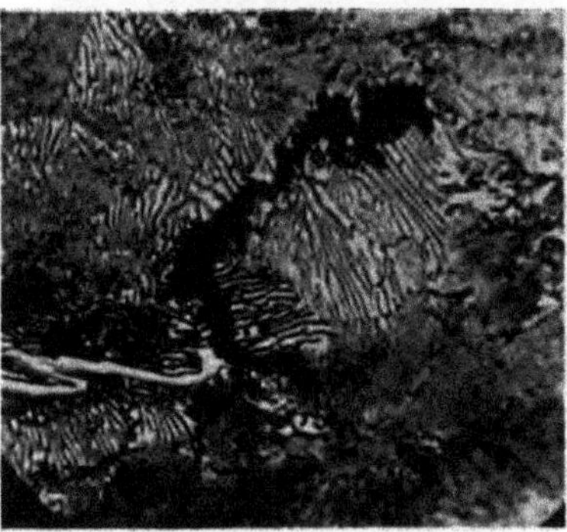

Fig. 29.—Graphitisation of Hypereutectoid Cementite in absence of manganese sulphide. 2-mm. oil-immersion objective. Etched in picric acid. × 1200.

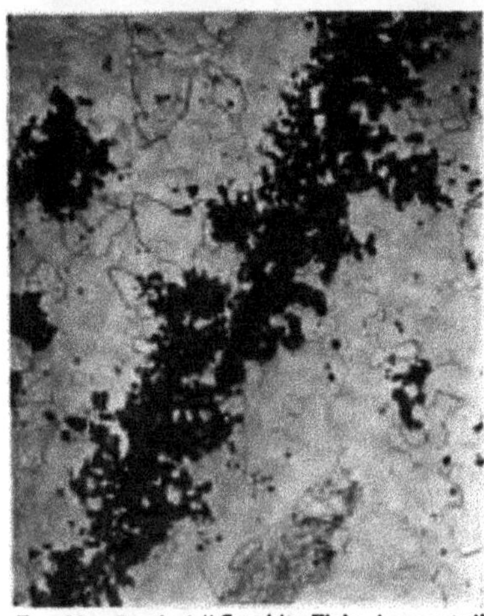

Fig 30.—Band of "Graphite-Flake-Aggregate" Temper-Carbon Nodules in sample of blackheart malleable cast iron. 16-mm. dry objective. Etched in 5% nitric acid in alcohol. × 150.

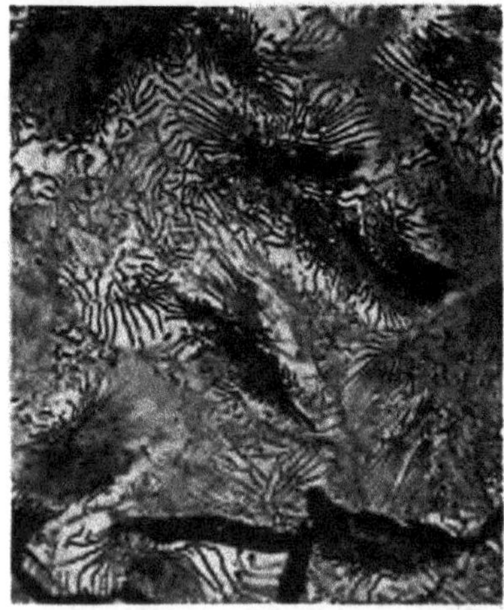

Fig. 31.—Intimate Occurrence of Fine and Coarse Pearlite in iron containing iron sulphide. Fine pearlite is the darkly etched constituent. 4-mm. dry objective. Etched in picric acid. × 600.

(Micrographs reduced to four-fifths linear in reproduction.)

[*Morrogh.*

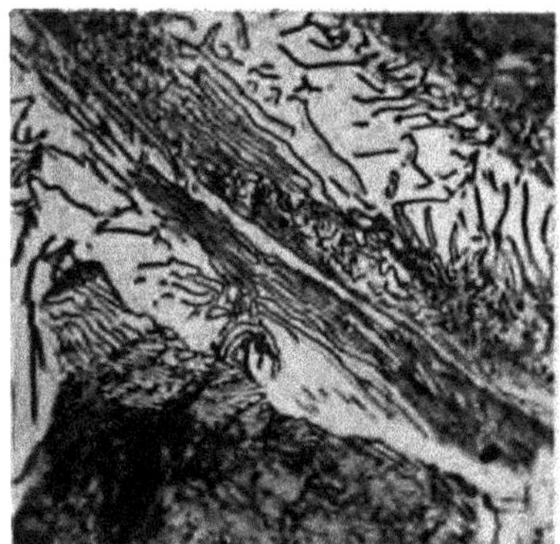

FIG. 32.—Fine Pearlite Area at high magnification. Thin band of hypereutectoid cementite running through fine pearlite area. 2-mm. oil-immersion objective. Etched in picric acid. × 2000.

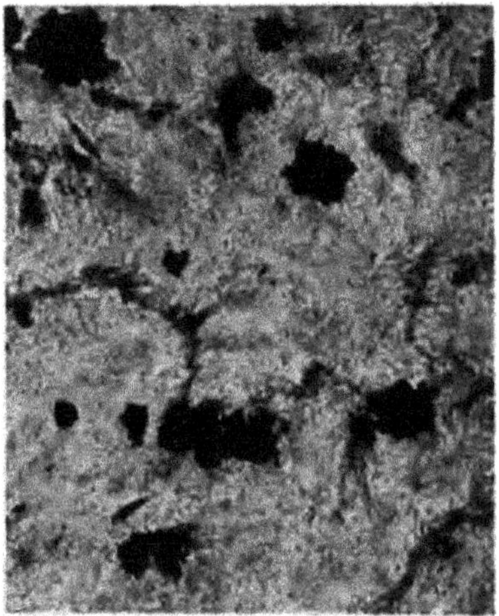

FIG. 33.– Network of very Fine Pearlite surrounding coarsely lamellar pearlite. Fine pearlite is the dark-etching constituent. 16-mm. objective. Etched in picric acid. × 60.

FIG. 34.—Area of Fine Pearlite in Fig. 33 at higher magnification. Hypereutectoid cementite revealed. 4-mm. dry objective. Etched in picric acid. × 600.

FIG. 35.—Manganese Sulphide associated with titanium cyano-nitride. × 1200.

FIG. 36.—Small Titanium Cyano-Nitride Cube

Fig. 38. Idiomorphic Blue-Pink Inclusion in pig iron. Inclusion on left is an allotriomorphic inclusion of the same colour, &c. 2-mm. oil-immersion objective. Unetched. × 1200.

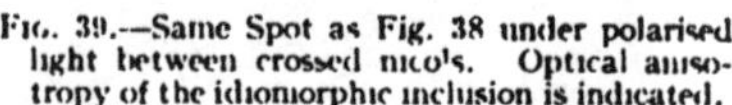

Fig. 39.—Same Spot as Fig. 38 under polarised light between crossed nicols. Optical anisotropy of the idiomorphic inclusion is indicated.

Fig. 40. Group of Blue-Pink Allotriomorphic Inclusions in blast furnace bear. 8 mm dry objective. Unetched × 250.

Fig. 41 Discontinuity of Allotriomorphic Lamellar Inclusions 2-mm. oil-immersion objective. Unetched × 2000.

Fig. 42.—Group of Blue-Pink Inclusions associated with small amount of grey khaki iron *sulphide* containing titanium. 2 mm. oil *immersion objective* *Unetched.* × 1200.

Fig. 43. Blue-Pink Inclusion surrounded by large amount of grey-khaki constituent. 2-mm. oil-immersion objective. Unetched. × 1200.

(*Micrographs* reduced to four-fifths linear in reproduction.) [*Morrogh.*

Fig. 44.—Typical Group of Anhedral Titanium Sulphide Inclusions. 2-mm. oil-immersion objective Unetched. × 1200.

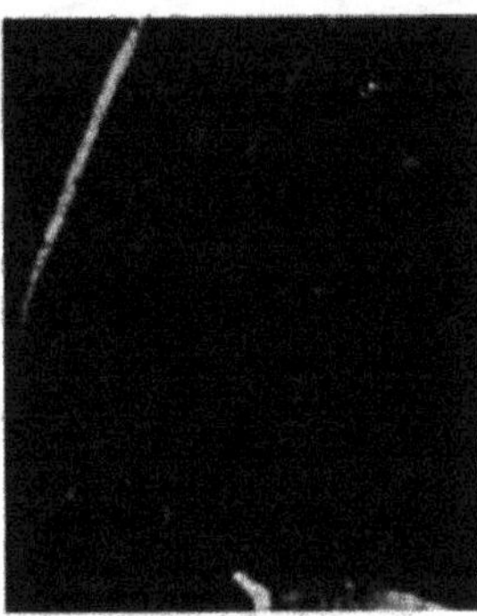

Fig. 45.—Same Spot as Fig. 44 under polarised light between crossed nicols. × 1200.

Fig. 47.—Titanium Carbide associated with Allotriomorphic Titanium Sulphide. 2-mm. oil-immersion objective. Unetched. × 1500.

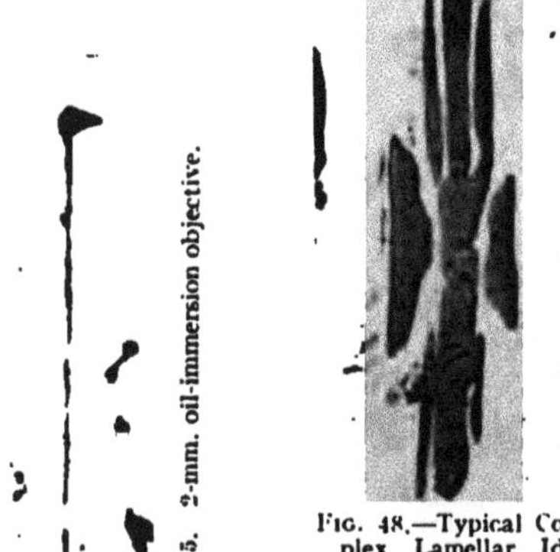

Fig. 46.—Very Long Lamella of Allotriomorphic Titanium Sulphide. From melt No. 15. 2-mm. oil-immersion objective. Unetched. × 1500.

Fig. 48.—Typical Complex Lamellar Idiomorphic Titanium Sulphide Crystal.

Fig. 49.—Idiomorphic Titanium Sulphide associated with Manganese Sulphide and Graphite.

Fig. 50.—Hexagonal Titanium Sulphide associated with Manganese Sulphide and Cubic Crystal of Titanium Carbide.

Figs. 48–50.—2-mm. oil-immersion objective. Unetched. × 1200.

(Micrographs reduced to four-fifths linear in reproduction.)

[Morrogh.

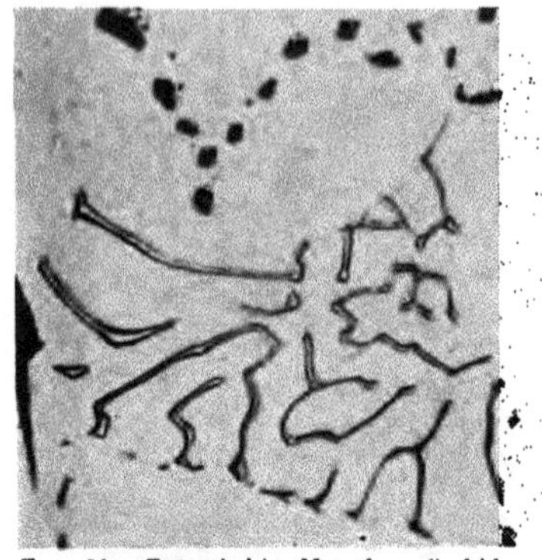

Fig. 51.—Eutectic-like Vanadium-Carbide and Iron-Sulphide-Rich Inclusion. 2-mm. oil-immersion objective. Unetched. × 2000.

Fig. 52.—Vanadium Carbide with Manganese-Sulphide and Iron-Sulphide-Rich Constituent. Oil-immersion objective. Unetched. × 2000.

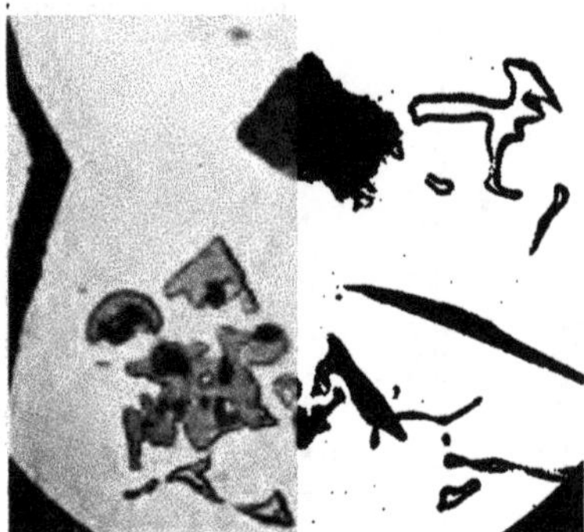

Fig. 53.—Titanium Sulphide associated with Manganese Sulphide, Iron Sulphide and Vanadium Carbide. 2-mm. oil-immersion objective. Unetched. × 2000.

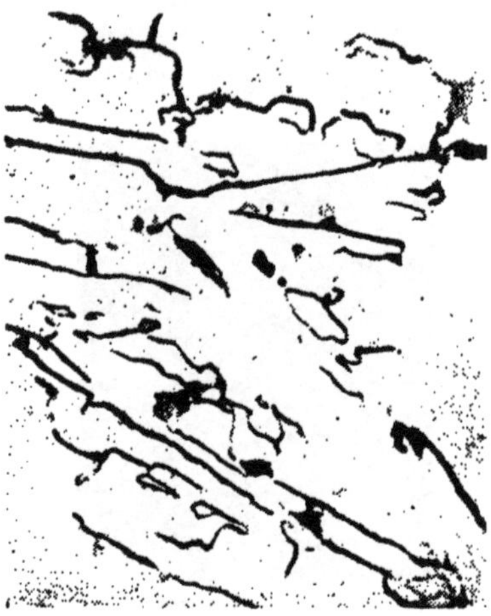

Fig. 54.—Graphite Formations in Melt No. 17. 4-mm. dry objective. Unetched. × 500.

Fig. 55.—Formation of Graphite around Cementite in Melt No. 17. 2-mm. oil-immersion objective. Etched in picric acid. × 2000.

Fig. 56.—Typical Lamellar Titanium Sulphide Crystal in a hot-poured 15% silicon iron Crystals of titanium carbide on left. 2-mm oil-immersion objective. Unetched. × 1200

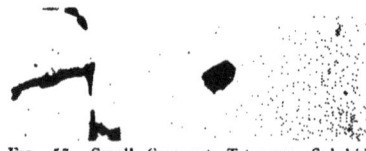

Fig. 57.—Small Compact Titanium Sulphid Crystal in cold-poured 15% silicon iron. 2 mm. oil-immersion objective. Unetchec × 1200.

(*Micrographs reduced to four-fifths linear in reproduction.*)

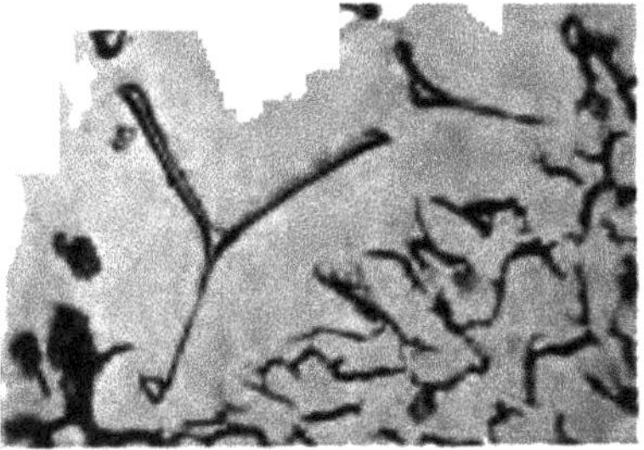

Fig. 58.—Allotriomorphic Form of Titanium Carbide in pig iron. 2 mm. oil-immersion objective. Unetched. × 1000.

Fig. 59.—Another Allotriomorphic Form of Titanium Carbide in same pig as Fig. 58. 2-mm oil-immersion objective. Unetched. × 1000.

Fig. 60.—Typical Orange - Yellow to Grey Inclusion in bar *Z2*.

Fig. 61.—Combination of Two Lamellar Orange - Yellow to Grey Zirconium Inclusions.

Fig. 62.—Orange-Yellow to Grey Inclusion through Titanium Carbide Crystal.

Fig. 63.—Zirconium Cyano-Nitride Crystal with cored structure.

Fig. 64.—Titanium Carbide Crystal with centre of zirconium carbide.

Figs. 60–64.—2-mm. oil-immersion objective. Unetched. × 1200.

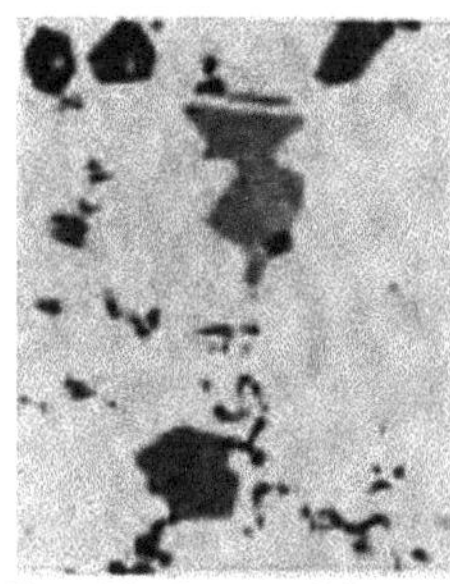

Fig. 65.—2-mm. oil-immersion objective. Unetched. × 1200.

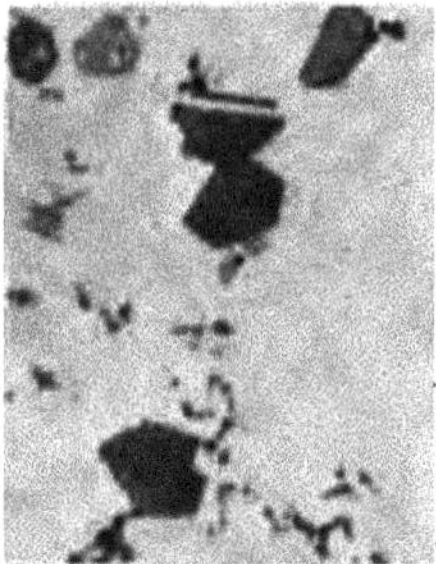

Fig. 66.—Same spot as Fig. 65 after rotation of stage through 90°.

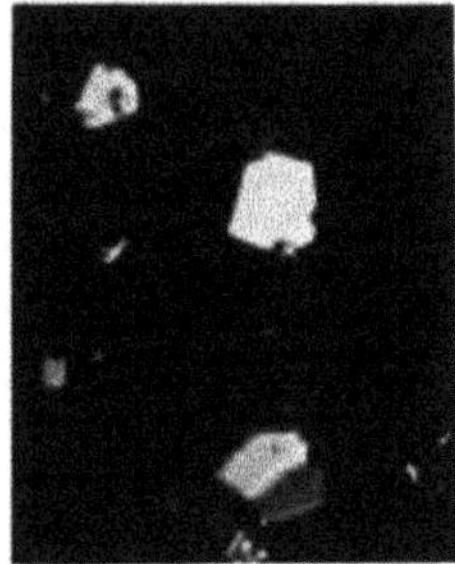

Fig. 67.—Same spot as Figs. 65 and 66 between crossed nicols.

Figs 65–67.— Orange-Yellow to Grey Inclusions surrounded by segregation of blue-grey zirconium cubes.

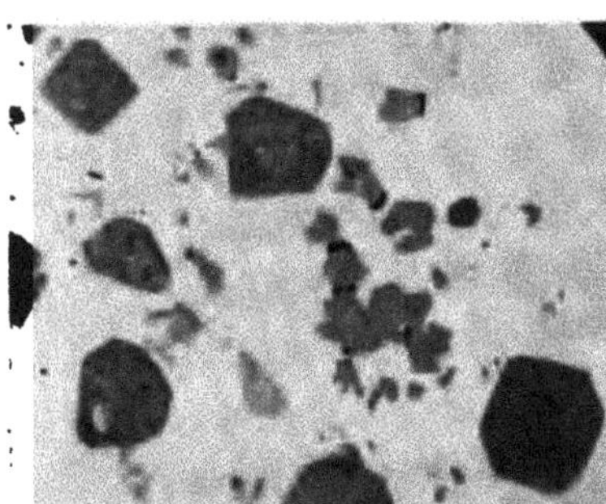

Fig. 68.—Group of Orange-Yellow to Grey and Blue-Grey Inclusions. 2-mm. oil-immersion objective. Unetched. × 1200.

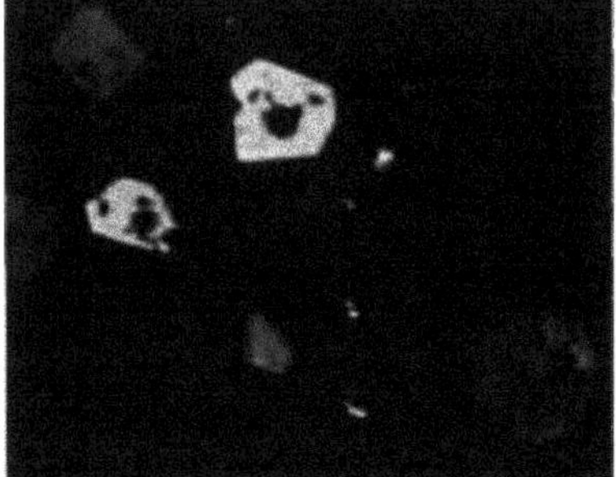

Fig. 69.—Same Spot as Fig. 68 between crossed nicols. Dark spots in lit-up inclusions are blue-grey carbide inclusions.

(Figs. 58, 59, 65–69 reduced to four-fifths linear in reproduction. Figs. 60–64 reproduced at same size.)

[Morrogh.

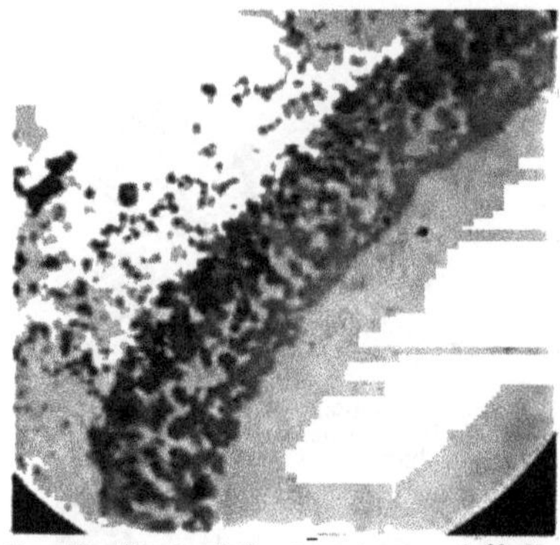

FIG. 70.—Bands of Constituents around Undissolved Particle of Ferro-Silicon-Zirconium. 2-mm. oil-immersion objective. Unetched. × 1200.

FIG. 71.—Same Spot as Fig. 70 under polarised light between crossed nicols.

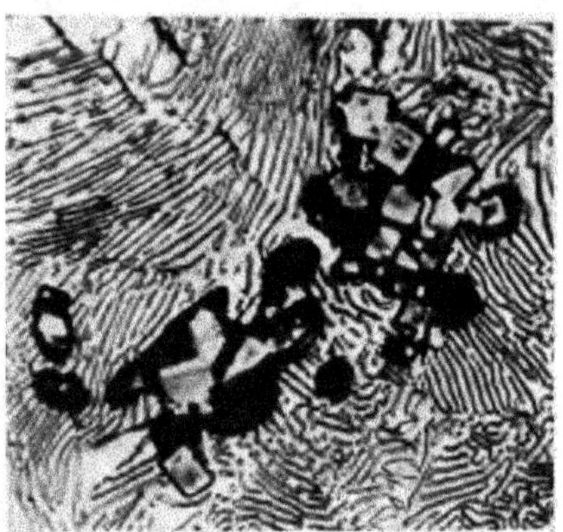

FIG. 72.—Group of Lemon-Yellow Zirconium Nitride Cubes associated with Blue-Grey Constituent. 2-mm. oil-immersion objective. Etched in picric acid. × 2000.

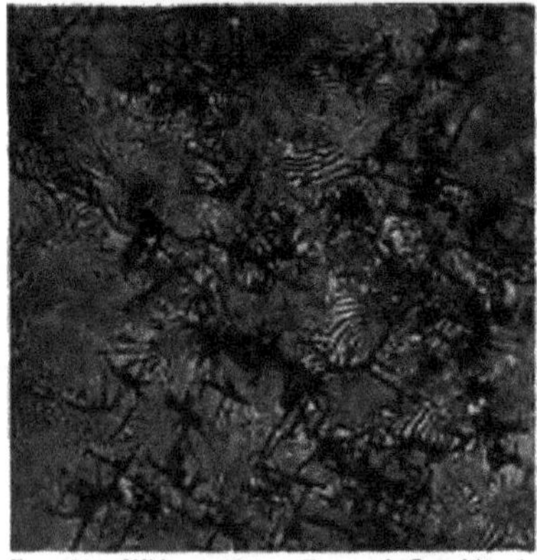

FIG. 73.—Widmanstätten Form of Graphite in bar Z13/4. 4-mm. dry objective. Etched in picric acid. × 500.

FIG. 74.—Ring of Graphite formed round a blue-grey inclusion in bar Z13/4. 2-mm. oil-immersion objective. × 2000.

(Micrographs reduced to four-fifths linear in reproduction.)

[*Morrogh.*

[*To face p.* 225 P.

particularly well in Fig. 40, which was taken from a blast-furnace bear sample having the following analysis :

Manganese.	Sulphur.	Titanium.
0·11%	0·024%	0·07%

When the lamellar allotriomorphic blue-pink inclusions in the pig-iron sample were examined at a very high magnification, it was observed that the lamellæ appeared to be in process of breaking up and that they were not always continuous. This is illustrated in Fig. 41.

The occurrence of these inclusions in titaniferous irons containing insufficient manganese to balance the sulphur suggested them to be probably titanium sulphide, and it is useful at this stage to consider the available literature pertaining to the presence of titanium sulphide in ferrous alloys. Portevin and Castro [45] observed that additions of titanium (0·5–1·0%) caused the sulphide phase in steels to change appreciably, giving irregular particles, harder and less deformable than manganese sulphide. These inclusions, the investigators say, are pale pink in colour and from this point of view are difficult to distinguish from titanium cyanonitride, but they are anisotropic. The analyses of the steel samples used by Portevin and Castro were :

	No. 2.	No. 7.
Carbon	0·080%	0·35%
Silicon	0·28%	0·45%
Manganese	0·30%	0·16%
Titanium	0·9%	0·24%
Sulphur	0·016%	0·015%
Phosphorus	0·020%	0·016%

It is seen that there was insufficient manganese in these steels to ensure complete neutralisation of the sulphur in the form of manganese sulphide.

Blair and Shimer,[46] when investigating cast iron prepared from titaniferous materials, observed a sulphide which they isolated chemically. The product had a bronze colour and gave the following analysis :

Titanium.	Iron.	Carbon.	Sulphur.
62·82%	1·82%	9·82%	22·64%

Portevin and Castro suggest that this substance may be regarded as a titanium sulphide, more or less contaminated with carbide.

Hanemann and Schrader [47] (*loc. cit.*, No. 4, Table 31) have recorded a "rare constituent," similar in morphology to the allotriomorphic blue-pink inclusions previously described, which they suggest might be titanium sulphide. It is interesting to note that the iron containing the inclusion shown by Hanemann and Schrader has just insufficient manganese (actually 0·44%) to balance the sulphur (actually 0·10%) according to the formula: Mn% = (1·7 × S%) + 0·3, and that the iron contains 0·02% of titanium.

Urban and Chipman[48] found that the addition of 1·26% of titanium to a synthetic alloy containing 0·15% of sulphur removed all the iron sulphide.

According to Mellor[49] titanium sulphide, TiS, is a reddish solid with a metallic appearance greatly resembling bismuth.

Since the facts available in the literature seemed to confirm the supposition that the blue-pink inclusions were probably titanium sulphides, it was decided to attempt to produce these inclusions artificially.

Melt No. 1.—For this purpose Armco iron was heated with sugar charcoal until it melted and gave a white iron of approximately 3% total carbon content. The iron was then treated while molten with iron sulphide to give approximately 0·5% of sulphur. This base metal was cast into small pigs and broken up for subsequent melting. A quantity of 150 g. of the base iron was then melted in a small Silit-rod crucible furnace and as much ferro-carbon-titanium as possible was dissolved in the melt. Any excess ferro-carbon-titanium was removed from the surface of the metal by skimming; the crucible was then removed from the furnace and allowed to cool in air. The small ingot so obtained was fractured from top to bottom. The fracture was white with a few specks of grey. The composition of the product was sulphur 0·465% and titanium 0·41%. The microscopic examination of the ingot showed the presence of a large number of allotriomorphic blue-pink inclusions, which exhibited exactly the same optical properties as those inclusions described previously. Generally, the pink inclusions, which may be tentatively termed titanium sulphide, were surrounded by a constituent resembling iron sulphide, but which actually was slightly darker, almost grey in colour, and which did not light up under polarised light between crossed nicols.

It would seem that this latter constituent is iron sulphide containing dissolved titanium (sulphide), and its etching properties were compared with those of iron sulphide in the following manner: Micro-specimens from the high-sulphur carburised Armco-iron pigs and from the titanium-treated crucible melt were etched together in a 1% aqueous oxalic acid solution.[50] After 1½ min. immersion in this solution at a temperature of 18·5° C., the iron sulphide in the blank specimen was completely etched away, while the greyish constituent in the titaniferous melt was completely unattacked. Even after 6 min. immersion in this etching reagent, neither the pink nor the grey constituents were attacked.

Further examination of the specimen at this stage gave the impression that all the iron-sulphide-rich component was not of the same colour. Some had a khaki tinge, while the rest had a grey coloration, and it was found that etching in a solution of ethyl alcohol saturated with stannous chloride [48] for 20 min. heightened the contrast between the grey and khaki components. By heat-tinting it was not found possible to differentiate between the grey and khaki parts—they both went deep purple, typical of iron-sulphide-rich inclusions. Micrographs showing typical formations of the allotriomorphic blue-pink and the grey-khaki inclusions are given in Figs. 42 and 43.

Now that it was known that the blue-pink inclusions could be produced synthetically, it was decided to determine the effect of increasing the manganese content of an iron originally containing these inclusions. To do this the following experiment was carried out.

Melt No. 2.—Sixty pounds of Swedish white iron were melted in an indirect rocking arc furnace. When the metal was molten, 5¼ oz. of iron sulphide were added, stirred in, and then 1·2 in. dia. bar No. 1 was poured; 14½ oz. of ferro-carbon-titanium were added, the metal was reheated and then bar No. 2 was poured. This operation was repeated for additions of 1 lb. 3½ oz. and 2 lb. 7 oz. of ferro-carbon-titanium, and for additions of 2¾ oz., 5½ oz. and 11 oz. of 80% ferro-manganese, giving seven 1·2 in. dia. bars in all. The chemical analyses and the types of inclusions found are indicated in Table II.

TABLE II.—*Effect of Titanium and Manganese Additions to an Iron Containing Iron Sulphide.*

Bar No.	Mn. %.	S. %.	Ti. %.	Inclusions Present.
1	...	...	...	Iron sulphide.
2	...	0·172	0·05	Iron sulphide and titanium sulphide detected.
3	...	0·162	0·06	Iron sulphide and trace of allotriomorphic titanium sulphide.
4	...	0·166	0·16	Iron sulphide and large amounts of allotriomorphic and iodomorphic titanium sulphide.
5	0·39	0·160	0·15	Large amounts of manganese sulphide and titanium sulphide, as No. 4.
6	0·96	0·145	0·13	Fairly large amounts of manganese sulphide. Trace of titanium carbide.
7	1·89	0·071	0·10	Manganese sulphide. Trace of titanium cyano-nitride.

The results given definitely indicate that additions of titanium to an iron containing iron sulphide results in the formation of a titanium sulphide and that manganese has a greater affinity for sulphur than has titanium. Additions of manganese to an iron containing titanium sulphide cause the disappearance of this constituent, until, when sufficient manganese is present to balance the sulphur, only manganese sulphide occurs. The fact that with increasing titanium contents in the above series, the allotriomorphic form of titanium sulphide occurred first and was followed by the idiomorphic form in the bar with the highest titanium content, suggests that the former probably remained dissolved until at least some of the metal was solid, and that the latter is a primary form which is deposited from the liquid.

Having determined the mode of occurrence of titanium sulphide formed by the addition of titanium to irons containing free iron sulphide, it was decided to try to prepare this constituent by the addition of iron sulphide to titaniferous pig irons. Two pig irons were chosen for this and a series of melts was made for each pig iron with increasing additions of iron sulphide.

TABLE III.—*Effect of Iron Sulphide*

Melt No.	Charge.	Mn. %.	S. %.	Ti. %.	V. %.	Fracture.	Manganese Sulphide.	
3	100 g. pig iron, 0·55 g. FeS	0·79	0·145	0·29	0·53	Very fine grey	Crystals	Pig
4	93 g. pig iron, 1·5 g. FeS	...	...	...	...	Slightly coarser than No. 3	Crystals	
5	140 g. pig iron, 3 g. FeS	0·79	0·452	0·13	0·52	Slightly coarser than No. 4	Large amounts	
6	150 g. pig iron, 4 g. FeS	...	...	...	...	Similar to No. 5	Do.	
7	150 g. pig iron, 5 g. FeS	0·57	0·501	0·10	0·49	Patches of coarse and fine grey	Large amounts in dendrites	
8	150 g. pig iron, 6 g. FeS	...	...	...	...	Similar to No. 7	Large amounts	
9	150 g. pig iron, 7 g. FeS	0·47	0·494	0·04	0·41	Similar to Nos. 7 and 8	Do.	
10	150 g. pig iron 10 g. FeS	0·33	0·331	0·08	0·43	Coarse grey	Fairly large amount	
11	126 g. pig iron, 12·6 g. FeS	0·14	0·947	0·08	0·33	White with grey rim	Not present	
12	146 g. pig iron, 0·5 g. FeS	1·40	0·183	0·10	...	Fine grey	Fair amounts	Pig
13	140 g. pig iron, 1 g. FeS	...	...	...	...	Do.	More than No. 12	
14	158 g. pig iron, 5 g. FeS	0·76	0·329	0·06	...	Slightly coarser	As No. 13	
15	148 g. pig iron, 6 g. FeS	...	...	...	...	Coarse grey	Do.	
16	149 g. pig iron, 7 g. FeS	0·56	0·281	0·03	...	Do.	Less than No. 15	
17	145 g. pig iron, 9 g. FeS	0·52	0·391	0·06	...	Mottled	Not present	

Melts Nos. 3–17.—The two pig irons used for these series had the following compositions :

	T.C. %	Si. %	Mn. %	S. %	P. %	Ti. %	V. %
Pig iron *A* .	4·00	2·60	1·10	...	...	0·60	0·50
Pig iron *B* .	3·6	2·06	1·57	0·058	1·04	0·23	0·07

For each melt, 100–150 g. of the pig iron were melted, and when molten the iron sulphide was added. The furnace employed was the Silit-rod furnace, as used for melt No. 1. After the iron sulphide addition, the metal was vigorously stirred with a silica tube; the

Additions to Two Pig Irons.

Titanium Carbide.	Vanadium Carbide.	Titanium Sulphide, Allotriomorphic.	Titanium Sulphide, Idiomorphic.	Iron Sulphide, Titanium Sulphide.
Iron A.				
Large amounts	Eutectic-like groups	Not present	Not present	Not present
Do.	Do.	Trace	Do.	Do.
Few crystals	Do.	Eutectic-like groups of spots and lamellæ	Large crystals	Do.
Do.	Do.	Do.	Lamellar idiomorphic crystals	Do.
Traces	Do.	Do.	Do.	Do.
Do.	Do.	Do.	Do.	Do.
None present	Do.	Do.	Do.	Do.
Do.	Do.	Do.	Do.	Traces
Do.	Not present	Not present	Not present	Large amounts
Iron B.				
Large cubic crystals	Trace	Not present	Not present	Not present
Do.	Do.	Do.	Do.	Do.
Do.	Do.	Trace	Do.	Do.
Less than No. 14	Do.	Fair amount	Fair amount	Do.
Very small amount	Do.	As No. 15	As No. 15	Do.
Not present	Not present	Not present	Not present	Present in round blobs

metal was then allowed to stand in the furnace for a few minutes before the crucible was withdrawn and cooled in the air. No attempt was made to control the temperature. Immediately the iron sulphide was added to the metal it became covered with a thick solid manganese sulphide slag, which was not removed. When cold, the resulting ingots were fractured from top to bottom. One vertical section was used for microscopical examination and the bottom part of the other was drilled for chemical analyses where and when required. The ingots were drilled at the bottom, because the microscopical examination indicated segregation to have occurred at the tops of the ingots, owing to the upward settling of manganese and titanium sulphides. It was not considered necessary to have every ingot chemically analysed and only a selection was used for this purpose. However, the chemical analyses given illustrate the general trend of the changes in composition with the increasing iron sulphide additions (Table III.).

The following general remarks can be made about these series of melts. With increasing additions of iron sulphide, the manganese content gradually drops, owing to the upward settling of manganese sulphide. At the same time there is a corresponding drop in the titanium contents, while, in the case of pig iron *A*, the vanadium contents drop only very slightly from the first to the last melt in this series. While there is sufficient manganese to balance the sulphur content the only inclusions present are manganese sulphide, titanium carbide and vanadium carbide. As soon as the manganese content is lower than the value of ($S\% \times 1{\cdot}7 + 0{\cdot}3$), titanium sulphide appears, first in the allotriomorphic form, then in the idiomorphic form. After this point, the titanium carbide is successively replaced by the idiomorphic titanium sulphide, which floats to the surface of the metal. This upward settling of titanium sulphide is not nearly so marked as that of manganese sulphide. The change in the fracture of the ingots from fine to coarse was accompanied by a corresponding change in the microstructure, from supercooled graphite to coarse flake graphite. This is very interesting in view of the known effects of titanium on the production of supercooled graphite, but it is impossible to tell whether the coarsening effect in this case was due to the gradual removal of the titanium or to the conversion of the titanium carbide to titanium sulphide.

A typical example of the allotriomorphic titanium sulphide in these series is shown in Figs. 44 and 45, actually taken from melt No. 10. In these micrographs it would appear that the titanium sulphide was originally lamellar, but at some later time spheroidisation took place. This idea is further emphasised by the composite micrograph in Fig. 46, taken from melt No. 15, showing a very long streak of the allotriomorphic titanium sulphide. Spheroidisation appears to have occurred in several places. In the early stages of the removal of titanium carbide, this constituent frequently occurs associated with allotriomorphic titanium sulphide. This is illustrated in Fig. 47, taken from melt No. 14.

In the case of the idiomorphic titanium sulphide, it was possible to observe its morphology and optical properties much more easily than previously. Large, well-formed crystals of the sulphide were present in melts Nos. 5 to 10, 15 and 16. These crystals have either a complex lamellar form or a hexagonal outline. Typical lamellar forms are shown in Figs. 48 and 49, taken from melt No. 10. The hexagonal form is illustrated in Fig. 50. It is possible that this latter form is merely a cross-section of the lamellar form.

So far in this paper, the colour of the titanium sulphide inclusions has been referred to as blue-pink. Actually, this constituent is pleochroic, and its colour varies from blue to pink, according to its position relative to the plane of polarisation of the incident light. There is no need to use a polariser in the microscope train to obtain the effect. The reflection of light from the glass-slip vertical illuminator produces enough polarised light to give an easily discernible change in colour on rotating the stage of the microscope. The use of a polariser merely increases the pleochroic colour contrasts. When a lamellar idiomorphic titanium sulphide inclusion is arranged parallel to the plane of polarisation of the incident light it appears pink, and when it is arranged at right angles to the plane of polarisation of the incident light it appears light blue. Intermediate positions show intermediate colorations, but the pink tends to predominate.

When a lamellar idiomorphic titanium sulphide inclusion is examined under polarised light between crossed nicols, it lights up and extinguishes four times for a complete revolution of the microscope stage. Extinction occurs when the length of the inclusion is parallel to either the plane of polarisation of the polariser or to the plane of polarisation of the analyser nicol. The inclusion lights up in positions at 45° to these directions. The behaviour of titanium sulphide when viewed under plane polarised light and under polarised light between crossed nicols is summarised in the diagram shown in Fig. 2.

The planes of polarisation of the polariser and analyser nicols are represented by the diametric lines PP' and AA', respectively. The behaviour of the inclusion when rotated between crossed nicols is given in the outer circle. For instance, when the inclusion lies parallel to PP' or AA', it is in positions of extinction, but when in the position ZZ' or XX' it lights up. The next circle refers to the pleochroic colorations obtained under plane polarised light or normal vertical illumination. When the inclusion is parallel to PP' it appears pink, and when at right angles to this it appears blue. The curved arrows in the inner circle refer to movements of the analyser nicol from the crossed position, and the letters by them refer to the colorations produced in the inclusion by such a movement of the analyser; the letter B indicates a blue coloration, and the letter *P* a pink coloration. For instance, when the inclusion lies in its position of maximum brightness, parallel to *XX′*, a slight clockwise rotation of the analyser gives a pink coloration, and an anti-clockwise rotation a blue coloration. If the stage is now rotated through 180°, exactly the same colorations are obtained for the same movements of the analyser. If, however, the stage is rotated through 90°, so that the inclusion now lies parallel to *ZZ′*, then a slight clockwise rotation gives a blue coloration, while an anti-clockwise rotation gives a pink coloration. When the inclusion lies parallel to the plane of polarisation of the incident light, *PP′*, then both clockwise and anti-clockwise rotations of the analyser give pink colorations, and when it lies parallel to the plane of polarisation of the analyser blue colorations are

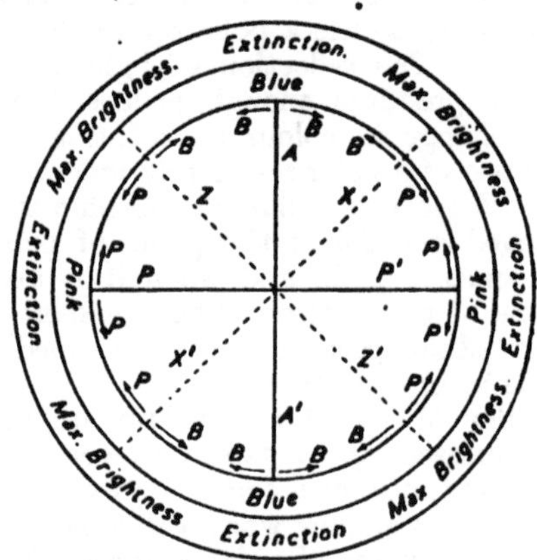

FIG. 2.—Diagrammatic Representation of the Optical Properties of Titanium Sulphide Inclusions.

obtained. All these optical effects can be observed much more distinctly with an oil-immersion objective than with a dry objective.

The optical properties of the titanium sulphide described above appear to constitute a very delicate test for this inclusion, and, as far as is known at present, they have never been described in the literature before.

When an idiomorphic titanium sulphide crystal with a hexagonal outline is examined under polarised light between crossed nicols, it remains dark for all positions of the stage, and under plane polarised light no pleochroic effect can be observed—it appears pink for all positions of the inclusion relative to the plane of polarisation of the incident light. These facts further support the view given above, that the hexagonal crystals are probably cross-sectional views of the idiomorphic lamellar titanium sulphide, since tetragonal and hexagonal crystals are singly refractive in the directions of their principal crystallographic axes, and rhombic, monoclinic and triclinic crystals are singly refractive in one or the other of the two optical axes. Of course, these facts only apply to transparent sections, but the implications of these observations are that the idiomorphic titanium sulphide crystals are prisms of hexagonal form, and the hexagonal nature of the crystals is only revealed in the comparatively rare cases when the polished surface of the specimen cuts cross-sections of the prisms.

Melt No. 12 showed complex formations of vanadium carbide, titanium sulphide and iron sulphide rich in titanium and manganese sulphide. The vanadium carbide in this sample frequently occurred in eutectic-like patches adjacent to eutectic-like patches of the iron-sulphide-rich constituent. A typical example of this is shown in Fig. 51. Fig. 52 and Fig. 53 illustrate the intimate occurrence of vanadium carbide with manganese sulphide, and of titanium sulphide with manganese sulphide. Melts Nos. 11 and 17 showed large amounts of an iron-sulphide-rich constituent which was not attacked by etching for 5 min. in a 1% aqueous oxalic acid solution, this indicating that this constituent was essentially the same as the khaki-grey constituent present in melt No. 1. Melt No. 17, which gave a mottled fracture, had some very interesting graphite formations. The graphite was arranged in very thin lamellæ around the lakes of cementite, also suggesting that it was the product of a reaction in the solid state. Typical micrographs of these graphite structures are shown in Figs. 54 and 55. Wells,[51] when studying graphitisation in high-purity iron-carbon alloys, found graphite precipitated round cementite and also as a Widmanstätten graphite pattern in cementite in annealed white irons. This would seem to imply that melt No. 17 solidified white and that some of the cementite decomposed after solidification.

Although the results obtained from the melts so far described indicate that if there is sufficient manganese to balance the sulphur content no titanium sulphide will be visible, one example has been found where this was not the case; this was a 15% silicon iron. A charge of 15 lb. of steel, 15 lb. of 10% ferro-silicon and 20 lb. of 30% ferro-silicon was melted in a salamander crucible, using a coke-fired, forced-draught furnace. When the charge was molten, 1 lb. of ferro-silicon-titanium was added, and when this was dissolved, carbon dioxide was bubbled through the melt for 2 min. After this treatment, four test blocks, each measuring 8 × 4½ × 1¼ in., were cast on to a chill in an open mould. The purpose of the titanium and carbon-dioxide treatment[52] was to give the iron a supercooled graphite structure. The test blocks were cast at successively lower temperatures, which were approximately 1450°, 1375°, 1300° and 1250° C. Chemical analysis of the block poured at the lowest temperature gave the following figures:

Manganese.	Sulphur.	Titanium.
0·54%	0·045%	0·52%

It is seen that there is ample manganese to balance the sulphur in this analysis, when the formula Mn% = (S% × 1·7 + 0·3) is used, yet microscopical examination revealed the complete absence of manganese sulphide; the only inclusions present were titanium sulphide and titanium carbide. There are two possible reasons which should be considered in trying to explain this anomaly:

(1) It is probable that the reaction between titanium sulphide and manganese is reversible and can be expressed as follows:

$$TiS + Mn \rightleftharpoons MnS + Ti.$$

This means that with relatively high titanium and low manganese contents, the reaction will proceed towards the left, whereas with relatively low titanium and high manganese contents the reaction will tend to proceed towards the right. Now, in the chemical analysis given above, while there is an excess of manganese over that required to balance the sulphur, the excess is not great. It should also be noticed that the titanium content of 0·52% is considerably higher than any encountered in the melts Nos. 2 to 17. These facts should, if the above reversible reaction is assumed, tend to cause the reaction to proceed towards the left, that is, encourage the formation of titanium sulphide.

(2) The sample in question is abnormal in the sense that it is a 15% silicon iron, and it is just possible that high silicon contents tend to alter the relationships between manganese, sulphur and titanium.

At the present time it is impossible to tell which of these factors was the cause of the occurrence of titanium sulphide.

In these samples it was possible to observe the effect of pouring temperature on the morphology of the titanium sulphide crystals. The sample poured at the highest temperature had typically lamellar idiomorphic crystals (Fig. 56) and the cold-poured sample had much smaller and very compact crystals (Fig. 57). There were many more crystals of the sulphide in the cold-poured sample than in the hot-poured one. In order to determine this definitely, the samples were examined at a magnification of 1500, the image being projected on to a ground-glass screen (the area visible in this way is referred to as a unit field), and the number of inclusions per unit field was counted as the specimen was slowly moved under the objective. Two hundred unit fields were counted and the average numbers of inclusions per unit field were calculated for both the hot- and the cold-poured sample, giving a value of 0·074 for the former and 0·470 for the latter. These results still further support the view that the idiomorphic titanium sulphide is a primary form which is precipitated from the liquid iron, since Norbury [38] has clearly demonstrated that primary dendrites in cast iron become smaller and more numerous with decreasing pouring temperature.

Group III.—White or Faint Grey Inclusions.

(*a*) *Idiomorphic Inclusions, and* (*b*) *Allotriomorphic Inclusions.*

The inclusions belonging to these sub-groups appear in all titaniferous irons and in irons containing vanadium. Hofmann and Schrader[53] observed that alloying ferro-titanium with cast iron produced a hard, cubic, greyish-white structural "constituent." This was separated from the alloy by solution of the iron in cold dilute hydrochloric acid and recognised as titanium carbide by X-ray analysis. Usually, titanium carbide occurs in the form of cubic crystals (Figs. 50 and 56), but occasionally it occurs in an allotriomorphic form (Figs. 58 and 59). This latter form occurs most frequently in titaniferous irons containing very small amounts of vanadium. For instance, the pig iron from which the micrographs in Figs. 58 and 59 were taken had the following analysis:

Total carbon	3·8%	Phosphorus	0·038%
Silicon	1·64%	Titanium	0·15%
Manganese	0·76%	Vanadium	0·02%
Sulphur	0·033%		

Both titanium carbide and vanadium carbide[34] have cubic crystal structures, and in irons containing both titanium and vanadium these two carbides appear to crystallise isomorphously or to yield solid solutions. The pig iron used for melts Nos. 3 to 11, in the previous section of this paper, contained 0·60% of titanium and 0·50% of vanadium, and its microstructure showed hard cubic crystals apparently all of the same type.

The number of titanium carbide crystals present in irons of different cross-sectional dimensions varies inversely as the cross-section for irons with the same titanium content. This was proved by the following experiment: A charge of 45 lb. of pig iron and 15 lb. of steel was melted in a salamander crucible, using a coke-fired, forced-draught furnace. The pig iron had the following composition:

Total carbon	4·04%	Sulphur	0·010%
Silicon	2·19%	Phosphorus	0·048%
Manganese	1·10%	Titanium	0·06%

When the charge was molten, a set of standard test bars, 0·6 in., 0·875 in., 1·2 in., 1·6 in. and 2·1 in. in dia., were cast, having the lengths 10 in., 14 in., 21 in., 21 in. and 27 in., respectively. Each of these bars was broken in transverse, and metallographic samples were taken from the bars at the points of fracture, that is, approximately half-way along the original bars. The micro-specimens were then examined with a 2-mm. oil-immersion objective, using a

10× eye-piece at a magnification of 1676 diameters. The number of titanium carbide and titanium nitride inclusions per unit field was counted for two hundred fields and the average number of these inclusions per unit field was then calculated for each bar. The titanium content of these bars was 0·05%. The results of the inclusion counts are given in Table IV. and in Fig. 3.

TABLE IV.—*Effect of Test-Bar Diameter on the Number of Titanium Carbide and Titanium Cyano-Nitride Inclusions.*

Diameter of Test Bar:	2·1 in.	1·6 in.	1·2 in.	0·875 in.	0·6 in.
Number of titanium carbide inclusions per unit field	0·385	0·515	0·645	0·925	1·195
Number of titanium cyano-nitride inclusions per unit field . . .	0·130	0·165	0·230	0·170	0·315
Total number of titanium inclusions per unit field	0·515	0·680	0·875	1·095	1·510

This variation of the number of titanium carbide crystals with the test-bar diameter may seem strange at first, but it should be remembered that such a variation is the general rule for the constituents of the cast-iron microstructure. For instance, in grey irons smaller sections show finer, and hence more numerous, graphite flakes, and for a given area of pearlite the number of cementite lamellæ will tend to increase. It is interesting to note that variations in test-bar diameter have much less effect on the number of titanium cyano-nitride crystals than on the number of titanium carbide crystals.

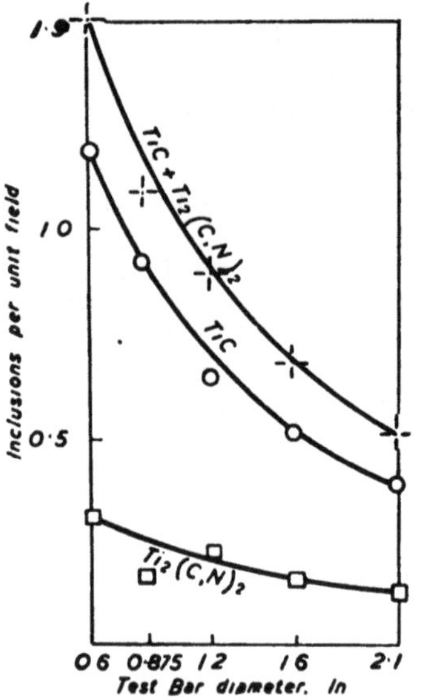

FIG. 3.—Effect of Test-Bar Diameter on the Number of Titanium Carbide and Titanium Cyano-Nitride Inclusions in Iron containing 0·05% of Titanium.

With increasing titanium contents the number of titanium carbide crystals increases, as also does the size of the individual crystals. The results of inclusion counts illustrating the combined effects of variations in test-bar diameter and titanium content on the number of titanium carbide and cyano-nitride crystals are given in Table V. and are illustrated graphically in Figs. 4 to 6.

Table V.—*The Combined Effects of Test-Bar Diameter and Titanium Content on the Number of Titanium Carbide and Titanium Cyano-Nitride Inclusions.*

Titanium:	0·05%.			0·07%.		
Test-Bar Diameter:	2·1 in.	1·6 in.	1·2 in.	2·1 in.	1·6 in.	1·2 in.
Number of titanium cyano-nitride inclusions	0·130	0·165	0·230	0·57	0·605	0·56
Number of titanium carbide inclusions	0·385	0·515	0·645	0·96	1·145	1·305
Total number of titanium inclusions	0·515	0·680	0·875	1·53	1·750	1·865
Titanium:	0·095%.			0·105%.		
Test-Bar Diameter:	2·1 in.	1·6 in.	1·2 in.	2·1 in.	1·6 in.	1·2 in.
Number of titanium cyano-nitride inclusions	0·76	0·68	1·21	0·505	0·425	0·22
Number of titanium carbide inclusions	1·32	1·595	2·47	1·915	2·405	3·61
Total number of titanium inclusions	2·08	2·275	3·68	2·42	2·83	3·83

The higher titanium contents were obtained by the addition of 4 oz., 6 oz. and 8 oz. of ferro-silicon-titanium to charges of pig iron and steel, as used for the first member of the series. It is seen that the number of titanium cyano-nitride crystals does not vary uniformly, as does the number of titanium carbide crystals. This is to be expected, since, for a given titanium content, the number of titanium cyano-nitride crystals must be a function of the nitrogen content of the metal, and it was not possible to control this factor during these experiments.

The equilibrium diagrams given by Tofaute and Büttinghaus [54] for the ternary system iron-carbon-titanium indicate titanium carbide to be appreciably soluble in austenite and, for hypo-eutectic alloys, completely soluble in molten iron. The following experiment was carried out in an attempt to determine whether the solubility of titanium carbide in austenite could be detected by inclusion counts for the temperature range 800–1000° C. To do this, sections, each 2 cm. long, were cut from a 0·875-in. dia. bar, containing 0·06% of titanium, and each section was annealed for 1 hr. at a constant temperature and quenched in water. The samples were cross-sectioned and polished, and the number of titanium carbide and titanium cyano-nitride crystals was counted, using the same method as for the other experiments in this part of the paper.

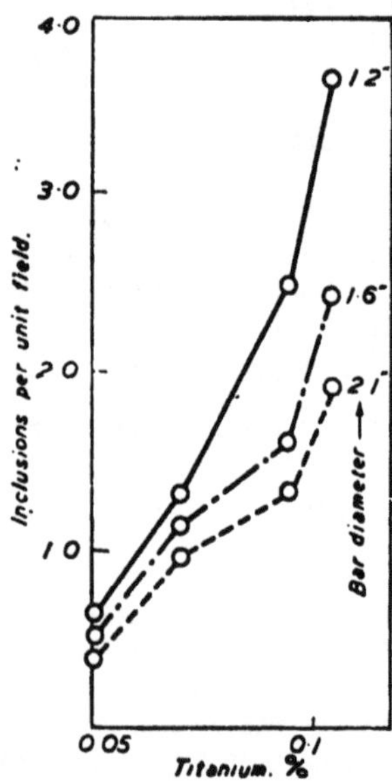

Fig. 4.—Combined Effects of Variations in Test-Bar Diameter and Titanium Content on the Number of Titanium Carbide Crystals.

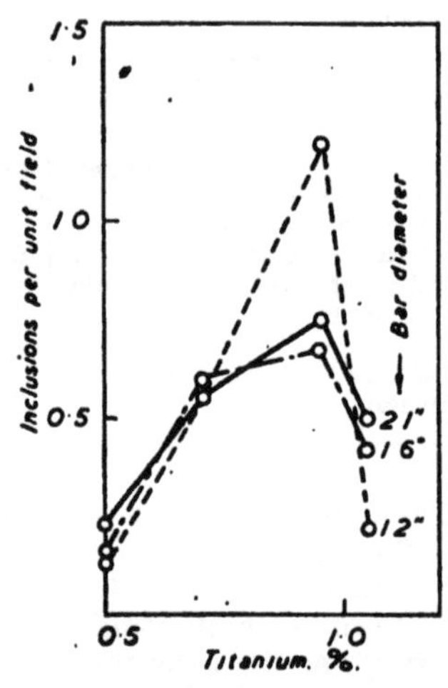

Fig. 5.—Combined Effects of Variations in Test-Bar Diameter and Titanium Content on the Number of Titanium Cyano-Nitride Inclusions.

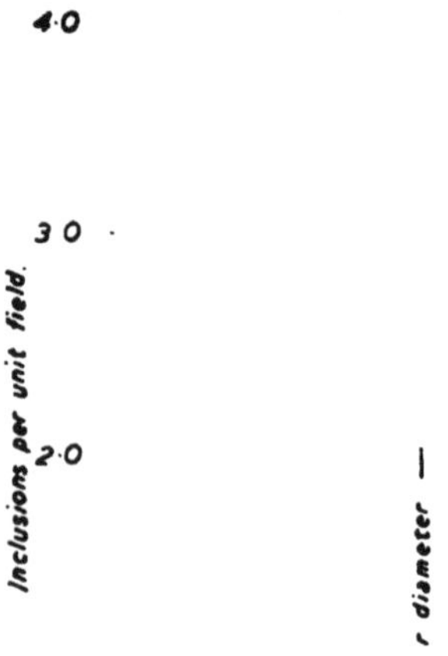

4·0
3·0
Inclusions per unit field.
2·0
r diameter

Fig. 6.—Combined Effects of Variations in Test-Bar Diameter and Titanium Content on the Total Number of Titanium Inclusions.

The results obtained are given in Table VI. and graphically in Fig. 7. While it is impossible to deduce anything very definite from these results, they do seem to indicate a slight decrease in the number of titanium carbide crystals on raising the temperature from 825° to 1000° C.

TABLE VI.—*Effect of Quenching Temperature on the Number of Titanium Carbide and Titanium Cyano-Nitride Inclusions.*

Temp. ° C.	Average Number of Inclusions per Unit Field.		
	TiC.	$Ti_2(C,N)_2$.	TiC + $Ti_2(C,N)_2$.
1000	0·855	0·145	1·00
975	0·62	0·31	0·93
950	0·67	0·295	0·965
925	1·055	0·19	1·245
900	1·02	0·16	1·18
875	1·16	0·225	1·385
850	1·27	0·225	1·495
825	1·22	0·145	1·365

It would appear that the number of titanium cyano-nitride crystals remains constant over this temperature range. The solu-

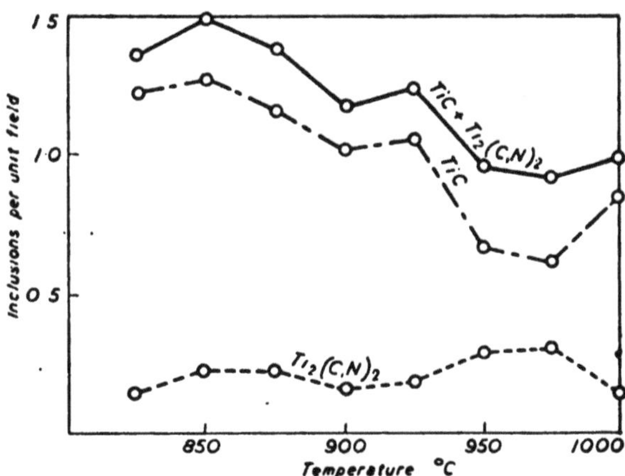

FIG. 7.—Detection of Solubility of Titanium Carbide in Austenite by Inclusion Counts.

tion of the titanium carbide would be very difficult to observe with higher titanium contents, because the decrease in size of the crystals consequent on the solution of the carbide would not be sufficient to effect a measurable decrease in the number of crystals.

The results of the inclusion counts presented so far in this paper must be recognised as having little, if any, quantitative significance; they indicate only the general tendencies with regard to the behaviour of titanium carbide.

Group IV.—Inclusions Produced by Alloying with Zirconium.

Zirconium is rarely added to cast iron, but its effects can be conveniently dealt with here, since the element itself shows considerable chemical similarity to titanium. Furthermore, the inclusions introduced into cast iron by zirconium in many ways resemble those introduced by the presence of titanium and, in at least one case, the two elements combine together to give a complex constituent. Nearly all cast irons contain small amounts of titanium, and so the effects of zirconium additions to cast irons are indissolubly associated with the effects of any titanium which is present.

There is practically no information in the literature on the precise effect of zirconium on the microstructure and inclusions of cast iron. However, there is some published information on the effect of this element on the inclusions in steels, and it is useful here to consider a brief review of this work. Feild [55] found that when deoxidising steel with silicon-zirconium, a considerable proportion of the zirconium was lost—the average yield of zirconium was 59·7%, while that of silicon from the same additions was 97%.

This investigator claimed that zirconium combined with oxygen, nitrogen and sulphur in the order given, and he was able to show that zirconium additions resulted in the formation of a bright lemon-yellow nitride. In another paper, Feild [56] points out that zirconium forms a sulphide very similar in colour to manganese sulphide, and he found that additions of zirconium removed red-shortness from steel when the finished product had a ratio of zirconium to sulphur equal to 1·41 : 1. Furthermore, he claimed that this ratio corresponded to the formation of the compound ZrS_2.

Urban and Chipman [48] found that additions of zirconium to sulphurised iron resulted in the formation of inclusions varying in colour from tan to grey. They found the tan inclusions to be anisotropic and the grey inclusions isotropic, and suggested that the sulphides observed by Field were manganese sulphide, perhaps contaminated with zirconium. These inclusions, they say, were angular or crystalline in outline and were not affected by annealing or forging.

The findings of Urban and Chipman were substantially confirmed by Portevin and Castro,[45] who found that the addition of 0·89% of zirconium to remelted Swedish iron gave yellowish-grey sulphide inclusions which were markedly anisotropic. They suggest that this phase replaces the sulphide of manganese (and zirconium) when the percentage of zirconium is sufficiently high. These investigators also describe the occurrence of blue-grey zirconium carbide, lemon-yellow zirconium nitride and zirconium cyano-nitrides ranging in colour from lemon-yellow or orange-brown to purple, violet and blue in one crystal. Both zirconium carbide [57] and zirconium nitride [58] have cubic crystal structures, and Portevin and Castro suggest that the solutions involved in the zirconium cyano-nitride crystals are continuous solid solutions of the system ZrN–ZrC.

It would thus appear from this review that additions of zirconium to cast irons might cause the appearance of zirconium sulphide, nitride, carbide and cyano-nitride phases. To determine the effect of zirconium on inclusions in cast iron, a series of irons having increasing zirconium contents was examined. These irons were prepared by melting a hematite base iron in a 250-lb. oil-fired crucible furnace. When molten, the requisite amount of metal was tapped into a handshank, and the alloy additions were made while the metal was in the shank. The zirconium was added in the form of ferro-silicon-zirconium having the following analysis:

Zirconium.	Silicon.	Iron.
38·8%	50·00%	8·25%

Bars 1·2 in. in dia. from eleven such irons were examined, having zirconium contents ranging from 0·04% to 0·54%. These bars have been numbered *Z*1 to *Z*11. Bars *Z*1, *Z*2, *Z*3 and *Z*9 were made from a base iron having 3·5% of total carbon and 1·4% of silicon, bars *Z*4, *Z*5, *Z*8 and *Z*11 from a base iron having 2·9% of total carbon and 2·2% of silicon, and bars *Z*6, *Z*7 and *Z*10 from a base iron having 3·3% of total carbon and 1·7% of silicon. In every case the addition was either ½ lb., 1 lb., 1½ lb. or 2 lb., which was added to about 28 lb. of the base iron. The analyses and alloy additions used are given in Table VII., which also indicates the types of inclusions met with in these bars. A more detailed description of the inclusions is given below.

Bar Z1, Zirconium 0·04%.—Titanium carbide cubes and manganese sulphide crystals are the only inclusions present in this bar.

Bar Z2, Zirconium 0·05%.—This bar contained titanium carbide and a small amount of manganese sulphide, as in bar *Z*1. Also present were a number of orange-yellow to grey lamellar inclusions, a typical example of which is shown in Fig. 60. These inclusions are presumably the same as those described by Urban and Chipman and by Portevin and Castro. Close examination revealed these inclusions to be pleochroic, which explains why their colour varied from orange-yellow to grey. When such an inclusion lies parallel to the plane of polarisation of the incident light it appears orange-yellow, and when, after a rotation of the microscope stage, it lies at right angles to the plane of polarisation of the incident light it appears light grey. The plane polarised

TABLE VII.—*Effect of Zirconium on the Inclusions in Cast Iron.*

• The presence of a constituent is indicated thus: ×.

Bar No.	Chemical Analysis.							Weight of Alloy Addition. Lb.	TiC, White-Grey.	(Ti,Zr)C, Pinkish.	MnS, Dove-Grey.	ZrC, Blue-Grey.	Orange-Yellow to Grey Inclusion.		Cored Cyano-Nitride.
	T.C. %.	Si. %.	Mn. %.	S. %.	P. %.	Ti. %.	Zr. %.						Lamellar.	Polygonal.	
Z1	3·39	1·45	0·80	0·021	0·061	0·06	0·04	½	×	...	×	...	...	...	...
Z2	3·51	1·53					0·05	1	×	...	×	...	×	...	...
Z3	3·42	2·05					0·09	1¼	×	...	...	...	×	...	...
Z4	2·91	2·24	0·81	0·017	0·035	0·08	0·12	¾	×	...	...	...	×	...	...
Z5	2·87	2·60					0·13	1	×	...	...	×	×	...	...
Z6	3·21	1·79	0·55	0·030	0·048	0·07	0·15	½	×	...	...	×	×	×	×
Z7	3·23	1·95					0·16	1	×	...	...	×	×	×	×
Z8	2·67	2·74					0·16	1½	×	...	...	×	×	×	×
Z9	3·45	2·47					0·25	2	×	×	...	×	...	×	...
Z10	3·17	2·45					0·29	1½	...	×	...	×	...	×	...
Z11	2·74	3·51					0·54	3	...	×	...	×	...	×	...

light produced by reflection from the glass slip of the vertical illuminator is quite sufficient to enable the easy detection of this

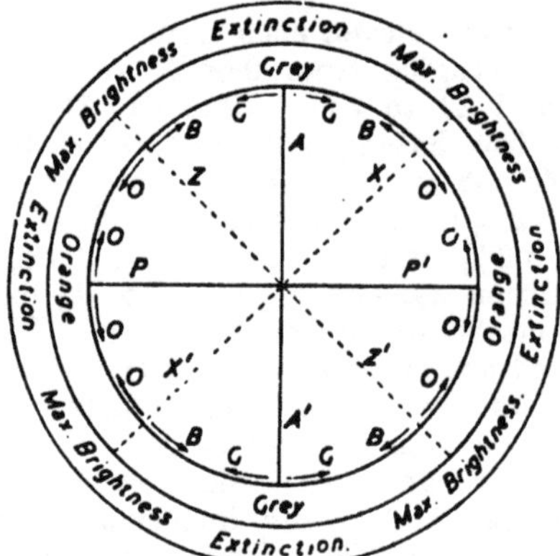

FIG. 8.—Optical Properties of the Orange-Yellow to Grey Zirconium Inclusion.

reflex pleochroism, the use of a polariser merely heightening the colour contrast in these two positions. The presence of the pleochroic effect in these inclusions has not been detected by previous investigators, and it explains why Portevin and Castro gave the colour of them as "yellowish-grey." When examined under polarised light between crossed nicols, they were found to be markedly anisotropic. The optical properties of this orange-yellow to grey inclusion are given diagrammatically in Fig. 8, using the same convention as in Fig. 2 for titanium sulphide; letter *O* indicates an orange-yellow coloration, *B* a blue, and *G* a grey coloration. In general, these optical phenomena can be detected more easily with an immersion objective than with a dry objective, although the use of the latter tends to exaggerate the pleochroic effects.

The orange-yellow to grey inclusions, while frequently lamellar themselves, have other lamellæ growing out of them, forming re-entrant angles (Fig. 61), and occasionally they occur cutting straight across titanium carbide crystals (Fig. 62).

Bar Z3, Zirconium 0·09%.—This bar contains titanium carbide as in the two previous bars, but no manganese sulphide is present. This constituent appears to have been completely replaced by the lamellar orange-yellow to grey zirconium inclusion.

Bar Z4, Zirconium 0·12%.—The inclusions in this bar are essentially similar to those in Z3.

Bar Z5, Zirconium 0·13%.—This bar contains titanium carbide as in the previous bars, but there is much more of the orange-yellow to grey inclusion, the lamellar form of which is partly replaced by a polygonal form. Also present are some blue-grey cubic inclusions which are isotropic. These latter appear to be the same as those described by Portevin and Castro as zirconium carbide.

Bar Z6, Zirconium 0·15%.—This bar is very similar to Z5, but the polygonal orange-yellow to grey inclusion is more in evidence, as is the cubic carbide inclusion. Titanium carbide is still present. A few crystals of the zirconium cyano-nitride are present in this sample. These crystals usually have a series of concentric rings ranging in colour from red-pink to deep purple. A typical crystal showing the concentric rings is illustrated in Fig. 63; the main body of the crystal is red-pink and the two dark rings are deep purple.

Bar Z7, Zirconium 0·16%, and Bar Z8, Zirconium 0·16%.—With regard to inclusions, these bars are identical. They show titanium carbide, polygonal orange-yellow to grey zirconium inclusions with a trace of the lamellar form, the blue-grey zirconium carbide and traces of the cored cyano-nitride.

Bar Z9, Zirconium 0·25%.—In this bar the lamellar orange-yellow to grey inclusion has been completely replaced by its polygonal form. The titanium carbide present has a definite pinkish appearance, and occasionally crystals of this phase can be seen to have crystallised around zirconium carbide cubes. This effect makes the centres of the titanium carbide crystals appear softer. An attempt has been made in Fig. 64 to show a titanium carbide crystal with a darkish centre of the zirconium carbide. The free zirconium carbide crystals tend to occur in large segregations. The individual crystals of this phase are usually very small and difficult to photograph. Fig. 65 shows a group of fairly large orange-yellow to grey inclusions surrounded by a segregation of the blue-grey zirconium carbide cubes. The apparently dark polygonal crystals were actually orange-yellow and the light crystal was grey in the particular stage position used. Fig. 66 depicts the same spot after rotating the stage through 90°; the inclusion which was formerly light (grey) is now dark (orange-yellow), while some of the inclusions which were formerly dark (orange-yellow) are now light (grey). Fig. 67 shows the same spot under polarised light between crossed nicols. The anisotropic character of the polygonal orange-yellow to grey inclusions is clearly seen, while the small carbide cubes have remained dark.

Bar Z10, Zirconium 0·29%.—This bar is essentially similar to Z9, but all the titanium carbide has been replaced by the faint pinkish phase, which is presumably a carbide of the type $(Ti,Zr)_2C_2$.

Bar Z11, Zirconium 0·54%.—This contains a few undissolved particles of ferro-silicon-zirconium, which probably account for the relatively high zirconium content. Large, well-developed crystals of the orange-yellow to grey inclusion and the cubic grey inclusion are present. Sometimes the cubic grey carbide inclusions occur inside the orange-yellow to grey inclusions. Typical examples of this mode of occurrence are given in Figs. 68 and 69. The very large crystals shown in these micrographs are the orange-yellow to grey ones and the crystals of a smaller order of size are the cubic blue-grey zirconium carbide inclusions. The undissolved particles of ferro-silicon-zirconium in this bar provide some interesting features. Each particle is surrounded by layers of three different phases. The innermost layer bordering on the alloy addition is an almost continuous band of a pinkish phase; the next layer is an almost complete ring of orange-yellow to grey inclusions; the outer layer consists of a band of dispersed blue-grey zirconium carbide crystals. Figs. 70 and 71 illustrate this mode of occurrence of the undissolved additions. Fig. 71, taken with polarised light between crossed nicols, enables the recognition of the band of orange-yellow to grey inclusions, which have lit up owing to their anisotropic character; the phases in the inner and outer bands have remained dark. A summary of the inclusions present in this series is given in Table VII. The crosses indicate the presence of inclusions.

From the examination of this series of irons, the following general conclusions can be drawn:

(*a*) Additions of zirconium up to 0·04% have no effect on the inclusions present in a titaniferous cast iron.

(*b*) Additions of zirconium above 0·04% result in the appearance of a lamellar orange-yellow to grey inclusion which completely replaces the manganese sulphide at 0·09% zirconium.

(*c*) At 0·13% zirconium, small crystals, blue-grey in colour and cubic in outline, are formed. These are probably zirconium carbide, ZrC. They gradually increase in number with increasing zirconium content.

(*d*) With 0·15% zirconium, the lamellar orange-yellow to grey inclusions become replaced by the polygonal form of the same phase, until, at 0·25% zirconium, the lamellar form has completely disappeared.

(*e*) With 0·25% zirconium, the titanium carbide begins to be replaced by a slightly softer pinkish constituent which is probably a complex titanium-zirconium carbide. At 0·29% zirconium all the titanium carbide in this series was replaced by this new phase.

From the results obtained thus far it was impossible to say definitely whether the orange-yellow to grey inclusion was a zirconium sulphide or a manganese-zirconium sulphide. With the object of determining more precisely the constitution of this inclusion, the following melt was made.

Melt Z12.—Sixty pounds of Swedish white iron were melted in a salamander crucible, using a coke-fired forced-draught furnace. When molten, ½ lb. of 14% ferro-silicon-zirconium was added and the metal was re-heated. The ferro-silicon-zirconium used had the following composition: Zirconium, 13·49%; silicon, 39·32%; titanium, 0·20%; manganese, sulphur and vanadium, nil. The addition was finely ground and packed in a sheet-steel container, which was dropped into the crucible and held under the surface of the metal by a plunger. It was considered necessary to take these precautions in view of the extremely low yields of zirconium in the previous series. When the crucible and its contents had been re-heated for a further 10 min., the crucible was removed from the furnace and a 1·2-in. dia. bar was cast directly from the crucible; 5¼ oz. of iron sulphide were then added to the metal and another 1·2-in. dia. bar was poured. Then 1 lb. of 14% ferro-silicon-zirconium was added in a similar manner to the first addition, and the crucible was again returned to the furnace to be reheated for 20 min. After this a third 1·2-in. dia. bar was poured; 5½ oz. of ferro-manganese were then added to the melt, and the fourth bar was poured. The bars were fractured in transverse and micro-sections were taken at the fractured ends. The analyses are given in Table VIII.

TABLE VIII.—*Neutralisation of Sulphur by Zirconium.*

Bar No.	Si. %.	Mn. %.	S. %.	Zr. %.	Fracture.
Z12/1	0·35	0·03	0·029	0·094	Grey with mottle
2	0·33	0·03	0·187	0·083	White
3	1·18	0·04	0·196	0·175	Grey with mottle
4	1·07	0·73	0·197	0·210	Grey

Bar Z12/1 showed small amounts of the orange-yellow to grey inclusion and traces of manganese sulphide.

Bar Z12/2 revealed only fairly large amounts of iron sulphide.

Bar Z12/3 contained iron sulphide and a bluish inclusion, having the form of iron sulphide. There was a complete range of colours, ranging from the khaki of iron sulphide to the bluish-grey of the other inclusion. The bluish inclusion, whilst having the same form as the iron sulphide, was completely isotropic when viewed under polarised light between crossed nicols. It would appear that these bluish inclusions are either iron sulphide containing zirconium or zirconium sulphide.

Bar Z12/4 had fairly large amounts of manganese sulphide and crystals of the orange-yellow to grey inclusion.

Melt Z13.—Since it appeared that there was insufficient zirconium completely to neutralise the sulphur in the bars of this melt, it was decided to carry out a similar experiment with larger additions of alloy. For this purpose a small Birlec 25-kW. rocking arc furnace was used. It was hoped by this means to obtain hotter metal and hence an improved yield of zirconium. A charge of 47 lb. of Swedish white iron and 23 lb. of steel was melted, and 5¼ oz. of iron sulphide were added to the molten metal. Then 2 lb. of 14% ferro-silicon-zirconium were added, the charge was reheated, and 1·2-in. dia. bar No. Z13/1 was poured. After this a further addition of 1 lb. of alloy was made, the metal was again reheated, and a second 1·2-in. dia. bar, Z13/2, was poured. This latter operation was repeated three more times, giving bars Z13/3, Z13/4 and Z13/5, and then 5 oz. of 80% ferro-manganese were added, and bar Z13/6 was poured. The analyses and fractures of these bars are given in Table IX.

TABLE IX.—*Neutralisation of Sulphur by Zirconium in Rocking Arc Furnace Melt.*

Bar No.	T.C. %.	Si. %.	Mn. %.	S. %.	Zr. %.	Fracture.
Z13/1	2·63	0·98	0·17	0·172	0·074	White
2	2·56	1·38	0·15	0·155	0·195	White
3	2·69	1·92	0·16	0·163	0·141	White
4	2·55	2·71	0·13	0·156	0·136	Mottled
5	2·53	3·91	0·18	0·152	0·121	Grey
6	2·50	3·98	1·68	0·143	0·016	Grey

It is seen that the yield of silicon from the additions of the ferro-silicon-zirconium was almost theoretical, while the yield of zirconium was very poor. Bars Z13/1 to Z13/5 all have very similar inclusions. They show a blue-grey constituent resembling manganese sulphide in colour, an iron-sulphide-rich constituent ranging in colour from khaki to blue-grey, and fairly large amounts of brilliant lemon-yellow zirconium nitride cubes. Bar *Z*13/6 has large amounts of manganese sulphide and also the brilliant lemon-yellow zirconium nitride cubes. Thus, it appears that the bars of this series are characterised by the presence of the zirconium nitride. The occurrence of this compound in the particular series is put down to the fact that the melting operation was carried out in an indirect rocking arc furnace, which exposes a large surface of the molten metal to the atmosphere. This has been confirmed by a number of other melts made in the furnace—every one containing zirconium showed large quantities of the lemon-yellow nitride.

A group of the cubic lemon-yellow nitride inclusions associated with the blue-grey constituent is shown in Fig. 72. When examined under polarised light between crossed nicols, the nitride inclusions give a bright yellow coloration which persists in all positions of the stage, while the blue-grey inclusions remain perfectly dark. While these latter inclusions resemble manganese sulphide in colour, they have apparently allotriomorphic outlines and do not show any translucency effects. It is thought that they were probably a complex sulphide of iron, manganese and zirconium.

Bar *Z*13/3 showed a small amount of graphite deposited in a very angular form. This effect was even more pronounced in bar *Z*13/4, which contained a fairly large amount of graphite. In this bar some of the graphite was deposited around the cementite in a similar manner to that described for melt No. 17 and illustrated in Figs. 54 and 55. The rest of the graphite was deposited in what appeared to be a Widmanstätten form and a small amount was deposited in a thin band around some of the blue-grey inclusions. A typical Widmanstätten arrangement is illustrated in Fig. 73, and a ring of graphite formed round a blue-grey inclusion in Fig. 74. A similar Widmanstätten form of graphite has been depicted by Hanemann and Schrader [47] (*loc. cit.*, No. 3, Table 22). However, the Widmanstätten form shown by these investigators consisted of hypereutectoid graphite which had crystallised on to the eutectic graphite. The example illustrated here is actually eutectic graphite which has probably formed by the decomposition of the solid eutectic cementite. It would thus appear that in irons which contain insufficient quantities of the elements manganese, titanium or zirconium, the sulphur acts as a much more potent carbide stabiliser at high temperatures than at low temperatures, owing, perhaps, to the reversibility of the reactions of iron sulphide with titanium, manganese and zirconium. The reversibility of these reactions would tend to give more iron sulphide at high temperatures than at low temperatures. This has definitely been proved for the reaction between manganese and iron sulphide, and has already been discussed with regard to the inclusions of Group I. The deposition of graphite round the blue-grey inclusions has important implications, but it is possible that this phenomenon is not due to particular properties of this inclusion, but rather to the fact that graphitisation will begin at any available solid-solid interface under suitable conditions.

Bars Z13/5 and Z13/6 have normal medium-sized eutectic flake graphite. The very big drop in zirconium content from bar Z13/5 to bar Z13/6 can only be explained by reference to the extreme oxidising power of the atmosphere in the melting unit employed.

Melts Z14 and Z15.—Having roughly determined the type of inclusions introduced into cast iron by the addition of zirconium, it was decided to investigate the effect of pouring temperature on these inclusions. To do this, two melts were carried out in a salamander crucible, using a coke-fired, forced-draught furnace. The charge for the first melt consisted of 36 lb. of Swedish white iron, 9 lb. of steel and 4½ oz. of 80% ferro-manganese, while that for the second melt consisted of 26 lb. of Swedish white iron, 9 lb. of steel and 9 oz. of ferro-manganese. These two melts have been given the numbers Z14 and Z15, respectively. In each case 1½ lb. of 14% ferro-silicon-zirconium was added to the molten charge, and four 1·2-in. dia. bars were poured at the following approximate temperatures: 1450°, 1350°, 1300° and 1225° C. The temperatures were taken with a platinum/platinum-rhodium thermocouple of the immersion type and are necessarily approximate, since a slight delay occurred between the taking of the readings and the pouring of the bars. The microscopical examination indicated that the pouring temperature only affected the number and size of the orange-yellow to grey inclusions. All the bars contained orange-yellow to grey inclusions in well-formed crystals and large segregations of the grey cubic zirconium carbide with a faint trace of the double titanium-zirconium carbide. In the bars poured at the lower temperatures, the orange-yellow to grey inclusions were fewer and larger than in the bars poured at higher temperatures. This was particularly noticeable in the bars poured at 1225° C., in which it could easily be seen that the large orange-yellow to grey inclusions were actually aggregations of smaller crystals of random orientation. To illustrate the effect of pouring temperature on the number of these inclusions, an inclusion count was carried out on all the bars, using the method described for counting the titanium carbide crystals. The results of these inclusion counts are given, along with the analyses, in Table X.

Melt Z16.—Melt Z13 had indicated the great affinity of zirconium for nitrogen and that a large proportion of the zirconium was fixed as zirconium nitride by melting in an indirect rocking arc furnace. With this in view an attempt was made to discover whether this inclusion could be introduced into crucible-melted

TABLE X.—*The Effect of Pouring Temperature on the Number of Orange-Yellow to Grey Zirconium Inclusions.*

Melt No.	Pouring Temp. °C.	Analysis.					Number of Inclusions per Unit Field.
		T.C. %.	Si. %.	Mn. %.	S. %.	Zr. %.	
Z14	1450	3·28	1·19	0·53	0·023	0·547	1·73
	1350						1·27
	1300						0·885
	1225						0·63
Z15	1450	3·44	1·25	0·94	0·23	0·496	1·56
	1350						1·115
	1300						1·085
	1225						0·605

iron by treating the molten metal with nitrogen. To do this a charge of 36 lb. of pig iron and 9 lb. of steel was melted in a salamander crucible, using a coke-fired, forced-draught furnace. The pig iron had the following composition:

Total carbon	4·04%	Sulphur	0·010%
Silicon	2·19%	Phosphorus	0·048%
Manganese	1·10%	Titanium	0·06%

When the charge was molten, 1½ lb. of 14% ferro-silicon-zirconium was added and a standard 1·2-in. dia. bar was poured. The crucible was then returned to the furnace and nitrogen was bubbled through the metal for 2 min., after which a second standard 1·2-in. dia. bar was poured. This melt has been numbered Z16. The analyses of the two bars are given in Table XI.

TABLE XI.—*Nitrogen-Treated Melt.*

Bar No.	T.C. %.	Si. %.	Mn. %.	S. %.	Zr. %.
Z16/1	3·15	2·74	0·90	0·038	0·32
2			0·90	0·040	0·25

Both bars contained identical inclusions—a number of orange-yellow to grey crystals, segregations of blue-grey zirconium carbide and a few crystals of the pinkish titanium-zirconium carbide. It would seem from these results that, under these conditions, zirconium either does not combine with nitrogen at all or does so only very slowly.

Group V.—Inclusions Produced by Alloying with Aluminium.

It has not been possible, so far, to find in the literature any reference to the effect of alloying cast iron with aluminium on the types of inclusions present in this alloy. For the purpose of this paper, several sets of irons having aluminium contents ranging from 0 to 10% were examined. These irons all contained enough manganese to balance the sulphur contents. The actual manganese contents ranged from 0·45% to 0·75% and the sulphur contents were all below 0·05%. In all the series examined, the same changes in the inclusions occurred with increases in aluminium content. Up to about 2% of aluminium, the only inclusions present in these series were normal idiomorphic manganese sulphide crystals. From 2% to 3·7% of aluminium the manganese sulphide crystals appeared to become much softer and were easily damaged during the polishing operation. Nevertheless, they still retained their characteristic idiomorphic outline. With aluminium contents above 4% the idiomorphic manganese sulphide was replaced by a phase of a similar colour but which was always perfectly spherical in outline. From 4% to 10% of aluminium the spheroids became larger and softer, being at the same time proportionally more difficult to polish. These spheroidal inclusions were optically isotropic and exhibited translucency effects similar to those of manganese sulphide.

Conclusion.

So far, this paper has been concerned with the metallography of small particles in the microstructure of cast iron, and throughout the paper these particles have been more or less continuously referred to as "inclusions." There appears to be a fairly general agreement that the terms "inclusion," "slag inclusion" and "non-metallic inclusion" are not very satisfactory for the collective description of these particles. Nevertheless, this has not prevented a liberal use of the terms, and the literature of ferrous physical metallurgy demonstrates a chaos of ideas on the subject, brought about in the first place by the lack of enforcement of exacting definition. In quite a large number of references, the particles are dealt with as a separate group of constituents entirely different from the major phases of the metallic matrix, with the implication that they are insoluble in either the solid or the liquid metal. In some extreme cases "inclusions" have, by analogy, been endowed with a misleading biological significance.

The chemical determination of oxides and silicates in steels and cast irons has proceeded apace in the last few years, and quite frequently this branch of activity is referred to as the analysis of non-metallic inclusions, &c. Thus there are two points of view on the question which are now actively developing—the chemical and the metallographical. At this point the question arises: Are the chemist and the metallographist talking about the same thing? In the case of steel, these two lines of investigation are probably correlative, but in the case of ordinary cast irons this is not so at the moment. The only "inclusions" seen in ordinary cast irons are manganese sulphide, titanium carbide, titanium cyano-nitride and perhaps iron sulphide. So far, the metallographist has not been able to detect oxides or silicates of iron, manganese or aluminium in ordinary commercial materials. From the point of view of the microscope, they do not exist, are completely soluble in the metal, or exist in such a finely divided form that they are not resolved. It is seen that the chemist and the metallographist have been tempted to use the same group of words to describe what are very different constituents—different in chemical composition, in mode of distribution and in order of size.

Portevin and Castro have pointed out that the term "inclusion" connotes a substance essentially foreign to the constitution of the alloy, perhaps detrimental to it, and more or less insoluble in the metal. With regard to such constituents as mechanically trapped furnace slag and moulding sand, it is obvious that the term "inclusion" is a good one, and it is still further improved by compounding—furnace-slag inclusion and moulding-sand inclusion. However, in the case of the particles described in this paper it is obvious that the terminology is not all that is to be desired. Perhaps with the exception of the cyano-nitrides, all the particles dealt with are at least partially soluble in molten metal at the normal pouring temperatures.

Portevin and Castro have also shown how difficult it is to subdivide "inclusions" into metallic and non-metallic groups, and they have suggested that the following arbitrary classification could be made:

Inclusions: Silicates, aluminates, oxides and sulphides.

Metallic Constituents: Nitrides, carbides (borides and perhaps phosphides).

This grouping would be very difficult to apply to cast iron, since the only metallographic inclusions would be the sulphides. Now, the blue-pink "inclusion" described in Group II. of this paper is largely titanium sulphide, but it probably also contains a fair amount of dissolved titanium carbide which is presumably a metallic constituent. In this context, it is interesting to note that Portevin [59] has made the very important point that the idea of an impurity or detrimental element is entirely relative and variable, and it would seem possible to extend this conception to the term "inclusion."

Inclusions have been stated to be the cause of a number of obscure phenomena exhibited by cast iron and it has been implied that they have properties peculiar to themselves. A careful metallographic consideration of all the particles dealt with indicates no properties, and, what is more important, no modes of formation peculiar to these particles which are not exhibited in an entirely analogous way by the major phases of the cast-iron microstructure. It would seem, then, that the essential difference between these particles and the remainder of the metallic phases is the fact that the former occur in much smaller quantities. With this point in view, it is suggested here that these particles could be called "minor phases" to great advantage. This terminology has not so many misleading implications as the one already in use, and is likely to prevent the formation of prejudices and biased opinions when the properties of these small particles are being investigated or discussed.

The author wishes to thank the Council of the British Cast Iron Research Association for permission to publish this paper.

REFERENCES.

(1) Von Keil, Mitsche, Legat and Trenkler: *Archiv für das Eisenhüttenwesen*, 1934, vol. 7, p. 579.

(2) Mitsche: *Iron and Steel Institute, Carnegie Scholarship Memoirs*, 1934, vol. 23, p. 65; 1936, vol. 25, p. 41.

(3) Bolton: *Transactions of the American Foundrymen's Association*, 1937, vol. 45, p. 467.

(4) Morrogh: *Bulletin of the British Cast Iron Research Association*, 1939, vol. 5, p. 296.

(5) Scott and Joseph: *Metals and Alloys*, 1938, vol. 9, p. 299.

(6) Levy: *Iron and Steel Institute, Carnegie Scholarship Memoirs*, 1911, vol. 3, p. 260.

(7) Stead: *Journal of the Iron and Steel Institute*, 1892, No. II., p. 240; 1893, No. I., p. 71.

(8) Baykoff: *Metallurgie*, 1909, vol. 6, p. 103.

(9) Herty and True: *Transactions of the American Institute of Mining and Metallurgical Engineers*, 1925, vol. 71, p. 540.

(10) Levy: *Proceedings of the Staffordshire Iron and Steel Institute*, 1909, vol. 25, p. 81 (*see* Hailstone's discussion, p. 96).

(11) McCance: *Journal of the Iron and Steel Institute*, 1918, No. I., p. 239.

(12) Houghton: *Proceedings of the British Foundrymen's Association*, 1906, p. 50.

(13) Heike: *Stahl und Eisen*, 1913, vol. 33, pp. 765 and 811.

(14) Shibata: *Technology Reports of the Tôhoku Imperial University*, 1928, vol. 7, p. 279.

(15) Allison: *Foundry Trade Journal*, 1928, vol. 38, p. 295.

(16) Herty and Gaines: *Transactions of the American Institute of Mining and Metallurgical Engineers*, 1927, vol. 75, p. 434.

(17) Herty and Gaines: *Blast Furnace and Steel Plant*, 1927, vol. 15, p. 467.

(18) Wohrman: *Transactions of the American Society for Steel Treating*, 1928, vol. 14, pp. 87, 255, 385 and 539.

(19) Andrew and Binney: *Journal of the Iron and Steel Institute*, 1929, No. I., p. 346.

(20) Andrew, Maddocks, and Fowler: *Journal of the Iron and Steel Institute*, 1931, No. II., p. 283.

(21) Bardenheuer and Geller: *Mitteilungen aus dem Kaiser-Wilhelm-Institut für Eisenforschung*, 1934, vol. 16, p. 77.

(22) Benedicks and Löfquist: "Non-Metallic Inclusions in Iron and Steel." London, 1930: Chapman and Hall, Ltd.

(23) Hanemann and Schildkötter: *Archiv für das Eisenhüttenwesen*, 1929, vol. 3, p. 427.

(24) Satô: *Technology Reports of the Tôhoku Imperial University*, 1931, vol. 10, p. 453.

(25) Vogel and De Vries: *Archiv für das Eisenhüttenwesen*, 1931, vol. 4, p. 613.

(26) Meyer and Schulte: *Archiv für das Eisenhüttenwesen*, 1934, vol. 8, p. 187.

(27) Vogel and Baur: *Archiv für das Eisenhüttenwesen*, 1933, vol. 6, p. 495.

(28) VOGEL and BAUR: *Archiv für das Eisenhüttenwesen*, 1937, vol. 11, p. 41.
(29) WENTRUP: *Iron and Steel Institute, Carnegie Scholarship Memoirs*, 1935, vol. 24, p. 103.
(30) SATÔ: *Technology Reports of the Tôhoku Imperial University*, 1934, vol. 11, p. 234.
(31) DAEVES: *Stahl und Eisen*, 1931, vol. 51, p. 202.
(32) JOSEPH and HOLBROOK: United States Bureau of Mines, 1934, *Report of Investigations No.* 3240, p. 13.
(33) WHITELEY: Seventh Report on the Heterogeneity of Steel Ingots, Section IIIA, *Iron and Steel Institute*, 1937, *Special Report No.* 16.
(34) WYCKOFF: "The Structure of Crystals." New York, 1931: Chemical Catalog Company Inc.
(35) SCHAFMEISTER and MOLL: *Archiv für das Eisenhüttenwesen*, 1936, vol. 10, p. 155.
(36) DAYTON: *Metals Technology*, 1935, vol. 2, No. 1, *Technical Publication* No. 593.
(37) PORTEVIN and CASTRO: *Journal of the Iron and Steel Institute*, 1935, No. II., p. 237.
(38) NORBURY: *Journal of the Iron and Steel Institute*, 1939, No. II., p. 161P.
(39) ALLEN: *Transactions of the American Foundrymen's Association*, 1931, vol. 39, p. 733.
(40) SCHWARTZ: *Metals and Alloys*, 1935, vol. 6, p. 328.
(41) SCHWARTZ and JUNGE: *Transactions of the American Foundrymen's Association*, 1934, vol. 42, p. 94.
(42) SCHWARTZ: *Journal of the Iron and Steel Institute*, 1938, No. II., p. 205P.
(43) BECKER: "Hochschmelzende Hartstoffe und ihre technische Anwendung." Berlin, 1933: Verlag Chemie, G.m.b.H.
(44) NORBURY: *Proceedings of the Institute of British Foundrymen*, 1928–1929, vol. 22, p. 151.
(45) PORTEVIN and CASTRO: *Journal of the Iron and Steel Institute*, 1937, No. II., p. 223.
(46) BLAIR and SHIMER: *Transactions of the American Institute of Mining and Metallurgical Engineers*, 1901, vol. 31, p. 748.
(47) HANEMANN and SCHRADER: "Atlas Metallographicus," vol. 2. Berlin, 1937: Gebrüder Borntraeger.
(48) URBAN and CHIPMAN: *Transactions of the American Society for Metals*, 1935, vol. 23, p. 645.
(49) MELLOR: "Comprehensive Treatise on Inorganic and Theoretical Chemistry," vol. 7, p. 90. London, 1927: Longmans, Green & Co., Ltd.
(50) BERGLUND: "Metallographer's Handbook of Etching." London, 1931: Sir Isaac Pitman and Sons, Ltd.
(51) WELLS: *Transactions of the American Society for Metals*, 1938, vol. 26, p. 289.
(52) NORBURY and MORGAN: *Journal of the Iron and Steel Institute*, 1936, No. II., p. 327P.
(53) HOFMANN and SCHRADER: *Archiv für das Eisenhüttenwesen*, 1936, vol. 10, p. 65.
(54) TOFAUTE and BÜTTINGHAUS: *Archiv für das Eisenhüttenwesen*, 1938, vol. 12, p. 33.
(55) FEILD: *Transactions of the American Institute of Mining and Metallurgical Engineers*, 1923, vol. 69, p. 848.
(56) FEILD: *Transactions of the American Institute of Mining and Metallurgical Engineers*, 1924, vol. 70, p. 201.
(57) FRIEDRICH and SITTIG: *Zeitschrift für anorganische und allgemeine Chemie*, 1925, vol. 144, p. 71.
(58) BECKER and EBERT: *Zeitschrift für Physik*, 1925, vol. 31, p. 268.
(59) PORTEVIN: *Bulletin de la Société des Ingénieurs-Soudeurs*, 1935, vol. 6, p. 1914.

[This paper was discussed jointly with the preceding one by H. Morrogh on "The Polishing of Cast-Iron Micro-Specimens and the Metallography of Graphite Flakes."]

JOINT DISCUSSION.

The two papers by H. Morrogh on "The Polishing of Cast-Iron Micro-Specimens and the Metallography of Graphite Flakes" and on "The Metallography of Inclusions in Cast Irons and Pig Irons" were discussed jointly.

Dr. J. E. Hurst (Messrs. Bradley and Foster, Ltd., Darlaston) welcomed the opportunity of speaking on the two papers presented by Mr. Morrogh, as he had been able to see some of the author's work at first hand, and would like, as would all cast iron metallurgists, publicly to congratulate him on the work he had done. The most important feature of that work was probably the systematic and organised way in which he had carried it out, and the mode of presentation in the two papers was also most praiseworthy. Most metallurgists had had the opportunity of making many observations of the type that the author had made, and they had always been perplexed by the difficulty of recording them in some systematic manner.

So far as cast iron metallurgy was concerned, the author had established a scientific method of studying the question of inclusions; he had shown rare experimental skill in the preparation and examination of his specimens and in the use of the modern microscope—he had used the rotating stage and the polarising method of examination very ably. Finally, the British Cast Iron Research Association, with which the author was associated, also deserved a share in the congratulations.

He had only one remark of a critical nature to make, and it concerned the Conclusion in the paper on the polishing of cast-iron micro-specimens, where it was stated that specimens could be prepared in fifteen minutes. Although the method of preparing the specimens was set out in very great detail, he would find it impossible to prepare a specimen himself and get the results which the author had obtained in such a short space of time as fifteen minutes.

In the same paper, he had been very fascinated by Figs. 19 to 22, in which the author was able to disclose—and it was probably the first time that this had been done—where the hypereutectoid and eutectoid graphite went to. People had talked very glibly in the past about hypereutectoid and eutectoid graphite, and might have surmised where it had finally deposited itself, but he did not think that they had ever before been shown so clearly exactly where the hypereutectoid graphite was to be found.

Dealing still with the same paper, it would be interesting to know whether the author took any precautions in the polishing of the graphite structure itself to be sure that the last traces of polishing powder were eliminated. It seemed possible for particles of polishing powder to remain embedded in the surface of such a soft material as a graphite flake. In the case of the large graphite flake shown in Fig. 4, there was a small white spot about half an inch from the bottom, and it would be of interest to know whether that happened to be a little particle of Diamantine powder. The author was obviously very accomplished in the microscopic examination of metals, and it would be interesting to know whether he had considered the method of electrolytic polishing, and whether there was any hope that that might be a royal road to polishing cast-iron specimens, thus saving people the trouble of trying to do what the author would like them to do in fifteen minutes.

In the other paper, on the metallography of inclusions in iron, Figs. 22 to 25 were especially fascinating, and particularly the one with the inclusion of sulphide in the centre. The author's work in this connection had brought out in a very striking manner the importance of the sulphur compounds, of both manganese and iron, in determining the structural characteristics at least of malleable cast iron, if not of ordinary grey iron; a good many cast iron metallurgists were convinced that the importance and significance of those two constituents had not yet been exhausted. One of the most startling things that the author had revealed was the absence of oxides and silicates, and still more startling was the fact that he found no oxide inclusions in the samples of aluminium-alloy cast iron, even when containing as much as 10% of aluminium. Personally, he was concerned with making castings which did not contain nearly as much aluminium as that, but which contained something of the order of 1·5% to 1·75% of aluminium for nitrogen-hardening cast-iron, and he felt sure that they were concerned with what they called oxide inclusions. Whether they were visible on micro-examination he did not know, but it was astonishing that the author was unable to see such oxide inclusions in his aluminium-alloy cast-irons. One wondered whether the samples which the author examined were treated by processes using chlorinated wax; it would be interesting to know.

At the present time, when the methods and the instruments of the physicist were occupying the attention of metallurgists so much, it was very refreshing to see that the trusted weapon of the metallurgist, the microscope, could be used so effectively.

Mr. T. Henry Turner (London and North Eastern Railway Company, Doncaster) suggested that the British Cast Iron Research Association should seriously consider linking up with the Scientific Instrument Research Association and the microscope makers and carrying the work in question still further, perhaps linking up with The Iron and Steel Institute as well, because in Great Britain there was not that real co-operation between the various sides in regard to metallography which was found abroad. On the now dark Continent there had been a marvellous collaboration between the microscope makers and the universities, but that had never been the case here. Now that the author had, so to speak, opened up the possibility of recommended methods of metallography, there should really be a team not only of cast iron men but also of scientific instrument men to deal with the problem, as well as to go on to the next stage.

The author's method of polishing had its limitations, and it was pointed out in the penultimate paragraph of Part I. of the paper on polishing cast-iron micro-specimens that phosphoric irons did not react to it so well as hematite irons. That was unfortunate, because in the particular work with which he personally was concerned the phosphorus always ran up to the 1½% range, and it stood up proudly when polishing was done by the method used by the author. It was probably correct to say that the author recommended a hard pressure working slowly, whereas ordinarily at Doncaster they used a soft pressure working quickly. That might be too concise a summary of the difference, but it roughly represented it; and with the high-phosphoric irons, after trying the author's method, they still thought that it suited them best to use the high-speed method. With regard to the speed, it would be useful if, when the paper was printed in its final form, the author would give the speed in feet per second rather than revolutions per minute, because when wheels could be of different sizes the latter did not convey very much. The author's work must certainly be continued.

Mr. J. G. Pearce (The British Cast Iron Research Association, Birmingham) remarked that it might not be altogether appropriate for him to congratulate the author, but it was interesting to recall that when the French metallurgical mission visited the laboratories of the British Cast Iron Research Association in March, 1940, the leader, Professor Portevin, was very interested in the author's work and expressed his very great admiration for it; indeed, he offered to arrange for its publication.

With reference to the paper on polishing, the speaker thought it would be agreed that in no country were finer or more informative reproductions of these structures being produced. It might reasonably be hoped that the full explanation of the technique would result in an improvement in many of the micrographs of these alloys published, although for ordinary routine works examinations simpler procedure might suffice.

So far as inclusions were concerned, it was now known what inclusions pig iron and cast iron contained, although microscopic evidence could give no quantitative estimation. He had little to add to what he had said on that matter in the Third Report of the Oxygen Sub-Committee,[1] in which he had endeavoured to indicate a few points which might assist in reconciling the apparent divergence between the results of Mr. Morrogh's work and that of Mr. Taylor-Austin; and in that connection he had been most interested in the

[1] *Journal of the Iron and Steel Institute*, 1941, No. I., p. 295 P (this volume).

comments made that afternoon on sub-microscopic inclusions in steels. He always felt that metallurgists must suggest with very great reluctance the possible causes of effects on lines which relieved them of the necessity for providing any direct evidence, and that was why the existence of sub-microscopic particles and inclusions of that kind required to be postulated with very great care.

The author had not dealt with inclusions of alloying elements other than those arising from zirconium and aluminium, which it was desirable, for the reasons pointed out by the author, to treat along with those of titanium. Titanium inclusions were now regarded as perfectly normal, but, with reference to what had been said by Mr. Turner, it should be mentioned that the work on the inclusions arising from alloying elements was being continued.

The author did not deal with the phosphide eutectic in either the binary or the ternary form, because it was not an inclusion; but a great deal was still obscure about that particular conglomerate, and he would like to recapitulate a few points which might tend to confuse what to some metallurgists was a perfectly simple matter, the binary or ternary phosphide eutectic. In the original chemical residue method for determining inclusions, the alcoholic iodine solvent as it was then used failed, because it did not dissolve the phosphide. The aqueous iodine solvent in its final form (*i.e.*, with various subsequent treatments) was much more successful, and it was shown in the Second Report of the Oxygen Sub-Committee [1] that with the aqueous iodine solvent, prior to the treatments, phosphorus remained in the residue after the aqueous iodine solution, and the quantity of it was similar to the amount of iron phosphide in the sample as recorded by Stead's method, details of which were given in classical papers to the Institute. Stead declared that dilute nitric acid separated iron phosphide from phosphorus in solution. In the iron-carbon alloys under consideration—pig irons and cast irons—metallurgists to-day did not believe that phosphorus was appreciably in solution, but the chemical evidence did suggest that there must be at least two forms in which phosphorus was present. About a quarter of it was soluble in aqueous iodine, and appeared to correspond with that stated by Stead to be in solid solution, and the large bulk of the remainder appeared to correspond with Stead's iron phosphide. Even so, there was still a small amount left in the residue after the treatments which followed the use of aqueous iodine, and which was apparently wholly removed when titanium was wholly removed. That suggested a connection between titanium and phosphorus. It was known microscopically that there was evidence of association between iron phosphide and iron carbide, which, with iron, made up the ternary phosphide eutectic. There was chemical evidence to support that association, because in pig and cast irons containing carbide the joint removal

[1] *See* Eighth Report on the Heterogeneity of Steel Ingots, *Iron and Steel Institute*, 1939, *Special Report No.* 25.

of the carbide and phosphide was a very much more difficult operation than the removal of phosphide in the absence of carbide.

Those points dealt with what could not be regarded strictly as an inclusion, but referred to a subject which he thought must be tackled sooner or later if metallurgists were to understand more thoroughly the structures and properties of the particular range of materials in question.

Mr. VERNON HARBORD (Messrs. Riley, Harbord and Law, London) said the point that he wished to make was in no way a criticism of the papers, but was a suggestion which might enable the author to carry his very valuable work a stage further. He might be accused of going back to almost prehistoric methods, but he thought that, in view of the improved technique in polishing, the old method of heat-tinting developed by Stead might possibly give very valuable information about the complex constituents in cast iron which were in question. He did not say that entirely without evidence. A few years ago, he was interested in certain constituents of cast iron, and he went to a very great deal of trouble in polishing specimens, rather on the lines outlined in the paper; and he found that, when heat-tinted, they were capable of examination under the highest power of the microscope without the general blurring effect which was obtained in the old days of ordinary polishing, and which definitely prescribed a limit to Stead's very valuable work at that time. It was not then found generally possible to examine any heat-tinted specimen at magnifications much over 100 diameters. He did not know whether anything had been done on those lines, but certainly a properly polished specimen was capable of examination under high power when heat-tinted, and it might be worth the author's while to consider that method as a further means of throwing light on a subject on which he had already thrown so much.

Mr. A. ALLISON (J. J. Habershon & Sons, Ltd., Sheffield) said that the two papers presented by Mr. Morrogh constituted a valuable record of the metallography of the constituents of cast iron from the three aspects of preparation of the specimens, examination of the matters found and verification of their character.

The improved technique had enabled better records to be made of sulphides and graphite than hitherto, and had shed further light on the composition and treatment of cast iron.

With regard to the preparation of specimens to give an accurate presentation of graphite, he (Mr. Allison) had outlined a method [1] which he was glad to find the author had also used and developed further. Apart from the method, he felt that a certain amount of personal skill was required to produce really satisfactory results, and the present micrographs confirmed this view.

[1] A. Allison, *British Cast Iron Research Association*, 1929, *Bulletin No.* 23.

On p. 222 P of the paper on the metallography of the inclusions, referring to Fig. 31 showing mixed pearlite, he was not sure if the author had made sufficient allowance for the varying angle of section of the pearlite grains, particularly as Figs. 28 and 29 appeared to be similar. However, Fig. 32 appeared to support the author's suggestion that a very small amount of iron sulphide was sufficient to produce a band of cementite running through a pearlite area under certain conditions not stated, and it was certainly of importance that this structure could be related to poor mechanical properties. The analysis quoted on p. 222 P might be considered in conjunction with the results of Taylor's experiments,[1] from which it appeared that the best mechanical properties of malleable iron were associated with a balanced ratio of manganese and sulphur. Presumably the formula $Mn\% = (S\% \times 1{\cdot}7) + 0{\cdot}3$ was based upon the effects upon the depth of chill, and might be a measure of another physical property of iron in the cast state.

It might also be expected that a lengthy heat treatment such as malleable-annealing would enable small proportions of dissolved elements to separate and assume a stable phase.

Since crystallisation occurred more easily in the presence of nuclei, there was little cause for surprise that MnS crystals should frequently be seen to contain a nucleus, nor that the crystals themselves should act as a starting point for graphitisation, and the author's excellent micrographs illustrated this point very efficiently.

In his opinion the paper generally derived much of its value from being based upon a large number of routine specimens, because he felt that phenomena observed in numerous specimens of widely different origin afforded cumulative evidence of a convincing character in support of the views expressed.

There was such a thing as bulk research or statistical research. In the case of many castings—chilled-iron rolls, for instance—it was not feasible to cut off a test-piece and make physical tests which could be related to service, as was done with forgings. Therefore it was necessary to employ the method of bulk or statistical research, based on records of a large number of cases.

He agreed with the author's conclusion as to the unsatisfactory terminology of constituents in a naturally heterogeneous material like cast iron and the suggestion that such things as manganese sulphide, &c., should be regarded as minor phases. Also he had sometimes wondered if the exhaustive researches into contained oxides and silicates were not explorations of a barren field, so far as the ultimate properties of the material were concerned.

Dr. T. SWINDEN (Member of Council; The United Steel Companies Ltd., Central Research Department, Stocksbridge, near Sheffield) associated himself with previous speakers in congratulating

[1] E. R. Taylor, *Iron and Steel Institute, Carnegie Scholarship Memoirs*, 1926, vol. 15, p. 381.

the author on a classical piece of work, and expressed his interest in two points which had already been made, the first being the possible use of anodic etching, of which there had already been some experience in the case of steel, and the second being that made by Mr. Pearce, namely, that, having developed this technique, the author could not be allowed to stop until he had revealed the at present sub-microscopic inclusions.

The PRESIDENT (Mr. John Craig, C.B.E., Messrs. Colvilles, Ltd., Glasgow) congratulated the author on, while still a young man, presenting papers of such importance. He was glad to think, he added, that young men were coming into the industry who early in life were offering such serious contributions to the solution of great problems.

JOINT CORRESPONDENCE.

Mr. W. B. PARKER (British Thomson-Houston Co., Ltd., Rugby) wrote that from the Introduction to the paper on the metallography of inclusions in cast irons and pig irons it was readily seen that Mr. Morrogh possessed a very clear and useful grip of the whole subject of " inclusions " in the metallic materials with which he dealt. The paper is one which should incite all young foundrymen and foundry chemists to take more and more interest in this subject, as it was directly of importance to their business.

What was required was still more practical work on the subject, which was unique in its combination of pure and applied metallurgy. For successful pursuit of the subject good microscopic equipment was required, also the capacity of close observation, resourcefulness and clever personal manipulation.

One of the most pleasing and useful features of the paper was Mr. Morrogh's careful descriptions of the metallographic specimens and the manner in which the individual micrographs were prepared. The latter made the paper really useful, since comparisons could be made with other work, personal or published. The micrographic work was really excellent.

In the Introduction the author dealt with the intricate subject of nomenclature, which still required more study. If the author's term " minor phases " were accepted it seemed to follow that metallurgically these minor phases should always have been formed by reactions between the normal constituents of *molten* pig irons, cast-iron scrap and steel scrap, also any special additions that went to form the *molten bath of metal*. It would cut out " inclusions " formed by added fluxes and the refractories of the furnaces.

The majority of the examples dealt with by the author were of his " minor-phase " type, but it was dangerous to limit the whole possibility of " inclusions " in cast iron and pig iron to this definition. On the second page it was stated that "oxides and silicates . . . are virtually non-existent from the point of view of the metallography of cast iron." This finding closely agreed with the present writer's, but, as admitted later by Mr. Morrogh, silicates could sometimes be found in cast iron. Experience showed that this most frequently occurred when the charges had been rapidly melted, tapped and used very soon after tapping. In these cases the silicates were "slags" from the cupola, and it was doubtful whether such substances should be termed "inclusions" or more justly "dross." Old foundrymen always permitted a "standing" period between filling the ladle and pouring the castings and frequently skimmed the metal twice during this period. This resulted in clean castings. Superheating in melting was probably desirable, but superheat in *poured* cast iron was seldom desirable.

With further reference to nomenclature it was noted that Bolton's suggestion to classify inclusions in cast iron as (1) foreign and (2) inherent was probably more practical, but the word "inherent" was rather too specific, since it appeared to have an implication which made it scientifically unsound. This term tended to give the idea of inevitability or unavoidability. Was this ever truly the case in foundry practice? The writer was of the opinion that it was not so with *many* "inclusions," and that foundry practice would be at fault if daily examinations seemed to indicate it to be so.

On the fourth page, "Metallography of Individual Inclusion Groups," it was noticed that Group I., sulphides, did not include nickel sulphide or molybdenum sulphide. These seemed desirable additions to the scheme, now that nickel-alloy cast irons were so common and molybdenum was increasing in employment. Molybdenum had a strong affinity for sulphur.

In considering the curves of Fig. 1 (*b*) it should be noticed that blast-furnace slag was used by Joseph and Holbrook for these determinations, but without the full analysis of that slag it could not be ascertained whether the slag was of strong, medium or weak basic reaction. In their original work Joseph and Holbrook gave considerable attention to the basicity of the slags that they employed, and Mr. Morrogh should supply the necessary data as a footnote.

On p. 219 P the author described *iron sulphide* and stated that (according to the collected data of Wyckoff) this substance had a hexagonal structure—and, by inference, that this was the form of iron sulphide found in cast iron and pig iron. This point was by no means certain, since there was a good chance that in *metals* one did not get only one sulphide of iron, and this was especially so in malleable cast iron. It should also be noted that the determinations quoted by Wyckoff (*per* de Jong and helpers) dated back to 1925–1927 and did not refer to the material isolated from cast iron or pig iron, but to minerals. This point concerning the form of iron sulphide was raised by the present writer in a report to the British Cast Iron Research Association in its early days (1922). Judging from the colour of the sulphide in the malleable cast iron *then reported upon* there was a probability that it was pyrrhotite (Fe_6S_7) This formula was variously written (6FeS + S) and (FeS + S_x), which showed that at the present time there existed considerable uncertainty as to its nature. In fact, Wyckoff stated[1]:

> "A striking characteristic of many of these substances (pyrrhotite-like crystals) is the ease and extent to which they can dissolve an excess of one or the other of their constituent elements."

In ordinary every-day cast iron and pig iron it was highly probable that the iron sulphide (if present) was not a simple substance of fixed crystalline type; so precise statements concerning the crystallographic structure of any compound should be accompanied by an analysis of it and the calculated formula. It seemed that at one end of the scale of iron sulphides one could get pyrites (cubic), FeS_2 (or FeS + S), or marcasite (rhombic), FeS_2, troilite (cubic or hexagonal), FeS, and pyrrhotite (hexagonal), Fe_6S_7. The matter could possibly be settled so far as cast iron and pig iron were concerned by isolation of the substance, followed by a chemical and X-ray analysis.

On p. 220 P the author made the statement that "iron sulphide can act as a *nucleus* for the temper carbon." Now, apart from the author's observations on p. 221 P, this raised a rather important question that so far had not been cleared up in metallurgical subjects, namely, could inorganic substances of different crystallographic systems induce crystallisation in one another when in the *molten* condition—*i.e.*, could they act as "nuclei" to each other *in a true sense?* For instance, could graphite (hexagonal) be induced to crystallise by means of pyrites (cubic) or by manganese sulphide, MnS (cubic)?

It might be surmised as very probable that similar crystallographic substances might or could induce crystallisation in one another; for instance, pyrrhotite (hexagonal) and graphitic carbon (hexagonal), or manganese sulphide (cubic) and iron (ferrite) (cubic). But one would like to know just what was understood or implied by the term "nucleisation"? Did the author intend this term to have the same significance as physical-crystallographers embodied in their term "germ-crystal"? If not, exactly what was the significance?

It must not be overlooked that for possibly a thousand years it had been known that crystals of inorganic salts from their saturated

[1] "The Structure of Crystals" (American Chemical Society Monograph Series, No. 19), Second Edition, p. 217. New York, 1931: Chemical Catalog Co., Inc.

aqueous solutions grew upon the sides of their container and upon anything suspended in the solution. For example, clean rods of iron suspended vertically in saturated aqueous borax solutions became covered by large beautiful crystals of borax (mono-symmetric system) in a comparatively short time, if the conditions were correct.

This type of crystallisation was, no doubt, mainly coupled with temperature gradients—the rod abstracted heat from the vat liquor. It seemed worth while to look more deeply into the so-called nucleisation from this point of view—it seemed impossible for it to be similar to the germ-crystal "nucleus." Its study would necessitate determinations as to *which* of the constituents involved solidified first in the liquid metallic bath, and whether it possessed the same crystal system throughout the complete solidification of the bath. It might not do so!

The author's discussion on pp. 220 P–223 P was very interesting and suggestive of more work in this and other directions. It was with much pleasure that the writer congratulated the author upon his paper.

Mr. G. F. Comstock (The Titanium Alloy Manufacturing Company, Niagara Falls, N.Y., U.S.A.) wrote that, after reading so much during the past five or ten years about the oxide and silicate inclusions in cast iron, which could only be imagined and were never actually seen with the microscope, it was refreshing to read this excellent paper on the metallography of inclusions in cast and pig irons, which was confined to a realistic view of the situation. The writer, who had been looking at the inclusions in polished sections of titanium-treated steel and cast iron through a microscope for many years, found nothing of importance to criticise in the paper, and appreciated this opportunity to congratulate the Institute and the author on the publication of so many facts pertaining to the metallography of cast iron, which had apparently escaped recognition in some quarters in the past. The author might perhaps be interested in some comparatively crude illustrations of some of the "minor constituents" which he had described, which appeared under the title "Titanium Nitride in Steel" in *Metallurgical and Chemical Engineering*, 1914, vol. 12, Sept., No. 9, p. 577.

There was one point on which some further explanation from the author would be helpful. He reported correctly that titanium carbide and nitride form a series of solid solutions, and that the so-called cyano-nitride was not a definite compound but merged gradually into the carbide when the nitrogen content decreased. This involved a fading-out of the pink or orange colour, so that in cast iron, where there was always plenty of carbon but not of nitrogen, typical angular titaniferous crystals of all shades, varying from distinct pink or orange to white or pale grey, were commonly seen. With this gradual merging of the two types, how was it possible for the author to report definite numbers of crystals per unit field of each type separately, as in Table IV., for instance? Was every crystal with the slightest trace of pinkish colour counted as cyano-nitride? It would seem that this would be a very difficult distinction to make with any degree of certainty, and it was remarkable that counts made in this way could check each other, even if made by the same observer at different times.

The evidence for the higher solid solubility of titanium carbide than of the cyano-nitride was extremely interesting to the writer, and more definite data on this point would be highly desirable.

Mr. J. A. Roadley (British Thomson-Houston Co., Ltd., Rugby) wrote that the polishing technique of grey cast irons was a subject which had never been deeply studied until the present paper by Mr. Morrogh. It was a technique which had aroused considerable interest at the B.T.H., Rugby, and had been studied very carefully by the writer and his colleagues. Although their study had perhaps not been quite so intensive as the author's, they had nevertheless arrived at practically the same conclusions, *i.e.*, rubbing with a series of emery papers and finally polishing with Diamantine powder upon a wheel covered with Selvyt cloth. There were a few slight differences in their technique, however, namely:

(1) They used four emery papers, grades 1*G*, 1*M*, 0 and 00.

(2) Instead of using a thick paste of Diamantine, they soaked the Selvyt cloth with water, allowed surplus water to be driven off centrifugally and then rubbed dry Diamantine powder well into the pile with the fingers. By this means, the amount of Diamantine powder used was cut down to a minimum. This was an important matter in industrial work.

(3) After etching, all further repolishing was done by hand, using magnesium oxide upon a wet Selvyt cloth, and not by machine. They were inclined to think that, no matter how much care was taken when polishing by machine, a perfect finish was not easily obtained. When the very final polishing process was done by hand the skill of the worker had full play, and the necessary frequent inspections of the surface were made as a matter of course.

It was presumed that when Mr. Morrogh spoke of emery paper grades 1*G*, &c., he meant the well-known brand of paper, namely, French Hubert emery paper. This should have been mentioned, as no doubt all metallurgists found it impossible to obtain this brand of paper just now, and the writer's experience was that the grades of the papers now obtainable did not correspond with the same grades of the Hubert brand. Also, it was now impossible to obtain Diamantine powder. Had Mr. Morrogh found a good substitute for this besides the magnesium oxide mentioned in his paper?

It might interest many other investigators to know that the B.T.H. Research Department did not develop this technique upon grey cast irons but on steels, from which it was applied to graphitic irons. They found the method of alternate polishing and etching an ideal way to prepare a micro-specimen of steel to show the exact nature and amount of non-metallic inclusions in it. These inclusions, of course, were not as soft and friable as graphite and therefore did not need etching and repolishing so many times as a graphitic iron, but they had found it a very good plan to etch and repolish several times all steels containing various non-metallic inclusions, especially when these were in very fine particles widely dispersed.

In his conclusion, Mr. Morrogh stated that this type of technique should be adopted in the preparation of specimens taken from nearly every type of ferrous and non-ferrous alloy. This was a statement which they at Rugby very heartily endorsed, as the results obtained were well worth the little extra time and trouble involved.

Part II. of Mr. Morrogh's first paper was certainly a very strong recommendation to Part I., as it was hard to imagine anyone producing better micrographs of grey cast iron than those obtained by the author from his specimens prepared by using his new technique of polishing.

Mr. Morrogh's remarks upon the structure of the graphite flakes were indeed very enlightening, and he was to be congratulated upon the efficient way in which he had dealt with the subject.

It was of very great interest to the writer that once more the use of polarised light had been introduced into a metallographic investigation, as it was his belief that polarised light could be much more widely employed in this branch of science than it had been to date, especially in the study of inclusions in steels and also in the study of intermetallic constituents in non-ferrous alloys such as in the aluminium light-alloy type of metals.

Referring to the paper on the metallography of inclusions Mr. Roadley wrote that the Introduction showed that the author had made a very wide and comprehensive study of all the information available upon the nature and character of inclusions or " minor phases " in cast iron and also steel.

The arbitrary groups into which the inclusions had been classified by the author were based upon the more evident of their characteristics, and, although, as stated by the author, they did not solve the ultimate identity of the inclusions, yet they did form a very useful primary test.

It was Group II., pink inclusions, in which the writer was most directly interested. It had been the writer's great privilege to visit Mr. Morrogh at the British Cast Iron Research Association laboratories at Birmingham during the course of this work and to collaborate with him in some small way upon the question of titanium inclusions. Although the writer was studying the titanium inclusions in a very complex alloy steel and the author was solely engaged upon cast iron, close agreement was obtained in the majority of cases.

During this short visit to Birmingham the writer noticed that the author was carrying out his work with a Vickers projection microscope and making his observations through the visual part of the apparatus. The writer was, during much of this investigation, employing the same model of Vickers microscope but with a proper graduated eye-piece analyser on loan from Mr. W. B. Parker. This eye-piece differed slightly from that used by Mr. Morrogh in that it was fitted with a nicol prism for the analyser as contrasted with the "film" type used by the B.C.I.R.A.

When the same specimens of steel were examined at Rugby, less definite results were obtained than with the "film-type" analyser. This was more noticeable in the case of the titanium sulphide inclusions. The change in colour from blue to pink or *vice versa* described by the author was very indecisive indeed and in many specimens which had definitely shown the change at Birmingham it was doubtful whether the change did take place at all at Rugby. It appeared to be partly a personal equation, since some of the writer's colleagues who were consulted stated there was a change and others that there was none. It must be remembered here that the inclusions in the alloy steel examined by the writer were very minute indeed and could in no way be compared in size with those obtained by Mr. Morrogh in his specially prepared cast irons, but, nevertheless, a definite uncertainty occurred in the results obtained upon the same micro-specimen of steel using the two different instruments.

Later on the subject was discussed with a representative of the makers of the Vickers projection microscope and the question of the eye-piece for use with polarised light was specially dealt with. The writer was then informed that when the direct visual part of the instrument was used, the apparatus was very definitely not suitable for work with polarised light, owing to the fact that a glass reflector had to be used to reflect the light into the eye-piece. The writer now had at his disposal an older micrescope of the Watson series which had recently been fitted with a nicol prism polariser, and an analyser, to continue the work. Unfortunately pressure of other work had prevented him from doing much as yet. Had Mr. Morrogh made any modifications upon his microscope to overcome the above difficulty of the reflector?

The method of preparation of the B.C.I.R.A. samples containing these pink to blue inclusions left no doubt that they were titanium sulphide and that their formation was due to insufficient manganese being added to the melt to desulphurise the metal.

In the highly alloyed wrought steel examined by the writer the inclusions found were:

(1) Grey cubic crystals of titanium carbide.
(2) Salmon-pink cubic crystals of titanium cyano-nitride.
(3) Blue to pink needle-like crystals of titanium sulphide.
(4) Golden-yellow cubic crystals.

The last-named crystals seemed to agree with the properties of pure titanium nitride, and yet it seemed improbable that free titanium carbide and titanium nitride would occur with titanium cyano-nitride (a solid solution of the two) in the same metal. Had the author ever encountered this golden-yellow crystal, and, if so, what were his views upon the subject?

On p. 237 P the author quoted Tofaute and Büttinghaus to the effect that titanium carbide was appreciably soluble in austenite. These authors were investigating the pure iron-titanium alloys with and without carbon. The temperature required to get the γ phase in such materials was high, and the point to note was that the observations only concerned rather high temperatures.

Mr. Morrogh then mentioned the results of an experiment carried out at the B.C.I.R.A. to determine the solubility of titanium carbide (TiC) in austenite at temperatures between 800° and 1000° C. Again, in this experiment, the austenite was of the pure or unstable variety. Such conditions were seldom met with in commercial titanium-containing wrought steels, and it seemed to the writer that titanium carbide was scarcely soluble in alloyed or stabilised austenite at room temperature.

The wrought alloy steels investigated at the British Thomson-Houston Co. all contained considerable percentages of chromium and nickel—as was seen from the typical analysis given below:

Typical Analysis of a Wrought Alloy Steel Examined.

Carbon . .	0·08%	Phosphorus .	Trace	Molybdenum .	4·31%
Silicon . .	0·28%	Titanium .	0·51%	Copper . .	4·47%
Manganese .	0·24%	Nickel . .	17·00%	Aluminium .	0·02%
Sulphur . .	0·02%	Chromium .	13·87%		

Taking this analysis, if all the carbon reported was present as titanium carbide, there would be 0·4% of TiC by weight. Since there was probably some carbon dissolved in the austenite *per se*, the true content of titanium carbide would be less than 0·4%, and, judged by the microstructure and the specific gravity of TiC, the amount of it precipitated out of solution was estimated as *over* 0·3% of TiC by weight, indicating that titanium carbide was not very soluble in alloyed (stabilised) austenite.

It therefore behoved all writers upon the subject to quote the analysis of the specimens discussed and thus to leave no doubt upon such important points as the type of austenite investigated.

With regard to the general characteristics of titanium sulphide, the writer found those in the steel, which was in the hot-forged condition, to be in perfect agreement with those found by the author in the cast irons. They were in the form of long thin needles (there was not sufficient present to form groups of parallel lamellæ), tending to become bulbous at the ends and also, in many cases, appearing to be in process of breaking up.

It would be seen that in this steel there was insufficient manganese to neutralise completely the sulphur in the form of manganese sulphide, as the product of Mn% × S% was only 0·0048, which was very much lower than the value of 0·03 found by Herty and Gaines.

On p. 225 P, the author mentioned that Blair and Shimer observed a sulphide with a bronze colour which was suggested by Portevin and Castro to be titanium sulphide more or less contaminated with carbide. Could Mr. Morrogh reproduce this sulphide? The steel examined by the writer contained an abundance of titanium carbide inclusions, but there was no sign of this bronze sulphide. Mr. Morrogh also showed micrographs of titanium carbide associated with titanium sulphide, but each was present as a separate constituent, *see* Figs. 47 and 50.

In Group IV., inclusions produced by alloying with zirconium, the author obtained close agreement in his observations with Urban and Chipman, and with Portevin and Castro, but in one detail it seemed that a rather important fact mentioned by Urban and Chipman had been ignored by Portevin and Castro and also by Mr. Morrogh. The additions of *zirconium* were brought about by adding ferro-silicon-zirconium (iron 8·25%, silicon 50%, zirconium 38·8%) and therefore more silicon was added than zirconium, but when the nature of the inclusions produced was studied, the action of this silicon was not mentioned. Might not these inclusions be something more than simple zirconium sulphide and carbide? Had one proof that rather complex compounds associated with the silicon were not formed, or that zircon (silicate of zirconium) was not present?

In alloy steels one was now also concerned with the presence of niobium (columbium)—and so the work became more and more complex and interesting.

In conclusion, the writer would like to congratulate Mr. Morrogh very heartily upon his remarkable photomicrography and on the vast amount of very useful information included in his papers.

Mr. J. V. Hardwick (British Thomson-Houston Co., Ltd., Rugby), referring to the paper on the polishing of cast-iron micro-specimens, wrote that Mr. Morrogh was to be congratulated upon the technique that he had developed in the preparation of his samples for micro-examination. The high quality of the micrographs clearly showed the value of his method.

The method used by Mr. Morrogh had been in regular use in the B.T.H. Research Laboratory for several years with slight modifications for the different types of steel and non-ferrous alloys met with in industry. The preparation of micro-specimens of ferrous and non-ferrous metals and alloys required a certain degree of skill and experience if misleading results were to be avoided.

Probably the most difficult type of steel to prepare for microscopical examination was the medium-carbon type in the normalised condition with a carbon content of approximately 0·40% and manganese 0·6–0·8%. In the writer's experience the repeated or alternate polishing and etching method of finishing off micro-sections of this class of steel in particular was essential if a true picture of the quantity and the type of non-metallic impurities was to be obtained. However carefully the specimen was prepared some surface "flow" of the soft ferritic portion of the structure took place and tended to smear over the very minute slag particles which were usually present in the ferrite. After one or two treatments by the alternate polishing and light etching treatment (using 1½% nitric acid in alcohol, followed by hand-polishing on Selvyt impregnated with well-ignited heavy-magnesia/water paste) it was usually found that very minute slag particles were revealed which were previously covered up. The edges of the larger visible slag grains were also noticeably cleared up by this treatment, and a perfect surface resulted. This final polished surface was more easily photographed and the subsequent etching of the structure more uniform than on the surface taken directly from the revolving polishing block using Diamantine.

At the present time when Diamantine powder was almost unobtainable the alternate polishing and etching method was more necessary than ever. More surface "flow" took place when using substitute powders for Diamantine, since they did not appear to have the cutting ability without scratching associated with Diamantine. This might possibly be overcome by heating the oxide (of aluminium) to a *high sintering* temperature, thus making it *crystalline* Al_2O_3.

In the above type of steel, carefully prepared, the condition of the surface was similar to that shown by the cast irons prepared by Mr. Morrogh, that was to say, the inclusions were below the general level of the surface as produced by the higher-speed polishing by wheels, and the alternate polishing and etching lowered the general surface and so revealed the minute slag grains.

A modification of the above method had been found necessary when dealing with certain types of alloy steels, *e.g.*, the stable austenitic type showing hard free carbides. When the surface of such a steel was examined directly from the polishing block it was usually found that these hard carbides were in relief to the general surface.

Before the alternate polishing and etching treatment was adopted, therefore, the writer had been in the habit of very lightly and carefully regrinding the specimen on a well-used Hubert 00 emery paper and then repolishing on the block with Diamantine. This operation levelled down the hard carbides, and then the surface was ready for the alternate polishing and etching treatment. The harder types of steels were much less troublesome to prepare, but they were all finally treated by the alternate polishing and etching method. Alloy cast irons were similar to the alloy steels when free carbides were present.

A ferrous specimen so treated would respond much more readily to a carbide etching reagent than one which had been etched directly from the polishing block. The removal of the microscopical "flowed" layer ("amorphous layer") from the surface of the specimen no doubt explained this action, and very minute carbides could readily be seen.

In the writer's opinion the preparation of microsections should never be rushed, especially in the first operation of cutting the section from the piece. Very careful wet-grinding or filing of the cut surface to be prepared should be done before the emery papers were used, to ensure complete removal of the distorted layer formed during the cutting of the section. The retention of any part of this layer made the specimen useless for microscopical examination in research work.

The best possible surface finish should be the aim in every case, since very misleading results were so very easily produced.

Dr. H. A. SCHWARTZ (The National Malleable and Steel Castings Company, Cleveland, Ohio, U.S.A.) wrote that in his paper on the metallography of inclusions the author dealt in detail with many types of inclusions. This discussion was limited to those of which the present writer had special knowledge. The relation of sulphide inclusions to the form and location of graphite nuclei was interesting and important to those studying the graphitising process. In this field the author presented some beautiful micrographs and some interesting observations.

The author's discussion directed attention to the possibility that all graphite might be nucleated by sulphides. This assumption was in no wise contradictory to the writer's well-supported observation which was much more than a "claim" that nodules always formed first at austenite-cementite interfaces. It was entirely likely that sulphides should form predominantly in such locations, which would completely harmonise the views of the author and of this commentator.

Carbon nodules of the "spherulite" type were habitually observed in the blackheart graphitising process if that process were conducted throughout below A_1 or if the white cast iron had been prequenched before graphitisation. The writer suspected that the two forms of nodules were related to the graphitising rate and conditions and only indirectly to the type of sulphide. It was, of course, well known that iron of a composition to precipitate iron sulphide graphitised but relatively slowly.

It was suggested that the relation between the crystallisation rate of graphite and the migratory rate of carbon under given conditions might be significant in this respect.

This commentator was not prepared to accept any assertions involving the graphitisation *in situ* of cementite until some explanation was offered as to how the resulting carbon, which occupied only about one-fifth of the volume of the original carbide, could form a pseudomorph of cementite. Indeed, there was no apparent reason to regard the graphite particle in Fig. 28 as pseudomorphic with cementite and hence to conclude that the sulphide was within, rather than on, the surface of the latter.

The coarsely divorced pearlite discussed by the writer was not unfamiliar to those who had examined difficultly "annealable" blackheart malleable. It sometimes occurred under conditions of high sulphur, but also under other conditions suspected to be related to an abnormal oxygen content.

The author's paper dealing with the polishing technique which produced his micrographs of graphite should prove an important contribution in a field in which progress had been rapid in the last few years.

If the author's observations regarding other inclusions were as thought-provoking as those regarding the sulphides, his paper might prove a milestone from which much progress would be measured.

Dr. A. L. Norbury (Research Department, Woolwich) wrote that the excellent and informative micrographs in the two papers spoke for themselves. In congratulating Mr. Morrogh on a first-class piece of work, it was appropriate also to congratulate Mr. L. W. Bolton, who, as Mr. Morrogh's predecessor in micrographical work at the British Cast Iron Research Association, had advanced the work to the position from which further advances had been so ably made by Mr. Morrogh.

Markings in graphite flakes had been attributed by Hanemann and Schrader to strain effects, but this explanation was not well established. Had Mr. Morrogh tried straining, at a high temperature, a piece of cast iron containing graphite flakes free from markings?

The markings might, alternatively, result from the manner in which the graphite flakes were formed. Certain "kish"-graphite flakes had been found by chemical analysis to contain as much as about 30% of iron, which possibly resulted from liquid entrapped during the aggregation of smaller flakes.

The different temper-carbon structures in Figs. 22 and 23 of the "inclusions" paper were attributed by Mr. Morrogh to the presence of iron sulphide in the former and to manganese sulphide in the latter. It should be ascertained that the different structures were not the result of differences in the stability of the carbide as affected also by the silicon content and the chemical composition of the white iron as a whole. The type of pearlite shown in Figs. 31 to 34 of the same paper was, he (the writer) suggested, better described as carbide-rich than as fine.

Mr. Morrogh's work had advanced the study of inclusions in cast iron very considerably and had taken it to the stage at which information on oxide-containing inclusions was now urgently required. There was a good deal of evidence that such inclusions were formed in molten cast iron under certain conditions. For example, the oxidation products of aluminium in molten cast iron could make the metal pasty, like honey, at temperatures well above the metal's freezing point. Silicates also were, presumably, being continuously generated in molten cast iron in certain industrial processes.

There was also evidence that the above oxidation products rapidly aggregated, possibly from a very fine initial condition, and tended to float out of the molten metal.

It was to be hoped that work would be continued on the deliberate synthesis and study of such oxygen-containing inclusions in compositions of iron free, in the early experiments, from other inclusions and from graphite.

Mr. J. C. Booth (The Lancashire Steel Corporation, Ltd., Irlam, Manchester), discussing the metallography of inclusions in cast iron, wrote that the construction of a suitable and elastic framework into which the known and the hitherto unknown inclusions could be fitted was difficult. It was complicated by the facility, possessed by most inclusion-forming materials, with which solid solutions of many components were formed the composition of which made indisputable identification almost impossible. The petrologist's classification did not offer much substantial help, for, whereas his structures were the result of heat, pressure and crystallisation over a period of time measured in geological units, the mineral inclusions of ferrous metallography formed in minutes or, at the most, hours, little time being allowed for differential crystallisation. Whether a given inclusion would ultimately occur as an allotriomorphic or idiomorphic crystal or merely as a glass depended upon the conditions prevailing at the time of formation. Since the development of characteristic external symmetry was governed largely by the purity, the temperature of formation, the mechanical strength of the growing crystal, the interfacial tension between the embryo crystal and the melt, and the conditions of crystallisation of the surrounding matrix, a change of external form was to be anticipated when any of the above conditions were varied. The internal symmetry depended, among other things, upon the speed of cooling—that was, whether opportunity was given for nucleation and crystal growth.

Therefore, the optical reaction of an inclusion might be affected by the rate at which it had cooled and its identity, based on the use of polarised light alone, uncertain. The effect of cooling speed on crystallisation was regularly seen in sections taken from various parts of steel ingots. The skin sections often showed a glassy type of inclusion, whilst those taken from more central regions indicated a progressive state of crystallisation.

Dark-ground (conical) illumination could be employed to supplement the use of polarised light. Under certain conditions manganese-rich sulphide showed a reflex the colour of which varied from a light brownish-yellow to a blood red, whereas iron-rich sulphide was always dark. Inclusions with the red coloration had been found most abundant in steels and certain irons with a high copper content.

One was left in some doubt as to whether the manganese-rich sulphide dendrites seen in cast iron had really crystallised from low-temperature solution as suggested; this would seem to imply a greater solubility in the iron than was at present accepted. Was it not possible that movement of semi-solid sulphide had been directed by the growing primary austenite dendrites and finally trapped and consolidated in the interstices, where, by virtue of their position, they appeared as dendrites? Some support for this mechanism was afforded by the statement in the last paragraph of p. 216 P, continued on p. 217 P, where it appeared that the external form of the sulphide was governed by the size and shape of the primary austenite.

The theory that iron sulphide nucleated the temper carbon in whiteheart malleable iron was interesting. Assuming, as stated, that the sulphide was in solution at the annealing temperature, what explanation did the author offer for the mechanism, since, according to modern X-ray work, when a compound molecule dissolved in a solvent its identity was lost? In many high-sulphur, low-manganese irons, the free sulphide was present in the carbide. It was conceivable that graphitisation might be induced by nucleisation at the perphery of a carbide globule and crystallisation proceed towards the centre. If the globule contained sulphide in solution this would be concentrated and, on cooling, precipitated near the centre of the sphere. If the seeding took place at the boundary of the carbide, the final structure of the nodule would depend upon the number of nuclei formed per unit of time and the linear velocity of crystallisation. A measure of support for this view was given in Figs. 22 and 24, where there was a difference in the number and size of the radial crystallites of temper carbon which could be ascribed to different rates of nucleisation and growth. Experimental evidence of this view could be obtained by so altering the annealing conditions as to induce changes in the rate of nucleisation.

Mr. H. H. SHEPHERD (Messrs. Crane, Ltd., London) offered his congratulations to Mr. Morrogh, not only on the excellent work which these two papers revealed from the point of view of the subject in general, but also on account of the very clear and concise way in which the subject matter had been set down. He was sure that everyone would agree that the micrographs were of the very highest order; perhaps the best tribute would be to say that they represented splendid examples of the metallographists' art.

These remarks were in themselves a tribute to the great care and attention which the author had given to the development of high-class technique for the polishing of cast-iron specimens, as set out in the first paper.

With regard to the metallography of inclusions in cast irons and pig irons, in the Introduction the author stated that "while oxides and silicates are very important inclusions in steels, they are virtually non-existent from the point of view of the metallography of cast iron." This, to the writer, seemed to be somewhat dogmatic and tended to contradict views which had been expressed from time to time, that hard "blowing" or excessive air supply to a cupola melting cast iron was apt to give rise to oxide inclusions in cast irons, especially those of the low-silicon and moderately low-manganese type, such as the white irons for the production of malleable castings. Again, from a consideration of the machineability of ordinary grey cast irons, there would appear to be some support for these views.

As to the appearance of iron oxide in cast iron, the writer had, in his firm's metallography laboratory, from time to time examined portions of cast-iron chills which had been burnt as the result of repeated use, and he thought they were able to identify iron oxide by means of polarised light, when it appeared an orange-yellow colour. The writer would appreciate Mr. Morrogh's observations on this.

On p. 209 P reference was made to "border-line compositions . . . where particles of moulding sand have been included at the edge of the castings, the metal in the immediate vicinity of the inclusions tends to solidify grey." This phenomenon had been repeatedly observed by the writer in connection with malleable castings. He had also found, as had other investigators, that iron oxide had an appreciable influence in this direction; for example, in pouring chill test blocks for the purpose of controlling the fracture of white iron produced by cupolas for malleable castings, if the pourer used a dirty ladle—that was to say, a ladle with a heavy iron-oxide coating or scale—then the chill test would almost invariably show up much more mottled, to even a heavy grey fracture, than would be the case if the metal were poured from a clean ladle. The writer believed it was Boegehold, the American investigator, who showed very clearly that iron oxide had an important influence in the direction of inoculating—as it were—a white iron to give a more mottled iron.

On p. 219 P it was stated that at least two different forms of temper-carbon nodules could exist in malleable-iron samples, namely spherulitic and also what may be termed "aggregated." This was particularly interesting, because the writer a considerable time ago satisfied himself that high manganese or excess manganese over the sulphur balance gave this flake-like aggregated temper carbon and was of course associated with low mechanical properties in the malleable cast iron.

Finally, with regard to the statement on p. 220 P that "when the white iron used for the manufacture of whiteheart malleable is poured at a fairly high temperature, the resultant malleable iron frequently contains groups of spherulitic temper carbon joined together forming an almost continuous and straight band of graphite," the writer had also observed this in connection with the manufacture of blackheart malleable cast iron and had long associated it with excessive pouring temperatures. It was, of course, much more dangerous to have this type of temper carbon—so well illustrated in Fig. 25—present in blackheart malleable iron than in whiteheart.

AUTHOR'S REPLY.

(Figs. A to J — Plates XLVIIA and XLVIIB.)

Mr. Morrogh wrote that, before replying to the points raised in the discussion, he would like to thank the various contributors for their kind appreciation of the work dealt with in these two papers.

Dealing with the comments made by Dr. Hurst, it could be said very definitely that in the polishing process described the polishing powder was in no way embedded in the graphite flakes. If by chance any fine particles of polishing powder were left on the specimen, then it would not etch correctly and a deep stain was produced, which obliterated the true structure. The small white spot inside a graphite flake in Fig. 4 of the first paper was a particle of metal and not a grain of polishing powder. Particles of metal quite frequently occurred entrapped in graphite flakes, particularly in coarse-graphite irons.

Some preliminary experiments on the application of electrolytic polishing to cast iron had revealed that grey cast iron might be one of the most difficult of the common alloys to polish by this method. The reagents used in this process appeared to penetrate the graphite flakes, causing swelling of the graphite crystals. While some polishing of the metallic matrix occurred, the graphite structure might be completely distorted. It must be emphasised, however, that these remarks on electrolytic polishing were based on only a few preliminary experiments, and at this stage it would be unwise to dismiss the possibility of electrolytically polishing grey-cast-iron microspecimens.

The samples used for examining the effect of aluminium on cast iron were prepared by the addition of the base iron to the requisite amount of molten aluminium contained in a ladle—no chlorinated wax was used.

Mr. Turner's remarks on the polishing of phosphoric cast irons were very interesting, but the author did not believe that the speed of the polishing pad had any great effect on the extent to which phosphide eutectic would stand up in relief from the polished surface. Many details in any polishing operation could be varied to suit the individuals concerned, and, as long as a certain routine was adhered to, good results could be achieved with a method which was fundamentally good. In this context, the speed of rotation of the polishing pad could be varied within certain limits without materially affecting the results obtained. Ternary phosphide eutectic tended to stand up in relief to a much greater extent than the pseudo-binary form. This relief effect could be reduced to some extent by a number of alternate polishings and etchings, using alkaline sodium picrate as the etching reagent for irons containing the ternary phosphide eutectic and alkaline potassium ferricyanide for irons containing the pseudo-binary form. This procedure was somewhat tedious, and the results obtained were not always comparable with the time required. To reduce the relief of the phosphide eutectic to a minimum, it was essential to carry out the rubbing-down operation as carefully as possible. The emery papers must be absolutely free from grit and not worn. Old emery papers caused the graphite to be burnished over, and thus more polishing and etching operations were required to remove this burnish—every re-etching and polishing increased the relief effect.

The expression of the speed of the polishing pad in feet per second was very difficult unless one particular zone of the pad was used. For instance, the specimen might be polished on the outer regions of the pad at the beginning of the operation and then gradually moved to the centre as the scratches disappeared. At other times, small awkwardly shaped specimens were polished near the centre, while large specimens were polished near the periphery of the disc. The best position for the polishing of any particular specimen depended on the skill of the operator. Thus, the speed of the polishing pad when expressed in feet per second could have very little meaning. Specimens used and described in the papers under discussion were prepared on a polishing disc 8 in. in dia., but no fixed zone of the polishing disc was used.

Mr. Turner's remarks regarding the standardisation of metallographic technique and the need for co-operation between the various scientific societies and the microscope manufacturers would be warmly endorsed by metallographists. There was still a considerable gap in Britain between the designer and manufacturer on the one hand and the user—the metallographist—on the other. In the past there had been a tendency for metallographic photomicrographic equipment to be produced merely by the modification of certain existing designs used by the biologist and petrologist. Metallographic technique had developed very special requirements, and the designer was not always fully acquainted with these. For his part, however, the metallurgist must realise that it was impossible to design photomicrographic equipment such that no knowledge of microscopy was necessary for its manipulation. Good photomicrographs could only be produced by having the basic principles of microscopy constantly in mind.

The author appreciated the query made by Mr. Allison regarding the occurrence of areas of fine pearlite intimately associated with areas of coarse pearlite in grey or mottled irons containing free iron sulphide. This structure was difficult to illustrate in any one micrograph, and it was merely the natural limits of photomicrographical illustration which caused any resemblance between Figs. 28 and 29 and Fig. 31 of the paper on inclusions. Actually this structure was strikingly different from that caused by the apparent variation in lamination of the pearlite due to the varying angle of section of the pearlite grains. The fine pearlite areas in samples containing iron sulphide tended to be arranged in a network, whereas the fine pearlite grains in a hematite iron containing sufficient manganese to balance the sulphur were statistically distributed at random.

The author thanked Mr. Harbord for his suggestions regarding the heat-tinting of cast-iron microspecimens. Actually, this method had been used considerably for the examination of the internal structure of graphite flakes and for the recognition of minor phases. Fig. 31 in the paper on polishing illustrated the use of heat-tinting for the examination of graphite flakes. It was impossible faithfully to reproduce a heat-tinted structure by photography in monochrome, but heat-tinting was very useful for increasing the contrast between the various structural features of graphite. It could, in fact, be regarded as an etching method for revealing the structure of graphite.

It was interesting to find that both Mr. Roadley and Mr. Hardwick agreed with the general principle that the preparation of metallographic specimens should be completed by alternate etching and polishing. The descriptions of their polishing technique illustrated admirably how the details of the polishing operation could be varied to meet special requirements. With regard to emery papers available at the present time, the author had found the American-made "Durex" emery papers to be excellent substitutes for the French emery papers. While these papers had a character slightly different from the French papers, grades 1, 0 and 00 Durex papers had been found to be approximately equivalent to grades 1*G*, 1*F* and 00, respectively, of the French papers. It was regrettable that as yet there appeared to be no single substitute polishing powder of the same quality as Diamantine. Many different types of alumina had been tried, including very finely divided brands, but none of these seemed capable of rapidly polishing without scratching, which was the unique characteristic of Diamantine. An investigation into the properties which conferred these qualities on this brand of alumina was urgently needed. At the moment the best alternative to the use of Diamantine was to remove the emery-paper scratches with rouge and then to carry out the alternate polishing and etching on another pad, using magnesium oxide. Magnesium oxide would give a better polished surface than Diamantine, but it polished very much more slowly.

The construction and assembly of a metallurgical polarising microscope was fraught with many difficulties, and so far no perfect design had been produced. For most of the author's work a Vicker's projection microscope was used. The polariser was a nicol prism, placed in the illuminating train so that its plane of polarisation was parallel to that of the partially plane-polarised light which would be produced by the reflection of unpolarised light at the glass-slip vertical illuminator. For the majority of cases a ½-in. dia. disc of Polaroid inserted in an 8× eye-piece was used for the visual work, while a nicol prism clamped over the projection eye-piece was used for the photographical recording of the observations made with polarised light and crossed nicols. From time to time the visual observations made with the Polaroid disc were checked, using instead a nicol prism clamped over the eye-piece. All the effects produced when using the Polaroid analyser were reproduced when the nicol prism analyser was employed. The Polaroid screen was preferred for general visual work, because the nicol prism caused the visible field to be considerably restricted. In contrast to Mr. Roadley's observation, the author found the various colour effects to be even more pronounced with the nicol prism than with the Polaroid analyser. The nicol prism also gave a darker metallic background, owing probably to the fact that it was capable of plane-polarising the light more completely. All the work under polarised light between crossed nicols was carried out with the objective stopped down as completely as possible, in order to reduce the amount of elliptically polarised light produced by reflection from the metallic surfaces.

In the Vickers projection microscope used by the author, the light was reflected into the visual tube by means of a metallic reflector, and not by a glass reflector as Mr. Roadley suggested. It had been possible to examine some of the titanium and zirconium minor phases under polarised light with an ordinary bench microscope, and all the observations which were made with the Vickers microscope were confirmed. It would thus appear that the metallic reflectors used in this microscope had little effect on these observations.

The particles present in the samples of steel which had been jointly examined by Mr. Roadley and the author were very small indeed and were probably standing up in slight relief. On a subject of this type it was very difficult to make any definite statements, since the elliptical-polarisation effects produced by a small particle in relief were frequently sufficient to mask the other effects. These samples were also highly alloyed steel, and therefore, since the present work was carried out on unalloyed cast irons, great caution was necessary in making comparisons. However, the author was inclined to the opinion that these particles were the titanium-sulphide constituent. The analysis of one of the steels in question was:

Total carbon . .	0·24%	Phosphorus . .	0·010%
Silicon . . .	1·09%	Nickel . . .	8·28%
Manganese . .	0·060%	Chromium . .	20·14%
Sulphur. . .	0·006%	Titanium . .	1·8%

This sample contained large amounts of titanium carbide and streaks of the pink constituent. It would be seen from the analysis that, while there was sufficient manganese to neutralise the sulphur as manganese sulphide, there was a large excess of titanium. This titanium content was greatly in excess of that in any of the cast-iron samples discussed. It had been pointed out on p. 234 P of the paper on inclusions that the reaction between titanium and manganese sulphide was probably reversible:

$$MnS + Ti \rightleftharpoons Mn + TiS.$$

The implication of this was that, even when manganese was present in sufficient amount to give manganese sulphide (under normal conditions), the simultaneous presence of a relatively large excess of titanium would cause the reaction to proceed to the right. If this was so, then the steel in question might be expected to contain titanium sulphide.

It had been possible to throw some light on this subject by means of an experiment carried out after the original paper was written: 3000 g. of a pig iron (total carbon 3·46%; silicon 1·16%; manganese 0·67%; sulphur 0·069%; phosphorus 0·46%) were melted in a high-frequency induction furnace and superheated to about 1650° C. Iron sulphide (15 g.) was then added, and this was quickly followed by an addition of 75 g. of titanium metal (2½% titanium) wrapped in a sheet steel container. When the additions were dissolved, half of the metal was cast in the form of a short 1·2-in. dia. bar; 56 g. of ferro-manganese (3% manganese) were then added to the remainder of the metal and another bar was cast. Examination under the microscope showed the first bar to contain well-formed crystals of titanium sulphide, while the second bar contained only large amounts of manganese sulphide. This experiment did lend colour to the idea that there was a reversible reaction between manganese sulphide and titanium. Nevertheless, the author felt that it was not out of place to emphasise again the necessity for the utmost caution when attempting to draw conclusions about the minor phases in steels on the basis of experiments carried out on cast irons.

In the author's experience the golden-yellow crystals mentioned by Mr. Roadley had never been found in cast irons, although seen quite frequently in steels. These particles were usually referred to as titanium nitride. Since titanium nitride formed a solid solution with titanium carbide, it might seem strange that the nitride, the cyano-nitride and the carbide could exist together in a ferrous alloy. Mr. Comstock also raised a pertinent question bearing on this point. The titanium nitride would be expected to distribute itself uniformly through the carbide to give a titanium cyano-nitride. Actually, this condition appeared to be rarely, if ever, realised in either cast irons or steels. In the inclusion counts reported in this paper, every titaniferous particle with a pink coloration was counted as the cyano-nitride. As Mr. Comstock pointed out, this was a difficult distinction to make, and to some extent this was borne out by the erratic nature of the curves shown in Fig. 5 of the second paper. It would appear that the carbide and nitride were either incapable of realising complete miscibility in ferrous alloys or that under normal conditions the reactions between carbon, nitrogen and titanium never reached equilibrium. This was a matter on which further information was required.

It should be remembered that the titanium sulphide-carbide isolated by Blair and Shimer and referred to in the discussion by Mr. Roadley was only described as having a bronze colour after chemical separation. Unfortunately, the metallographic data given by Blair and Shimer about this substance were very scanty, but it was possible that it would have had a somewhat different colour when examined as part of the microstructure of the pig iron. It was felt that the compound isolated by these investigators was the same as that described as titanium sulphide in the paper under discussion. No evidence had been obtained that this blue-pink phase contained titanium carbide, but, in view of the results obtained by Blair and Shimer, it would be wise to keep this possibility in mind.

When Mr. Roadley mentioned zirconium sulphide he presumably referred to the orange-yellow to grey phase. It should, however, be noted that no claim was made that this phase was simple zirconium sulphide. On p. 245 P of the paper on inclusions, it was pointed out to be impossible, on the basis of the experiments carried out, to say whether these particles were a zirconium sulphide or a manganese-zirconium sulphide. For this reason the phase was referred to throughout the paper as the "orange-yellow to grey phase," so that no assumptions were made in regard of its composition. However, since with increasing amounts of zirconium this phase gradually replaced the manganese sulphide, the weight of the evidence was behind the idea that it at least contained a sulphide. As Mr. Roadley pointed out, the possibility of the presence of zirconium silicates must not be overlooked. No definite evidence was available on this point, but it should be noted that the orange-yellow to grey phase was quite different from that which Portevin and Castro suggested to be a zirconium silicate. The possibility of the complete absence of microscopically visible zirconium oxides (or silicates) was suggested by the results obtained when irons containing zirconium were remelted. For instance, an iron of the following composition was remelted in a Salamander crucible, using a coke-fired forced-draught furnace: Total carbon 3·09%; silicon 2·09%; manganese 0·93%; sulphur 0·018%; titanium 0·04%; zirconium 0·23%. The analysis of this iron after remelting was: Total carbon 3·06%; silicon 2·15%; manganese 0·82%; sulphur 0·034%; titanium 0·08%; zirconium 0·04%.

Before remelting, the iron contained fair amounts of the orange-yellow to grey phase and large numbers of the grey cubic particles. After remelting, only manganese sulphide was visible. One would expect the decrease in the zirconium content to be an oxidation loss. Now, if the orange-yellow to grey and the grey cubic phases were oxides or silicates, why should they disappear on oxidation of a melt containing zirconium? If visible zirconium oxides or silicates could occur at all in cast irons, one might expect them to occur under these conditions. The fact that the small grey cubes occurred isomorphously associated with titanium carbide suggested very definitely that they were a zirconium carbide.

Mr. Parker pointed out the desirability of avoiding the term "inherent" with regard to the classification of inclusions. This could not be too strongly laboured. The borrowing of the conception of heredity from the biologist was scientifically unsound and open to many objections. Such terms tended to shroud the problem with mystery rather than to provide a path for the elucidation of the effects of these small particles. Quoting Wyckoff that compounds with the nickel-arsenide structure were characterised by their ability to dissolve an excess of one or other of their constituent elements, Mr. Parker suggested that the ratio of sulphur to iron in the iron sulphide in cast iron was higher than that given in the formula FeS. There was no doubt that certain naturally occurring and synthetic pyrrhotites could be more accurately represented by $FeS + S_x$, but this applied to the compound formed generally in the presence of excess sulphur. In ferrous alloys there was a huge excess of iron, and, leaving aside the other sulphide-forming elements and considerations of the crystal structure, it would seem more probable that the ratio of sulphur to iron would be lower than that given in the formula FeS. There was, however, no direct evidence to support this view. While the khaki-coloured sulphide present in ferrous alloys was usually referred to for the sake of convenience as iron sulphide, it should be remembered that, if there was any manganese, titanium, copper, aluminium or chromium in the alloy, then the compound was an iron-sulphide-rich complex containing any combination of these elements in solid solution.

Consideration of the sulphides of nickel and molybdenum had been omitted from the paper on inclusions, because it was hoped in a short time to present a paper in the form of a Carnegie Scholarship Memoir on the neutralisation of sulphur in cast irons by the alloying elements nickel, copper, chromium, molybdenum and aluminium.

The application of the term " nucleus " and its derivatives with regard to the isothermal decomposition of iron carbide during the malleablising of cast iron could scarcely have the same significance as the term " germ-crystal " when used in connection with crystallisation from a liquid. At this point it would be useful to consider the main features of the decomposition of iron carbide. The isothermal decomposition of iron carbide in equilibrium with austenite was a complex reaction, our knowledge of which was very limited. After a period of " incubation," during which the graphitising rate was very small, graphite nodules started to form and the rate of graphitisation rapidly increased. It was possible that the austenite provided the means for the diffusion of the carbon, either before or after the breakdown of the iron carbide, to the centres of graphite formation. The complete change was probably accompanied by a change in the type of austenite and in the composition of the austenite. While a small amount of the iron carbide might decompose without appreciable diffusion of the carbon, most of the carbon diffused to a small number of selected spots from which the temper nodule grew. Dr. Schwartz maintained that graphite nodules first appeared at austenite/cementite interfaces, and thus the point at which the nodule first began to form on any particular interface could be regarded as the nucleus. The term " nucleus " was thus used to define a chemical or physical condition which favoured the deposition of graphite at a particular point, and with the present state of our knowledge could have little precise meaning. To substantiate his view that temper carbon originated at austenite/cementite interfaces, Dr. Schwartz had shown low-power micrographs of partially graphitised white iron in which the temper carbon tended to border on the cementite areas. Since the sulphides present in cast iron tended to segregate in the extra-dendritic regions, it seemed likely that they might constitute the particular points on the austenite/cementite interfaces at which graphitisation could occur. Such an explanation was, and no doubt Dr. Schwartz would concur with this view, extremely empirical and required much modification and supplementing to bring it into line with our knowledge of the many factors, such as heating rate, casting conditions and prequenching treatment, which affected the nodule number.

The term " nodule " might itself be misleading when applied to the temper carbon in some malleable irons. The use of the improved polishing technique for malleable irons had revealed that the graphite crystals composing the "graphite-flake-aggregate" temper carbon could exist in various degrees of dispersion. At the one extreme the temper carbon existed in the form of compact masses, approximately spheroidal in shape, consisting of a number of apparently haphazardly orientated graphite crystals. This form was quite different from the spherulitic type of temper carbon. At the other extreme, the graphite crystals comprising the nodule were not welded together, but were frequently partially separated by metal. The concept of a nucleus for the formation of a temper carbon nodule might tend to suggest the gradual deposition of graphite around a given point, the result being a compact mass of graphite. While this point of view might be tenable for the initial stages of the decomposition of graphite, it would appear that after a time a considerable mass of metal around the original nucleus was in a condition capable of permitting the deposition of graphite. At the present time it was impossible to give an account of the precise factors which affected the degree of dispersion of the graphite crystals, but it could be said that, in general, for irons containing only manganese sulphide the dispersion increased with higher annealing temperatures and with smaller nodule numbers.

The author found it impossible to agree with Dr. Schwartz's statement that "carbon nodules of the 'spherulite' type were habitually observed in the blackheart graphitising process if that process were conducted throughout below A_1 or if the white cast iron had been prequenched before graphitisation." Micrographic evidence was presented here to support this disagreement.

The accompanying Fig. A showed the temper carbon distribution in a malleable iron of the following composition: Total carbon 3·2%; silicon 1·26%; manganese 0·49%; sulphur 0·109%; phosphorus 0·06%; nickel 1·32%; chromium 0·52%. This sample had been annealed at 900° C. for 40 hr. and then air-cooled. Fig. B illustrated the refinement of the temper carbon produced by prequenching the same material before giving it this heat treatment. To carry out the prequenching treatment, the white iron sample was immersed in a salt bath for 10 min. and then quenched in water. Fig. C showed a typical "graphite-flake-aggregate" temper carbon nodule in the plain annealed iron, and Fig. D illustrated a temper carbon nodule in the prequenched material. It was obvious from this latter micrograph that the prequenching had not caused the formation of spherulitic temper carbon. The characteristics of the temper carbon present in such prequenched samples were further illustrated in Fig. G, a prequenched sample of the same material after only 16 hr. at 900° C.

It had been found for certain irons that if the time of soaking prior to the quenching treatment was increased, then a less drastic refinement of the temper carbon might occur. In such cases the individual temper-carbon nodules appeared smaller and more compact than those in the plain annealed irons, but they were still not of the spherulitic type. Figs. E and F illustrated this point with respect to an iron of the following composition: Total carbon 3·01%; silicon 0·87%; manganese 0·39%; sulphur 0·092%; phosphorus 0·03%. Fig. E showed a typical temper-carbon nodule in a sample of this iron which had been annealed at 900° C. for 48 hr. and then cooled in the furnace. Fig. F depicted the temper carbon in a sample of the same material soaked at 950° C. for 1 hr. prior to prequenching and then annealed for 48 hr. at 900° C.

The statement made by Dr. Schwartz apparently had no reservation with regard to chemical analysis, and all the samples cited above had the sulphur in the form of manganese sulphide. It had, however, been possible to investigate the effect of prequenching an iron containing iron sulphide. Fig. H showed a typical spherulitic temper-carbon nodule in an annealed iron of the following composition: Total carbon 3·22%; silicon 0·80%; manganese 0·18%; sulphur 0·466%; nickel 1·77%. This iron was annealed for 16 hr. at 900° C. and then cooled in air. The temper carbon nodule was seen to have formed around a duplex particle of iron and manganese sulphide. A sample of the same white iron was then prequenched from 900° C. after soaking in a salt bath for 15 min. After prequenching, the specimen was annealed at 900° C. for 16 hr. Fig. I showed the minute spherulitic temper-carbon nodules found in the prequenched sample.

As a result of experiments, such as those described above, the author had arrived at the conclusion that prequenching did not alter the type of temper carbon formed, but only affected the number and size of the temper-carbon nodules.

It was interesting to read that both Dr. Norbury and Dr. Schwartz were inclined to the opinion that the different types of temper carbon were not necessarily due to the presence of the different sulphides *per se*, but rather to the effect of the sulphides on the stability of the carbides and hence on the rate of graphitisation. It was impossible to gainsay these suggestions, but in this context it was interesting to consider the irons illustrated in Figs. H and I. These samples contained large amounts of iron sulphide, but on account of the presence of nickel the graphitising rate was fairly rapid (the eutectic carbide was completely broken down after 16 hr. at 900° C.). Both of these samples contained spherulitic temper carbon, and this fact must tend to disprove the suggestions made by Dr. Schwartz and Dr. Norbury. It seemed also that Dr. Schwartz's suggestion tended to contradict his claim that prequenching produced spherulitic temper carbon. He suggested that spherulitic temper carbon was the result of the slow rate of graphitisation produced by the presence of iron sulphide, but he had shown elsewhere that prequenching increased the rate of first-stage graphitisation.

With regard to graphitisation below the critical range, the author had examined several irons annealed at sub-critical tem-

FIG. A.—Temper Carbon produced by annealing white iron at 900° C. for 40 hr. Unetched. × 60.

FIG. B. Very Fine Temper Carbon in pre-quenched sample. Unetched. × 60.

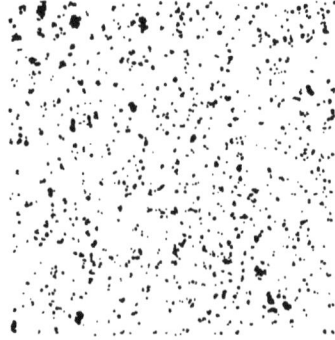

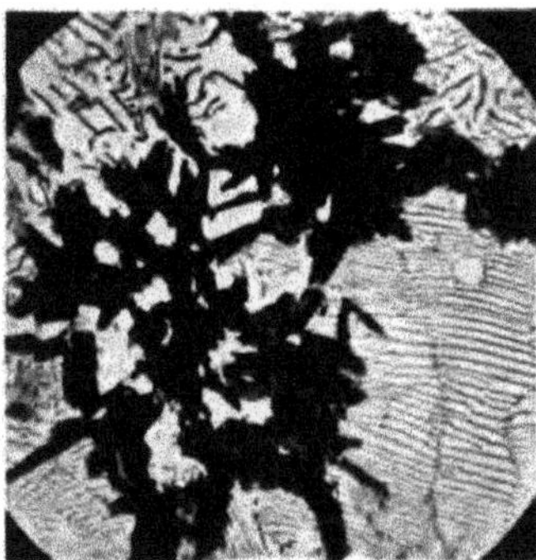
FIG. C.—Typical Graphite-Flake-Aggregate Temper Carbon in plain annealed iron. Etched in picric acid. × 1200.

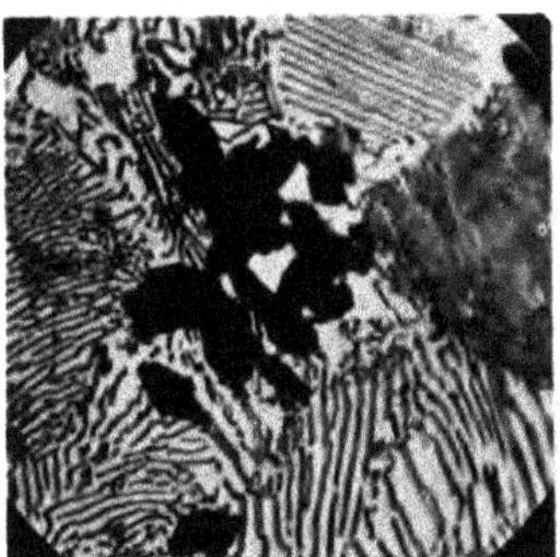
FIG. D.—Typical Temper Carbon in pre-quenched sample. Etched in picric acid. × 2000.

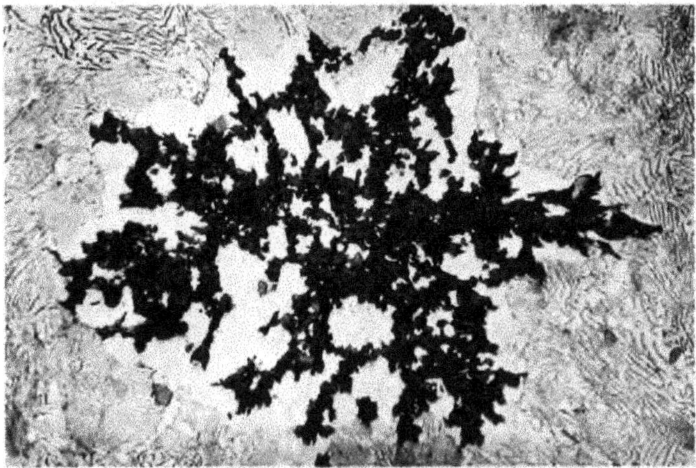
FIG. E.—Dispersed Nature of Graphite Crystals of Temper Carbon in plain annealed sample. Etched in picric acid. × 500.

(Micrographs reduced to two-thirds linear in reproduction.)

[Morrogh ; author's reply.
[To face p. 284 P.

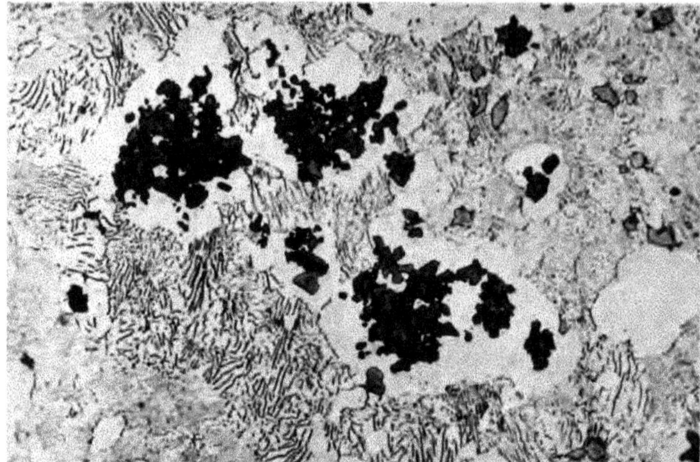

Fig. F. Compact Graphite-Flake-Aggregate Temper Carbon produced by prequenching. Etched in picric acid. × 500.

Fig. G.—Temper Carbon in incompletely annealed prequenched sample. Etched in picric acid. × 2000.

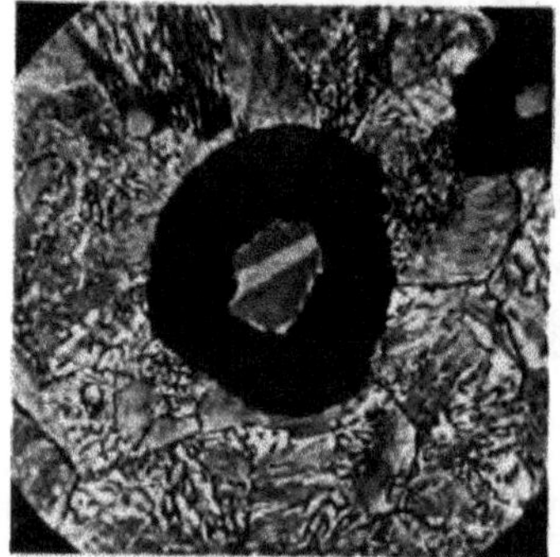

Fig. H.—Spherulitic Temper Carbon around a duplex MnS–FeS particle. Etched in picric acid. × 1200.

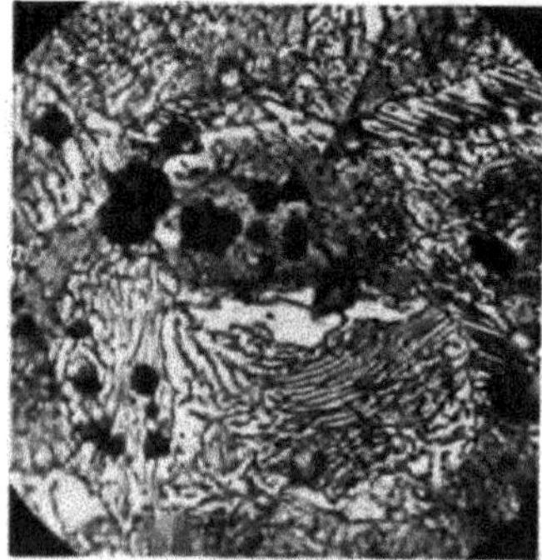

Fig. I.—Minute Spherulitic Temper-Carbon Nodules in prequenched sample. Etched in picric acid. × 2000.

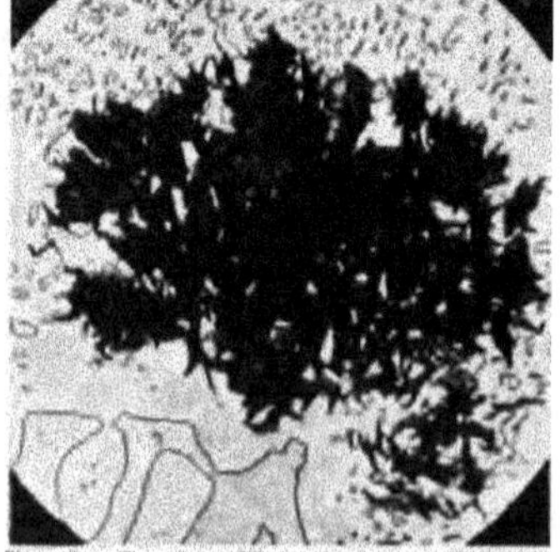

Fig. J.—Graphite-Flake-Aggregate Temper Carbon formed by annealing below the critical range. Etched in picric acid. × 1200.

(Micrographs reduced to two-thirds linear in reproduction.)

[Morrogh; author's reply.

[To face p. 285 P.

peratures, in which all the sulphur was in the form of manganese sulphide. None of these irons contained spherulitic temper carbon. Fig. J showed graphite-flake-aggregate temper carbon in a white iron which had been annealed for 4 days at 700° C.

Hanemann and Schrader's theory that the markings in graphite flakes were due to strain effects was, as Dr. Norbury pointed out, not well established. A grey cast iron which had been strained at about 900° C. had been examined, but no difference was observable between the graphite flakes in this material and those in the as-cast material—both samples showed many graphite flakes with complex structures and a few flakes free from structural complications. The experimental difficulties were greater than Dr. Norbury's question implied, since no grey iron had all its graphite flakes free from structural complications, and so it was necessary to strain a particular graphite flake containing no markings in order to throw some light on this point.

It appeared that Mr. Booth had misinterpreted the statements made on pp. 216 P and 217 P with regard to the effect of pouring temperature on the size of manganese sulphide. Actually, it was pointed out that Norbury had shown primary austenite dendrites to decrease in length and become more compact with decreasing pouring temperature. It was also stated that manganese sulphide dendrites behaved in a similar manner. This effect of pouring temperature was characteristic of all primary dendrites, whether they were ferrous, non-ferrous or non-metallic. The terms "hot" and "cold" applied to pouring temperature were purely relative, and were apt to be misleading. It would be better to consider not the actual pouring temperature, but rather the difference between the pouring temperature and the freezing point. Thus, an iron with a total carbon content of 3% would show a pronounced columnar primary structure, while an iron with a total carbon content of 2% would show an equi-axed structure when poured at, say, 1400° C. Now, if Mr. Booth's suggestion that the manganese sulphide had its shape determined by the arrangement of the primary dendrites were correct, then one would expect compact sulphide crystals to accompany the equi-axed low-total-carbon structures and manifestly dendritic sulphide crystals to accompany the higher-total-carbon structures. However, this was not the case —the morphology of manganese sulphide in cast irons was a function of the actual pouring temperature and not of the difference between the pouring temperature and the freezing point of the metallic alloy. The "anchor-shaped" dendritic manganese sulphide crystals were quite a frequent occurrence in commercial low-carbon irons, of which the pouring temperatures were relatively high and the differences between the freezing point of the alloys and the pouring temperatures were low. If manganese sulphide did become entrapped and consolidated between the austenite dendrites as Mr. Booth implied, it would be difficult to explain its pronounced idiomorphic appearance. In the author's experience, it appeared that the size and morphology of the individual sulphide crystals were determined by the pouring temperature and the distribution of the sulphide particles on the primary metallic dendrites—that was, on the pouring temperature and the total carbon content.

With regard to the nucleation of temper carbon by iron sulphide, Mr. Booth should note that the author did not suggest that all the iron sulphide was in solution in manganese sulphide. In fact, in the first paragraph on p. 221 P it was stated that "the manganese sulphide inclusions rich in iron sulphide do not act as nuclei for graphite, but the relatively small amount of iron sulphide remaining undissolved does act in this manner." Thus the problem of how dissolved iron sulphide could preserve its ability to produce a certain type of temper carbon did not arise.

Mr. Shepherd's remarks on the presence of oxides in cast iron and their relation to melting practice touched on the very heart of the problem of "inclusions" in cast iron. When the author stated that oxides and silicates were virtually non-existent from the point of view of the metallography of cast iron, it was not suggested that these particles were entirely absent from the metal—the chemical evidence proved their presence. The problem had resolved itself into the determination of the precise mode of occurrence of the extremely low concentration of oxides detected by chemical means. They might be in solution or in some other state of sub-microscopic dispersion in either the metallic or non-metallic phases of the cast-iron microstructure. Mr. Shepherd had instanced the effect of melting conditions and cleanliness of ladles on the resulting cast irons. The author agreed with these observations, but wished to point out that as yet they were not accompanied by any observations indicating changes in concentration or distribution of oxides or silicates. In the paper under consideration the author had attempted to deal only with those particles visible under the microscope.

It was not considered that the products of the oxidation or corrosion of solid cast iron could be classed as "inclusions," and so they had been entirely omitted from the paper dealing with these particles. In the initial stages of the oxidation of grey cast iron, a dark translucent phase was sometimes seen, which gave orange colorations under polarised light between crossed nicols. This might correspond to the constituent mentioned by Mr. Shepherd, although its composition was unknown to the author. Further oxidation gave a blue-grey iron oxide, probably FeO, which remained dark under polarised light for all positions of the stage. Drastic oxidation could produce a yellowish-coloured phase, probably Fe_2O_3, which lit up under polarised light. The mechanism of the oxidation of cast iron and the constitution of its products were a subject about which there was little precise information and which was still a matter for conjecture. Its very nature insisted that it should be approached from a different angle from "oxide or silicate inclusions."

ATMOSPHERIC EXPOSURE TESTS ON COPPER-BEARING AND OTHER IRONS AND STEELS IN THE UNITED STATES.[1]

By EWART S. TAYLERSON (Pittsburgh, Pa., U.S.A.).

Paper No. 4/1940 of the Corrosion Committee (communicated by Dr. W. H. Hatfield, F.R.S.).

SUMMARY.

The results of atmospheric exposure tests of twelve different irons and steels for a period of five years at three locations in the United States of America are reported. These include three steels, containing 0·03%, 0·2% and 0·5% of copper, respectively, tested by the Corrosion Committee, six steels ranging in copper content from very low to 0·5%, a copper-bearing wrought iron, and two low-alloy steels. The last nine materials were of American origin and were selected to illustrate the large difference in corrosion rate that can be obtained owing to variation in analysis. The locations included an industrial district on marine marshes, an inland industrial district and a rural district. The results illustrate the great influence of copper and the even greater protective value of higher percentages of alloying elements.

The comparative pollution of the atmosphere at these three locations was evaluated by exposure of the Corrosion Committee's standard pollution samples for a period of two years.

Introduction.

During a discussion with Dr. U. R. Evans when he was visiting the United States of America the suggestion was made that it would be interesting to compare corrosion rates of steels in the States with those obtained at the various exposure stations of the Corrosion Committee. In 1931 several specimens of the Committee's steels *X*, *Y* and *Z* were received from Dr. J. C. Hudson, the Committee's Official Investigator. At that time the Company with which the author is associated started an extensive investigation on a large number of other materials, so that it was possible to include the British steels in exposure tests at three locations. While these tests were intended to supply information for the commercial development of low-alloy steels, the results for a number of the American materials have been included in this paper as a matter of general interest.

Materials Referred to in this Paper.

The chemical compositions of the materials covered in this paper are given in Table I., which has a footnote reference to the Corrosion Committee's Report as to the exact source of the Committee's steels *X*, *Y* and *Z*. These were received in the form of 15 × 10 × ⅜-in.

[1] Received October 22, 1940.

flat bars, which were hot-rolled in packs to 0·030 in. and annealed. This material was then pickled, and all samples were sheared to 4 × 6 in.

TABLE I.—*Chemical Analysis of the Materials.*

Material.	C. %.	Mn. %.	P. %.	S. %.	Si. %.	Cu. %.	Ni. %.	Cr. %.
Committee's steel *X* * (0·03% Cu)	0·18	0·61	0·041	0·043	0·008	0·03	0·05	0·04
Committee's steel *Y* * (0·26% Cu)	0·19	0·60	0·040	0·043	0·009	0·26	0·05	0·05
Committee's steel *Z* * (0·54% Cu)	0·20	0·62	0·038	0·043	0·010	0·54	0·05	0·05
Committee's pollution specimens (ingot iron)	0·02	0·05	0·013	0·044	0·008	0·07	0·05	0·001
Experimental heat, low Mn and Cu	0·015	0·02	0·004	0·038	0·004	0·02	0·006	0·003
Experimental heat, medium Mn, low Cu	0·040	0·24	0·062	0·032	0·008	0·01	0·002	0·009
Open-hearth iron, very low Cu	0·018	0·025	0·005	0·019	0·002	0·008	0·002	0·01
Steel *S* (0·05% Cu)	0·040	0·49	0·074	0·042	0·003	0·05	0·006	0·09
Wrought iron (0·29% Cu)	0·018	0·095	0·096	0·028	0·14	0·29	...	0·006
Steel *K* (0·27% Cu)	0·026	0·47	0·066	0·045	0·003	0·27	0·03	0·03
Bessemer steel (0·5% Cu)	0·10	0·40	0·107	0·057	0·007	0·52	0·002	0·02
3·5% Ni steel	0·043	0·46	0·014	0·027	0·004	0·06	3·52	0·16
Cr–Si–Cu steel	0·11	0·62	0·012	0·006	1·25	1·03	0·07	3·88

* *See* First Report of the Corrosion Committee, *Iron and Steel Institute*, 1931, *Special Report No.* 1, pp. 97, 98. The original 15 × 10 × ½-in. specimens used for rolling the sheets were: *X*110813, *X*110814, *X*110910, *X*110911; *Y*111413, *Y*111414, *Y*111510, *Y*111511; *Z*112013, *Z*112014, *Z*112110, *Z*112111.

The American materials, selected to cover a large variation in corrosion rate, consisted of nine heats: (*a*) Three steels of low copper content, two of which were experimental crucible heats and the third was a commercial open-hearth iron. (*b*) Two steels, marked *S* and *K*, from the same rephosphorised open-hearth heats that were exposed by Committee A–5 of the American Society for Testing Materials [1] at five different American locations. (*c*) A copper-bearing American wrought iron. (*d*) An acid Bessemer copper steel. (*e*) Two low-alloy steels, namely, a 3½% nickel and an experimental chromium-silicon-copper steel.

The specimens were supported on insulators at an angle of 30° to the horizontal.[2] Pickled and weighed specimens of all materials were exposed simultaneously. After various intervals of time complete sets were removed, electrolytically cleaned and reweighed.

Exposure Locations.

The three exposure locations selected were: (*a*) Kearny, New Jersey, about three miles from New York City on reclaimed marine

[1] "Report of Sub-Committee VIII. on Field Tests of Metallic Coatings," *Proceedings of the American Society for Testing Materials*, 1938, vol. 38, Part I., p. 89, Table V.

[2] H. S. Rawdon, "Atmospheric Corrosion Testing," American Society for Testing Materials Symposium on Corrosion Testing Procedures, 1937, p. 42, Fig. 1.

marshes, surrounded by many industrial plants, where the atmosphere is foggy but does not deposit much soot or cinders; (b) Vandergrift, Pennsylvania, twenty-five miles north-east of Pittsburgh in a river valley on steel-mill property, an inland industrial atmosphere contaminated by soot and cinders; and (c) South Bend, Pennsylvania, a rural site, twelve miles from the nearest industry, where the atmosphere is free from much pollution owing to favourable prevailing winds.

The specimens were exposed at all three locations in October, 1931.

A rough comparison of the corrosiveness of the atmospheres at the three American locations and at the testing stations of the Corrosion Committee is rendered possible by the results of tests on some of the Committee's open-hearth iron pollution specimens, which were forwarded by Dr. Hudson in 1936. Standard 4 × 2 × ⅛-in. specimens of this material were exposed vertically in April, 1936, at the three locations. The losses in weight of these specimens after one year's exposure are given in Table II., together with similar figures for the Corrosion Committee's stations at Sheffield, Woolwich and Llanwrtyd Wells.[1] It would seem from the data that the American atmospheres are rather less corrosive than the British ones, which have been selected to correspond to them as closely as possible.

Table II.—*Comparison of the Corrosiveness of American and British Atmospheres.*

Locality.	Type of Atmosphere.	Rate of Corrosion of Standard Ingot-Iron Specimens (One Year's Exposure).	
		Loss in Weight. G. per 4 × 2 × ⅛-in. Specimen.	Equivalent Loss of Metal. In. × 10^{-3}.
United States.			
Kearny, New Jersey	Industrial, marine.	7·75	3·45
Vandergrift, Pa.	Industrial, inland.	8·34	3·71
South Bend, Pa.	Rural.	4·27	1·90
Great Britain.			
Sheffield	Industrial, inland.	11·30	5·03
Woolwich	Industrial, inland.	8·40	3·74
Llanwrtyd Wells.	Rural.	4·87	2·17

The data given in Table II. refer to specimens exposed for a period of one year, but other sets of specimens were exposed for two years. In addition, for comparison purposes in connection

[1] Fifth Report of the Corrosion Committee, *Iron and Steel Institute*, 1938, *Special Report No.* 21, p. 67. The results for the British stations are mean values for several consecutive periods of one year's exposure.

with other tests, duplicate specimens of the two materials *S* and *K* were exposed vertically and in the 30° to the horizontal position.[1] The results of these tests, over one and two years' exposure, are given in Table III.; they indicate that less corrosion occurred on the inclined specimens. It should be emphasised that these two-year tests were made at a later date than the main test and were begun in the spring, so that the two sets of results are not strictly comparable.

TABLE III.—*Comparison of the Corrosion Committee's Ingot-Iron Pollution Specimens with Steels* S *and* K *of the American Society for Testing Materials. Effect of the Angle of Exposure.*

Locality and Position.	Loss in Weight. G. per 4 × 2-in. Specimen.*					
	Ingot Iron, Corrosion Committee.		Steel *S* (0·05% Cu), A.S.T.M.		Steel *K* (0·27% Cu), A.S.T.M.	
	1 Year.	2 Years.	1 Year.	2 Years.	1 Year.	2 Years.
Kearny, New Jersey.						
Vertical . .	7·75	11·20	6·58	9·29	5·54	7·76
30° to horizontal .	...	...	5·46	7·42	4·59	6·22
Vandergrift, Pa.						
Vertical . .	8·34	14·04	6·41	10·21	5·54	9·07
30° to horizontal .	...	...	5·93	8·41	5·37	7·20
South Bend, Pa.						
Vertical . .	4·27	6·44	3·77	5·07	3·40	4·90
30° to horizontal .	...	...	3·14	4·41	2·97	4·09

* Mean values for two or three specimens throughout.

Discussion of the Results.

The results shown in Table IV., which relate to five years' exposure, have been compiled for the British steels *X*, *Y* and *Z* in accordance with the Committee's method of calculating the rates of corrosion. In this connection, the curves shown in Figs. 1, 2 and 3 indicate the wide variation with exposure period of atmospheric corrosion rates. It is believed that this is due to the progressive formation of rust coatings which differ in their protective value It is interesting to note that after an interval of two to three years, within the limitations of these tests, the corrosion losses appear to be proportional to time and the rate constant. The actual losses in weight of all twelve steels, which were exposed at 30° to the horizontal, are given in Table V.

[1] Rawdon, *loc. cit.*

TABLE IV.—*Results of Atmospheric Corrosion Tests on Committee Steels* X, Y *and* Z.

Specimens 6 × 4 × 0·030 in., exposed at 30° to horizontal. Tests begun October, 1931; duration of exposure, 5 years.

Original Specimen Number.	Loss in Weight (5 Years). G. per Specimen.	Ratio.	Equivalent Annual Corrosion Rate.	
			In. × 10^{-3}.	Oz. per sq. ft.
Kearny, New Jersey (an industrial marine atmosphere).				
X110813 } Cu 0·03%	53·3			
X110814 } approx.	60·8			
	Mean 57·1	100	1·86	1·15
Y111413 } Cu 0·26%	32·9			
Y111414 } approx.	34·2			
	Mean 33·6	59	1·09	0·68
Z112013 } Cu 0·54%	29·3			
Z112014 } approx.	29·5			
	Mean 29·4	51	0·96	0·59
Vandergrift, Pennsylvania (an industrial atmosphere).				
X110813	48·2			
X110814	53·0			
	Mean 50·6	100	1·64	1·02
Y111413	35·0			
Y111414	38·6			
	Mean 36·8	73	1·20	0·74
Z112013	30·6			
Z112014	32·7			
	Mean 31·7	63	1·03	0·64
South Bend, Pennsylvania (a rural atmosphere).				
X110813	18·6			
X110814	20·8			
	Mean 19·7	100	0·64	0·40
Y111413	15·5			
Y111414	17·4			
	Mean 16·5	84	0·54	0·33
Z112013	15·0			
Z112014	14·8			
	Mean 14·9	76	0·48	0·30

The results of the tests at the three locations indicate that the marine industrial atmosphere at Kearny, New Jersey, where the specimens remained comparatively clean, was considerably more corrosive than that at Vandergrift, Pennsylvania, where the specimens were subjected to deposits of soot and cinders. This increased corrosion rate may have been due to the high humidity and the

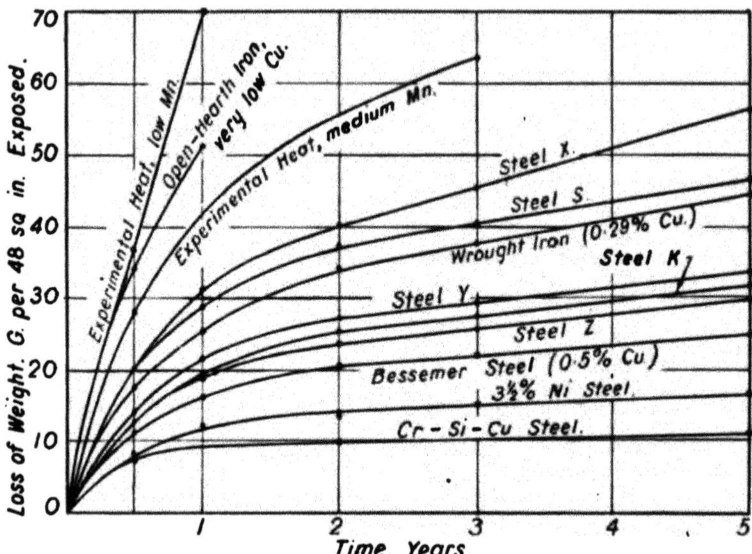

FIG. 1.—Results of Atmospheric Corrosion Tests at Kearny, New Jersey (industrial marine atmosphere).

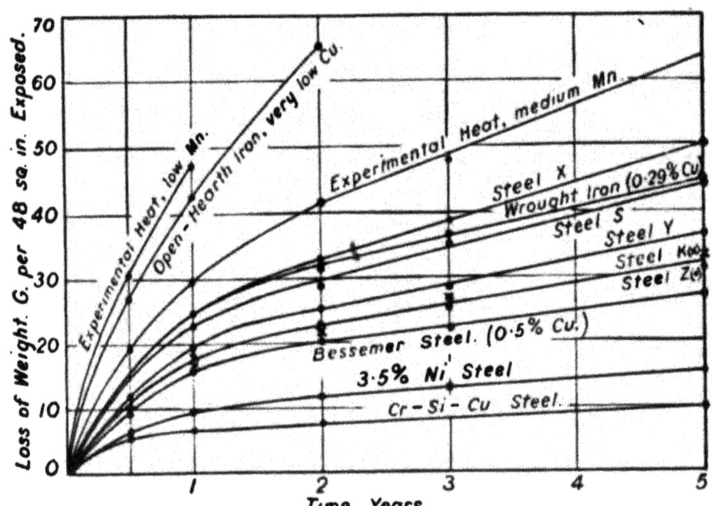

FIG. 2.—Results of Atmospheric Corrosion Tests at Vandergrift, Pa. (industrial atmosphere).

TABLE V.—*Results of the Tests.*

Material.*	Loss in Weight. G. per 6 × 4 × 0·030-in. Specimen.													
	Kearny, N.J. Industrial Marine Atmosphere. Duration of Exposure, Years—					Vandergrift, Pa. Industrial Atmosphere. Duration of Exposure, Years—					South Bend, Pa. Rural Atmosphere. Duration of Exposure, Years—			
	0·5.	1.	2.	3.	5.	0·5.	1.	2.	3.	5.	0·5.	2.	3.	5.
Committee's steel *X* (0·03% Cu)	20·1	31·1	40·2	44·9	56·6	15·0	24·8	32·9	38·5	50·6	4·1	11·3	14·6	19·7
Committee's steel *Y* (0·26% Cu)	14·2	21·8	27·1	29·1	33·6	11·3	19·7	24·9	28·3	36·8	3·9	10·0	12·8	16·5
Committee's steel *Z* (0·54% Cu)	12·7	19·7	23·5	25·9	29·4	11·8	18·1	22·5	25·6	31·7	3·9	9·6	12·0	14·9
Experimental heat, low Mn and Cu	36·7	70·2	...	...	...	30·8	47·0	...	...	...	8·7	19·5	25·8	37·2
Experimental heat, medium Mn, low Cu	27·8	42·0	55·4	63·7	66·6†	19·0	29·3	41·9	47·5	64·0	6·3	14·2	18·2	23·9
Open-hearth iron, very low Cu	34·0	51·0	...	...	...	26·9	42·7	65·5	...	...	5·6	14·7	19·7	28·3
Steel *S* (0·05% Cu)	20·1	28·6	37·2	40·2	46·4	14·7	22·6	27·7	35·2	44·3	4·4	10·7	14·4	17·9
Wrought iron (0·29% Cu)	17·9	25·0	33·9	37·3	44·5	14·4	24·6	31·0	36·2	44·7	5·6	12·9	15·8	19·8
Steel *K* (0·27% Cu)	12·9	19·3	25·4	27·2	31·8	10·3	16·9	21·8	26·9	33·0	4·4	9·7	12·0	14·7
Bessemer steel (0·5% Cu)	11·3	16·3	20·6	22·1	24·9	9·5	15·9	20·1	21·6	27·4	3·8	8·3	10·6	11·5
3·5% Ni steel	8·2	12·4	13·8	15·0	16·6	6·7	9·8	12·0	13·6	15·7	3·9	8·3	9·5	11·2
Cr–Si–Cu steel	7·9	9·0	9·9	...	10·9	5·9	7·0	7·6	...	10·1	3·5	5·3	5·8	6·9

* The results for the Committee's steels *X*, *Y* and *Z* are mean values for two sheet specimens, each prepared from one of the original flat bar specimens (*cf.* Table I., footnote, and Table IV.).

† This loss is for 42 months' exposure.

contaminated marine marsh atmosphere. The corrosion rates at South Bend, Pennsylvania, in the rural atmosphere are very much lower, as shown in Fig. 3, and it was necessary to plot the results of several steels in a single band to avoid confusion.

It is interesting to note that at these locations the effect of the addition of copper in the case of the Committee's steels was much more pronounced than that shown in the results published by the Committee for other locations. It is believed that this difference

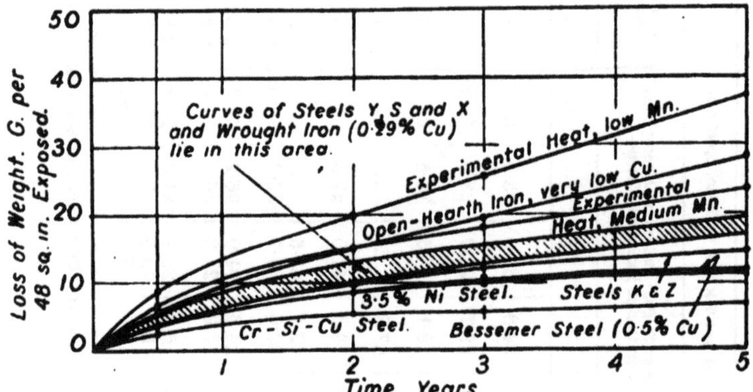

Fig. 3.—Results of Atmospheric Corrosion Tests at South Bend, Pa. (rural atmosphere).

may be due to the specific nature of these industrial atmospheres, which seem to have a pronounced tendency to form harder and more adherent scales on the high-copper steels. The differences noted are considerably less under rural conditions.

The results of the whole test confirm those of other investigators, in that very pure materials low in copper have high corrosion rates, which, however, are decreased by the slight additions of manganese and phosphorus. Among the commercial steels it is apparent that the effect of phosphorus is beneficial when combined with copper. The two higher-alloy steels show a significant increase in corrosion resistance.

THIRD REPORT OF THE OXYGEN SUB-COMMITTEE

OF THE COMMITTEE ON THE HETEROGENEITY OF STEEL INGOTS.[1]

(Section II., Part A: Fig. 1 = Plate XLVIII.)
(Section III., Part A: Fig. 3 = Plate XLVIII.)
(Section IX., Part A: Fig. 10 = Plate XLIX.)

Paper No. 5/1941 *of the Committee on the Heterogeneity of Steel Ingots (submitted by the Oxygen Sub-Committee).*

CONTENTS.

---

[1] A Joint Committee of The Iron and Steel Institute and The British Iron and Steel Federation, reporting to The Iron and Steel Industrial Research Council.

SECTION I.—INTRODUCTION.

THE Second Report of the Oxygen Sub-Committee appeared as Section VI. of the Eighth Report on the Heterogeneity of Steel Ingots.[1]

Dr. Sykes and Dr. Jay have joined the Sub-Committee meantime, and Dr. Bramley resigned on leaving Sheffield University to take up a post in industry. The present personnel is as follows:

Dr. T. Swinden (*Chairman*) .	The United Steel Companies, Ltd.
Professor J. H. Andrew .	The University, Sheffield.
Mr. E. W. Colbeck . .	I.C.I. (Alkali), Ltd., Northwich.

[1] *Iron and Steel Institute*, 1939, *Special Report No.* 25, pp. 37–234.

Mr. S. W. Craven .	. I.C.I. (Alkali), Ltd., Northwich.
Dr. C. H. Desch, F.R.S.	. Technical Advisor to the Iron and Steel Industrial Research Council.
Dr. H. A. Dickie .	. Messrs. Stewarts and Lloyds, Ltd.
Mr. N. Gray . .	. Guest Keen Baldwins Iron and Steel Co., Ltd.
Dr. A. H. Jay . .	. The United Steel Companies, Ltd.
Dr. W. R. Maddocks .	. The University, Sheffield.
Dr. W. C. Newell .	. The Brown-Firth Research Laboratories.
Mr. J. G. Pearce .	. The British Cast Iron Research Association.
Mr. T. E. Rooney .	. The National Physical Laboratory.
Mr. H. A. Sloman .	. The National Physical Laboratory.
Mr. W. W. Stevenson	. The United Steel Companies, Ltd.
Dr. C. Sykes . .	. The National Physical Laboratory.
Mr. E. Taylor-Austin .	. The British Cast Iron Research Association.
Mr. A. E. Chattin (*Secretary*)	The Iron and Steel Institute.

The incidence of the war has interfered considerably with the prosecution of the work of the Sub-Committee, but, nevertheless, sufficient has been done in the interim to warrant the presentation of this Report. As formerly, the Report is sectionalised under the various methods investigated, and the Chairman takes this opportunity of again paying testimony to the generous collaboration between the various laboratories which has enabled this to be done. It is evidence of the virility of the Sub-Committee that certain opinions expressed in the individual contributions are not necessarily held unanimously by the members, and the vigorous discussions in the Chemists' Panel and Sub-Committee have led to the construction of the forward programme.

Since the publication of the Second Report, the work of the co-operating members has been concerned mainly with a more detailed study of existing methods, rather than with developing new ones, and the additional knowledge presented is a definite advance on previous conceptions of those methods. The respective authors have endeavoured to describe as concisely as possible the essential items on which advance has been made, and to state the possibilities and limitations of the respective methods as they are practised to-day.

It is hoped that this will prove of service not only to those who for many years have worked on the subject of determining the gases (oxygen, hydrogen and nitrogen) in steel, but particularly to those laboratories who desire to take up this work.

The necessity for these determinations cannot now be questioned, and, although the problem of accurately determining quantitatively the respective oxides in all circumstances still demands further work, this Report will, it is hoped, clarify the position as to the relative merits and demerits of the respective methods used.

SECTION II.—The Vacuum Fusion and Vacuum Heating Methods.

Part A.—*The Determination of Oxygen, Nitrogen and Hydrogen in Steel. A Survey of the Vacuum Fusion Method, with a Note on the Solid Solubility of Oxygen in High-Purity Iron.**

By H. A. Sloman, M.A., A.I.C. (National Physical Laboratory, Teddington).

(Fig. 1 = Plate XLVIII.)

Synopsis.

The work of the Oxygen Sub-Committee on the vacuum fusion method is reviewed. A comparison of the forms of apparatus in use in the co-operating laboratories and a survey of the fundamental work which has been carried out in a critical examination of the method at the National Physical Laboratory are made. The present position with regard to the determination of oxygen, nitrogen and hydrogen in irons and steels is summarised and shown to be completely satisfactory in the cases of the first two but open to some doubt in the case of hydrogen. A note on the solid solubility of oxygen in high-purity iron is included, and it is suggested that the limit is between 0·006% and 0·003% of oxygen.

Prior to 1933, no serious attempt to examine the vacuum fusion method had been made in Great Britain, although work had been carried out in several European and American laboratories for some years previously. Doubt, however, still existed as to the validity of the method and as to its universal applicability to all classes of steels and irons.

The attention of the steel industry and of research institutions in Britain was being increasingly directed towards the desirability of establishing a reliable method for the determination of oxygen in steel, and accordingly, in 1933, an investigation was commenced at the National Physical Laboratory on the vacuum fusion method in an attempt to examine critically its possibilities and limitations. Simultaneously, the University of Sheffield took similar action, and shortly afterwards apparatus were set up by two industrial research laboratories, namely, the Central Research Department of The United Steel Companies, Ltd., and the Brown-Firth Research Laboratories. For some time all these laboratories worked more or less independently, but the formation of the Oxygen Sub-Committee of the Committee on the Heterogeneity of Steel Ingots in 1935 drew them much closer together and inaugurated the first experiment in co-operative research on a large scale. This experiment has proved completely successful; overlapping has been avoided, the results obtained in one laboratory are freely disclosed

* Received March 6, 1941.

to the others long before they are made generally available by publication, and the work of each investigator greatly benefits by the criticisms and comments of fellow-members of the Sub-Committee. Progress has been rapid, and our knowledge of the method to-day is undoubtedly far more comprehensive than would otherwise have been possible.

Principles of the Method.

The vacuum fusion method depends fundamentally on the affinity of carbon for oxygen and its ability to decompose oxides at high temperatures with the formation of carbon monoxide. These reactions are of the balanced equilibrium type, and it is accordingly necessary not only to carry out the determination at a high temperature but also to remove the products (*i.e.*, the carbon monoxide) rapidly and completely if total reduction of the oxides is to occur. From the quantity of carbon monoxide produced and the weight of sample taken the oxygen content of the steel is determined.

Apparatus.

In order partly to attain the temperature necessary for the rapid reduction of most oxides and partly to bring carbon into intimate contact with them, the steel or iron sample is held in the liquid state in a graphite crucible, whereupon solution of carbon from the crucible and its diffusion throughout the melt take place very rapidly. Essentially, therefore, the apparatus consists of a closed system—with the sample or samples in a convenient side tube—containing a graphite crucible heated by suitable means and into which, after degassing, the sample can be introduced from outside by means of a magnet. The system should be connected to a vacuum pump capable of producing a high degree of vacuum and having a fast pumping speed. This pump is usually of the mercury-vapour diffusion type and should be attached to the system by short wide connections in order that full advantage may be taken of its pumping speed and a high vacuum maintained above the surface of the melt during the evolution of the carbon monoxide, which normally takes 15–20 min. The gas is allowed to collect on the " backing " side of the pump, usually in some form of Toepler pump, from which it can be transferred to a gas-analysis apparatus, and in order that the rise in pressure on the backing side of the diffusion pump, due to the accumulation of gas, shall not reach a value such as to prevent its efficient working, it is desirable that the pump shall be of the multi-stage type capable of operating against a high backing pressure.

Recent Modifications.

The apparatus set up in the four above-mentioned laboratories, while they differ in many details, satisfy all the above essential requirements, and full details of their construction and operation have already appeared in previous Reports.[1,2] Since then no drastic changes have been found necessary, but some modifications have been introduced in two of the laboratories, particularly in the direction of increasing the maximum temperatures at which degassing can be carried out and thus lowering the "blanks" or alternatively shortening the times necessary to attain the previous blank values. At the National Physical Laboratory, where graphite powder is used as a thermal insulator, further experience has been gained in the method of packing the powder round the crucible, and the degassing temperature has been raised from about 2400° C. to about 2600° C. The blank has been correspondingly lowered from about 0·006 ml. per hr. to about 0·004 ml. per hr. This represents the lower limit of blank value in the particular apparatus at the National Physical Laboratory; because there exists a constant blank for the mercury-vapour pump of about 0·003 ml. per hr. The blank for the furnace parts is thus only about 0·001 ml. per hr., but since the pump forms part of the gas collecting system its own blank cannot be ignored. The reason for the gas evolution from the pump has not been found. It has been shown that it does not occur when power is not applied to the mercury boiler and also that it continues indefinitely at a constant rate when the boiler is hot. It would appear to be due to some slight porosity in the metal case which is only noticeable when hot. The total blank is, however, so small as to be completely negligible.

In the Brown-Firth apparatus, the molybdenum radiation screen has been replaced by a hollow graphite one filled with powdered graphite. It is constructed of two thin graphite cylinders differing in diameter by about one inch. The smaller cylinder is inserted in the other and the space between is filled with graphite filings passing a 300-mesh sieve in conformity with the experience at the National Physical Laboratory. The use of this screen has raised the temperature of the furnace for the same power input by several hundred degrees and, unlike the molybdenum one, it has an almost indefinite life. Owing to limitations of space inside the furnace, the four supporting pillars for the head of the furnace have been made thinner in diameter to allow for the extra space taken up by the graphite screen. The water-cooling has also been made more effective by soldering the inner water-cooling ring to the lower connecting plate. This plate is thereby kept much cooler and a waxed vacuum joint has been dispensed with.* An open full-length mercury manometer has also been added to the equipment. It is connected to the mercury of the Toepler pump, so that on compression of the gas in this pump a rapid indication

* Complete elimination of joints in the various forms of apparatus is almost impossible, but a reduction in their number to the lowest possible minimum is, in the author's opinion, very essential. This applies particularly to greased or waxed joints in positions liable to become heated, when gases may be evolved.

of the amount of gas present is obtained. This has been found particularly useful at times when the quantity of gas given off is too small to be transferred accurately to the analysis apparatus. In such cases, the quantity of gas can be measured by means of the manometer. The use of this manometer serves a purpose similar to that of the McLeod gauge which has always formed part of the apparatus at the National Physical Laboratory.

In Table I. are set out for comparison the essential details of the four sets of apparatus.

The "Blank."

One of the major considerations in the operation of the method is that the blank for the apparatus at the operating temperature should be as low as possible. This is particularly true in one primarily designed for research on all types of materials, including those containing very low percentages of oxygen, hydrogen or nitrogen, such, for instance, as oxygen-free iron. In this respect the apparatus at the National Physical Laboratory is considerably in advance of the others, in spite of the presence of the large mass of fine graphite powder, which might be expected to contain large quantities of gas, the removal of which would present difficulties. It would appear, therefore, that the exceptionally low blank is due to the very high temperature reached during the degassing period. The low power consumption for this temperature should also be noted as illustrating the excellent thermal insulating properties of the powder, a point already noted above in the modification recently effected in the apparatus at the Brown-Firth Research Laboratories. The use of the ball stopper covering the hole in the lid of the crucible in two laboratories may also be noted. This not only prevents splashes of molten metal from being ejected from the crucible but also considerably reduces the radiation losses from the open end. At the highest temperatures these losses may lower the maximum attainable by as much as 200° C.

Fundamental Examination of the Method.

Turning now to the possibilities of the method, practically all the fundamental work has been carried out at the National Physical Laboratory. It has always been felt by the Oxygen Sub-Committee that this type of work constitutes the best method by which the unique facilities of the National Physical Laboratory can contribute to the co-operative investigation. It was first shown that the results on many different commercial steels were reproducible. Where this was not so, it could always be traced to segregation—such, for instance, as occurs in rimming steels, where the result depends on the position in the specimen from which the sample is taken—or to trapped slag. With one reservation to be discussed below, it was then found that several (six or eight) samples, either

pertaining to each individual design of apparatus, and it would be desirable for each investigator to check this point before making determinations on this type of steel.

From the reproducibility of these early results it became apparent that they had some definite meaning, even if they did not necessarily represent the true oxygen contents. At this time, no other reliable method existed for the determination of total oxygen in irons and steels. A comparison was therefore impossible. The behaviour of synthetically prepared simple and complex oxides was accordingly examined. This work has already been described by the author in previous Reports.[1, 2] Weighed quantities of oxide were inserted in capsules of a steel of known oxygen content, and it was found that within quite narrow limits of accuracy all those oxides, whether simple or complex, likely to be present in all types of steel were completely reduced by carbon at sufficiently high temperatures. As was expected, alumina and the complex oxides containing alumina were the most difficult to reduce rapidly and completely. However, with a high vacuum above the surface of the melt, a temperature of about 1550° C. was found to be sufficiently high to complete reduction of these oxides in 15–20 min.

The effect of metallic constituents was next investigated in a similar manner. In the case of manganese, quantitative results were obtained in capsules containing up to 12–15% of manganese, but above this percentage the results tended to be low. For aluminium, no variations in oxygen content were detected with up to 2·5% of that metal, the maximum so far investigated. In view of the easily reducible nature of the oxides of all the other metals likely to be found in steels, it is unlikely that the presence of these metals themselves would affect the determinations, since in no case do problems connected with volatility and film formation exist to anything like the extent that occurs with manganese. The examination of the effect of additions of these metals has, therefore, not been undertaken.

Present Position of the Method.

On the assumption that the synthetically introduced oxides and metals behave in a manner similar to those actually present in steel, a temperature of 1550–1600° C. is satisfactory, and the method gives true values for total oxygen for all irons and steels within the composition ranges investigated by the capsule method. The usual time for complete reduction of a sample is 20 min. or less, although in isolated cases 25 min. or even 30 min. may be required. This extra time is usually associated with the presence of aluminium. It may be noted that the other three laboratories prefer an operating temperature of 1650° C. as a measure of safety, but it is not apparent from co-operative work on the same steels that the higher temperature has any advantage in either the speed or completeness of reduction.

The co-operative examination of both commercial and specially prepared materials has formed an important feature of the work of the Oxygen Sub-Committee ever since its formation and large numbers of steels, cast irons and pig irons have been examined by the vacuum fusion method, in some cases by all four laboratories and in others by two or three. So far the attention of the Sub-Committee as a body has been mainly directed to carbon steels, including a series made up to contain ½% of different alloying elements, and no difficulty has been experienced in obtaining reliable results on any of the specimens examined. Individual members have obtained good results on all classes of alloy steels and there seems to be no reason to suppose that the alloy steels which form part of the Sub-Committee's future programme will present any unexpected problems.

The reproducibility of the results on different samples of the same iron or steel is very striking, and is of the same order no matter whether the different samples are reduced in the same apparatus at any one laboratory or divided between two or more. The range of total oxygen content with which one is normally concerned is from about 0·002% to about 0·03%, and provided that the weights of samples are adjusted to give reasonable volumes of gas for subsequent analysis, the order of error is about ± 0·0005% to about ± 0·001% over the above range of oxygen contents. In the case of the special work on oxygen-free high-purity iron carried out at the National Physical Laboratory, duplicates of the type 0·0004 ± 0·00005% of oxygen have been obtained.

Comparison with the Aluminium Reduction Method.

As pointed out above, at the time when this investigation was commenced no other reliable method existed. Within the last year or so this situation has changed with the introduction of the aluminium reduction method by Gray and Sanders. This method was originally described in a previous Report.[3] Certain modifications, recently introduced, are fully discussed in the present Report (Section III., Part A).

Essentially the method is as follows: The oxides are converted into alumina by melting the steel sample in the presence of excess aluminium at about 1150° C. and estimating the oxygen chemically as alumina. It is thus entirely independent, and, in general, gives results in remarkably good agreement with the vacuum fusion method. In some cases where discrepancies have been noted, they have been traced to unsuspected factors influencing the results by the aluminium reduction method, and modifications to overcome them have led to the values by the two methods coming into line. Some uncertainty still appears to exist in the values by the aluminium reduction method in the case of certain cast irons, owing to the interference of large amounts of silica, but the general agreement between the two methods is very gratifying and considerably strengthens the case for the reliability of each. In many cases this agreement is of the same order as that existing between different laboratories operating the vacuum fusion method itself.

In a comparison of the merits of the two methods it should be noted that the aluminium reduction method gives total oxygen only, whereas the vacuum fusion method gives not only oxygen but hydrogen and nitrogen as well. The fractional form of the latter method may eventually be proved to go even further, and split, with reasonable accuracy, the total oxygen into its constituent components. It seems probable that the aluminium reduction method is somewhat more rapid than the vacuum fusion method, but against this must be set the fact that only one analysis can be made in the former as against a large number in the latter. This disadvantage is, however, largely off-set by the relative cheapness and simplicity of the equipment required in the former method, which would permit several furnace tubes to be set up and run concurrently if required.

Fractional Form of the Vacuum Fusion Method.

It may be convenient here to discuss briefly the present position of the fractional form of the vacuum fusion method. Most of the earlier work on this elaboration of the ordinary method was carried out in the United States of America by Reeve,[4] Hoyt and Scheil,[5] Motok,[6] &c. More recently the method has received the attention of the Oxygen Sub-Committee, and the first paper by Swinden, Stevenson and Speight appeared in a previous Report.[7] The results of their experience since that time are described in the present Report (Section II., Part B).

The method is based on the fact that the constituent oxides in steel are not all equally readily reduced by carbon, and therefore, by carrying out a series of partial reductions at different temperatures from about 1080° C. upwards, it is hoped that the oxides will be reduced separately in the order FeO, MnO, SiO_2 and Al_2O_3. The first point of interest lies in the fact that the sum total of the oxygen calculated from the four fractions agrees, within experimental limits, with the total oxygen as determined by the ordinary vacuum fusion method. This applies equally to hydrogen and nitrogen. The only point, therefore, upon which doubt can be thrown is whether the separate fractions really represent the oxygen associated with the particular oxides. Two possible reasons for this doubt exist. First, interaction may occur between the oxides and metals during the melting with tin in the presence of carbon, tending to increase the low-temperature fractions at the expense of the higher. Secondly, it may be impossible to reduce separately the various oxides when they occur together or in combinations. These points were discussed by Swinden and his co-authors in their earlier paper, in which attention was drawn to the fact that the true base pressure of the system between fractions could not be attained even after an excessively long time taken over the collection of each fraction. Curves were given illustrating the rapid evolution of the bulk of the gas at the commencement accompanied by a sharp drop in pressure, followed by an excessively slow decrease in pressure, which in some cases had not reached the true base value even after 4 hr. These facts can only be explained on the assumption of a gradual reduction of more stable oxides during the collection of each fraction, but there is no evidence to show which of the above hypotheses offers the better reason for their occurrence. In order, therefore, to obtain more satisfactory results and to cut down the time during which these unwanted oxides have an opportunity to reduce, the investigators decided to cease the collection of each fraction as soon as the first rapid drop in pressure had occurred. There is an obvious justification for this decision, in that the rate of reduction of the next higher oxide must vary with the pressure above the melt, and if this is not permitted to fall below a certain—undetermined—value, reduction will be inhibited. At the same time, of course, there will be a tendency for the wanted oxide to be incompletely reduced at this higher pressure, and, in the limit, the balance of the fractions may be considerably in error in the opposite direction, the high-temperature fractions being increased at the expense of the lower. The basis for the work on the method is thus seen to be still somewhat empirical, the results depending largely on the particular technique employed. The only basis of comparison available at the moment is with the chemical residue methods, which in some cases are themselves open to doubt as regards not only the total oxygen but also the various fractions. Nevertheless, the investigators are to be congratulated not only on the very valuable contribution that they have made to our knowledge of the method but also on the way in which they have brought forward the difficulties which confront workers in this field. Further investigation of a fundamental nature is urgently required, and the Oxygen Sub-Committee have this in hand as part of their future programme.

One disadvantage of the method which it would seem impossible to overcome is that only one sample can be reduced each time the apparatus is assembled. The lengthy processes of dismantling, cleaning, assembling and degassing are necessary after every single determination. This arises from the volatility of the tin present, which if taken much above 1300° C. behaves in the same way as manganese or aluminium. One result of this is that subsequent temperature measurements are rendered valueless. Thus, having once taken the melt to about 1600° C. to obtain the Al_2O_3 fraction, it will be difficult to drop the temperature with the necessary accuracy to the required point for the first (FeO) fraction from a second sample. This difficulty can largely be overcome if the Al_2O_3 fraction is not collected—that is, if the melt is never taken above about 1300° C. Under these conditions it may be possible to obtain all the fractions up to and including the SiO_2 on several samples one after the other in the same assembly. Subsequent total oxygen determinations on similar samples by the ordinary vacuum fusion process would give the Al_2O_3 by difference.

Type of Sample.—Surface Oxygen.

It should be noted that the samples employed in both forms of the vacuum fusion method are solid cylinders, usually about 1 in. long and ⅝ in. in dia., weighing 15–20 g. Such samples have a small surface area per unit weight, and loss of accuracy due to interference by surface oxygen is thus reduced to a minimum. The use of millings or other types of sample having a large surface area is to be avoided not only in the vacuum fusion method but in all others, and work showing the magnitudes of the errors involved by their use has been carried out at the National Physical Laboratory. The author,[2] using the vacuum fusion method, determined the surface oxygen on pure iron sheet under various conditions and found that for freshly abraded surfaces it amounts to $1{\cdot}5 \times 10^{-6}$ g. per sq. cm., which for the particular thickness of sheet used (0·004 in.) is equivalent to 0·004% of oxygen. Rooney,[8] using the alcoholic iodine method on similar pure iron, found that freshly prepared millings gave an oxygen value about 0·006% higher than that obtained with a solid sample. The agreement between these results is striking in view of the possibility that the degree of surface oxidation produced during the preparation of sheet is not necessarily comparable with that which occurs during milling. Further comprehensive investigation of surface oxygen as it affects not only pure iron but different types of steel and cast iron is being undertaken. For this purpose use will be made of the vacuum fusion method and also, where practicable, the alcoholic iodine method.

Turning now to elements other than oxygen, both nitrogen and hydrogen can be determined by the vacuum fusion method.

Nitrogen.

A comparison between the results obtained by the vacuum fusion and chemical methods has nearly always shown excellent agreement in the materials so far examined.[2] The evidence strongly suggests that nitrogen is present in iron and steel in a combined form. A few cases have been found where the values by the former method are somewhat higher than those by the latter. These have always occurred in molybdenum or titanium steels containing stable nitrides, the decomposition of which is not necessarily completely effected by the usual chemical method.[2] It is recognised, however, that these exceptional steels yield reliable results by the chemical method, provided that appropriate modifications, leading to complete decomposition of the nitrides, are made, and it is suggested that there is good evidence that the difficulty arises from the stability to the action of hydrochloric acid of the nitrides present in these steels.

A contrast to the above is provided by the recent experience at The United Steel Companies' laboratory in the cases of one or two alloy steels which tend to give lower results by the vacuum fusion method than by the chemical. It was found that from these steels the nitrogen is evolved only with difficulty and that to obtain results comparable with those given by the chemical method prolonged evacuation at 1600° C. is required. It is possible that this difficulty is connected with the lowest pressure that it is possible to attain in the particular apparatus, which, as will be seen from Table I., is somewhat higher than in some of the others. Further examination of these steels is awaited.

The total nitrogen as determined by the fractional vacuum fusion method agrees well with the results obtained by the ordinary vacuum fusion method, the bulk of the gas being evolved in the final (Al_2O_3) fraction.* It would thus appear that the form of combination of the nitrogen is such that a high temperature is required for its decomposition, and it is suggested that it probably occurs as aluminium nitride.

The position, then, with regard to nitrogen may be considered satisfactory. By the use of the vacuum fusion method attention has been drawn to possible errors in the chemical method normally employed as applied to certain classes of steel, and modifications have been suggested.

Hydrogen.

It has been known for some time that hydrogen is evolved from samples during the vacuum fusion process and that the results are reproducible. Once again, difficulties in the way of comparison arose owing to the lack of any alternative reliable method of analysis. Results obtained at the National Physical Laboratory led the Oxygen Sub-Committee, soon after their appointment, to extend their field of interest to include the determination of hydrogen, and, as a result, opportunity was first provided to demonstrate quantitatively the progressive lowering of the values for hydrogen in the same steel due to increasing amounts of hot-working. This first led the author to suggest that an alternative method for the determination of hydrogen in irons and steels might be to hold the sample at between 600° and 900° C. in a high vacuum, whereupon the hydrogen should be completely evolved. This experiment was carried out on samples of many different steels, and it was found that in all cases evolution of hydrogen did in fact occur. This ceased after 1–2 hr., depending on the particular steel and size of sample, and the quantities obtained were in agreement with

* Swinden, Stevenson and Speight, *see* reference No. 7 and this Report, Section II., Part B.

those by the vacuum fusion method.[2] Newell has constructed an apparatus especially for the purpose of estimating hydrogen in steels based on this vacuum-heating extraction method. The apparatus is very neat and simple, and full details have been published.[9] It will be noted that he has confirmed the author's results in every case and has applied this simplified method with success to many steels outside the range originally examined at the National Physical Laboratory. A similar apparatus is now being installed in The United Steel Companies' laboratory, in which it is proposed to incorporate a special tap of the type described by Willems.[10] This permits samples to be introduced into an evacuated system at any stage in the course of an experiment and without contamination by tap lubricant. Where the hydrogen content only is required, the vacuum heating method is considerably more simple and rapid than the vacuum fusion method, but it is not, of course, applicable to the determination of oxygen and nitrogen.

As indicated above, results by the fractional vacuum fusion method are in agreement with those by the ordinary vacuum fusion method. It should be noted that in the former method, the whole of the hydrogen is invariably obtained with the first fraction,* thus strengthening the underlying basis of the low-temperature vacuum heating procedure.

The agreement between laboratories on hydrogen is not as good as on the other elements, particularly where the hydrogen content is low. This may possibly be due to variability in the hydrogen from sample to sample or to uncertainties arising out of the blanks for the various systems. In some cases these blanks are relatively large and contain considerable percentages of hydrogen.

While from the foregoing it might appear that the general position with regard to hydrogen is satisfactory, actually it is far from being so when the work of Chaudron[11] and his collaborators is considered. Using ionic bombardment of the sample in the cold, they found volumes of hydrogen many times larger than those given by any other method, not only in steels but also in other metals, such as aluminium. The results are in fact so different as to be quite irreconcilable, and one is forced to conclude either that there is, in metals, a large quantity of hydrogen which cannot be extracted by any form of heating or melting *in vacuo*, or alternatively that Chaudron's results are subject to some unsuspected interference from outside. Recent work by Schmid and von Schweinitz[12] appears to confirm the latter view. The matter is of such importance that it cannot be allowed to rest, and the Oxygen Sub-Committee have arranged for an apparatus similar to that used by Chaudron to be set up in an attempt to throw further light on the differences. Meanwhile the view, not only of the author, but of the other members of the Sub-Committee operating the vacuum

* Swinden, Stevenson and Speight, *loc. cit.*

fusion method, is that the evidence is largely in favour of true figures being obtained by this method.

Other Uses of the Method.

In conclusion, it may be pointed out that, in addition to the continued use of the vacuum fusion method for its usual purpose of determining the oxygen, nitrogen and hydrogen in cast iron and steel, its scope is gradually being extended to other problems. At the National Physical Laboratory it has, as already indicated, been applied successfully to the determination of surface oxygen on high-purity iron sheet and the work is now being expanded to include the surface oxygen on millings. Recently, other ferrous materials have been examined, and preliminary work on non-ferrous metals such as aluminium, cobalt, &c., is in progress.

Solubility of Oxygen in High-Purity Iron.

In the course of the production of high-purity low-oxygen iron, a combination of microscopic examination and the vacuum fusion method has incidentally resulted in a more accurate knowledge of the solubility limit of oxygen in iron. Tritton and Hanson [13] put the limit at about 0·05% of oxygen, but this figure was probably subject to errors owing to experimental difficulties existent at the time. Micrographs in Fig. 1 show the microstructures of four samples of iron after slow cooling and containing 0·15%, 0·0059%, 0·0028% and 0·0004% of oxygen, respectively. The first sample was saturated with FeO at the melting point, and the others were produced at various stages during the deoxidation of the iron by hydrogen. The amount of oxygen in solution in relatively pure iron after slow cooling to room temperature thus appears to lie between 0·006 and 0·003%. It is hoped that opportunity will enable these limits to be even more closely defined in the near future.

The author wishes to acknowledge the assistance o those members of the Oxygen Sub-Committee to whose work reference is made in the paper, which is published by permission of the Director, National Physical Laboratory.

REFERENCES.

(1) SLOMAN: Seventh Report on the Heterogeneity of Steel Ingots, p. 82, *Iron and Steel Institute*, 1937, *Special Report No.* 16.
RAINE and VICKERS: *Ibid.*, p. 100.
STEVENSON and SPEIGHT: *Ibid.*, p. 65.

(2) SLOMAN: Eighth Report on the Heterogeneity of Steel Ingots, p. 42, *Iron and Steel Institute*, 1939, *Special Report No.* 25.
BRAMLEY and RAINE: *Ibid.*, p. 86.
NEWELL: *Ibid.*, p. 97.

(3) GRAY and SANDERS: Eighth Report on the Heterogeneity of Steel Ingots, p. 103, *Iron and Steel Institute*, 1939, *Special Report No.* 25.

(4) Reeve: *Transactions of the American Institute of Mining and Metallurgical Engineers*, 1934, *Iron and Steel Division*, vol. 113, p. 82.
(5) Hoyt and Scheil: *Transactions of the American Institute of Mining and Metallurgical Engineers*, 1937, *Iron and Steel Division*, vol. 125, p. 313.
(6) Motok: *Transactions of the American Society for Metals*, 1937, vol. 25, p. 466.
(7) Swinden, Stevenson and Speight: Eighth Report on the Heterogeneity of Steel Ingots, p. 63, *Iron and Steel Institute*, 1939, *Special Report No.* 25.
(8) Rooney: Eighth Report on the Heterogeneity of Steel Ingots, p. 141, *Iron and Steel Institute*, 1939, *Special Report No.* 25.
(9) Newell: *Journal of the Iron and Steel Institute*, 1940, No. I., p. 243P. This Report, Section II., Part C.
(10) Willems: *Archiv für das Eisenhüttenwesen*, 1938, vol. 11, p. 627.
(11) Portevin, Chaudron and Moreau: *Comptes rendus*, 1935, vol. 201, p. 212; 1937, vol. 204, p. 1252; 1938, vol. 207, p. 235.
Chaudron: *Foundry Trade Journal*, 1937, vol. 56, p. 509.
(12) Schmid and von Schweinitz: *Aluminium*, 1939, vol. 21, p. 772.
(13) Tritton and Hanson: *Journal of the Iron and Steel Institute*, 1924, No. II., p. 85.

Part B.—*The Fractional Vacuum Fusion Method for the Separation of Oxides and Gases in Steel. Further Practice and Typical Results.*[1]

By T. SWINDEN, D.Met., W. W. STEVENSON, A.I.C., and G. E. SPEIGHT, B.Sc., A.I.C. (Stocksbridge, near Sheffield).

Synopsis.

Typical results obtained by the fractional vacuum fusion procedure for several types of steel are presented and compared in some instances with results obtained by the alcoholic iodine method. The usefulness and limitations of the fractional technique are discussed.

Since the publication of the authors' paper on this subject,[2] the fractional vacuum fusion method has been constantly applied in the Stocksbridge Laboratory to various types of material and a large amount of data has been accumulated. In many cases, the materials have been examined by other methods, notably the alcoholic iodine and the direct total vacuum fusion methods; some of the results obtained are reported in Tables II. to VIII., to indicate the scope and usefulness of the fractional determination.

The process is carried out essentially as described in the previous paper, and provides a rapid method whereby easily reducible oxides present in a steel are distinguished from the less easily reducible compounds. The method has been criticised on the grounds that

[1] A communication from The United Steel Companies, Ltd., Central Research Department, Stocksbridge, near Sheffield, received February 7, 1941.
[2] Eighth Report on the Heterogeneity of Steel Ingots, p. 63, *Iron and Steel Institute*, 1939, *Special Report No.* 25.

TABLE II.—*Determination of Oxygen by Fractional Vacuum Fusion.*

Cast *X*4890.

Manufacture: Basic open-hearth.

Analysis:	C. %.	Mn. %.	Si. %.	S. %.	P. %.	Cr. %.
	0·09	0·35	0·08	0·022	0·012	0·03

	Oxygen. %.	Oxide. %.	Hydrogen.		Nitrogen. %.
			Ml. per 100 g.	%.	
		Cast X4890A.			
FeO . .	0·0008	$0{\cdot}003_5$	0·4	$0{\cdot}00003_5$	...
MnO . .	0·0044	$0{\cdot}019_5$	...	...	0·0002
SiO_2 . .	0·0047	0·009	...	...	0·0004
Al_2O_3 . .	0·0024	0·005	...	...	0·0042
Total . .	$0{\cdot}012_5$	0·037	0·4	$0{\cdot}00003_5$	0·005
Direct total .	0·0125	...	0·2	...	0·004
		Cast X4890C, *aluminium-treated.*			
FeO . .	0·0001	$0{\cdot}000_5$	$0{\cdot}2_5$	0·00002	...
MnO . .	$0{\cdot}0001_5$	$0{\cdot}000_5$	$0{\cdot}0_5$	$0{\cdot}00000_5$	...
SiO_2 . .	0·0011	0·002	...	...	0·0002
Al_2O_3 . .	0·0055	$0{\cdot}011_5$	...	...	0·0027
Total . .	0·007	0·015	0·3	$0{\cdot}00002_5$	0·003
Direct total .	0·006	...	0·3	...	$0{\cdot}002_5$

the separate constituents of complex non-metallic inclusions cannot be expected to respond to complete fractional separation, and the authors acknowledge that there is justification for this criticism. If an absolute criterion of the distribution of oxygen in steels were known, it would be possible to investigate with certainty the degree of accuracy of the fractional separation, but no such standard is available. Hence, we are dependent upon less absolute methods of assessing the validity of new developments of this type, and the agreement of certain fractions with residue methods is sufficiently good to warrant confidence in the fractional technique. Practical difficulties in the method of fractionally reducing synthetic mixtures of complex oxides and silicates have been observed,[1] but it is planned to proceed, forthwith, upon an investigation into the response of certain definite compounds, such as fayalite, spinels, &c., to fractional reduction.

The separation of iron and manganese oxides is admitted to be a difficult operation, owing to the small interval between the tempera-

[1] Reeve, *Transactions of the American Institute of Mining and Metallurgical Engineers*, 1934, Iron and Steel Division, vol. 113, p. 82.

TABLE III.—*Determination of Oxygen in Boiler-Plate Material.*

Manufacture: Basic open-hearth.

Analysis:	C. %.	Mn. %.	Si. %.	S. %.	P. %.
Sample 433 . . .	0·16	0·58	Trace	0·041	0·051
,, *M*5 . . .	0·13	0·55	0·08	0·049	0·036

Sample.	Alcoholic Iodine.			Fractional Vacuum Fusion.							
	Constituent.	Oxygen. %.	Oxide. %.	Oxygen. %.		Oxide. %.		Hydrogen. Ml. per 100 g.		Nitrogen. %.	
433	Total residue.	...	0·070								
	FeO.	0·003	0·014	0·001		0·004$_5$		0·2$_5$		...	
	MnO.	0·003	0·013	0·005$_5$		0·024		0·1		0·0001$_5$	
	SiO_2.	0·006	0·011	0·004		0·007$_5$		...		0·0001$_5$	
	Al_2O_3.	0·003$_5$	0·007$_5$	0·001$_5$		0·003$_5$		...		0·0024$_5$	
	TiO_2, Cr_2O_3, P_2O_5.	...	0·017$_5$								
	Total.	0·015$_5$	...	0·012				0·3$_5$		0·002$_5$	
	Direct total by vacuum fusion.			0·012				0·6		0·003	
*M*5	Total residue.	...	0·051								
				(*a*)	(*b*)	(*a*)	(*b*)	(*a*)	(*b*)	(*a*)	(*b*)
	FeO.	0·001$_5$	0·007	0·001$_5$	0·001$_5$	0·006$_5$	0·007	0·3$_5$	0·3	...	...
	MnO.	0·003	0·013	0·007	0·006$_5$	0·030$_5$	0·029	0·1$_5$	0·1$_5$	...	0·0001$_5$
	SiO_2.	0·009	0·016$_5$	0·001$_5$	0·001$_5$	0·003	0·003	...	...	0·0001$_5$	...
	Al_2O_3.	Trace	0·000$_5$	0·002	0·002	0·004	0·004	...	...	0·0036	0·003$_5$
	TiO_2, Cr_2O_3, P_2O_5.	...	0·007$_5$								
	Total.	0·013$_5$	...	0·012	0·011$_5$	...	...	0·5	0·4$_5$	0·003$_5$	0·003$_5$
	Direct total by vacuum fusion.			0·012				0·3$_5$		0·003$_5$	

Table IV.—*Determination of Oxygen in Boiler-Plate Material.*

Manufacture: Basic open-hearth.

Analysis:	C. %.	Mn. %.	Si. %.	S. %.	P. %.
Top plate . . .	0·14	0·63	0·014	0·046	0·035
Bottom plate . . .	0·15	0·63	0·009	0·049	0·038

Sample.	Constituent.	Alcoholic Iodine.				Fractional Vacuum Fusion.				Chlorine Method.*	
		Oxygen. %.		Oxide. %.		Oxygen. %.	Oxide. %.	Hydrogen. Ml. per 100 g.	Nitrogen. %.	Oxygen. %.	Oxide. %.
Top plate.	Total residue	...	...	0·100	0·092						
	FeO.	0·0024	0·0020	0·011	0·009	0·0017	0·007$_5$	0·1$_5$	...	0·0023	0·010$_5$
	MnO.	0·0074	0·0072	0·033	0·032	0·0075	0·033	0·1$_5$	0·0001	0·0074	0·033
	SiO_2.	0·0138	0·0125	0·026	0·023$_5$	0·0118	0·022	...	0·0003$_5$	0·0117	0·022
	Al_2O_3.	0·0068	0·0064	0·014$_5$	0·013$_5$	0·0062	0·013	...	0·0028	0·0060	0·013
	TiO_2, Cr_2O_3, P_2O_5.	...	...	0·010$_5$	0·010$_5$	...	...	...	...	...	...
	Total.	0·030$_5$	0·028	...	...	0·027	...	0·3	0·003	0·027$_5$	
	Direct total by vacuum fusion.					0·027	...	0·1	0·003		
						0·027	...	...	0·003$_5$		
Bottom plate.	Total residue.	...	...	0·083	0·098						
	FeO.	0·0019	0·0023	0·008$_5$	0·010	0·0025	0·011	0·3	0·0002		
	MnO.	0·0066	0·0070	0·029	0·031	0·0078	0·035	...	...		
	SiO_2.	0·0078	0·0087	0·014$_5$	0·016$_5$	0·0107	0·020	...	0·0002$_5$		
	Al_2O_3.	0·0063	0·0070	0·013$_5$	0·015	0·0054	0·011$_5$	...	0·0023$_5$		
	TiO_2, Cr_2O_3, P_2O_5.	...	...	0·013$_5$	0·016						
	Total.	0·022$_5$	0·025	...	...	0·026$_5$	...	0·3	0·003		
	Direct total by vacuum fusion.					0·024$_5$	...	0·1	0·003		
						0·024$_5$	...	0·1	0·004		

* Chlorine results by I.C.I. (Alkali), Northwich.

Table V.—*Fractional Vacuum Fusion Determination of Oxygen in 0·[illegible]% Molybdenum Steel Plates.*

Manufacture.	Basic Open-Hearth.			Electric Arc.	
Mark.	*H*140.	*H*141.	*H*142.	⅜-in. Plate.	1-in. Plate.
Analysis:					
Carbon. % . . .	0·23	0·22	0·23	0·25	0·22
Manganese. % . . .	0·55	0·53	0·51	0·50	0·46
Silicon. % . . .	0·13	0·09	0·09$_5$	0·16	0·17$_5$
Sulphur. % . . .	0·043	0·030	0·030	0·019	0·017
Phosphorus. % . . .	0·043	0·044	0·037	0·031	0·031
Nickel. % . . .	0·04	...	...	0·11	0·15
Chromium. % . . .	0·07	...	...	0·21	0·20
Molybdenum. % . . .	0·51	0·59	0·59	0·56	0·54
Copper. % . . .	0·02	...	...	0·10	...

Sample.	Constituent.	Oxygen.		Hydrogen.		Nitrogen. %.
		Oxygen. %.	Oxide. %.	Ml. per 100 g.	%.	
*H*140.	FeO fraction.	0·0006	0·003	0·2	0·00002	Nil
	MnO "	0·0035	0·015$_5$	...	...	0·0002
	SiO_2 "	0·0057	0·011	...	...	0·0002
	Al_2O_3 "	0·0037	0·008	...	...	0·0035
	Total.	0·013$_5$	0·037$_5$	0·2	0·00002	0·004
	Direct determination.	0·013	...	...	...	0·004
*H*141.	*First Determination.*					
	FeO fraction.	0·0009	0·004	0·2$_5$	0·00002	Nil
	MnO "	0·0043	0·019	0·0$_5$	0·00000$_5$	0·0001
	SiO_2 "	0·0101	0·019	...	...	0·0004
	Al_2O_3 "	0·0113	0·024	...	...	0·0028
	Total.	0·026$_5$	0·066	0·3	0·00002$_5$	0·003$_5$
	Second Determination.					
	FeO fraction.	0·0008	0·003$_5$	0·1	0·00001	Nil
	MnO "	0·0043	0·019	...	...	0·0001
	SiO_2 "	0·0109	0·020$_5$	...	...	0·0004
	Al_2O_3 "	0·0123	0·026	...	...	0·0035
	Total.	0·028$_5$	0·069	0·1	0·00001	0·004
	Direct determination.	0·023	...	Nil	...	0·003$_5$
*H*142.	FeO fraction.	0·0009	0·004	0·1$_5$	0·00001$_5$	0·0002
	MnO "	0·0036	0·016	...	...	0·0003
	SiO_2 "	0·0070	0·013	...	...	0·0003
	Al_2O_3 "	0·0035	0·007$_5$	...	...	0·0031
	Total.	0·015	0·040$_5$	0·1$_5$	0·00001$_5$	0·004
	Direct determination.	0·015	...	Nil	...	0·004
⅜-in. Plate.	FeO fraction. MnO "	0·0003	0·001$_5$	0·2$_5$	0·00002	...
	SiO_2 "	0·0012	0·002$_5$	...	...	0·0009
	Al_2O_3 "	0·0007	0·001$_5$	...	...	0·0071
	Total.	0·002$_5$	0·005$_5$	0·2$_5$	0·00002	0·008
1-in. Plate.	FeO fraction. MnO "	0·0005	0·002	0·3	0·00002$_5$	0·0006
	SiO_2 "	0·0020	0·004	...	...	0·0010
	Al_2O_3 "	0·0012	0·002$_5$	...	...	0·0064
	Total.	0·003$_5$	0·008$_5$	0·3	0·00002$_5$	0·008

Table VI.—*Determination of Oxygen by Fractional Vacuum Fusion on Plate Material.*

Manufacture: Basic Bessemer.

Analysis:	*A2.*	*B3.*	*E2.*
Carbon. %	0·20	0·14	0·15
Manganese. %	0·49	1·08	0·55
Silicon. %	Trace	Trace	0·01

Analysis:	*A2.*	*B3.*	*E2.*
Sulphur. %	0·064	0·063	0·052
Phosphorus. %	0·054	0·063	0·060

Sample.	Constituent.	Oxygen. Oxygen. %.		Oxygen. Oxide. %.		Hydrogen. %:		Hydrogen. Ml. per 100 g.		Nitrogen. %.	
Plate *A2.*	$FeO + MnO$.	0·0075		$0·033_5$		0·00010		$1·1_5$		0·0006	
	SiO_2.	Nil		Nil		...		...		0·0006	
	Al_2O_3.	0·0006		$0·001_5$		...		...		0·0100	
	Total.	0·008		0·035		0·00010		$1·1_5$		0·011	
	Direct total determination.	0·008		...		0·00016		$1·7_5$		0·011	
Plate *B3.*	$FeO + MnO$.	0·0089		0·039		0·00016		$1·7_5$		0·0004	
	SiO_2.	Nil		Nil		...		...		0·0008	
	Al_2O_3.	0·0013		0·003		...		...		0·0091	
	Total.	0·010		0·042		0·00016		$1·7_5$		$0·010_5$	
	Direct total determination.	0·010				0·00018		$2·0_5$		0·012	
		1st Test.	2nd Test.	1st Test.	2nd Test.	1st Test.	2nd Test.	1st Test.	2nd Test.	1st Test.	2nd Test.
Plate *E2.*	$FeO + MnO$.	0·0064	0·0064	0·028	0·029	0·00009	0·00009	$1·0_5$	$1·0_5$	0·0003	0·0002
	SiO_2.	0·0061	0·0058	$0·011_5$	0·011	...	...	...	...	0·0013	0·0011
	Al_2O_3.	0·0029	0·0025	0·006	$0·005_5$	...	...	...	...	0·0094	0·0094
	Total.	$0·015_5$	$0·014_5$	$0·045_5$	$0·045_5$	0·00009	0·00009	$1·0_5$	$1·0_5$	0·011	$0·010_5$
	Direct total determination.	0·014				0·00014		1·5		$0·011_5$	

Table VII.—*Determination of Oxide Inclusions by the Fractional Vacuum Fusion Method.*

	Cast 52363.	Cast 52308.
Manufacture:	Basic open-hearth.	
Ladle additions:		
Carbon	80 lb.	180 lb.
Ferro-manganese	750 lb.	1000 lb.
Silico-manganese	250 lb.	100 lb.
Mould addition: Alsimin	6 lb.	2½ lb.
Pit analysis:		
Carbon. %	0·14	0·18
Manganese. %	0·49	0·70*
Silicon. %	0·009	...
Sulphur. %	0·042	0·045
Phosphorus. %	0·037	0·051

Sample.	Constituents.	Alcoholic Iodine Method.		Fractional Vacuum Fusion Method.			
		Oxygen. %.	Oxide. %.	Oxygen. %.	Oxide. %.	Hydrogen. Ml. per 100 g.	Nitrogen. %.
Cast 52363.	Total residue.	...	{0·065, 0·066				
Plate No. 27, top of ingot.	FeO.	0·001	0·006$_5$	0·001	0·005	0·5$_5$	0·0002
	MnO.	0·003$_5$	0·014$_5$	0·006	0·027	0·1	0·0001
	SiO_2.	0·008	0·015	0·009	0·017	...	Nil
	Al_2O_3.	0·010	0·021	0·003$_5$	0·008	...	0·0024
	TiO_2.	...	0·001				
	Cr_2O_3.	...	0·002				
	P_2O_5.	...	0·0035				
	Total.	0·023	...	0·019$_5$	...	0·6$_5$	0·002$_5$
	Total by direct vacuum fusion.			0·018	...	0·4	0·003
Cast 52363.	Total residue.	...	0·079				
Plate No. 29, bottom of ingot.	FeO.	0·001$_5$	0·006$_5$	0·001$_5$	0·007	0·5	0·0002$_5$
	MnO.	0·006	0·027	0·005	0·023$_5$	...	0·0001
	SiO_2.	0·007$_5$	0·014	0·009$_5$	0·017$_5$	...	0·0002$_5$
	Al_2O_3.	0·010	0·021	0·004	0·009	...	0·0021
	TiO_2.	...	0·000$_5$				
	Cr_2O_3.	...	0·001$_5$				
	P_2O_5.	...	0·003				
	Total.	0·025	...	0·020	...	0·5	0·002$_5$
	Total by direct vacuum fusion.			0·019$_5$	...	0·3$_5$	0·0025
Cast 52308.	Total residue.	...	{0·058, 0·057				
Top of Ingot.	FeO.	0·001$_5$	0·007$_5$	0·001	0·004	0·3$_5$	0·0001
	MnO.	0·004$_5$	0·022	0·004$_5$	0·019	0·0$_5$	0·0001
	SiO_2.	0·006	0·011	0·004	0·007$_5$	...	0·0003
	Al_2O_3.	0·003	0·006$_5$	0·001$_5$	0·003$_5$	...	0·0032
	TiO_2.	...	0·000$_5$				
	Cr_2O_3.	...	0·002				
	P_2O_5.	...	0·005				
	Total.	0·015	...	0·011	...	0·4	0·003$_5$
	Total by direct vacuum fusion.			0·010$_5$	...	0·9	0·003$_5$
Cast 52308.	Total residue.	...	{0·080, 0·082				
Bottom of ingot.	FeO.	0·001$_5$	0·006$_5$	0·001$_5$	0·007$_5$	0·3$_5$	0·0001$_5$
	MnO.	0·006	0·027	0·006	0·025$_5$	...	0·0002$_5$
	SiO_2.	0·007	0·018	0·010$_5$	0·019$_5$	...	0·0004
	Al_2O_3.	0·009	0·019$_5$	0·003	0·007	...	0·0024
	TiO_2.	...	0·001$_5$				
	Cr_2O_3.	...	0·002				
	P_2O_5.	...	0·007				
	Total.	0·023$_5$	...	0·021	...	0·3$_5$	0·003$_5$
	Total by direct vacuum fusion.			0·019$_5$	...	0·4	0·003$_5$

TABLE VIII.—*Determination of Oxygen in Steel by the Fractional Vacuum Fusion Method.*

Cast 20/9556.

Manufacture: Acid open-hearth.

Pit sample analysis:

C. %.	Mn. %.	Si. %.	S. %.	P. %.	Ni. %.	Cr. %.	Mo. %.
0·28	0·67	0·28$_5$	0·031	0·031	2·34	0·67	0·42

Analysis of stage test mould samples (not treated with aluminium) taken during manufacture:

	C. %.	Mn. %.	Si. %.	Ni. %.	Cr. %.	Mo. %.
No. 5	1·40	0·14	0·02	2·29	0·16	0·43
No. 6	0·84	0·26	0·21	2·30	0·19	0·50
No. 8	0·38	0·20	0·16	2·35	0·18	0·49

Sample.	Oxygen.		Hydrogen.		Nitrogen.	
	Oxygen. %.	Oxide. %.	Ml. per 100 g.	%.	Ml. per 100 g.	%.
No. 5 unkilled.						
FeO	0·0040$_5$	0·018	0·7$_5$	0·00007	0·15	0·0002
MnO	0·0029	0·013	0·3$_5$	0·00003	0·07	0·0000$_5$
SiO_2	0·0007$_5$	0·001$_5$	...	...	0·15	0·0002
Al_2O_3	0·0009	0·002	...	...	1·34	0·0017
Total	0·008$_5$	0·034$_5$	1·1	0·00010	1·71	0·002
Direct total oxygen determination	0·006$_5$					
No. 6 unkilled.						
FeO	0·0023	0·010$_5$	0·5	0·00004$_5$	...	...
MnO	0·0047$_5$	0·021	0·1$_5$	0·00001$_5$	0·11	0·0001$_5$
SiO_2	0·0013$_5$	0·002$_5$	...	...	0·06	0·0000$_5$
Al_2O_3	0·0018	0·004	...	...	0·96	0·0012
Total	0·010	0·038	0·6$_5$	0·00006	1·13	0·001$_5$
Direct total oxygen determination	0·009$_5$					
No. 8 unkilled.						
FeO	0·0018	0·008	0·4	0·00003$_5$	...	...
MnO	0·0071	0·031$_5$	0·2$_5$	0·00002	0·19	0·0002$_5$
SiO_2	0·0033$_5$	0·006$_5$	...	...	...	...
Al_2O_3	0·0053$_5$	0·011$_5$	...	...	0·89	0·0011
Total	0·017$_5$	0·057$_5$	0·6$_5$	0·00005$_5$	1·08	0·001$_5$
Direct total oxygen determination	0·016					

tures at which iron and manganese oxides are reduced respectively. The lower temperature is limited by the melting point of the alloy with tin, and attempts to obtain a substantial decrease in this temperature, avoiding the use of easily volatilised metals, have not met with success up to the present time.

The fractional vacuum fusion method has been particularly successful in the examination of unkilled steels of the rimming type, where the oxygen is believed to be present as basic oxides not combined with any substantial amount of silica. With the present technique of the residue methods, the values obtained for FeO and MnO in such materials are of doubtful value, being usually low.

Moreover, if oxygen exists "in solution in the steel" it is possible that the residue methods may fail to separate oxygen existing in this form.

Brief Comments on the Results.

Taking the figures in Table IV., where fractional vacuum fusion results and alcoholic iodine residues are given for the top and bottom plates rolled from an ingot of basic open-hearth steel, agreement between the two methods is quite satisfactory. Detailed comments on the significance of the fractional results were made in the authors' previous publication (*loc. cit.*), but the following points are reiterated:

(1) The fractional result for iron oxide is usually lower than that of the residue methods in steels containing a preponderance of silicate non-metallics; in unkilled steels of the rimming type, the fractional result is usually higher in iron oxide than that obtained in separated residues.

(2) A higher figure for manganese oxide is usually obtained by the fractional method than by residue procedures.

(3) Steels containing complex silicates tend to give slightly low silica results by the fractional method, which fairly readily recognises such non-metallics by their comparative slowness of reduction at the temperature of the silica fraction.

(4) Fractional alumina results are usually lower than those obtained by residue methods. This may be due to the separation by the residue methods of compounds of aluminium other than oxides, and further investigation of this possibility is in hand by the Sub-Committee.

PART C.—*The Determination of Hydrogen by Vacuum-Heating.*[1]

BY W. C. NEWELL, PH.D., D.I.C. (BROWN-FIRTH RESEARCH LABORATORIES, SHEFFIELD).

IN the *Journal* of the Institute [2] the author has already published a paper giving details of an apparatus which he designed and constructed for the determination of hydrogen in steel by heating in a high vacuum at a temperature of only 600° C. The rate of evolution of hydrogen from steel over a range of temperature from 400° to 900° C. was explored, and the total hydrogen given off was shown to be independent of the temperature. A careful comparison was made of the results obtained by this method and those by the author's vacuum-fusion apparatus, and a number of results were given for different types of steel, showing that there is a good agreement.

[1] Received February 7, 1941.

[2] *Journal of the Iron and Steel Institute*, 1940, No. I., p. 243 P.

Since the publication of the above paper, to which reference should be made for further details, the single-stage mercury-vapour pump has been replaced by a similar two-stage pump. Whilst the pumping rate is thereby improved and the degree of vacuum attained is presumably higher, no difference has been observable in the performance of the apparatus, which has continued to be used in a highly satisfactory manner for fundamental research work upon the hydrogen-iron system.

SECTION III.—The Aluminium Reduction Method.

Part A.—*The Development and Comparison of Two Procedures for the Aluminium Reduction Method for Determining Oxygen in Steel. General Summary Showing Applicability of the Aluminium Reduction Method.*[1]

By N. GRAY, M.Met., and M. C. SANDERS (Port Talbot).

(Fig. 3 = Plate XLVIII.)

Synopsis.

Recent developments in the design and manipulation of the apparatus for the aluminium reduction method using an atmosphere of hydrogen are described, with particular reference to the simplification of the blank estimation.

The development is traced of a modification in the procedure whereby the atmosphere of hydrogen is dispensed with and the determination carried out in a tube from which the atmosphere has been evacuated to a pressure of 8 mm. of mercury by means of a pulsometer pump.

Results obtained on various steels on which the oxygen content had previously been determined by vacuum fusion and by the aluminium reduction method in hydrogen are given. The oxygen content in known weights of various oxides and compounds has been recovered in this way. The summary calls attention to the possible application of the method in other directions.

Determination of Oxygen in Hydrogen Atmosphere.

No alterations in the design of the apparatus or in the procedure for determining oxygen by the aluminium reduction method in an atmosphere of hydrogen have been made since the publication of the Second Report of the Oxygen Sub-Committee, but one or two modifications had been made just previously and were only lightly touched upon, and are referred to in greater detail below.

Modification of the Original Apparatus.—The Drechsel washing bulb containing sulphuric acid has been removed.

The opening of the silica tube after, or in restarting, an experiment has been simplified by the insertion of a three-way capillary

[1] Received March 10, 1941.

glass stopper, connected to the atmosphere through a Drechsel washing bulb containing syrupy phosphoric acid, between the soda-asbestos tube and the phosphoric anhydride tube. It is considered that this arrangement reduces the possibility of explosions to a minimum.

The present arrangement is shown in Fig. 2.

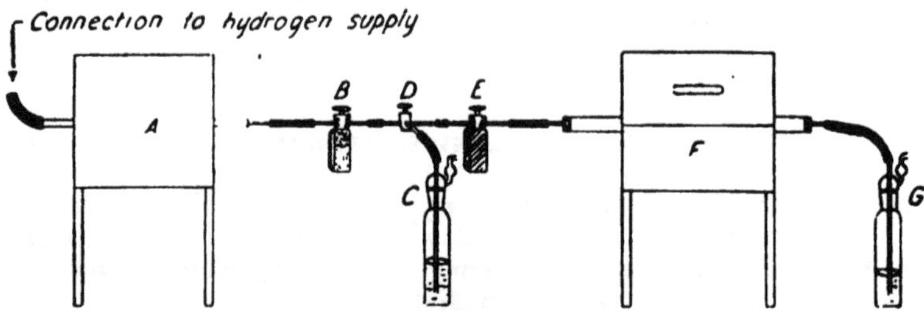

A. Preheater furnace for tube containing platinum-asbestos.
B. Nesbitt bulb containing soda-asbestos.
C. Drechsel scrubber containing syrupy phosphoric acid, sp. gr. 1·75.
D. Capillary glass tube, T-form, with 3-way stopcock.
E. Nesbitt bulb containing phosphorus pentoxide suspended on glass wool.
F. High-temperature furnace for melting of metal in graphite boat.
G. Drechsel scrubber containing concentrated sulphuric acid.

FIG. 2.—Present Arrangement of the Apparatus for the Aluminium Reduction Method.

The Blank Estimation.—As carried out in the original method, the determination of the blank called for considerable care, and even so the fused sample in the boat was usually covered with a film of white furry material, which was thought at the time to be aluminium carbide but which has since been found to be aluminium sulphide. This film was only soluble with great difficulty in the acid mixture prescribed and hence tended to give a slightly high blank upon ignition of the final alumina residue. For this reason, scrubbing of the hydrogen with sulphuric acid was discontinued, and it has also been found a safeguard to use black pressure tubing containing no sulphur compounds for making connections. As a result of these precautions the film has disappeared.

When the method is applied to steel samples the contamination of the hydrogen with traces of sulphur compounds does not interfere with the determination of oxygen, since the presence of iron prevents the formation of these small amounts of slowly soluble sulphur compounds.

The fused aluminium invariably stuck to the boat and had to be prised away, bringing with it, usually, small fragments of graphite which contaminated the precipitate until burned away in the final ignition.

The solution of the fused aluminium also usually occupied some hours, and it was felt that this was rather a weak point, as the blank and the determination of a steel sample could not be said to be carried out in identical conditions.

The addition of 0·2 g. of a steel of low oxygen content increased the speed of solution to about 35 min., which is much more in line with an actual determination. It was also found to prevent the adherence of the fused aluminium to the boat, and in the view of the authors has simplified considerably the carrying-out of the blank determination. Provided that a new boat is used which has been heated in the tube to 1100° C. in an atmosphere of hydrogen for 1 hr. and afterwards kept in a desiccator till required and if care is taken in the preparation of the aluminium by filing and degreasing, there should be no difficulty in obtaining a blank of a sufficiently low order.

The quantities of steel and aluminium used for a determination of oxygen have been reduced to 5 g. and 7 g., respectively, but if the steel is found to have a very low oxygen content, it is considered desirable to use the full amounts.

Determination of Oxygen in Vacuo.

The authors recognise that the manipulation of material in an atmosphere of hydrogen calls for some little care and attention to detail if explosions are to be avoided.

They themselves have never experienced this trouble since the inception of the method, but it was thought that if the reduction could be carried out *in vacuo* the method would perhaps be more widely adopted in some laboratories where the vacuum technique is thoroughly understood.

The following work *in vacuo* was carried out therefore simultaneously with that of Stevenson and Speight, described in Section III., Part B of this Report.

Apparatus and Procedure.—The experiments were conducted in a 30-in. Vitreosil tube of 1 in. bore, heated by means of a split electric furnace as used for the determination of hydrogen. The only advantage of this type is that of quick cooling. One end of the tube was sealed with a tight-fitting rubber bung and collodion flexile, the other end being used for the insertion of the samples and the evacuation of the tube.

The evacuation was carried out by means of a pulsometer pump, which was capable of reducing the pressure to 8 mm. of mercury. Apart from this being the only apparatus available, it was feared that the use of a very high vacuum might cause a collapse of the silica tube at the temperature of the experiments.

The blank or sample, placed in a graphite boat, made up as previously described for the determination in a hydrogen atmosphere, was introduced into the tube cold, and it was found that the pressure could be reduced to 8 mm. of mercury in about 7 min.

In all the experiments, however, evacuation was continued for ¾ hr. in the cold. The tube was then raised to 1100° C. in ½ hr., this temperature being maintained for 1½ hr., after which the furnace was switched off and the tube allowed to cool. The pump was kept running during the whole period. When cold, the boat was removed from the tube and the alumina content of the sample determined analytically as previously described. Fig. 3 shows the graphite boat containing the iron-aluminium alloy as produced at 1100° C. *in vacuo*.

The Blank Estimation.—It was anticipated that, on account of the inability to obtain a lower pressure in the tube than 8 mm., the blank would be rather high. Repeated trials, however, on both 7 g. and 14 g. of aluminium (to which had been added 0·2 g. of a steel of low oxygen content) gave blanks of 0·0004 g. and 0·0008 g. of alumina, respectively. This compares very favourably with the blank obtained in an atmosphere of hydrogen, and it was decided to carry out a series of experiments on steels of known oxygen content as determined by the vacuum fusion method and by the aluminium reduction method in hydrogen.

In addition, experiments were also made on the recovery of the oxygen contained in small added amounts of pure silica, aluminium silicate, ferric oxide and manganese oxide. These were added to the ordinary determination of a steel containing 0·008% of oxygen. The results are shown in Tables IX. and X.

TABLE IX.—*Comparison of Results obtained by Vacuum Fusion and Aluminium Reduction.*

Sample No.	Description of Sample.	Oxygen, %, by—		
		Vacuum Fusion.	Aluminium Reduction—	
			In Hydrogen.	*In Vacuo.*
1	Low-carbon rimming steel.	0·018	0·020	0·0164
4	Low-carbon killed steel.	0·002	0·004	0·004
7	Open-hearth ingot iron.	0·106	0·105	0·103
A	Semi-killed basic open-hearth steel.	0·014	0·016	0·015 0·017

From these experiments it would seem that the aluminium reduction method carried out *in vacuo*, when applied to ordinary carbon steels or to those oxides and combinations of oxides normally found in plain carbon steels, gives results which compare favourably with those obtained in an atmosphere of hydrogen.

Summary.

Since its introduction, the aluminium reduction method has been applied to a considerable number of plain carbon and simple

TABLE X.—*Check Tests on Added Compounds.*

Aluminium reduction *in vacuo.*

Compound Tested.	Weight Taken. G.	Equivalent Weight of Oxygen. G.	Total Weight of Alumina Found. G.	Weight of Alumina due to Blank and Oxygen in Steel. G.	Weight of Alumina Recovered. G.	Equivalent Weight of Oxygen Recovered. G.
Silica, SiO_2.	0·01	0·00533	0·0134 0·0132	0·00125 0·00125	0·01215 0·01195	0·0057 0·0056
Aluminium silicate, $Al_2O_3.2SiO_2$.	0·01	0·00528	0·0117 0·0114	0·00125 0·00125	0·01045 0·01015	0·00491 0·00477
Ferric oxide, Fe_2O_3.	0·02	0·006	0·0136 0·0134	0·00125 0·00125	0·01235 0·01215	0·00581 0·00571
Manganese oxide, MnO.	0·02	0·0045	0·0108 0·0110	0·00125 0·00125	0·00955 0·00975	0·00449 0·00458

alloy steels of various compositions and of known oxygen content as determined by the vacuum fusion method. With a little practice it is comparatively simple to operate, and the introduction of the modified method *in vacuo* should encourage its use as a rapid control method in steel manufacture and testing.

Section III., Part C of this Report deals with its application to pig irons and cast iron in the laboratory of the British Cast Iron Research Association. Much painstaking work has been carried out by this co-operator, some of which has been done in collaboration with the authors. Whilst agreement has not always been reached, and considerable work remains to be done, particularly with respect to the effect of varying silicon contents, there is reason to believe that these difficulties may be overcome in the future and the method established on a satisfactory basis.

The method has been applied to the determination of the oxygen in ferro-silicon and ferro-manganese, although it is not certain at this stage what value can be ascribed to the information so obtained. It has also been used to a very limited extent for non-ferrous materials, chiefly nickel and copper. Here, its application ought to be of considerable value, and will repay further study.

The experiments *in vacuo* were first discussed with Dr. B. Jones, University College, Cardiff, and were carried out with his collaboration in the Department of Metallurgy. Acknowledgments are due to Dr. Jones for the photograph shown in Fig. 3 and to Professor W. R. D. Jones for his interest during the work.

Part B.—*A Description of the Aluminium Reduction Method as Operated at Stocksbridge.*[1]

By W. W. Stevenson, A.I.C., and G. E. Speight, B.Sc., A.I.C.
(Stocksbridge, near Sheffield).
(Under the direction of T. Swinden, D.Met.)

Synopsis.

A description is given of the procedure adopted for the aluminium reduction method operated *in vacuo*. Typical results are presented and comparison is made with results obtained by the vacuum fusion method. Close agreement is recorded on many types of ferrous materials.

The method of Gray and Sanders[2] consists of the heating, at 1150° C., of a steel specimen placed between two pieces of aluminium in an atmosphere of hydrogen, with subsequent acid solution of the iron-aluminium melt and chemical determination of the alumina formed by reduction of the oxides in the steel. Under these conditions, the present authors obtained blank values which were unusually high and compared unfavourably with those reported by Gray and Sanders, the high blanks being due to imperfect purification of the hydrogen used. At this stage in the work, Gray and Sanders[3] indicated their interest in vacuum technique instead of operation in hydrogen; the present authors, therefore, abandoned the hydrogen method in favour of carrying out the aluminium reduction *in vacuo*. This has the advantage that it permits of the use of a tube closed at one end and a single furnace instead of the split furnace originally used by Gray and Sanders. A comparatively high blank persisted in early experiments on the vacuum method, owing to air leaking back through the mechanical oil pump, but this was overcome by incorporating a tap between the exit tube from the furnace and the pump. The equipment and method employed are thus as follows:

The reaction chamber (Fig. 4) consists of a silica tube, 24 in. long by 1 in. in dia., closed at one end, in which is placed the graphite boat containing the sample and 14 g. of aluminium strip. The other end of the silica tube is closed by a rubber bung fitted with a tube leading to a glass tap and then to a mechanical oil pump. The boat, 4 in. in length, is machined from a half section of 1-in. round bar of pure Acheson graphite. The steel sample consists of five discs, $\frac{1}{16}$ in. thick, cut from $\frac{1}{2}$-in. dia. bar; these are sandwiched between strips $2\frac{3}{4}$ in. long by $\frac{5}{16}$ in. wide by $\frac{1}{8}$ in. thick (approx.) of freshly filed aluminium. Evacuation of the silica tube is carried out for 20–30 min. at room temperature, and then the tube is introduced

[1] A communication from The United Steel Companies, Ltd., Central Research Department, Stocksbridge, near Sheffield, received February 7, 1941.

[2] *Journal of the Iron and Steel Institute*, 1938, No. I., p. 348 P.

[3] Private communication.

into a Silit-rod furnace at 1150° C. When the temperature of the tube has attained 1000° C.—after about 4 min.—the tap is closed, and the gases evolved from the graphite are allowed to accumulate in the tube. The pressure thus created prevents the premature collapse of the tube. The silica tube attains a temperature of 1150° C. after 10–15 min. and is maintained at this temperature for 1 hr.

The tube is then allowed to cool to atmospheric temperature and the iron-aluminium melt is detached from the boat and transferred

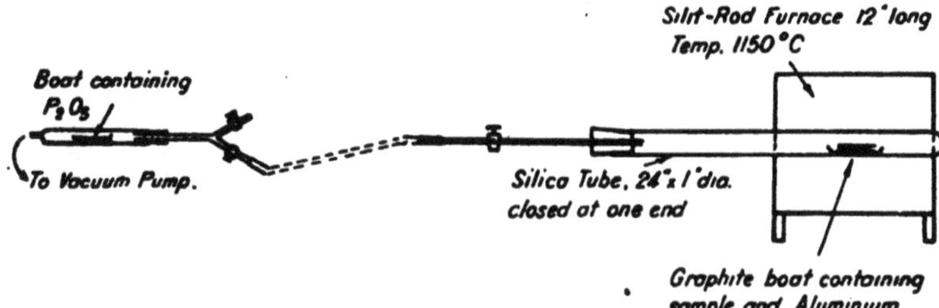

Fig. 4.—Diagrammatic Sketch of Apparatus for the Aluminium Reduction Method (Vacuum Technique).

to a beaker. No difficulty is experienced in dissolving the melt in hydrochloric acid (1 : 1), followed by oxidation with nitric acid (sp. gr. 1·42), after which the alumina is filtered on fine ashless paper or a pulp filter. The filter is washed with water and warm hydrochloric acid (1 : 1), then with hot water and hot sodium carbonate solution (3%), and finally with water. The residue is ignited in a platinum crucible at 1000° C., treated with a few drops of sulphuric and hydrofluoric acids, re-ignited and weighed.

The results obtained on a number of samples by the above method are given in Table XI.

Comments.

The aluminium reduction method operated *in vacuo* gives a blank value of reasonably low proportions. The results shown in Table XI. obtained on eight samples of plain carbon steels and one low-alloy steel are in fair agreement with the total-oxygen values reported by vacuum fusion, the aluminium reduction results being all slightly on the low side after applying the blank correction. When no blank correction is made, agreement with the vacuum fusion method, on the steels examined, is very close. The gases evolved from the graphite boat during the early heating of the furnace and after closing the exit tap have not been examined, but in view of the fact that high results compared with the vacuum fusion method results are not obtained, they are apparently not of an oxidising nature.

Table XI.—*Results Obtained by the Aluminium Reduction Method.*

Aluminium Blank (obtained by testing steel of very low oxygen content).

Steel Sample.	Weight of Aluminium.	Weight of Steel.	Oxygen Content (by Vacuum Fusion).	Al_2O_3 found.	Blank (14 g. of Aluminium).
*H*3716 (*a*)	14·71 g.	8·75 g.	0·0025% (= 0·0005 g. Al_2O_3)	0·0009 g.	0·0004 g. Al_2O_3
„ (*b*)	13·99 g.	9·35 g.	0·0025% (= 0·0005 g. Al_2O_3)	0·0009 g.	0·0004 g. Al_2O_3

This blank of 0·0004 g. of Al_2O_3 is equivalent to 0·002% of oxygen per 10 g. of steel.

Sample.	Analysis.								Oxygen (by Vacuum Fusion). %.	Aluminium Reduction Method.	
	C. %.	Mn. %.	Si. %.	S. %.	P. %.	Ni. %.	Cr. %.	Mo. %.		Oxygen. %. (No Blank Deducted).	Oxygen. %. (Deduction, 0·002% Oxygen).
*G*1808	0·13	0·43	0·24	...	...	4·25	1·21	0·24	$0·009_8$	0·010	0·008
1168	0·11	1·10	...	0·185	0·053	...	...	...	$0·009_8$	0·0010	0·008
										0·009	0·007
										0·008	0·006
*SG*2356	0·10	0·40	0·07	0·040	0·048	...	...	...	0·023	0·021	0·019
*SG*2616 *	0·02	Nil	0·33	0·012	0·049	...	...	...	0·072	0·069	0·067
*SG*1101 *	0·02	Nil	0·01	0·013	0·099	...	...	...	0·175	0·194	0·192
532/1	0·41	0·73	0·18	0·024	0·042	...	...	...	$0·010_8$	0·011	0·009
										0·009	0·007
532/2									$0·007_8$	0·008	0·006
532/3									...	0·009	0·007
532/Billet									...	0·009	0·007

* For details of preparation *see* Eighth Report on the Heterogeneity of Steel Ingots, p. 196, *Iron and Steel Institute*, 1939, *Special Report No.* 25.

Part C.—*The Determination of Total Oxygen in Pig Iron by the Aluminium Reduction Method.*[1]

By E. TAYLOR-AUSTIN, F.I.C. (British Cast Iron Research Association, Birmingham).

Synopsis.

An account is given of the application of the aluminium reduction method for the determination of total oxygen to a series of pig-iron samples. Modification has been found necessary to overcome the interference set up by silicon in certain irons. The final modified procedure is described in detail, and results obtained are compared with those given by the aqueous iodine and other methods.

The method adopted for this series of experiments was that described by Gray and Sanders in the Second Report of the Oxygen Sub-Committee.[2] A thorough investigation of the application of this process to the determination of total oxygen in pig iron was

[1] Received February 22, 1941.

[2] Eighth Report on the Heterogeneity of Steel Ingots, pp. 103–108, *Iron and Steel Institute*, 1939, *Special Report No.* 25.

made and it was found necessary to modify the original procedure in order to avoid interference from silicon.

The following method has been successfully applied to a series of grey pig irons, and a comparison of the results obtained with those by other processes is given in Table XII.

TABLE XII.—*Total Oxygen Contents of Pig Irons obtained by Various Methods.*

No.	Type of Iron.	Total Oxygen, %, by—		
		Modified Aqueous Iodine.	Aluminium Reduction.	Vacuum Fusion.
13	Armco iron	0·086	0·085	...
14	Swedish iron	0·017	0·018	...
16	Pig irons	0·015	...	0·019
19	Pig irons	0·010	0·011	0·011
31	Pig irons	0·009	0·006	0·005
21	Pig irons	0·010	0·007	0·008
107	Pig irons	0·0063	0·0050	...
101	Pig irons	0·0053	0·0047	...
18	Pig irons	0·016	0·015	...
73	Pig irons	0·013	0·018	...
25	Pig irons	0·0091	0·0084	...
86	Pig irons	0·0084	0·010	...
111	Pig irons	0·021	0·019	...
84	Pig irons	0·0074	0·0077	...
85	Pig irons	0·013	0·013	...
55	Pig irons	0·013	0·011	...
23	Pig irons	0·016	0·017	...
103	Pig irons	0·011	0·012	...
38	Pig irons	0·011	0·013	...
79	Pig irons	0·0077	0·0071	...

Modified Aluminium Reduction Method.

Apparatus.—This is as described by Gray and Sanders,[1] except that the silica combustion tube of the main furnace is heated by a winding of Kanthal resistance wire instead of Silit-rod elements.

Preparation of Samples.—Samples should weigh between 7 and 10 g. and should be in the form of strips not greater than $\frac{1}{16}$ in. thick taken from the centre of a complete cross-section of the pig.

Determination of the "Blank." —The "blank" on the pure aluminium sheet used is determined by fusing 14 g. of the material together with 0·20 g. of a steel of low oxygen content exactly as described by Gray and Sanders. In the author's experiments a blank value of 0·0008 g. was obtained.

Procedure.—Fuse 7–10 g. of sample with 14 g. of pure aluminium sheet at 1100° C. exactly as described in the original process. At the conclusion of the heat withdraw the boat from the cold tube and transfer the alloy to a 1500-ml. tall-form beaker. Add 5 g.

[1] Eighth Report on the Heterogeneity of Steel Ingots, pp. 103–108, *Iron and Steel Institute*, 1939, *Special Report No.* 25.

each of citric and tartaric acids, followed by 200 ml. of nitric acid (sp. gr. 1·20). Warm gently until the citric and tartaric acids have dissolved, and add 60 ml. of concentrated hydrochloric acid, carefully, in small amounts. When the violent reaction subsides, heat to boiling, and keep hot until solution is complete (about 30 min.). Cool somewhat, add 10 g. of ammonium persulphate and boil gently for 1 hr. Dilute with an equal volume of hot water, boil and filter on an ashless paper-pulp pad.

Wash the residue with hot 5% hydrochloric acid and hot water. Transfer the precipitate and pad to a 300-ml. "carbon dioxide" flask, add 120 ml. of a 5% sodium carbonate, 10% sodium citrate solution and stir mechanically for 1 hr. at 80° C., as described in the procedure for the aqueous iodine method (Section VI., Part A).

Filter the solution through a 4·25-cm. Whatman No. 42 filter paper contained in a small Büchner funnel; wash well with hot water, followed by hot 5% hydrochloric acid and finally with hot water. Ignite the residue to constant weight in a platinum crucible. Remove any traces of silica remaining in the precipitate by volatilisation with hydrofluoric acid in the presence of sulphuric acid in the usual manner. The residue so obtained should be pure alumina, and the weight after deduction of the blank is used for the calculation of the oxygen content of the material under examination.

Development of the Modified Procedure.

At the beginning of 1940 the necessary apparatus and equipment were installed and experiments were begun. After certain initial difficulties in technique had been overcome, results were obtained which were in close agreement with those of Gray and Sanders working on joint samples of pig iron.

After these preliminary tests the method was applied to the series of pig irons already examined by the modified aqueous iodine method, and it was found that in all but a few cases good agreement was obtained between the total oxygen results by the two processes. Where agreement was not obtained, the results by the aluminium reduction method were of a considerably lower order than those by the aqueous iodine process.

Further investigation of these discrepancies revealed the fact that the silica figures obtained by the aqueous iodine method were high for reasons given in Section VI., Part A of this Report. Redeterminations by the iodine process, suitably modified, on samples which had previously yielded checking results by both the methods under review revealed the fact that some of these silica figures were also in error.

This led to a thorough examination of the application of the original Gray and Sanders procedure to pig-iron samples, and it was found that this process sometimes yielded results which were greatly in excess of those recorded by the aqueous iodine method. It was subsequently revealed that this error was caused by the presence of excessive amounts of silica in the final ignited residue obtained after solution of the alloy in mineral acids; the removal of this silica by volatilisation with hydrofluoric acid made no appreciable difference to the amount of alumina found by chemical analysis of the residue.

Further experiments showed that the extent of this error increased as the silicon contents of the original pig irons became greater.

The author was therefore of the opinion that the presence of this silica was due to hydrolysis during the solution of the melt in acid and that this silicic acid adsorbed aluminium salts from the solution, thus causing excessive amounts of alumina in the final residue.

Accordingly the sodium carbonate-citrate treatment, already used in the aqueous iodine method, was applied to unignited residues from the aluminium reduction method. It was found that this procedure successfully removed the hydrolysed silicic acid, and the results for total oxygen fell considerably. Further, the final figures were in good agreement with those obtained by the latest aqueous iodine process.

The aluminium reduction method was therefore modified to incorporate this treatment, and in this latter form was applied to the series of pig irons already mentioned. The results obtained for total oxygen by both procedures were now found to be in excellent agreement and no further discrepancies have so far arisen.

The application of the aluminium reduction method to pig iron is, at the present time, limited somewhat by the difficulty of obtaining test-pieces from samples of white and other hard irons. This difficulty does not arise if samples can be cut with a hack-saw, but if it is necessary to use a carborundum wheel to obtain specimens, considerable oxidation appears to take place on the surfaces, and high results are recorded. Attempts to use drillings or crushings enclosed in an aluminium envelope have so far failed, owing to the difficulty of removing trapped oxygen by passing hydrogen through the furnace. This difficulty would probably not arise if fusion were carried out *in vacuo* instead of in hydrogen, but the author has had no experience of the former procedure, although it has been examined by other workers (*see* Part B of this Section). The problem of obtaining suitable samples from hard materials is receiving further attention.

Conclusions.

The experimental work so far carried out, has led to the following conclusions :

(1) The modification introduced successfully obviates any interference arising from the presence of silicon in the original material.

(2) The modified aluminium reduction method can be applied to the determination of the total oxygen content of pig irons generally, provided that a suitable sample can be obtained.

SECTION IV.—The Chlorine Method.

Part A.—*Description of the Procedure now Adopted for the Chlorine Method.*[1]

By E. W. COLBECK, M.A., S. W. CRAVEN, A.M.C.T., and W. MURRAY, A.M.C.T., A.I.C. (Northwich).

Synopsis.

Details of the present procedure for the determination of non-metallic inclusions in steel by the chlorine method are given.

A scheme of analysis applicable to chlorine residues containing appreciable amounts of phosphorus pentoxide is set out.

An analytical procedure involving the use of a photometer is outlined.

Introduction.

Since the Seventh Report on the Heterogeneity of Steel Ingots was published in 1937, a number of improvements have been incorporated in the method adopted for the determination of non-metallic inclusions in steels by the chlorine method. It appears, therefore, desirable to describe the present procedure as operated by I.C.I. (Alkali), Ltd., in their Winnington Research Laboratory.

Description of the Apparatus.

A sketch of the apparatus is given in Fig. 5. The chlorine, contained in a 70-lb. gas cylinder, is prepared electrolytically and purified by a liquefaction process. The first 50 litres of gas from each

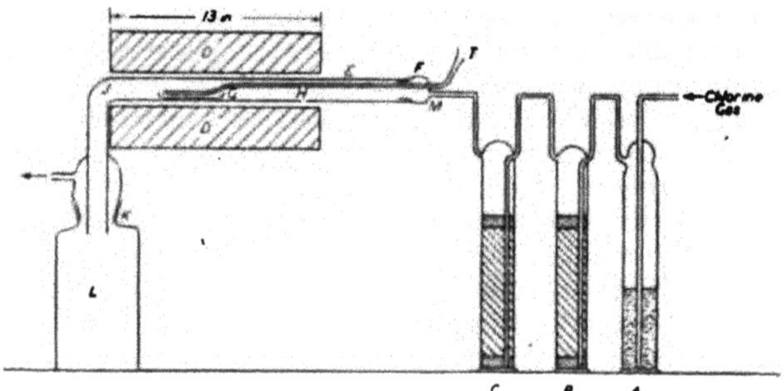

Fig. 5.—Apparatus for the Determination of Oxide Inclusions in Steel by the Chlorine Method.

new cylinder are passed to the atmosphere, since the bulk of any gaseous impurities in the chlorine is contained in this fraction. The gas is led from the cylinder through concentrated sulphuric acid in the wash-bottle *A* and calcium chloride and phosphorus pentoxide

[1] Received February 11, 1941.

in the drying towers *B* and *C*, respectively. The rate of flow of chlorine is approximately 15 litres per hr.

The steel sample, contained in a fused-silica boat *G*, is chlorinated in the Pyrex glass reaction tube *E*. This tube has a ground-glass stopper *F*, fitted with an inlet for gas and a thin Pyrex tube *H* carrying a chromel-alumel thermocouple *T*. In order to make the temperature measurement as accurate as possible the tube *H* is arranged so that its end is right in the boat and in immediate contact with the steel. The length of the tube is such that the end of the thermocouple can be placed at any point along the boat *G* when in position.

The wiring of the electrically-heated furnace *D* is arranged so that it is evenly heated throughout its whole length. This ensures that all the volatile chlorides are carried away to sublime in the collecting bottle *L*, which is connected to the reaction tube *E* by means of the ground-glass joint *K*. The bottle *L* is fitted with an exit tube for excess chlorine gas.

The bend *J* in the reaction tube fits as closely to the furnace *D* as is practicable, so that the chlorides fall into the bottle *L* and do not block the reaction tube itself.

Form of Sample Used.

Solid test-pieces of steel are always used. As is well known, the reaction between iron and chlorine is strongly exothermic, and is accompanied by a rise in temperature of the specimen. With solid test-pieces, the initial rate of attack is slow, so that the amount of heat evolved in a given time is much smaller than is the case when drillings are used. Further, the heat evolved is absorbed into the mass of metal without causing intense local heating.

Experiments with solid test-pieces showed that it was possible to control the exothermic reaction between iron and chlorine so that the temperature of the specimen never exceeded 300° C.

Two methods of obtaining a sample for chlorination are employed. In the first, the bar is skimmed up in a lathe until all surface marks are removed, and thin cross-sectional slips are then cut from it, each approximately 0·2 in. long. Sufficient of these slips to make 10–15 g. weight of sample are taken for each determination. With bars of ⅜ in. dia. or less the slips fit into the boat in such a way that it is possible to bring the tip of the sheathed thermocouple close to them by making the couple dip down at the front of the boat and then run parallel just above the specimens to the other end of the boat. If the bars are from ⅜ to ¾ in. in dia. each slip is carefully folded over in a vice. The samples are dried in alcohol before weighing and chlorinated immediately.

In the second method, one solid test-piece cut from the skimmed bar may be used, provided that it will fit well down into the boat. It has been found possible, for example, to use one test-piece cut from a bar of ⅜ in. dia.

Procedure.

The reaction tube *E*, the collecting bottle *L* and the ground-glass stopper *F* carrying the thermocouple sheath are rinsed with methylated spirits and dried in an air oven at 110° C. overnight. The apparatus is then assembled as shown in Fig. 5, but the drying train is not connected to the reaction tube. The furnace *D* is switched on and the temperature raised to 300° C. This temperature is maintained for about 10 min., the furnace is switched off and the reaction tube is brought rapidly to room temperature by means of a stream of dry compressed air introduced at *M*.

The boat *G* is dried by heating it in a Bunsen flame to a dull red heat; it is allowed to cool in a desiccator. The weighed sample is then placed in the boat and introduced into the reaction tube *E*. The drying train is connected up at *M* and the flow of chlorine started. The boat *G* is not placed in the centre of the furnace *D*, but nearer to the bend *J*, so that the volatile chlorides have a shorter distance to travel.

When chlorine has been passing through the apparatus for ½ hr., the furnace *D* is switched on and the temperature is raised slowly to between 150° and 160° C. The furnace is then switched off, as the residual heat is sufficient to start the reaction between the chlorine and the sample. The exothermic reaction raises the temperature of the steel, but by switching off the furnace at 150–160° C., the temperature should never exceed 300° C. As soon as the temperature begins to fall the furnace is switched on and maintained at 350° C. throughout the test. When no trace of volatile chlorides can be seen entering the receiving bottle, chlorination is complete, but the stream of chlorine is continued for a further 15 min.

The furnace is then switched off and the reaction tube cooled to room temperature, the passage of chlorine being continued. The chlorine is shut off, the drying train *A*, *B* and *C* is stoppered to prevent the inlet of air, and the boat and its contents are removed from the reaction tube immediately.

The contents of the boat are washed out with water at once, and the residue is filtered through a 5·5-cm. No. 42 Whatman filter paper.

Test Run, using a Standard Steel.

Experience over the past three years has indicated that before commencing work on a fresh series of steels it is necessary to carry out a preliminary test, using a steel the oxygen content of which is known. Also, this test run is always carried out when alterations have to be made to any part of the apparatus or when a fresh cylinder of chlorine is brought into operation. The standard steel chosen is one on which a large number of determinations have been carried out by the chlorine method and on which the percentage of total ignited residue has regularly fallen between 0·041 and 0·044%;

this represents a total oxygen content of 0·013–0·014%. The total oxygen figures obtained by vacuum fusion on this steel lie between 0·0135% and 0·014%. Work on the new steels is not commenced until a chlorination of the standard yields a total ignited residue within the limits mentioned above, namely, 0·041 to 0·044%.

An interesting instance occurred recently, a description of which will serve to draw attention to the importance of the use of a standard. After a period during which the apparatus for chlorination had been standing unused for some weeks, a sample of the standard steel was chlorinated before commencing work on a number of new steels. The first run gave a total ignited residue of 0·10%, a figure well outside the acceptance limits; a repeat test gave even higher results. Since the chlorinations appeared to have proceeded in a normal manner the chlorine supply was suspected of being impure. A further cylinder of chlorine was obtained, and the first test run with this again gave a very high result for the total ignited residue. A very careful examination was then carried out on all parts of the apparatus, when a fine line was observed in the reaction tube in a position just above where the test specimen is located during chlorination. It was suspected that this might be a minute crack, which would be expected to widen when the tube was heated to 350° C. An attempt was made to repair the tube and a further test run carried out; this gave a total ignited residue appreciably lower than the previous figures, but it was still in excess of the maximum acceptance figure of 0·044%. The suspected portion of the reaction tube was cut out and a fresh piece of Pyrex glass inserted. The first chlorination carried out after this alteration gave a total ignited residue of 0·044% on the standard steel, and subsequent chlorinations in the repaired apparatus behaved normally.

Analysis of the Residue.

In the Seventh Report on the Heterogeneity of Steel Ingots, a scheme of analysis for chlorine residues was set out. Two modifications to this scheme—the colorimetric determinations of phosphorus and chromium—were given in the Eighth Ingot Report. All the steels examined yielded residues containing only traces of P_2O_5, and it is believed that the procedure by which the Al_2O_3 fractions were obtained as "difference" figures after the precipitation of the hydroxides of Group III. with ammonia and the colorimetric determination of Fe_2O_3, Cr_2O_3, P_2O_5, &c., was accurate. At the same time it was realised that the direct determination of Al_2O_3 would be much more preferable.

Since the publication of the Eighth Ingot Report the possibility of the retention of larger amounts of P_2O_5 (*e.g.*, of the order of 0·002–0·003 g.) in chlorine residues has been considered. When phosphorus is present in the residues in such amounts it is impossible to obtain accurate SiO_2 figures by direct volatilisation with hydrofluoric acid in the presence of sulphuric acid, and the difference figures for Al_2O_3

are uncertain. Modifications would have to be made to the usual scheme of analysis, and details of a new procedure are given below. This procedure is regarded as an intermediate stage in the development of a satisfactory analytical scheme, and it is hoped that further advances will be made when work with a photometer is completed. An outline of the scheme proposed, using such an instrument, is appended.

Modified Analytical Procedure (Applicable to Chlorine Residues containing Appreciable Amounts of P_2O_5).

Ignition of the Non-Metallic Residue.—The residue is filtered on a Whatman No. 42 paper (5·5 cm.) and washed with cold water. The washed residue is transferred to a tared platinum crucible, ignited at a low temperature until all carbonaceous matter is destroyed and then ignited at a temperature of 1000° C. for a further 10 min.

The crucible and contents are transferred to a desiccator, allowed to cool and weighed. The ignition at 1000° C. is repeated for another 10 min., and the crucible is cooled and reweighed. This weight should be constant before proceeding with the analysis.

Determination of Silica.—The residue is carefully fused with 0·5 g. of dehydrated "Analar" sodium bisulphate, the melt allowed to cool and extracted with dilute hydrochloric acid. The insoluble matter is filtered on a 5·5-cm. Whatman No. 42 filter paper and washed alternately with hot dilute hydrochloric acid and hot water. The filter paper and contents are transferred to a tared platinum crucible and ignited to constant weight. The residue is treated with one drop of sulphuric acid (sp. gr. 1·84) and a few drops of hydrofluoric acid, evaporated to dryness and reweighed. The loss in weight gives the weight of silica.

Any residue remaining after the volatilisation of silica is fused with 0·3 g. of sodium bisulphate, the melt extracted with a little dilute hydrochloric acid and added to the main solution.

Precipitation of the Group III. Hydroxides.—Three to four grammes of "Analar" ammonium chloride are added to the main solution, the volume of which is adjusted to approximately 100 ml. The solution is boiled and oxidation is ensured by the addition of four drops of concentrated nitric acid. The mixed hydroxides of Group III. are precipitated by the careful addition of 1 : 1 ammonium hydroxide until a slight excess is obtained. The solution is then boiled until the vapours no longer smell of ammonia. (When the operation is carried out correctly, the solution is neutral to methyl orange.) The solution is filtered on a 5·5-cm. No. 40 Whatman filter paper and washed with a hot 2% ammonium nitrate solution which is neutral to methyl orange. The filtrate is reserved for the determination of manganese.

Separation of Iron and Titanium.—The mixed hydroxides of Group III. are dissolved from the filter paper with a little hot dilute

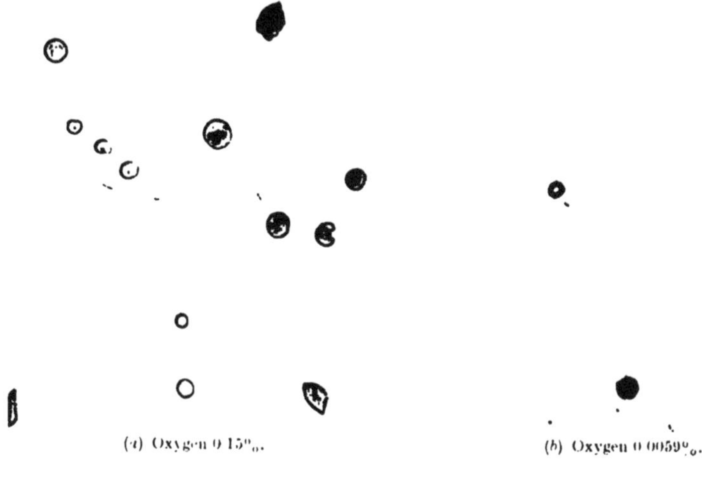

(a) Oxygen 0·15%. (b) Oxygen 0·0059%.

(c) Oxygen 0·028%. (d) Oxygen 0·0004%.

FIG. 1.—Oxygen in Iron. × 750 (reduced to two-thirds linear in reproduction)

SECTION III., PART A.

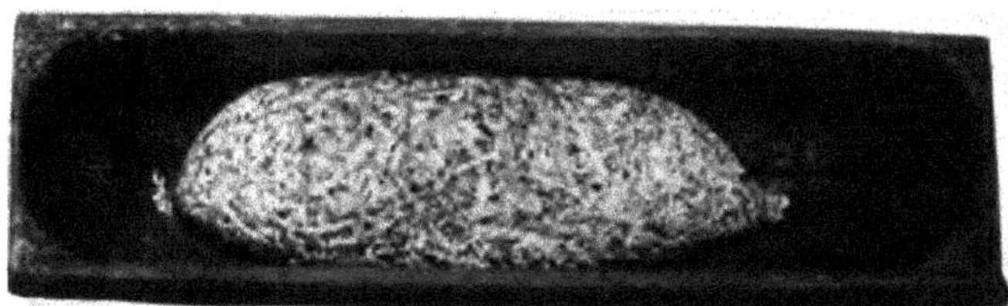

FIG. 3.—The *Graphite Boat* containing Iron Aluminium Alloy as produced at 1100° C. in vacuo.

[Third Report of the Oxygen Sub-Committee.
[To face p. 330 P.

FIG. 10.—Sulphur Print of Basic Bessemer Rimming Steel Sample No. 8706F. × 1.

hydrochloric acid, and iron and titanium are separated by the procedure described by Berman, Chap and Taylor.[1]

The solution is transferred to a 75-ml. separating funnel, its volume adjusted to approximately 50 ml. and hydrochloric acid added so that the acidity is not less than 10%. A 6% aqueous solution of cupferron is added until the reagent produces a momentary white precipitate at the point of contact with the acid solution; 10 ml. of chloroform are added and the separator is shaken vigorously for 15 seconds. The layers are allowed to separate, and then more cupferron is added until the white precipitate reappears. The shaking-out process is repeated without removing the chloroform layer until, after alternate treatment with reagent and shaking-out, the aqueous layer is quite clear of any turbidity, except that due to suspended chloroform. The clear chloroform layer is drawn off into a second separator, and the aqueous solution in the first separator is extracted with small portions of chloroform until the solvent remains colourless. These extracts are added to that in the second separator. The combined chloroform extracts are washed with about 10 ml. of water. The washed chloroform extract is drawn off into a tared platinum crucible and the solvent evaporated on a steam bath. The crucible is then cautiously heated at a low temperature until all fuming ceases, and is finally ignited to constant weight.

The oxides are then fused with 0·5 g. of Analar sodium bisulphate and extracted with 10% sulphuric acid. Iron and titanium are determined colorimetrically, using thioglycollic acid and hydrogen peroxide, respectively.

Separation of Aluminium and Chromium.—The volume of the combined water extracts and washings from the separation of iron and titanium are reduced to approximately 50 ml. by boiling, and aluminium and chromium are precipitated free from phosphorus by the method suggested by Schoeller and Webb.[2]

One gramme of tartaric acid is dissolved in the solution followed by 0·5 g. of tannic acid. The solution is then made just ammoniacal to litmus and 2·0 g. of ammonium acetate are added. The complex is then boiled for 2 min., and the precipitate is filtered on a 7-cm. Whatman No. 41 filter paper and washed thoroughly with a hot neutral solution containing 2% of ammonium chloride, a little tannic acid and a little tartaric acid. The wet filter paper and contents are transferred to a tared platinum crucible and ignited to constant weight (Al_2O_3 and Cr_2O_3).

The ignited residue is then examined for chromium. It is fused with a little sodium carbonate, the melt is extracted with water, about twenty drops of hydrogen peroxide are added and the liquid is boiled until oxygen effervescence ceases. Chromium is then

[1] Berman, Chap and Taylor, *Journal of the Association of Official Agricultural Chemists*, 1937, vol. 20, p. 635.

[2] Schoeller and Webb, *Analyst*, 1929, vol. 54, p. 704.

determined colorimetrically, using diphenylcarbazide. If chromium is present, the weight of Cr_2O_3 is deducted from the original weight of residue to give the amount of Al_2O_3.

Separation of Manganese.—The filtrate from the separation of the mixed hydroxides is treated with 5 ml. of bromine water and made alkaline with ammonia. The solution, measuring not more than 100 ml., is brought to the boil, stirring vigorously. It is allowed to boil for about two or three minutes and is then filtered on to a 5·5-cm. No. 42 Whatman filter paper. After washing with hot water, the precipitate is ignited in a tared platinum crucible and weighed as Mn_3O_4. The manganese content of the latter is checked colorimetrically.

Results of some Test Analyses made to Check the above Procedure.

A number of test analyses were made to check the accuracy of the analytical procedure described above. In all cases the amounts of added oxides were unknown to the operator. Some typical results are given in Table XIII.; this Table also contains figures obtained

TABLE XIII.—*Results Obtained on Synthetic Solutions Using the Modified Analytical Procedure.*

Test Solution.	Fe_2O_3.		TiO_2.		Al_2O_3.		Cr_2O_3.		P_2O_5 added. G.	$NaHSO_4$ added. G.
	Added. G.	Found. G.	Added. G.	Found. G.	Added. G.	Found. G.	Added. G.	Found. G.		
1	0·0008	0·0008	...	...	...	...	...	...	...	...
2	0·0016	0·0015	...	...	...	...	...	...	...	...
3	0·0014	0·0014	...	...	...	...	...	...	0·0002	...
4	0·0009	0·0008	...	...	...	...	...	...	0·0002	...
5	0·0008	0·0008	...	...	...	...	...	...	0·0020	...
6	0·0012	0·0013	...	...	...	...	...	...	0·0030	...
7	...	...	0·0012	0·0011	...	...	...	...	...	...
8	...	...	0·0020	0·0020	...	...	...	...	...	...
9	...	...	0·0010	0·0010	...	...	...	...	0·0002	...
10	...	...	0·0016	0·0015	...	...	...	...	0·0020	...
11	...	...	...	...	0·0061	0·0062	...	...	...	...
12	...	...	...	...	0·0013	0·0012	...	...	...	...
13	...	...	...	...	0·0011	0·0010	...	...	0·0002	...
14	...	...	...	...	0·0014	0·0014	...	...	0·0030	...
15	...	...	...	...	0·0048	0·0046	...	...	0·0030	...
16	...	...	...	...	...	...	0·0017	0·0016	...	...
17	...	...	...	...	...	...	0·0011	0·0010	0·0020	...
18	0·0010	0·0010	0·0005	0·0005	0·0012	0·0012	0·00025	0·00025	0·0001	2·0
19	0·0013	0·0014	0·0003	0·0003	0·00096	0·0010	0·00038	0·00038	0·0001	2·0
20	0·0010	0·0010	0·0006	0·0006	0·00145	0·0014	0·0005	0·0005	0·0020	2·0
21	0·0015	0·0014	0·0005	0·0004	0·0012	0·0013	0·00038	0·0004	0·0030	2·0

on two synthetic solutions containing iron, titanium, aluminium, chromium and phosphorus in proportions approximating to those in an average non-metallic residue, and on two others containing appreciable amounts of P_2O_5. Sodium bisulphate was present in all four solutions.

Outline of the Proposed Scheme of Analysis when a Photometer is Available.

It is believed that the use of a photometer will allow of a much simpler scheme of analysis for chlorine residues than the modified procedure given above. Such a scheme has already been described by Stevenson [1] in the Eighth Ingot Report.

In the analysis of chlorine residues up to the present, the determination of silica has always followed after the residue has been ignited and weighed. Silica could be separated quantitatively and an ignition avoided if a wet oxidation with perchloric acid were carried out on the residue (and filter paper) after filtration. Preliminary experiments have shown that such a procedure can be readily carried out.

After the determination of silica, it is proposed to separate iron and titanium from the filtrate by means of cupferron and chloroform, to ignite them to the oxides and weigh, and to determine these elements colorimetrically.

In the water extract from the cupferron-chloroform separation it is thought that aluminium, chromium, phosphorus and manganese can all be determined by direct colorimetric methods as suggested by Stevenson (*loc. cit.*). The use of the procedure proposed by Lampitt and Sylvester [2] for the stabilisation of the colour developed by the addition of aurintricarboxylic acid to aluminium solutions is under consideration. Work on these lines has been successfully carried out, using a Lovibond tintometer.

PART B.—*General Summary Showing the Applicability and Utility of the Chlorine Method.*[3]

BY E. W. COLBECK, M.A., S. W. CRAVEN, A.M.C.T., AND W. MURRAY, A.M.C.T., A.I.C. (NORTHWICH).

SYNOPSIS.

The applicability of the chlorine method to various types of steel is discussed. Good agreement in the total-oxygen figures obtained by the chlorine method and the vacuum fusion method on fully killed plain carbon steels was obtained. The influence of sulphur and phosphorus in steels of this type was investigated.

Possible reasons why the total oxygen result determined by the chlorine method on a rimming steel is lower than that by the vacuum fusion method are put forward.

Results obtained by the chlorine method on various types of weld metal are given.

---

[1] Stevenson, Eighth Report on the Heterogeneity of Steel Ingots, Section VI., Part C, *Iron and Steel Institute*, 1939, *Special Report No.* 25.

[2] Lampitt and Sylvester, *Analyst*, 1932, vol. 57, p. 418.

[3] Received February 11, 1941.

Future work on the effect of increased chlorination temperatures on the recovery of non-metallic inclusions is mentioned, particularly in connection with the proposed examination of a series of alloy steels.

SINCE the publication of the Eighth Report on the Heterogeneity of Steel Ingots no further work in collaboration with other Members of the Oxygen Sub-Committee has been completed, except the examination of a basic Bessemer rimming steel ingot, details of which are given in Section IX., Part A of the present Report. In general, therefore, the position of the chlorine method remains the same as set out in the conclusions at the end of Section VI., Part 4 of the Eighth Report. Some additional materials have been examined, the results of which are given later (Table XVI.), but corroborative figures by the alcoholic iodine and the vacuum fusion methods are not available.

No further work has been done by the chlorine method on pig and cast iron.

Carbon Steels.

(1) *Killed Steels.*—The results obtained by the chlorine method on killed steels of the type mentioned in the Eighth Report (*i.e.*, samples *SG*1267, *SG*1268, *SG*1269 and *SG*2694) show:

(*a*) That good agreement for total oxygen with the vacuum fusion method is obtained;

(*b*) that the amounts of individual oxides are in reasonable agreement with those obtained by the modified alcoholic iodine method used at the National Physical Laboratory;

(*c*) that the chlorine method is capable of giving closely reproducible results;

(*d*) that the interference from sulphur is not appreciable (sample *SG*1267 contained 0·084% of sulphur);

(*e*) that the amount of phosphorus retained in the residues is small. For example, 0·002% of P_2O_5 was present in the residue from sample *SG*1268, which contained 0·094% of phosphorus;

(*f*) that the presence of large amounts of sulphur and phosphorus existing together in a steel do not appear to affect the results (*see* results for samples *SG*1269 and *SG*2694).

These conclusions are illustrated by the results shown in Table XV., obtained on the steels the chemical analyses of which are given in Table XIV.

(2) *Rimming Steels.*—In Section IX., Part A of the present Report the examination of a basic Bessemer rimming steel ingot is described. The results obtained for total oxygen by the chlorine method on the core and envelope of this ingot are much lower than those returned by the vacuum fusion method. The reasons for

TABLE XIV.—*Chemical Analyses of Steels* SG1267, SG1268, SG1269 *and* SG2694.

High-frequency electric steels made in sillimanite crucibles.

Steel Sample Number.	Chemical Analysis.					
	C. %.	Si. %.	S. %.	P. %.	Mn. %.	Ti. %.
*SG*1267	0·32	0·22	0·084	0·009	0·68	0·002
*SG*1268	0·33	0·25	0·017	0·094	0·67	0·007
*SG*1269	0·31	0·24	0·083	0·086	0·68	0·012
*SG*2694	0·30	0·26	0·092	0·163	0·62	0·001

this are not yet fully understood. The following suggestions are, however, put forward on the limited evidence at present available :

(*a*) The inclusions present in rimming steel may be susceptible to attack by chlorine at 350° C., perhaps because of the fact that in this type of steel they consist largely of FeO and MnO, present as such and not linked with appreciable amounts of SiO_2 or Al_2O_3 as in killed steels.

(*b*) There is considerable controversy as to whether oxygen exists in " solid solution " in steel. If it is present in rimming steels in such a form it is probable that it would not be retained in chlorine residues, although it would be extracted by the vacuum fusion procedure.

The above explanations should be considered as tentative; it is clear that a much larger number of samples must be examined before any definite conclusions can be put forward.

(3) *Miscellaneous Materials Examined.*—A few preliminary experiments have been carried out on some types of steel that have not yet been examined by the Oxygen Sub-Committee on a co-operative basis. The results obtained by the chlorine method are shown in Table XVI. Since, however, no comparative figures by either the alcoholic iodine or the vacuum fusion methods are available, the results must be treated with reserve.

Sample *C*6 is an example of a 0·65% carbon steel which has been successfully examined by the chlorine method. In this case the retention of phosphorus in the chlorine residue is of a low order.

In the silicon-killed example *SG*2616, the silica fraction of the chlorine residue predominates. Similarly, the alumina fractions of the residues from the aluminium-killed steels *TC*4 and *TC*8 are noticeably higher than in the steels *TC*2 and *TC*6, which had no aluminium additions.

Samples *TC*6 and *TC*8 are instances of steels containing 1·5% of manganese which have been chlorinated without difficulty at 350° C.

The oxygen contents of the weld-metal deposits made by the arc

TABLE XV.—*Comparison of Oxygen Results Obtained by the Chlorine, the Modified Alcoholic Iodine and the Vacuum Fusion Methods.*

Steel Sample Number.	Co-operating Laboratory.	Method.	Form of Sample Used.	Total Ignited Residue. %.	Residue Analysis.										Total Oxygen. %.
					SiO_2.		FeO.		MnO.		Al_2O_3.		P_2O_5. %.	TiO_2. %.	
					SiO_2. %.	O_2. %.	FeO. %.	O_2. %.	MnO. %.	O_2. %.	Al_2O_3. %.	O_2. %.			
*SG*1267	I.C.I. (Alkali), Ltd.	Chlorine at 350° C. for 4 hr.	Discs.	0·044	0·007	0·004	0·014	0·003	0·006$_5$	0·001$_5$	0·012	0·005$_5$	0·001	Nil	0·014
				0·043$_5$	0·006$_5$	0·003$_5$	0·012$_5$	0·002$_5$	0·006$_5$	0·001$_5$	0·012$_5$	0·006	...	Nil	0·013$_5$
				0·041	0·007	0·004	0·012	0·002$_5$	0·006$_5$	0·001$_5$	0·012$_5$	0·006	0·001	Nil	0·014
	National Physical Laboratory.	Modified alcoholic iodine.	Thin discs.	0·043	0·013$_5$	0·007	0·004	0·001	0·007$_5$	0·001$_5$	0·009	0·004$_5$	0·001	0·005	0·014
		Vacuum fusion.	Solid piece.	...	...	...	...	...	...	...	...	...	...	...	0·014
*SG*1268	I.C.I. (Alkali), Ltd.	Chlorine at 350° C. for 4 hr.	Discs.	0·041$_5$	0·007$_5$	0·004	0·010$_5$	0·002$_5$	0·006	0·001$_5$	0·012	0·005$_5$	0·002	Nil	0·013$_5$
				0·045	0·006	0·003	0·012$_5$	0·003	0·005$_5$	0·001	0·014	0·006$_5$	0·002	Nil	0·013$_5$
				0·044	0·007$_5$	0·004	0·010	0·002	0·006	0·001$_5$	0·012$_5$	0·006	0·002	Nil	0·013$_5$
	National Physical Laboratory	Modified alcoholic iodine.	Thin discs.	0·065	0·010	0·005	0·005$_5$	0·001	0·011	0·002$_5$	0·009$_5$	0·004$_5$	0·014$_5$	0·013$_5$	0·013
		Vacuum fusion.	Solid piece.	...	...	...	...	...	...	...	...	...	...	...	0·013
*SG*1269	I.C.I. (Alkali), Ltd.	Chlorine at 350° C. for 4 hr.	Discs.	0·043	0·006	0·003$_5$	0·012$_5$	0·003	0·003	0·000$_5$	0·012	0·005$_5$	0·002	Nil	0·012
				0·043	0·006	0·003$_5$	0·013	0·003	0·003	0·000$_5$	0·012$_5$	0·006	...	Nil	0·013
	National Physical Laboratory.	Modified alcoholic iodine.	Thin discs.	...	0·005	0·002$_5$	0·009$_5$	0·002	0·017	0·004	0·010	0·004$_5$	0·013	0·020$_5$	0·013
				0·073	0·005$_5$	0·003	0·006	0·001$_5$	0·012	0·002$_5$	0·010$_5$	0·005	0·013	0·022$_5$	0·012
		Vacuum fusion.	Solid piece.	...	...	...	...	...	...	...	...	...	...	...	0·012
*SG*2694	I.C.I. (Alkali), Ltd.	Chlorine at 350° C. for 4 hr.	Discs.	0·048	0·011	0·006	0·009$_5$	0·002	0·010	0·002	0·011	0·005$_5$	0·002$_5$	Nil	0·015$_5$
				0·048$_5$	0·012	0·006$_5$	0·009$_5$	0·002	0·010	0·002	0·010	0·005	0·003	Nil	0·015$_5$
	National Physical Laboratory.	Modified alcoholic iodine.	Thin discs.	0·068	0·018	0·009$_5$	0·008$_5$	0·002	0·010	0·002	0·008	0·004	0·017	0·002$_5$	0·017$_5$
		Vacuum fusion.	Solid piece.	...	...	...	...	...	...	...	...	...	...	...	0·013

TABLE XVI.—*Oxygen Results by the Chlorine Method on Miscellaneous Materials.*

Sample Number.	Description of Sample.	Chemical Analysis.	Residue Analysis.									Total Oxygen. %.
			SiO_2.		FeO.		MnO.		Al_2O_3.		P_2O_5. %.	
			SiO_2. %.	O_2. %.	FeO. %.	O_2. %.	MnO. %.	O_2. %.	Al_2O_3. %.	O_2. %.		
C6	Carbon tool steel.	C 0·65% Si 0·27% S 0·035% P 0·029% Mn 0·39%	<0·000$_5$	Nil	0·012	0·002$_5$	<0·000$_5$	Nil	0·010	0·004$_5$	0·001	0·007
SG2616	Remelted electrolytic iron (silicon-killed).	C 0·02% Si 0·33% S 0·012% P 0·049% Mn Nil	0·134 0·137	0·071$_5$ 0·073	0·006 0·006	0·001$_5$ 0·001$_5$	<0·000$_5$ <0·000$_5$	Nil Nil	0·010 0·009$_5$	0·004$_5$ 0·004$_5$	0·001 0·001	0·077$_5$* 0·079*
TC2	Tropenas converter steel (acid lining).	C 0·12% Si 0·34% S 0·058% P 0·057% Mn 0·75%	0·027	0·014$_5$	0·007	0·001$_5$	0·033	0·007$_5$	0·007$_5$	0·003$_5$	0·001	0·027
TC4	Tropenas converter steel (acid lining) killed with 0·07% of aluminium.		<0·000$_5$	Nil	0·010	0·002	<0·000$_5$	Nil	0·044	0·021	0·001	0·023
TC6	Tropenas converter steel (acid lining).	C 0·19% Si 0·45% S 0·053% P 0·059% Mn 1·5%	0·019	0·010	0·008	0·002	0·033	0·007	0·007$_5$	0·003	0·001	0·022$_5$
TC8	Tropenas converter steel (acid lining) killed with 0·07% of aluminium.		<0·000$_5$	Nil	0·010	0·002	<0·000$_5$	Nil	0·050	0·023	0·001	0·025
W1 *W2*	All-weld-metal test-pieces made by the Union Melt process.		0·102 0·104	0·054 0·055	0·008 0·016	0·002 0·004	0·027 0·102	0·006 0·023	0·006 0·014	0·003 0·007	0·001 0·001	0·065 0·089
W3	All-weld-metal test-piece. Electrode had slag-shielded TiO_2 alloy coating, containing ferro-manganese and silicates.		0·095	0·050$_5$	0·013	0·003	0·077	0·016	0·003	0·001$_5$	0·001	0·071
W4	All-weld-metal test-piece. Electrode was gas-shielded. Coating was organic of the cellulose type.		0·122	0·065	0·037	0·008	0·065	0·014$_5$	0·014	0·006$_5$	0·001	0·094
W5 *W6* *W7* *W8*	All-weld-metal test-piece made by the oxy-acetylene process with a neutral flame.		0·011 <0·000$_5$ <0·000$_5$ <0·000$_5$	0·006 Nil Nil Nil	0·011 0·007 0·004 0·004$_5$	0·002$_5$ 0·001$_5$ 0·001 0·001	0·013 <0·000$_5$ <0·000$_5$ <0·000$_5$	0·003 Nil Nil Nil	0·003 0·015 0·014 0·008$_5$	0·001$_5$ 0·007 0·006$_5$ 0·004	0·001 0·001 0·001 0·001	0·013 0·008$_5$ 0·007$_5$ 0·005

* Total oxygen by fractional vacuum fusion method (United Steel Companies, Ltd.) = 0·072%.

process are rather high; this, in general, appears to be due to the presence of entrapped silicates, which is in good agreement with what is normally found when such welds are examined under the microscope. The oxygen content of the oxy-acetylene welds is of a very much lower order, owing to the almost complete absence of silicates in the residues.

Alloy Steels.

No further co-operative work has been done on alloy steels since the publication of the Eighth Report. In that Report [1] results for total oxygen on a series of steels containing amounts of alloying elements of the order of 0·5% were given. The agreement between the chlorine and vacuum fusion methods was fair in all cases.

Preliminary experiments have indicated that as the chromium content of steels rises the resistance of the alloys to attack by chlorine increases. It is, therefore, probable that chlorination temperatures above 350° C. will have to be employed when the percentage of chromium rises above a certain limiting figure. Before the determination of non-metallic inclusions by the chlorine method on a new series of alloy steels (particularly those containing chromium) is proceeded with, it seems essential that the effect on the inclusions of increasing chlorination temperatures should be explored.

SECTION V.—The Alcoholic Iodine Method.

Part A.—*Present Position, Limitations and Possibilities of the Alcoholic Iodine Method, with a Note on Factors Affecting the Presence of Phosphorus in the Residue.*[2]

By T. E. ROONEY, A.M.S.T., F.I.C. (National Physical Laboratory, Teddington).

Synopsis.

The investigations on the alcoholic iodine method carried out by the laboratories collaborating in the work of the Oxygen Sub-Committee are summarised and references to the Reports where more complete details can be found are given; brief mention is made of the method which has become standard practice and of a modification which is convenient for routine determinations. The types of steel that can be successfully examined are indicated, and a modification that may be necessary for steels of the rimming type is described. The limitations of the method and the possibilities of extending its use to alloy and other steels are discussed, and a note on factors affecting the presence of phosphorus in the residue is appended. The successful use of alcoholic iodine solution for stripping oxide films from iron is mentioned.

[1] Eighth Report on the Heterogeneity of Steel Ingots, Section VI., Part 4, sub-section *B*, p. 111, *Iron and Steel Institute*, 1939, *Special Report No.* 25.

[2] Received February 25, 1941.

Present Position of the Alcoholic Iodine Method.

The Method.—Since the publication of the Second Report of the Oxygen Sub-Committee the stirring apparatus devised at the National Physical Laboratory has been tested and adopted by other laboratories. This apparatus is similar to that described in the First Report,[1] but includes certain minor modifications which are given in the Second Report.[2] The stirring method (Second Report, *loc. cit.*, p. 146) has become standard practice. The stirring is carried out at 60–65° C. in order to ensure complete decomposition of sulphides. Solid samples of thin section are used instead of the millings employed in the older procedure in order to minimise the effects of surface oxide films, and the alcoholic iodine solution is deoxidised by passing a stream of nitrogen through the boiling liquid.

A recent modification of the above method has been devised at Stocksbridge (this Section, Part B), and as modified is similar to the "boiling method" (Second Report, *loc cit.*, p. 142). Rubber stoppers were used instead of ground-glass joints in the preliminary tests, but were a source of weakness in the method. They were replaced by glass joints so arranged that the greased surfaces are outside the reaction flask. The solution after reaction has ceased is allowed to cool to room temperature and is filtered through a paper-pulp filter previously washed with dry alcohol. This method simplifies the procedure to some extent, but the standard practice above noted is adhered to, *viz.*: (1) Exclusion of oxygen and water vapour from the atmosphere of the reaction vessel. (2) Agitation of the solution. (3) Maintenance of the reaction solution at a fixed temperature. (4) Use of solid samples. (5) Deoxidation of the alcoholic iodine solution.

Application.—The method can be used for the determination of oxides in low- and medium-carbon steels not containing appreciable amounts of elements which form stable carbides or insoluble iodides. Such steels represent a large proportion of the total tonnage of manufactured steel.

In Table XVII. results are given for materials to which the method has been successfully applied. It is not suggested that this list exhausts the applicability of the method, and it is expected that copper and nickel steels as well as other materials will yield reliable results.

It is well known that most analytical methods can be varied or modified to suit the material under consideration. In Table XVIII. the results of some experiments on a rimming steel are given in order to illustrate how a simple modification can be made to suit

[1] Seventh Report on the Heterogeneity of Steel Ingots, p. 113, *Iron and Steel Institute*, 1937, *Special Report No.* 16.

[2] Eighth Report on the Heterogeneity of Steel Ingots, p. 142, *Iron and Steel Institute*, 1939, *Special Report No.* 25.

TABLE XVII.—*Analyses of Steel Samples Examined.*

Mark.	Material, and Method of Manufacture.	Analysis. C. %.	Si. %.	S. %.	P. %.	Mn. %.	Other Elements. %.	Total Oxygen, %, by— Alcoholic Iodine.	Vacuum Fusion.	Co-operating Laboratory.
*SG*1267	Steel, crucible.	0·32	0·22	0·084	0·009	0·68	Ti. 0·002	0·014	0·014	National Physical Laboratory.
*SG*1268	Steel, crucible.	0·33	0·25	0·017	0·094	0·67	Ti. 0·007	0·013	0·013	National Physical Laboratory.
*SG*1269	Steel, crucible.	0·31	0·24	0·083	0·086	0·68	Ti. 0·012	0·012	0·012	National Physical Laboratory.
*SG*2694	Steel, crucible.	0·30	0·26	0·092	0·163	0·62	Ti. 0·001	0·017	0·013	National Physical Laboratory.
*PI*3	Pure iron rolled sheet.	0·006	0·0005	...	0·001	...	Ni. 0·023	0·0018	0·0017	National Physical Laboratory.
*EU*3	Pure iron ingot.	0·01	0·0004	0·001	0·003	...	Ni. 0·02	0·0012$_5$	0·0012	National Physical Laboratory.
*PO*5	Pure iron + phosphorus.	As *EU*3	above.	...	0·053	...		0·0028	0·0033	National Physical Laboratory.
4	Steel, Al-killed.	0·17	0·09	0·029	0·014	0·65	Al. 0·027; Cr. 0·008; V. 0·001	0·005$_5$	0·002	National Physical Laboratory.
5	Steel, Si-killed.	0·22	0·14	0·042	0·020	0·45	Al. ...; Cr. 0·020; V. 0·001	0·011	0·010	National Physical Laboratory.
6	Steel, Si-killed.	0·43	0·20	0·027	0·014	0·47	Al. ...; Cr. 0·012; V. 0·001	0·008	0·005	National Physical Laboratory.
*SG*2616	Steel, Si-killed.	0·02	0·33	0·012	0·049	Nil	Al. ...; Cr. ...; V. ...	0·077	0·072	National Physical Laboratory.
	Steel, boiler plate:									
27	Top of ingot.	0·14	0·01	0·043	0·037	0·49	...	0·023	0·019$_5$ *	Central Research Department, The United Steel Co.'s, Ltd.
29	Bottom of ingot.							0·025	0·020 *	Central Research Department, The United Steel Co.'s, Ltd.
31	Top of ingot.	0·18	Trace	0·045	0·051	0·70	...	0·015	0·011 *	Central Research Department, The United Steel Co.'s, Ltd.
35	Bottom of ingot.							0·023$_5$	0·021 *	Central Research Department, The United Steel Co.'s, Ltd.
*A*1802	Steel, weld metal.	...	...	...	...	...	...	0·076	0·073$_5$ *	Central Research Department, The United Steel Co.'s, Ltd.
*A*1803	Steel, weld metal.	...	...	...	...	...	...	0·094	0·104 *	Central Research Department, The United Steel Co.'s, Ltd.

* Simplified modification of the method.

TABLE. XVIII.—*Oxygen Determinations on Rimming Steel.*

Composition of steel : C 0·03% ; Si 0·002% ; Mn 0·31% ; S 0·036% ; P 0·011% ; Cr 0·004% ; V 0·001%.

Samples in form of discs.
Total oxygen by vacuum fusion, 0·020%.

Weight of Sample. G.	Time of Stirring and Max. Temp.	Total Ignited Residue. %.	SiO_2. SiO_2. %.	O_2. %.	FeO. FeO. %.	O_2. %.	MnO. MnO. %.	O_2. %.	Al_2O_3. Al_2O_3. %.	O_2. %.	Cr_2O_3. %.	P_2O_5. %.	Total. %.	Total Oxygen. %.
9·85	3½ hr., 64° C.	0·026	0·0018	0·0010	0·015	0·0033	0·0054	0·0012	0·002	0·001	0·0005	0·003	103	(1) 0·006$_5$
9·86	5 hr., 35° C.	0·041	0·0021	0·0011	0·008	0·0018	0·024	0·005	0·002	0·001	0·0005	0·004	105	(2) 0·009
9·31	4 hr. 40 min., 33° C.	0·126	0·0018	0·0010	0·060	0·013	0·046	0·010	0·001	0·0005	0·0005	0·005	99·2	(3) 0·024$_5$

this class of material. This steel is described as of the low-carbon rimming type and its analysis is as follows:

Carbon.	Silicon.	Manganese.	Sulphur.	Phosphorus.	Chromium.	Vanadium.
0·03%	0·002%	0·31%	0·036%	0·011%	0·004%	0·001%

In the first experiment (1) the standard procedure was used, in the second (2) a maximum temperature of 55° C. of the mixture was reached in about 45 min., falling to 40° C. towards the end of the experiment, and in the third (3) no external heat was applied to the iodine solution and a maximum temperature of 33° C. was reached in about 30 min., falling to 25° C. towards the end of the experiment.

These experiments tend to show that oxides of iron and manganese contained in some types of rimming steel are soluble in alcoholic iodine solution if the temperature exceeds 33° C. If the temperature is controlled in this region a reasonable result can be obtained on this material.

In the paper by Swinden and Stevenson (Section IX., Part A) an example is given of a basic Bessemer rimming steel which, in a similar manner to the rimming steel noted above, yields lower calculated total oxygen values by the chlorine and iodine methods than the total oxygen by the vacuum fusion method. It is probable in this case also that the use of a lower reaction temperature will result in a total oxygen value by the iodine method more in line with that by the vacuum fusion method.

Although outside the scope of the work of the Sub-Committee, it may not be out of place to refer to an application of the method other than for the determination of oxide inclusions. Reference has been made elsewhere to the successful use of alcoholic iodine solution for stripping oxide films.[1] The work illustrates the accuracy which is possible in isolating iron-oxide films.

Limitations of the Method.

(1) Difficulties have been experienced in dealing with aluminium-killed steels and some to which aluminium has been added, mainly of the medium- and high-carbon type. Table XIX. gives the analyses of the steels and Table XX. the results of iodine determinations. The difficulty probably arises owing to the retention of undecomposed aluminium carbide and/or aluminium nitride [2] in the oxide residue after treatment with the iodine solution.

(2) The low solubility of iron carbide in the iodine solution is another factor which limits the applicability of the method. In Table XVII. results are given for steels containing up to 0·43% of carbon

[1] Vernon, Wormwell and Nurse, *Journal of the Chemical Society*, 1939, April, p. 621.

[2] Klinger and Koch, *Technische Mitteilungen Krupp, Forschungsberichte*, 1938, May, No. 3, p. 49.

TABLE XIX.—*Analyses of Aluminium-Killed Steels.*

Co-operating laboratory, National Physical Laboratory.

Mark.	Material.	Analysis.					
		C. %.	Si. %.	Mn. %.	P. %.	Al. %.	N. %.
B	Pure iron + carbon *	0·70	...	...	...	N.D.†	0·0007
1	Basic electric Al-killed	0·13	0·23	0·56	0·029	0·005	0·006
3	Basic electric Al-killed	0·39	0·18	0·51	0·027	0·018	0·006
5	Basic electric Al-killed	0·58	0·26	0·57	0·025	0·013	0·007

* Aluminium accidentally introduced from refractories during preparation.
† N.D. = not determined.

TABLE XX.—*Oxygen Determinations on Aluminium-Killed Steels.*

Co-operating Laboratory, National Physical Laboratory.

Sample.	Time of Stirring and Max. Temp.	Total Ignited Residue. %.	Residue Analysis.								P_2O_5. %.	Total Oxygen, %, by—	
			SiO_2.		FeO.		MnO.		Al_2O_3.			Alcoholic Iodine.	Vacuum Fusion.
			SiO_2. %.	O_2. %.	FeO. %.	O_2. %.	MnO. %.	O_2. %.	Al_2O_3. %.	O_2. %.			
B	7½ hr., 64° C.	0·075	0·0016	0·0009	0·0376	0·0084	...	...	0·033	0·016	...	0·025	0·0029
1	4½ hr., 65° C.	0·044	0·013	0·007	0·004	0·001	0·010	0·002	0·012	0·006	0·004	0·016	0·0097
3	6 hr., 61° C.	0·046	0·0014	0·001	0·007	0·002	0·011	0·003	0·006	0·003	0·014	0·009	0·0035
3	8 hr., 65° C.	0·036	0·0016	0·001	0·009	0·002	0·0016	0·0004	0·007	0·003	0·012	0·006	0·0035
5	6½ hr., 65° C.	0·043	0·0011	0·001	0·011	0·002	0·0005	0·0001	0·013	0·006	0·015	0·009	0·0023

to which the method has successfully been applied. It is probable that steels containing up to 0·6% of carbon and low aluminium and nitrogen contents can be examined by the method. Work has been in progress at Sheffield University (Section IX., Part B) on a series of carbon steels in order to determine the solubility of iron carbide in alcoholic iodine solution.

(3) Alloy steels containing appreciable amounts of elements such as chromium, vanadium, titanium, &c., which form stable carbides, or elements such as titanium and molybdenum the iodides of which are insoluble in alcohol, when treated by the standard procedure yield residues which contain an undue proportion of material other than oxides. The magnitude of this effect may be illustrated by quoting the results obtained on two commerical alloy steels which have been examined at the National Physical Laboratory. A molybdenum-titanium steel (carbon 0·13%, silicon 0·03%, manganese 0·71%, molybdenum 0·52% and titanium 0·24%) yielded a total ignited residue of 0·6% containing both titanium and molybdenum. This is to be attributed to the insoluble character of the iodides of these elements, and it is possible that their carbides may not be easily decomposed by the iodine solution. A portion of the molybdenum was volatilised during the ignition of the residue, but practically a quantitative recovery of titanium was obtained, *viz.*, 0·23%.

A nickel-chromium steel (carbon 0·5%, nickel 2·0%, chromium 2·0%) yielded a residue before ignition of 7–9% which consisted principally of chromium carbide. The vacuum fusion value for total oxygen was 0·003%.

The presence of large amounts of stable carbides in the residues obtained from alloy steels is a serious complication and may make impossible the correct evaluation of the oxides present.

(4) Certain steels have been examined which yield variable results for the MnO fraction. By prolonging the stirring operation after the steel was dissolved normal values for the MnO were obtained. Examples may be quoted from the Second Report (p. 154), *viz.*, steels *SG*1269 and *SG*2694. Both these steels contained 0·3% of carbon, were melted in the induction furnace and were killed steels.

The results recorded in Table XXI., obtained on steel *SG*2694, illustrate this type of effect.

Steel *SG*2694 contained 0·163% of phosphorus, and the usual procedure, *viz.*, stirring at just over 60° C. until the sample was dissolved, gave very high values for the MnO fraction. In the first three experiments in Table XXI. the stirring was continued until the samples were just dissolved. In experiment 4 stirring was continued for about 2 hr. after the sample was dissolved. The magnitude of the MnO fraction was considerably reduced and the total oxygen obtained was then in reasonable agreement with the vacuum fusion result.

TABLE XXI.—*Oxygen Determinations on Steel* SG2694.

Composition of steel: C 0·30%; Si 0·26%; S 0·092%; P 0·163%; Mn 0·62%.

Total oxygen by vacuum fusion, 0·013%.

Experiment.	Weight of Sample. G.	Number of Discs.	Time of Stirring and Max. Temp.	Total Ignited Residue. %.	SiO₂.		FeO.		MnO.		Al₂O₃.		TiO_2. %.	Cr_2O_3. %.	P_2O_5. %.	Total. %.	Total Oxygen. %.
					SiO_2. %.	O_2. %.	FeO. %.	O_2. %.	MnO. %.	O_2. %.	Al_2O_3. %.	O_2. %.					
1	10·44	32	1½ hr., 65° C.	0·088	0·015	$0·007_5$	0·008	0·002	0·031	0·007	0·008	0·004	0·002	0·001	0·016	97·8	0·021
2	10·15	33	2 hr., 65° C.	0·102	0·017	0·009	0·008	0·002	0·042	$0·009_5$	0·008	0·004	0·002	0·001	0·018	97·4	0·024
3	10·26	19	4½ hr., 64° C.	0·11	0·017	0·009	N.D.†	...	0·039*	...	N.D.†	...	N.D.†	N.D.†	0·016	...	...
4	10·23	20	7 hr., 65° C.	0·068	0·018	$0·009_5$	$0·008_5$	0·003	0·010	0·002	0·008	0·004	0·002	0·002	0·017	100	0·017

* MnO was determined directly with bismuthate after removal of SiO_2.
† N.D. = not determined.

TABLE XXII.—*Oxygen Determinations on an Alloy of Pure Iron + 0·053% of Phosphorus.*

Total oxygen by vacuum fusion, 0·0033%.

Weight of Sample. G.	Form of Sample.	Time of Stirring and Max. Temp.	Total Ignited Residue. %.	SiO₂.		FeO.		Al₂O₃.		P_2O_5. %.	Total. %.	Total Oxygen. %.
				SiO_2. %.	O_2. %.	FeO. %.	O_2. %.	Al_2O_3. %.	O_2. %.			
10·25	Thin slices.	1 hr. 20 min., 63° C.	0·0067	0·0007	0·0004	0·0041	0·0009	0·0032	0·0015	Trace	106	0·0028
10·77	Millings.	1 hr. 15 min., 62° C.	0·027	0·0009	0·0005	0·0205	0·0046	0·0027	0·0013	0·0012	102	0·0064

The explanation for the variation in the MnO content with stirring time is not yet clear.

Possibilities of Further Improving or Modifying the Method.

(*a*) It may be possible in the case of aluminium-killed steels to determine the amount of aluminium carbide and nitride in the residue directly.

(*b*) In the case of carbon and alloy steels it may be possible to isolate the carbides from the oxide portion of the residue by similar means to those employed in electrolytic methods, by magnetic treatment during filtration (percolation method, Second Report, p. 144), or by treatment such as that used in the aqueous iodine method (this Report, Section VI., Part A).

(*c*) As noted above, difficulties connected with the MnO fraction are receiving further attention.

Note on Factors Affecting the Presence of Phosphorus in the Residue.

The presence of phosphorus in the residue has always been an elusive subject to investigate, probably because it is affected by a number of different factors. Its influence on the analysis of the residue was demonstrated in the Second Report (p. 145). Phosphorus contained in a steel had been regarded as being completely soluble in the iodine solution, but it was found that a small quantity was retained in the residue. This small quantity forms P_2O_5 on ignition and to a greater or less extent affects the accuracy of the analysis of the residue. In the ammonia method of estimating the alumina some P_2O_5 is retained in the precipitate, even after reprecipitation, and, being registered as alumina, has an appreciable effect on the calculation of the total oxygen.

Recently it has been noted in the analysis of chlorine residues that the presence of excessive amounts of P_2O_5 interferes with the direct determination of silica by the hydrofluoric acid method (Section IV., Part B). This observation agrees with those of other members of the Oxygen Sub-Committee using the alcoholic iodine method, who have advocated indirect methods for the determination of silica (First Report, p. 118, and Second Report, p. 180). The retention of appreciable amounts of phosphorus in chlorine residues is commented on in Section IV., Part A of the present Report, and is worth emphasising here that, as hitherto, the amount retained has been much lower in chlorine than in iodine residues, and it may be interesting to examine the effects with the iodine method.

Effect of Oxygen on the Phosphorus Content of the Residue.—Some experiments were made at the National Physical Laboratory on an alloy of pure iron and phosphorus. It was hoped by the use of this material to reduce the number of factors affecting the influence of phosphorus in the alcoholic iodine method.

The alloy contained 0·053% of phosphorus. The iodine solution was deoxidised twice with nitrogen, the stirring method was used and a temperature of just over 60° C. was reached in about ½ hr. and maintained until the end of the experiment.

One experiment carried out with thin discs, in order to minimise the effect of surface oxygen, showed that the phosphorus does not interfere with the method when applied to iron containing phosphorus and only small amounts of oxygen. A second experiment with millings indicated that surface oxidation introduces sufficient oxygen to have a small effect on the phosphorus content of the residue. The results are recorded in Table XXII.

The next step appears to be the examination of an alloy of pure iron, phosphorus and about 0·01% of oxygen, in order to determine whether, in the presence of iron oxide ("body oxygen"), P_2O_5 is retained in the residue.

In conclusion, the author wishes to acknowledge the assistance of the members of the Oxygen Sub-Committee whose work is referred to in the paper, and for permission to publish acknowledgment is made to the Director of the National Physical Laboratory.

Part B.—*A Simplification of the Alcoholic Iodine Method for the Determination of Oxide Residues in Steel.*[1]

By W. W. STEVENSON, A.I.C., and G. E. SPEIGHT, B.Sc., A.I.C. (Stocksbridge, near Sheffield).

(Under the direction of T. Swinden, D.Met.)

Synopsis.

The conditions necessary for the successful operation of the standard alcoholic iodine method for the determination of oxide inclusions in iron and steel are discussed and the development and operation of a simplified method is described. Some possible errors are reviewed and typical results by the simplified procedure on various materials are given.

The use of a solution of iodine in absolute alcohol for the determination of oxides in steel was recommended by Willems,[2] who claimed that the reaction should be carried out in a neutral atmosphere and in the absence of water. These conditions were covered in an apparatus designed by B. A. Bannister [3] at Sheffield University, and also in a modified apparatus used by Rooney and Stapleton.[4] The method consisted of dissolving 7–8 g. of millings in a solution of 70 g. of pure iodine in 600 ml. of specially dried methyl alcohol, in

[1] A communication from The United Steel Companies, Ltd., Central Research Department, Stocksbridge, near Sheffield, received February 10, 1941.

[2] Willems, *Archiv für das Eisenhüttenwesen*, 1927–28, vol. 1, p. 655.

[3] Bannister, Sheffield University, Ph.D. Thesis, 1933.

[4] Rooney and Stapleton, *Journal of the Iron and Steel Institute*, 1935, No. I., p. 249.

an enclosed vessel contained in a mechanically rotated box. The vessel had been previously filled with nitrogen, and all filtrations, &c., were also carried out in an atmosphere of purified nitrogen; a non-metallic residue of a few milligrammes in weight resulted after solution and filtration of the iron.

The later developments of the alcoholic iodine method are described in the previous Reports of the Heterogeneity of Steel Ingots Committee,[1] and have taken the form of (*a*) modified apparatus, *i.e.*, means of "churning," "stirring" and "boiling," to maintain constant the temperature of the reaction, (*b*) de-aeration of the alcohol solvent, (*c*) the use of piece samples as a precaution against surface oxide content of millings, and (*d*) modified methods of analysis of the residues. The possibility of applying the method to carbon steels, low-alloy steels, and pig and cast irons has been studied and a comparison of the results with those obtained from other methods has been made.

Our knowledge of the alcoholic iodine method indicates that comparable results are obtained only by rigid adherence to certain conditions, which may be enumerated as follows:

(1) *The exclusion of oxygen and water vapour from the atmosphere of the reaction vessel.* This is attained in the standard method by extensive purification of synthetic methanol by successive fluxing and distillation over metallic calcium, and by the use of a suitable apparatus whereby carefully purified nitrogen is admitted to the evacuated reaction vessel.

(2) *Agitation of the solution during reaction.* The reaction proceeds very slowly in the absence of movement, and mechanical shaking was first adopted. The standard method now uses an electrically-driven stirrer which is inserted in the reaction vessel.

(3) *Maintenance of the solution at a fixed temperature.* The reaction between iodine and iron evolves a large amount of heat, which serves to increase the temperature of the solution. Much of the early work was vitiated, owing to the variable temperature attained by the iodine solution, which thus reacted in an indefinite manner towards the sulphides present in the sample.

(4) *The use of a sample of small surface area* is essential, since the risk of error due to surface oxidation of the steel during the preparation of the sample is minimised. The ideal material size would be one piece of appropriate weight, but since solution of this would be too slow, a compromise is effected by using thin discs about 2 mm. in thickness.

[1] Andrew, Raine and Vickers, Sixth Report on the Heterogeneity of Steel Ingots, *Iron and Steel Institute*, 1935, *Special Report No.* 9, p. 50. Rooney, Stevenson and Raine, Seventh Report, 1937, *Special Report No.* 16, p. 109. Rooney, Eighth Report, 1939, *Special Report No.* 25, p. 141. Taylor-Austin, Eighth Report, 1939, *Special Report No.* 25, p. 159.

(5) *De-aeration of the alcohol solvent* is recommended. Alcohol is capable of dissolving a large volume of oxygen, and since great care is taken to remove oxygen from the atmosphere of the reaction vessel, it is advisable to remove any air possibly dissolved in the liquid solvent. In the standard method this is carried out by distillation of the alcohol while maintaining a passage of purified nitrogen.

The standard method as described in the Eighth Report on the Heterogeneity of Steel Ingots satisfactorily complies with the above conditions, but requires the use of a specially designed complicated apparatus, which at first examination restricts the widespread use of the method to the field of research. It has now been found that a simpler form of apparatus, more adaptable to ordinary chemical laboratory conditions, may be used. In addition, in the apparatus used in the standard method several greased joints are liable to contaminate the extracted residue and thereby introduce complications. The modified apparatus, which satisfies the above conditions, also avoids the possibility of grease contamination. A description of the apparatus follows:

The early attempts at simplification of the iodine method were carried out with the usual chemical laboratory ware shown in Fig. 6. This arrangement consists of an R.P. flask (1000 ml. capacity) fitted with a rubber bung and a straight 8-in. Liebig condenser. The upper end of the condenser is closed by a bung, carrying a long delivery tube (4 mm. in dia.), reaching to within a fraction of an inch of the bottom of the R.P. flask, and a shorter fine-capillary outlet tube. The long delivery tube is connected to a supply of nitrogen, purified as described in a previous Report of the Oxygen Sub-Committee [1]; thus, an atmosphere of purified nitrogen is maintained in the apparatus.

The sample, in the form of pieces about 12–13 mm. in dia. and 2 mm. in thickness, and of total weight approximately 10 g., is cleaned in benzene and ether, and immediately weighed and placed in the clean and dried R.P. flask. The alcohol-iodine solvent, prepared as described on p. 109 of the Seventh Ingot Report (70 g. of A.R. iodine in 600 ml. of purified methyl alcohol), is filtered through a fine filter paper, previously washed with alcohol, into the R.P. flask containing the sample. The condenser and delivery tube are attached to the flask, which is supported on a sand tray or electrically-heated plate, and before applying heat the supply of pure nitrogen is commenced.

After passing a stream of nitrogen for 1 hr., heat is supplied to the solution and in about 15–20 min. the solvent boils gently. The nitrogen flow is continued throughout the test and solution of the sample commences and is usually complete in 5–6 hr., although an occasional sample may require longer; completeness of solution is

[1] Andrew, Raine and Vickers, *loc. cit.*

revealed quite easily by inspection of the flask and contents. The source of heat is removed and the reaction flask allowed to cool while still maintaining the flow of nitrogen (usually this is continued overnight). When cold, the solution is filtered through a tightly packed paper-pulp filter (previously washed with alcohol), and the residue is washed rapidly with purified methyl alcohol, dried, ignited and weighed in the customary manner. The time interval between disconnecting the nitrogen flow and completion of the washing must be very short and care must be taken to avoid "creeping" of the iron-iodide/iodine solution on the walls of the funnel. If the latter point is overlooked, danger exists in alcohol evaporating and leaving a dried residue of iron iodide, which may decompose, leaving a basic iron residue on subsequent washing with alcohol. Periodic washing

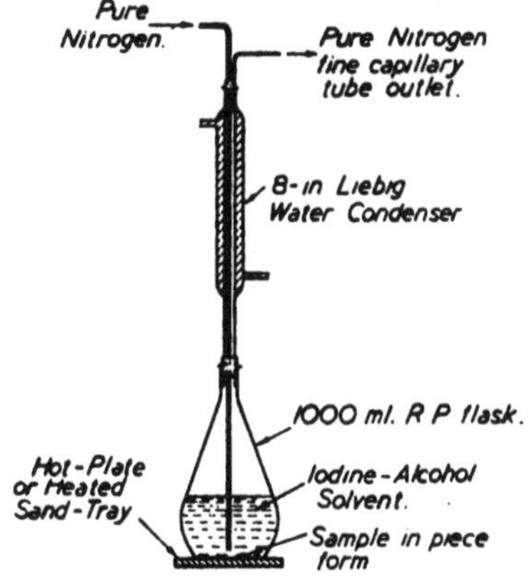

FIG. 6.—Original Apparatus for the Modified Alcoholic Iodine Method.

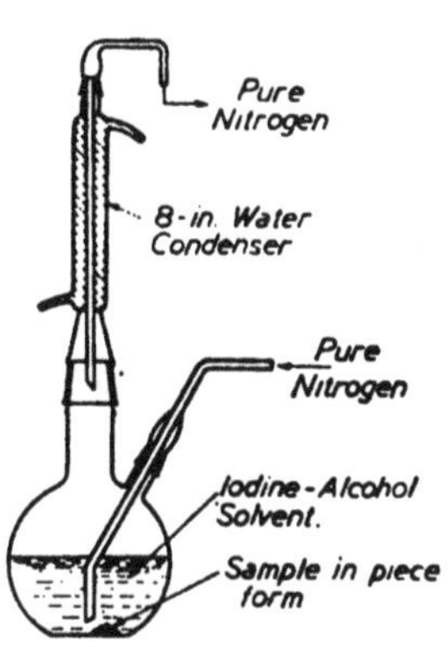

FIG. 7.—Later Apparatus Incorporating Ground Joints.

with drops of alcohol from a dropping bottle while the filtration is proceeding overcomes this possibility.

It will be seen that the above procedure is very simple and fulfils the essential conditions of the alcoholic iodine method. Oxygen and water vapour are excluded by the continued passage of purified nitrogen; agitation and the fixed temperature of the solution are obtained by the bubbling of the nitrogen through the gently boiling solution. Also, the passing of nitrogen through the heated solvent is equivalent essentially to the de-oxygenation as carried out at the National Physical Laboratory. In addition, the residue and iodine solvent do not come into contact with any greased joints and hence contamination is avoided. The modified method also permits of a number of samples being dissolved in a battery, since four such apparatus have been connected in series and tests conducted simultaneously.

An apparent objection lies in the fact that the solution after reaction is filtered in the open atmosphere. However, the results indicate that there is little or no danger in exposing the iron-iodide/iodine solution, after reaction, to the atmosphere, provided that the solution is cold and filtration is carried out rapidly, *i.e.*, in about 10–15 min. The black residues on the filter show no evidence of basic iron contamination and analysis for iron oxide has revealed the presence of normal quantities. The chief source of extraneous iron oxide in the ignited residue is imperfect washing of the filter with alcohol, which penetrates the core of the filter with difficulty. Hence, whilst filtration of the iodide solution must be speedily

TABLE XXIII.—*Determination of Non-Metallic Inclusions by the Modified Alcoholic Iodine Method. Weld-Metal Deposits.*

	Weld Metal *A*1802.		Weld Metal *A*1803.	
	Test 1.	Test 2.	Test 1.	Test 2.
Total ignited residue. %	0·247	0·250	0·284	0·289
Silica (SiO_2). %	$0{\cdot}090_5$	0·090	$0{\cdot}106_5$	0·105
Iron oxide (FeO). %	0·007	$0{\cdot}006_5$	$0{\cdot}023_5$	$0{\cdot}022_5$
Titania (TiO_2). %	0·020	$0{\cdot}020_5$	Nil	Nil
Alumina (Al_2O_3). %	0·005	0·005	0·003	0·003
Manganese oxide (MnO). %	$0{\cdot}108_5$	$0{\cdot}110_5$	$0{\cdot}138_5$	$0{\cdot}139_5$
Chromium oxide (Cr_2O_3). %	$0{\cdot}000_5$	$0{\cdot}000_5$	$0{\cdot}001_5$	0·001
Total.* %	$0{\cdot}231_5$	0·233	0·273	0·271
P_2O_5 not determined.				
Total oxygen. %.				
(*a*) Calculated from the sum of SiO_2, FeO, Al_2O_3 and MnO	0·076	$0{\cdot}076_5$	0·094	$0{\cdot}093_5$
(*b*) By the vacuum fusion method	$0{\cdot}073_5$		$0{\cdot}104_5$	

* The oxides FeO and MnO are possibly changed during ignition to Fe_2O_3 and Mn_3O_4.

carried out, the subsequent washing must under no circumstances be hurried.

The preliminary tests with the above apparatus were very satisfactory, but for the early observation that the use of rubber bungs was a weakness in the method. Accordingly, the apparatus illustrated in Fig. 7, in which the rubber bungs are replaced by standard ground joints, so arranged that the greased surfaces are outside of the reaction flask, was designed. This apparatus is used identically as described previously and also enables several determinations to be conducted simultaneously.

The above method has been adopted for the determination of oxide residues in killed steels in the Central Research Department of The United Steel Companies, Ltd., and Tables XXIII. to XXV.

TABLE XXIV.—*Determination of Non-Metallic Inclusions by the Modified Alcoholic Iodine Method. Boiler Plate Steels.*

Method of manufacture : Basic open-hearth.

	Cast 52363.	Cast 52308.
Ladle additions :		
Carbon	80 lb.	180 lb.
Ferro-manganese . . .	750 lb.	1000 lb.
Silico-manganese . . .	250 lb.	100 lb.
Mould addition : Alsimin . .	6 lb.	2½ lb.
Pit Analysis :		
Carbon. %	0·14	0·18
Manganese. %	0·49	0·70
Silicon. %	0·009	...
Sulphur. %	0·042	0·045
Phosphorus. %	0·037	0·051

	Cast 52363.		Cast 52308.	
	Plate No. 27, Top of Ingot.	Plate No. 29, Bottom of Ingot.	Plate No. 31, Top of Ingot.	Plate No. 35, Bottom of Ingot.
Total ignited residue. % . .	{0·065 0·066}	0·079	{0·058 0·057}	0·080 0·082
Iron oxide (FeO). % . . .	$0·006_5$	$0·006_5$	$0·007_5$	$0·006_5$
Manganese oxide (MnO). % .	$0·014_5$	0·027	0·022	0·027
Silica (SiO_2). % . . .	0·015	0·014	0·011	0·013
Alumina (Al_2O_3). % . .	0·021	0·021	$0·006_5$	$0·019_5$
Titania (TiO_2). % . . .	0·001	$0·000_5$	$0·000_5$	$0·001_5$
Chromium oxide (Cr_2O_3). % .	0·002	$0·001_5$	0·002	0·002
Phosphoric oxide (P_2O_5). % .	$0·003_5$	0·003	0·005	0·007
Total.* %	$0·063_5$	$0·073_5$	$0·054_5$	$0·076_5$
Total oxygen. %.				
(*a*) Calculated from the sum of SiO_2, FeO, Al_2O_3 and MnO	0·023	0·025	0·015	$0·023_5$
(*b*) By the fractional method † .	$0·019_5$	0·020	0·011	0·021
(*c*) By the direct total oxygen method	0·018	$0·019_5$	$0·010_5$	$0·019_5$

* The oxides FeO and MnO are possibly changed during ignition to Fe_2O_3 and Mn_3O_4.

† Results by the fractional method are given in Section II., Part B of this Report.

include a number of the results obtained. In many cases oxygen values by the vacuum fusion technique are also recorded, and comparison with the standard procedure is shown in the results for transformer steels, of which sample *K*446 gives good agreement, indicating the absence of hydrolysis during filtration, since the modified

TABLE XXV.—*Determination of Non-Metallic Inclusions in Transformer Steels by the Alcoholic Iodine Method.*

Analysis of materials :	Cast K446.	"Stalloy."	American-manufacture Sheet.	Belgian-manufacture Sheet.
Carbon. %	0·05	0·23	0·037	0·035
Manganese. %	0·06	$0·05_5$	0·032	0·06
Silicon. %	4·01	4·18	4·08	4·30
Sulphur. %	0·026	0·024	0·01	0·003
Phosphorus. %	0·012	0·023	0·01	0·005
Tin. %	0·030	...	0·028	...

Sample.	Total Residue. %.	SiO_2. %.	FeO. %.	Method.
K446	0·090 0·098	0·042 0·045	0·013 0·010	Standard procedure.
American	0·066	0·038	0·004	Modified procedure.
Belgian	0·025	0·007	0·003	Modified procedure.
"Stalloy"	0·051	0·004	0·002 (TiO_2 $0·036_5$)	Modified procedure.

method gives slightly the lower iron oxide value. It is necessary, however, to carry out the filtration cold and speedily, *i.e.*, in about 10–15 min.

SECTION VI.—THE AQUEOUS IODINE METHOD.

PART A.—*Recent Developments in the Determination of Oxide Inclusions in Pig Iron by the Modified Aqueous Iodine Method.*[1]

BY E. TAYLOR-AUSTIN, F.I.C. (BRITISH CAST IRON RESEARCH ASSOCIATION, BIRMINGHAM).

SYNOPSIS.

Since the publication of the original modified aqueous iodine method, the wider application of the procedure has led to the introduction of certain modifications. A study of the variations in hydrogen-ion concentration during the decomposition of pig iron samples has provided an explanation for the apparently anomalous behaviour of synthetic manganous oxide towards the iodine solvent and has also demonstrated the futility of carrying out experiments on synthetically prepared oxides, sulphides, &c., in the absence of iron.

A study has been made of the behaviour of iron carbide during the process, and it has been shown that no interference is set up by this compound. The behaviour of titanium carbide and manganese sulphide has also been examined, and there is evidence that the small traces of phosphorus hitherto found in the final non-

[1] Received February 22, 1941.

metallic residues is in some way connected with titanium and not, as was at one time thought, with calcium and/or magnesium.

The revised procedure, which has now been successfully applied to a series of over forty pig irons, is described in detail.

In a paper entitled "The Determination of Non-Metallic Inclusions in Pig and Cast Iron by a Modified Aqueous Iodine Method,"[1] the author described a procedure upon which it was hoped to build a reliable method for the determination of oxide inclusions in pig and cast iron. A number of problems were, however, outstanding at the time of publication of the original investigation; these have now been explained and certain modifications have been made to increase the range of application of the method to cover all types of pig iron other than alloy varieties. The following procedure has been successfully applied to a series of more than forty typical pig irons and a selection of the results obtained is compared with those by other methods in Table XXVI.

Table XXVI.—*Inclusions in Pig Iron Determined by Various Methods.*

No.	Type of Iron.	Residue Analysis (as percentage of Metal) by Modified Aqueous Iodine Method.					Total Oxygen. %.		
		Total Oxides (SiO_2+ FeO+ Al_2O_3+ MnO).	SiO_2.	FeO.	MnO.	Al_2O_3.	Calc. from Residue Analysis.	By Aluminium Reduction.	By Vacuum Fusion.
13	Armco iron	0·350	0·013	0·279	0·018	0·040	0·086	0·085	...
14	Swedish iron	0·054	0·009	0·037	Nil	0·008	0·017	0·018	...
16	Pig irons	0·028	0·008	0·010	0·002	0·008	0·015	...	0·019
19		0·028	0·008	0·014	0·001	0·005	0·010	0·011	0·011
31		0·026	0·006	0·014	Nil	0·006	0·009	0·006	0·005
21		0·029	0·005	0·015	0·001	0·008	0·010	0·007	0·008
107		0·017	0·005	0·009	Nil	0·003	0·0063	0·0050	...
101		0·019	0·003	0·014	Nil	0·002	0·0053	0·0047	...
18		0·039	0·020	0·014	0·001	0·004	0·016	0·015	...
73		0·043	0·009	0·031	Nil	0·003	0·013	0·018	...
25		0·025	0·008	0·012	0·001	0·004	0·0091	0·0084	...
86		0·023	0·007	0·012	0·001	0·003	0·0084	0·010	...
111		0·046	0·020	0·023	0·001	0·002	0·021	0·019	...
84		0·021	0·005	0·013	Nil	0·003	0·0074	0·0077	...
85		0·038	0·011	0·023	Nil	0·004	0·013	0·013	...
55		0·040	0·010	0·024	0·001	0·006	0·013	0·011	...
23		0·046	0·012	0·025	0·001	0·008	0·016	0·017	...
103		0·032	0·011	0·017	0·002	0·002	0·011	0·012	...
38		0·031	0·010	0·017	Nil	0·004	0·011	0·013	...
79		0·021	0·005	0·010	Nil	0·006	0·0077	0·0071	...

Revised Aqueous Iodine Method.

Reagents.

Iodine/Potassium-Iodide Solution.—Dissolve 30 g. of pure resublimed iodine and 30 g. of potassium iodide in a minimum quantity

[1] Eighth Report on the Heterogeneity of Steel Ingots, pp. 121–138, *Iron and Steel Institute*, 1939, *Special Report No.* 25.

of water. When solution is complete, filter and dilute to 120 ml. This solution is sufficient to treat 5 g. of iron.

Sodium Carbonate-Citrate Solution.—Dissolve 6 g. of anhydrous sodium carbonate and 12 g. of sodium citrate in 120 ml. of water.

Ammonium Citrate Solution.—Dissolve 10 g. of ammonium citrate in 100 ml. of water.

Peroxide-Citrate Solution.—To 170 ml. of 10% ammonium citrate solution add 30 ml. of 30-volume hydrogen peroxide (stabilised with sulphuric acid).

Procedure.

Transfer 120 ml. of aqueous iodine/potassium-iodide to a 300-ml. "carbon dioxide" flask fitted with a mechanical stirrer and nitrogen tubes. Immerse the flask in cold running water and stir for 30 min. with nitrogen passing through the solution. At the end of this period, rapidly introduce 5 g. of sample in the form of drillings. Continue to pass nitrogen and stir for 3 hr., keeping the flask immersed in running water throughout. Filter the solution through a Whatman No. 42 filter paper (4·25-cm.) contained in a small Büchner funnel, applying gentle suction, and wash free from iodine with 5% potassium iodide solution and cold water. Transfer the paper and precipitate back to the original flask, add 120 ml. of sodium carbonate-citrate solution, place the flask in a water-bath at 80° C. and stir for 30 min. Filter the solution as before and wash well with hot 2% sodium citrate and hot water. Place the paper and precipitate in the original flask, add 200 ml. of peroxide-citrate solution and again stir for 2 hr. at 80° C. Filter on a Büchner funnel as previously, washing with hot 2% ammonium citrate and water. Next stir the residue as before with 200 ml. of ammonium citrate solution for 2 hr. Filter, wash with hot dilute ammonium citrate solution and water and again stir with citrate reagent for 30 min. Filter the solution and observe whether or not the filtrate is free from yellow coloration. If this is the case, wash with hot 2% ammonium citrate and hot water and finally stir the residue with 120 ml. of carbonate-citrate solution for 30 min. Filter, wash as before, and transfer the pad and precipitate to a 150-ml. beaker. If the filtrate is yellow in colour, the residue should be stirred in it for a further 30 min. (making one hour in all). Repeat the filtration and washing and treat for a further 30 min. with citrate reagent; again filter, and examine the filtrate. This series of operations should be repeated until the final filtrate is free from yellow coloration. The residue is finally treated with carbonate-citrate solution as previously described.[1]

To the residue in the beaker add 10 ml. of concentrated sulphuric acid and 10 ml. of concentrated nitric acid and heat gently until

[1] *Note.*—For all the pig-iron samples so far examined, a period of 2 hr. with ammonium citrate, after the peroxide-citrate treatment, has proved sufficient to yield a colourless filtrate on further testing.

sulphur trioxide fumes are evolved. Allow to cool and add 2 or 3 ml. of fuming nitric acid from a spotting bottle and again evaporate to fumes. Repeat these additions of fuming nitric acid until all graphite and other carbonaceous matter have been removed. Cool, dilute to 50 ml., boil, and filter off silica on a Whatman No. 42 filter paper; wash the precipitate with hot 5% sulphuric acid and cold water. Ignite the silica, &c., in a platinum crucible and treat it with a few drops of 20% sulphuric acid; heat to expel excess sulphuric acid and ignite to constant weight. This procedure ensures that any elements forming stable sulphates are weighed in this form after ignition. Remove silica by volatilisation with hydrofluoric acid in the presence of sulphuric acid in the usual manner.

Fuse any residue remaining after the removal of silica with 0·5 g. of potassium bisulphate and extract the melt in the main solution.

The remainder of the analysis may be outlined briefly as follows:

(1) Iron, titanium, vanadium and zirconium (if present) are precipitated with cupferron. The residue is ignited, fused with bisulphate and the constituents determined colorimetrically.

(2) The excess cupferron in the filtrate from the above separation is destroyed with sulphuric and nitric acids and the resulting solution used for the determination of manganese, aluminium and phosphorus. Manganese is determined by the periodate, aluminium by the aurintricarboxylate, and phosphorus by the ammonium-molybdate/stannous-chloride method.

Notes on the Revised Aqueous Iodine Method.

(*a*) Throughout the development of the above procedure, a careful check was kept on the impurities introduced from chemicals and glass-ware. Recent determinations of blank values have shown that, although further treatments have been incorporated, no appreciable increase has taken place. The average total blank, including ash from filter papers, is as follows:

SiO_2. %.	Fe_2O_3. %.	Al_2O_3. %.	MnO. %.
0·00040	0·00018	0·00008	Nil

(*b*) All analytical determinations, with the exception of that of silica, are now made colorimetrically, a Lovibond tintometer being used for colour measurement.

(*c*) Attempts were made to apply colorimetric methods to the determination of silica. Two methods were examined, *viz.*, one based upon the yellow silico-molybdate colour, and the other on the blue colour obtained by reducing the latter compound. Neither procedure was found to be sufficiently sensitive for the accurate determination of the small amounts of silica present in non-metallic residues.

Development of the Revised Procedure.

The original modified aqueous iodine procedure [1] yielded results for total oxygen which were, generally speaking, in good agreement with those obtained by the vacuum fusion method on the same materials. There were, however, as indicated at the time, problems which called for further investigation. Some preliminary work on the behaviour of manganous oxide was also reported previously, and it was stated that reliable results for this oxide could be obtained by the procedure if the washing of the residue with ammonium citrate solution were omitted, since this latter reagent appeared to dissolve manganous oxide. Subsequent work showed this statement to be erroneous, since, whilst synthetically prepared manganous oxide is but slightly soluble in aqueous iodine/potassium-iodide, it is readily dissolved by this solution in the presence of iron.

The Behaviour of Manganous Oxide.—A recent study of the variations in hydrogen-ion concentration during the decomposition of iron by aqueous iodine/potassium-iodide revealed the fact that on adding the iron to the iodine solution, which has an initial pH value of 6·1, the temperature rises very rapidly, in spite of water-cooling, and the pH value falls equally rapidly to about 1·0; the actual fall varies slightly with the particular type of iron concerned. Thus, the acidity of the solution approaches that of $N/10$ acid about 15 min. after adding the iron sample. These experiments have provided an explanation for the apparent differences in the behaviour of synthetic manganous oxide to which reference has already been made.[1] It has been stated by Klinger and Koch [2] that manganous oxide is soluble in solutions having a pH value below 5·0, so that it is to be expected that it will readily dissolve in the aqueous iodine solution if iron is present. This was confirmed by experiments using synthetically prepared manganous oxide. It was also shown that the oxide is completely soluble in hot ammonium citrate solution. Attempts to bring about the decomposition of the iron in iodine solutions of controlled pH value (5·0–6·0) have so far failed owing to hydrolysis of the iron; if citrates or tartrates are added to avoid the latter, manganous oxide is dissolved. The conclusion has therefore been reached that the aqueous iodine method does not yield results for manganous oxide, existing as such, but from a consideration of the chemical properties of manganese silicate, the author is of the opinion that, if it exists combined with silica in this form, it remains in the final non-metallic residue. Thus, the small amounts of manganese oxide found in pig-iron residues appear to have existed originally as silicate.

[1] Eighth Report on the Heterogeneity of Steel Ingots, pp. 121–138, *Iron and Steel Institute*, 1939, *Special Report No.* 25.

[2] Klingor and Koch, *Technische Mitteilungen Krupp, Forschungsberichte*, 1938, No. 3, May, pp. 49–65.

The study of variations in *p*H values and in temperature suggests that the decomposition of the iron by the iodine solution is complete in a comparatively short time (15–20 min.) and that the additional time required to obtain a residue suitable for further treatment is occupied by the conversion of iron phosphide to phosphate.

The Behaviour of Manganese Sulphide.—Experiments carried out with synthetically prepared samples of manganese sulphide, along similar lines to those used in examining the behaviour of manganous oxide, showed that the sulphide is not completely decomposed by aqueous iodine in the presence of iron; approximately 90% passes into solution. It is, however, completely soluble in hot ammonium citrate solution, so that it is concluded that manganese sulphide causes no interference in the modified process.

It appears probable, in the light of this recent work, that the manganese obtained in the residues referred to above when omitting the citrate treatment was due, not to oxide, but to small amounts of undecomposed sulphide.

The Behaviour of Iron Carbide.—A few years ago, Bihet and Willems [1] extracted iron carbide from a sample of Swedish white iron and examined its solubility in a variety of solvents; they reported that the carbide is most soluble in cuprous ammonium chloride, less soluble in alcoholic iodine, and least soluble in aqueous iodine solutions.

More recently, Bramley, Maddocks and Tateson [2] suggested that the iron carbide present in steel specimens containing more than 0·6% of carbon causes serious interference when the alcoholic iodine method is employed to separate the non-metallic residue from such materials.

It appeared probable, therefore, that the aqueous iodine method would be subject to such interference from iron carbide, especially in view of the work of Bihet and Willems. The preliminary work on samples of grey iron gave no indication that such interference was occurring, but the problem was investigated by an examination of synthetically prepared white irons having a combined carbon content of 3·2% and a silicon content of 2%. Silicon was added in order to reduce the amount of ferrous oxide in the material.

When this material was examined by the standard procedure, it was found that the initial residues contained unusually large amounts of iron and, moreover, consistent results could not be obtained. After the application of the ammonium citrate treatment, however, the amounts of iron found were consistent and of a much lower order than previously. It was therefore concluded

[1] Bihet and Willems, *Archiv für das Eisenhüttenwesen*, 1937, vol. 11, Sept., pp. 125–130.

[2] Bramley, Maddocks and Tateson, Eighth Report on the Heterogeneity of Steel Ingots, pp. 11–25, *Iron and Steel Institute*, 1939, *Special Report No.* 25.

that iron carbide causes no interference in the modified aqueous iodine process.

From the results obtained on these synthetic white irons it would appear that iron carbide is only partially decomposed during the initial attack with aqueous iodine but that the small amount remaining is completely removed by treatment with hot ammonium citrate solution.

The Behaviour of Iron Oxide.—The general results obtained by the original procedure suggested that, in spite of the acidity of the final iodine solution, oxides other than that of manganese and silicates are unattacked. This is to be expected from the known chemical properties of such oxides as those of silicon and aluminium, although the behaviour of iron oxide is less certain.

In the original work the procedure developed was applied to a very limited number of irons, and when in subsequent tests this range was greatly extended it was found that, in certain cases, abnormally high figures were recorded for iron oxide; in these cases it was observed that a second treatment with ammonium citrate yielded a filtrate which was still yellow in colour. It was subsequently found that if citrate treatments were applied until a colourless filtrate was obtained, consistent figures for iron oxide, of a much lower order than previously, were obtained. This modification was therefore incorporated in the new standard procedure. The application of this latter method to a series of over forty irons has given evidence that ferrous oxide is not appreciably attacked during the decomposition of the material by aqueous iodine/potassium-iodide. The total oxygen results calculated from the aqueous iodine figures are in good agreement with those obtained directly by other methods, particularly the modified aluminium reduction procedure, to which reference is made elsewhere (Section III., Part C.).

The Determination of Silica.—The application of the original procedure to a wider variety of materials revealed the fact that in certain cases the recommended treatment with sodium carbonate-citrate solution failed to remove the whole of the silicic acid present. Evidence of this was provided by the fact that the oxygen equivalent of the recorded silica percentage was in excess of the *total* oxygen given by the modified aluminium reduction process. It was found necessary to repeat the carbonate-citrate treatment after the application of the other treatments. When this was done no further trouble was experienced, and this procedure was also adopted as standard. It was also found that the results for silica were improved by substituting wet oxidation for ignition as a means for removing carbon, &c.

The Behaviour of Titanium.—From the commencement of the work on the aqueous iodine method it was observed that the whole of the titanium present in the iron remains in the final non-metallic residue.

Recent results have indicated that the amounts of oxide inclusions in pig iron are of a much lower order than was at first assumed. Further, none of the treatments so far devised had had the slightest effect upon the titanium content of residues, and analysis was therefore being carried out in the presence of comparatively large amounts of this element. It occurred to the author that this titanium might be removed from unignited residues by treatment with hydrogen peroxide.

Preliminary tests showed that titanium carbide was soluble in ammonium citrate solutions containing hydrogen peroxide, but experiments with neutralised hydrogen peroxide alone failed owing to hydrolysis of the titanium.

Accordingly hydrogen peroxide was added to the ammonium citrate solution used for treatments, and it was found that in most cases this removed the whole of the titanium from the residue without appreciable influence on the oxides present. In a few cases, *viz.*, those of irons containing 0·3% or more of titanium, a trace (0·01%) remained in the final residue, but was insufficient to complicate the analysis. Further, this application of peroxide-citrate solution instead of citrate alone halved the time required for the decomposition of iron carbide. This change was therefore made in the standard procedure.

The Behaviour of Phosphorus.—In the original work it was shown that, whilst iron phosphide is relatively insoluble in ammonium citrate solution, iron phosphate is soluble under the conditions laid down in the standard procedure. It was also observed that in spite of this solubility of the phosphate in the reagent, it was impossible to remove every trace of phosphorus from the residues, even by very prolonged treatment with hot citrate solution; the amount remaining was equivalent to about 0·01% of P_2O_5.

In the more recent work it was found that when treatments were carried out with citrate solutions containing hydrogen peroxide, not only was titanium removed from the residue but the small traces of phosphorus hitherto found also disappeared. In every case where the titanium content of the residue was nil, phosphorus could not be detected by the sensitive colorimetric method employed.

It would therefore appear that this final trace of phosphorus is in some way connected with titanium, and is not combined with calcium and magnesium as was originally suggested. In this connection it is noteworthy that the results of all tests so far made for calcium and magnesium have been negative in character in spite of improved analytical technique.

Conclusions.

From the work carried out since the publication of the original aqueous iodine method the following conclusions may be drawn:

(1) The solution of manganous oxide, as such, during the

decomposition of samples by aqueous iodine/potassium-iodide cannot be prevented, and hence results for this oxide cannot be obtained at present.

(2) Iron carbide, iron phosphide, manganese sulphide and titanium carbide cause no interference in the new procedure.

(3) The modifications introduced surmount the difficulties which had arisen during the more general application of the method originally suggested.

(4) The present procedure has been successfully applied to a series of forty-five different types of pig iron without further complications; it is therefore believed that the process may be employed for the determination of oxide inclusions in all types of pig iron, excluding alloy irons, and that results for SiO_2, FeO, MnO (existing as manganese silicate) and Al_2O_3 are reliable.

Part B.—*The Present Position of the Determination of Oxide Inclusions in Pig Iron and Cast Iron.*[1]

By J. G. PEARCE, M.Sc., F.Inst.P., M.I.Mech.E., M.I.E.E. (British Cast Iron Research Association, Birmingham).

Synopsis.

The progress of the estimation of oxides in pig iron and cast iron by chemical methods is indicated, with reasons for the adoption of the aqueous iodine and aluminium reduction methods. The range of oxides in forty-five representative British pig irons is given. Reference is made to the disparities between chemical evidence and microscopic evidence of oxide inclusions, and it is suggested that the contradictions are more apparent than real.

Inclusions in cast iron and pig iron are of as much interest metallurgically as are those in steels, but not entirely for the same reasons. Present consideration is confined to solid and to inherent, as distinct from foreign, particles, although a sharp distinction between them is difficult to draw.

Knowledge of the amount, nature, size and mode of distribution of these particles may throw light on many still obscure points, foremost among which is whether small particles of silica or alumina or complexes into which they enter are in any way associated with the precipitation of graphite, and some investigators have not hesitated to postulate the nucleating effect of sub-microscopic particles of iron silicate. Furthermore, it is necessary to find out whether they offer any explanation of other points, such as the so-called hereditary properties of pig iron (persistence of properties through subsequent remelting), and the difference between hot- and

[1] Received February 10, 1941.

cold-blast or between coke and charcoal pig irons. Do they throw any light on why pig irons of different origins but similar chemical analyses behave differently in practice during the manufacture of cast or wrought steel or iron? Do they change as between ordinary and desulphurised pig irons, between pig irons and their remelted forms, refined pig irons and cast steels and irons, or between coarse-graphite and supercooled-graphite irons? Does superheating or agitating the melt have any influence on them? Whether such particles ultimately prove to have any such effects or not, it is necessary to find out, so that unprofitable speculation may be avoided.

When the British Cast Iron Research Association began to study this subject in conjunction with the Oxygen Sub-Committee, the alcoholic iodine method was naturally adopted, as it permits the quantity of the individual oxides to be determined, although the vacuum fusion method is now acknowledged as unrivalled for the determination of total oxygen. No special difficulty was anticipated, but after exhaustive trial the alcoholic iodine method was abandoned for pig and cast irons on account of interference set up by phosphorus (*i.e.*, the phosphide compound is only partially attacked by the solvent and hence some is left in the residue), as described in the Second Report.[1] The chlorine method was then tried and also abandoned for pig and cast irons, this time on account of interference set up by manganese.[2] The older acid methods were re-examined, but work with both strong and dilute acids indicated that these procedures have a very limited application to pig and cast iron. With dilute acids, solution of the metal occupies considerable time, and such large amounts of silicic acid separate, that filtration of the final residue is very difficult; with strong acids, the silica results are considerably lower than those given by aqueous iodine, and it would therefore appear that these solvents decompose some of the silicates present in the residue. Results for alumina are in good agreement with those by aqueous iodine. It was therefore concluded that, as far as pig and cast iron are concerned, only alumina figures are obtainable from acid methods, which therefore have only a limited application to these materials.

A method based on the use of aqueous iodine as a solvent was then evolved and adopted, and was described in the Second Report[3] as the foundation of a reliable method, although various problems remained to be solved. In the meantime, the publication in the same Report of the aluminium reduction method by Gray and Sanders[4] led to the trial of this in order to provide a total-oxygen

[1] Taylor-Austin, Second Report of the Oxygen Sub-Committee, *Iron and Steel Institute*, 1939, *Special Report No.* 25, p. 159.

[2] Colbeck, Craven and Murray, Second Report of the Oxygen Sub-Committee, *Iron and Steel Institute*, 1939, *Special Report No.* 25, p. 109.

[3] Taylor-Austin, Second Report of the Oxygen Sub-Committee, *Iron and Steel Institute*, 1939, *Special Report No.* 25, p. 121.

[4] Gray and Sanders, Second Report of the Oxygen Sub-Committee, *Iron and Steel Institute*, 1939, *Special Report No.* 25, p. 103.

check on the aqueous iodine by a method less expensive than vacuum fusion. Very little adaptation was needed, and the aqueous iodine and aluminium reduction methods as now used for pig and cast iron are described in the present Report,[1] with some results from both methods, and from vacuum fusion where these figures are available. The accord is very satisfactory, and it is felt that the two methods are now in reasonably final form.

The acquisition of the necessary technique enabled the methods to be applied to the examination of a series of representative British pig irons, of which thirty, supplied by the Blast-Furnace Committee of the British Iron and Steel Federation, were on the 1940 programme. To date, about 45 have been examined by both methods, although the analysis of the results is not yet complete. Furthermore, very little has been done on cast irons. The preliminary figures indicate the following ranges for the various oxides in the above pig irons :

Silica (SiO_2)	0·003–0·020%
Ferrous oxide (FeO) . . .	0·009–0·045%
Manganous oxide (MnO) . . .	Nil–0·002%
Alumina (Al_2O_3)	Nil–0·008%
Total oxides	0·017–0·060%
Total oxygen	0·005–0·020%

The oxides as a whole are similar to those found in steel, but not necessarily of the same order. It must not be assumed that they are present as simple oxides. They are expressed in this form because the residue representing inclusions separated from the metal has to be ignited or subjected to chemical oxidation to get rid of graphite. They are almost certainly present in a more complex form, and thermodynamic considerations suggest that inclusions to be expected in pig and cast iron, apart from sulphides, are alumina, silicates of iron and manganese and, under certain conditions, free FeO, MnO or SiO_2. The calculation of the FeO and MnO content of the residue to silicates, allowing for the FeO in solution, checks well with the determined values of SiO_2 in the residue, but this is based on equilibrium conditions.

As the work on chemical methods progressed two points became evident. First, the amount of inclusions present proved to be very small, compared, for example, with the sulphides present. (Sulphides are decomposed in the modified aqueous iodine procedure and hence do not appear in the final residue. Sulphur is determined by direct chemical analysis, although this does not show any division between iron and manganese. Sulphides as such can be separated from iron by electrolytic methods, although little work has so far been done on such methods. But it may be noted that just as efforts can be made microscopically to see what is determined by chemical methods, so, chemically, efforts can be made to separate what is visible microscopically and quantitative figures obtained.)

[1] Taylor-Austin, this Report, Section III., Part C, and Section VI., Part A.

Indeed, the oxide inclusions in pig iron appear to be less than are found in many steels, and form a slender basis on which to erect a superstructure of theory. Secondly, there appeared to be a lack of relation between inclusions chemically determined and those visible microscopically.

The laboratories of the British Cast Iron Research Association have examined many thousands of samples of pig and cast iron in all their forms, and the author initiated a more systematic examination from the point of view of inclusions, primarily to see whether any light would be thrown on the way in which the chemically determined oxides were actually present. The results of this examination are now available,[1] and, even acknowledging that the microscope can give no evidence of particles in solution or of submicroscopic size, the results are startling.

Apart from sulphides, the inclusions visible in pig and cast iron are due to titanium, and these are so common that their presence must be regarded as normal. Incidentally, experience does not suggest that inclusions microscopically visible in pig iron differ materially in quantity or type from those in cast iron. Most striking of all, however, is the complete absence of any microscopic evidence of the presence in ordinary commercial materials of oxides or silicates of iron, manganese or aluminium.

Various suggestions can be put forward to assist in resolving the apparent contradiction between the chemical and microscopic methods of approach, but the author does not propose at present to go beyond recording his conviction that the results obtained by both methods, as far as they go, are correct, and that the resolution of the contradiction is facilitated by a clear statement and is bound to lead to further advance. The work is being continued from both aspects.

That the contradiction is more apparent than real is shown by the fact that, excluding sulphides for the reason mentioned above, the principal visible inclusions are those of titanium. Omitting the latest treatment in the aqueous iodine method, these can be frequently found in the chemical residue as TiO_2, but as there is ground for belief that they are actually present in the metal as carbide or cyano-nitride, they are not considered among the oxide inclusions mentioned above. The absence of the oxide as such is supported by chemical evidence. Chromium occasionally appears as Cr_2O_3 in the residue, but the inclusions in alloy pig and cast irons have not yet been dealt with. So far as FeO, SiO_2, MnO and Al_2O_3 are concerned, the form in which they exist may well prove too small in quantity for microscopic identification.

The problem of nomenclature has been dealt with (Morrogh, *loc. cit.*). The use of the terms "metallic" and "non-metallic" can only be defended by special interpretations of them. Certain

[1] Morrogh, "The Metallography of Inclusions in Cast Irons and Pig Irons," *Journal of the Iron and Steel Institute*, 1941, No. I., p. 207 P (this volume).

constituents of the metal other than ferrite, pearlite and cementite, particularly graphite and phosphide, are non-metallic, while certain carbides, *e.g.*, titanium carbide, may partake of the nature of inclusions. It is also desirable to distinguish between inherent inclusions and foreign inclusions due to the pick-up of material from furnace linings, slags or mould material, heterogeneous in their distribution and influenced by operating technique. As there are likely to be similarities between them, the foreign inclusions of one melt may become the inherent inclusions of subsequent melts.

As far as pig-iron samples are concerned, an adequate number are available for further work, but samples would be appreciated of both pig irons and cast or wrought material made therefrom where unexpected behaviour has been encountered in practice, apparently due to the raw materials used, and not explained by chemical analysis or other factors.

SECTION VII.—The Hydrogen Reduction Method.[1]

Submitted by W. W. STEVENSON, A.I.C. (Stocksbridge, near Sheffield).

Experimental work has been proceeding at Stocksbridge on the suitability of the hydrogen reduction method for determining that part of the oxygen content of steel which corresponds to the low-temperature fraction of the fractional vacuum fusion method.

A number of refractory materials has been employed for the furnace tube—translucent silica, porcelain, fused alumina, &c.—without obtaining a sufficiently low blank value to render the procedure scientifically attractive. For example, at 950° C. translucent silica gives a blank value of the order of 0·0004 g. of water vapour per hr. in contact with hydrogen, passed at 2 litres per hr. at atmospheric pressure. This value is far too high to permit of placing any reliance on oxygen results obtained on samples of steel by this method. It has recently been stated, in committee, that transparent silica is far more stable towards reduction by hydrogen than any of the refractory tubes on which experiments have yet been made in this connection.

It is therefore planned to continue the work on transparent silica, the results of which will be communicated later, together with the negative results already obtained.

[1] A communication from The United Steel Companies, Ltd., Central Research Department, Stocksbridge, near Sheffield, received February 10, 1941.

SECTION VIII.—The Analysis of Non-Metallic Residues Extracted by the Alcoholic Iodine Method.[1]

By G. E. SPEIGHT, B.Sc., A.I.C. (Stocksbridge, near Sheffield).

Synopsis.

The improvements in the analysis of non-metallic residues effected since the publication of the Second Report of the Oxygen Sub-Committee are given, and the progress towards complete colorimetric determination is indicated. Details of a colorimetric method for small quantities of alumina and a brief outline of the present complete scheme of analysis of non-metallic residues are given.

In the Eighth Report on the Heterogeneity of Steel Ingots comprehensive details of the analysis of extracted non-metallic residues are given. These methods include alternative procedures for both colorimetric and gravimetric determinations of the major part of the constituents in the ignited residue. Since the publication of that Report, progress in analysis has been towards the almost complete adoption of colorimetric methods, using for the comparison one of several well-known instruments, amongst which are the Pulfrich photometer used by the Central Research Department (United Steel Companies, Ltd.) and Sheffield University, the Lovibond tintometer used by the British Cast Iron Research Association, and the Bausch and Lomb colorimeter used by the National Physical Laboratory.

The colorimetric method for the determination of small concentrations of elements in solution is one of the most accurate and easily conducted processes in analytical chemistry, and satisfactory colour measurements are available for all the constituents with the exception of silica. Attempts have been made to apply colorimetric measurement in this instance; two well-known methods, one based on the yellow colour of silico-molybdate and the other on the blue colour of the reduced compound, were not found sufficiently sensitive for the small amounts of silica occurring in the extracted non-metallic residues. This oxide is still determined gravimetrically by the loss in weight after treatment with hydrofluoric and sulphuric acids of the silica precipitate obtained by decomposition of the ignited residue with strong acids or potassium or sodium bisulphate fusion.

Colorimetric Determination of Alumina.

The colorimetric determination of alumina is the least satisfactory of the tests for the remaining oxides. The method, which

[1] A communication from The United Steel Companies, Ltd., Central Research Department, Stocksbridge, near Sheffield, received February 10, 1941.

is based on the colour reaction with aurintricarboxylate, is subject to interference by other elements, and slight variations in the conditions of the solution cause rapid coagulation of the lake and apparent fading of the coloured solution. The procedure has been modified by Sheffield University to enable measurement by the Pulfrich photometer, and an account of the method is given below.

After reviewing the literature dealing with this method, certain modifications mentioned by Lampitt and Sylvester [1] were investigated and found to ensure greater uniformity of results. These modifications were :

(*a*) The addition of glycerin to increase the stability of the lake.

(*b*) The heating of the solution on a water-bath during the formation of the lake.

(*c*) The thorough cooling of the solution before the addition of the ammonia/ammonium-carbonate solution.

The effect of the addition of glycerin on the depth of the colour of the resulting lake is shown in Fig. 8, which also gives some idea of

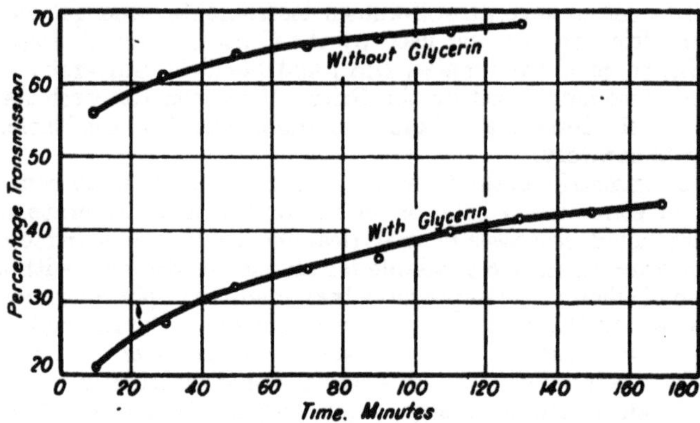

FIG. 8.—The Effect of Glycerin on the Depth of Colour of the Aluminium Lake with Ammonium Aurintricarboxylate.

the rate of fading of the aluminium lake over a period of approximately 3 hr. The concentration of aluminium in each of these tests was the same (0·1 mg. of Al_2O_3 per 100 ml.), the only difference between the two being that for the upper curve no glycerin was added to the solution, whilst for the lower one 20 c.c. of a 50% solution of glycerin were added to the solution before the addition of the " aluminon " reagent.

The method finally adopted for the determination is described

[1] *Analyst*, 1932, vol. 57, p. 418.

below, together with the strengths of the various solutions required. It should be pointed out that, whilst it is still necessary to standardise the conditions closely and to ensure that all the reagent solutions are made up as accurately as possible, it is no longer necessary to prepare a set of standards each time a determination is carried out, which economises both time and also the quantity of reagent used. It is essential, however, to test thoroughly each different supply of "aluminon" reagent, as variations in lake-formation capacity have been observed.

The Solutions Required.

The following are the solutions necessary for the method :

Hydrochloric acid	10% solution.
Ammonium acetate	25% solution.
Glycerin	50% solution.

"*Aluminon*" *reagent.*—0·1 g. of "aluminon" is dissolved in water and then made slightly ammoniacal, after which the solution is boiled to drive off excess ammonia. The solution is then made up to exactly 100 c.c.

Ammonium carbonate.—60 c.c. of ammonia (sp. gr. 0·880) are added to 940 c.c. of a saturated solution of ammonium carbonate.

The Method.

A suitable fraction (*e.g.*, one-tenth) of the aluminium-chromium solution obtained during the general separation of the oxides described later is diluted with water to 30 c.c. in a 100-c.c. standard flask. This solution is then made just alkaline with dilute ammonia and then just acid with 10% hydrochloric acid, after which an excess of 5 c.c. of this acid is added. To this solution are then added 5 c.c. of 25% ammonium acetate solution, 20 c.c. of 50% glycerin solution and 5 c.c. of the solution of "aluminon." The flask containing the solution is then immersed in boiling water for exactly 5 min., after which it is thoroughly cooled for a standard time; when quite cold, 10 c.c. of the ammonium carbonate solution are added. The solution is then diluted to 100 c.c. and allowed to stand for exactly 15 min. at a temperature of approximately 15° C.

The extinction (E) of the solution is then measured against water, using either the 5-cm. or the 2-cm. cells, according to the depth of colour developed. Filter *S*53 is used in the photometer in conjunction with the Pulfrich filament lamp.

A standard calibration curve obtained under the above standardised conditions is shown in Fig. 9, in which the extinction coefficient (K; *i.e.*, the extinction per centimetre of solution) is plotted against the concentration.

General Scheme for the Analysis of Ignited Non-Metallic Residues.

The methods of analysis which have now been adopted for ignited non-metallic residues may be summarised briefly as follows :

Silica is determined by digestion of the residue with strong acids or fusion with potassium or sodium bisulphate, followed by filtration, ignition and treatment with hydrofluoric and sulphuric acids.

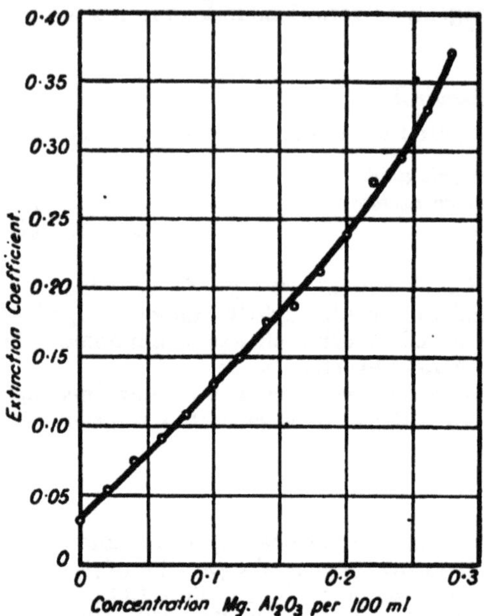

Fig. 9.—Standard Calibration Curve for the Colorimetric Determination of Aluminium in the Pulfrich Photometer.

The oxides of iron, titanium, vanadium and zirconium (if present) are precipitated by cupferron in the acid filtrate from the silica precipitate. The residue is ignited, fused with potassium bisulphate and extracted in dilute sulphuric acid. Aliquot portions of the resulting solution are taken for the colorimetric determination of :

Iron, by the thioglycollic acid procedure.
Titanium, by hydrogen peroxide.
Vanadium, by hydrogen peroxide.

A test for zirconium is carried out by precipitation of its phosphate in dilute sulphuric acid solution with ammonium phosphate.

The cupferron in the filtrate is destroyed, and aluminium and chromium are precipitated by 8-hydroxyquinoline in buffered solution. The precipitate is redissolved, and portions of the resulting solution are taken for the determination of aluminium by aurintricarboxylate, and chromium by diphenyl carbazide. Manganese and phosphorus are determined in the filtrate (after destroying the 8-hydroxyquinoline by nitric and sulphuric acids) by the colour reactions with potassium periodate and ammonium-molybdate/stannous-chloride, respectively. In cases of high phosphorus-pentoxide contents, it is sometimes preferable to use the gravimetric lead-molybdate procedure.

SECTION IX.—Reports on Materials Examined.

Part A.—*An Examination of the Oxygen Content of a Basic Bessemer Rimming Steel.*[1]

By T. SWINDEN, D.Met., and W. W. STEVENSON, A.I.C.
(Stocksbridge, near Sheffield).

(Fig. 10 = Plate XLIX.)

Synopsis.

The results of an examination of the oxygen content of a basic Bessemer rimming steel are presented. This examination comprises (a) microscopical examination, (b) chlorine extraction, (c) alcoholic iodine extraction, and (d) vacuum fusion determination, by both the total and the fractional procedure.

The results are discussed and show that loss of oxygen occurs in the residue methods as operated at present. It is suggested that the lower oxygen result is due to the presence of uncombined iron and manganese oxides, which are attacked by the halogen reagent.

A billet sample of basic Bessemer rimming steel (blow No. 8706*F*) was subjected to careful examination of its oxygen content by vacuum fusion and residue methods, by two co-operating laboratories, the Central Research Department of The United Steel Companies, Ltd., and I.C.I. (Alkali), Ltd., Northwich. The present paper submits a summary of the work on this material.

Pit Sample Analysis.

The analysis of the pit sample was as follows :

Carbon.	Manganese.	Silicon.	Sulphur.	Phosphorus.
0·04%	0·45%	Trace	0·036%	0·049%

Samples.—A section of a rolled slab, 7½ in. × 3¼ in., obtained from the middle of an ingot was machined to give two bars, dia. ½ in., one from a position midway between the central axis and the junction of rim and core, designated " core," and the other from midway between the outer edge and the junction of the rim and

[1] A communication from The United Steel Companies, Ltd., Central Research Department, Stocksbridge, near Sheffield, received February 10, 1941.

core, designated "rim." Later work necessitated further samples, which were obtained from the remaining parts of the slab section. The disposition of the samples is shown in Fig. 11 and is described later.

Programme of Tests.

(1) *A Microscopical Examination and Sulphur Print* were made on the entire slab section. The sulphur print is illustrated in Fig. 10.

(2) *Chlorine Method Extractions* were carried out by I.C.I. (Alkali), Ltd., Northwich, on ½-in. dia. bars machined as described above. Details of the method employed are given in the Seventh Report on the Heterogeneity of Steel Ingots.[1]

(3) *Alcoholic Iodine Method Extractions* were made on ½-in. dia. bars machined as above. Details of the method as employed by

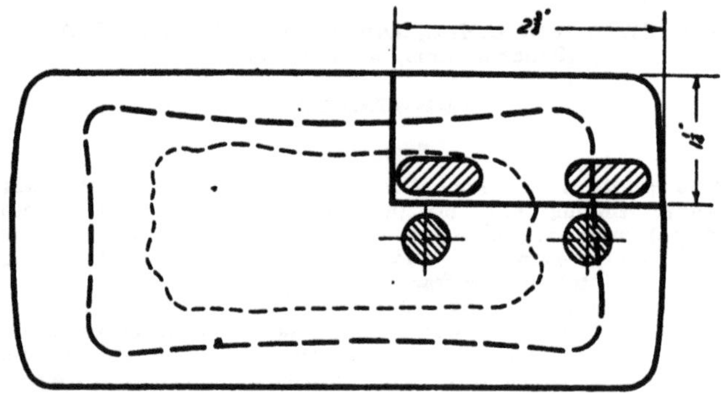

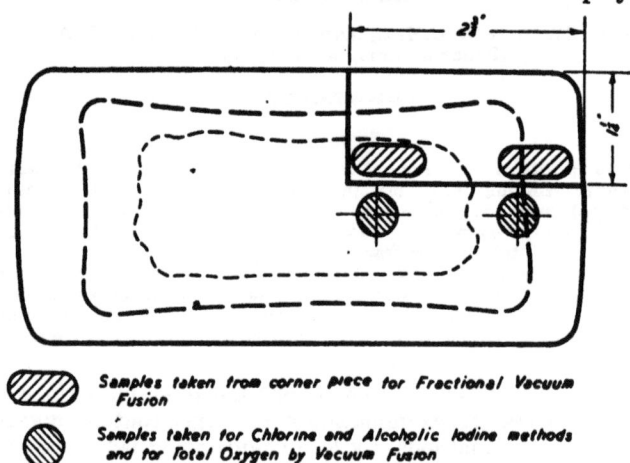

Samples taken from corner piece for Fractional Vacuum Fusion

Samples taken for Chlorine and Alcoholic Iodine methods and for Total Oxygen by Vacuum Fusion

Fig. 11.—Disposition of Samples Examined.

the Central Research Department (The United Steel Companies, Ltd.) are given in the Seventh and Eighth Reports on the Heterogeneity of Steel Ingots.[2]

(4) *The Total Oxygen by the Vacuum Fusion Method* was determined on ½-in. dia. bars obtained as above. This method, as employed by The United Steel Companies, Ltd., is described in the Seventh Report on the Heterogeneity of Steel Ingots (*loc. cit.*, p. 65).

(5) *Oxygen Determinations by the Fractional Vacuum Fusion Method* were carried out on samples obtained from the core and rim of the entire slab section, as shown in Fig. 11. These determinations

[1] Colbeck, Craven and Murray, *Iron and Steel Institute*, 1937, *Special Report No.* 16, p. 124.

[2] Rooney, Stevenson and Raine, *Iron and Steel Institute*, 1937, *Special Report No.* 16, p. 112. Rooney, *Iron and Steel Institute*, 1939, *Special Report No.* 25, p. 142.

were carried out after the ½-in. dia. bars prepared for the investigation had been exhausted; only a small piece of the original billet section was available. For this reason, the oxygen value of the rim test by the fractional method is somewhat higher than the direct determination on the ½-in. rods, as discussed later.

Results of the Examination.

Microscopical Examination.—There were fairly numerous inclusions throughout the rim, consisting, *at the outside edge of the rim*, of:

(*a*) Numerous small globular inclusions of silica, and silica mixed with iron-manganese oxide.

(*b*) A few elongated inclusions of iron-manganese oxide with probably a little sulphide.

(*c*) Occasional large segregates of iron-manganese silicate.

(*d*) Comparatively few sulphide inclusions.

The globular inclusions decreased towards the centre of the rim and the elongated mixed oxides with sulphide were larger and more numerous. A marked increase in size and number occurred at the junction of the rim and core. The inclusions throughout the core were numerous, and typical of rimming steel, consisting of elongated and massive monophased sulphides (occasionally heavily segregated) and also sulphides admixed with iron-manganese oxide.

Oxygen, Nitrogen and Hydrogen Results.—Table XXVII. gives the

TABLE XXVII.—*Fractional Vacuum Fusion Results on Basic Bessemer Material.*

Blow No. 8706*F*.

Constituent.	Oxygen. %.	Oxide. %.	Hydrogen. Ml. per 100 g.	Nitrogen. %.
		Core.		
FeO . .	0·0101	0·0452	2·15	0·0010
MnO . .	0·0070	0·0309	...	0·0006
SiO_2 . .	Nil	Nil	...	0·0015
Al_2O_3 . .	0·0001₅	0·0003	...	0·0163
Total . .	0·017	...	2·15 (0·00019₅%)	0·019₅
		Rim.		
FeO . .	0·0137	0·0617	0·75	0·0004₅
MnO . .	0·0107	0·0475	0·20	0·0003
SiO_2 . .	0·0011	0·0021	...	0·0004₅
Al_2O_3 . .	0·0022	0·0046	...	0·0089
Total . .	0·028	...	0·95 (0·00008₅%)	0·010

results obtained by the fractional vacuum fusion determinations, whilst the entire results of the examination by vacuum fusion and by the residue methods are shown in Table XXVIII.

TABLE XXVIII.—*Basic Bessemer Rimming Steel* (*Core and Rim*). *Analy*

Blow No.

Co-operating Laboratory.	Method.	Weight of Steel taken. G.	Total Ignited Residue. %.	Residue Analysis, with each Constituent					
				SiO_2.		FeO.		MnO.	
				SiO_2.	O_2.	FeO.	O_2.	MnO.	O_2.
I.C.I. (Alkali), Ltd.	Chlorine.	11·29	0·027	N.d.	...	0·011	0·0025	Trace	...
Central Research Department, The United Steel Companies, Ltd.	Alcoholic iodine.	11·75	0·039	0·001	0·0004	0·027	0·0060	0·002	0·0005
	Vacuum fusion.	...	...	...	...	...	...	...	...
	Fractional.	...	...	Nil	Nil	0·045	0·0101	0·031	0·0070
I.C.I. (Alkali), Ltd.	Chlorine.	10·71	0·017	N.d.	...	0·009	0·0020	Trace	
Central Research Department, The United Steel Companies, Ltd.	Alcoholic iodine.	12·06	0·056	<0·001	0·0004	0·041	0·0092	0·004	
	Vacuum fusion.	...	...	...	...	...	...	...	
	Fractional.	...	...	0·002	0·0011	0·061$_8$	0·0137	0·047$_8$	

N.d. = not

* In the residue methods, this figure is the sum of the oxygen contents of the FeO, MnO, SiO_2 and Al_2O_3.

Comments on the Results Obtained.—(1) Substantial agreement between the fractional and total vacuum fusion tests was obtained in the case of hydrogen and nitrogen, and it is noted that the segregation of these two elements in the rim and core followed that of the carbon, manganese, sulphur and phosphorus.[1] Furthermore, it will be seen that the hydrogen content of the samples was evolved in the lower-temperature fractions only; this agrees with the principle of the vacuum heating method described by Sloman [2] and also by Newell.[3] As has been observed in every fractional vacuum fusion determination yet carried out by the authors, the major part of the nitrogen was found in the final fraction.

(2) The total and fractional vacuum fusion figures for total oxygen agree in the case of the core, but the fractional total is higher in the rim. This is explained by the different positions in the billet of the specimens used for the two tests. The segregation of oxygen

[1] Swinden, Seventh Report on the Heterogeneity of Steel Ingots, *loc. cit.*, p. 15.

[2] Sloman, Eighth Report on the Heterogeneity of Steel Ingots, *loc. cit.*, p. 50.

[3] Newell, *Journal of the Iron and Steel Institute*, 1940, No. I., p. 243 P.

in a rimming steel is somewhat different from that of the other elements, and the extent of the variation is dependent on the carbon content. The oxygen content decreases from the chill across the

l Results by Chlorine, Alcoholic Iodine and Vacuum Fusion Methods.

IF.

...essed as a Percentage of the Steel.							Total Components, as % of Ignited Residue.	Total Oxygen (including that from Cr_2O_3).	Total Oxygen.* %.	Nitrogen. %.	Hydrogen. Ml. per 100 g.
Al_2O_3.		TiO_2. %.	V_2O_5. %.	ZrO_2. %.	Cr_2O_3. %.	P_2O_5. %.					
O_2.	O_2.										
05	0·0023	N.d.	N.d.	...	0·008	0·001$_5$	...	0·007	0·005	...	...
02	0·0008	Nil	...	...	0·001	0·003	99·8	...	0·008	0·018†	...
.. 00$_5$	... 0·0001$_5$	... ...	... ...	... ...	... ...	... ...	... ...	... ...	{0·016$_5$ {0·016$_5$ 0·017	0·019 0·019$_5$ 0·019$_5$	2·5 2·6 2·15
05	0·0022	N.d.	N.d.	...	0·002	0·001	...	0·005	0·004	...	...
02	0·0008	Nil	...	...	0·001	0·003	99·4	...	0·011	0·010†	...
.. 04$_5$	... 0·0022	... ...	... ...	... ...	... ...	... ...	... ...	... ...	{0·020 {0·018 0·028	0·010 0·010 0·010	1·35 1·3 0·95

...rmined.

† By chemical solution method.

rim, rises towards the junction of rim and core and maintains a fairly constant value across the core. Thus, all positions in the core may be expected to give similar oxygen figures, whereas rim positions will be higher or lower in oxygen, depending on their proximity or otherwise to the outer edge. The test-pieces for fractional vacuum fusion contained a greater proportion of the outer parts of the rim than the ½-in. rod from which the total vacuum fusion specimens were machined. In comparing the results given by the residue methods with the individual vacuum fusion fractions on the rim, it should be borne in mind that the total of the separate fractions is 0·028% of oxygen, against 0·020% of oxygen by the direct total vacuum fusion determination on the sample comparable to that used for the residue methods.

(3) The total oxygen contents in both the rim and the core calculated from the analyses of the chlorine and iodine residues are lower than the figures obtained by the vacuum fusion method.

(4) The greatest discrepancies between the fractional vacuum fusion and the residue analyses are in the iron and manganese oxides. The residue methods show very low contents of manganous oxide and considerably less iron oxide than the fractional vacuum

fusion method. The solubility of manganese oxide in alcoholic iodine solution under the conditions of the iodine method has been shown in a previous Report,[1] and it is possible that iron oxide is affected similarly. It was hoped that the chlorine method would give more satisfactory recoveries of iron and manganese oxides, but the results of both sets of determinations show an even lower content of these oxides.

(5) The small amount of silica found in the rim by fractional vacuum fusion, as against no silica found by the alcoholic iodine and the chlorine methods is explained by the disposition of the respective samples. Microscopical examination showed that silica and silicates were found only at the outside edge of the rim, none of which was included in the ½-in. rod used for the residue methods.

(6) If the fractional vacuum fusion figures be accepted as representing the contents of the various oxides constituting the total-oxygen figure, then it follows that the chlorine and alcoholic iodine methods, as at present operated, fail to give satisfactory residues from materials of the rimming type, in which the oxides are of a highly basic character. Alternatively, if this type of material contains " oxygen in solution " in the iron, it may be suggested that oxygen existing in this form is not recovered by residue methods.

PART B.—*A Note on the Examination of a Series of Carbon Steels.*[2]

SUBMITTED BY W. R. MADDOCKS, B.Sc., Ph.D. (SHEFFIELD).

AN investigation has been undertaken to ascertain the influence of carbon on the alcoholic iodine extraction method for the determination of oxygen in steel. A series of carbon steels containing from 0·10% to 1·13% of carbon was chosen. Oxygen determinations were carried out on these steels in the normalised, quenched and spheroidised states.

The results obtained so far indicate that the amount and state of the carbon present have an influence on the iron-oxide fraction of the residue. Normalised high-carbon steels give iron-oxide fractions which are much greater than would be expected. All the steels with more than 0·45% of carbon, examined in the quenched state, give higher iron-oxide fractions than in the normalised condition. Steels with less than 0·45% of carbon appear to give the highest value for the iron-oxide fraction after spheroidisation.

[1] Taylor-Austin, Eighth Report on the Heterogeneity of Steel Ingots, *loc. cit.*, p. 167.

[2] A communication from the Metallurgy Department, Sheffield University, received February 10, 1941.

An interesting feature of the work is that the P_2O_5 fraction of the residue is always greater in the quenched steels than in those that are normalised. The lowest values were always obtained from the spheroidised samples.

It is felt that misleading conclusions would be drawn from the work if it were published in its present form.

SECTION X.—General Summary.

By T. SWINDEN, D.Met., Chairman of the Oxygen Sub-Committee.

As stated in the Introduction, the work of the members since the Second Report was published has consisted mainly of a more detailed study of existing methods, and this Report is intended to present a concise statement concerning their possibilities and also their limitations as we see them to-day.

In Section II., Part A Mr. Sloman describes the developments in vacuum fusion equipment and restates the view that this represents our accepted method for the accurate determination of total oxygen. The importance of recognising the presence of " surface oxidation " is emphasised, it being shown that this may account for as much as 0·004–0·006% of oxygen on pure iron. Nitrogen results obtained by the vacuum fusion method have provided some interesting new data when studied in conjunction with those obtained by chemical methods, whereby some indication is given of the form of existence of the nitrogen in the steel. The hydrogen content of steel is being regarded with increasing interest, and determination by vacuum *heating* (as distinct from vacuum *fusion*) is being extended.

In Section II., Part C Dr. Newell summarises briefly his further data on this subject, following the useful paper which he contributed to The Iron and Steel Institute last year.

A significant though short section of Mr. Sloman's paper refers to the solubility of oxygen in high-purity iron. More will be said on this subject later, but, meantime, taking advantage of the availability of appropiate samples, it is indicated that the cold solubility of oxygen in pure iron lies between 0·006 and 0·003%.

The position of the fractional vacuum fusion method is fairly stated, and this subject is further elaborated by Dr. Swinden, Mr. Stevenson and Mr. Speight in Section II., Part B, in which further typical data by this method are given and are compared with results obtained by the alcoholic iodine and chlorine methods. The fractional vacuum fusion method has been particularly successful in the examination of unkilled steels of the rimming type, in which, it is suggested, the oxygen is present essentially as basic oxides.

As an alternative method to vacuum fusion for the determination of total oxygen alone, attention is directed to the aluminium reduction method introduced by Mr. Gray and Mr. Sanders. In Section III., Part A these authors detail the improvements introduced in the method since it was originally described, notably whereby the atmosphere of hydrogen is dispensed with and the determination carried out under reduced pressure. Results are given indicating excellent agreement with the vacuum fusion method when applied to carbon steels, and check tests on added compounds containing oxygen show excellent recovery. A further study of the method on a wide variety of materials is being undertaken. This method has been in regular use for some time in the Central Research Department of The United Steel Companies, Ltd., and in Section III., Part B Mr. Stevenson and Mr. Speight contribute a note describing their equipment and their experience on a series of steels, which confirms very good agreement with the vacuum fusion method.

Further work on the aluminium reduction method is dealt with in Section III., Part C, where Mr. Taylor-Austin describes a modified method applied to the determination of total oxygen in pig iron. The high silicon content necessitates a modification of the original procedure as described, and evidence is submitted that this method can be satisfactorily applied to pig iron generally, provided that a suitable sample can be obtained.

We turn now to the "residue" methods, wherein the non-metallic inclusions are separated chemically from the iron and are submitted to chemical analysis, leading in turn to a calculated content of oxygen. The necessity for further work on all these methods has been stressed in earlier Reports, and it is gratifying to be able to state that definite progress has been made in elucidating some of the earlier difficulties and discrepancies.

In Section IV., Part A Mr. Colbeck, Mr. Craven and Mr. Murray describe their further work on the chlorine method. The latest type of equipment and procedure is carefully explained as well as an improved technique in the analysis of the residue. In Section IV., Part B the same authors show the satisfactory application of the method to a wide variety of steels. Good agreement is forthcoming with the vacuum fusion method when testing fully killed steel. In the case of rimming steel, lower results are obtained by the chlorine method and explanations are submitted for these. In their earlier work the authors showed that the chlorine method was very promising for certain alloy steels which contained elements known to interfere seriously with the iodine method, and this Part refers to the further work which is being pursued in the application of the chlorine method to alloy steels.

In Section V., Part A Mr. Rooney has summarised concisely the present position of the alcoholic iodine method for which he has been so largely responsible. The Part covers the work both at the National Physical Laboratory and in the collaborating laboratories. The method as now practised and herein described includes the various modifications introduced since the last Report and can now be regarded as proved and accepted for the accurate determination of the oxide-containing inclusions in low and medium straight carbon steels. The interference either proved or anticipated from elements forming stable carbides or insoluble iodides is recognised and is being made the subject of further study. As Mr. Rooney points out, it is not unusual for an accepted method to be modified to suit different types of material, and certain modifications are proposed for the application of the alcoholic iodine method to rimming steel, to which further reference is made in Section IX.

Incidentally, note is made of the successful use of alcoholic iodine solution for stripping oxide films on iron. The work since the last Report has naturally been directed mainly towards overcoming the limitations of the method, and this work has given rise to some interesting thoughts on the form of existence of certain elements in the steel, *e.g.*, aluminium and phosphorus. Sheffield University has done considerable research on the interference by carbon and has indicated the limits beyond which carbon interferes with the accuracy of results. Phosphorus has also had considerable attention and this has led to the necessity for thoroughly drying the alcohol.

In Section V., Part B Mr. Stevenson and Mr. Speight describe a simplification of the alcoholic iodine method which nevertheless fulfils all the essentials and appears to produce results of sufficient accuracy for all practical purposes, subject, of course, to the limitations of the alcoholic iodine method previously referred to. The British Cast Iron Research Association has collaborated in investigating particularly the use of the aqueous iodine method, and in Section VI., Part A Mr. Taylor-Austin describes a revised aqueous iodine method, surveying its possibilities and limitations with particular reference to the determination of oxide inclusions in pig iron. In Section VI., Part B Mr. Pearce presents a general review of the present position of the determination of oxide inclusions in pig iron and cast iron.

The hydrogen reduction method (one of the earliest ever proposed for determining oxygen in steel) has been the subject of intermittent attention for many years. Interest has been renewed in the possibility of accurately determining the more readily reducible oxides, *e.g.*, iron oxide and manganese oxide, and in Section VII. Mr. Stevenson briefly states the result of a considerable amount of work which unfortunately up to the present has not led to a satisfactory adaptation of this method. The work, however, is proceeding.

The technique of the analysis of residues is obviously important and was dealt with at considerable length jointly by the members of the Chemists' Panel in the last Report. In Section VIII. Mr. Speight suggests some further improvements in the direction of the almost complete adoption of colorimetric methods.

In Section V., Part A Mr. Rooney referred at length to rimming steel which, as then stated, led him to suggest some modification of the alcoholic iodine method when examining steel of that type. In Section IX., Part A Dr. Swinden and Mr. Stevenson describe work on a billet of rimming steel examined by the chlorine, alcoholic iodine and vacuum fusion (both total and fractional) methods. These results confirm the view that the residue methods tend to give low results, which, it is suggested, are due to the attack of the halogen reagent on uncombined iron and manganese oxides. It would appear that the fractional extraction method, as previously mentioned, is well suited for the fractional determination of oxygen in this type of steel.

Finally, in Section IX., Part B Dr. Maddocks presents a short note, which in itself does not reflect the large amount of work which this entailed, summarising briefly the results of an examination of a series of carbon steels by the alcoholic iodine method. As intimated in other Sections, this work again brings out interesting points concerning the constitution of steel, with particular reference to the form of existence, in this case, of the phosphorus.

SECTION XI.—Work in Hand and Future Programme.

Although in the various Sections indications have been given of certain work in hand, considerations of space render it impossible to set out a detailed programme. It is, nevertheless, desirable that this should be referred to briefly, if only for the purpose of soliciting criticism and comment on the work of the Oxygen Sub-Committee. In both of the preceding Reports a word of warning has been given against the too hasty adoption of any method without a full understanding of the limitations thereof, in the absence of which results are misleading and erroneous deductions are inevitable. A good example may be quoted from the present Report, where more specific attention is given to the influence of surface oxygen and the importance of the preparation of the sample used for the determination.

It is gratifying to be able to confirm that the vacuum fusion method does in fact determine the total oxygen content of the steel and that the technique is now well established. The outstanding feature of the present Report as regards total oxygen is the increased confidence which is being obtained in the aluminium reduction method, having in mind the simpler equipment necessary. Further work on the possible limitations and on improvements of the aluminium reduction method form a definite feature of the forward programme.

It so often happens, however, that the total oxygen, interesting though it may be, represents only a small part of the story, and work must continue until we are able to determine with complete accuracy the oxygen in its various forms of occurrence in the inclusions or in fact in the steel itself. It would seem sometimes that the residue methods employed to achieve this end tend to become more and more complicated and to be relegated to a position in the research laboratories rather than in the routine steelworks laboratory. It may be hoped, however, that in due course, when the limitations are more fully explored, simple modifications of the alcoholic iodine and/or chlorine methods will result in the present difficulties being eliminated. Even if the ultimate conclusion is that these methods must be regarded as being confined to the work of the research chemist and therefore of somewhat limited application, the work would still remain of very great value. The protagonists of these methods are still hopeful of their more general application in due course. In this direction it is hoped that further use will be made of the simplified alcoholic iodine method described in Section V., Part B.

The criticisms that have been made of the fractional extraction method are fully recognised, and the future programme will explore the merits of this method with particular reference to the breakdown of oxygen-containing minerals. One drawback attending all this work on the determination of oxygen is that there is no ultimate and infallible check on the accuracy of results, and this fact has led to the necessity for cross-checking in the various laboratories by the same and by alternative methods. The Sub-Committee are now fortunate, however, in having secured a quantity of iron substantially free from oxygen, which has been prepared for them by the National Physical Laboratory. This is a very important contribution to their work and will allow the members to proceed forthwith on the programme already planned, whereby known quantities of various types of slags and minerals can be incorporated with the oxygen-free iron as a base. In a number of Sections reference has been made to the possible interference of alloying elements. This becomes increasingly important as the content of small amounts of various elements as "residuals" in carbon steels tends to increase. The Sub-Committee have therefore obtained a very comprehensive series of commercial alloy steels, which have now been processed into suitable form and distributed to the various laboratories for work by their respective methods on the interference of alloying elements.

Mr. Colbeck and his collaborators propose to continue their work on the effect of increased chlorination temperatures, which will be of particular interest and service in this direction.

It has, it is hoped, been noted with interest that, while the work is directed essentially towards the accurate determination of oxygen in iron and steel, a number of very significant observations arise, leading one to speculate on the form of existence of certain constituents, particularly aluminium and phosphorus, and these features will not be lost sight of in the forward programme. Further, although this body is called the Oxygen Sub-Committee, it has naturally embraced two other elements, namely, nitrogen and hydrogen, because of the facility provided for determining these in the vacuum fusion method. This in turn has led to the more popular use of the vacuum heating method for determining hydrogen, and here again the work promises to be of the greatest importance in linking up with important research into certain steel defects, which, according to certain schools of thought, are attributed essentially to hydrogen.

The trend of the work must inevitably arouse serious thought on the question of the solubility of oxygen in iron at normal temperature. Mr. Sloman has given further evidence of a definite, if small, solubility of oxygen in iron, and we must eventually be able to determine quite definitely the form of existence and the amount of this oxygen, which, it is reasonable to believe, has a profound influence upon the physical properties of the iron itself. By this is meant an influence distinct from that of the non-metallic inclusions, the bulk of which contain oxygen and are, of course, recorded in the total oxygen determination. Whether or not the oxygen or iron oxide in solid solution is determined by the residue methods remains an important part of the future programme. Rimming steel appears to provide a very interesting subject for the further study of this problem, particularly since, as indicated in Section IX., the amended residue methods give a lower total oxygen value than that obtained by vacuum fusion.

It is hoped to proceed with certain work of a more fundamental nature such as the X-ray examination of residues extracted by the iodine and chlorine methods as well as by vacuum fusion determinations of the oxygen in the residues. There are, however, certain practical difficulties which require to be overcome in this section of the work.

On the other hand, the directly practical aspect of oxygen determination is not lost sight of. While the foregoing work relates entirely to the determination of oxygen (hydrogen and nitrogen) in the final steel, a good deal of work has in fact been done in the individual steelworks laboratories on samples taken throughout various steelmaking processes with quite interesting results. Contemplation of the value which could be attached to these results, if we were quite certain of the absolute accuracy of the various fractions, urges one in the furtherance of this task of finalising rapid methods of making such determinations. Moreover, the initial problem of taking reliable and representative samples of liquid metal has not been overlooked and research is being carried out in this direction as a first essential. We may look forward to the day when the reactive oxygen as the principal chemical reagent in steelmaking processes will be determined directly in the stage laboratory by methods approaching in simplicity those at present used for carbon as an alternative to the present "iron in slag" determination, which at best can only be an approximation and then only under prescribed conditions.

The foregoing refers essentially to steel. It is hoped and anticipated, however, that the happy collaboration with the British Cast Iron Research Association will continue and that further light will be shed upon the question of non-metallic inclusions in pig and cast irons. The extremely fine work by Mr. Morrogh on the metallography of these inclusions raises certain very interesting problems which require elucidation in the light of the work of the Sub-Committee.

The war conditions under which a good deal of this work has been carried out naturally have had their effect on the progress of these researches. It is, indeed, gratifying to be able to present a Report at all under these circumstances, but it is in the certain belief that this work is of vital importance to the fuller understanding of the physical chemistry of steelmaking and all that this means in avoiding trouble and improving quality, that one is inspired to find the necessary time to pursue what might otherwise be regarded as an academic piece of work suitable only for peace-time pursuit.

DISCUSSION.

Dr. W. H. Hatfield, F.R.S. (Vice-President; Brown-Firth Research Laboratories, Sheffield), Chairman of the Committee on the Heterogeneity of Steel Ingots (the Parent Committee of the Oxygen Sub-Committee) had hoped to open the discussion. Unfortunately, he was not able to remain to deliver a contribution in person, but he had intimated to Dr. Swinden that, in his view, the work of the Sub-Committee had carried our knowledge to a very practical stage, and that the Sub-Committee was to be congratulated on the work they had done. It would appear, in his view, that they would be able to handle the question of oxygen in a very effective manner in the near future. Dr. Hatfield had promised to contribute a fuller statement in writing (*see* Correspondence below).

Mr. Vernon Harbord (Messrs. Riley, Harbord and Law, London) said he fully appreciated the very valuable work which the Oxygen Sub-Committee had been doing in trying to determine by chemical methods the oxygen in steel, but it seemed to him that the question was not only one of chemical oxide determination but also one of the distribution of the oxygen in the steel. In the case of the many steels which came to him for examination, it had been found that those which were regarded with most suspicion from an oxidation point of view were not those which showed a vast amount of inclusions under the ordinary microscopic examination but those which suggested an almost sub-microscopic distribution of oxide. Therefore, although he appreciated the value of the total-oxygen determination, he thought the distribution of the oxide was of rather more importance than the total oxygen content of the steel, and he would like to put forward that view to the Sub-Committee for their comments.

Mr. R. Percival Smith (Messrs. Steel, Peech and Tozer, Sheffield) said he did not propose to discuss the Report in a technical or scientific sense, but he would like to say something about it from the steelmaker's point of view.

In the first place, he wished to say that a number of his immediate associates had accompanied him to the meeting in order to be present when the Bessemer Gold Medal was presented to Dr. Swinden, because they all knew that they had greatly benefited by their association with him. In presenting the Medal the President had referred to the work that Dr. Swinden had done as the leader of a great research department, but, because one man's own contribution, however distinguished, was small when viewed in the light of the combined contributions of a close-knit association of men, he ventured to suggest that Dr. Swinden's services to the industry as the leader of a research department were greater than his individual efforts could possibly be. He was convinced that it was true to say that the collaboration to-day between those engaged in research and those engaged in steelmaking was greater than it had ever been before, and certainly the works were benefiting greatly from that association. Perhaps only those in close touch with the day-to-day work of productive departments could fully realise what that meant. It seemed to him that Britain, if it pursued the course it was now on, was bound to do better than it had ever done before, and he thought he was right in saying that even now there were signs that in certain directions this country was fully equal to if not ahead of the other great steel-producing countries.

The works of The United Steel Companies had the benefit of a personal partnership with men such as Dr. Chesters, Dr. Jay and Mr. Stevenson of Dr. Swinden's staff, and the whole steel trade was reaping the benefits of the work already done or in the course of being done by Committees dealing with refractories, liquid steel temperatures, methods of analysis, the grading and use of dolomite, and oxygen determination in steel, in all of which activities Dr. Swinden was taking a prominent part, and much of which he had inspired and was leading. These all worked towards one common end and would, with the companion activities in workshops and works laboratories, within measurable time effect great changes in shop outlook and practice, and, as Dr. Swinden had said, give the trade vast benefits in better steel and economy in time, material and effort.

There was only one thing that he was anxious to know and perhaps an answer could not be given yet, namely, this: Would steelmakers soon have means at their disposal to determine the oxygen content of the steel on the furnace stage before tapping?

Dr. R. J. SARJANT (Messrs. Hadfields, Ltd., Sheffield) said that, as one connected with a works that was attempting to use modern methods of physical chemistry in steelmaking and to apply them to developments in steel manufacture, it gave him great pleasure to convey to Dr. Swinden and the members of his Sub-Committee very hearty congratulations on the standard of the work which they put forward in their Report. He thought that the Report was really a milestone in the application of science to steelmaking, which was often described as an art, and that it would be of great value, because now it would not be long, he thought, before there was actually a method of determining oxygen on the stage. When that happened, steelmaking would become more of a science rather than an art.

He had read the Report with great interest. Whilst the Sub-Committee were attempting to determine a method for the estimation of oxygen in steel, he wondered whether they could give their attention to finding a quick method for the determination of gases.

That might seem a very far-reaching suggestion, but he did not think that it was. Not only was trouble experienced in steelmaking due to oxygen, but in the case of many puzzling things which occurred one could say " That cannot be oxygen; it must be gas." If, therefore, the Sub-Committee could give their attention to the gas content of the steel at the same time as to the oxygen, he thought their contribution would be very welcome to the industry. The quick determination of the gases was not going to be done by any of the known methods of extraction, but he thought that it should be possible, by correlating the work that the Sub-Committee were doing with certain *ad hoc* observations on the steel specimens in the furnace, ultimately to have a technique which would make it possible not only to judge the oxygen rapidly but also to judge the gases.

He had himself applied the Schenck method of determining the oxygen content from the rate of carbon drop, and he had been astounded to find, when one had an accurate method of determining the oxygen in the steel, how closely the data which were determinable from the rate of carbon drop could be correlated with the actual oxygen as determined by the most modern methods which had been described. The importance of the gas question could not be overrated. The Report itself contained many interesting points of which as time went on the value would become clearer. In particular, it had now been ascertained once and for all by the work of the Sub-Committee that the segregation of the gases hydrogen and nitrogen followed the segregation of the constituents such as carbon, silicon and manganese. That was going to be of extreme value in connection with other work which was being done by the Iron and Steel Research Council.

One other point which struck him as significant in the detailed technique that had been developed was in the extraction method, where certain difficulties had been found in connection with the porosity of the mercury boiler when it became hot, and that was ascribed to a certain amount of gas leakage through the body of the metal. That was, he thought, in keeping with what one would expect. Many people did not realise that if metals were heated they became more permeable as the temperature was raised, and that was no doubt the case with the metal of the mercury boiler.

Those who were dealing with the matter from the point of view of works investigation would like to be quite clear as to exactly what were the limits of error that might be expected to be due to experimental method in the laboratory, and what were the so-called sampling errors, due to the method of taking the sample, whether it was taken from part of an ingot or from part of a casting, or whether the ingot was rolled down to a billet and one ultimately sampled the billet or the final bar, or whether an attempt was made to take the sample in the furnace. He would like to see the Sub-Committee devote their attention particularly to specifying what they considered to be, in the most accurate methods which they described, the limits of error, as they saw the matter to-day; and he thought that they had taken the subject far enough to be able to say.

Mr. D. A. OLIVER (Messrs. William Jessop & Sons, Ltd., Sheffield), having expressed his admiration of the Report, said that he was going to discuss the methods purely from the point of view of steelmaking. He regarded the extended work on the fractional method of determination as being of outstanding importance, because, as had already been pointed out in the discussion, there seemed to be accumulating evidence in favour of inclusions which were sub-microscopic. That view, he believed, was put forward first by Dr. McCance in his classical paper in 1925, and was a view which personally he regarded as being extremely sound and of great practical importance. If one determined the total oxygen in steel, one measured the oxygen in the different phases in which the inclusions occurred, and one should, of course, determine as part of the total the oxygen in those sub-microscopic particles. By the fractional method, some information was given as to how equilibrium had been set up during the cooling process, and to a large extent how the steel had been deoxidised. His own view was that these methods, which had been so admirably developed, would be of great value in the laboratory as a check on steelmaking processes. He did not suppose that any steelmaker would contemplate analysing fully every cast of steel, but if he cared to standardise his methods in steelmaking and to analyse a number of typical casts, there would probably be growing evidence in favour of the contention that such a campaign would give most valuable guidance in the better control of steelmaking.

There seemed to be increasing evidence that oxygen by itself was not a bad thing. In other words, heats which were allowed to solidify on the final silicon and manganese present (*i.e.*, without the addition of further deoxidisers) did not necessarily have bad mechanical properties—in some cases they could be most excellent—but steel melted as nearly as possible in the same way and fully deoxidised with a more powerful deoxidiser, such as aluminium or zirconium, would often give inferior mechanical properties, particularly ductility, on account of the fine subdivision of the inclusions. From that point of view he felt that the fractional method, supplemented by either the vacuum fusion total-oxygen method or the new, simple and attractive Gray and Sanders aluminium method, would go a long way towards throwing light on this problem. It was becoming increasingly important to aim at reproducing the same state of affairs in repeated casts of the same steel. If some reproducible deoxidation procedure was followed, it was evident that within certain practical limits it was possible to achieve a high degree of uniformity. In his own works they had put such a practice into operation over the last two years, and the success to date had been over 98%. In the light of those results, he suggested that the new methods which had been so admirably developed in the last Report of the Sub-Committee, if applied to a rigid regime in melting procedure, would result in a great stride forward in the quality of steel products in industry.

CORRESPONDENCE.

Dr. W. H. Hatfield, F.R.S. (Vice-President; Brown-Firth Research Laboratories, Sheffield), wrote that the present Report, although dealing with investigations carried out under war-time conditions, showed unabated enthusiasm and success on the part of Dr. Swinden and the members of the Oxygen Sub-Committee in developing the technique of this valuable study. Indeed, it was clear that, in spite of the war, the work of the Committee had made very considerable progress since the previous Report in regard both to improved and simplified technique, and to the more complete understanding of the significance of the results obtained. Considerable further work would no doubt be necessary before a full knowledge of the facts was attained, but the simplified iodine determination and the work on the aluminium reduction method were notable developments in the direction of the simplification procedure so desirable for routine testing purposes.

The work of the Sub-Committee and its members on the vacuum fusion method had been shown to provide an accurate measure of the total oxygen present, as well as the nitrogen and hydrogen contents, but the realisation of the full value of oxygen determinations towards a proper understanding of the physical chemistry of steelmaking, as well as in regard to the properties of the final steel, must depend on an accurate knowledge of the quantitative distribution of this oxygen among the various elements with which it was combined. The results obtained by the fractional vacuum heating method unfortunately did not so far give figures on which much reliance could be placed, thereby emphasising the importance of the residue method. Thus further developments in the accuracy and simplicity of the chemical extraction methods were of the utmost importance.

In the consideration of the chemical and physical state of the oxygen present it was also important to know the extent to which it existed in solid solution in the metal and the microscopic evidence for a limit of solubility between 0·003 and 0·006% in pure iron was interesting in this connection. It must, however, be borne in mind that the values obtained at normal temperature might not correctly represent the state at the melting point, since some change might occur on cooling, even in the solid state. Moreover, the solubility might be appreciably different in the presence of the other elements occurring in commercial irons and steels. Further study in this direction was very desirable, particularly in view of the doubt as to how far such dissolved oxygen was revealed by the residue methods.

Regarding alternatives to the alcoholic iodine method of extraction, there would appear to be a distinct possibility that the aqueous iodine method could give high oxygen figures from hydrolysis of dissolved iron. The apparent contradiction between the results given by this method for cast iron, and the absence of any indication of oxygen in the micrographic study, might be explained in this way, but the amount found in the residue was small, and other explanations such as complete solubility of the oxygen in the metal or masking of oxide inclusions by the other inclusions present, were equally possible. Care in interpreting results by this method was evidently desirable until such points of doubt had been finally settled.

The chlorine method of extraction had been very fully studied and was capable of giving results similar to those by the alcoholic iodine method, but the latter still appeared to be the most satisfactory residue procedure so far devised, particularly with regard to the development of reliable and speedy routine testing.

In conclusion it could be said that here we had a most complex subject very difficult to handle experimentally which was being developed by Dr. Swinden's Committee in a masterly manner. Already, the subject had largely been cleared, and the work in the near future bids fair to completely achieve results which could not fail to be of immeasurable value to those interested in the scientific aspects of steel metallurgy.

Dr. A. McCance (Member of Council; Messrs. Colvilles, Ltd., Glasgow) wrote that this Report recorded a further stage in the development of the technique for making oxygen determinations of practical aid to steelmakers in the works. The progress made since the Oxygen Sub-Committee took the matter in hand had been quite marked, and the work of Dr. Swinden and his colleagues was greatly appreciated by everyone in the industry.

In spite of the progress made with the vacuum fusion method, it still unfortunately remained essentially a method for the laboratory, requiring careful and, above all, intelligent handling and skilled control.

Its chief drawback was the length of time required for degassing. While the oxygen determination itself could be done quickly, the period of 2–3 hr. necessary to reduce the blank to a manageable size was a distinct disadvantage. The possibility of substantially reducing this time was, no doubt, occupying the attention of the Sub-Committee, and it was to be hoped that they would be successful during the next stages of their work in dealing with this admittedly difficult problem. Only when this had been solved could the development of more robust apparatus suitable for handling in the steelworks be proceeded with.

Even in its present form, of course, there was still much useful investigation work to be done. The two main problems of determining and identifying the gas content in steels and the fractional separation of inclusions received adequate attention in the Report.

In connection with the latter, it might be necessary to change the outlook on the problem and consider less the total oxygen given off at any temperature than the rate of evolution at different temperatures.

In the case of a pure oxide mixed with excess carbon, the rate of reduction would depend on the reaction pressure of the evolved gases, which would be a mixture of CO and CO_2, with the CO predominating.

If the rate of evolution was determined at several temperatures, the curve connecting these two quantities would be characteristic of the oxide concerned, and it should be possible to identify not only the pure oxides in this way but also to some extent mixtures of two oxides. At any rate, oxide inclusions could be classified, and perhaps identified, where there were not too many classes involved.

Another method that might be considered was to determine the actual pressure attained at any given temperature, and, using the well-known relation between the logarithm of this pressure and the temperature, to identify the reactants. Some modification of the existing apparatus would probably be required for this purpose, as means for introducing a back pressure of CO would be necessary and the pump no longer required, but these were details to which the Sub-Committee would doubtless give thought as opportunity arose.

The possibilities of useful work were limitless, and the Sub-Committee could be congratulated not only on what they had already achieved but on the inviting prospect of future work which lay before them.

Mr. W. A. Baker (British Non-Ferrous Metals Research Association, London) wrote that in Section II., Part A, reference was made to errors arising from surface oxidation and further work on this point was contemplated. He would suggest that attention should also be given to another source of error arising from the use of samples with a high surface-area/volume ratio, namely, that due to adsorbed moisture, &c. Adsorbed gases might be removed in the initial out-gassing of the vacuum fusion apparatus, but other methods of analysis might be liable to serious error from this source. The magnitude of the possible error arising from this cause was indicated by the work of Schmid and von Schweinitz. These workers used vacuum fusion and ionic bombardment methods and demonstrated that the large quantities of gas extracted from samples of aluminium derived mainly from moisture, &c., adsorbed on the surface of the sample. In the course of the British Non-Ferrous Metals Research Association's investigations on welding, similar results were obtained, again on aluminium samples, using the apparatus which the writer recently described [1] for the estimation of oxygen in metals by hydrogen reduction. The procedure for estimating oxygen consisted in heating the sample in a closed system, initially evacuated, containing a small excess of hydrogen, the resultant water vapour being condensed during the reduction and subsequently vaporised, and its amount determined by observing its vapour pressure (hydrogen could be equally well determined by using instead a small excess of oxygen). In using this method he had not observed any appreciable blank due to reduction of the silica combustion tube by hydrogen, and he thought it was likely that the large blank attributed to this source in Section VII. of the Report was due rather to traces of oxygen remaining in the hydrogen prior to its entry to the combustion tube. By using an amount of hydrogen only moderately in excess (2 to 3 times) of that required to combine with the oxygen in the sample the blank from the latter source was reduced to negligible proportions, *i.e.*, less than 0·00002 g. of oxygen per reduction.

Mr. J. H. Whiteley (Consett Iron Co., Ltd.) wrote that the series of papers comprising this Report contained much information of service to those engaged on the difficult task of separating and estimating the various non-metallic inclusions that occurred in iron and steel. The work, taken as a whole, marked a definite step forward, and the development of a reliable means of determining different inclusions other than sulphides now seemed to be within the bounds of possibility. If this could be accomplished the identification of these particles under the microscope would be greatly facilitated as would the study of their origin and effects. The attempt to solve the problem by fractional vacuum fusion, described in Part B of Section II., was of particular interest, since the method would be a valuable alternative to the extraction process. Unfortunately, there was as yet no criterion that the distribution of the oxygen in the inclusions as thus indicated was really trustworthy. This, in fact, was recognised in the Report, for in Section XI. (p. 385 P) it was remarked that there was "no ultimate or infallible check on the accuracy of the results." The statement then followed that an endeavour would be made to meet this need by incorporating known amounts of slag and minerals with oxygen-free iron. Whether this plan would be entirely successful was open to question, for it could not be taken for granted that some reduction of both MnO and SiO_2 by the iron would not take place.

[1] "The Estimation of Oxygen in Metals," *Journal of the Institute of Metals*, 1939, vol. 65, p. 345.

In considering this matter it occurred to the writer that another way in which the fractional vacuum fusion SiO_2 and Al_2O_3 results might be checked would be to take as the standard of comparison the figures obtained by the alcoholic iodine method, for, whatever the shortcomings of this method might be, it should at any rate extract completely both SiO_2 and Al_2O_3, no matter in what form they existed as particles. As far as he could see, the only likely sources of error to be guarded against were the following:

(1) The presence in the residue of aluminium carbide or nitride as suspected by Rooney (p. 347 P). This could, of course, be avoided by using steels to which no aluminium had been added. The possible presence of silicon carbide should also not be overlooked, and to ensure that no error was introduced from this cause it might be preferable to kill the metal with phosphorus instead of silicon.

(2) The retention of silicates or aluminates in solid solution in the MnS inclusions. As to the extent of such solubility there was little available evidence; it was probably negligible but, as a safeguard, steels with a very low sulphur content could be used.

Taking these precautions, and given a uniform distribution of the inclusions, the results for SiO_2 and Al_2O_3 yielded by the alcoholic iodine method should be correct, thus enabling the accuracy of the fractional vacuum fusion method for these two substances to be measured.

Some support for this suggestion appeared from an examination of the results presented in Tables III., IV. and VII., where determinations both by the alcoholic iodine and by the fractional vacuum methods were recorded. In reviewing these results it was assumed that aluminium had been added to sample 433 and to that belonging to Table IV. but not to sample *M*5, and the supposition was then

TABLE A.—*Comparison of Oxygen Results obtained by the Alcoholic Iodine and the Fractional Vacuum Fusion Methods.*

Sample.	Oxygen, %, obtained by—					
	Alcoholic Iodine Method.			Fractional Vacuum Fusion Method.		
	SiO_2.	Al_2O_3.	Total.	SiO_2.	Al_2O_3.	Total.
Table III., sample 433	0·006	Nil	0.006	0·004	0·0015	0·0055
Table IV. Top plate	0·0138	0·0033	0·0171	0·0118	0·0062	0·018
Table IV. Top plate	0·0125	0·0054	0·0179	0·0118	0·0062	0·018
Table IV. Bottom Plate	0·0078	0·0063	0·0141	0·0107	0·0054	0·0161
Table IV. Bottom Plate	0·0087	0·0065	0·0152	0·0107	0·0054	0·0161
Table VII. Cast 52363 Top	0·008	0·005	0·013	0·009	0·0035	0·0125
Table VII. Cast 52363 Bottom	0·0075	0·0045	0·012	0·0095	0·004	0·0135
Table VII. Cast 52308 Top	0·006	Nil	0·006	0·004	0·0015	0·0055
Table VII. Cast 52308 Bottom	0·007	0·005	0·012	0·0105	0·003	0·0135

made that the difference between the total oxygen obtained by the iodine extraction and that found by ordinary vacuum fusion, which in nearly every instance was the lesser of the two, was due to the presence of aluminium carbide and/or nitride. This difference was accordingly deducted from the oxygen percentage in the Al_2O_3 as estimated by the iodine method, and the amounts of SiO_2 and altered Al_2O_3 were then compared with those estimated by the other method. It was recognised, of course, that in taking the whole of the difference from the Al_2O_3 an excessive deduction might have been made; nevertheless, as shown in Table A, the total oxygen figures for SiO_2 and Al_2O_3 by the two methods now agreed remarkably well, and, moreover, the proportionate amounts of the two substances were roughly similar, whereas in the case of Plates No. 27 and 29 and Cast 52308 (bottom), Table VII., they were reversed.

SUB-COMMITTEE'S REPLY.

On behalf of the Sub-Committee, Dr. Swinden, the Chairman, thanked the various contributors for the encouraging manner in which the Report had been received. In spite of war-time conditions, definite additional progress had been made, and appreciation of that progress was reflected in the pressing requests that further necessary work still required to be done by the Sub-Committee.

It was realised fully that the very accurate determination of total oxygen, nitrogen and hydrogen, by the vacuum fusion method, provided figures which could not be interpreted with the desired completeness in the absence of certain knowledge as to the distribution of the oxygen present. In this connection, two angles to the problem presented themselves, namely, (*a*) the post-mortem examination of non-metallic inclusions, and (*b*) the more important determination of the FeO-MnO fraction on the steelmaking stage, raised by Mr. Percival Smith and by Dr. Sarjant. The latter determination was one of the definite objectives of the Sub-Committee.

At the present time, there were two methods of approach apparently possible, namely, conversion of the oxygen present into an insoluble form with chemical extraction, or vacuum fusion in a comparatively robust equipment, incorporating a Willems tap. So far as the Sub-Committee were aware, the latter method had not yet been practised, but there were possibilities, when certain practical difficulties had been overcome. In any furnace-stage procedure, however, the question of sampling was as important, if not more important, than the actual extraction of the oxygen present, and it was in this preliminary operation that initial research should be directed. In regard to the examination of non-metallic inclusions, several contributors had discussed the various residue methods and the fractional vacuum fusion procedure.

Mr. Vernon Harbord and Mr. Oliver mentioned the occurrence of sub-microscopic inclusions, and this subject was discussed fully in regard to pig and cast irons by Mr. Pearce in Section VI., Part B of the Report.

Of the two suggestions to explain the apparent contradiction between the chemical and micrographic results on pig and cast iron, both were possible, and the next step was to consider the solubility of oxides and silicates in such particles as manganese sulphide. While agreeing with Mr. Whiteley that the solubility of silicates and aluminates in manganese sulphide might be low, there might be sufficient manganese sulphide present in pig and cast irons to effect appreciable solution of the total proportion of oxides, silicates or aluminates present, thereby masking these inclusions against visual observation.

Dr. Hatfield's criticism of the chlorine method was not justified on consideration of the results obtained up to the present time on solid steels. The Sub-Committee regarded the chlorine method as being definitely valuable, particularly for materials containing carbide-forming elements, which vitiated the alcoholic iodine method.

With reference to Dr. Hatfield's suggestion that iron might be hydrolised in the aqueous iodine procedure, this possibility had been recognised by all workers on the method. This hydrolysis and consequent formation of ferric hydroxide and of hydriodic acid was the probable reason for the increase in *p*H value to which reference was made on p. 362 P of the Report. In the treatments accompanying the aqueous iodine method, the ammonium citrate dissolved the hydroxide. Tables XXXII. to XXXV. inclusive of the Sub-Committee's Second Report [1] showed that prolongation of the citrate treatment resulted in a constant oxide value, which implied that the whole of the hydroxide had been removed, and hence was not available to increase the oxygen content of the residue. Evidence was also given in the present Report of checks between the aqueous iodine, aluminium reduction and vacuum fusion methods.

The Sub-Committee were particularly happy to have Mr. Baker's contribution in regard to high hydrogen figures, resulting from adsorbed moisture on the surface of samples submitted to ionic bombardment. In Mr. Sloman's paper, Section II., Part A of the Report, reference was made to the work of Schmid and von Schweinitz, in which criticism was made of the work of Chaudron on ionic bombardment, for this reason. As stated in Mr. Sloman's paper, an apparatus had been constructed at the National Physical Laboratory to check the discrepancy in the hydrogen results by the two methods. The Sub-Committee did not consider that the blank in the hydrogen reduction method was caused by oxygen in the hydrogen, as suggested by Mr. Baker, as this would be a constant

[1] Eighth Report on the Heterogeneity of Steel Ingots, pp. 37–234, *Iron and Steel Institute*, 1939, *Special Report No.* 25.

value. They considered reduction of the refractory tube to have taken place, as the value of the blank increased with increasing temperature.

The Chairman was particularly gratified at the continued interest shown in the fractional vacuum fusion method, and the operators of this method were indebted to Dr. McCance and Mr. Whiteley for their very constructive proposals for future work. The Stocksbridge laboratory had repeatedly stated that the actual values of the fractions were not an infallible measure of the separate non-metallics present in all classes of steel. Rates of reduction of the different fractions had often been utilised in interpreting the results of fractional analyses. Dr. McCance's suggestion that the rate of evolution of gas from the various fractions and Mr. Whiteley's proposal that alcoholic iodine silica and alumina values be used as "standards" for comparison were readily accepted in connection with future research on the method. It should be noted, however, that the Sub-Committee suspected that the alumina figures from the alcoholic iodine method in certain steels might be high, owing to possible carbide interference. Mr. Rooney was actually working on a series of aluminium steels, made specifically to check this and other factors, in connection with alumina in the alcoholic iodine method. Dr. Hatfield's suggestion that the results obtained by the fractional method were unreliable was not justified by the results obtained, many of which were in very close agreement with those of residue methods. The time-rate of reducibility of the fractions, as stated earlier, was also a very useful feature of the method.

Dr. Sarjant raised the question of the relative accuracy of the various methods. In the Sub-Committee's opinion, the vacuum fusion method when applied to homogeneous billet or bar samples was at least as accurate as other standard methods of chemical analysis. Under similar conditions of sampling, the various operators of the residue methods claimed almost equal accuracy, with few exceptions, most of which were known.

Mr. Oliver's study of steelmaking methods and deoxidation control was very interesting, being similar to other studies made by the Chairman for the same purpose. Unfortunately much of the early work prior to the formation of the Sub-Committee had been valueless, because the early oxygen methods gave unreliable results. In consequence, the early data had to be discarded and built up again, when a truer realisation of the significance of the various methods had been achieved.

The Sub-Committee were continuing their work in the sure belief that the data being obtained as a result thereof would contribute very materially to a fuller understanding of the physical chemistry of steelmaking.

THE FORMATION AND PROPERTIES OF MARTENSITE ON THE SURFACE OF ROPE WIRE.

BY E. M. TRENT, M.MET., PH.D. (THE ROYAL SCHOOL OF MINES, LONDON).

(Figs. 1 to 23—Plates L. to LIII.)

SUMMARY.

The occurrence of thin layers of martensite on the worn surfaces of wires in wire ropes used for mining is described. These layers are very easily cracked, and the cracks lead to a rapid failure of the wires through fatigue. This produces a very dangerous form of deterioration, particularly in mining haulage ropes.

During friction the rope may frequently seize or weld locally on to the object against which it is rubbing, and a thin layer of the object may be torn away and remain adherent to the wire surface.

The metallurgical structures of these thin layers are described. Martensitic surfaces similar to those found in service were reproduced on wires in the laboratory by a number of methods, such as striking the wire a glancing blow with a hardened steel tool or rubbing the wire with a steel tool under heavy pressure. These layers are examined metallurgically and their effect on the mechanical properties of the wire is examined briefly. The cause of a certain type of corrosion pitting known as " chain pitting," which occurs in service on the worn surfaces of wires in haulage ropes, is traced to a localisation of corrosion at cracks in a martensitic surface.

INTRODUCTION.

WIRE ropes used in mining, in particular haulage ropes, are frequently subject to severe wear in service. This wear, by reducing the cross-sectional area of the outer wires, gradually reduces the strength of the rope. A more serious effect where severe wear takes place is the formation of a brittle surface on the wire. Cracks are formed in this brittle surface and these can lead to the rapid failure of the outer wires before they have been seriously reduced in cross-section. This type of deterioration is more difficult to detect by inspection, and is therefore more dangerous. In a recent case examined by the Safety in Mines Research Board a haulage rope failed after only six months' service. Examination on the day before failure had revealed no defect. Investigation after failure showed that the outer wires had been reduced in diameter by only 11%, while the general recommendation of the Safety in Mines Research Board with regard to wear is that a round strand rope becomes unsafe and " should not be allowed to remain in service after 40% of the thickness of the outer wires has been worn away at the worst point."

In this case the deterioration had been caused by the formation of a brittle layer on the worn surface. Examination of sections of the wires under the microscope showed the presence of a layer of martensite, of maximum thickness approximately 0·003 in., covering a large area of the worn surface. Cracks formed in this martensitic layer, which was very brittle, caused the failure of twenty out of the thirty-six outer wires in the rope at the point of fracture.

The presence of such martensitic layers, caused by a rapid heating and cooling of the immediate surface of the wires during friction, has been known for some time. It was described by Atkins in a paper to the Iron and Steel Institute in 1927,[1] and a more detailed description is given in the Annual Report of the Safety in Mines Research Board for 1934.[2] The importance of these martensitic layers as a factor in the deterioration of wire ropes, however, has been underestimated. The following figures indicate the frequency with which martensite occurs and its importance as a cause of failure in mining haulage ropes. Of the last thirty-five haulage-rope breakages examined by the Safety in Mines Research Board, two were capel failures. Of the remaining thirty-three, twenty-four, or 72%, showed the presence of a layer of martensite on the worn surfaces, while in the case of thirteen, or 39%, the martensite was definitely an important contributory cause of failure. Because of its importance in this respect further investigations into the mechanism of the formation and the properties of the martensitic layers were carried out.

Structure of the Martensitic Layers.

The area covered by any one layer of martensite on the worn surface is comparatively small and in the form of a narrow strip. This strip extends along the worn surface of the wire in the direction of rubbing, that is, in the direction of the axis of the rope. In the case of a Lang's-lay rope the strip extends diagonally across the worn surface of the wire. If the worn surface is polished and etched the strips of martensite show up white against a dark background, as illustrated in Fig. 1. When the wire is sectioned through the worn surface, polished and etched in 1% nitric acid in alcohol, the martensite appears as a white or slightly yellow structureless layer sharply separated from the sorbite of the normal wire (Fig. 2). This layer normally varies in thickness from approximately 0·0001 in. to 0·0035 in. In cross-section the strips are lenticular in shape, being thickest towards the centre and decreasing in thickness towards the sides, the total width of the strip being usually not more than 0·03 in. The small size of the martensitic areas indicates that the area of contact between the wire and the rubbing agent is very small and thus the local pressure may be very high.

In many cases the wire surface has been subjected to friction a great many times. The layers of martensite thus formed frequently overlap, and a typical structure is seen in Fig. 3. This is a longitudinal section through the worn surface of a wire. The first layer of martensite that was formed is the white area furthest to the right in the micrograph; it originally covered almost the whole area of the surface shown here in section, and it gradually decreased in thickness towards the right. The second layer was caused by further friction on the surface already covered with martensite; this is the white area towards the left of the micrograph. The two layers are separated by a dark band; this was the result of a tempering of the first layer of martensite, which, at this point, was reheated by the friction to a temperature below 700° C. In some cases several layers of martensite are produced one on top of the other, the last layer formed in each case tempering the layer immediately preceding it, so that the layers are separated by dark bands.

The temperature gradient in the wire caused by the friction is undoubtedly very steep. A rough estimation of the steepness of this gradient can be made from the tempered layers described above. These tempered zones are usually not more than 0·002 in. thick. Supposing that the minimum temperature required to form martensite is 700° C. and the minimum temperature to cause a visible darkening of the martensite is 400° C., then there is a temperature gradient of 150° C. per 0·001 in. of the tempered martensite. It is possible that a higher temperature than 400° C. might be required to produce a visible darkening of the martensite, owing to the extremely short time of tempering, but the figure given above may be regarded as a rough estimate. It may be assumed that the temperature gradient is at least as steep in the martensite—*i.e.*, in that depth of the steel that was converted to austenite—so that in a layer of martensite 0·003 in. thick it is probable that the surface reached a temperature of at least 1150° C.

The heating and cooling of the surface layer must take place in a very short space of time—a matter of a few seconds if not less. The transformation, on heating, from sorbite to austenite must proceed very rapidly, and this is facilitated by the fact that in rope wire the carbide is very finely dispersed. The structure of the sorbite is either not resolvable or only just resolvable at a magnification of 1000 diameters. A coarser structure would require either a longer time or a higher temperature for the complete transformation to austenite. Where there are ferrite streaks at the grain boundaries of the steel these are frequently not dissolved into the austenite on heating, and are visible in the martensite. This is probably due to the fact that, although they are thin, these streaks of ferrite are many times thicker than the cementite laminæ of the sorbite, and hence would take longer to dissolve. The very rapid cooling of the surface layer is due to the continuous contact of the very thin heated layer with the main body of the wire, which remains cold; the cooling is mainly due to conduction through the wire.

The martensite etches either white or slightly yellow in colour, depending on the time of etching, but no acicular structure is developed, probably because of the extremely small grain size produced by the rapid heating and cooling. When a layer of martensite is subjected to a second period of friction, the very small grain size and the even finer distribution of the carbide again facilitate its rapid transformation to austenite, while the martensite which is reheated to a temerature below the critical point is tempered, with the precipitation of very finely dispersed carbide particles. This finely dispersed carbide etches very rapidly in 1% nitric acid in alcohol, quickly becoming jet black.

So far, only martensitic layers formed from the steel of the wire itself have been considered. This is the main cause of surface brittleness in wires from haulage ropes. If the worn surfaces of the wires are examined at a low magnification they are frequently seen to be rough, as if there had been actual seizing of the wire on to the object against which it was rubbing. If a wire is sectioned through these rough areas, the extreme outer layer of the martensite is seen to be by no means homogeneous, as shown in Figs. 4, 5 and 6. At least a part of this material is apparently iron or steel which was welded on to the wire from the surface against which the wire was rubbing under conditions of high pressure and temperature. These welded-on layers are usually extremely thin in haulage ropes (less than 0·0005 in.) and for this reason the structures are difficult to interpret. Further consideration will be given to these structures in the Section on experimental work below; from the latter it was concluded that the welded-on material shown in Figs. 5 and 6 is a fairly high-carbon steel or cast iron, while that shown in Fig. 4 is probably mild steel or wrought iron.

On the worn surfaces of two winding ropes, much heavier layers of extraneous material were found welded on to the wire. Fig. 7 is a section through the surface layers of one of these. The extraneous material here consists of ferrite with numerous slag particles, welded on to a partially tempered layer of martensite formed from the steel of the wire. The welded-on material is probably either wrought iron or the surface layers of a weld. In the second case, the welded-on material was obviously a mild steel. It had been welded on to the wire from a mild-steel plug in a capel, against which the wire seized when the rope failed at the capel. Examination of the plug showed that its surface layers also had been heated to a high temperature and that thin flakes of martensitic steel (0·0005 in. thick) from the wire had been welded on to the plug in some places.

It is thus demonstrated that, under the conditions of friction found in service, the rope can seize or weld locally on to the object against which it is rubbing. When such welding takes place the welded materials subsequently break apart, either (*a*) at the weld, or (*b*) by tearing away portions of the wire or the other object. Which of these occurs depends on the strength of the weld (*i.e.*, on the temperature and pressure of welding, &c.) or on the strength of the two materials and the presence in them of cracks or other sources of weakness. In any case this must lead to rapid wear of the object against which the rope is rubbing, and probably also of the rope itself. Where such seizing occurs the coefficient of friction will greatly increase, and consequently the heat generated will also be much greater and martensite is more likely to be formed.

The Cracking of Martensitic Layers.

The thin layers of martensite on the surface of the wire are very brittle and can withstand very little, if any, plastic deformation. Thus it is usual to find that the martensite produced on rope wires in service contains numerous cracks. Two cracks are seen in Fig. 3 and one in Fig. 5. An examination of the worn surfaces at a low magnification frequently reveals a whole series of cracks running across the wire, as shown in Fig. 1. Once a crack is formed in the martensite it acts as a stress-raiser. The bending stress induced in the wire—for instance, when the rope passes over a pulley—is concentrated at the base of the crack, which is rapidly deepened by fatigue until the wire breaks. The wire illustrated in Fig. 7 thus failed in fatigue, and the micrograph shows a secondary fatigue crack starting from the martensite. In wires that have been fractured by a single application of a bending stress at a cross-section containing an embrittled surface, it is frequently found that the fracture started from a crack in the worn surface and proceeded for a certain distance in shear. Thus a crack in a martensitic layer constitutes an important source of weakness in rope wire.

These cracks are most frequently found at or near a point where a layer of martensite has been reheated by the overlapping of a second layer, as in the case of the two cracks shown in Fig. 3. The stresses set up by the heavy pressure and friction and those caused by the steep temperature gradients involved are probably the cause of the cracking of the martensite in most cases. It is possible that the pressure of the rope on the object against which it is rubbing, by causing a local bending of the wire, would itself be sufficient to propagate the cracks into the sorbite. The crack shown in Fig. 8 may have been formed in this way. Thus, sufficiently high rubbing stresses may, by themselves, cause failure, even though the actual loading stresses on the rope are safe and the pulley diameters sufficiently large.

The cracks found in martensite are not always normal to the wire surface but are often bent over, as shown in Fig. 5. Here the crack was originally formed in a direction normal to the surface in the first layer of martensite. It was then bent over to the left by the dragging of the metal involved in the formation of the second layer. The crack was finally covered over by the layer of welded-on material, so that it does not reach the surface of the wire. This bending-over of the crack was possible because, during the time of friction, this layer of the metal was austenitic and at a temperature of over 700° C. and hence was highly plastic. In this case it is obvious that the direction of friction has been from right to left. In Fig. 9 the crack has been dragged by friction in both directions, indicating that the friction occurred while the rope was travelling both to the left and to the right.

Experimental Work: Laboratory Production of Martensitic Layers.

In one rope, a wire which had been damaged by a blow from a sharp object, such as a chisel, was found to have a layer of martensite on the cut surface. Wire cut by hand clippers sometimes had a layer of martensite on the clipped end, as shown in Fig. 10. It was found that on crushing a wire on an anvil by means of a blow with a heavy hammer, the wire tended to split longitudinally on a plane at approximately 45° to the direction of the blow. A layer of martensite was formed along this plane when sufficient deformation had taken place; Fig. 11 depicts a cross-section through such a wire.

These preliminary experiments suggested the possibility of reproducing in the laboratory the martensitic layers on the wire surface by striking the wire a glancing blow.

The wire used for these experiments was of "plough steel" grade—a grade of hard-drawn, sorbitic wire of a type used for wire ropes. The wire was 0·1 in. in dia., with a tensile strength of 109 tons per sq. in. In the reverse-bend test over a 5-mm. radius it required 9 or 10 bends through 180° to fracture it. The number of twists to failure in the torsion test was 41 over a length of 100 diameters. The composition of the wire was as follows:

Carbon.	Manganese.	Silicon.	Sulphur.	Phosphorus.
0·61%	0·70%	0·112%	0·035%	0·042%

An attachment was made by which a hardened steel tool, rounded at the end, was fastened to the pendulum of an Izod impact testing machine. The wire was held by means of a special grip in the vice of the machine; the axis of the wire was in the plane of swing of the pendulum and the wire was slightly bowed. The tool thus struck the wire a glancing blow and the energy absorbed was measured in foot-pounds on the scale of the Izod machine. It was found that when the energy absorbed was below 8 ft.lb. no martensite was formed and the surface was merely scored. When the tool was lowered so that a deeper "cut" was taken, the wire was very little scored and a layer of martensite was formed, which increased in thickness as the energy absorbed increased.

The tool was deeply nicked by the blow and that part of the wire on which martensite was found had increased in diameter by 0·001 in. This suggested that material from the tool had been welded on to the wire. This was confirmed by weighing the tool and the wire before and after the operation. The wire gained 0·0021 g. in weight while the tool lost 0·0039 g. on the average.

Micro-examination revealed two layers in the martensite, the outer layer being formed from the tool and the inner one from the steel of the wire itself (Fig. 12). The inner layer consisted of the normal slightly yellow-coloured martensite. The outer layer was much whiter, but had brown irregular markings running vertically through it. It was found that where the outer layer was cold-worked after the formation of martensite, as, for instance, where the wire was cut with clippers, the brown markings became an interlacing network of lines and the whole of the outer layer etched a yellow-brown colour. Several of the wires were tested in fatigue, and Fig. 13 shows a fatigue crack starting from the martensite. Where the fatigue crack passes through the outer layer, this layer has been cold-worked and the brown markings have become a mass of interlacing lines. Under high magnification these brown markings are seen to be a series of very fine needles, (Fig. 14). This microscopical evidence indicates that the majority of the outer layer at first consisted of austenite (which etched white), with a few series of needles of martensite. The application of cold-work rapidly converted this austenite to martensite. The tool from which the austenitic layer was formed was made from a plain carbon tool steel containing approximately 1·2% of carbon. The retention of so much austenite in a plain carbon steel is most unusual and must have been caused by the exceptionally rapid cooling of this very thin layer.

These structures very closely resemble the structures described above which are found in the outer layers of martensite on some rope wires from service—compare the structures in Figs. 12 and 13 with those in Figs. 5 and 6. In Fig. 5 the material above the crack has apparently been cold-worked and transformed entirely to martensite. In these cases the layers are thinner and it is not possible to show up the needle structure of the martensite formed in the austenite. It must be deduced from these structures that the rope had been rubbing against some steel high in carbon, or a cast iron.

Properties of the Martensitic Layers.

The martensitic layers produced in the laboratory were easily cracked by bending the wire through a small angle, *e.g.*, through an angle of 20° over a radius of approximately ⅛ in. In the reverse-bend test over a 5-mm. radius, a wire with martensite on the surface broke after 7 bends, while a wire without martensite withstood 9 or 10 bends. Cracks were formed during the first bend, and Fig. 15 shows the crack formed after two bends. In this severe test the fracture was propagated mainly by shear, with large tears accompanying each bend to produce a very rough fracture. Fig. 16 illustrates some of the cracks produced by 12 bends through approximately 20° over a 5-mm. radius. With this small angle of bend the cracks produced were much more smooth and more like fatigue cracks, the amount of tearing accompanying each bend being much smaller. Under these conditions a wire with a layer of martensite contained a crack penetrating to one-third of the diameter after 12 bends, while a wire without martensite required 45 bends to produce the same result.

It has not been possible to carry out a full series of fatigue tests for the comparison of wires with and without martensitic coatings. This is because, in producing the martensitic coating as described above, the tool slightly nicks the wire at the point where it first comes in contact with the wire. The wires tend to fail at this damaged part, and thus the results of the fatigue tests cannot be regarded as demonstrating conclusively the effects of martensite on the fatigue properties of the wire. The results, however, indicate that a martensitic coating, starting without cracks, has little influence on the fatigue limit or on the number of reversals to failure up to a stress of approximately 45 tons per sq. in. This might be expected, since the tensile strength of martensite is certainly not lower than that of the sorbite of the drawn wire. However, if the martensite were cracked before the test it would be expected that the fatigue limit would be greatly lowered, since these cracks would act as stress raisers. Also, if the reversed stress were raised to a point where the plastic deformation of the wire itself was comparatively large, the martensite, being unable to deform plastically, would be cracked during the first few reversals of stress and the wire would fail rapidly. Thus one would expect that when the stress was raised above a certain point the wires with a martensitic coating would prove themselves markedly inferior to the normal wire. Such a tendency is indicated by the result of the reverse-bend test through a small angle described above.

The martensitic coating appeared to have little effect on the tensile strength or on the torsion-test results.

The above method of forming martensite by means of a rapid glancing blow probably does not correspond closely to the conditions under which it is formed in rope wire; here the martensite is formed by a slower friction with heavy pressure. In order to reproduce these conditions a wire was fixed to the travelling table of a planing machine, and was worn by lowering on to it a tool with a rounded end. At first a tool tipped with a polished piece of tungsten carbide was used. The wire was very rapidly worn,

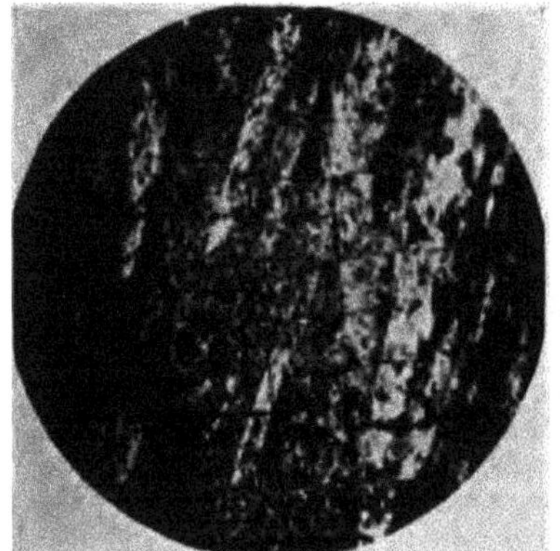

FIG. 1.—Worn Surface of Wire Showing Martensite Streaks and Cracks. × 50.

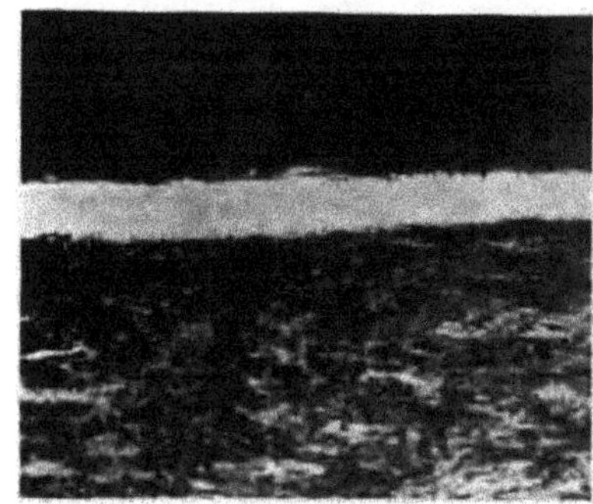

FIG. 2.—Section through Martensitic Layer. × 200.

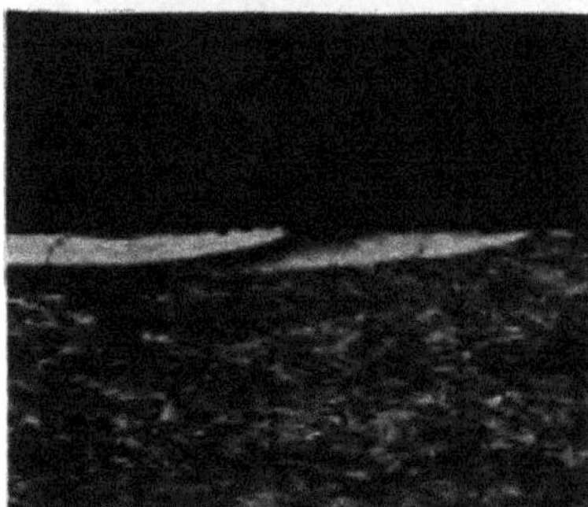

FIG. 3.—Two Martensitic Layers Overlapping. × 150.

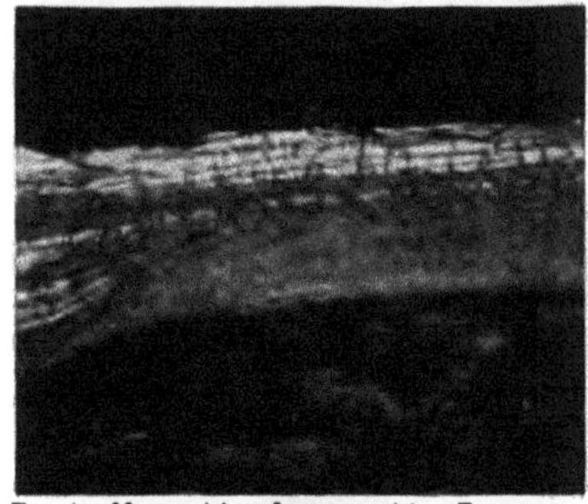

FIG. 4.—Martensitic Layer with Extraneous Material Welded on. × 500.

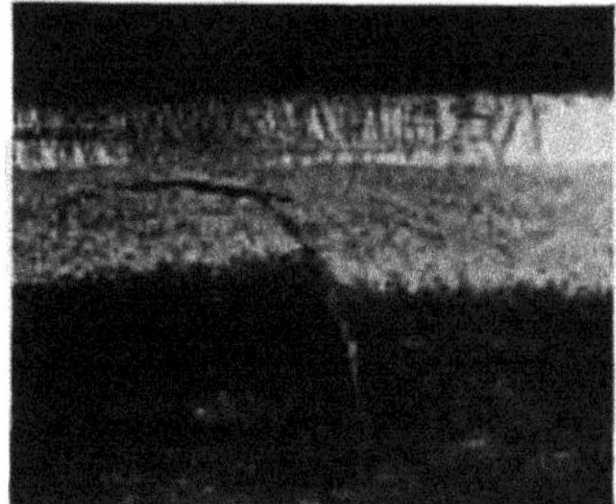

FIG. 5.—As Fig. 4, showing crack in martensite. × 600.

FIG. 6.—As Fig. 4, at higher magnification. × 1000.

(Micrographs reduced to two-thirds linear in reproduction.)

[*Trent.*

[*To face p.* 408 P.

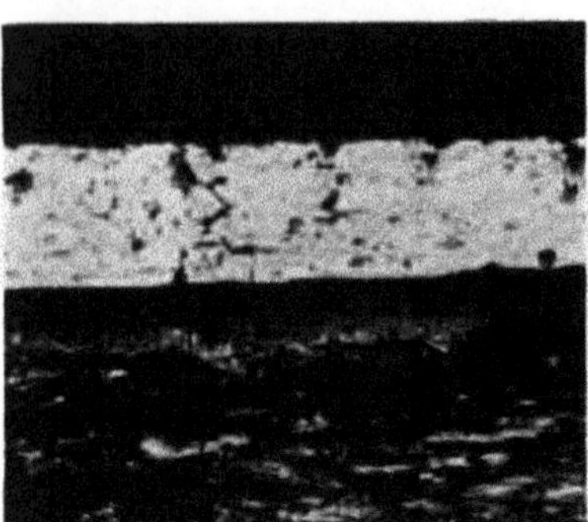

FIG. 7.—As Fig. 4, showing fatigue crack. × 200.

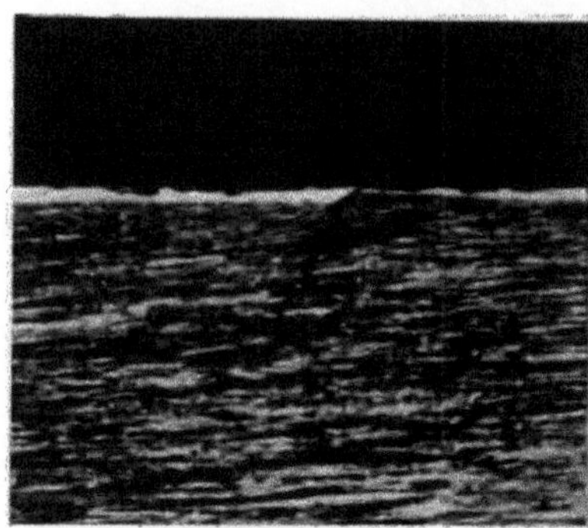

FIG. 8.—Crack Starting from Thin Martensitic Layer. × 180.

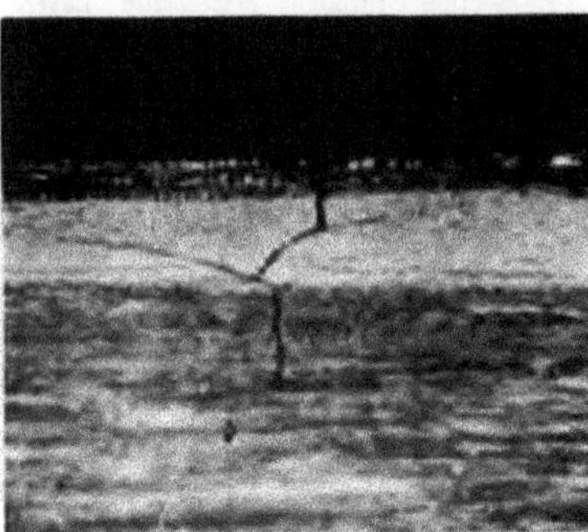

FIG. 9.—Crack in Martensitic Layer. × 500.

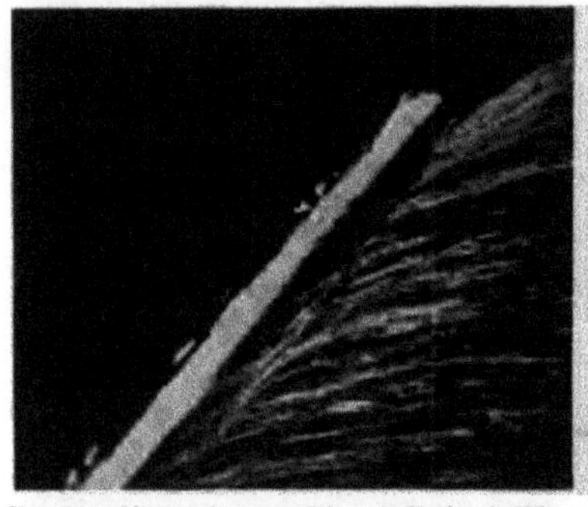

FIG. 10.—Martensite on Clipped End of Wire. × 430.

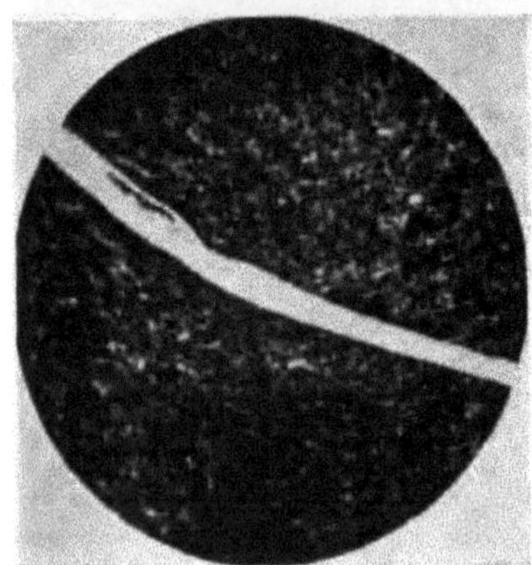

FIG. 11.—Martensite Band in Wire Crushed by Hammer Blow. × 150.

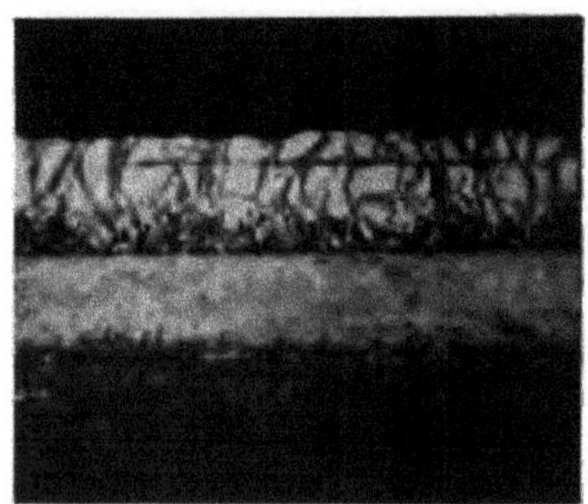

FIG. 12.—Martensite and Austenite-Martensite Layers produced on wire in laboratory. × 500.

(Micrographs reduced to two-thirds linear in reproduction).

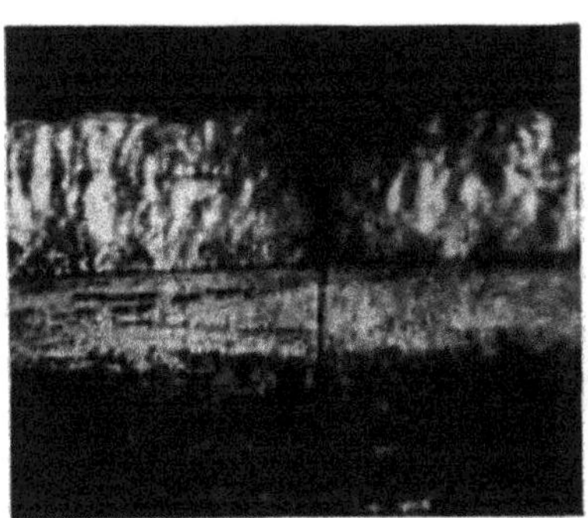
FIG. 13.—As Fig. 12, with fatigue crack. × 500.

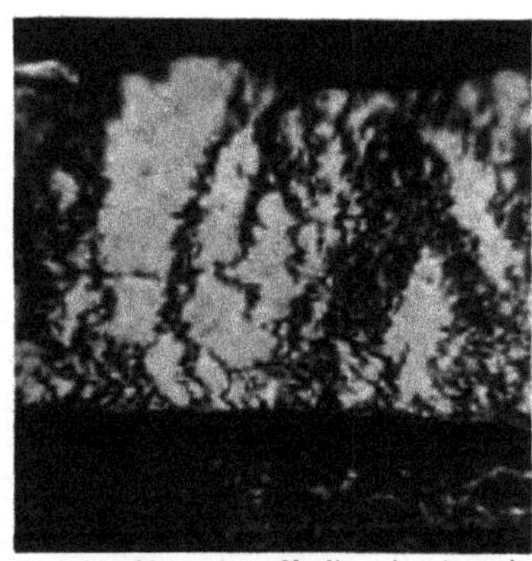
FIG. 14.—Martensitic Needles in Austenite. × 1200.

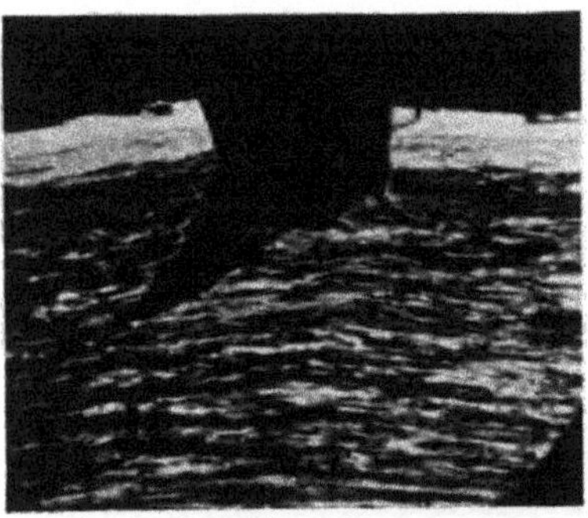
FIG. 15.—Crack Starting in Martensite, produced by two reversed bends. × 150.

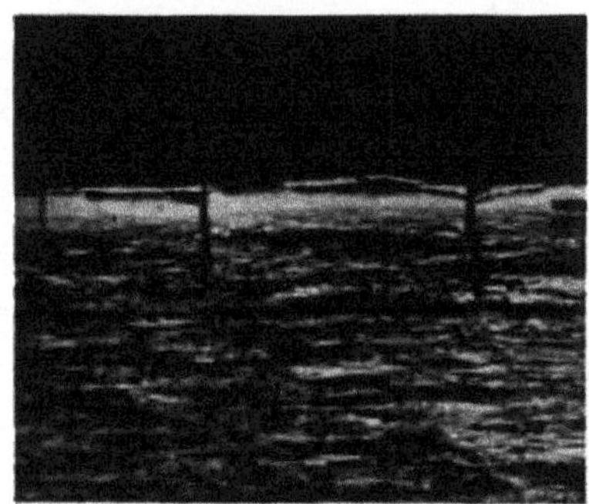
FIG. 16.—Cracks Starting in Martensite, produced by 12 bends through 20°. × 150.

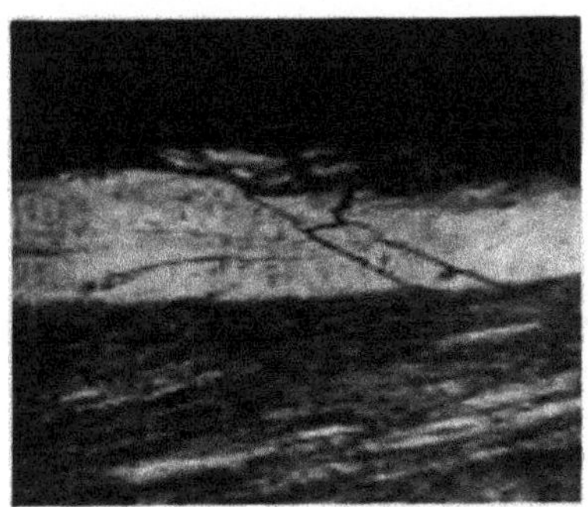
FIG. 17.—Martensite Produced in Laboratory by Rubbing. × 500.

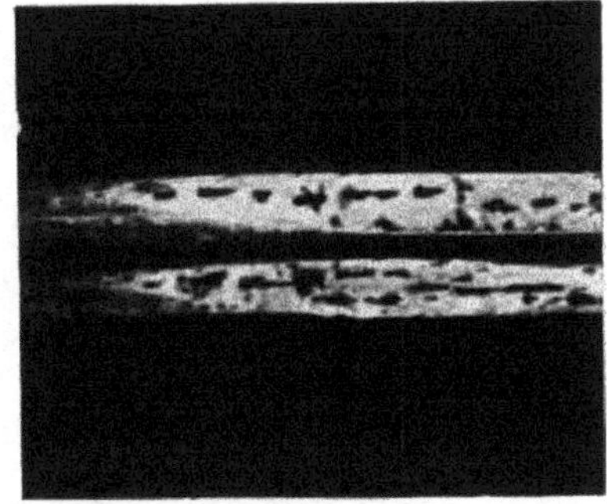
FIG. 18.—"Chain Pitting" on Worn Surface of Wires. × 4.

(Micrographs reduced to two-thirds linear in reproduction.)

[*Trent.*

FIG. 19.—Surface of Pitted Wire, showing large corrosion pits and martensitic surface with cracks. × 50.

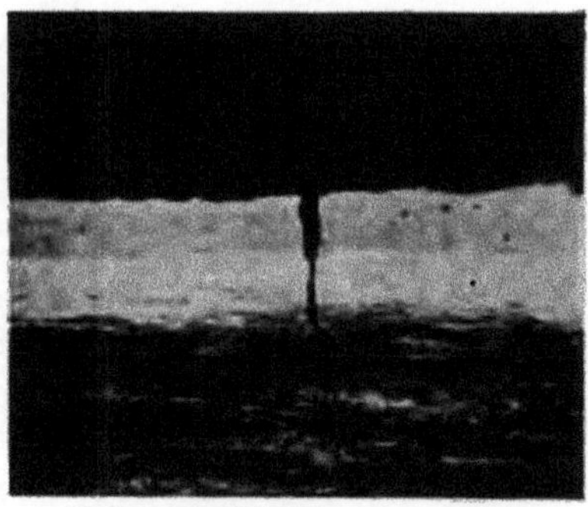

FIG. 20.—Crack in Martensite. × 500.

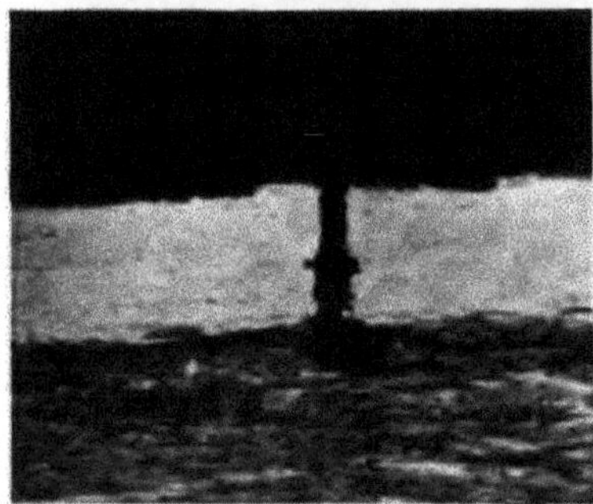

FIG. 21.—As Fig. 20, after 2 hr. in 1% sulphuric acid. × 500.

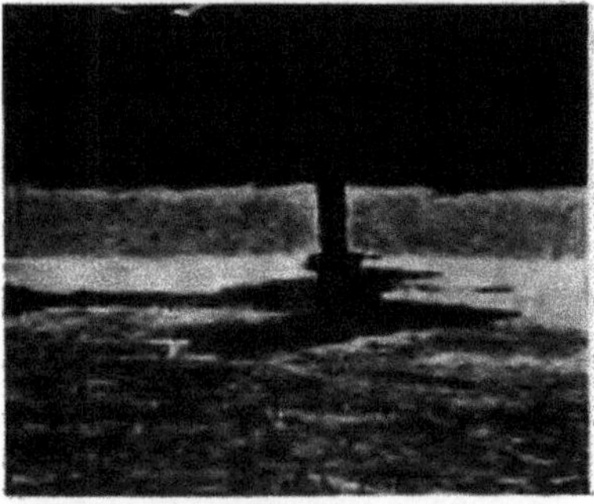

FIG. 22.—As Fig. 20, after 6 hr. in 1% sulphuric acid. × 500.

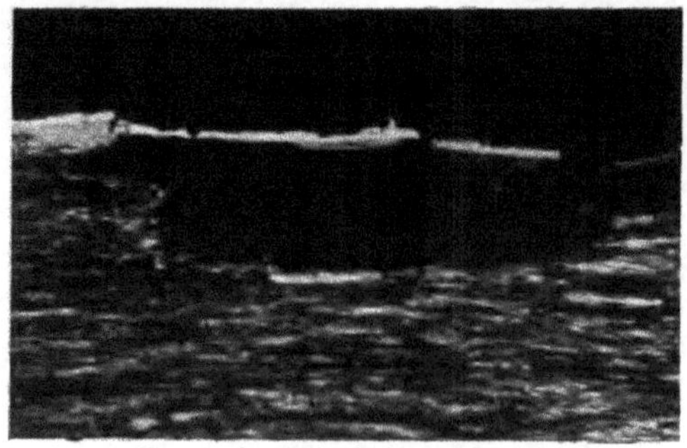

FIG. 23.—As Fig. 20, after 14 hr. in 1% sulphuric acid. × 200.

(Micrographs reduced to two-thirds linear in reproduction.)

a polished surface being produced, while the tool itself was not marked. The worn surface, however, was not excessively brittle, and no martensite could be detected under the microscope. A layer of martensite was produced, however, by first wearing the wire with a tungsten-carbide tool to produce a flat surface, and then wearing it with a hardened steel tool, which tore and roughened the surface to some extent, although the tool itself was hardly marked. From the appearance of the wire and the tool it seems that some seizing took place, which did not occur with the tungsten-carbide tool. Fig. 17 shows a section through the martensite thus produced. The speed of travel of the wire relative to the tool was approximately 0·4 ft. per sec., which is less than one-tenth of the maximum speed of a haulage rope in service.

Discussion.

The work of Bowden and his co-workers[3, 4] has shown that during friction there is a considerable rise in temperature at the interface. For pairs of metals such as lead and steel, Wood's metal and steel, constantan and steel, &c., it was found that the surface temperature achieved was directly proportional to the relative velocity of the two surfaces and to the load, until the melting point of the lower-melting metal was reached, after which no increase in surface temperature occurred. It was also found that in most cases the sliding of one metal over the other took place in jerks, the surfaces sticking together, then slipping a short distance, then sticking again. The coefficient of friction was found to attain very high values when two similar metals were rubbed together, and in this case seizing readily occurred and both surfaces were torn.

The present work on martensite is in agreement with these results. Since the pressures associated with the wear of haulage ropes are much greater than those used by Bowden, one would expect to find higher surface temperatures than those recorded in his work; also the depth of the high-temperature zone is much greater. The seizing of the wire with the object against which it was rubbing is also in agreement with the process of sliding as described by Bowden, but, once again, the pressure being much higher, the actual area of seizing, or welding-together, is probably considerably greater.

Since the quantity of heat liberated during friction is proportional to the coefficient of friction, and since the coefficient of friction is highest for two similar metals, one would expect martensite to be formed most readily when steel rubbed against steel. As described above, when a wire was rubbed with a tungsten-carbide tool no visible layer of martensite was formed, while a heavy layer was produced with a hardened carbon-steel tool. The higher coefficient of friction and the higher surface temperatures attained between two similar metals are undoubtedly associated with the fact that seizure and welding would occur more readily between two similar than between two dissimilar metals.

A white layer a few thousandths of an inch in thickness is formed on the bore of machine-gun barrels near the breech, and a recent paper by Snair and Wood[(5)] describes this layer. The conclusion is reached that this layer consists of "nitrides in solution and not martensite." The conditions of formation of this layer, however—*i.e.*, the presence of carbon monoxide and dioxide, methane, nitrogen, hydrogen and moisture under a pressure of 51,000 lb. per sq. in.—are so different from those involved in the formation of the layer on rope wire that the results of this work seem to have little bearing on the present research. The evidence presented of the tempering of the layers on rope wire, &c., shows that in this case it is a layer of martensite and not a layer of nitrides in solution that is being dealt with.

A Note on Corrosion Pitting on Worn Surfaces of Haulage Ropes.

A distinctive form of corrosion pitting is sometimes found on the worn surfaces of haulage ropes. This takes the form of chains of pits running down the centre of the worn surface, as shown in Fig. 18. It was noticed that this pitting was frequently associated with the presence of martensite. Examination of the pitted surface at low magnification after polishing and etching revealed in nearly every case layers of martensite on the unpitted part of the surface with cracks running into the pits, as shown in Fig. 19, which is a plan view of the worn and pitted surface. Micro-examination of sections through the pi e surface showed that in many cases the corrosion pits undermined the martensitic surface. This suggested that the pitting might be caused by a localisation of corrosion at the cracks in the martensite. It is difficult to prove this by the examination of pitted wires, since the process of pitting proceeds simultaneously with the processes of wear and martensite formation, so that evidence of the original cracks from which the pits were formed is destroyed in the process. However, the presence of cracks and the undermining of the martensite by corrosion indicate the probability that this is the cause of pitting.

An attempt was made to reproduce this pitting in the laboratory. Wires coated with martensite by striking a glancing blow, as described above, were bent so as to crack the martensite. The wires were then immersed in a 1% solution of sulphuric acid for various periods of time at room temperature (approximately 20° C.). After 6 hr. small areas of the martensite between the cracks began to flake off, being completely undermined by corrosion. After 14 hr. comparatively large areas of martensite had flaked off, exposing corrosion pits of considerable depth. Figs. 20 to 23 show the progress of corrosion. Fig. 20 illustrates a crack in the martensite before corrosion. Fig. 21 depicts the widening of the crack and the formation of a pit in the sorbite below the martensite after 2 hr. corrosion. Fig. 22, taken after 6 hr. corrosion, reveals a considerable deepening and extension of the pit, so that it undermines a much greater area of the martensite. Fig. 23 shows a deep pit formed after 14 hr. corrosion, with some flakes of martensite still remaining above it.

Although these tests deal only with one corrosive medium, they show that corrosion pits can be formed owing to cracks in the martensitic surface. This pitting may be due either to an electrochemical effect caused by a difference in potential between the martensite and the sorbite (enhanced in this case by the fact that the outer layer of the martensite was of a different composition from that of the wire itself) or to differential æration, causing localisation of corrosion at the base of the crack. In the latter case it would be expected that corrosion would occur at any crack in the steel, whether martensite were present or not. There is some indication that this may be so, but in any case the importance of martensite in causing pitting is not diminished, since the presence of martensite seems to be the main cause of cracking of the worn surface.

Conclusions.

(1) The formation of martensite on rope wire is an important cause of deterioration and failure in haulage ropes, and can also contribute to the deterioration of winding ropes.

(2) The presence of martensite on the worn surfaces of the wires may lead to rapid failure of the rope by fatigue before inspection reveals that the rope is in a dangerous condition. Care should be taken to prevent the rope from rubbing against any metal object, such as seized pulleys or rollers, rails, steel sleepers, roof girders, &c.

(3) In rubbing against such objects the rope wire may seize on to them, and this leads to the rapid wear both of the rope and of the object against which it is rubbing. Thin layers of metal may be welded on to the rope wire, owing to the high pressures and temperatures during friction, and microscopical examination may be able to give an indication of the nature of the material against which the rope was rubbing.

(4) Under certain conditions of corrosion the presence of martensite containing cracks leads to the formation of corrosion pits on the worn surface of wire. When these pits take the form of " chain pitting " they indicate that martensite is probably present.

Acknowledgments.

This investigation was carried out as part of the work of the Safety in Mines Research Board on Wire Ropes. The author wishes to thank the Board for permission to publish this paper, and also Mr. A. E. McClelland, Mr. W. J. Allum, and other members of the Board for their assistance in carrying out the practical work.

REFERENCES.

(1) E. A. Atkins: *Journal of the Iron and Steel Institute*, 1927, No. I., p. 466.
(2) "Thirteenth Annual Report of the Safety in Mines Research Board," 1934, pp. 115–116.
(3) F. P. Bowden and K. E. W. Ridler: *Proceedings of the Royal Society*, 1936, vol. 154, pp. 640–656.
(4) F. P. Bowden and D. Tabor: *Proceedings of the Royal Society*, 1939, vol. 169, p. 391.
(5) W. H. Snair and W. P. Wood: *Transactions of the American Society for Metals*, 1939, vol. 27, No. 3, pp. 608–620.

CORRESPONDENCE.

(Fig. A — Plate LIIIA.)

Dr. T. Fairley (Park Gate Iron and Steel Co., Ltd., Rotherham) wrote that the hardness of cold-drawn wires was by no means constant across the cross-section. The material within a few thousandths of an inch of the skin was usually very much softer than the core. Having investigated the changes in hardness throughout the cross-sections of various steel wires, he was very interested in Dr. Trent's paper, particularly in his experiments designed to reproduce the martensitic layers similar to those found in service.

The method which he (the writer) had used was to mount the wire at a slope of 1 in 10 in a specially made steel holder, then to grind down the specimen until an elliptical section was obtained, and to measure, by means of the Vickers diamond hardness testing machine, the hardness at every succeeding hundredth of an inch along the major axis of the ellipse. Thus the hardness at every succeeding thousandth of an inch along the radius of a wire was obtained. Accurate micrographs of the edges of cross-sections of the wires were made in order to measure the depth of observable decarburisation. On comparing the hardness figures with the micrographs, it was found that the softer outside extended to a depth of approximately twice the depth of the observable decarburisation. Dr. Trent has reached his conclusions solely from micrographical inspection, and the writer felt that a hardness survey of his specimens would have given valuable information.

He had investigated the effect of progressive cold-drawing on the internal hardness and structure of a 0·66% carbon steel wire. The wire was cold-drawn from a diameter of 0·324 in. to a diameter of 0·080 in. in nineteen passes, and the hardness was investigated at various stages of the cold-drawing process. The analysis of the wire was :

Carbon.	Manganese.	Sulphur.	Phosphorus.
0·66%	0·56%	0·037%	0·031%

and the results were as follows :

Diameter.	Total Reduction of Area.	Internal Hardness.	Surface Hardness.	Depth of Soft Zone.
0·116 in.	87·2%	430	340	0·007 in.
0·080 in.	93·9%	485	410	0·005 in.

This wire might be compared with that used by Dr. Trent, and it was important to note that the depth of softening was approximately 0·005 in., whilst the total depth of martensite and austenite-martensite layers produced in Dr. Trent's experiment was less than 0·003 in. (Fig. 12). Owing to wear, the outsides of the wires used in the ropes in service were removed and the material where the martensitic layers were formed would be truly representative of the wire, whereas in the laboratory experiments it appeared that none of the outside material of the wire was removed and it was probable that the carbon content would increase rapidly from the surface to a depth of a few thousandths of an inch. Also, such decarburised material would etch white and might give rise to some doubt in the interpretation of the micrographs.

Mr. A. Scholes (Messrs. Pattinson and Stead, Middlesbrough) wrote that the formation of layers of martensite caused by friction of the outer wires of wire ropes, and its attendant dangers, had been known by investigators of wire-rope failures at a much earlier period than was indicated by the author.

Dr. Stead, in a lecture before the West of Scotland Iron and Steel Institute at the beginning of 1912,[1] referred to this type of failure. An illustration of martensite formation on the surface of a wire taken from a winding rope and another showing the cracking of this brittle layer on bending were included in the account of the lecture given in the *Journal* of the above Institute (Figs. 15 (*a*) and 15 (*b*)).

In the course of investigations of wire-rope failures in the laboratory of Messrs. Pattinson and Stead over a period of more than thirty years, numerous examples of the formation of martensite, or "frictional hardenite" as it was often called, had been found, and the writer's experience was in agreement with that of the author as to the frequency with which martensite was the cause, or a contributory cause, of failure in haulage ropes.

He could also confirm the author's statement in his conclusions with regard to the formation of martensite on the surface wires of winding ropes, although, as would be expected, it was not a common cause of failure in these ropes, as in the case of haulage ropes.

He had met with one or two rare examples of martensite formation in the drawing of certain types of rope wire. The martensite was probably formed by the temporary seizing of the wire in the die, due to lack of efficient lubrication. In view of the results of experimental work by the author, martensite formation from this cause was much less likely to happen with the use of the modern tungsten-carbide die than with the older type of steel die.

There was one point in connection with the formation of martensite on steel-wire surfaces which was not referred to by the author, and that was the effect of the composition of the steel on the formation of the martensite layer. For example, under similar frictional conditions martensite layers would be more liable to form on a steel wire of high manganese content than on a similar wire of low manganese content.

The author gave some very interesting examples of the welding on the surface of the wire of material from the object with which

[1] J. E. Stead, *Journal of the West of Scotland Iron and Steel Institute*, 1911–12, vol. 19, p. 169.

it came in contact. The writer had met with a few examples of a welded-on layer of extraneous material, but in the great majority of wire-rope failures due to martensite formation which he had examined, either welding-on did not take place, or the weld must have broken at its junction with the martensite layer.

In conclusion, he would like to congratulate Dr. Trent on the very thorough manner in which he had investigated the problem of the formation of martensite on the surface of rope wire. From the practical standpoint he had done a great service in pointing out to rope users, particularly haulage-rope users, the importance of reducing friction on the surface of the rope to a minimum in order to avoid excessive surface overheating from this cause.

Dr. R. Schnurmann (London, Midland and Scottish Railway Company, Research Laboratory, Derby) wrote that Dr. H. O'Neill had drawn his attention to the author's paper in which the suggestion was made that under the conditions of friction found in service a wire rope could seize or weld locally on to the object against which it was rubbing.

"Seizure" was a descriptive term which covered either of two observations: The relative motion between two solid bodies in frictional contact might be found to *cease* when the tangential force required for the maintenance of sliding increased and exceeded the force exerted by the propelling mechanism, or else under conditions of a sufficiently large frictional grip at the true boundary of the two solids sliding might be found to continue by *tearing* if the shear strength of the surface layers of one of the friction elements were smaller than the external force. In both cases Amontons' law, according to which the tangential force of friction should be a constant proportion of the normal force and independent of the area of contact, would not be obeyed. A deviation from this law in the direction of a more than proportionate increase of the force of friction with the area of contact would not afford any criterion as to the mechanism of "seizure" beyond the indication that the surfaces in frictional contact suffered appreciable damage. Since "seizure" occurred whenever the force of friction at the true boundary reached a value which balanced the external tangential force, it was obvious that it depended upon the operative conditions whether bodily seizure would be experienced. For instance, a steel pin and ring might be assembled in a machine which was able to exert an external tangential force of any value between zero and a maximum value T_1. It might happen that the pin bodily seized in the ring when the force of friction had risen to a value $F_x = T_1 - X$ where $0 < X < T_1$. An increase of the external force from $T_1 - X$ to T_1 might not suffice to initiate further relative motion of the two friction elements, so that it would be necessary to transfer them to a larger machine which would be capable of exerting an external tangential force $T_2 > T_1$ of sufficient magnitude to overcome the seizure experienced at values smaller than T_2. If, on the other hand, the melting point of the lower-melting material had been reached at the contact area during rubbing between *non-porous* solids, the deviation from Amontons' law would be in the opposite direction to that mentioned above, because the molten material would act as a lubricant—as was well known, for instance, from the slipperiness of ice near its melting point. The force of friction would decrease with the advent of thick-film lubrication, and the speed of sliding would increase correspondingly if the external force could be maintained at a constant value throughout, so that lubrication by the molten surface layers would persist, and seizure could not occur.

If local welding were due to melting, as distinct from a slow diffusion process in the solid phase, very large values of both the force of friction and the speed of sliding would be required. It was quite true to say that as soon as the surfaces in frictional contact suffered damage the particles plucked out from at least one of them would act as cutting tools, and at a uniform rate of propulsion the force of friction would increase, so that the amount of frictional heat generated over a given path of sliding would increase correspondingly. But it ought to be borne in mind that at every stage the temperature of the surface irregularities was determined by the equilibrium between the generated and the dissipated amount of heat, and that the dissipation of heat by conduction, radiation and convection would also increase with the temperature of the surface irregularities, so that, as Bowden and Ridler [1] had shown experimentally, *large* values of both the normal load and the speed of sliding were required to achieve an appreciable temperature rise. And this was in agreement with Herbert's [2] measurements of the temperature generated on a tool-steel testing machine. He found a temperature of 700° C. when he used, for instance, a Stellite tool cutting dry under standard conditions at the high cutting speed of 200 ft. per min. The consideration of thermal equilibrium in the case of Bowden and Leben's [3] experiments on jerky motion would lead to the result that the temperature rise per jerk was only about 1° C. when, for instance, a silver sphere was held against a uniformly propelled steel plate, and each jerk, covering a path of sliding of 0·0017 in., was completed within 0·001 sec.

It was unfortunate that in the author's only experiment with dissimilar friction elements, when a steel wire was rubbed with a tungsten-carbide tool, no visible layer of martensite was formed,

[1] F. P. Bowden and K. E. W. Ridler, *Proceedings of the Royal Society*, 1936, A, vol. 154, pp. 640–656.

[2] E. G. Herbert, *Proceedings of the Institution of Mechanical Engineers*, 1926, vol. 1, pp. 289-308.

[3] F. P. Bowden and L. Leben, *Proceedings of the Royal Society*, 1939, A, vol. 169, pp. 371–391.

whereas a heavy layer was produced with a hardened carbon-steel tool. The author had thus been led to accept Bowden and Leben's statement that the coefficient of friction would be " highest for two similar metals," although this statement did not cover the results obtained by these authors with steel sliding on steel. The " coefficient of friction " was a function of the method by which it was determined and of accidental surface contamination rather than a physical constant of any given material.[1] The process of sliding as described by Bowden had been discussed elsewhere,[2] where the experimental result had been mentioned that the coefficient of static friction between *naked* metal surfaces cleaned by volatilisation in a high vacuum assumed the order of unity at room temperature.[3] Layers of adsorbed and condensed matter between the irregularities of two metal surfaces in frictional contact were known to alter drastically the coefficient of friction, and with the same friction elements values between almost zero and infinity could be obtained with a given external tangential force when both the thickness and the nature of the film of foreign matter were varied.[4] Particularly if these films were extremely thin, they would exert adhesive properties and might be very tenacious, so that there was no necessity to assume that welding would take place during seizure. For example, a small lead cylinder was cast on to a horizontal glass plate in air at atmospheric pressure and temperature and was found to stick to the plate so as to carry a nominal normal stress of 330 lb. per sq. in. The coefficient of static friction determined by inclining the glass plate was infinity in this case, although there could certainly be no suggestion of welding. The profound influence of extremely thin films of foreign matter upon the frictional behaviour between a steel pin and ring had been studied[5] with a " press-fit " technique at a sufficiently small rate of propulsion that sliding proceeded by a series of heavy jerks if the surfaces did not suffer appreciable tearing, whereas at the same small rate of propulsion sliding proceeded smoothly if the conditions of the experiment were such that the surfaces suffered severe tearing. In these latter cases marked deviations from Amontons' law were observed, and bodily seizure of the pin and ring was experienced when the force of friction rose to a value which exceeded the capacity of the 15-ton spring-balanced recording Buckton machine in which the friction elements were assembled. In one of these cases a longitudinal section of the seized elements was prepared and examined under high-power magnification. This showed that particles had been plucked out from the surfaces and

[1] R. Schnurmann, *Journal of Applied Physics*, 1940, vol. 11, p. 624.
[2] R. Schnurmann, *Engineer*, 1939, vol. 168, p. 278.
[3] R. Schnurmann, *Proceedings of the Physical Society*, in course of publication.
[4] R. Schnurmann, *Nature*, 1940, vol. 145, p. 553.
[5] R. Schnurmann, *Engineering*, 1940, vol. 150, pp. 236–237.

dragged along, but surfaces of separation were clearly visible. There was no evidence of any appreciable discoloration of the contact area, as would have been expected if the temperature had been high enough for local welding to occur.

Direct experimental evidence for the tenacity of, for instance, monomolecular patches of condensed water vapour was found when a mercury column was made to slide in an evacuated glass capillary of about 1·5-mm. bore. Slow sliding proceeded by jerks when at room temperature the column which compressed the residual gas and vapour in the closed capillary reversibly condensed water vapour at a pressure of 6×10^{-3} mm. of mercury on the stable adsorption layer which at room temperature adhered to the glass wall (nominal area, 1·3 sq. in.), whereas sliding proceeded smoothly when the capillary had been connected to a liquid-air trap which removed the water vapour before the mercury entered the capillary.

Dr. C. H. Desch, F.R.S. (Vice-President; Iron and Steel Industrial Research Council, London), wrote that the paper gave a very clear account of the embrittlement of rope wires by the formation of layers of martensite. An early observation of such hard layers, formed by friction on the outer wires, was made by Stead in 1917 [1] on a South African mining rope used on an incline. The crowns of the wires were stated to be difficult to scratch with a steel point. On bending, cracks were formed in the hard layer. No martensite was found, and the effect was attributed to cold-working, but it was probable that martensite was really present. The effect was essentially the same as that found on the surface of rails.

It was clear that very high temperatures could be attained on rubbing. The author's observation that the same effect could be produced by impact, with a pendulum machine or by a blow with a chisel, was very interesting. There was room for much more work on the effects of deformation at high speeds. The deformation then being virtually adiabatic, very high temperatures might be reached locally, and the subsequent cooling would be equivalent to a severe quenching.

The welding-on of foreign material described on p. 404 P could also occur within a rope. The writer had described [2] a winding rope of Lang's-lay construction, which broke after only ten weeks' use. Some of the wires were deeply groved, whilst their neighbours were built up to a corresponding extent. Two of the built-up ridges were shown in Fig. A. These ridges were very hard, and occurred at intervals corresponding with the lay of the strand and mainly well below the surface of the rope. In attempts to reproduce the structure in the laboratory by causing wires to rub over one

[1] Quoted in "Wire Ropes for Hoisting," *South African Institution of Engineers*, 1920, p. 272.

[2] *Transactions of the Institution of Mining Engineers*, 1928, vol. 75, p. 19.

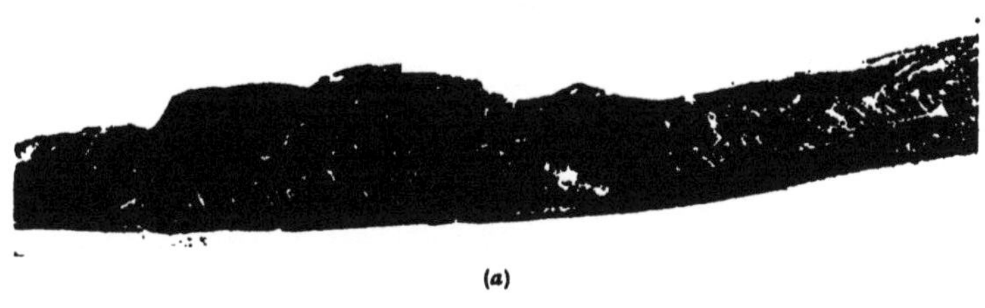

(*a*)

(*b*)

Fig. A.—Built-up Ridges on Internal Rope Wires. × 5.
(*See* C. H. Desch's contribution.)

another obliquely while under load, only very small ridges could be formed, the bulk of the material dislodged in grooving being removed as powder. He had also observed extensive pitting of internal wires, caused by the action of acid waters, which were often found in mines. Such attack had been found in the interior wires of locked-coil ropes, whilst the outer wires were uncorroded and well lubricated.

It was perhaps premature to exclude the formation of nitride in the hard layer. Heavy polishing of a metallographic steel specimen on a dry pad increased the nitrogen in the surface layer, as found by microchemical analysis, from 0·01 to 0·097%,[1] and the rubbed surfaces of several steels in service showed a similar increase in nitrogen.[2] A worn manganese-steel spring showed an increase from 0·006 to 0·09%. The micrographs were very similar to those given by the author, and fatigue cracks were found to start from the hard patches. Under the high local stresses demonstrated by Bowden there was little doubt that activation of atmospheric nitrogen could occur.

It is regretted that, owing to illness, the author's reply has not been received at the time of going to press.

[1] H. J. Wiester, *Archiv für das Eisenhüttenwesen*, 1936, vol. 9, p. 525.
[2] H. Schottky and H. Hiltenkamp, *Stahl und Eisen*, 1936, vol. 56, p. 444.

THE PRACTICAL SIDE OF BLAST-FURNACE MANAGEMENT, WITH ESPECIAL REFERENCE TO SOUTH AFRICAN CONDITIONS.

By R. R. F. WALTON (Pretoria, South Africa).

This paper was discussed at the Autumn Meeting held in Sheffield on November 12, 1940; in the author's absence it was presented by Mr. J. E. Holgate. It will be found, together with the discussion and correspondence to which it gave rise, in the *Journal of The Iron and Steel Institute*, 1940, No. II., p. 13P. The author's reply was not received in time for inclusion with the paper itself, and is printed below.

AUTHOR'S REPLY.

Mr. Walton wrote, in reply, that he wished to thank Mr. Holgate for introducing the paper, and all those who took part in the discussion.

At the time when Mr. Holgate left South Africa, 300-ton furnaces were suggested for the Iscor plant, but it was decided later to adopt a bolder policy and to install a 500-ton furnace. This was blown-in in March, 1934, and was followed by No. 2 furnace, of the same capacity, in November, 1936. The policy mentioned had been amply vindicated by the performance of the plant, and, as stated later, the output had exceeded 600 tons per day per furnace. It should be noted that the South African ton equals 2000 lb., and this unit is used throughout.

As pointed out by Mr. Holgate, Transvaal coals made a comparatively inferior coke, which could be improved chemically and physically by blending with Natal coals. It had been found as economical, however, to use 100% Transvaal coal and to increase the proportion of rich Thabazimbi ore in the burden, so that about 85% of this ore was used in the normal basic burden.

Regarding the two types of gas-cleaning plant, the wet-washing system was installed along with No. 2 furnace, because of lower first cost, and operating and maintenance charges. This was referred to again, in reply to Mr. Brown. The two furnaces were on a common gas main and delivered to a common clean-gas main, with suitable arrangements for isolating either furnace and adjusting the gas volume delivered to each cleaner, but there was no attempt to use the cleaners in series, as this was unnecessary. The gas from the Lodge Cottrell plant averaged about 0·004 grains of dust per cu. ft., and from the Theisen plant 0·006 grains.

Referring to ferro-manganese production at Newcastle, since the paper was prepared a further run on this alloy had been made, with the following results:

Average daily output . .	75·93 tons
Average grade	78·5% of manganese
Coke	4246 lb. per ton of alloy
Coal	169 „ „ „
Ore	3852 „ „ „
Dolomite	1410 „ „ „

As would be seen from a comparison with previous results, the improvement, in output and consumption, was due to the use of an ore lower in silica, illustrating the importance of a low slag volume when making ferro-manganese.

In reply to Mr. Clements, the following data gave average figures for the Pretoria practice, making basic iron for the open-hearth plant:

Output per Furnace: 610 tons per day.

Analysis of Metal:

C. %.	Si. %.	S. %.	P. %.	Mn. %.
4·15	0·70	0·03	0·13	1·50

By increasing the slag volume to about 1000 lb., the sulphur could be kept very low, as illustrated by the following averages for both furnaces over February and March, 1941:

Sulphur . .	0·015%	Manganese . .	1·79%

Weight of Slag: 875 lb. per ton of metal.

Slag Analysis:

SiO_2.	M_2O_3.	CaO.	MgO.
32%	16%	36%	13·5%

Consumptions of Raw Materials:

Thaba ore . . .	2510 lb.	
Pretoria ore . .	470 „	
Manganese ore . .	100 „	
Dolomite . . .	580 „	
Limestone . . .	300 „	
Coke	1530 „	= 1250 lb. of carbon.

No outside scrap was used.

Analysis of Coke:

Ash.	Volatile.	Sulphur.	Moisture.
16·1%	1·1%	0·8%	0·3%

Blast Volume: 40,000 cu. ft. per min. under Pretoria conditions.
Average Blast Temperature: 1200° F.
Number of Tuyeres: Nine 5½-in. and one 4-in. over tap-hole.
Carbon Charged: Per sq. ft. of hearth per hr., 158 lb.
Per sq. ft. of bosh per hr., 106 lb.
Dust Made: 100 lb. per ton of metal.
At present this material was dumped.

Analysis of Dust:

Fe. %.	C. %.	CaO. %.	MgO. %.	SiO_2. %.	H_2O. %.
42·5	14·1	3·5	1·0	12·5	2·7

Gas Analysis (by volume):

CO_2. %.	CO. %.	H_2. %.	N_2. %.
14·2	26·0	1·7	58·1

Top-Gas Temperature: 350° F.

The best month's output for one furnace was 19,979 tons, with a coke consumption of 1493 lb.

Replying to Dr. Marshall, it was true that a higher blast temperature could be carried when the breeze was removed from the coke, but, for the purposes of the tests detailed in Table I., the blast temperature was held as constant as possible, and the burden adjusted to keep the metal regular in each period. The average blast temperature was around 1200° F., although, as mentioned in the paper, the stoves could maintain a temperature of over 1600° F. The author preferred, however, to handle the furnaces as described, keeping this extra heat in reserve to overcome emergencies.

The points raised by Mr. Mitchell and Mr. Gerber were covered by the figures given in reply to Mr. Clements.

Concerning Mr. Fisher's query as to the use of the breeze screened out at the furnaces, about 2,000 tons per month of −½-in. breeze were burnt at the power station, as an adjunct to the normal blast-furnace gas-firing. It was burnt on chain grates, together with discard from the coal washery. About 400 tons per month of the larger size were used for lime-burning in the calcining kilns, while a small amount was sold for domestic use. At present the balance was stocked.

Replying to Mr. Brown, the various figures given for the coke consumption arose in the following manner. The rate of 1530 lb. per ton of metal was an average year-in and year-out, while the figures in Table I. were obtained in the summer months, when the humidity was high, and were further increased by the fact that the equivalent of all cast house and pig machine scrap was deducted from the output, so that the tests should be exactly comparable for each period. The lowest coke consumption over any one month had been 1432 lb.

The operating and maintenance cost of the Theisen plant was about two-thirds of that of the Lodge Cottrell plant, made up as follows:

Gas-Cleaning Costs: Six Months' Average.

	Wet Cleaning.	Electrostatic.
Production salaries and wages	0·019*d*	0·053*d*
Electricity	0·047*d*	0·033*d*
Steam	...	0·027*d*
Water	0·012*d*	0·008*d*
Operating stores	0·001*d*	0·004*d*
Maintenance	0·022*d*	0·043*d*
General services	0·019*d*	0·021*d*
	0·120*d*	0·189*d*

The gas from either plant was amply clean enough for modern stoves and coke-ovens.

With regard to the tuyère life, this could be attributed to (*a*) exceptional freedom from hanging and slipping, due to correct furnace lines and regularity of materials and conditions, and (*b*) an ample supply of clean water, with attention to strainers. No coolers and only two flat plates had been lost since the plant started up in March, 1934. All coppers were of the usual cast standard—about 98·5% copper.

Admittedly the resistence of the burden was partly responsible for the necessity of pulling the gas back through a stove while tuyering, but the Baer explosion valves on the bleeders definitely baffled the gas. This was evidenced by the fact that when drying out the furnace with gas, after relining, the bleeders would not take it all and burnt gas was forced down into the dust-catcher.

THE MANUFACTURE OF STEEL BY THE PERRIN PROCESS.

By B. YANESKE (Chemical, Metallurgical and Research Department, The Tata Iron and Steel Company, Ltd., Jamshedpur, India).

This paper was discussed at the Autumn Meeting held in Sheffield on November 12, 1940; in the author's absence it was presented by Dr. T. Swinden. It will be found, together with the discussion and correspondence to which it gave rise, in the *Journal of the Iron and Steel Institute*, 1940, No. II., p. 35P. The author's reply was not received in time for inclusion with the paper itself, and is printed below.

AUTHOR'S REPLY.

Mr. Yaneske wrote that, before replying to the discussion, he desired to express his grateful thanks to Dr. Swinden for so kindly presenting the paper on his behalf.

With reference to Dr. Swinden's remarks in opening the discussion, it was quite evident to the author that he had a very clear understanding of the necessary requirements for the successful operation of the process. As stated by Dr. Swinden, the slag must be sufficiently fluid and at a sufficiently high temperature. In the author's experiments, notwithstanding the exothermic nature of the reaction between the metal and slag, it was found preferable to have the slag at a temperature 50° C. higher than that of the metal to be dephosphorised. With regard to the most suitable pouring height, the experiments showed that the more violent the agitation of the metal and slag, the more intensive was the dephosphorisation which followed, as would be expected. With the presence of a fair amount of carbon in the metal—say, 0·30% or more—the agitation in the ladle became very violent, so that a lower pouring height was necessary to effect the same degree of dephosphorisation than was the case with metal containing 0·10% of carbon or less. Mr. Perrin's original idea was to pour the metal and slag into a mechanical shaker with the object of well churning them together. That might be possible when the carbon in the metal was very low—that was, 0·05% or less—but would result in the contents being ejected from the shaker with the extremely violent agitation that ensued when a considerable amount of carbon was present in the metal. The author and his colleagues appreciated very much the complimentary remarks contained in the remainder of Dr. Swinden's contribution.

The author agreed with Mr. Whiteley that the experiments seemed to indicate that the Fe_2O_3 in the synthetic slag was responsible for accomplishing most of the oxidation of the phosphorus in the metal. Also, the presence of a much higher percentage of Fe_2O_3 in the synthetic slag as compared with the basic open-hearth slag appeared to explain the increased capacity of the synthetic slag to effect dephosphorisation; nevertheless, it was observed during the experiments that whenever the acid content of the dephosphorising slag was abnormally high—say, over 10% of P_2O_5 + SiO_2—the dephosphorisation was less intensive than when the slag contained much less than 10% of these acids, with approximately the same percentage of Fe_2O_3 in the slag. In reply to Mr. Whiteley's query as to whether the author had tried to dephosphorise metal straight from the blast-furnace, this was attempted without much success in one of the early experiments. As explained by Dr. Swinden in his reply to Mr. Whiteley, the obvious difficulty was the high silicon content of the metal. However, as the result of the experiment might be of interest, particulars were given below :

Heat No. P4. *12th November*, 1935.

Weight of blast-furnace iron used in the experiment . .	21 tons 16 cwt.
Weight of synthetic slag poured in ladle	3 tons 10 cwt.
Proportion of slag used	16%

Analysis of Iron.

	Before Mixing.	After Mixing.
C. % . .	3·95	3·84
Mn. % . .	0·54	0·60
S. % . .	...	0·016
P. % . .	0·280	0·208
Si % .	0·92	0·45

Analysis of Slag.

	Before Mixing.	After Mixing.
SiO_2. % .	8·12	11·20
Al_2O_3. % .	1·60	1·67
FeO. % .	3·99	18·44
Fe_2O_3. % .	27·80	9·00
MnO. % .	7·60	11·40
CaO. % .	46·60	44·20
MgO. % .	3·45	3·02
P_2O_5. % .	0·50	1·17
TiO_2. % .	Nil	0·38
S. % . .	Trace	0·07

As will be seen from the above results, the proportion of silicon removed from the pig iron was double that of the phosphorus.

Since the conclusion of the experiments referred to in the paper, further work had been done by the author and his colleagues with the object of eliminating some of the silicon from the molten pig iron as it was being tapped from the blast-furnace. This was accomplished by the use of iron ore from dust to pea size, in the following manner : Iron ore was screened to pass through ¼-in. mesh. About 2% of the screened iron ore, containing from 60 to 66% of iron, was used, calculated on the expected weight of pig iron, half the quantity of the iron ore being placed in the ladle bottom prior to the tapping of the blast-furnace, whilst the other half of the iron ore was fed into the stream of molten pig iron running from the blast-furnace into the ladle. It was found in numerous experiments that from 20 to 50% of the silicon content of the iron could be removed in this simple manner. Mill scale or blast-furnace flue-dust could be substituted for the iron ore, the quantity to be added depending on their oxide of iron contents. The following were particulars of one of these experiments :

1600 lb. of iron ore, passed through ⅜-in. mesh sieve, were added to 38⅜ tons of molten pig iron, half the ore being placed in the ladle bottom and half fed into the stream. The following was the analysis of the iron before and after treatment with iron ore :

	Before Treatment.	After Treatment.
C. %	4·27	4·06
Mn. %	0·44	0·32
Si. %	0·97	0·48

It would be observed that some carbon and manganese were also removed from the pig iron by this treatment.

Such molten pig iron containing less than 0·50% of silicon had been successfully dephosphorised from 0·30% to less than 0·05% of phosphorus (and also desiliconised) by pouring it very slowly from the blast-furnace ladle held at a height of about 25 ft. into another ladle containing not less than 16% of synthetic basic oxidising slag. The treated metal, containing about 3% of carbon, with manganese, silicon, phosphorus and sulphur below 0·05%, was by this means available for conversion into steel by transferring it either to a basic or acid open-hearth or an electric furnace for the removal of the carbon to the desired extent, thus eliminating the Bessemer converter from the process.

Mr. Whiteley asked whether the Perrin process would be equally efficient in dephosphorising metal containing about 1·5% of phosphorus. The author had never tried to dephosphorise metal with such a high phosphorus content, but was of the opinion that the phosphorus could be removed successfully by two or three successive treatments with synthetic slag, this being more effective than attempting to dephosphorise the metal with a very large volume of slag in one operation.

Mr. Percival Smith had called attention to the repeated remarks of the author that the steel produced was satisfactory and had raised the question "satisfactory for what purpose?" By the description "satisfactory" the author meant that the steel rolled well and the yield of first-class material was normal as compared with similar grade and section produced by the basic open-hearth and duplex processes. The author agreed with Mr. Percival Smith that much could be done by the co-operation of steelmakers in large-scale research work with the object of improving the present so-called standard methods of steelmaking, which, although well-known to be very inefficient, had remained unaltered for many decades.

In reply to the questions raised by Mr. Robinson, the author had not found any difficulty in melting the synthetic slag in an ordinary basic open-hearth tilting furnace. Dolomite or magnesite had proved to be a satisfactory lining to stand up against the highly basic synthetic slag. With regard to the temperatures of the steel and slag before and after treatment, provided that the temperature of the dephosphorising slag was not less than that of the steel to be treated, the net result of the mixing was an increase in the temperature of the steel due to the exothermic reaction that took place, so that usually no heavy metal skulls were formed. In each of the experiments described in the paper the weight of the ladle skull was given, from which it was seen that, despite the necessity to re-ladle the metal in the experiments, the ladle skull formed was never excessive. By avoiding re-ladling, skulls should be the exception rather than the rule.

The author was in agreement with Mr. Mitchell in that Mr. Perrin's theories had proved to be practically sound in regard to both his deoxidising and dephosphorising methods. With reference to the latter, however, Mr. Perrin had not, prior to the author's experiments, worked out a reliable method for the direct manufacture of steel to chemical specification from the dephosphorised blown metal without the necessity of transferring it to a finishing furnace. Mr. Mitchell was correct in his statement that a suitable pouring height to ensure intimate mixing of the metal and slag was an essential practical detail, which might be the main difficulty in the application of the Perrin process in an existing plant, just as was found to be the case in the existing duplex plant at Jamshedpur.

Mr. Hock also referred to the effect of the temperature of the slag. As stated previously, the most satisfactory results were obtained when the temperature of the slag was higher than that of the metal to be dephosphorised. It was absolutely essential that the slag was sufficiently fluid, and greater fluidity was obtainable with a high temperature.

In his reply to Mr. Gerber at the Meeting, Dr. Swinden had clearly explained the difference between the Perrin process and the Aston-Byers process. The author was interested to learn that Mr. Gerber had carried out some experiments on similar lines to the Perrin process which had resulted in the production of practically killed high-carbon dephosphorised steel. With reference to Mr. Gerber's query as to whether the metal could be bottom-poured from a lower height than 20 ft. without materially affecting the results, the author had tried this by pouring the metal into the slag through a 3-in. dia. nozzle from a height of 10 ft., the ladle being moved about during the pouring. The result of the dephosphorisation was found to be not so good as by lip-pouring from double this height, and, moreover, a heavy metal skull was formed, owing to the re-ladling involved and the length of time lost in pouring the metal through the nozzle. In the author's opinion, re-ladling should never be resorted to in any steelmaking process, and in the new plant being erected at Jamshedpur, the blown metal would be slowly poured direct from the Bessemer converter from a height of not less than 20 ft. into the dephosphorising slag contained in the casting ladle, so that re-ladling would thus be avoided.

Dr. Hatfield had confined his remarks mainly to Mr. Perrin's deoxidising process. The author had been privileged to witness at Ugine many experimental heats which were thoroughly deoxidised by tapping the metal into a suitable molten acid slag. It appeared to the author that the chief value of Mr. Perrin's deoxidising process lay in the fact that basic Bessemer steel could be converted into as good a quality as acid Bessemer steel and basic open-hearth steel into as good a quality as acid open-hearth steel or electric-furnace steel by this process, thus eliminating the necessity for low-phosphorus pig iron and scrap. With regard to Dr. Hatfield's question concerning the cost of the Perrin process, this had been aptly dealt with in Mr. Mather's written contribution to the discussion.

In replying to Mr. Whiteley the author had covered the point raised by Mr. Manterfield regarding the greater efficiency of the synthetic slag as compared with the basic open-hearth slag used for dephosphorising. Concerning the composition of the slag of heat No. *P*8 after mixing, this was considered to be a "freak," as in all the other experiments in which a synthetic slag was employed, the ferric-oxide content decreased considerably while the ferrous-oxide content increased considerably after the mixing.

The author desired to express his appreciation of the tribute paid to himself and to the Tata Company by the President in his contribution to the discussion. Mr. Craig's remarks were full of wisdom regarding the lines along which further research work might be conducted with the object of advancing the art of steel-making, which had been allowed to lie dormant for so many years, probably because the metallurgical world had taken it for granted that no worthwhile improvements could be effected. In this respect the courage of Mr. Perrin in attempting to revolutionise the art was deserving of very high praise indeed.

With reference to the communication from Mr. Russell, the silicon content of the hot metal used in the experiments was intentional in order to obtain the necessary casting temperature. Basic pig iron with a silicon content of not more than 0·90% was also made at the Tata Works for the straight basic open-hearth process. Mr. Russell was correct in assuming that the scrap available in India was limited, which fact was, of course, taken into consideration when it was decided by the Tata Company to adopt both the duplex and Perrin processes at Jamshedpur. Mr. Russell requested information as to the nature of the lining which would be adopted for the open-hearth finishing furnaces. In the complete scheme there would be, in addition to two basic-lined slag-melting furnaces, two basic-lined steel-melting furnaces and one acid-lined steel-melting furnace. No trouble was anticipated from pouring the hot dephosphorised metal on to the silica lining. Regarding the linings of the dephosphorising ladles in the new plant, the experiments had proved that a firebrick lining coated with a basic cement was satisfactory, although experiments were at present being carried out with various compositions of bricks made from Indian raw materials with the object of providing a lining that would not require a basic-cement coating.

Mr. Harbord's written contribution was full of interesting observations. Regarding his inquiry as to whether the author had carried out any experiments with synthetic slag in which the FeO was largely in excess of the Fe_2O_3, the author desired to state that he had not purposely done so, as this condition was present in all the basic open-hearth slags used in the experiments.

The author was particularly interested in Mr. Harbord's description of the experiments on the washing of pig iron recorded by Sir Lowthian Bell as far back as 1878.

Mr. Harbord was correct in his surmise that dephosphorisation could be effected with a synthetic slag having a lower lime content and a higher oxide of iron content than the slags used in the experiments recorded in the paper, for in several other experiments, not recorded, dephosphorisation to the desired extent had been accomplished with such slags, of which the lowest lime content was 35·45% with a total oxide of iron content of 34·05%.

With regard to Mr. Harbord's query as to the necessity for desiliconising before dephosphorising, the author had already discussed this point in his reply to Mr. Whiteley.

Mr. Harbord's reference to the experiments carried out by Sir Lowthian Bell recalled the fact, not generally known, that Mr. Benjamin Talbot obtained a patent in 1892 relating to "Improvements in the Treatment of Iron and Basic Slag and in Extracting Silicon and Phosphorus." This patent appeared to be an almost complete anticipation of the Perrin process, as would be seen from the following abstract from the patent specification :

> "This invention has for its object the obtaining of iron which has in it silicon and phosphorus and in extracting the silicon and phosphorus therefrom by an operation carried on outside of the furnace and without the employment of additional fuel for the purpose. To this end the molten iron is subjected to a process of filtration through liquid basic slag, the effect of which is to cause the union of the metalloids and other impurities with the slag, leaving the iron in a condition of purity. The reaction which occurs is very violent and results in the maintenance of a suitable temperature without the employment of fuel or the application of heat from an external source. The preferred practice is to draw the molten

slag in a deep body or column into a retaining vessel lined with highly refractory material and thereafter to pour the molten iron into the slag through which it sinks. The purified iron is then tapped away at the bottom of the vessel and additional iron passed through the slag, the operation being carried on either continuously or intermittently until the slag becomes inert. The delivery of slag into and from the vessel may be continuous or intermittent.

"Good results may be obtained by pouring the molten metal and molten slag into the vessel simultaneously. If complete purification be desired so that practically all the impurities will be removed, it is found that the best results are obtained by subdividing or breaking up the fluid iron into a series of small streams and allowing it to fall into and through the slag.

"The invention is also applicable in the manufacture of steel which shall contain an exceedingly small percentage of phosphorus. By this process metal is used which contains one per cent. of phosphorus, and the phosphorus is eliminated to three-tenths of one per cent. It is preferred to install the process as follows: The iron is blown in the usual acid converter as is done when making ordinary acid steel. The silicon is eliminated and practically all the carbon, and as soon as the flame is perceived to be dropping, which indicates that the metal is nearly decarbonised, the converter is turned down and the blast taken off. A crane is arranged which commands the Bessemer converter and on this crane a cylindrical or other vessel is suspended into which a stated quantity of liquid basic slag has been put. The crane is placed under the converter and the latter emptied of its metal into the filter which contains the slag, care being taken that no siliceous slag is emptied from the converter into the basic slag. The metal descends through the body of liquid basic slag and the remaining traces of silicon and practically all the carbon and phosphorus are expelled, and the metal is a bath of pure metallic iron. The crane of the filter which contains the pure metal is swung over the casting ladle, and the metal is tapped out preferably at the bottom. As the metal is running out, the ferro-manganese or spiegel is introduced into the casting ladle so that the metal becomes deoxidised and is then poured from the casting ladle into ingots. The manganese is introduced after the filtering process, as the iron-oxidising basic slag would have a tendency to expel manganese from the metal. If Bessemer metal is taken which contains a high percentage of phosphorus, the vessel is turned down a few seconds earlier than is done when the metal contains a smaller percentage of phosphorus. This leaves a little more carbon in the metal, causing a much more active reaction between the slag and molten metal and giving a better chance for the expulsion of the phosphorus.

"When acid steel is made in the Siemens open-hearth furnace, the course of procedure is exactly the same as when it is made in the converter. That is, the metal is tapped from the furnace, filtered, the phosphorus reduced and then deoxidised in the casting ladle."

The author felt sure that Members would wonder, as the author had done, why Mr. Talbot's ideas of half a century ago were not developed.

The author was very pleased that Mr. Mather had contributed to the discussion, because it was chiefly through his encouragement that the experiments on dephosphorisation had been carried out to a successful conclusion. Mr. Mather was noted for his initiative and shrewdness, and it was due to his guidance as Technical Director that the Tata Company had finally decided to erect a new plant at Jamshedpur to operate the Perrin steelmaking process on a large commercial scale.

On the other hand, the author was disappointed, as he felt sure other Members would be, that owing to the international situation Mr. Perrin had been prevented from joining in the discussion of his process.

In conclusion, the author wished to take this opportunity of thanking Dr. Swinden for so ably replying to the various points raised in the discussion at the Sheffield Meeting.

SECTION II.

A SURVEY OF LITERATURE ON THE MANUFACTURE AND PROPERTIES OF IRON AND STEEL, AND KINDRED SUBJECTS.

CONTENTS.

The Editor has been assisted in the preparation of this survey by R. A. Ronnebeck.

REFRACTORY MATERIALS

The Effect of Hydrocarbon Gases on Refractory Materials. Part IV. A Study of the Effect of Ethylene on Refractory Materials. E. Rowden. (Transactions of the British Ceramic Society, 1940, vol. 39, Sept., pp. 266–268). The author reports on his experiments the object of which was to determine whether an unsaturated hydrocarbon gas might attack firebricks more quickly and at a lower temperature than methane, a saturated hydrocarbon. Ethylene was the gas selected for the tests. He found that when this gas was passed over refractory specimens maintained at 600° and 800° C., the only effect was to cause discoloration throughout the material. The gas itself polymerised and condensed, forming tar, other liquids and solid hydrocarbons, and at the higher temperature a larger amount of the heavier hydrocarbons was produced. The experiments were discontinued, as apparently decomposition of the attacking gas with the formation of carbon deposits is required in order to bring about disintegration of firebricks.

The Effect of Hydrocarbon Gases on Refractory Materials. Part V. A Further Study of the Effect of Coal Gas on Refractory Materials. E. Rowden. (Transactions of the British Ceramic Society, 1940, vol. 39, Sept., pp. 269–278). The author reports on some further experiments in the study of the effect of coal gas on refractory materials maintained at high temperatures. The particular objects in this case were to determine: (*a*) The effect of the presence of carbon dioxide in the coal gas; and (*b*) the minimum temperature at which disintegration of the firebricks and silica bricks occurred. The general conclusions are as follows: (1) Dry coal gas direct from the mains produced no effect, or very little effect, on specimens at 500°, 800° and 900° C.; (2) moist coal gas produced no effect on specimens maintained at 900° C.; (3) dry gas from which the carbon dioxide had been removed led to the cracking and disintegration of specimens maintained at 800° and 900° C.; (4) the minimum temperature at which attack took place was approximately 800° C.; (5) of the seven specimens subjected to the action of the gas, two silica and one firebrick specimens were never affected, whilst in the remaining four firebrick specimens, cracking, disintegration and the formation of large carbon spots resulted; (6) the "iron spots" in the fireclay brick appeared to be the foci of the carbon deposition from which resulted the brick disintegration; and (7) it was thought that methane, of which there was 24% by volume in the coal gas, was the constituent that caused the disintegration.

The Measurement of Apparent Porosity. H. H. Macey. (Transactions of the British Ceramic Society, 1940, vol. 39, Sept., pp. 279–288). The author describes an investigation the object of which was to make a critical comparison of different methods of determining porosity, one of the important properties of refractory bricks. The methods compared included two air-expansion methods, two displacement methods (one with water and one with paraffin) and four modified displacement methods (two with water and two with paraffin). The author found that the most common method—that of boiling and subsequent evacuation—has the faults of having both the largest inherent and largest random error. (The "inherent" error is that inherent in the method; the "random" error is the experimental error and includes errors due to both the method and the observer.) In his opinion the modified displacement method using water is the best one for determining porosity, as it is very accurate and requires little time.

Use of Aluminium Metal in the Ceramic Industry. I. Properties of Refractories Produced with Mixtures of Aluminium, Fire Clay and Grog. H. G. Schurecht and H. I. Sephton. (Journal of the American Ceramic Society, 1940, vol. 23, Sept., pp. 259–264). The authors found that the addition of aluminium powder to fireclay-grog mixtures greatly increases the strength of the fired brick as a result of a reaction at high temperature between the metal and the silica present in the clay and grog. This reaction takes place at 930° C. and causes the temperature to rise rapidly, so that these refractories need be heated to only 930° C. to produce hard and well-fired bricks. The authors studied the properties of bricks produced from a variety of mixtures of grog, fireclay and aluminium powder, and give data for their drying and firing shrinkage, porosity, thermal behaviour, compressive strength, load-carrying capacity and resistance to spalling.

Refractories for High-Frequency Induction Furnaces. L. F. Keeley. (Metallurgia, 1940, vol. 22, Sept., pp. 157–158: Refractories Journal, 1940, vol. 16, Oct., pp. 411–414). The author discusses in an elementary manner the refractories used for high-frequency furnaces, giving particular attention to the technique of ramming and patching refractory linings.

FUEL

Safety Fuel Engineering. (Iron and Steel Engineer, 1940, vol. 17, Aug., pp. 40–52). A number of papers on safety precautions in the use of gaseous and liquid fuels in large industrial plants are presented. These papers have been prepared by members of the fuel department of the Bethlehem Steel Co. The subjects dealt with are:

Methods of Purging Gas Lines, Holders, Tanks, &c., by G. J. Campbell.

Carbon Monoxide and its Detection, by J. S. Morris.

Operation of Gas Safety Regulators, by D. G. Hisley.

Safety and the Burning of Blast-Furnace Gas, by C. E. Duffy.

Safety Procedure in Gas Fired Furnaces, by J. F. Black.

Precautions in the Burning of Premixed Gases, by H. L. Halstead.

Safety Procedure in Heavy Oil Fired Furnaces, by E. C. Davis.

Boiler Combustion Control in the Steel Mill. M. J. Boho. (Blast Furnace and Steel Plant, 1940, vol. 28, Aug., pp. 806–809). The author reviews the development of the boiler plants of large steelworks during the last twenty years, and describes, as an example, a plant consisting of five boilers with a total steam capacity of about 1,450,000 lb. per hr. Three of the boilers are fired with a combination of blast-furnace gas and coal and two with blast-furnace gas alone, and the author describes the complicated automatic control system which regulates the steam output.

The Net Calorific Value of Coals. T. Evans. (Analyst, 1940, vol. 65, June, pp. 352–353: Fuel, 1940, vol. 19, Sept., pp. 181–183). The author found that it is possible to calculate the net calorific value of coals from most of the British coalfields by the use of data available in the proximate analysis and without the additional determination of the hydrogen content. His method of calculating the net calorific value does not, however, seem to be applicable to coals from overseas.

Coal-Screening Plant at Primrose Hill Colliery. (Mechanical Handling and Conveying, 1940, vol. 27, Oct., pp. 235–236). A new coal-screening plant erected at Primrose Hill Colliery by Goodall, Clayton & Co., Ltd., is described. It is designed for screening 160 tons per hr. into five different sizes. It consists of the following parts: An automatic tippler, a feeder conveyor, a jigging screen, three picking belts fitted with pulsating loaders, cross conveyors for dressings and a longitudinal conveyor for chippings and crushed dressings.

Gravity Coal Washing. A. D. Cummings. (National Association of Colliery Managers: Iron and Coal Trades Review, 1940, vol. 141, Oct. 11, pp. 360–361; Oct. 18, pp. 383–384). The author reviews the principal processes of gravity separation of coal, dividing them into three classes, *viz.*, those using (*a*) heavy liquids, (*b*) salt solutions and (*c*) suspensions of insoluble material in water. He states that the first two classes are, at the present time, quite unimportant compared with the last. He describes at some length the most frequently used processes of this last class, which are the Chance, Tromp, Loess and Barvoys processes, pointing out that too few results have as yet been published to make a careful comparison of the relative merits of these processes possible.

A Medium-Sized Coke-Oven Plant. (Coke and Smokeless Fuel Age, 1940, vol. 2, Aug., pp. 175–179). An illustrated description is given of a new coke-oven plant in the north of England, designed to carbonise 300–400 tons of coal per day. The plant was designed and constructed by Simon-Carves, Ltd.

Friability, Grindability, Chemical Analyses and High- and Low-Temperature Carbonization Assays of Alabama Coals. E. S. Hertzog, J. R. Cudworth, W. A. Selvig and W. H. Ode. (United States Bureau of Mines, 1940, Technical Paper No. 611). The authors present the results of a study of some physical and chemical properties of Alabama coals. The research was carried out under a co-operative agreement between the Bureau of Mines and the School of Mines of the Alabama University.

Light Oil Recovery from Coke-Oven Gas. W. Tiddy and M. J. Miller. (American Gas Journal, 1940, vol. 153, Sept., pp. 7–10, 46). After a brief review of the development, since 1860, of the recovery of light oil from coke-oven gas, the authors describe the three processes which are employed for this purpose and which are designated as: (*a*) The cooling and compression system; (*b*) the adsorption system; and (*c*) the absorption or wash-oil system. In conclusion they consider the advantages and disadvantages of the elimination of light oil, the main components of which are benzene, toluene, xylene and naphthalene.

Blast-Furnace Gas Cleaning. (Iron and Steel, 1940, vol. 13, Sept., pp. 458–459). An illustrated description is given of a complete gas-cooling and cleaning plant designed for a throughput of 6·5 million cu. ft. per hr. of blast-furnace gas. The plant consists of three units, each comprising a precooler, a disintegrator and a spray separator. The plant effluent is dealt with by two Dorr thickeners. The degree of gas-cleanliness of not exceeding 0·02 g. per cu. m. as specified by the contractors has easily been maintained.

Developments in Electrical Precipitation in Steel Plants. C. W. Hedberg and L. M. Roberts. (Iron and Steel Engineer, 1940, vol. 17, July, pp. 18–23). The authors describe recent developments in the use of electrical precipitators for cleaning blast-furnace and coke-oven gas at American steelworks. To prevent the building-up of a solid precipitate on the electrodes, a system of vertical tubular collecting electrodes has been developed, the upper ends of which are fitted into a header plate; the top side of the header plate is divided into ponds by vertical metal strips, and a water supply is so connected that controlled amounts of water are supplied to each pond, whence it overflows uniformly and runs down the inner surface of each tubular electrode. An improvement which reduces the consumption of electric power is the use of the " half-wave " electrical set; with this equipment two instead of one bank of electrodes are connected to a single rectifier, and a power impulse is supplied to each bank on alternate half-cycles instead of on consecutive half-cycles. Experience has shown that the electric power consumption for blast-furnace gas cleaning averages between 0·4 and 1·0 kWh. per 100,000 cu. ft. of gas cleaned, and that the total water consumption, including overflow for the collecting electrodes and the amount required for flushing, averages 150–500 gal. for the same quantity of gas. The authors also describe some tests made first with a pilot plant and then with a 60,000-cu.-ft.-per-min. precipitator for cleaning gas from blast-furnaces producing ferro-manganese and high-silicon iron. In conclusion they give an account of some successful results achieved in the detarring of coke-oven gas by electrical precipitation.

The Use and Application of Town's Gas as an Industrial Fuel. J. E. White. (Fuel Economy Review, 1940, vol. 19, pp. 6–13, 32). The author discusses the advantages of the use of town gas as an industrial fuel, comparing it with producer gas and referring in particular to wartime requirements. He deals mainly with the application of gas in the metallurgical industry and gives a brief illustrated description of numerous modern types of gas-fired salt baths and furnaces, including reverberatory, tilting, reheating and annealing furnaces.

PRODUCTION OF IRON

Continuous Pig Casting. R. Trautschold. (Steel, 1940, vol. 107, Sept. 9, pp. 50, 72). An illustrated description is given of a continuous casting machine at one of the blast-furnaces of Canadian Furnace, Ltd., Port Colborne, Ontario. The conveyor carries moulds between two endless chains. The machine has worked satisfactorily for long periods, producing 45-lb. pigs of cast iron at the rate of 70 tons per hr.

Production of Pig Iron in the Electric Furnace. C. Hart. (American Institute of Mining and Metallurgical Engineers, Technical Publication No. 1230: Metals Technology, 1940, vol. 7, Sept.). The author describes and illustrates the three best-known types of electric furnace used for the production of pig iron. These are: (1) The Swedish Elektrometall high-shaft furnace; (2) the Siemens Halske low-shaft furnace; and (3) the Tysland furnace as improved by Hole and tried out by Christiania Spigerwerk in 1925. He presents data comparing their consumptions and outputs and, in conclusion, reviews the economic possibilities of the smelting of iron ores in electric furnaces in Sweden, Norway, France, Italy, Finland, the United States, Canada and South America.

Ironfounding in England 1490–1603. R. Jenkins. (Transactions of the Newcomen Society, 1938–39, vol. 19, pp. 35–48). The author gives a brief historical account of iron-founding in

England from its commencement in Sussex in 1490 up to 1603, when large numbers of guns were being cast, not only for English ships, but also for export to the Continent. He includes an account of the method then used for preparing the moulds for guns.

FOUNDRY PRACTICE

Moisture Control of Cupola Air. J. L. Brooks. (Foundry, 1940, vol. 68, Sept., pp. 38–39, 110). The author describes a plant, used by a foundry manufacturing piston rings, for removing the moisture from the cupola blast. The absorbent used is a solution of lithium chloride. The ability of this solution to absorb moisture depends on the temperature at which it is held and its concentration. In the plant described, which operates on what is known as the Kathabar system, the concentration of the solution is held constant and its temperature is varied so that the required amount of moisture is absorbed.

Sulphur in the Cupola. C. D. Abell. (Foundry Trade Journal, 1940, vol. 63, Oct. 3, p. 219). The author discusses a number of methods of eliminating sulphur during the melting of iron in a cupola. These methods include: (1) The use of ferro-manganese; (2) the production of a special slag; (3) the addition of lime or lime-washed coke, blast-furnace slag, soda ash or common rock salt; and (4) increasing the amount of slag produced by the cupola, altering its composition and running the furnace more slowly.

New *SMZ* Alloy Transforms White Cast Iron to Strong Grey Iron. (Steel, 1940, vol. 107, Sept. 23, p. 57). Some particulars are given respecting the use of a graphitising alloy known as *SMZ* (it contains silicon, manganese and zirconium) as a ladle addition to cast iron. It is claimed that this addition will convert a normally hard white iron into a high-duty grey iron.

Pearlitic Malleable from the Electric Furnace. E. F. Cone. (Metals and Alloys, 1940, vol. 12, Aug., pp. 150–153). The author reviews the properties and gives a brief illustrated account of the production, at the Belle City Malleable Iron Co., Racine, Wisconsin, of Belmalloy and Belectromal, the former being a pearlitic, the latter a high-strength malleable cast iron. Both alloys are made from a base metal, the composition of which is not stated, by melting with certain alloy additions in an electric-arc furnace with subsequent annealing in a Dressler tunnel kiln.

Investigates Problem of Cooling Sand. F. H. Amos. (Foundry, 1940, vol. 68, Aug., pp. 34–35, 92). The author considers the problem of the cooling of moulding sand and presents a number of graphs and data respecting the heat losses during the various stages of founding as the iron cools from tapping temperature to room temperature.

The Running of Castings. F. Henon. (Foundry Trade Journal, 1940, vol. 63, Oct. 10, pp. 239–242). The author discusses the theory and practice of metal-pouring with special reference to the type of runner, the number and dimensions of the gates and risers, and the speed at which the metal fills the mould. He develops a number of formulæ for calculating the velocity of the metal entering the mould under different conditions of pouring.

The Importance of Tellurium in Manufacturing. C. C. Drake. (Mines Magazine, 1940, vol. 30, Sept., pp. 498–500, 516, 519). The author outlines numerous improvements in the methods of melting, casting and cooling employed in the manufacture of chill-cast railway wagon wheels in the United States. He refers in particular to a method of controlling the depth of the chilled and transition zones by additions of silicon and chromium to the ladle, and states that a more recent development is the simultaneous use of the graphitising and stabilising agents, graphite and tellurium. This technique permits of a greater measure of control of the depth of chill and produces a softer and stronger metal which supports the chilled tread and flange of the wheel.

Causes of Gray Iron Casting Defects. W. B. McFerrin. (Iron Age, 1940, vol. 146, Sept. 5, pp. 44–47). The author presents in tabular form a list of thirty-two types of defects found in grey-iron castings and their causes, the latter being classified according to their degree of probability.

PRODUCTION OF STEEL

Weirton Steel Company. C. Longenecker. (Blast Furnace and Steel Plant, 1940, vol. 28, Aug., pp. 773–790, 826). After outlining the development of the Weirton Steel Co. since its foundation 35 years ago, the author describes the plant in detail, under the following headings: Coke-ovens; blast-furnaces; open-hearth and Bessemer department; soaking pits; 40-in. blooming mill; bar mills; 35-in. breakdown mill; 23-in./29-in. mill for structural steel; sheet mill; Weirton tinplate mill; strip mills; finishing department for wide-strip mill; finishing department for 10- and 16-in. strip mills; Steubenville tinplate mill; power production; boilers; metallurgical department; mechanical department; and accident prevention.

Youngstown Sheet and Tube Company, Chicago District. T. J. Ess and J. D. Kelly. (Iron and Steel Engineer, 1940, vol. 17, Sept., pp. Y-1–Y-18). The authors present a detailed account of the activities and plant of the Youngstown Sheet and Tube Co., at Indiana Harbor. This plant includes two batteries of coke-ovens, two blast-furnaces, eight open-hearth furnaces, two converters, a blooming mill, billet and bar mills, a skelp mill, merchant mills, pipe production facilities and a modern wide-strip mill.

Wisconsin Steel Company. T. J. Ess. (Iron and Steel Engineer, 1940, vol. 17, Sept., pp. W-2–W-15). The author gives a detailed and illustrated description of the works of the Wisconsin Steel Co., near Chicago. This plant includes three batteries of coke-ovens, three blast-furnaces, nine open-hearth furnaces, bloom and billet mills, and section and bar mills capable of producing 480,000 tons of rolled steel per annum.

Granite City Steel Company—Steel Pioneers in the West. T. J. Ess and J. D. Kelly. (Iron and Steel Engineer, 1940, vol. 17, July, pp. 1-GC–10-GC). A description with numerous illustrations is given of the development and present plant of the Granite City Steel Co., at Granite City, Illinois. The company now has an annual ingot capacity of 400,000 tons from ten 60-ton basic open-hearth furnaces. The final product is mainly steel sheet and strip, for which a 90-in. hot strip mill and cold-reduction mills were installed in 1936.

Open-Hearth Trends. W. J. Reagan. (Steel, 1940, vol. 107, July 22, pp. 62–65, 75; July 29, pp. 58–62, 72). In his discussion of the trends of open-hearth practice in the United States the author gives reasons for his opinion that the capacity of these furnaces will not increase beyond about 150 tons. Other points worthy of note are: (*a*) Sloping back walls are becoming standard practice; (*b*) insulation is now largely used below the floor level, and its use above this point is being extended; (*c*) the increased use of basic refractories, especially chrome and chrome-magnesite bricks, for the furnace hearth; (*d*) the increased use of instruments, particularly for temperature measurement; and (*e*) the extension of the duplex process to very large plants, as exemplified by a works now under construction which will include three 1200-ton mixers, five 60-ton basic Bessemer converters and six 150-ton open-hearth tilting furnaces.

Handling by Magnet Cranes in the Steel and Engineering Industries. (Iron and Steel, 1940, vol. 13, Sept., pp. 446–449). Electro-magnets of different types for lifting steel scrap, ingots, tubes, bars and angles in steelworks, foundries and engineering shops are described.

Statistical Analysis of Metallurgical Problems. E. M. Schrock. (Metal Progress, 1940, vol. 38, Aug., pp. 153–158). The author outlines some successful applications of statistical analysis to metallurgical problems connected with the control of materials and the production of steel at the works of the Jones and Laughlin Steel Corporation.

REHEATING FURNACES

Billet Heating. C. F. Herington. (Iron Age, 1940, vol. 146, Aug. 29, pp. 26–29). The author describes, with numerous illustrations, a modern furnace fired with pulverised coal for the reheating of billets and bars. The capacity of the furnace is about 120 tons of 15-ft. bars per 7½-hr. day, with a fuel consumption of about 210 lb. of coal per ton of steel. The furnace design provides for four streams of gases moving from a single bank of burners at one end of the furnace; these divide at the centre of the furnace and move towards each end of the hearth, passing both above and below the steel. A high degree of automatic control has been incorporated.

New Type Billet Heating Furnace for Non-Ferrous Metals and Steel. (Iron and Steel, 1940, vol. 13, Sept., pp. 469–470). A brief description is given of an automatic, gas-fired furnace for heating billets for subsequent pressing or forging operations. The furnace is designed to take 17 insulated trays 5 ft. by 2 ft. on which the billets are loaded; these trays are moved through the furnace by hydraulic pushers, and the discharging mechanism is so arranged that when the press operator wants a heated billet, he pushes a button, which causes one billet to be discharged on to a conveyor leading to the press. The heating and soaking time is 1½–2 hr.

FORGING, STAMPING AND DRAWING

Valve Forging. (Automobile Engineer, 1940, vol. 30, Aug., pp. 235–238). An illustrated description is presented of the machinery used and the sequence of operations in the manufacture of valves for internal-combustion engines at the works of Guest, Keen and Nettlefold, Ltd. Special completely automatic electric upsetting machines have been developed for mass production and over ten million valves have been produced during the five years ending 1938.

Power Economy in Hydraulic Press Forging. V. Tatarinoff. (Heat Treating and Forging, 1940, vol. 26, June, pp. 277–280). The author describes some examples of the uneconomic operation of hydraulic presses and discusses some recommendations for securing more economic working. He recommends that the losses of pressure water due to leakage should be recorded at least twice a week, and if the leakage exceeds 5 gal. per min. the source should be traced and repairs made. Good co-ordination between the press operators and the pump-room staff is also necessary, and a system of signalling, whereby the former can indicate to the latter whether the class of work to be undertaken is light, medium or heavy, will also assist in the more economical use of the pumping plant.

Progress in Metal Stamping Tools and Equipment. C. L. Szalanczy. (Heat Treating and Forging, 1940, vol. 26, July, pp. 340–343). The author describes and illustrates some recent developments in the design and construction of metal dies for the mass production of stampings.

ROLLING-MILL PRACTICE

Quality Problems in the Operation of Wide Strip Mills. G. D. Tranter. (Iron and Steel Engineer, 1940, vol. 17, July, pp. 38–44). The author describes the technique employed in the rolling of wide steel strip at the Middletown works of the American Rolling Mill Co. He stresses the co-ordination which exists between the open-hearth, slabbing-mill and strip-mill departments. These rolling mills have been described in a recent article entitled " New Slabber-Edger Setup " (*see* Journ. I. and S.I., 1940, No. II., p. 13 A).

Tension Control for Skin Pass Mills. C. P. Croco. (Steel, 1940, vol. 107, Sept. 16, pp. 68–72, 149). The author describes the development, at first on an experimental scale, and then on a full scale for service in a rolling mill, of a tensometer for automatically maintaining the desired tension on steel strip as it passes from one mill-stand to the next.

Strip Plant Finishing Equipment. D. A. McArthur. (Iron and Steel Engineer, 1940, vol. 17, Aug., pp. 22–30). The author describes and illustrates the processes and equipment used for the finishing operations of continuously rolled steel strip in accordance with modern American practice.

The Push Bench Process of Seamless Tube Manufacture. J. Porteous. (B. H. P. Review, 1940, vol. 17, Aug., pp. 8–12). The author describes the plant and process used in the production of seamless steel tubes at the works of Stewarts and Lloyds (Australia) Pty., Ltd., Newcastle, New South Wales. The push-bench process, which is a development of the Ehrhardt process, is used. In this, the billet is heated and pierced to form a " bottle," which is a short length of thick tube closed at one end; the bottle is then reheated and pushed through a series of roller dies of gradually decreasing diameter. The closed end is cut off, and, after being reheated, the rough tube passes through a finishing mill, which reduces it to the required dimensions.

Electrical Equipment for Armco Slabbing Mill. A. F. Kenyon. (Iron and Steel Engineer, 1940, vol. 17, Aug., pp. 32–38). An illustrated description is given of the electrical equipment and circuits for the drive of the reversing slabbing mill of the American Rolling Mill Co., Middletown, Ohio.

D.C. Mill Motor Standards. G. R. Carroll. (Iron and Steel Engineer, 1940, vol. 17, Aug., pp. 54–57). The author, who is chairman of one of the standardisation committees of the Association of Iron and Steel Engineers, presents the findings of his committee after its examination of the question of standard ratings and dimensions for D.C. motors for rolling mills.

Electron Tube Control for Steel-Mill Application. H. L. Palmer and H. W. Poole. (Iron and Steel Engineer, 1940, vol. 17, July, pp. 46–59). After explaining the theory of electric relays operated by a photo-electric cell in conjunction with electron valves and amplifiers, the authors describe some applications of these instruments to the control of certain rolling-mill equipment, such as automatic sheet catchers, rod-mill shears, run-out tables, pyrometers, pin-hole detectors for tinplate, strip-mill coilers, electric resistance welders and timing devices.

PYROMETRY

Optical Pyrometry. W. E. Forsythe. (Journal of Applied Physics, 1940, vol. 11, June, pp. 408–419). The author explains the principles and construction of optical pyrometers, methods of calibrating and testing them, and the calculations involved in their design and application.

Practical Applications of Temperature Control and Measurement. W. M. Barrat. (Journal of the Junior Institution of Engineers, 1940, vol. 51, Oct., pp. 14–23). The author describes in an elementary way the use of thermocouples in pyrometers and in instruments installed for the automatic temperature control of gas-fired furnaces, referring in particular to the importance of temperature control in the heat treatment of steel.

A High Sensitivity Radiation Pyrometer. N. E. Dobbins, K. W. Gee and W. J. Rees. (Transactions of the British Ceramic Society, 1940, vol. 39, Aug., pp. 253–257). The authors have developed a pyrometer which consists of a lens and mirror system throwing an image of the surface of a small specimen inside a furnace on to a screen in which is cut an elliptical hole of such dimensions that the light forming the centre part of the image is allowed to pass and strike a photo-electric cell. A calibration curve for the pyrometer was obtained by setting it up in front of a surface the temperature of which had been determined by means of an optical pyrometer. The authors claim that their pyrometer is much more sensitive than the usual type of optical or radiation pyrometer, an alteration in temperature of 1° C. at 2000° C. being easily detected, and, with certain precautions, even a change of only 0·5° C. A further advantage is that the target area needed for observation is very small.

HEAT TREATMENT

Principles of Gas Carburizing. J. A. Dow. (Industrial Heating, 1940, vol. 7, June, pp. 491–494, 502; July, pp. 594–596). The author discusses the principles of gas carburising in an elementary way, dealing mainly with the importance of the composition and uniform distribution of the carburising gas and emphasising the deleterious effect of the presence of carbon dioxide and water vapour. He briefly describes a method of measuring the carburising strength of a gas mixture. In conclusion he reports on his own experiments which indicated that: (*a*) The results obtained by gas-carburising in a muffle and in a radiant-tube furnace are equally satisfactory; and (*b*) the case depths obtained under identical conditions of treatment are approximately the same for steels of different compositions.

Salt Solution Ejector. (Machine Shop Magazine, 1940, vol., June, p. 71). A brief description is given of a simple device which facilitates the removal of spent salt solution from a salt-bath furnace. The device consists of an inverted U-tube one leg of which is immersed in the bath and the other projects into a catch-pit situated at a lower level than the bath. A motor-driven suction pump is connected to the tube, and, on switching on the motor, the liquid is drawn over from the salt bath to the catch-pit. A 15-ton bath can be completely emptied in about 20 min. The control unit is manufactured in a number of standard sizes, the largest having a maximum continuous discharge capacity of 25 tons per hr.

Some Metallurgical Phases of Flame Hardening. R. O. Day. (Metals and Alloys, 1940, vol. 12, Aug., pp. 167–171). The author considers flame-hardening to be a simple process of heating and quenching in accordance with recognised theories of hardening. He points out that the speed of operation, whilst fast, is not greater than the speed of reaction of hardenable steel of normal pearlitic or sorbitic structure; and he discusses the phenomena involved in the flame-hardening of alloy steels, cast iron and malleable iron. In conclusion he reports on his study of the hardness and microstructure of flame-hardened cases, which revealed that, although the temperature gradient during treatment is of necessity steep, there is never an abrupt change from the case to the untreated core, but there is always a definite transition zone in which the hardened case gradually changes in both hardness and structure to the untreated base material, this transition zone being, as a rule, shallower under thin cases and in alloy steels.

Flame and Electrical Surface Hardening. (Mechanical World, 1940, vol. 108, Oct. 11, pp. 265–267). Descriptions are given of the principles and equipment used for the surface-hardening of steel by the Shorter and Tocco processes. The former process is a well-known one. In the latter process the heat is applied electrically by surrounding the part with an inductor block which carries a H.F. current of the order of 9000 amp. at a frequency of about 2250 cycles per sec. The inductor block also incorporates a number of quenching jets, and mechanism is provided by means of which the current is automatically switched on and off and the water is turned on at the correct time interval. As special inductor blocks are usually required for each particular job, the Tocco process is most economical for mass-production work such as the hardening of the journals of automobile crankshafts.

Nitriding of Steels. F. W. Haywood. (Wild-Barfield Heat-Treatment Journal, 1940, vol. 4, Sept., pp. 11–16). The author describes and illustrates the theory and practice of nitriding as applied to steels containing one or more elements such as aluminium, chromium or vanadium, which readily form nitrides, and 0·2–0·5% of carbon. He enumerates the advantages of nitriding as follows: (1) The process is clean, cheap and simple; (2) there is no distortion of the parts, owing to the low temperature of the treatment; (3) as quenching is not required, the risk of quenching cracks is avoided; (4) slight growth is experienced, but this is much smaller than with carburising; (5) extremely hard cases are produced and the greatest degree of hardness is obtained 0·001–0·003 in. below the surface, from which depth the hardness gradually tapers off into the core, so that the risk of spalling is eliminated.

The Metallurgy of High Speed Steel. J. G. Ritchie. (Metals Treatment Society of Victoria and the Melbourne University Metallurgical Society: Australasian Engineer, 1940, vol. 40, Aug. 7, pp. 16–20, 35). The author points out the theoretical basis of the heat treatment for high-speed tool steel (containing tungsten 18%, chromium 4% and vanadium 1%) and explains the practical significance of each theoretical point. As the alloying elements in high-speed steel have many similar effects on the normal iron-carbon diagram, it is possible as an approximation to consider this steel as a ternary iron-carbon-tungsten alloy, and the author accordingly presents a section of this ternary diagram at 20% tungsten. After describing the casting and forging of ingots of high-speed steel, the author goes on to discuss the structural changes which occur during the heat treatment and tempering of this steel with reference to the diagram mentioned above. He concludes with some observations on the control of the atmosphere and the temperature of heat-treatment furnaces.

Selection of Steel and Heat Treatment for Spur Gears. H. B. Knowlton and H. E. Snyder. (Transactions of the American Society for Metals, 1940, vol. 28, Sept., pp. 687–701). The authors discuss the selection of steel and the correct heat treatment for the manufacture of case-hardened spur gears on the basis of both service experience and laboratory tests.

Heat Treatment—Large Batch Treatment. W. H. Grain. (Machine Shop Magazine, 1940, June, pp. 84–87). The author describes some modern heat-treatment furnaces for treating material in batches, and stresses some points which require particular attention when purchasing such a furnace.

Heat Treating the Track of a Tractor. H. J. Gregg. (Industrial Heating, 1940, vol. 7, July, pp. 587–592). The author describes the heat treatment of the shoes, links and driving sprockets of tractor tracks, and briefly touches on their manufacture as practised at the Caterpillar Tractor Co., Peoria, Illinois.

Flue Gases from Heat-Treatment Furnaces. O. G. Pamely-Evans. (Metallurgia, 1940, vol. 22, Sept., pp. 159–160). The author outlines the methods for the physical and chemical examination and for the conditioning of flue gases in heat-treatment furnaces, making a sharp division between muffle furnaces and those furnaces in which the combustion products are in direct contact with the charge to be heated. He suggests a combination of instruments which would allow close watch to be kept on the flue gases, so that even varying weather conditions might be indicated. He points out, however, that such instrumentation would be expensive and the return on capital outlay only small; whereas the installation of part of the items suggested might prove desirable.

Controlled Atmospheres for Modern Furnaces. H. M. Heyn. (Industrial Heating, 1940, vol. 7, Aug., pp. 683–688, 692). The author enumerates the gases and gas mixtures forming the more commonly used protective atmospheres of heat-treatment furnaces. He discusses the production of these atmospheres and their selection for different purposes.

Bright Annealing of Chromium Steels. O. Dahl and F. Pawlek. (Iron and Coal Trades Review, 1940, vol. 141, July 26, pp. 85–86; Aug. 2, pp. 113–114). An abridged English translation is presented of the authors' paper, in which they discuss the properties essential to a protective atmosphere for the successful bright annealing of 18/8 stainless steels. The paper originally appeared in Stahl und Eisen, 1940, vol. 60, Feb. 15, pp. 137–142. (*See* Journ I. and S.I., 1940, No. I., p. 242 A).

Continuous Annealing. R. J. Wean. (Iron and Steel Engineer, 1940, vol. 17, Sept., pp. 59–62). The author describes and illustrates a continuous annealing plant for annealing tinplate strip. The principal feature of this equipment is that the heating and cooling chambers are in the vertical, instead of the horizontal, position. The furnace is 30 ft. high, and is gas-fired by means of horizontal radiant tubes. This plant will anneal tinplate strip up to 38 in. wide at speeds between 75 and 300 ft. per min., according to the gauge. The advantages claimed for this type of plant are: (1) Economy of floor space; (2) ease of control; (3) uniformity in the properties of the product; (4) elimination of the conventional electrolytic cleaning; and (5) fast operation.

Continuous Heat Treating Machine Quickly Built for Emergency Service. H. C. Knerr. (Metal Progress, 1940, vol. 38, Aug., pp. 147–152). The author describes an automatic heat-treatment plant which was built in two months for the heat treatment of forgings, castings and rough-machined parts of uniform size up to about 5 in. in dia. and 2 ft. in length. The whole unit consists of three parts, a preheating and soaking furnace, a quenching device and a tempering furnace. Its overall length is 45 ft. Both furnaces are fired with propane, the former working at up to 1600° F. and the latter at up to 1300° F. The mechanical devices passing the material through the unit enable a throughput of 1000 lb. per hr. to be maintained.

Strand Heat Treating Cold Strip. N. P. Goss and C. H. Vaughan. (Steel, 1940, vol. 107, Sept. 9, pp. 52–57, 72). The authors describe a continuous annealing furnace which was installed at the works of Cold Metal Process Co., Youngstown, Ohio, in 1937, for the heat treatment of wide cold-rolled steel strip. The heating chamber of this furnace is 3 ft. 8 in. wide by 30 ft. long, and its capacity is 40 tons of low-carbon steel strip in widths up to 36 in. in 24 hr. The strip passes over the roller-hearth of the furnace, through an insulated tunnel into a cooling chamber 72 ft. long in which the cooling is done by a helical coil of tubing carrying cooling water. The grain size of the finished strip can be controlled by adjusting the speed at which it passes through the furnace. In conclusion the authors discuss the hardness, grain size and deep-drawing properties of strip which has been heat treated in this furnace.

How Thermal Stress-Relieving Reduces Residual Stresses. H. Lawrence. (Welding Engineer, 1940, vol. 25, Apr., pp. 19–21, 23). The author explains how residual stresses are set up in welded structures and the principles involved in the methods of relieving them.

A Stress-Relieving Furnace for Welded Structures. L. G. A. Leonard. (Metallurgia, 1940, vol. 22, Sept., pp. 137–138). The author describes a stress-relieving furnace recently installed at the works of Daniel Adamson & Co., Ltd. It is of the car-bottom type and of large size, being 36 ft. long, 14 ft. wide and 14 ft. high. It has been designed to take pressure vessels weighing from 25 to 30 tons, and can be used for normalising at 920° C., as well as for stress-relieving at 600–650° C. It is heated by 36 low-pressure gas-burners arranged in three zones, which in turn are subdivided into four groups of burners. The temperature of the zones is controlled independently by thermocouples inserted through the roof.

Additional Furnace and Quenching Facilities Installed at Carnegie-Illinois South Works. (Industrial Heating, 1940, vol. 7, Aug., pp. 704–709). An electrically-heated bell-type heat-treatment furnace and a new quenching tank are described which were recently installed at the South Works plant of the Carnegie-Illinois Steel Corporation, Chicago. The plant now consists of three furnaces and two quenching tanks equipped to operate under fully automatic control. The new furnace is designed to handle charges of up to 18,000 lb., taking bars up to 10 in. in dia. and 30 ft. long, or plates up to 52 in. wide. The capacity of the new quenching tank is 18,000 gal.

Bright Hardening of Tool Steels Without Decarburization or Distortion. J. R. Gier and H. Scott. (Transactions of the American Society for Metals, 1940, vol. 28, Sept., pp. 671–684). The authors describe a furnace and heat-treatment procedure for the hardening of tool steel without the risk of surface damage by oxidation and decarburisation. The method is particularly suitable for alloy steels of the air-hardening type used for blanking dies. A furnace with an all-metal muffle and a protective atmosphere of dissociated ammonia are used.

Heat Treating Small Arms. (Industrial Heating, 1940, vol. 7, Aug., pp. 690–692). A brief illustrated description is given of the heat-treatment and blueing departments of Colt's Patent Fire Arms Manufacturing Co., Hartford, Connecticut.

WELDING AND CUTTING

Minerals Used in Welding. O. C. Ralston and M. W. von Bernewitz. (United States Bureau of Mines, 1940, June, Information Circular 7121). The authors present a general review of the subject of welding with particular reference to the minerals used in the fluxes and electrode coatings.

The Nature of the Welding Arc. A. A. Alov. (Welding Industry, 1940, vol. 8, Sept., pp. 213–217). The author studies the forces operative within the arc in electric welding and the transfer of metal from the electrode to the parent metal.

Advantages of Bronze Rods in Oxy-Acetylene Welding of Cast Iron. (Welding Engineer, 1940, vol. 25, Apr., pp. 24–26). The technique and advantages of using bronze rods for the oxy-acetylene welding of cast iron are described.

Oxyacetylene Welding of Carbon-Molybdenum Steel Piping for High-Pressure, High-Temperature Service. Part I. Laboratory Development. R. M. Rooke and F. C. Saacke. (International Acetylene Association: Welding Journal, 1940, vol. 19, Aug., pp. 553–562). The authors describe and discuss the welding and heat-treatment procedure for satisfactory welding and stress-relieving by the oxy-acetylene process in the making of joints in molybdenum steel tubes for service at high temperatures and pressures. The American Standards Association Code and the American Society of Mechanical Engineers Boiler Construction

Code for welded joints in pipe lines and pressure vessels are also considered.

Oxyacetylene Welding of Carbon-Molybdenum Steel Pipe. Part II. Field Testing and Application. A. N. Kugler. (International Acetylene Association: Welding Journal, 1940, vol. 19, Aug., pp. 562–567). The author discusses some results obtained in field tests on oxy-acetylene-welded joints in high-pressure steel tubes and presents a suggested specification entitled "Process Specification for Oxy-Acetylene Fusion Welding of Carbon-Molybdenum Steel Pressure Pipe Lines."

Brazing with Salt Bath Furnaces. A. E. Ballis. (Industrial Heating, 1940, vol. 7, June, pp. 496–498). The author describes the use for brazing of an electrically-heated salt-bath furnace, which has been developed during the last ten years, and which is also suitable for heat-treating operations, *e.g.*, annealing, normalising, hardening and case-hardening. The composition of the salt bath is so selected that it is also effective as a flux. Further advantages of the salt-bath method of brazing, enumerated by the author, are: (1) The equipment is simple and inexpensive; (2) the furnace works at a constant temperature and need not be reheated after each load; (3) the light racks used to hold the work absorb little heat; (4) chromium is not oxidised, so that chromium alloys can be brazed; (5) the treated parts can be quenched quickly or cooled slowly, as required, while the protective coating of the salt prevents oxidation or scaling.

Welding Metallurgy. Part IX. Rate of Cooling. O. H. Henry and G. E. Claussen. (Welding Journal, 1940, vol. 19, Aug., pp. 571–575). Continuation of a series of articles (*see* Journ. I. and S.I., 1940, No. II., p. 221 A). The authors apply the iron-carbon diagram to a consideration of how the rate of cooling of a welded joint affects its physical properties.

Some Notes on Weldability. E. D. Lacy. (Metal Treatment, 1940, vol. 6, Autumn Issue, pp. 128–130). The author describes the research programme of the Institute of Welding now in hand on the weldability of mild steel, high-tensile structural steel, steel for pressure vessels, alloy steels, stainless steels and cast iron.

Factors Affecting Residual Stresses in Welds. C. T. Gayley and J. G. Willis. (Welding Journal, 1940, vol. 19, Aug., pp. 303-S–306-S). The authors give an account of an investigation in which the influence of a number of factors on the degree of distortion produced in electrically welded steel plates was studied. The factors in question were: (1) The design of the joint; (2) the number of layers per inch thickness of weld metal; (3) the rate of weaving during the deposition of the weld metal; (4) the flow of heat between successive layers of deposited metal; (5) the arc energy; (6) the diameter of the electrode; and (7) the amount of penetration into the parent metal. The data obtained are given in tables. The general conclusions from the results obtained are: (*a*) The fewer the beads making up the weld, the less will be the distortion; (*b*) the steeper the temperature gradient during cooling, the greater will be the residual stresses and the distortion; (*c*) in making V butt joints in plates up to $\frac{3}{4}$ in. thick each plate edge should be bevelled at an angle of about 22°; (*d*) welding with a high weaving rate causes less distortion than a low weaving rate; (*e*) the welded joints with the greatest penetration also exhibit the greatest distortion; and (*f*) the interpass temperature, which governs the flow of heat between successive beads, and the rate of weaving are closely related. When a joint is welded with a few beads the temperature of the parent metal is raised to a higher point than when a large number of beads are deposited, because the beads are thinner and require less time for deposition. This raising of the temperature with a few heavy beads has a pronounced preheating effect and reduces the distortion very considerably.

Second Progress Report—Joint Investigation of Continuous Welded Rails. H. F. Moore, H. R. Thomas and R. E. Cramer. (Welding Journal, 1940, vol. 19, Aug., pp. 293-S–302-S). This is the Second Progress Report on an investigation of welded joints in rails commenced in 1937 and sponsored by the Association of American Railroads and the Engineering Experiment Station of the University of Illinois. (An abstract of the First Progress Report appeared in Journ. I. and S.I., 1939, No. II., p. 343 A). The work reported in the first report has been continued. Sixteen rolling-load tests have been completed, and further bend tests and more microscopical studies have been made. Plans have been made for the systematic reporting of welded-rail failures in service. The following papers are included in the report:

Tests of Welded Joints Under Repeated Wheel Load, by N. J. Alleman and H. F. Moore.

Metallographic Tests, by R. E. Cramer and E. C. Bast.

Mechanical Tests of Specimens from Welded Joints, by S. W. Lyon.

Flame-Cutting Steel for Forging or Machining. F. Collins. (Machinery, 1940, vol. 57, Oct. 17, p. 79). The author points out that an increasing use is being made of oxy-acetylene flame-cutting machines as a substitute for sawing in the preparation of bars and billets for forging, and surveys the differences between flame-cutting and sawing. An advantage of the former process is that it requires only half as much time as the latter, and the loss of metal is smaller. The author then briefly considers the possibilities of removing metal from blanks by flame-cutting in order to reduce the machining time. In conclusion he deals with the effect of flame-cutting on the surface hardness, pointing out that it may cause serious brittleness in the more hardenable steels, and he gives a list of carbon and nickel steels which can be flame-cut, and of certain alloy steels which, owing to the formation of cracks where the flame impinges, are unsuitable for flame-cutting.

CLEANING AND PICKLING OF METALS

Speed and Economy are Features of the Flame-Gouging Process. (Welding Engineer, 1940, vol. 25, Mar., pp. 29–32). The technique and several applications of the flame-gouging process are discussed, and the tools and nozzles used are described and illustrated.

Notes on the Conditioning of Semi-Finished Steel. (Metal Progress, 1940, vol. 38, Aug., pp. 168–170). Some methods of conditioning steel blooms, billets and slabs are described and illustrated. These methods include the application of water at high pressure, scarfing with the oxy-acetylene flame, mechanical chipping, grinding, and sand- and shot-blasting.

Autogenous Deseaming for Removing Surface Defects from Ingots and Billets. J. Aversten. (Jernkontorets Annaler, 1940, vol. 124, No. 5, pp. 212–224). (In Swedish). A description is given of the theory and practice of deseaming ingots and billets with a special oxy-acetylene torch. This process has been in use for some time in the United States, and has recently been introduced at a number of steelworks in Sweden.

Efficient Trichlorethylene Degreasing. E. E. Halls. (Metal Treatment, 1940, vol. 6, Autumn Issue, pp. 131–133). The author discusses methods of promoting the efficiency of trichlorethylene degreasing plants. He points out that in the presence of light, or of traces of moisture, or a combination of the two, this solvent decomposes to a very small degree with the formation of hydrochloric acid; it is therefore frequently subjected to a stabilising treatment to suppress this decomposition. He presents a table showing the extent of the decomposition of trichlorethylene under different conditions of exposure, and graphs demonstrating how its boiling point is raised by increasing contamination with oil and wax. He recommends that the following criteria be fulfilled for the efficient operation of a degreasing plant: (1) Only the best stabilised solvent should be used; (2) the design of the plant should have regard to the mode of heating, the recovery of the solvent, the escape of vapour and adequacy of cooling; (3) the vapour or liquor or both should be selected according to the class of work; (4) the plant should be operated at full capacity so that the inevitable solvent losses are associated with maximum throughput; (5) the mode of packing the work should be carefully studied; and (6) in deciding the location of a plant, attention should be paid to the loss of solvent which can be caused by draughts and the need for protecting the operators from fumes.

Heating Open Tanks by Submerged Combustion. W. G. See. (Industrial Heating, 1940, vol. 7, July, pp. 604–608). The author briefly describes a submerged combustion system developed and patented by the Submerged Combustion Co. of America, Hammond, Indiana. The system is designed for the heating of pickling tanks and is said to have several advantages as compared with heating by open steam jets.

Data on the Use of Stannous Chloride as a Pickling Inhibitor, T. P. Hoar and S. Baier. (Sheet Metal Industries, 1940, vol. 84. Sept., pp. 947–948). The authors present the results of a series of tests with pickling baths containing as a pickling solution 7% (by weight) of sulphuric acid at 80° C. to which different inhibitors were added. The object of the experiments was to determine: (*a*) The inhibitory power of stannous chloride as compared with other inhibitors; (*b*) the comparative rate of deterioration of the bath when in use and while standing idle; (*c*) the possibility of regenerating the spent bath; (*d*) the possibility of using materials other than flour to reduce the amount of hydrogen taken up by the steel. The data obtained with fifteen different inhibitors are compared in a table. It was found that the most satisfactory bath inhibited with stannous chloride contained 0·05% of stannous chloride ($SnCl_2.2H_2O$), 0·025% of gelatin and 0·05% of cresolsulphonic acid. The gelatin gave a cleaner solution and better inhibition than the flour originally proposed, and the cresolsulphonic acid prevented oxidation.

A Critical Discussion of the Data Provided on the de Lattre Processes. (Sheet Metal Industries, 1940, vol. 14, Aug., pp. 829–830). In this article some criticisms are put forward with regard to the claims made for the de Lattre process of pickling described in a recent paper (*see* Journ. I. and S.I., 1940, No. II., p. 189 A). In the description of the de Lattre process it is stated that no appreciable quantity of hydrochloric acid is added during working and that it is the sulphuric acid which is consumed; on the other hand, reasons are given in the present article why it is more probable that it is mainly the hydrochloric acid which does the pickling, and it therefore appears that the process is nothing more than the regeneration of hydrochloric acid by means of sulphuric acid by the reaction of the latter with the additional ferrous chloride formed. It is also disputed that the large savings of acid consumption claimed are really due to the adoption of the de Lattre process, for it is more likely to be attributable to the use of an inhibitor where none was previously used. In conclusion it is stated that the following points should be taken into account in considering the de Lattre process as an alternative to straight hydrochloric acid pickling: (1) The suitability of the existing materials of construction for use with mixed acids at slightly higher temperatures; (2) the present cost of waste disposal compared with the capital and operating cost of a recovery plant; (3) the possibility of obtaining a regular market for copperas ($FeSO_4.7H_2O$) slightly contaminated by chloride; and (4) the greater chemical control necessary to maintain the correct proportion of the two acids.

The Electrolytic Polishing of Stainless Steels. H. H. Uhlig. (Electrochemical Society, Oct., 1940, Preprint No. 25). The author developed the use of glycerine/phosphoric-acid mixtures for the electrolytic polishing of 18/8 stainless steel. He determined the optimum conditions to produce polish with the aid of a photoelectric spectrophotometer and found that the best polish was obtained with an electrolyte consisting of approximately 42% of phosphoric acid, 47% of glycerine and 11% of water, used at 100° C. or higher with an anodic c.d. of at least 0·1 amp. per sq. in. Some other organic additions to phosphoric acid also proved useful. These had, in general, high boiling points, contained one or more hydroxyl groups and were soluble in phosphoric acid. These organic-solution/phosphoric-acid electrolytes could also be used to polish mild steel, chromium steel and 18/8 steel containing molybdenum. The author assumes that the organic additions to phosphoric acid affect the electrolytic-polishing process by two independent mechanisms, the first being a change in the conductivity of the electrolyte and the anodic phosphate film which produces a high polish, and the second being a considerable retardation of localised pickling of the anode by acid.

Electrolytic Cleaning, Pickling and Polishing of Wire and Wire Products. C. L. Mantell. (Wire and Wire Products, 1940, vol. 15, Aug., pp. 413–415). The author describes the general principles of electrolytic cleaning as applied to wire and reviews some recent American patents relating to this process.

COATING OF METALS

Control of Spangles. W. G. Imhoff. (Steel, 1940, vol. 107, Sept. 23, pp. 54–56). The author discusses the factors controlling the formation of spangle in the galvanising of steel sheets. The principal factors are the period of immersion and the temperature of the bath. Galvanising at low temperature produces a soft and ductile coating, whilst high-temperature coatings are hard and brittle, and the spangles become smaller as the temperature of the bath increases until at 945° F. with 30 sec. immersion there is no spangle at all, because the coating is " burned ".

Improvements in Galvanizing Plants and Practice. N. E. Cook. (Iron and Steel Engineer, 1940, vol. 17, July, pp. 28–35). The author gives an account of the improvements in the technique of galvanising which have taken place in the last ten years as a result of applying scientific methods of research and control.

High-Production Cleaning, Pickling and Plating Line. (Steel, 1940, vol. 107, Sept. 23, pp. 50–52, 84). A description is given of the plant at an American works for the cleaning, pickling and galvanising of 4000 lb. of small parts per hr. The plant is operated by only seven men, and a high degree of mechanisation contributes materially to the maintenance of this high output.

Addition Agents in the Electrodeposition of Zinc. J. L. Bray and F. R. Morral. (Electrochemical Society, Oct., 1940, Preprint No. 15). The authors studied the electrodeposition of zinc on patented steel wire in the presence of various amines. They found that amines with a low number of carbon alkyl radicals had a favourable influence on the brightness and ductility of the electrodeposits. Very small amounts of these amines proved to be effective, and the required concentration was dependent on the amine nitrogen present.

X-Ray Analysis of Hot-Galvanized Heat-Treated Coatings. F. R. Morral and E. P. Miller. (American Institute of Mining and Metallurgical Engineers, Technical Paper No. 1224: Metals Technology, 1940, vol. 7, Sept.). The authors report on an investigation by X-ray diffraction analysis to determine the phases present in hot-dipped, heat-treated zinc coatings on steel sheets and wire. The results showed that the zinc-iron alloys present consisted mainly of the ζ-phase ($FeZn_{13}$), which corresponds to that found by Schueler in 1925. Small amounts of Fe_5Zn_{21} and δ_1 phase were also present in the coating.

Surface Cladding. G. E. Gude, jun. (Steel, 1940, vol. 107, Aug. 26, pp. 44–46). The author describes a technique which has been developed for welding panels of stainless steel on to carbon-steel plate for the fabrication of corrosion-resistant vessels. The stainless-steel sheet generally used for this purpose is 12–16 gauge, and it has been found economical to cut it into panels about 3 ft. long and 4–6 in. wide. These panels are placed in position on the carbon-steel plate with sufficient space between them for a welder to deposit one bead from a high-alloy welding electrode in such a way that it joins two adjacent panels together and simultaneously welds both edges to the carbon-steel plate. A stress-relieving treatment is necessary after lining a vessel in this way.

Reducing the Cost of Metal Spraying. E. T. Parkinson. (Steel, 1940, vol. 107, Sept. 9, pp. 46–49). The author gives some particulars of the improved performance of some of the latest metal-spraying pistols which use oxygen in conjunction with acetylene, propane or natural gas. With this equipment the following quantities of metals can be deposited per hour: (1) 8–10 lb. of steel; (2) 7–9 lb. of aluminium; (3) 15–20 lb. of bronze; and (4) 30–35 lb. of zinc.

Phosphatising Aids Rapid Finishing. E. Armstrong. (Metal Treatment, 1940, vol. 6, Autumn Issue, pp. 105–110, 118). The author describes in detail the process and equipment used for phosphatising steel sheets as a treatment preparatory to enamelling. The results of numerous salt-spray exposure tests are presented in tables, from which a comparison of the corrosion-resisting properties of phosphatised and non-phosphatised enamelled steel specimens can be made.

Determination of Thickness of Acid-Resistant Portion of Vitreous Enamel Coatings. W. N. Harrison and L. Shartsis. (Journal of

Research of the National Bureau of Standards, 1940, vol. 25, July, pp. 71–74). A method of preparing and testing an oblique section of enamelled iron is described, the object of which is to enable the thickness of the outer acid-resistant coating to be determined.

Recent Developments in the Application of Asphalt-Mastic Coating to Both New and Operating Pipe Lines. W. W. Colley. (Proceedings of the Natural Gas Section of the American Gas Association, 1940, pp. 81–84). The author describes the latest technique for the preparation and application of asphalt-mastic composition for the protection of gas mains.

Protective Coating for Steel Water Lines. D. Bronson. (Journal of the American Water Works Association, 1940, vol. 32, Aug., pp. 1385–1393). The author describes several recognised methods of protecting the inner and outer surfaces of steel water-mains in accordance with American practice. Experience indicates that the most efficient form of protection for both the inside and outside of steel pipes is afforded by coal-tar enamel. For the inside a priming coat of liquid coal-tar is applied cold, after which the coal-tar enamel is applied hot from a retractable feed line, while the pipe is spun at a fairly high speed. For the outside, after the coal-tar enamel has been applied a coal-tar-impregnated asbestos-felt wrapper is wound round the pipe.

Protective Coatings for Aircraft. (Canadian Metals and Metallurgical Industries, 1940, vol. 3, July, pp. 174–175). Some particulars are given of the numerous paints, varnishes and enamels used for the protection of steel and non-ferrous metal parts of aircraft.

Rust Problems. R. W. Baker. (Steel, 1940, vol. 107, July 1, pp. 62–64). After briefly reviewing methods of combating the rusting of steel, the author describes the properties of a clear blend of natural oils which has many advantages as a priming coat. This oil can be applied to steel which has already rusted ; it has a penetrating action and it gets into the spaces between the steel and the rust particles ; it also hardens into an elastic film, the hardening commencing at the bottom so that moisture and air are gradually forced out of it. When the top surface has become a dry film a surface coat of paint can be applied. No indication of the cost of the oil is given, but it is claimed to be extensively used for maintenance work in the field, especially for surfaces which cannot easily be cleaned before painting.

PROPERTIES AND TESTS

Modulus of Elasticity of Alloys. (Metallurgist, 1940, vol. 12, Oct., pp. 129–131). This is an abridged translation of a paper published by L. Guillet, jun., in Revue de Métallurgie, Mémoires, 1939, vol. 36, pp. 497–521. (*See* Journ. I. and S.I., 1940, No. I., p. 295 A.)

800-Ton Hydraulic Testing Machine. (Engineering, 1940, vol. 150, Sept. 13, pp. 201–202). A description, with several photographs and diagrams, of an 800-ton hydraulic testing machine by A. J. Amsler and Co. is presented. The machine is intended mainly for the testing of special pit-cage wire ropes for deep coal mines. The machine has an overall length of 103 ft. and the movable head, which has a stroke of 6 ft. 6 in., is operated by two identical cylinders supplied with oil from a pump driven by an 11-h.p. electric motor.

Chevenard's Micromachine for the Determination of the Mechanical Properties of Metals. P. Sederholm. (Jernkontorets Annaler, 1940, vol. 124, No. 7, pp. 313–316). A detailed description is given of a Chevenard testing machine purchased by Jernkontoret for their laboratories in 1939. The machine is intended for making tensile, bend and shear tests on very small specimens. The specimens for tensile tests are 15 mm. long and 3 mm. in dia. Some particulars of a similar machine have been given previously (*see* Journ. I and S.I., 1935, No. I., p. 457).

The Tensile Testing Laboratory of the Royal Technical College [Kungl. Tekniska Högskola] and Some of the Work Carried Out There. F. Odqvist. (Teknisk Tidskrift, 1940, vol. 70, Aug. 17, Mekanik, pp. 81–91). (In Swedish). The author describes many of the tensile- and fatigue-testing machines and some of the research work that has been done with them at the Royal Technical College, Stockholm. The machine of particular interest is an Amsler hydraulic pulsating testing machine of recent design capable of applying stresses in the ranges 0–30 tons and − 15 to + 15 tons per sq. in ; the mechanism of this machine is described in detail with diagrams. Improved apparatus for long-time creep tests and some investigations in progress are also described. Details are also given of a new type of fatigue test on ring-shaped test-pieces vibrated at their natural frequency. Finally, some fatigue tests on crane hooks and models of crankshafts are reported and comparisons are made with the results of rotating bend tests on the same materials.

A Simple Method of Finding Poisson's Ratio. Mary D. Waller. (Proceedings of the Physical Society, 1940, vol. 52, Sept. 1, pp. 710–713). The authoress describes a method of measuring Poisson's ratio in terms of the ratio of two specified natural frequencies of a freely vibrating square plate, and gives results for steel, brass, copper, aluminium and glass. In conclusion she suggests that her method of estimating Poisson's ratio might be of value in the mechanical testing of metals and alloys.

The Interrelation Between Stress and Strain in the Tensile Test. E. J. Janitzky and M. Baeyertz. (Transactions of the American Society for Metals, 1940, vol. 28, Sept., pp. 714–723). Three empirical equations which express the relation between stress and strain in the tensile test, as determined on round specimens with a gauge length of 2 in. and a diameter of 0·505 in., are presented. The validity of these equations is demonstrated by means of experimental data obtained on steels of very different chemical compositions and after various forms of heat treatment, such as annealing, normalising, and quenching followed by tempering.

Laboratory Tensile Tests Shed Light on Fish-Eye Fractures in Weld Metal. Mildred Ferguson. (Welding Engineer, 1940, vol. 25, Apr., pp. 22–23). The authoress describes a certain type of fracture in tensile test specimens of weld metal. These fractures are distinguished by shallow circular cavities with a deep pinhole in the centre, and they are therefore sometimes called "fish-eye fractures." A microscopical examination of sections has revealed successive stages of the development of this type of fracture, and has shown that it is centered on an inclusion and that the metal first separates from the inclusion in the direction of stress and this separation becomes large before the transverse rupture forming the shallow cavity occurs.

Stress-Strain Analysis from Crack Formations in Brittle Lacquer Coatings. (Machinist, 1940, vol. 84, Sept. 21, pp. 322 E–323 E ; Sept. 28, pp. 328 E–330 E). The use of brittle lacquers as developed by the Massachusetts Institute of Technology for the detection of stress concentrations in steel parts and assemblies is described and illustrated. It has been found necessary to prepare a graduated series of eleven lacquers, one of which is selected according to the humidity and temperature of the atmosphere. The interpretation of the cracks appearing in the lacquer is explained, and it is pointed out that the cracks which form in the brittle lacquer during a static test appear at those points where fatigue cracks are most likely to form in the structure under the dynamic loading of actual operation.

Hydrogen, Flakes and Shatter Cracks. C. A. Zapffe and C. E. Sims. (Metals and Alloys, 1940, vol. 11, May, pp. 145–148 ; June, pp. 177–184 ; vol. 12, July, pp. 44–51 ; Aug., pp. 145–148). The authors survey the literature on the relation of faults, in particular flakes and shatter cracks, to the presence of hydrogen in steel. In the first part they deal with the solubility of hydrogen and methods of calculating the hydrogen pressure in steel, and the diffusion and occlusion of hydrogen in steel. In the second part they discuss theories concerning the mechanism whereby hydrogen may be retained in steel at ordinary temperatures. In the third part the influence of the presence of hydrogen on the formation of flakes and shatter cracks is dealt with. In the concluding part details are given of the furnace technique required to prevent the formation of flakes in steel. A bibliography of 239 references arranged in chronological order from 1863 to 1939 is presented.

The Relationship Between the Mechanical Properties of Materials and the Liability for Failure in Service. L. W. Schuster. (Transactions of the Liverpool Engineering Society, 1939–40, vol. 61, pp. 36–68). The author explains the use of various forms of mechanical tests applied to steel and the value of data on tensile strength, yield point, limit of proportionality, reduction of area, resistance to bending and impact, fatigue strength and impact strength. From his observations he draws the conclusion that, when service breakages under dynamic loading are the result of some inferior mechanical property, the inferiority must be sought in some property not usually measured by any of the tests previously mentioned. In conclusion he points out the importance of the properties at the exposed surfaces of parts and that, in testing, the direction of loading in relation to the grain of the material should correspond with that in service.

Fatigue Failure. G. Sachs. (Iron Age, 1940, vol. 146, Sept. 5, pp. 31–34; Sept. 12, pp. 36–40). The author surveys the literature on the effect of stress raisers on the fatigue strength of metals and alloys. He distinguishes between three types of stress raisers, *viz.*, machined notches, corrosion and chafing or fretting, and presents numerous curves which show how the fatigue strength of specimens is reduced by these factors. A bibliography with forty references is appended.

Fatigue Strength of Drilled and Notched Steel Test-Pieces under Alternating Tensile and Compression, Bending and Torsional Stress. (Metallurgia, 1940, vol. 22, Sept., pp. 139–140). A review is presented of the conclusions reached by Körber and Hempel as a result of their fatigue, tensile, compression, bend and torsion tests on plain, drilled and notched specimens of eight unalloyed and six chromium-nickel steels. A full report of this investigation was published in Mitteilungen aus dem Kaiser-Wilhelm-Institut für Eisenforschung, 1939, vol. 21, No. 1, pp. 1–19. (*See* Journ. I. and S.I., 1939, No. II., p. 33 A).

The Influence of Magnetic Fields on Damping Capacity. E. R. Parker. (Transactions of the American Society for Metals, 1940, vol. 28, Sept., pp. 661–667). The author reports on an investigation of the damping capacity of a 0·40% carbon steel and a steel containing 5% of chromium and 0·5% of molybdenum in constant and alternating magnetic fields. A Föppl-Pertz machine was used for determining the damping capacity, which in this case was measured in terms of the heat dissipated by a unit volume of metal during a completely reversed cycle of stress. The loss of energy from a specimen in a constant magnetic field was found to be less than half that with a zero field, whilst an alternating field reduced the energy loss even further. The endurance of fatigue specimens was found to be reduced by the presence of a constant magnetic field.

Steel Hardenability as Related to Physical Properties. G. T. Williams. (Steel, 1940, vol. 107, Aug. 19, pp. 49–54, 80). The author discusses the relation between the hardenability of steel and its ductility in a tensile test. He carried out a number of "end-quench" tests with low-alloy steels containing chromium and molybdenum, and others containing vanadium, the carbon being in the range 0·40–0·50%. The principal object was to determine the effect of the heat-treatment temperature and time on the resulting hardness and physical properties. The end-quench test is made by quenching the end instead of the whole of the specimen and making hardness determinations at increasing distances from the end. The author had presumed that if one steel had higher hardenability than another it would also have better structural uniformity and mechanical properties, but the data he obtained showed that whilst the hardenability was increased by heat-treating at 1700° F. instead of 1500° F., the mechanical properties deteriorated slightly, and that there was no direct relationship between the hardenability as demonstrated by the end-quench test and the ductility.

Magnetic Permeability of Some Austenitic Iron-Chromium-Nickel Alloys as Influenced by Heat Treatment and Cold Work. J. B. Austin and D. S. Miller. (Transactions of the American Society for Metals, 1940, vol. 28, Sept., pp. 743–755). The authors report on their investigation of the effect of different forms of heat treatment and cold-work on the magnetic permeability of 18/8, 18/12 and 25/12 chromium-nickel steels. The apparatus used is described in detail. They arrived at the following general conclusions: (1) The austenitic iron-chromium-nickel alloys, if free from ferrite, all have practically the same permeability, which, in the experiments described, was 1·003 under a magnetising force of 436 oersted; (2) the 18/8 and 18/12 groups become magnetic if sufficiently cold-worked, but the effect is very much greater in the former, whereas the permeability of the 25/12 group is practically unaffected by cold-work. In general, the higher the carbon and the higher the nickel/chromium ratio, the smaller is the effect of cold-work on permeability; (3) in none of the alloys was the increase in permeability directly proportional to the amount of cold-work; in fact, there appeared to be no simple relation between these quantities; (4) careful welding does not appreciably alter the permeability of these alloys; and (5) if it is essential to have a very low permeability then the nickel/chromium ratio should be as high as the other requirements of the alloy will permit.

Electric Furnace Cast Irons Containing Copper. T. E. Barlow. (Metals and Alloys, 1940, vol. 12, July, pp. 36–39; Aug., pp. 158–165). The author's paper is based upon unpublished progress reports by C. H. Lorig and E. C. Kron of the Batelle Memorial Institute, who studied the effects of copper, in amounts up to 3%, on four series of cast irons, *viz.*: (1) High-carbon cast irons with 3·30–3·44% of carbon and 0·63–2·50% of silicon; (2) medium-carbon cast iron with 3·07–3·20% of carbon and 0·68–2·61% of silicon; (3) low-carbon cast irons with 2·80–2·90% of carbon and 1·09–2·52% of silicon; and (4) alloy cast irons with 3·12–3·23% of carbon, 1·48–1·74% of silicon and various combinations of nickel, chromium, molybdenum and vanadium. The irons had been produced in an indirect-arc electric furnace, under carefully controlled conditions, and test bars of four different sections, 0·875, 1·2, 2·0 and 4·0 in. in dia., were prepared. Numerous tables and diagrams are reproduced which show the effect of the copper content of the bars on their tensile strength, transverse strength, transverse deflection, Brinell hardness, modulus of rupture, resilience and impact value. The results indicate the value of copper, either alone or in simple combination with other alloying elements, as an addition to cast iron. In general, the presence of 1–2% of copper resulted in an increase in the tensile strength, the transverse strength and the Brinell hardness; and the tendency of copper to lower the values for transverse deflection was only pronounced in the low-carbon specimens. In conclusion some experiments on the effect of copper on the chilling properties of cast iron are reviewed. These indicated that copper reduces both the chill and the shrinkage—a result which supports the contention that an addition of copper exerts a beneficial effect on the density of heavy sections of cast iron.

Glass Moulds Present Many Problems. R. M. Scafe. (Foundry, 1940, vol. 68, Aug., pp. 30–31, 103–105; Sept., pp. 36–37, 101–103). The author discusses the special properties which are required of the iron used for making moulds in which glass is cast and the peculiarities of a number of alloy cast irons which are used for this purpose.

Castings Must Pass Many Tests. M. D. Johnson. (Foundry, 1940, vol. 68, Aug., pp. 28–33, 108). The author describes the organisation of an inspection department of an iron foundry, and discusses the utility of a number of methods of testing castings.

Some Low-Alloy Steels. W. Ashcroft. (Metallurgia, 1940, vol. 22, Sept., pp. 161–163). The author reviews the compositions, properties and uses of some low-alloy steels, dealing separately with manganese, chromium, nickel, molybdenum and nickel-chromium-molybdenum steels.

Titanium and Some Properties of Cr-Mo Steel for Airplane Tubing. G. F. Comstock. (Metals and Alloys, 1940, vol. 12, July, pp. 21–26). The author investigated whether small additions of titanium to S.A.E. *X*4130 chromium-molybdenum steel would improve its ductility and weldability without impairing any of its other properties. He also investigated the effects of greater additions of manganese and copper, as it had been said that the presence of these elements counteracted a reduction in strength caused by titanium. The experiments led to the following conclusions: (1) In forged and normalised samples welded lengthwise, machined flat and bent cold with the weld outside, the presence of titanium improved the bending properties, also when additional manganese and 0·31% of copper were present. The best results were obtained with steels containing (*a*) 0·15% of titanium and (*b*) 0·09% of titanium and 0·31% of copper. (2) The presence of more than 0·1% of titanium produced a softer steel, but did not decrease the degree of hardening by welding. Steel with 0·85% of manganese and 0·093% of titanium was harder than that with the normal manganese content, but there was no greater hardening by welding than with ordinary S.A.E. *X*4130 steel. Micrographs showed a narrower hardened and coarse zone in the titanium steels. (3) The improved ductility, impact value and microstructure produced by titanium, or by titanium together with aluminium, are thus reflected in a corresponding improvement in the results of bend tests on welds. (4) Steel S.A.E. *X*4130 containing 0·09–0·10% of titanium and 0·85% of manganese is of satisfactory strength in the normalised condition, the hardening by welding is not excessive and the ductility after welding, as well as the resistance to impact, are definitely improved as compared with steel S.A.E. *X*4130 without titanium.

Combined Stress Experiments on a Nickel-Chrome-Molybdenum Steel. J. M. Lessels and C. W. MacGregor. (Journal of the Franklin Institute, 1940, vol. 230, Aug., pp. 163–181). The authors carried out experiments in which thin-walled tubes of a nickel-chromium-molybdenum steel were subjected to combined axial tension and internal pressure. The axial loading was applied by means of a 30-ton Baldwin-Southwark hydraulic testing machine, and the internal pressure was supplied by means of an Amsler pendulum dynamometer and pump of 17,000 lb. per sq. in. capacity. Tangential strains were measured by two hydraulic lateral extensometers, one of which was designed for use in axial tension tests and the other for use in axial compression tests. These extensometers, recently developed by the authors, possessed the advantage of measuring accurately the average of the tangential strains over a considerable length of the tubes instead of at a single cross-section only. Axial strains were measured with a Martens extensometer. The authors report also on some torsion tests on solid bars, carried out for comparative purposes. In conclusion the authors point out that their experimental results are in very satisfactory agreement with the theory of constant energy of distortion, developed independently by von Mises, Huber and Hencky.

The Properties of Cold-Reduced Tinplate. H. A. Stobbs. (Tin and Its Uses, 1940, Oct., pp. 2–4). The author points out that cold-reduced tinplate has the following advantages as compared with the product obtained by the older pack-rolling method: (1) Improved mechanical properties; (2) greater uniformity and the fact that it can be rolled to closer dimensional tolerances; (3) greater corrosion resistance; (4) better appearance; and (5) greater continuity and uniformity of the tin coating. He describes how these advantages, in spite of various difficulties, led to the rapid development of this method of manufacture.

Metals and Alloys in an Air-Conditioned Automobile. H. Chase. (Metals and Alloys, 1940, vol. 12, Aug., pp. 141–144). The author discusses the metals and alloys used in the construction of an air-conditioned motor-car developed by the Packard Motor Car Co. The metals used consist mainly of high-duty cast iron, malleable iron, hardened alloy steel, annealed copper, brass and bronze.

METALLOGRAPHY AND CONSTITUTION

Testing Steel With the Cutting Flame. J. R. Cady. (Transactions of the American Society for Metals, 1940, vol. 28, Sept., pp. 646–659). It has been found that certain characteristics of steel, such as its carbon content and hardenability, affect the structure of the surface which is left after cutting with an oxy-acetylene torch. In the present work the author describes a technique of flame-cutting and preparing metallographic specimens by which exaggerated effects are produced so that an examination of these specimens provides a means of determining some of the properties of the steel.

X-Ray and Electron Diffraction Studies. F. R. Morral. (American Zinc Institute, 1940, Apr., Preprint). The author gives an elementary account of the properties of X-rays and electron beams, and refers briefly to their use in metallographic research.

Analyzing Chromium Plate by X-Ray Diffraction. J. T. Wilson. (Steel, 1940, vol. 107, Aug. 26, pp. 38–42, 66–69). The author describes the apparatus and technique employed for studying differences in the quality of chromium-plating produced under different plating conditions. A study was made of the effect of increasing the concentration of trivalent chromium in the bath; the X-ray diffraction patterns obtained from specimens plated at varying concentrations seemed similar and showed that a high concentration of trivalent chromium had little, if any, effect on the nature of the coating produced, provided that the current densities were maintained at the same value and the temperatures were accurately controlled. The author is of the opinion that the presence of trivalent chromium increases the bath resistance and that changes in the throwing power of the bath occur as the percentage of trivalent chromium increases.

Non-Parallelism of Lattice Planes in Tin Coatings on Steel. B. Chalmers. (Nature, 1940, vol. 146, Oct. 12, p. 493). By X-ray and optical examination of the crystals of the tin layer on tinplate the author found that a continuous change of orientation takes place along such crystals in their direction of growth. The explanation he proposes for this phenomenon is that the conditions of formation of the layer cause a greater concentration of iron in the tin near the tin-iron interface than at the free surface, and that this causes a gradual contraction of the lattice as the interface is approached.

Analysis of the Cold-Rolling Texture of Iron. C. S. Barrett and L. H. Levenson. (American Institute of Mining and Metallurgical Engineers, Technical Publication No. 1233: Metals Technology, 1940, vol. 7, Sept.). The authors describe an investigation of the behaviour of large single crystals of iron when subjected to cold-rolling, the particular object being to determine the end orientation.

The initial diameter of the crystals was between 0·25 and 0·5 in., and reductions of 82–97% in thickness were obtained in 20 to 40 passes. The initial orientations of the crystals were obtained by the X-ray back-reflection technique or by optical reflection from pits formed by etching; the mean final orientations were determined by X-ray Laue photograms. Depending on the initial orientation of both the normal and rolling directions, some of the crystals maintained or rotated into a reasonably sharp single orientation during deformation; others rotated into two distinct orientations by the formation of fragments or of deformation bands, while a third category fragmented into a large number of major and minor orientations sometimes approaching the entire polycrystalline pole figure.

Influence of Austenitic Grain Size on the Critical Cooling Rate of High-Purity Iron-Carbon Alloys. T. G. Digges. (Journal of Research of the National Bureau of Standards, 1940, vol. 24, June, pp. 723–741). Continuing his investigation of the factors affecting the hardenability of high-purity iron-carbon alloys (*see* Journ. I. and S.I., 1938, No. II., p. 178 A), the author, in the present paper, examines the influence of the grain size of the austenite on the critical cooling rate of these alloys containing carbon in the range 0·23–1·21%. He describes the apparatus and experimental procedure for heating small specimens *in vacuo* and in dry nitrogen to different temperatures and quenching them in hydrogen. Cooling curves were obtained by a string galvanometer combined with a special photographic apparatus. The grain size established at temperatures ranging from 1425° to 1600° F. increased markedly with decrease in the rate of heating through the transformation range, whereas the grain size at 1800° F. was not so noticeably dependent on the rate of heating. The grain sizes at temperatures in the range 1425–2100° F. were also determined. For any selected temperatures within the range 1600–2100° F. and with the heating rate employed, all the alloys were found to have approximately the same average grain size, provided that the carbon was completely dissolved at that temperature. From this it appears that the carbon in solution is not effective in inhibiting the grain growth of the austenite in these high-purity iron-carbon alloys, but that the actual temperatures and the rate of heating through the transformation range are the dominant factors in controlling the grain size of the austenite. There is a narrow range of cooling rate within which the austenite transforms at Ar′ and a wide range of quenching rate in which the larger portion of the austenite transforms at Ar″. This is evidence that the transition from the unhardened to the hardened condition in high-purity iron-carbon alloys is brought about by a small change in the cooling rate. The critical cooling rate was taken as the average cooling rate between 1110° and 930° F. which produced in the quenched specimen a structure of martensite with nodular troostite in amounts estimated to be between 1% and 3%. The critical cooling rate decreased progressively (*i.e.*, the hardenability increased) as the austenitic grain size increased in the alloys containing 0·80–0·85% of carbon, and for each size of austenite grains investigated the critical cooling rate decreased continuously with increase in the carbon content. For complete solution and uniform distribution of the carbon at the time of quenching, the critical cooling rate may be approximately represented by the equation:

$$R = \frac{410N^{-4}}{C + 0{\cdot}2},$$

where R is the critical cooling rate in degrees Fahrenheit per second, N is the number of austenite grains per square inch at 100 diameters, and C is the percentage of carbon.

Effect of Composition and Steelmaking Practice on Graphitization Below the A_1 of Eighteen One Per Cent. Plain Carbon Steels. C. R. Austin and M. C. Fetzer. (American Institute of Mining and Metallurgical Engineers Technical Publication No. 1228: Metals Technology, 1940, vol. 7, Sept.). The authors report on an investigation of the correlation between the graphitising tendency at 760° C. of a number of carbon steels after quenching from various temperatures above the critical point, and (1) the minor variations in the analysis of the steels, and (2) the deoxidation practice employed in their manufacture. Eighteen steels containing 0·98–1·12% of carbon and very small percentages of chromium, nickel and copper were used; these were heat-treated at temperatures in the range 750–1000° C., then tempered at 200° C. for 1 hr. before annealing for 125 and 600 hr. at 670° C. in a lead bath covered with charcoal. Microscopical examinations were then made to study the distribution of the graphite. It was found that the graphitisation was closely associated with the deoxidation practice. Alumina formed during deoxidation or produced in the steel by the oxidation of residual aluminium appears to be the active nucleating agent when annealing is carried out in an oxidising atmosphere, but the alumina must be present in the steel in a certain physical state to be effective. These conditions are met when aluminium has been added in the mould or when alumina is formed by oxidation of the residual aluminium in the steel. In general, it was observed that an increase in any of the elements usually present in carbon steel (such as manganese, silicon, sulphur, phosphorus, chromium, nickel, copper and nitrogen) has no effect upon the graphitisation tendency; the steels highest in silicon have the least tendency and those low in silicon, manganese, phosphorus and nitrogen have a greater tendency to graphitise.

The Metallography of Inclusions in Cast Irons and Pig Irons. H. Morrogh. (Iron and Steel Institute, 1941, this Journal, Section I.). A preliminary scheme of classification for inclusions in cast irons and pig irons has been developed. Using this classification, the various inclusions are dealt with under the appropriate headings.

Various experiments have been performed to elucidate the nature and mode of occurrence of these particles.

The effect of pouring temperature on the morphology of manganese sulphide is discussed. Both manganese and iron sulphide were found to behave as nuclei for the formation of temper carbon in malleable iron. Manganese sulphide gives "graphite-flake-aggregate" temper carbon and iron sulphide gives spherulitic temper carbon.

A blue-pink inclusion has been observed in various cast irons containing titanium and insufficient manganese to neutralise all the sulphur as manganese sulphide. This constituent has been prepared in a number of melts and shown to be probably titanium sulphide. Two forms of the titanium sulphide inclusion occur, one allotriomorphic and one idiomorphic. The complicated optical properties of this inclusion, as revealed by the metallurgical polarising microscope, are described in detail.

The effects of test-bar diameter and titanium content on the number of titanium carbide and titanium cyano-nitride crystals were determined by means of inclusion counts. An attempt was made to determine whether the solubility of titanium carbide in austenite could be detected by the inclusion count method.

The effect of zirconium, in amounts up to about 0·5%, on the inclusions in cast irons was studied. With increasing zirconium content it was found that the manganese sulphide in the base iron was gradually replaced by an orange-yellow to grey inclusion. When all the manganese sulphide had been removed from the structure, blue-grey cubes of zirconium carbide appeared, which combined with the titanium carbide present to give a complex titanium-zirconium carbide. The optical properties of the orange-yellow to grey inclusion, as revealed by the polarising microscope, are given in detail. In melts carried out in a rocking arc furnace, the yield of zirconium from ferro-silicon-zirconium additions was very poor and most of the zirconium appeared to be fixed as lemon-yellow zirconium nitride. An attempt to introduce this inclusion into crucible-melted cast iron by bubbling nitrogen through the melt failed.

Very little analogy was found between the inclusions in steels and cast irons, the latter being characterised by the almost complete absence of visible oxides or silicates. In conclusion, it is suggested that the small particles referred to in the paper could be termed "minor phases" to great advantage with regard to definition.

CORROSION OF IRON AND STEEL

The Reaction between Iron and Water in the Absence of Oxygen. M. de K. Thompson. (Electrochemical Society, Oct., 1940, Preprint No. 24). In order to study the corrosion of iron by water in the absence of oxygen, the author left pure iron powder at 25° C. in oxygen-free distilled water in glass flasks through which hydrogen bubbled continuously. After three days a black deposit began to form on the flasks, at the water surface and at the tip of the inlet tube, which could be identified as magnetite. This was therefore the final product of corrosion in the absence of oxygen, as had been found by some previous investigators. The author points out that the reason why iron with a smooth surface does not react with oxygen-free water is that its overvoltage is higher than the voltage corresponding to the small free-energy decrease of the reaction. This does not apply to iron powder, which, owing to its rough surface, gives the lowest possible overvoltage.

Tarnish and Corrosion of Galvanised Sheets. F. R. Morral. (American Zinc Institute, Apr., 1940, Preprint). After an extensive review of the literature on the corrosion and tarnishing of zinc and galvanised sheet, and on the electron-diffraction analysis of zinc, the author reports on his examination, by electron and X-ray diffraction methods, of freshly galvanised sheets, and samples of sheets covered with various degrees of tarnish and white rust. He found that freshly hot-galvanised sheets have a transparent zinc oxide coating, approximately 1000 Å. thick, which is formed while the coating cools down from the molten state on leaving the galvanising pot. This oxide film is more protective against atmospheric corrosion than pure zinc. It can never be changed to zinc hydroxide, as is often assumed in the literature. The author found, however, that moisture causes the film to grow until it becomes visible as a yellow-brown tarnish, provided that carbon dioxide is absent. If it is present, basic zinc carbonate (white rust) is formed, which, under the same conditions, also develops on electro-galvanised sheets. Under normal atmospheric conditions a transparent bluish film of basic zinc carbonate was found to develop on galvanised sheet.

Corrosion of Metals Used in Aircraft. W. Mutchler. (Journal of Research of the National Bureau of Standards, 1940, vol. 25, July, pp. 75–81). The author reviews the work that has been done since 1925 on the investigation of the corrosion of metals used in aircraft construction. This work has been sponsored by the National Advisory Committee for Aeronautics, the Army Air Corps and the Bureau of Aeronautics of the Navy Department. Most of the tests were made on non-ferrous metals and alloys, but some investigations were also made with stainless steel. For this purpose panels of polished cold-rolled sheets of 18/8 and 18/12 chromium-nickel steel were exposed to attack in tidal waters and it was observed that such steels containing molybdenum were more resistant to attack than those containing additions of titanium or niobium.

Electrolytic Corrosion of Ship Structures. J. H. Paterson. (Transactions of the Institution of Engineers and Shipbuilders in Scotland, 1940, vol. 83, pp. 340–355). In his general discussion of the causes of the corrosion of ship structures the author explains the general electro-chemical theory of corrosion, the terms "*p*H value" and "electro-chemical series" and quotes some practical examples where the application of the results of scientific investigation has led to the prevention of the corrosion of ships. In dealing with localised corrosion the author points out that in steel in which the carbon is unevenly distributed, the higher-carbon areas will be anodic to the lower-carbon areas, and that steel which has been subject to stress and is in a condition of strain is anodic to less strained steel alongside it.

Marine Corrosion and Erosion. J. W. Donaldson. (Metal Treatment, 1940, vol. 6, Autumn Issue, pp. 119–127). The author reviews the results of recent investigations of the marine corrosion of steel, referring in particular to the reports published by the Corrosion Committee of the Iron and Steel Institute, the work of the Committee appointed by the Institution of Civil Engineers and that of the American Society for Testing Materials. In conclusion he deals with researches into the causes and methods of preventing the cavitation erosion of metals and alloys with special reference to Mousson's work (*see* Journ. I. and S.I., 1938, No. II., p. 50 A).

Influence of Cyclic Stress on Corrosion Pitting of Steels in Fresh Water, and Influence of Stress Corrosion on Fatigue Limit. D. J. McAdam, jun., and G. W. Geil. (Journal of Research of the National Bureau of Standards, 1940, vol. 24, June, pp. 685–714). The authors report on a further stage of their investigation of the combined effects of stress and corrosion on the fatigue limit of steel. Earlier stages of this investigation have been reported previously (*see* Journ. I. and S.I., 1931, No. I., p. 757). In the study which forms the subject of the present paper the influence of the degree of stress, its frequency and the period during which the corrosive medium (in this case, water) was applied on the size and shape of the corrosion pits was studied. Specimens were prepared from four types of steel which contained: (*a*) Carbon 0·46%, (*b*) nickel 3·7%, and chromium 0·26%, (*c*) nickel 1·51% and chromium 0·73%, and (*d*) nickel 3·26% and chromium 1·55%, respectively. A comparison of the specimens subjected to corrosion with and without fatigue stressing showed that the effect of fatigue stresses is to increase the size of the corrosion pits, to cause the pits to extend and merge in a direction transverse to the specimen and to increase in depth. When the combined influence of the stress, the frequency and the corrosion time is sufficiently great, transverse fissures appear at the equators of the rounded corrosion pits. When the combined influence of these three variables is somewhat less, transverse crevices are formed. When the combined influence is still less, the rounded pits may merely extend laterally and deepen. When the combined influence is small, the effect may be only to increase the size and depth of the pits. The application of fatigue stress, while increasing the depth of the pits apparently decreases the tendency to spread, and, while it may have little apparent effect on the size of most of the pits, it has a great effect on the size of a few of them.

Results Obtained With Cathodic Protection. F. A. Hough. (Proceedings of the Natural Gas Section of the American Gas Association, 1940, pp. 68–81). The author summarises the evidence collected from a number of gas companies in the United States respecting the experience gained, and the cost of installing and operating apparatus for the cathodic corrosion control of pipe lines carrying town gas. The principles of cathodic corrosion control have been described previously (*see* Journ. I. and S.I., 1940, No. II., p. 231 A).

The Corrosion of Underground Piping. C. H. S. Tupholme. (Mining Magazine, 1940, vol. 63, Oct., pp. 177–184). The author explains the mechanism of the corrosion of pipe lines, particularly with reference to pipes in mines. He describes several methods of protection, including the cathodic protection technique (*see* Journ. I. and S.I., 1940, No. II., p. 231 A), and the use of bitumens, tar enamels, impregnated wrappings, cement and mortar, cellulose nitrate plastic, pyralin and rubber.

BOOK NOTICES

MORLEY, ARTHUR. "*Strength of Materials.*" Ninth Edition, 8vo. Pp. x + 571. Illustrated. London, 1940. Longmans, Green and Co. (Price 15*s.*)

When a technical book reaches its ninth edition, and about fifty thousand copies have been sold, there is little that a reviewer can say about it, since it is obvious that the book must have established itself, in the first place because it fulfilled the object with which it was written, must have maintained its position over a long period because nothing definitely better has appeared, and must go on being sold and used for a long time as is the way of the technical classics. Morley's *Strength of Materials* has been in use for many years by students of engineering, metallurgy and other subjects who have to study this branch of technical science. The new editions and impressions published between 1908 and 1939 have been produced without resetting the type, so that substantial revision or enlargement of the text has been impossible. The need for resetting the type for the ninth edition has given an opportunity for more complete revision than was possible before, and the author has taken advantage of this to rewrite chapters or articles on those parts of the subject in which research or development has made substantial changes necessary. As would be expected, a considerable amount of space has had to be found for dealing with fatigue, and creep is also discussed, though very briefly. The chapter on the mechanical properties of metals is designed to give a short account of modern knowledge about this, but extensive consideration of the more definitely metallurgical side of the strength of materials is not to be looked for in a work which is strictly limited in size and is primarily concerned with stress calculations. The book was originally written for the use of students. The new edition is intended for the same purpose, and no attempt has been made to expand the text to rival some of the larger and newer works, or to cover the ground of several books. Of the seventeen chapters that the book contains, thirteen are devoted to elastic stress and strain, strain energy and strength, bending and beams, direct and bending stress, twisting, pipes, cylinders and discs, flat plates, and vibrations and critical speeds. Two of the remaining chapters deal with testing and testing machines, while the others are concerned with the mechanical properties of metals and with special materials such as concrete, stone, brick and timber.

J. M. ROBERTSON.

NEWITT, DUDLEY M. "*The Design of High-Pressure Plant and the Properties of Fluids at High Pressure.*" 8vo. Pp. viii + 491. Illustrated. London, 1940. Oxford University Press. (Price 35*s.*)

This book is essentially a survey of the specific effects of pressure on physical processes taking place in liquid and gaseous systems. It deals therefore with a highly specialised subject, and is mainly based on investigations that have been carried out in the past few years. As, however, work at high pressure requires the use of suitable plant, a part of the book is devoted to the design of plant for high-pressure service and to the nature and properties of the materials suitable or available for its construction. The five chapters which constitute this first part of the book deal with the properties of materials used in the construction of high-pressure plant and equipment, cylinders for the storage and transport of permanent and liquefiable gases, stresses and strains in the walls of cylinders subjected to internal and external pressure, the design of plant, and the measurement of high pressures. It is this part of the book that will be of most direct interest to metallurgists, and in view of the variety of uses of gas cylinders, the large number produced and the many different ways of making them, the second and third chapters are of special interest. The main part of the book is, however, that dealing with the properties of fluids at high pressures, and this occupies three hundred and fifty pages, as against one hundred and thirty occupied by the first part. Here the author is entirely concerned with physical phenomena such as the kinetic theory and the ideal fluid, equations of state, the compressibility of gaseous mixtures, the thermodynamic properties of gases, the liquefaction of gases, the compression and circulation of gases, the pressure-volume-temperature relationships of liquids, &c. As no other book on this important subject has yet been published, this one should be of great value to those concerned with high-pressure work, either experimentally or in industry.

J. M. ROBERTSON.

MINERAL RESOURCES

Synopsis of the Mineral Resources of Scotland. M. Macgregor. (Geological Survey of Great Britain, Special Reports on the Mineral Resources of Great Britain, 1940, vol. 33). A short account is presented of the nature, distribution and uses of the various economic minerals of Scotland, together with references to the sources from which more detailed information can be obtained. A short section on water supply has been added.

Chrome Ore and Chromium. R. Allen and G. E. Howling. (Imperial Institute, Mineral Resources Department, London, 1940). An account is given of the occurrences of chrome ore in the British Empire and foreign countries (*see* p. 81A).

Iron Deposits of the Steeprock Lake Area. M. W. Bartley. (Canadian Mining Journal, 1940, vol. 61, Sept., pp. 566–572). The author reports on a geological and geophysical survey of hematite deposits discovered in 1937 below the bed of Steeprock Lake in the province of Ontario, Canada. The deposits are overlain by 75–150 ft. of clay and 100–200 ft. of water. One deposit with a possible length of 3300 ft. and another of 700 ft. have been indicated, but information on their width and vertical extent is not yet available. Analyses of the hard hematite cores from the drillings show an average iron content of 58%. The principal deposit is situated four miles from the southern line of Canadian National Railways and good water transportation is available from the site to the railway. Electric power is also available from a plant on Steeprock Lake.

Iron-Ore Developments in Sierra Leone. W. S. Edwards. (G.E.C. Journal, 1940, vol. 11, Aug., pp. 96–102). The author describes the ore treatment, conveying and loading equipment of the Sierra Leone Development Co., Ltd., a company which is operating an iron-ore mine in Sierra Leone 300 ft. above sea level and 50 miles from the coast. No analysis of the ore is given, but one deposit is described as a soft, powdery, micaceous ore, mostly under $\frac{1}{16}$ in. in size. The first cargo of this ore was shipped to Glasgow in 1933. Numerous illustrations are given of the electric shovel, conveyors, truck tipper, screening plant and the travelling storage bridge.

Coal and Lignite Deposits in Siberia. (Iron and Coal Trades Review, 1940, vol. 141, Aug. 30, p. 215). A brief account is given of the distribution of coal and lignite deposits in Siberia, from which it is apparent that, whilst they are enormous in quantity, they are unevenly distributed over an area of 6,000,000 sq. miles, and it is questionable whether any substantial development will prove possible until better railway facilities are provided to expedite distribution, and until a considerable advance has been made in industrial evolution.

ORES—MINING AND TREATMENT

Iron-Ore Concentration and the Lake Erie Price. E. W. Davis. (American Institute of Mining and Metallurgical Engineers, Technical Publication No. 1202: Metals Technology, 1940, vol. 7, Sept.). A price for ore, known as "the Lake Erie price," is established every spring, and it determines the maximum price that will be paid during the coming season for iron ore of the Mesabi non-Bessemer grade; it is based on the iron and phosphorus contents of the ore. In the present paper the author derives a simple formula which enables the value of any ore to be calculated from the Lake Erie price.

The Separation of Apatite from Red Hematite Ores by Flotation. G. G. Bring. (Jernkontorets Annaler, 1940, vol. 124, No. 7, pp. 277–312). (In Swedish). The author reports the results of a laboratory investigation of the flotation process of separating the apatite present in Swedish red hematite ores, using hydrofluosilicic acid and waterglass as separating media.

REFRACTORY MATERIALS

(Continued from pp. 1 A–2 A)

The Action of Alkalies on Refractory Materials. Part VI.—The Action of the Vapour from a Potash-Silica Glass on Refractory Materials at 1200° C. F. H. Clews, H. M. Richardson, A. Chadeyron and A. T. Green. (Transactions of the British Ceramic Society, 1940, vol. 39, Oct., pp. 289–296).

The Action of Alkalies on Refractory Materials. Part VII. Some Aspects of the Action of Alkali Chloride Vapours on Refractory Materials at 1000° C. F. H. Clews, H. M. Richardson and A. T. Green. (Transactions of the British Ceramic Society, 1940, vol. 39, Oct., pp. 297–303). In the present investigation, which is an extension of that reported in Part V. (*see* Journ. I. and S.I., 1940, No. II., p. 87A), the action of sodium chloride vapour at temperatures between 700° and 1000° C., in the dry state, and in the presence of steam, on refractory materials was studied. The following conclusions were arrived at: (1) Sodium chloride vapour in dry air at 1000° C. behaves similarly to potassium chloride in causing expansion effects in refractory materials; (2) the presence of steam in the reaction vessel with which the tests were conducted does not appear appreciably to modify the expansion caused by potassium chloride vapour and dry air at 1000° C.; (3) alternations of temperature between 700° and 1000° C. tend to increase the rate of expansion caused by potassium chloride vapour; (4) disruption of a coarse-grained sillimanite material bonded with bentonite is caused by alkali chloride vapour treatment at 1000° C.; and (5) the transverse modulus of rupture of a fireclay material is considerably decreased by exposure to potassium chloride vapour at 1000° C.

The Action of Alkalies on Refractory Materials. Part VIII. Experiments on the Action of the Vapour from Sodium Aluminate at Various Temperatures. F. H. Clews, H. M. Richardson and A. T. Green. (Transactions of the British Ceramic Society, 1940, vol. 39, Oct., pp. 304–311). The nature of the action of alkali vapours on refractory materials is dependent to some extent on the compound used as a source of alkali. The results of experiments using vapours from molten potash, a potash-silica glass and potassium chloride have already been reported. In the present investigation data were obtained using sodium aluminates as the source of the alkali vapour. Sodium aluminates with a molecular ratio of $Na_2O : Al_2O_3$ of 1·27 : 1 and 1·09 : 1 were selected. Samples from eight types of commercial refractory bricks were tested by suspending them from sillimanite rods by platinum wires in sillimanite crucibles containing the powdered sodium aluminate. The crucibles were heated to temperatures of 1200°, 1100°, 1000° and 900° C., and the rate of increase in weight was noted. From the data obtained the authors made the following deductions: (1) The rate of increase in weight is greater, the higher the temperature and the greater the content of sodium oxide in the aluminate; (2) highly siliceous materials increase in weight more rapidly than fireclay, and fireclay more rapidly than sillimanite; (3) experiments with fireclay specimens of artificially increased porosity suggested that the chemical composition of the refractory is of more importance in determining the rate of absorption than the porosity; and (4) the rates of penetration of the alkali from sodium aluminate at 1000° C. into a silica, a fireclay and a sillimanite product decrease in that order.

Effects upon Furnace Refractories of Protective Gases for High Carbon Steels. J. H. Loux. (American Society for Metals, Oct., 1940, Preprint No. 24). The author considers the problem of the selection of the best type of refractory brick to resist the attack of the protective atmospheres used for the heat treatment of high-carbon steels. He suggests that the problem is similar to that of selecting bricks for blast-furnace linings. A number of illustrations are presented of five different qualities of American firebrick after a three-months test on the base of a removable-hood type of furnace in which an atmosphere containing about 20% of carbon monoxide was used. The results of this test showed that cheap firebricks glazed on five sides and resting on the unglazed side, and hard pressed, de-aired, Kentucky clay firebricks with a dense surface offered the greatest resistance to mechanical abrasion and to the action of the protective atmosphere.

Apparatus and Method for the Determination of Porosity in Refractory Materials. G. E. Seil, B. S. Tucker and H. A. Heiligman.

(Journal of the American Ceramic Society, 1940, vol. 23, Nov., pp. 330–333). The authors describe an apparatus and procedure for determining the porosity of refractory materials. The samples used consisted of 100-g. portions of grain or crushed brick between $\frac{1}{4}$ in. and $\frac{3}{8}$ in. in size. The true specific gravity of the sample was determined with a pycnometer, and the volume was determined by a mercury-displacement method, the apparatus for which is described in detail. The calculation of the porosity from the true specific gravity, the weight and the volume of the sample by means of a simple formula is explained. The results obtained are accurate within $\pm 0{\cdot}25\%$. The advantages of the method include the following: (1) A composite representative sample from a mass of crushed refractory material can be used; (2) samples from any portion of a brick can be tested and any variation in the porosity of the brick can thus be established; (3) it eliminates the inaccuracies inherent in methods in which free liquid has to be wiped away from the surface of the sample; and (4) it is adaptable to any type of ceramic material.

FUEL

(Continued from pp. 2 A–5 A)

The Place of Coal in the Steel Plant—Past, Present and Future. H. V. Flagg. (Mining and Metallurgy, 1940, vol. 21, Oct., pp. 457–461). The author refers to changes in the use of fuel at the works of the American Rolling Mill Co., Middletown, which have resulted in a very greatly reduced consumption of coal in the various processes of steelmaking, and compares the fuel-consumption data for the years 1927 and 1940. He discusses the increased use of natural gas in the United States and the advantages and disadvantages of pulverised coal as a fuel for soaking pits and heat-treatment furnaces.

Steam Turbine Practice in Steelworks. A. C. Hirst. (English Electric Journal, 1940, vol. 10, Nov., pp. 40–48). The author presents a short survey of the various applications of steam in steelworks practice, with particular reference to steam turbines for power generation and turbo-drives.

The Influence of Ash Constituents on the Reactivity of Coke. R. N. Golovati. (Journal of Applied Chemistry, U.S.S.R., 1939, vol. 12, pp. 1178–1186: Fuel in Science and Practice, 1940, vol. 19, Oct., pp. 206–211). The author presents the results of a study of the effect of the presence of chlorides of iron, aluminium, calcium, magnesium, manganese and potassium, and of silicic acid either separately or in mixtures, on the reactivity of two representative cokes from the Don Basin in the U.S.S.R. The reactivity was calculated in accordance with Bähr's formula. It was shown that iron chloride increases, and silicic acid reduces, the activity of coke, not only at 700° C., but also at the temperature of the initial reaction between carbon dioxide and the carbon of the coke. The influence of certain catalysts on the velocity of this reaction was also examined, and it was found that their effect depended on the nature of the coke. Good catalysts, in decreasing order of effectiveness, were iron, calcium and potassium.

Studies on Some Characteristics of Indian Coking Coals. R. K. Dutta Roy. (Records of the Geological Survey of India, 1940, vol. 75, May, Professional Paper No. 3). It has been established that, in India, good coking coals for the smelting of iron ores will only be available for a limited period. In the present report the author presents the results of his investigation, carried out at the request of the Coal Mining Committee set up by the Government of India, on the following characteristics of Indian coking coals: (*a*) The caking index; (*b*) the swelling properties; (*c*) decomposition and softening points; and (*d*) blending. From the information and data presented it is expected that some important economies in the utilisation of Indian coking coals can be effected.

PRODUCTION OF IRON

(Continued from pp. 5 A–6 A)

The Output of Blast-Furnace Dust. N. I. Manzyuk. (Metallurg, 1939, No. 10–11, pp. 35–43). (In Russian). Observations on the output of blast-furnace flue dust are important, as they enable some conclusions to be drawn regarding the working of the furnace, and, more particularly, they provide a guide as to the necessary proportioning of the constituents of the burden. In connection with the latter purpose, frequent determinations of the dust output are necessary in view of its possible appreciable and rapid variations. At the works of tho Kuznetskiy Metallurgical Group the amount of blast-furnace dust carried over is roughly estimated every two hours by emptying the contents of the primary dust-catchers into dump cars. A large number of observations on the total dust carried over and the amount of dust collected in the primary dust-catchers were made on three blast-furnaces. Curves plotted from the data obtained indicate that there is a relation between these two amounts, though a number of the results deviate appreciably. The latter were recorded during erratic working of the furnace as a result of scaffolding. Under such conditions a smaller proportion of the larger total amount of dust carried over is retained in the primary dust-catchers. In discussing the relation between the manner in which the blast-furnace is working and the output of dust, the author mentions some results which show that dust carried over during the erratic working of a furnace in a poor condition has an increased ferrous-oxide content, and that there is also some metallic iron present. The evidence suggests that under such conditions dust carried upwards settles and accumulates in the furnace, and is only carried over after further crushing. The large number of factors affecting dust output overlap to a certain extent and cannot, generally, be dealt with separately. A certain relationship, characteristic for each particular furnace, was found to exist between the average daily blast pressure and the daily dust output.

The Output of Dust and the Distribution of the Gas Stream in the Throat of the Blast-Furnace. D. A. Pospelov. (Metallurg, 1939, No. 10–11, pp. 44–53). (In Russian). In some introductory remarks, reference is made to the unsatisfactory results accompanying the excessive moistening of the ore charged into the blast-furnace in an attempt to reduce the evolution of dust. The distribution of the gas stream in the throat of the No. 3 blast-furnace of the Zaporozhstal works was investigated in view of its obvious connection with dust evolution. This was done by taking measurements of the carbon dioxide present at points along the radius of the throat in the space just below the lower edge of the distributor cone. It was observed that with an increase in the carbon-dioxide content of the gas near the periphery, and a decrease near the centre, the dust evolution increases and the working of the furnace becomes less steady. In order to get undisturbed working of the furnace a steady gas stream, either peripheral or axial, must be obtained ; a low dust evolution is, however, most frequently obtained with a steady peripheral gas stream. The level of the charge must also not be allowed to drop too low. The distribution of the gas stream may be controlled by varying the weight of the charges dropped into the furnace from the distributor. In the Zaporozhstal furnaces a decrease in the charge results in less ore reaching the centre, whilst an increase in the charge favours peripheral distribution of the blast. The way in which the above observations can be utilised in practice is exemplified by the methods adopted to control the working of the No. 3 blast-furnace during, and following, the blowing-in. It is concluded that one can either have a steadier working of the furnace and a lower production of dust with, at the same time, a slightly increased consumption of coke as a result of peripheral working, or an increase in the output of the furnace and a lower coke consumption with, at the same time, an increase in the evolution of dust.

On the Question of Running Blast-Furnaces on Iron-Coke. K. V. Messerle and G. S. Gerasimchuk. (Metallurg, 1939, No. 10–11, pp. 21–27). (In Russian). The authors discuss experience gained in Russia in the use of iron-coke in blast-furnaces. Iron-coke is obtained by adding blast-furnace flue dust to the coke-oven charge. Incidentally, it enables rich non-coking coals to be utilised.

The properties of the iron-coke used in a number of blast-furnaces are described. Some of them had unsatisfactory drum-test properties, and their sulphur content, generally speaking, was high. In several blast-furnaces the use of this coke was not successful. Experiments on the use of iron-coke in three blast-furnaces at the Stalin works were successful, owing to its satisfactory characteristics when employed, either alone or mixed with ordinary coke. The mean diameter of the iron-coke lumps was about 7% greater than that of ordinary coke. This, together with its much lower porosity, improved the distribution of the blast in the furnace and enabled a slightly lower blast pressure to be used. The coke had been prepared from a charge to which had been added 10% of blast-furnace dust, which figure is regarded as the optimum. The average iron content of the dust was 7·5%, of which up to 65% was metallic iron, and the sulphur content was 1·7%. Operating data for the furnaces showed that the use of this iron-coke resulted in a reduction of the fuel consumption and an improved utilisation of the volume of the furnace. While the sulphur content of the slag increased, that of the iron remained unchanged, owing to an increase of 40° C. in the temperature of the slag. When iron-coke is used the temperature of the blast should be raised.

Manufacture of Ferro-Silicon in the Blast-Furnace. (Foundry Trade Journal, 1940, vol. 63, Oct. 31, p. 284). The results of Russian investigations on the manufacture of ferro-silicon in the blast-furnace are briefly presented. (See Journ. I. and S.I., 1940, No. II., p. 211 A).

Manufacture of Sponge Iron at Nildih. (Tisco Review, 1940, vol. 8, Sept., p. 710). A brief account is given of the working of a small blast-furnace, 4 ft. 6 in. high, built of local clay at Nildih, on the river Suvarnarekha, India, in which about 18½ lb. of sponge iron were made in a day.

The Iron-Powder Situation. A. T. Fellows. (Metals and Alloys, 1940, vol. 12, Sept., pp. 288–291). The author considers the economic aspect of the development of the manufacture of iron powder in the United States. He estimates that if iron powder were available at a price of about 10 cents per lb., a market for 500–1000 tons per annum might be anticipated, and that if iron powder of better quality suitable for electrical purposes were available at prices ranging from 10 to 60 cents per lb., 300–500 tons per annum might be sold. He points out that the largest source of supply, Sweden, is now cut off from America owing to the war in Europe, and that it is quite possible that, after the war, iron powder of Swedish origin may be shipped to America at low controlled prices which do not reflect the cost of production.

Germany's Iron and Steel Position. G. Abrahamson. (Iron and Steel, 1940, vol. 14, Oct. 22, pp. 36–42). The author discusses the measures taken in Germany during the last few years to make the production of iron and steel less dependent upon foreign supplies of ore. In the field of raw-material supply it was the aim to intensify mining in Germany, and in the sphere of iron and steel production it was to bring technical conditions into line with the position created by the enforced changes in the supply of raw materials; quality entered into the question only in so far as changes were necessary in order to satisfy the consumer. The author then relates how the Czech and Polish steel industries have been incorporated into the German economic sphere; he also endeavours to assess the position created by the German control of the iron and steel industries of France, Belgium and Luxemburg. In summarising the present position of Germany's iron and steel industry, the author estimates that the output of steel has declined since the outbreak of war by at least one-quarter and is likely to fall further as a result of receding material supplies. The capture of vast quantities of armaments, transport and scrap from France and Belgium, while a factor of great importance, can only temporarily affect the situation.

Brazil's Iron and Steel Industry. (Iron and Steel, 1940, vol. 14, Oct. 22, pp. 49–51). After a brief history of the negotiations of the Brazilian Government first with the Itabira Iron Ore Company and then with the United States Steel Corporation for the establishment of a large iron and steel works in Brazil, a short account is given of the plan envisaged by the Executive Board of the National Siderurgical Plan for the development of the production of steel in Brazil. Some statistics are given relating to the production and consumption of pig iron, sheet iron and steel in Brazil, from which it is seen that while the annual consumption is about 5,500,000 tons, the production is only about 300,000 tons.

FOUNDRY PRACTICE

(Continued from pp. 6 A–7 A)

The Significance of Hydrogen in the Metallurgy of Malleable Cast Iron. H. A. Schwartz, G. M. Guiler and M. K. Barnett. (American Society for Metals, Oct., 1940, Preprint No. 3). The authors had intended to study the influence of hydrogen on the rate of graphitisation in the production of malleable iron, but they found this to be affected by a number of variables which required preliminary study. The present paper is an exploratory survey of the whole subject of the significance of hydrogen in the production of malleable iron. The results presented apply only to white cast irons of the compositions commonly encountered in malleable iron practice. The authors' conclusions may be summarised as follows: (1) White cast iron will dissolve approximately twice as much hydrogen from a given atmosphere as will pure iron at the same temperature; (2) no evidence was obtained that graphite adsorbs any appreciable amount of hydrogen; (3) commercially produced white cast irons may contain 0·0002–0·0010% of hydrogen by weight; (4) the hydrogen referred to in (3) originates from rust in the charge, from moisture in the air, and possibly from the fuel; (5) the oxidising reactions during melting reduce the hydrogen content, but, in general, the oxygen content of metal high in hydrogen is likely to be high; (6) the introduction of a small amount of almost any solid metallic addition into liquid white cast iron reduces the hydrogen content; and (7) after annealing, the hydrogen content of the iron approaches a constant value slightly above 0·0001%, irrespective of the initial hydrogen content.

Short-Cycle Annealing. R. M. Cherry. (Iron Age, 1940, vol. 146, Sept. 26, pp. 34–37). The author describes in detail the design of large and small modern electric furnaces for the annealing of malleable iron. Details of both batch and continuous furnaces are given and attention is drawn to the reduction in annealing time and labour, the improved conditions of working, the saving in floor space and the uniformity in the quality of the product as compared with the older annealing technique when the charges were placed in heavy cast iron pots.

New Custom Alloy Shop. G. R. Reiss. (Steel, 1940, vol. 107, Oct. 14, pp. 111–112, 139). A brief description is given of the development and present plant of the Youngstown Alloy Casting Corporation, which specialises in the manufacture of small alloy steel castings with high wear-resisting properties, such as mandrels and guide-shoes for tube mills.

Application of Controlled Directional Solidification to Large Steel Castings. J. A. Duma and S. W. Brinson. (Transactions of the American Foundrymen's Association, 1940, vol. 48, Dec., pp. 225–277). This paper has already been published in the Journal of the American Society of Naval Engineers, 1940, vol. 52, Feb., pp. 26–59, and an abstract appeared in Journ. I. and S.I., 1940, No. II., p. 10 A.

Coatings for Wood Patterns. F. C. Cech and V. J. Sedlon. (Transactions of the American Foundrymen's Association, 1940, vol. 48, Dec., pp. 369–392). The authors present and discuss the replies received to a questionnaire sent out to a large number of American varnish and paint manufacturers and pattern-makers in order to obtain information on suitable coatings for wood patterns. From the replies received it was found that: (1) 75% of those concerned prefer flake shellac to " cut " shellac (the latter is prepared by covering flake shellac with a solvent over-night and stirring and thinning out the solution on the following morning); (2) patented solvents are rapidly replacing wood alcohol; (3) black extract is superseding lampblack as a pigment; (4) vermilion is preferred to other shades; (5) yellow pigment is little used; (6) aluminium powder, mixed with shellac or varnish, is gaining in favour; (7)

oxalic acid should not be used for reducing the opacity of shellac; (8) earthenware containers for shellac are preferred; (9) the spraying of patterns is not practicable; and (10) the pattern coating of the future will probably consist of a priming coat of shellac with a top coat of lacquer.

Core Behaviour at Elevated Temperatures. H. W. Dietert. (Foundry, 1940, vol. 68, Oct., pp. 52, 112–113). The author discusses the relation between the tensile strength and hardness of cores and their expansion and rate of collapse at elevated temperatures. He shows that as the tensile strength of a core decreases, the expansion decreases rapidly, and that a reduction in the hardness also decreases the expansion. An adjustment of the oil/sand ratio of the core thus provides a method of controlling the expansion. The rate of collapse of a core as it is heated by molten metal may be varied, not only by the tensile strength, but also by the core permeability and changes in the clay content of the mixture.

Barrel-Type Valve Lifters Molded in Green Sand. E. C. Hoenicke. (Foundry, 1940, vol. 68, Oct., pp. 42–45, 107–109). The author describes the moulding technique which was devised for the production in large quantities of barrel-shaped valve tappets for internal combustion engines. Green sand was used for both the mould and the cores, and the tappets are now cast in the vertical plane on a single metal chill-plate for the entire mould. It was found that the cost of 32 patterns for each mould was very high, and it was eventually decided to use die-cast patterns of a zinc-base white metal.

Ford Cast-Steel Crankshafts Now Poured Continuously. E. F. Cone. (Metals and Alloys, 1940, vol. 12, Sept., pp. 267–270). The author describes the plant developed by the Ford Motor Co. for the production of steel for casting automobile crankshafts. The plant consists of two cupolas or stacks 25 ft. high and 36 in. in dia., each of which has a long fore-hearth or air-furnace at the base fired by pulverised coal. The flame from this fuel sweeps along the hearth and up the cupola where the charge is preheated. The charge is made up of 40% of returned scrap, 20% of basic iron and 40% of briquetted borings, and it is put in through the charging door 15 ft. above the foundry floor. The charge melts at the base of the cupola, from which it feeds continuously into the fore-hearth, where it is brought up to about 2700° F. The metal is then tapped off into an 8-ton transfer ladle, in which it is taken to a 15-ton electric-arc furnace for superheating to 2900° F. During the brief period it is in this furnace spectrographic analyses are taken and alloys are added to bring the metal to the desired composition. From the electric furnace the metal is conveyed in ladles to a rocking type of holding furnace also fired with pulverised coal. The steel then flows continuously into a pouring car, whence it pours simultaneously into two moulds of four crankshafts each. The composition of the cast metal is as follows: carbon 1·35–1·60%, manganese 0·70–

0·90%, silicon 0·85–1·10%, copper 1·50–2·00%, chromium 0·40–0·50%, phosphorus 0·10% max. and sulphur 0·08% max.

Centrifugally Cast Gear Blanks. (Machinery, 1940, vol. 57, Oct. 31, pp. 117–119). An illustrated description is given of the foundry practice at the Dearborn Works of the Ford Motor Co. for the centrifugal casting of gear blanks for automobiles and tractors. The gears are cast up to 13 in. in dia. The steel is melted in a 15-ton electric furnace, whence it is delivered to a 10-ton holding furnace; from the latter the metal is carried by ladles suspended on a monorail to the pouring-stations of four large turntables, each of which carries eighteen steel dies or moulds. As each mould approaches the pouring-station an electric motor beneath the mould is automatically started and rotates it during pouring and cooling. The temperature of the metal entering the mould is 2850–2900° F. After spinning for 2 min. the moulds reach the unloading station, where the casting is removed and new cores are inserted. The analysis and some of the mechanical properties of the finished castings after machining and heat treatment are given.

Scrap Salvage by Magnetic Separation. W. E. Box. (Foundry Trade Journal, 1940, vol. 63, Oct. 24, pp. 265–266). The author describes and illustrates a number of magnetic separating machines designed for recovering iron scrap in the form of brads, spillings, &c., from used moulding sand and for treating the waste going to tips and slag-banks, including sweepings from workshops, areas and yards, and for treating existing dumps and slag-banks.

PRODUCTION OF STEEL

(Continued from pp. 7 A–8 A)

The Acid Bessemer Process of 1940. H. W. Graham. (American Institute of Mining and Metallurgical Engineers, Technical Publication No. 1232: Metals Technology, 1940, vol. 7, Oct.). The author points out that the Bessemer process of steelmaking developed rapidly in America in its early history owing to economic reasons, but that it receded steadily from its peak for both economic and metallurgical reasons. He states, however, that the decline of the Bessemer process has gone somewhat further than is justified by present economic considerations and that the production of Bessemer steel is likely to remain on the same level, or perhaps even to increase in coming years. He compares the Bessemer and open-hearth processes at some length, and deals in separate sections with production, availability of scrap, investment costs, product losses, metallurgical history, inherent factors, basis for control, end-point determination, temperature control, design and maintenance, oxygen and nitrogen control, and physical properties of the products.

The Metallurgical Observer. N. H. Bacon. (Sheffield Society of Engineers and Metallurgists: Iron and Coal Trades Review, 1940, vol. 141, Nov. 8, pp. 467–468; Nov. 15, p. 493). The author describes the development at a large Sheffield steelworks of a team of metallurgical observers and its duties. The advantages of promptly supplying reports on the casting conditions to the rolling-mill staff, and of the collection and examination of records of the charging, melting, refining, teeming, rolling and dressing of the steel are noted. The changes in practice which led to an improvement in the quality of the billets included: (1) An extension of bottom-pouring practice; (2) controlling the teeming speed by a partially shut stopper; (3) lengthening the pouring nozzle from 7 to 10½ in.; and (4) boiling the nozzles in tar until saturated. In the concluding part of his paper the author discusses some factors affecting the life of ingot moulds, sand inclusions in tyre steel and corner segregation in octagonal alloy-steel ingots which were discovered by statistical analysis.

Tailor-Made Steels. T. Grey-Davies. (Sheet Metal Industries, 1940, vol. 14, Aug., pp. 827–828; Nov., pp. 1174–1175, 1178). Continuation of a series of articles (*see* Journ. I. and S.I., 1940, No. II., p. 139 A). In Parts XVI. and XVII. of this series the author explains the mechanism of the solidification of rimming-steel ingots and presents some illustrations showing the appearance of a section of an ingot at different heights. He also discusses the American practice of adding very small quantities of sodium fluoride when teeming in order to assist the rimming reactions. The addition of this compound at the commencement of pouring accelerates the reaction in the lower half of the ingot mould, thereby producing an ingot with a more uniform skin thickness. It also enables steels of higher carbon content to rim more readily, and causes a substantial improvement in the quality of the surface. He suggests that the powerful action of a fluoride such as fluorspar in increasing the fluidity of basic slags is a subject requiring greater study by steelmakers.

The Thermal Relations Between Ingot and Mould. T. F. Russell. (Iron and Steel Institute, 1941, this Journal, Section I.). Saitô's formulæ for the distribution of heat between ingot and mould are examined quantitatively. The practical variations from the ideal case are discussed, and it is thought that the possibility of deriving more exact formulæ than Saitô's is very remote.

Examples are worked out for circular and square ingots of the same cross-sectional area cast into moulds of four different thicknesses. Curves are drawn showing the temperature at different points in the ingot and moulds, the temperature distribution across a diameter and the total quantities of heat in the ingot and mould at different times. Curves are also drawn to show the effect of " mould ratio " on the temperature cycle occurring in the ingot near to the mould wall and on the time taken for the temperature at the centre of an ingot to fall certain amounts representing solidification. The latter show that solidification is accelerated by increasing the mould thickness until the mould ratio is about 0·8–1·0, but that a further increase in the mould thickness has an inappreciable effect.

Four sets of experimental results on the measurement of mould temperature are examined, and they show that the greatest difference between theory and practice is found at positions in the mould near to the inner face. This is attributed to the effect of the air-gap forming between the ingot and the mould.

ROLLING-MILL PRACTICE

(Continued from pp. 10 A–11 A)

Calibration Which Takes into Account Elastic Deformations and Their Experimental Investigation. Yu. M. Faynberg. (Metallurg, 1939, No. 10–11, pp. 141–148). (In Russian). The author studies methods of determining the elastic deformation of rolls. He points out the need for taking into account all elastic deformations when setting the roll opening, the accuracy of which, particularly in continuous rolling, is of considerable importance. In view of the difficulty of accounting for all the elastic deformations of the rolls and stand by calculation, and in order to avoid the tediousness of accounting for them by trial-and-error methods, the author has developed a method for indirectly measuring these deformations by recording oscillographically the variations in the power requirements of the screw-down motors as the rolls are pressed down on to one another. Allowances are made for the power losses in the screw-down mechanism and for the pressure of the metal on the rolls in actual rolling.

The Use of Super-Reductions in Rolling. S. A. Kushakevich. (Metallurg, 1939, No. 10–11, pp. 123–140). (In Russian). Super-reductions in rolling are defined as reductions per pass which, by the method of calculation developed by Kirchberg, involve a coefficient of increase in length greater than 2, *i.e.*, a reduction greater than 50%. The old theory that reductions greater than 50% will lead to cracking of the metal is shown to be wrong, as the increase in length of the metal in a pass is shown to be the result of triaxial compression rather than a process akin to drawing. The author goes on to give experimental data obtained in the rolling of several steels and non-ferrous alloys. Super-reductions in the hot-rolling of both ingots and semi-rolled materials were found to be feasible without detriment to the structure or mechanical properties. In conclusion, the author considers mathematically the question of pressure on the rolls, the gripping of the metal by the rolls when rolling with super-reductions, the effect of this method of rolling on the output capacity of rolling-mill units, and the saving in fuel costs as a result of the less frequent reheating required because of the reduced heat losses during rolling. An increase in the energy consumption is the one serious drawback to the process of rolling with super-reductions.

The Ratio of the Diameters of the Rolls of the Three-High Lauth Mill. B. Gol'denberg. (Stal, 1939, No. 10–11, pp. 52–53). (In Russian). After various tests with a three-high Lauth mill the author concludes that the optimum ratio of the diameter of the top and bottom rolls to that of the centre roll is 0·65.

Continuous 250-mm. Wire Mill of the Kirov Works. A. Novgorodtsev and N. Rymkevich. (Stal, 1939, No. 10–11, pp. 38–45). (In Russian). A detailed and illustrated description is given of the Morgan continuous wire rod mill and accessory equipment installed recently at the Kirov works in the U.S.S.R. The plant is designed for the rolling of wire 5 to 8 mm. in dia. from billets measuring 9140 × 54 × 54 mm., weighing 206 kg. A Morgan pusher-type preheating furnace fired with a mixture of blast-furnace and coke-oven gas with a capacity of 50 tons per hr. is employed. The mill has a total of nineteen stands in four groups. A table showing the rolling schedule and the speeds of rotation of the rolls is given. With a useful working time of 6940 hr. per year the annual output is planned to amount to 220,000 tons.

Specialisation of Rolling Mills in the U.S.S.R. and New Rolled Sections. V. Volobnev. (Stal, 1939, No. 10–11, pp. 30–37). (In Russian). The author discusses the demand in the U.S.S.R. for rolled-steel sections of special shape. He points out the need for special shapes in different industries, a need which should be satisfied as far as possible without any duplication. Enquiries addressed to thirty-one consumer organisations brought replies regarding the shapes of sections required. They included a number of new shapes for railway construction, railway rolling stock, automobile and tractor construction, machine tools, architectural uses, mining and chemical plant construction, electrical and power engineering and for miscellaneous purposes. Some special features, and the production and advantages in the application of some of these sections, are discussed and illustrated.

Rolling of Plates of *DS* Steel at the Petrovskiy Works. M. Galemin and A. Nazarenko. (Stal, 1939, No. 10–11, pp. 46–49). (In Russian). The authors investigated the effects of preheating, of changes in the rolling schedule and of different rates of cooling on the properties of 20-mm. thick steel plates rolled at the Petrovskiy Works in the U.S.S.R. The composition of the steel used was carbon 0·17–0·21%, manganese 0·71–0·84%, silicon 0·25–0·40%, chromium 0·50–0·60%, copper 0·50–0·60%, sulphur and phosphorus each 0·04% max. The plates were rolled from slabs measuring 1600 × 1200 × 200 mm. In order to produce the optimum mechanical properties it was found that rolling should be commenced and finished at temperatures of 1100–1150° C. and 830–870° C., respectively, and that the plates should be cooled rapidly from the latter temperature to 400–450° C. It was also found necessary to increase the manganese content to 0·85–1·20% in order to bring the steel up to the following specification: Tensile strength 52–62 kg. per sq. mm., yield point 36 kg. per sq. mm. and 18% elongation in the direction of rolling.

Experimental Rolling of Sheet from Large Stainless-Steel Ingots. D. Gurevich and S. Belorusov. (Stal, 1939, No. 10–11, pp. 49–52). (In Russian). The authors report on an investigation of the possibility of rolling sheets from 10-ton ingots of stainless steel. Two ingots of 18/8 stainless steel were prepared from two heats. Their weights were 9·5 and 10·3 tons, respectively. The ingots were first rolled into slabs and then down to sheet in a continuous sheet mill. Details are given regarding the surface state of the ingots, their preheating, and of the rolling in the slab and sheet mills. An examination of the macrostructure of the slabs was made. Chemical analysis of the slabs showed that segregation in the 10-ton stainless-steel ingots was negligible. It is concluded that the rolling of 10-ton ingots into sheet is feasible. Such ingots should be heated to 1170–1180° C. for rolling into slabs. The temperature at the end of the slab-rolling process should not be lower than 900° C. The slabs should be heated to 1200–1230° C. for rolling into sheet.

Cold Tube-Rolling Mills of the Rockwright Type. A. Grebenichenko. (Stal, 1939, No. 10–11, pp. 58–61). (In Russian). The author discusses the design of Rockwright mills for cold-rolling tubes, drawing attention to some unsatisfactory features and pointing out the differences between the American and the modified Russian types.

Welding Tubes. G. Kentis, jun. (Steel, 1940, vol. 107, Oct. 7, pp. 64–68, 77). The author describes a modern tube mill in which the skelp, after going through the shaping rolls, passes under shaped electrodes so that the seam is welded by the electric resistance process. Standard mills of this type are now available in five sizes for the production of steel tubes from ¼ in. in outside dia. with 0·02-in. walls to 5 in. in dia. with 0·209-in. walls from hot- or cold-rolled strip.

HEAT TREATMENT

(Continued from pp. 12 A–16 A)

Heat Treatment—Gas Carburising. D. McPherson. (Machine Shop Magazine, 1940, vol. 1, Oct., pp. 84–86). The author discusses the action of carbon monoxide and of hydrocarbon gases in gas-carburisation practice with particular reference to the suitability of methane. He also offers some general conclusions which have become evident from his observations of gas-carburising practice.

These are: (1) Unless the carburising gases are thoroughly mixed and uniformly distributed in the furnace or muffle, control over carburisation is impossible; (2) the carbon formed by the breakdown of the carburising gas must be atomic carbon and the subsequent formation of molecular carbon must be prevented; and (3) it is necessary to be able to control the temperature of the muffle within close limits, which must be determined for each particular plant and gas mixture.

Factors Affecting the Activity of Carburizing Compounds. M. Sutton and R. A. Ragatz. (American Society for Metals, Oct., 1940, Preprint No. 4). The authors report on an investigation of the factors affecting the energising action of the carbonates of barium, calcium and sodium when mixed with charcoal for the carburising of steel. The influence of temperature and of the proportion of the carbonate in the mixture were the principal factors studied. From the data obtained the authors came to the following conclusions: (1) Calcium carbonate is quite inert as an energiser with hardwood charcoal; (2) both sodium and barium carbonates are powerful energisers and give results of the same order of magnitude; the former appears to have a slightly greater effect when operating at the high temperatures normally used in carburising, and it is of interest that it has a definite energising action at such low temperatures as 650° and 700° C., whereas barium carbonate has practically no effect at these temperatures; and (3) increasing the concentration of either sodium or barium carbonate above 5% does not increase the energising effect. The effect of different carburisation temperatures upon the depth of case and the amount by which the weight of the specimens was increased is shown in a series of twenty graphs. The authors also examined the relative carburising effect of the following carbonaceous materials, using the previously named carbonates as energisers; (*a*) Hardwood charcoal; (*b*) preheated hardwood charcoal; (*c*) charcoal made from pure cane sugar; (*d*) graphite; (*e*) carbon black; (*f*) retort carbon; and (*g*) metallurgical coke. Their conclusions were as follows: (1) Without any energiser, the three charcoals are far superior to the other materials, the sugar charcoal possessing the lowest activity of the three; (2) calcium carbonate appears to be completely inert except when used in conjunction with metallurgical coke; in this case it is extremely effective, although not as good as the other two carbonates; (3) sodium carbonate is an extremely effective energiser with all of the substances tested; (4) barium carbonate is very effective except with retort carbon; in most instances, barium and sodium carbonates produce effects of about the same magnitude, the latter being usually slightly superior, but with metallurgical coke the barium carbonate is the better of the two; (5) some mixtures, notably sugar charcoal with sodium or barium carbonate, and carbon black with sodium carbonate, closely approach the activity of fresh charcoal energised with the corresponding carbonate; and (6) a definite correlation exists between the precipitation of graphite on the surfaces of the steel specimens and the degree of carburisation, for it was observed that those compounds producing a high degree of carburisation also caused the precipitation of graphite on the surfaces of the specimens.

Surface Saturation of Iron and its Alloys with Boron. I. E. Kontorovich and M. Ya. L'vovskiy. (Metallurg, 1939, No. 10–11, pp. 89–98). (In Russian). After reviewing earlier work on the absorption of boron by steel, the authors discuss the results they obtained by saturating the surface of Armco iron, plain carbon steel and several low- and high-alloy steels by heating at temperatures of 700–1000° C. for various periods of time. The microstructure and thickness of the diffusion layer were studied with reference to the effect of the alloy additions, and X-ray studies were also made of the phases present at different depths. The latter showed the predominance of the iron boride (Fe_4B_2) compound. The extreme hardness of the surface layers (1200–1450 Vickers-Brinell units) is ascribed to the presence of this compound. The depth of the layer formed depends on the composition of the steel and is directly proportional to the duration and temperature of the treatment. Elements which raise the A_3 point markedly reduce the thickness of the diffusion layer. Low-carbon alloys develop their maximum hardness on being treated with boron. The surface hardness is retained after repeated heating.

Diffusion and its Significance in the Heat-Treatment Shop. F. W. Haywood. (Wild-Barfield Heat-Treatment Journal, 1940, vol. 3, Mar., pp. 92–99). The author discusses, in simple language, the mechanism of diffusion in solids, and he emphasises its importance in carburising processes, considering separately the phenomena involved when applying solid, liquid and gaseous carburisers.

Surface Reactions and Diffusion. J. E. Dorn, J. T. Gier, L. M. K. Boelter and N. F. Ward. (American Society for Metals, Oct., 1940, Preprint No. 47). The authors have developed an equation relating the rate of the surface reaction during case-hardening treatment to the depth of the case. The equation is limited to linear diffusion where only two phases are present, *viz.*, the solid and either the gaseous or liquid phases; the formation of a second solid phase and its influence on the diffusion are not taken into consideration. The equation is applicable to nitriding and denitriding with cracked ammonia. No corresponding equation could be developed for the carburising process, as the experimental data for this process are too involved.

Heat Treating with Induction Heat. E. Blasko. (American Society for Metals, Oct., 1940, Preprint No. 48). The author reviews the development by the Ford Motor Co. of induction heating for heat-treatment operations, using 1920-, 960- and 60-cycle currents. He points out that originally the surface-heating properties of induction heating were over-emphasised. The possibility of heating parts in their full cross-section instead of only at the surface opened a wide field for both high- and low-frequency induction heating. Generators of lower frequency and smaller size could be used, so that the installation costs became smaller than those for surface heating. In addition, parts heated in their full cross-section proved to have better mechanical properties.

Inherent Characteristics of Induction Hardening. M. A. Tran and H. B. Osborn, jun. (American Society for Metals, Oct., 1940, Preprint No. 49). The authors enumerate the inherent characteristics of induction hardening as follows: (1) The unusual speed with which carbide solution takes place; (2) the intensity of this diffusion; (3) the very fine and homogeneous martensite produced; and (4) the resulting exceptionally high hardness. They report on their experimental study of induction hardening undertaken for the purpose of establishing plausible explanations for these inherent characteristics of the process. The investigation, which was carried out with an S.A.E. 1050 fine-grained steel and involved hardness tests and metallographic examination, led to the following conclusions: (*a*) Complete carbide solution can be accomplished by induction heating in as short a time as 0·2 sec. (*b*) Carbide diffusion in a pearlitic structure becomes essentially complete within the grain before any appreciable diffusion into the ferrite boundaries is discernible. (*c*) Induction heating produces finer, more homogeneous and more highly dispersed austenite than other methods of heating, resulting in a quenched martensite which possesses the same characteristics. (*d*) Owing to the great fineness and homogeneity of the austenite, a specimen of steel heated by induction will have a low critical cooling rate, and this may be the reason for the increased depth of the quenching effect in sections heated by induction. (*e*) The rapid carbide diffusion may be due to the existence, in certain of the micro-constituents, of temperatures which are not measurable but are presumably considerably higher than the observed average temperature.

Nitrided Stainless Steels. H. Drever. (Metals and Alloys, 1940, vol. 12, Sept., pp. 271–273). The author describes an improved nitriding process for the case-hardening of stainless steel. The parts to be treated by this process are placed in the muffle of a furnace which is kept at 1000–1100° F. The furnace atmosphere is derived from dry ammonia and it is used in two states. In the first stage of the process the ammonia is dissociated, and, after purification and activation, it is passed into the muffle so as to purge all the air out of it. In the second stage, the ammonia is ionised by a silent electrical discharge before it enters the nitriding chamber, where some of the nascent nitrogen is absorbed by the charge in the furnace; the hydrogen passes out of the muffle together with the balance of the nitrogen and some undissociated gas. Tests have shown that an extremely hard case can be obtained by this process.

The corrosion resistance of stainless steel is, however, slightly impaired by nitriding and the material is not recommended for use in contact with heavy acids such as nitric acid. Its resistance to corrosion by steam at high pressure, by petrol, paraffin, fuel oil and tap water remains unchanged. Salt-spray and long-time immersion tests are being carried out, the results of which are not yet available.

Furnace Atmosphere Generation. S. Tour. (American Society for Metals, Oct., 1940, Preprint No. 6). The author discusses the factors affecting the composition and properties of the products of combustion in the preparation of atmospheres for heat-treatment furnaces. He finds that in burning rich mixtures of fuel gas and air, the composition of the products of combusion for a given gas/air ratio are dependent upon the temperature of the combustion chamber. Heavy deposits of soot occur at low combustion-chamber temperatures, and the gaseous products are low in hydrogen and carbon monoxide and high in water vapour and volatile hydrocarbons. At high temperatures coke is deposited and the gaseous products are high in hydrogen and carbon monoxide and low in water vapour. In practice, an intermediate range of 1600–1900° F. may be used, where difficulties due to heavy soot or coke deposits do not develop; this range is wide enough to allow for a considerable variation in the gaseous products of combustion.

Water Vapor in Furnace Atmospheres. S. Tour. (American Society for Metals, Oct., 1940, Preprint No. 8). The author discusses the composition of atmospheres for heat-treatment furnaces, their preparation from the products of combustion and methods of determining the water-vapour content. He points out that standard methods of gas analysis when applied to furnace atmospheres give the composition of the mixture as cooled to room temperature. During this cooling the surplus water vapour is condensed to water, leaving the remaining gases in the saturated state at the temperature of the analysis. Assuming that no other gas reactions have taken place during the cooling, the water-vapour content can be recalculated so as to allow for this loss of water, provided that the original amount of water is known. He gives some examples of the calculation of the total water-vapour content for certain simple fuel gases and known air/gas ratios. The results show that, in some cases, the actual water-vapour content in the products of combustion of rich mixtures of gas and air as used for the atmospheres of heat-treatment furnaces may be as high as 18%. By cooling to remove the water and then reheating, the composition of the atmosphere may be changed, for the effect of this is to reduce the residual hydrogen and carbon dioxide and to increase the residual carbon monoxide. It is thus evident that the dehydration, reheating and recirculation through the furnace of the gaseous products of the combustion of rich mixtures of fuel gas and air provide a means of controlling the carburising or decarburising power of the atmosphere.

A Balanced Protective Atmosphere—Its Production and Control. J. R. Gier. (American Society for Metals, Oct., 1940, Preprint No. 5). The author describes the production on a laboratory scale of a balanced protective atmosphere for the heat treatment of steel, and a device by which the carburising power of such an atmosphere can be measured. The protective atmosphere is produced from a mixture of natural gas and air, and it is known as "endogas" because it is formed by an endothermic reaction between the hydrocarbons and the air in the presence of an electrically heated catalyst. The important feature of the method of production is that, by varying the air/gas ratio of the feed to the generator, the carburising power of the atmosphere can be adjusted to balance the carbon pressure of any steel. The device for measuring the carburising capacity of the atmosphere is known as a hot-wire carbon gauge. In principle, it involves the heating of a thin steel wire in the test gas until a carbon equilibrium is established between the gas and the wire; this requires only a few minutes, after which the wire is rapidly cooled to retain its carbon in solid solution as martensite. The electrical resistance and certain other physical properties of the wire after this treatment under standardised conditions can then be used as a measure of the carbon content of the wire, which, in turn, measures the carburising power of the gas.

Dimensional Changes on Hardening High Chromium Tool Steels. H. Scott and T. H. Gray. (American Society for Metals, Oct., 1940, Preprint No. 7). The authors report on an investigation of the factors which affect the changes in the dimensions of tool steels during hardening. The factors they examined were: (1) Segregation in the steel used; (2) incomplete hardening, particularly at the surfaces; (3) distortion of the specimen in the furnace under its own weight; (4) loss of metal by scaling; (5) release of stresses resulting from previous heat treatment; (6) changes in the rate of heating; (7) changes in the rate of quenching; (8) inherent increase in the specific volume of the specimen on hardening by quenching; (9) directional effects of the change in volume referred to in (8); and (10) decrease in the specific volume of hardened steel on tempering to a lower hardness. From the results of their tests, the authors came to the following conclusions: (*a*) The change of dimension due to (5) is negligible in tools machined from annealed bars that have not been deformed cold except by normal cutting operations; (*b*) preheating is unnecessary to control dimensional changes, even when the tool is of rather complicated shape; (*c*) distortion due to (7) can be effectively eliminated by using a gas as the quenching medium; (*d*) when tools are hardened all over to a hardness not less than Rockwell 64, an increase in volume takes place which may amount to 0·7% of the volume in the annealed state; (*e*) the volume increment resulting from hardening diminishes rapidly with increasing chromium content, being only 0·1% in an 11% chromium steel; (*f*) the expansion on hardening a steel containing 1% of carbon and

5% of chromium is the same in all directions, amounting to approximately 0·001 in. per inch; (*g*) on hardening a steel containing 1·5% of carbon and 11% of chromium, practically all the movement is in the direction of extension during hot-working and is equal to that of a 5%-chromium steel; (*h*) the size and shape of the test-piece or bar stock from which specimens are cut has no substantial effect on the dimensional changes; and (*i*) with the kinds of steel and the hardening practice adopted by the authors the changes in dimensions are predictable, even for complicated dies.

Distortion in the Heat Treating of Meehanite Metal. T. E. Eagan. (Foundry Trade Journal, 1940, vol. 63, Oct. 31, pp. 281–282). The author's paper on the limitation of the distortion which occurs in the heat treatment of Meehanite is reproduced. This appeared previously in Heat Treating and Forging, 1940, vol. 26, May, pp. 225–229 (*see* Journ. I. and S.I., 1940, No. II., p. 184 A).

Hardening Characteristics of Various Shapes. M. Asimow and M. A. Grossmann. (American Society for Metals, Oct., 1940, Preprint No. 32). The authors have previously suggested a method of determining the severity of quenching of round bars which also served as a criterion of hardenability (*see* Journ. I. and S.I., 1940, No. II., p. 61 A). In the present paper they report on the application of their method to shapes other than rounds. They developed a series of charts for flat plates which is entirely analogous to that previously set up for rounds, and the calculated values were found to be in close agreement with experimental data. The authors also propose an approximation for irregular shapes which do not permit of precise calculation. In this approximate method each point of the irregular shape is considered to behave like a corresponding point in a round bar of equivalent diameter.

Effects of Small Amounts of Alloying Elements on the Tempering of Pure Hypereutectoid Steels. C. R. Austin and B. S. Norris. (American Society for Metals, Oct., 1940, Preprint No. 36). The authors studied the softening characteristics of 1·1% carbon steels, made from electrolytic iron melted under hydrogen in an induction furnace and containing 0·5% or less of one of the following elements: Manganese, silicon, nickel, chromium, copper, aluminium, sulphur, phosphorus and tin. The steels were hardened and subsequently tempered for periods ranging from 30 min. to 125 hr. at five different temperatures, *viz.*, 550°, 590°, 630°, 670° and 710° C., and the effect of the heat treatment was investigated by hardness determinations and micrographic studies. The authors found that a linear relationship exists between the Rockwell B or Brinell hardness and the logarithm of the tempering time, so long as graphitisation does not occur. Heavy graphitisation occurred, however, in some of the aluminium steels when tempered at 550°, 590° and 630° C. After a tempering period of 1 hr., at any of the temperatures studied, hardness values from 10 to 30 Brinell units greater than those for plain carbon steel were obtained with alloys containing approximately

0·5% of manganese, silicon, aluminium or copper. For chromium the values were 70–100 units higher, whereas nickel had little effect. The influence of chromium was noteworthy even in amounts of 0·07–0·08%. The addition in such small amounts of the other elements studied had hardly any effect. In order to demonstrate the effectiveness of the various alloying elements on the resistance to softening, the authors present curves which show the variations in the time rate of softening as a function of the hardness of the steel tempered at 710° C. They also derived data to permit the presentation of curves illustrating the effect of temperature on the variation in softening rate, with change in hardness, of both the chromium and nickel alloy steels containing approximately 0·5% of these elements. At all temperatures investigated the relative rate of softening at any selected Brinell hardness was very much greater for the nickel than for the chromium steel.

WELDING AND CUTTING

(Continued from pp. 16 A–18 A)

Arc Welding Chrome-Vanadium Steel. I. Z. Kagan. (Welding Industry, 1940, vol. 8, Nov., pp. 276–277). The author reports on a comparative investigation of the merits of electric welding steel containing carbon 0·15–0·25%, chromium 0·80–1·10% and vanadium 0·15%, using electrodes of low-carbon steel and electrodes of steel similar to the parent metal coated with an ilmenite-base flux containing additions of ferro-chromium and ferro-vanadium. The results showed that welds of very high quality could be made with the latter electrodes, and that the arc-welding process could be satisfactorily employed for the fabrication of internal parts of plant for the production of synthetic ammonia.

The Butt Welding of Steel Tubes and Pipes. H. Harris, J. E. Jones and A. L. Skinner. (Transactions of the Institute of Welding, 1940, vol. 3, July, pp. 116–156). The authors present a report on an extensive investigation of the general problem of the butt welding of steel tubes and pipes. The defects found and reported are those which, in their magnitude and frequency of occurrence, are peculiar to the butt welding of pipes if adequate precautions are not taken; such defects as slag inclusions and lack of penetration are not dealt with. In the first part of their investigation, the authors examine the properties of joints made by different techniques, using different shapes of steel backing rings and no backing ring. The technique of welding with no backing ring was developed because of troubles experienced with the welding of thick-walled tubes of steel containing carbon 0·20–0·27% and molybdenum 0·5%. In this method the first run is deposited by autogenous welding and the remaining runs by the metallic-arc process; a special type of V-notch preparation is required and the minimum depth of the first run should be about ⅛ in. The results of a large number of tests have proved that, by proper appreciation of the factors involved, welds of a very high quality can be made without the necessity of using a backing ring and without any special preheating. The authors also point out that the reverse bend test is not entirely a test of the soundness of the weld, for in many cases reflects unduly the directional properties of the pipe material. In the next part, the authors discuss at length the factors influencing the formation of cracks in the base of welds, dealing in turn with the electrode, the contour of the surfaces to be welded, the temperature of the pipes being welded, the proficiency of the welder and the metallurgical character of the pipe steel. The authors then consider the necessity and methods of preheating, stress-relieving and normalising. In conclusion the different methods of examining butt welds in pipes are dealt with. The authors do not recommend X-ray testing for pipe joints but consider macro-etching to be the most informative method. Over 170 illustrations of joints and test-pieces accompany the report.

Temperature Distribution and Shrinkage Stresses in Arc Welding. D. Rosenthal and J. Zabrs. (Welding Journal, 1940, vol. 19, Sept., pp. 323-S–331-S). The authors report on their investigation of the relationship which exists in arc welding between shrinkage stresses and temperature distribution. A bead of weld metal was deposited along the edge of each of the specially prepared specimens of 0·47-in. steel plate in such a manner that the quantity of heat supplied to the plate per unit of time and the speed of advance of the electrode were kept constant. The distortion of the plate at different distances from the weld was measured by three different methods and the shrinkage stresses were calculated from these measurements. It was established that two zones were created, one of plastic deformation near the weld, and one of elastic deformation farther away. The authors conclude that the relation between the heat input to the weld and the maximum stress in each of these two zones is as follows: (1) Within the limits of the experiments, the maximum stress in the elastic zone increases in almost direct proportion with the increase in the power of the arc measured in watts; (2) increasing the speed of advance of the electrode has the opposite effect, but it is very slight; (3) the maximum stress in the plastic zone depends very little, if at all, on the temperature distribution; and (4) for the type of welding discussed, a low rate of heat input is in every respect preferable to a high rate of heat input.

MACHINING

Crankshaft Construction at Australian Iron & Steel Ltd., Kembla Works. J. B. Robinson. (B.H.P. Review, 1940, vol. 17, June, p. 18). The author describes the machine-shop technique employed at the Kembla Works of Australian Iron and Steel, Ltd., in the production of large crankshafts built up of forged steel webs, crank-pins and journals. The crankshafts produced are up to 25 ft. long with 21-in. dia. journals and a total weight of 30 tons.

Machining With Single Point Tools. M. Kronenberg. (Transactions of the American Society for Metals, 1940, vol. 28, Sept., pp. 725–742). One of the most important conclusions drawn from recent metal-cutting research is that two main relationships exist between the dimensions of the chip and the cutting speed. One of these is based on the life of the tool and the other on the capacity of the machine. The author discusses these relationships and presents tables of data and graphs with the aid of which cutting speeds and forces for various materials and tools can be calculated using simple formulæ.

Chip Formation, Friction and High Quality Machined Surfaces. H. Ernst and M. E. Merchant. (American Society for Metals, Oct., 1940, Preprint No. 53). The authors point out that the coefficient of friction between chip and cutting tool is usually very high, and that this is responsible for the formation of the so-called built-up edge, with the accompanying roughness of the finished surface, and, in general, for the inefficiency of all present-day metal-cutting processes. They report on their mathematical and experimental study of the friction phenomena involved in machining operations, and they show that the hardness of the metal surfaces in contact and the resistance to shear at the areas of contact are factors of great influence. The resistance to shear of ordinary dry metal surfaces in contact is decreased owing to the presence of firmly adsorbed surface films consisting of material of low shear strength. This leads to the conclusion that great advances in the field of metal cutting could be made after the development of cutting fluids which would provide and maintain material of low shear strength at the chip/tool interface, even at very high cutting speeds. They also consider the further development of free-cutting steels, the cutting of which is facilitated owing to the addition of substances, such as lead and sulphur, capable of producing films of low shear strength.

PROPERTIES AND TESTS

(Continued from pp. 23 A–29 A)

Investigation of the Internal Stresses in Cast-Iron Rolling-Mill Rolls in Relation to the Conditions of their Use. N. Krupnik. (Stal, 1939, No. 10–11, pp. 53–57). (In Russian). The author reports on an investigation of the internal stresses in four cast-iron rolls, the first of which had been cast in the usual way, the second had been given a stress-relieving anneal, the third had been in use in a rolling mill, and the fourth had been used under special conditions which involved internal heating. Disc-shaped specimens, 40 mm. thick, were cut from the central portions of the rolls and subjected to microscopic examination and chemical analysis. Preliminary experiments also showed that with this particular iron, which contained carbon 3·26%, silicon 0·7% and manganese 0·9%, the annealing treatment should not, for practical purposes, exceed 3–5 hr. at 500° C., in order to avoid graphitisation of the pearlite. The author discusses the theory of Sachs' method, which was used to measure tangential and radial stresses in the specimens. This involved progressively increasing the diameter of a hole in the disc and measuring the change in its external diameter. The method of measurement and the precautions against errors due to temperature changes are described. In the as-cast condition, the specimen was found to have tangential stresses greater than 11 kg. per sq. mm. and radial stresses up to 12 kg. per sq. mm. These had been reduced in the annealed specimen to 6 and 5·6 kg. per sq. mm. respectively. Stresses in the used roll were found to be slightly higher than those in the as-cast roll. Internal heating of the fourth roll appeared to have reduced the stresses somewhat and altered their distribution.

Effect of Surface Conditions on Fatigue Properties. O. J. Horger and H. R. Neifert. (American Society for Metals, Oct., 1940, Preprint No. 52). The authors studied the influence of surface condition on the fatigue strength of steel. They tested specimens varying from 1 to 11·5 in. in dia. in a rotating cantilever beam machine and compared the results with those obtained with specimens 0·3 in. in dia. in a Moore rotating-beam machine. They investigated the effect of surface finish, as produced by machining operations, for rough-turned, smooth-turned, polished and super-finished surfaces. They determined the roughness of the surface from profilograph records and correlated it with endurance limit values. Some of the tests were made on specimens with stress concentrations produced by a press-fitted wheel as a stress raiser. Observations were also made as to the increase in the formation of nitrides on the surface of specimens after fatigue tests in the region of the fitted wheel. In conclusion, the authors report on some tests comparing the increase in fatigue strength due to surface finish with that due to rolling, burnishing, metal spraying and flame-hardening.

Chafing Fatigue Strength of Some Metals and Alloys. G. Sachs and P. Stefan. (American Society for Metals, Oct., 1940, Preprint No. 38). The authors found endurance tests on cylindrical test bars to be a simple method for determining the effect of stress raisers on the fatigue strength. They point out that there are three types of stress raisers known, *viz.*, notches, corrosion and lateral pressure connected with a chafing on the surface of parts subjected to vibrating stresses. They report on their study of the effect on the fatigue strength of the last-named type of stress raiser, which is met with in press fits, bearing seats, propeller bosses and jaws. They used a Farmer rotating-beam type of machine, which produced a complete reversal of stress during each cycle and a uniform bending moment on the specimen. The speed of testing was approximately 3500 cycles per min. The grips, described at some length in the paper, were chosen so as to give a tight chafing press fit. In the first part of their paper the authors report on their study of the effects of alloying, heat treatment and cold-work on the chafing fatigue strength of a great number of wrought alloys. In the second part, which is abstracted from a thesis by E. J. Jory, the results of some chafing fatigue tests carried out under identical conditions with some cast and wrought alloys of similar compositions are compared. In general, the chafing fatigue strength was found to be closely related to the notch fatigue strength obtained with specimens having very sharp notches. Annealed wrought metals had a higher chafing fatigue strength than the same metals in the cold-worked, strain-hardened or precipitation-hardened state, and cast alloys had a higher chafing fatigue strength than wrought ones.

Damping Capacity, Endurance, Electrical and Thermal Conductivities of Some Gray Cast Irons. C. H. Lorig and V. H. Schnee. (Transactions of the American Foundrymen's Association, 1940, vol. 48, Dec., pp. 425–445). The authors investigated the effect of additions of up to 3·0% of copper on the damping capacity, endurance limit and the electrical and thermal conductivities of some low-, medium- and high-carbon cast irons. The apparatus with which the damping capacity was tested is described and illustrated. In this apparatus the specimens were submitted to torsional oscillations in which the material at the surface was subjected to stresses equal to 15–35% of the tensile strength of the iron. At stresses equal to 15% of the tensile strength the damping capacity appeared to increase with increasing copper content. At somewhere between 15% and 35% of the tensile strength this relationship was reversed. Fatigue tests on six types of medium-carbon grey iron with additions of copper in the range 0·0–3·0% demonstrated that the ratio of the endurance limit to the tensile strength decreased slightly with increasing copper content. With the range of copper content investigated, the copper had little, if any, effect on either the thermal or the electrical conductivity.

Fatigue and Damping Studies of Aircraft Sheet Materials: Duralumin Alloy 24*ST*, Alclad 24*ST* and Several 18/8 Type Stainless Steels. R. M. Brick and A. Phillips. (American Society for Metals, Oct., 1940, Preprint No. 39). The authors studied the relation between the fatigue and damping properties of various sheet metals for aircraft. The materials tested were the two wrought aluminium alloys 24*ST* and Alclad 24*ST*, and nine commercial 18/8 stainless steels, of low and medium carbon contents, in the annealed, cold-rolled, stabilised and aged conditions. For both fatigue and damping tests the same specimens were used, the preparation and dimensions of which are described in detail. The fatigue tests were carried out with a constant-deflection type of machine. The tests were started at high deflections and were continued at progressively lower ranges of stress. For the damping tests and the photographic reproduction of damping curves the authors developed a special apparatus which is described at some length. The endurance-limit/tensile-strength ratio of the stainless steels was found to vary between 30% and 70% for different commercial finishes, but to be relatively constant for each type of surface. The fatigue values obtained were considerably scattered. The authors think this to be due not only to errors inherent in the method of testing but also to work-hardening and minute surface defects, the latter resulting in crack formation and the former in increases of stress at constant deflection near the endurance limit. The damping capacity decreased almost continuously with increase in tensile strength. The study of the damping curves led to the following conclusions: (1) An increase of the carbon content from 0·07% to 0·11% decreased the damping capacity of the annealed 18/8 steels. (2) Damping was greatly decreased by cold-work. (3) Repetition of the damping test, with the maximum stress above the endurance limit, increased the damping capacity slightly, particularly at higher stresses. (4) Stressing which started below the endurance limit and was continued for a long time, reaching stresses above this limit, raised the damping capacity of annealed or slightly worked steels considerably, and also slightly increased that of stronger specimens. (5) Work-hardening accompanying under-stressing seemed to cause a slight increase in damping capacity. In conclusion the authors briefly report on additional tests by which the mechanical hysteresis loops formed by complete cycles of the load-bending test were determined for several of the 18/8 steels under investigation, before and after fatigue stressing near the endurance limit. The mechanical-hysteresis loops obtained gave a qualitative idea only of the relative damping capacities.

Pendulum Hardness Testing Machine. (Iron Age, 1940, vol. 146, Sept. 26, p. 38). A description is given of a pendulum type of indentation hardness-testing machine with which hardness tests can be made on specimens held at temperatures up to 900° C. This machine has been described previously by Cornelius and Trossen in Stahl und Eisen, 1940, vol. 60, Apr. 4, pp. 293–294 (*see* Journ. I. and S.I., 1940, No. II., p. 27 A).

Metals in Thin Layers—Their Microhardness. C. G. Peters and F. Knoop. (Metals and Alloys, 1940, vol. 12, Sept., pp. 292–297). The authors describe a diamond-pyramid hardness testing machine for use on very thin layers of metal and discuss some results obtained with it on ferrous and non-ferrous metals. The shape of the diamond in this machine is such that the length of the impression produced is seven times the width and thirty times the depth, and the sensitivity of the tool is such that a load of 0·5 kg. on hard steel makes an impression 100 μ long, which can be measured with an accuracy of 1 μ with a micrometer microscope. In making tests on the coating of chromium-plated steel it was observed that the hardness of the base metal did not affect the results obtained. To obtain consistent results it was found that the thickness of the coating should be at least fourteen times the depth of the impression. The results discussed include those from tests on numerous specimens of high-speed steel heat-treated and nitrided in various ways.

Effect of Deoxidation on Hardenability. G. V. Cash, T. W. Merrill and R. L. Stephenson. (American Society for Metals, Oct., 1940, Preprint No. 30). The authors studied the effect of deoxidation practice on the hardenability of steel, examining the behaviour of five different heats, specimens from which were prepared in the fine- and coarse-grained state. Their investigation indicated that the hardenability of any steel is influenced by its austenitic grain size at the time of quenching, and that the only effect of the deoxidation practice is to determine the temperature at which any given grain size can be produced. The authors conclude, therefore, that the hardenability is a linear function of the grain size at the temperature of heat treatment.

Hardenability Testing Tool Steels by Oil Quenching Small Cones. E. K. Spring and J. K. Desmond. (Steel, 1940, vol. 107, Oct. 28, pp. 58–62). The authors describe a method of testing the hardenability of tool steels by tests on cone-shaped specimens, and of classifying the steels by the results obtained. The specimens are 8 in. long, 2 in. of which are left cylindrical with a diameter of 0·625 in.; the other 6 in. are turned so as to taper from 0·625 in. down to 0·200 in. The specimens are heated to the desired temperature and held for the requisite time, and are then quenched in oil. Hardness explorations are made along the length of the cone and the hardenability of the steel is classified by the diameter of the cone at the point where a hardness of Rockwell C60 is obtained. Some results obtained with specimens of low-alloy steels containing nickel and chromium are discussed.

The Effect of Grain Size on Hardenability. M. A. Grossmann and R. L. Stephenson. (American Society for Metals, Oct., 1940, Preprint No. 31). The authors studied the effect of the grain size on the hardenabilities of steels of a wide range of composition. For each steel, a series of specimens with different grain sizes was produced by quenching from different temperatures. It was found that the greater the hardenability of a steel due to its chemical composition, the more was that hardenability affected by the grain size.

Age Hardening of Cold Reduced Strip. P. J. McKimm. (Steel, 1940, vol. 107, Sept. 30, pp. 44–49; Oct. 7, pp. 46–50, 78–79). The author presents and discusses data which show how the rolling temperature and the degree of reduction in the rolling of steel strip affect its hardness and tensile properties and its susceptibility to ageing. The economic aspect of producing non-ageing steel is also dealt with.

Measuring the Tendency of Some Cast Irons to Seize Under Sliding Friction. A. H. Dierker, B. Fried and H. H. Dawson. (Transactions of the American Foundrymen's Association, 1940, vol. 48, Dec., pp. 355–367). The authors describe an apparatus and testing procedure which were developed for determining the resistance to seizure of the surfaces of a number of cast irons when in sliding contact under conditions of boundary lubrication. A shaping machine was used to give a reciprocating motion to a cylindrical specimen which was moved over the surface of a block. Pressure was applied to the specimen by means of a hydraulic system and the procedure adopted was to increase the load by 2·6 lb. every 5 strokes until seizure occurred. The area of contact and the load at seizure were measured and the resistance to seizure of the material in question was thus determined in pounds per square inch. The results are given for thirteen plain and low-alloy cast irons and for one low-carbon steel. The specimen and the block were of the same material in each test. From the data obtained it is deduced that the load-carrying capacity of cast iron depends on two apparently independent characteristics, namely, its resistance to seizure and its " pliability," a term which the authors define as the ability of the material to give contact over a large area. Microscopical examinations revealed that, in some cases, there were considerable differences in structure in various parts of the same specimen. The authors were unable to establish any relation between either the microstructure or the hardness and the resistance to seizure, but their general conclusion was that large amounts of carbides or of ferrite are undesirable.

Correlation of High-Temperature Creep and Rupture Test Results. R. H. Thielemann. (American Society for Metals, Oct., 1940, Preprint No. 1). The author reports on a series of tensile tests at high temperature on specimens prepared from four alloy steels. These steels contained: (*a*) Nickel, chromium and molybdenum; (*b*) chromium, molybdenum and vanadium; (*c*) molybdenum and 0·5% of carbon; (*d*) molybdenum, 7% of chromium and 1% of silicon. At least six specimens of each steel were prepared, and different stresses were applied to each after it had reached the desired temperature, and measurements of the creep were made during the progress of the long-time tests. The results are shown by curves obtained by plotting the creep against the time on a logarithmic scale. It was observed that with steel (*a*) the creep curves were fairly straight up to about 2% elongation, whilst those for steel (*b*) indicated a rapid creep rate until the elongation was about 1%, whereas above this value the creep rate slowed down.

Effect of Nitrogen on the Case Hardness of Two Alloy Steels. S. W. Poole. (American Society for Metals, Oct., 1940, Preprint No. 33). The author studied the effect of nitrogen on the case hardness of two qualities of steel used in the manufacture of automobile gears, *viz.*, a medium-carbon nickel-chromium steel and a high-carbon chromium steel. He treated the samples in five different commercial salt baths, two of which, containing sodium cyanide and calcium cyanamide, respectively, were chosen so as to introduce considerable quantities of nitrogen into the case. This treatment was followed by oil-quenching and subsequent drawing in a temperature range of 150–315° C. The sodium cyanide bath gave the best results in so far as high surface case hardness was concerned, and it was found that the maximum surface case hardness was generally greater for the chromium steel than for the nickel-chromium steel. The maximum hardness of the former occurred at a lesser depth below the surface than with the latter, owing to the smaller amount of austenite formed on quenching. Nitrogen in the case proved to be valuable in inhibiting the softening effect of tempering. This was more clearly indicated in the chromium-nickel than in the chromium steel.

Phosphorus as an Alloying Element. N. Roy. (Tisco Review, 1940, vol. 8, Sept., pp. 706–709, 714). The author reviews the results of some experiments in which the effect of increasing the quantity of phosphorus in plain and alloy steels on their tensile strength and corrosion resistance was examined. The alloys tested included the iron-phosphorus series in combination with one of the following elements: (*a*) Silicon; (*b*) manganese; (*c*) sulphur; (*d*) copper; (*e*) chromium; (*f*) nickel; (*g*) molybdenum; and (*h*) vanadium. The results may be summarised as follows: (1) The tensile strength and the yield strength of plain and alloy steels are both increased by the addition of phosphorus; (2) with low-carbon steels, both plain and alloyed, this increase in strength can be secured with but little sacrifice in ductility; (3) the addition of phosphorus has a marked effect in reducing wear as measured by a Spindel machine; and (4) the resistance to atmospheric corrosion was materially improved with the progressive additions of phosphorus.

Effects of Sulphur on Electric Furnace Cast Iron. F. Holtby and R. L. Dowdell. (Transactions of the American Foundrymen's

Association, 1940, vol. 48, Dec., pp. 303–343). The authors report on an investigation of the effect of increasing the percentage of sulphur in cast iron on its properties. An indirect-arc furnace with a capacity of about 2 cwt. was used for preparing the melt, the charge for which was made up of pig iron, steel scrap and ferromanganese; the sulphur was adjusted by additions of iron sulphide. The iron produced contained approximately 3·50% of carbon, 2·70% of silicon, 0·66–1·09% of manganese and 0·12% of phosphorus. Over 40 casts were made under controlled conditions. The data obtained led to the following conclusions: (1) Increasing the sulphur content up to about 0·18% in electric-furnace cast iron lowers the transverse strength, tensile strength and hardness, and increasing the sulphur content beyond 0·18% improves these properties; (2) as the sulphur content is increased the manganese content is reduced, the reduction being more rapid when the sulphur content exceeds 0·18%; (3) an increase in the sulphur content does not affect the depth of chill obtainable if the manganese content is kept constant; (4) the flowability of the iron decreases slightly with increasing sulphur content up to about 0·18%, beyond which point it decreases very rapidly; (5) increasing the sulphur content up to 0·14% decreases the machinability; further increases up to 0·18% increase the machinability; above this point the machinability decreases rapidly; (6) the contraction of this class of iron increases with increasing sulphur content up to 0·14%, from 0·14% to 0·18% of sulphur it decreases, and above 0·18% it again increases; (7) when the sulphur content exceeds about 0·18%, blow-holes tend to form in the iron; and (8) the presence of up to about 0·18% of sulphur in the iron used in this investigation is not considered very detrimental to its properties.

Definition of Cast Iron. (Bulletin of the British Cast Iron Research Association, 1940, vol. 6, No. 6, pp. 145–146). Some definitions of cast iron are discussed, notably that of Norbury, which reads: "Alloys of iron and carbon with or without other elements, which contain carbide eutectic (white cast irons) or graphite eutectic (grey cast iron) or both carbide eutectic and graphite eutectic (mottled cast iron) in the structure." This definition has been modified by the United States Bureau of Standards, as follows: "Cast iron is a cast alloy of iron and carbon, with or without other elements, in which the carbon content exceeds the maximum limit of solid solubility, as determined at any temperature (which in plain cast iron is 1·7%), and hence contains eutectic carbon or graphite as a structural feature. It is not usefully forgeable at any temperature." A third definition in the form of a long descriptive statement suitable for elementary textbooks is also suggested.

Sand Affects Physical Properties of Gray Iron. H. W. Dietert and E. E. Woodliff. (Transactions of the American Foundrymen's Association, 1940, vol. 48, Dec., pp. 393–422). The authors report on an investigation of the effect of changes in the moisture content and permeability of moulding sand on the properties of grey iron castings. The conclusions they reached are as follows: (1) An increase of moisture in the sand reduces the fluidity of the metal; (2) the transverse strength and deflection of a grey iron of similar composition to that used in the investigation are reduced by an excess of moisture in the sand; (3) the fracture of a grey iron becomes finer and lighter in colour as the moisture content of the sand increases; (4) the graphitic carbon is refined by increasing the moisture content of the sand; (5) the areas of ferrite in grey iron are increased when the moisture content is high and this decreases the strength of the metal; (6) a high moisture content frequently causes gas inclusions in the cast metal; (7) the porosity of a casting can be reduced by increasing the permeability of the sand, provided that the mass of metal is not too great; (8) the fluidity of the metal can be reduced by using a sand of high permeability; (9) the depth of chill obtainable in a grey iron is increased as the moisture content of the sand is raised; (10) the fracture of a grey iron becomes finer and lighter in colour as the permeability of the sand increases; (11) the pearlitic areas are more completely developed when sands of low permeability are used; and (12) generally speaking, changes in the moisture content of the moulding sand have a greater influence on the physical properties of the casting than changes in the permeability.

Materials of Construction for Chemical Engineering Equipment. (Chemical and Metallurgical Engineering, Sept., 1940, Editorial Supplement). A comprehensive table is presented of the sources of supply, compositions and properties of 264 ferrous alloys and 102 non-ferrous alloys used in the construction of chemical works plant, all of which are produced in the United States. The alloys are listed in alphabetical order of their trade names, and in addition to the usual tensile and hardness properties, particulars are also given, in most cases, of their specific gravity, melting point, expansion, thermal conductivity, machinability, and resistance to heat, abrasion and to some acids and alkalis.

METALLOGRAPHY AND CONSTITUTION

(Continued from pp. 30 A–33 A)

Mounting Micro-Specimens. D. McPherson. (Machine Shop Magazine, 1940, vol. 1, Oct., pp. 54–56). The author describes a simple and easily constructed apparatus for the mounting of specimens in bakelite or other plastic material for microscopical examination. The body of the apparatus consists of a short length of steel bar about 5 in. in dia. and 3 in. long. It is bored through the centre to take a steel plug by which the pressure on the plastic material is exerted. Each face of the bar is provided with an annular recess and a cover-plate; each recess houses an electric heating element. A tray or base-plate is provided, on the centre of which the specimen is placed. The apparatus is placed between the rams of a press or testing machine; the current is switched on, and when the desired temperature is reached the centre hole is charged with the powdered plastic, the plug is inserted and the pressure is applied; after the "curing" period the specimen and mount can be removed.

The Technique of Microradiography and its Application to Metals. G. L. Clark and W. M. Shafer. (American Society for Metals, Oct., 1940, Preprint No. 21). The authors study the theory and practice of the production of microradiograms of metal specimens. They find that by using the Lippmann-Gevaert emulsion of an extremely fine grain size for the silver halide, an X-ray image on a photographic plate may be magnified up to 200 times without loss of detail. The authors also work out theoretical X-ray absorption equations and show how these can be used to determine the X-ray wave-length which will produce the maximum contrast between the constituents in a specimen such as an aluminium-copper alloy. A careful study of photographic emulsions and developers is presented. A camera for multiple exposures is described and some results obtained with it for steel and some non-ferrous alloys are reported. It is demonstrated that microradiography supplements microscopy and has great metallurgical usefulness.

Notes on the Interpretation of X-Ray Diffraction Diagrams and Evidence of Mosaic Structures. N. P. Goss. (American Society for Metals, Oct., 1940, Preprint No. 23). The author explains briefly the fundamental principles of the examination of metals by X-ray diffraction and cites some experiments which show that the appearance of Laue spots may be due either to radiation characteristic of the target element, or to continuous radiation which varies over a considerable range of wave-length, or to one of these superimposed on the other. He also presents and discusses evidence produced by X-rays, which shows that the grains in annealed metals are built up of crystal units smaller than the grains and differing slightly in orientation.

The Phenomenon of Inverse Segregation in Iron Alloys ("Hard Spots" and "Black Spots"). M. P. Slavinskiy. (Metallurg, 1939, No. 10–11, pp. 118–122). (In Russian). The phenomenon of inverse segregation in alloys is discussed with reference to the conditions under which it appears, its causes and consequences. Assuming that inverse segregation is a phenomenon appertaining to alloys of the solid solution type, the author points out that there is no apparent reason for the generally accepted view that the iron-carbon alloys do not suffer from this defect. A number of examples in which hard spots were encountered in the machining of cast, rolled and stamped parts are described and illustrated. The hard spots interfered with machining, while in some instances—*e.g.*, in milling—the hard spot or inclusion was torn out of the surface. Hard spots were encountered in an iron casting containing 3·2% of carbon, and in plain carbon steels with average carbon contents of 0·65%, 0·46% and 0·62%, respectively. In the case of the casting, the carbon content of the hard spot was found to be 2%. The hard spots are ascribed to the expulsion from the interior of still liquid alloy enriched with the low-melting-point constituents. The hard spots are sometimes referred to as black spots owing to their appearance on the light-coloured machined steel surface. Under the microscope the black spots frequently appear as stripes running in the same direction as the fibres. They have a weakening effect on the metal.

Dendritic Structure. C. H. Desch. (Sheffield Metallurgical Association : Iron and Coal Trades Review, 1940, vol. 141, Nov. 8, p. 474 ; Nov. 15, pp. 495, 492). The author discusses the influence of dendritic structure on the heterogeneity of alloys and methods of studying one of the four types of heterogeneity, namely, the existence of a composition gradient within individual dendrites. In steels, the austenite crystals showed composition gradients within the dendrites. This structure was important for practical purposes because it persisted during forging. The author observed that the proportions of the alloying elements in an alloy steel were not constant throughout the breadth of a dendrite. The different portions of a dendrite would therefore have different transformation points, and this might play an important part in the production of internal stresses and cracks. The author refers to eleven possible methods of determining the actual composition gradient within a dendrite ; they involve : (1) Chemical etching ; (2) contact printing ; (3) microchemical examination of shavings ; (4) spectrographic study of isolated spots ; (5) determination of electrolytic potential at points ; (6) measurement of the electrolytic potential as a whole ; (7) X-ray microscopy ; (8) X-ray transmission pictures ; (9) the thermo-magnetic method ; (10) measurement of electrical resistance ; and (11) observation of the broadening of X-ray lines. He discusses the advantages and disadvantages of these methods with references to work which has already been done by some of them and suggests that further work might usefully be done by methods (1), (7) and (9).

Structure and Properties of Some Iron-Nickel Alloys. G. Sachs and J. W. Spretnak. (American Institute of Mining and Metallurgical Engineers, Technical Publication No. 1246 : Metals Technology, 1940, vol. 7, Oct.). The authors studied the structure and physical properties of eleven iron-nickel alloys, ranging in composition from nil to approximately 100% of nickel, after subjection to a variety of heat-treatment procedures and, in some cases, to cold-working. The investigation led to a proposed new equilibrium diagram for the iron-nickel system, which differs from those proposed by Owen and Sully (Philosophical Magazine, 1939, vii, vol. 27, p. 614) and by Bradley and Goldschmidt (Journ. I. and S.I., 1939, No. II., p. 11 P) in that the α-phase is stable at room temperature up to very high nickel contents. By means of powder diagrams the authors observed structural changes towards equilibrium conditions in the 28% nickel alloy in both the cold-rolled and annealed conditions. This definitely established the heterogeneous (α + γ) equilibrium for this alloy for a certain temperature range below an upper critical temperature of 480° C. On rolling, the alloy tended to be completely transformed into the α-state, and this led to the conclusion that it consists of the pure α-phase when in the equilibrium condition at room temperature and at temperatures up to about 200° C. Hardness tests indicated that alloys containing more than 4% of nickel are, with increasing nickel content, increasingly affected by residual strain-hardening and transformation-hardening. The authors developed hypothetical hardness curves for ferrite and martensite and found that alloys containing 15–25% of nickel are generally in an almost completely martensitic condition. The alloy with 28% of nickel became strain-hardened to a particularly large extent by cold-rolling, owing to the simultaneous formation of carbon-free martensite, and the hardening of this alloy by cold-work exceeded that of any other iron-nickel alloy.

Structural Changes in Low Carbon Steels Produced by Hot and Cold Rolling. N. P. Goss. (American Society for Metals, Oct., 1940, Preprint No. 14). The author describes an investigation by X-ray diffraction methods of the structure of hot- and cold-rolled steel strip, the object of which was to determine the orientation of the crystals after heavy reduction by hot-rolling at a temperature between the A_1 and A_3 points, and the effect of subsequent cold-rolling upon this orientation. He found that the complex orientations could not be adequately determined by the Laue method alone, so this was supplemented by the production of surface diffraction diagrams by reflecting the X-ray beam from the surface, from the edge parallel to the rolling direction and from the edge transverse to the rolling direction. The data obtained proved that cold-rolling has a marked effect on the crystallographic orientation of hot-rolled steel strip. The author also investigated whether the directional properties imparted by the cold-rolling could be removed by heat treatment. He found that a normalising treatment at 1700° F. did not remove the directional properties entirely, although the grains were completely recrystallised and quite uniform in size, whilst annealing at 1300° F. was much less effective in removing the directional properties.

Surface Carbon Chemistry and Grain Size of 18/4/1 High Speed Steel. W. A. Schlegel. (American Society for Metals, Oct., 1940, Preprint No. 20). The author presents the results of an investigation of the effect of various heat treatments on the degree of carburisation and the grain size of 18/4/1 tungsten-chromium-vanadium high-speed steel. His general conclusions are: (1) Atmospheres which carburise this steel in short periods of heating may eventually cause decarburisation if the time at high temperature is sufficiently prolonged; (2) in the as-quenched condition the carburised zone cannot be distinguished under the microscope unless the steel has been tempered at 950–1150° F., in which case it will appear as partially retained and partially transformed austenite; (3) the retained austenite can be broken down by repeated tempering. The austenite produced on the surface by normal heat treatment can, as a rule, be broken down by a second tempering at 1050° F.; the rate at which it is transformed at any specific temperature depends upon the carbon concentration; (4) a study of the grain size and the characteristics of the fracture of this steel indicates that it is not susceptible to grain growth when treated at 2350° F. or lower; and (5) the composition of the furnace atmospheres used in heat-treatment furnaces for this high-speed steel does not affect its grain size or its fracture characteristics; the only factors which do affect the grain growth are the time and temperature of the treatment.

Effect of Rate of Heating through the Transformation Range on Austenitic Grain Size. S. J. Rosenberg and T. G. Digges. (American Society for Metals, Oct., 1940, Preprint No. 12: Journal of Research of the National Bureau of Standards, 1940, vol. 25, Aug., pp. 215–228). The authors determined the austenitic grain size at different temperatures with various rates of heating through the transformation range for a number of high-purity alloys of iron and carbon and for some steels with a wide range of carbon content. For the alloys containing 0·50% of carbon, the grain size at 1475°, 1500° and 1600° F. increased as the rate of heating decreased, whereas the grain size at 1800° F. was not as noticeably dependent upon the rate of heating. Coarser grains were produced at the lower temperatures with slow rates of heating than at higher temperatures with fast rates of heating. All the materials did not behave in the same manner, and it is therefore necessary, when making grain-size specifications, to consider the possible effects that the rate of heating through the transformation range may have upon the grain size of the steel.

Influence of Austenitic Grain Size on the Critical Cooling Rate of High Purity Iron-Carbon Alloys. T. G. Digges. (American Society for Metals, Oct., 1940, Preprint No. 10). This is a reproduction of the author's paper which appeared in the Journal of Research of the National Bureau of Standards, 1940, vol. 24, June, pp. 723–741 (*see* p. 31 A).

The Mechanism and Kinetics of the Crystallisation of Alloys of the Eutectic Type. Ya. V. Grechnyy. (Metallurg, 1939, No. 10–11, pp. 9–20). (In Russian). The author reports on a study of the structure of ledeburite prepared by melting electrolytic iron with electrode graphite in magnesite crucibles. The irons after superheating to 1300–1700° C., were cooled at rates varying from several tens to several thousands of degrees per second. In all cases microscopic and X-ray examination showed the cementite to be the leading phase in the crystallisation of the ledeburite, and accordingly changes in the shape of its crystals were studied. The cementite crystals were mainly lamellar in shape, though with more rapid cooling some typical dendritic crystals were observed. At very high rates of cooling, bent lamellæ joined at the ends were formed, presumably by the splitting of wider lamellæ. The eutectic austenite crystals, primarily of the acicular, but sometimes of the dendritic type, were observedtto have grown into the cementite lamellæ, and to be oriented parallel to their shorter edge forming a honeycomb type of structure. In addition some austenite separated out at the boundaries of the ledeburite grains. Typical ledeburite was not observed in metal cooled slowly. As the rate of cooling was increased, the shape of the ledeburite grains varied from crystalline to a dendritic and feathery structure. Some X-ray evidence indicated that the cementite in a grain of ledeburite is in the form of a single crystal. Some observations on the structure of the fayalite-wüstite eutectic are also reported.

Cementite Stability and Its Relation to Grain Size, Abnormality and Hardenability. C. R. Austin and M. C. Fetzer. (American Society for Metals, Oct., 1940, Preprint No. 29). In their previous paper on the " Effect of Composition and Steelmaking Practice on Graphitization Below the A_1 of Eighteen One Per Cent. Plain Carbon Steels " the authors had suggested that the presence of alumina might serve as a graphitiser during prolonged tempering (*see* p. 32 A). In the present paper they report on a further examination of these steels with a view to correlating the carbide stability, or tendency to graphitisation, with the grain size of the fracture at various temperatures, the McQuaid-Ehn grain size and abnormality, and the hardenability of the steels. The steels were tempered at 670° C. for periods up to 600 hr., after quenching from 1000° C. Ten steels proved to be completely free from graphitisation, whereas eight others showed graphitisation ranging from about 50% to 100%. In general, the results indicated that coarse-grained normal steels are stable whilst fine-grained abnormal steels graphitise. Alumina, which had proved to have such a profound effect on grain size and abnormality, was similarly effective in lowering the stability of the carbide. The data which the authors obtained from the various steels of common grain size did not permit of any attempt at correlation with carbide stability.

The Influence of Alloying Elements on the Position of the Martensite Point, the Amount of Retained Austenite and Its Stability on Tempering. V. I. Zyuzin, V. D. Sadovskiy and S. I. Baranchuk. (Metallurg, 1939, No. 10–11, pp. 75–80). (In Russian). The authors studied the effect of additions of silicon, manganese, nickel, cobalt, aluminium, copper, chromium, tungsten, vanadium and molybdenum, in amounts varying in some instances up to 8%, on the martensite point, on the amount of retained austenite and on its stability in 0·9–1% carbon steel. Manganese, chromium, molybdenum, nickel, copper, tungsten and vanadium lowered the martensite point by 55°, 15°, 30°, 17°, 10°, 12° and 35° C. respectively, for an addition of 1%. Cobalt and aluminium added to a 0·76% carbon steel raised the martensite point by 12° and 30° C. respectively, whilst silicon had no effect. All the elements except cobalt and aluminium raised the amount of retained austenite in the quenched steel, whilst cobalt and aluminium lowered it. Manganese, chromium and silicon had no marked stabilising effect on the retained austenite, but the other elements studied (in amounts up to 3%) had only a slight stabilising effect. The carbide-forming elements (chromium, molybdenum, tungsten and vanadium), which produce an intermediate stability zone at 400–600° C. in the isothermal transformation of primary austenite, give in a similar manner the same stability zone in the transformation of retained austenite.

The Influence of the Method of Quenching on the Amount of Retained Austenite in Structural Chromium-Nickel Steels. V. D. Sadovskiy and N. P. Chuparkova. (Metallurg, 1939, No. 10–11, pp. 80–89). (In Russian). The authors investigated the effects of different methods of quenching chromium-nickel steels on the amount of retained austenite. Basic open-hearth steels of the following compositions were used in the investigation:

	(1)	(2)	(3)
Carbon, %	0·38	0·40	0·30
Manganese, %	0·42	0·34	0·42
Silicon, %	0·34	0·16	...
Chromium, %	1·38	1·13	1·47
Nickel, %	3·07	3·20	3·6
Phosphorus, %	0·20	0·007	...

The amount of retained austenite was determined by a ballistic method, using as a standard a specimen quenched in ice-cold water and then cooled in liquid air to produce a minimum retained austenite content. Preliminary experiments showed that raising the maximum temperature from 825° to 950° C. had little effect on the amount of retained austenite. Quenching with incomplete cooling, followed by stepped tempering, resulted in a lowering of the martensite point of the retained austenite and when the tempering was sufficiently prolonged the martensite transformation on cooling to room temperature was completely eliminated. Quenching steel (1) from 830° C. in media with temperatures from 20° to 550° C. showed that two maxima in the retained austenite content were obtained at 200–250° C. and 325–400° C. with a sharp minimum at 300° C. The retained austenite formed at these two points behaved differently on subsequent cooling and tempering. The austenite in specimens quenched at 350° C. was much more resistant to tempering than that in those quenched at 200° C. Specimens of steel

(3) quenched in oil at 200° C. and in salt at 350° C. were studied with regard to the rates of decomposition of the retained austenite at tempering temperatures up to 700 °C. In specimens quenched at 200° C. the austenite may pass through the first zone of rapid transformation at 300–400 ° C. to undergo isothermal decomposition at 650–700° C. Evidence was also obtained regarding changes during secondary quenching, *i.e.*, cooling after high-temperature tempering.

The Effect of Molybdenum on the Isothermal, Subcritical Transformation of Austenite in Low and Medium Carbon Steels. J. R. Blanchard, R. M. Parke, and A. J. Herzig. (American Society for Metals, Oct., 1940, Preprint No. 35). The authors, in their study of the isothermal subcritical transformation of austenite in low- and medium-carbon steels with molybdenum contents from 0·15% to 0·75%, found molybdenum to be effective in promoting the hardenability of these steels. They carried out metallographic and dilatometric examinations and prepared S-curves from the isothermal data, which indicated that the influence of molybdenum on the decomposition of austenite is complex.

Transformation of Austenite on Continuous Cooling and its Relation to Transformation at Constant Temperature. R. A. Grange and J. M. Kiefer. (American Society for Metals, Oct., 1940, Preprint No. 9). The authors studied the transformation of the austenite in a steel containing carbon 0·42%, nickel 1·79%, chromium 0·80% and molybdenum 0·33% at seven different constant cooling rates, by the examination of more than a hundred specimens. From the data obtained they constructed a cooling transformation diagram analogous to the isothermal diagram. They discuss the general relationship of the transformation on continuous cooling to that at constant temperature, and propose a simple empirical method for estimating cooling transformation phenomena from data obtained at a constant temperature. This method, when applied to the above steel, was in excellent agreement with the cooling transformation diagram determined experimentally. When the isothermal data are known, the empirical method provides a means of predicting the course of the transformation of austenite for any given rate of cooling.

CORROSION OF IRON AND STEEL

(Continued from pp. 33 A–36 A)

The Passivation and Coloring of Stainless Steel. G. C. Kiefer. (American Society for Metals, Oct., 1940, Prepreprint No. 42). In the first part of his paper the author discusses in an elementary way the phenomenon of the passivation of metals, considering in particular the influence of certain alloy additions, *viz.*, chromium, nickel and molybdenum, from the point of view of their passivating influence. In conclusion he reports on his own experiments on rendering stainless steel less prone to pitting by treatment with a solution containing 4% hydrofluoric and 4% chromic acid. In the second part of his paper the author briefly reviews various methods of colouring stainless steel.

Pitting Type of Corrosion on the Surface of Stainless Steel Tubes and Methods of Eliminating it. V. Radchenko and N. Solonyy. (Stal, 1939, No. 10–11, pp. 62–63). (In Russian). An investigation into the cause of the pitting of stainless steel tubes during pickling in a mixture of nitric and hydrochloric acids is described. The factors studied included the effect of the concentrations of the two acids in the pickling bath, the effect of the proportion of nitric to hydrochloric acid, the effect of the temperature of the bath, and the effect of the heat-treatment temperature. It is concluded that, with insufficient nitric acid present, the passivating film will not have time to form, or, if formed, it will be destroyed by the hydrochloric acid. As the proportion of nitric acid is increased the passivating film is gradually rendered more stable, and eventually, only local breakdown will occur. At these points, pitting will take place, the corrosion being due to local concentration cells. Further increase in the proportion of nitric acid will render the passivating film resistant to hydrochloric acid. The above theory was confirmed by experiments which showed that, as expected, there was a critical range of nitric acid concentrations in the nitric-acid/hydrochloric-acid pickling bath within which a pitting type of corrosion is observed. This critical concentration range depends on the chromium content of the steel, and is also shifted by potassium dichromate (which stabilises the film) and by additions of salts of iron, chromium and nickel.

Corrosion Cracking. G. Sachs. (Iron Age, 1940, vol. 146, Oct. 3, pp. 21–28). The author reviews the literature on the causes and effects of residual stresses in steel, and on the corrosion cracking of ferrous and non-ferrous metals and alloys.

Corrosion Resistance of Tin Plate. Influence of Steel Base Composition on Service Life of Tin Plate Containers. R. R. Hartwell. (American Society for Metals, Oct. 1940, Preprint No. 44). The author studied the influence of the composition of the base metal on the corrosion resistance of tinplate containers. He did not apply the statistical method of Hoar, Morris and Adam (*see* Journ. I. and S.I., 1939, No. II., p. 55 P) but used samples of tinplate the composition and production of which had been carefully controlled in all stages, eliminating, as far as possible, all variables except those to be studied. Over a period of several years he made experimental packs of corrosive fruits in plain and enamelled cans made of different tinplate containing various amounts of carbon, phosphorus, sulphur, silicon, copper, manganese, nickel and chromium. Within the limits in which they are usually present in tinplate steel, carbon, sulphur and manganese did not prove to have any commercial significance. Phosphorus, silicon and copper had pronounced effects on the corrosion resistance of the cans, the first two being of greater influence in enamelled cans and the last in plain cans. Increasing amounts of phosphorus and silicon in the steel base were detrimental to the service life of the containers, whilst copper had a favourable or unfavourable effect, depending on the product packed. Since the detrimental effects of copper were usually more obvious than the beneficial ones, the author is of the opinion that a tinplate made from a steel base low in phosphorus (max. 0·015%), copper (max. 0·06%) and residual silicon is more desirable for strongly corrosive products. A few tests with two qualities of tinplate steel containing 0·19% of chromium and 0·13% of nickel respectively, seemed to indicate that the presence of these two elements may, in certain instances, be detrimental to corrosion resistance.

ANALYSIS

Manganese, Chromium and Nickel in 18/8 Alloy Steels. L. Silverman and O. Gates. (Industrial and Engineering Chemistry, Analytical Edition, 1940, vol. 12, Sept., pp. 518–519). The authors have adapted the method for the determination of manganese in steel suggested by Sandell, Kolthoff and Lingane (*see* Journ. I. and S.I., 1935, No. II., p. 529) to the analysis of 18/8 steels, in such a way that, after the titration of manganese, the same solution can be used for the titration of chromium and nickel, the former being determined by the ferrous sulphate and the latter by the cyanide method. Silicon, molybdenum, niobium and phosphorus do not interfere, and the authors describe, as an example, the procedure followed in the complete analysis of an 18/8 steel containing these elements.

The Determination of Phosphorus in Titanium Steels. A. T. Etheridge and D. H. Higgs. (Analyst, 1940, vol. 65, Sept., pp. 496–498). The authors have developed a method which allows the determination of phosphorus in titanium steels without preliminary removal of the titanium, provided that the amount of the latter does not exceed 1%. They suggest the following procedure: A 2-g. sample is dissolved in 45 ml. of nitric acid (sp. gr. 1·2) and oxidised with permanganate solution, the excess of which is destroyed with sodium or potassium nitrite. After adding 50 ml. of concentrated nitric acid (sp. gr. 1·42) the solution is boiled for 1 min. Then 50 ml. of a molybdate solution containing 55 g. of ammonium molybdate, 50 g. of ammonium nitrate and 40 ml. of ammonia (sp. gr. 0·95) per litre are added. The mixture is stirred vigorously for some minutes with a policeman rod, and after standing for about 30 min. the precipitate is filtered off and dealt with in the usual way.

The Spekker Steeloscope. (Wild-Barfield Heat-Treatment Journal, 1940, vol. 4, Sept., pp. 20–21). A brief description is given of a simplified form of spectroscope known as the " Spekker Steeloscope " which is intended for the rapid detection of such elements as nickel, chromium, molybdenum, manganese, tungsten, cobalt, cadmium and vanadium which are present in the more common alloy steels. The eye-piece is the only moving part of the instrument and the determination of the elements is still further facilitated by the scale engraved with the chemical symbols. The apparatus requires little skill and after a few days practice an operator can classify a specimen of steel in 30–60 sec.

A Semi-Micro Method for the Determination of Carbon and Hydrogen in Coals. R. Belcher and F. Smith. (Fuel, 1940, vol. 19, Sept., pp. 181–183). Friedrich's micro method for the determination of carbon and hydrogen in organic compounds (published in Mikrochemie, 1932, vol. 10, p. 329) has been adapted by the author for use as a semi-micro method for the determination of these two elements in coal. The method proved to be rapid and to give results comparable with those obtained by the standard method.

BOOK NOTICES

(Continued from pp. 36 A–37 A)

AMERICAN INSTITUTE OF MINING AND METALLURGICAL ENGINEERS. "*Transactions*, volume 140. *Iron and Steel Division*." 1940. 8vo. Pp. 511. New York : The Institute. (Price $5.00).

This volume contains papers with complete discussion presented to the Iron and Steel Division at meetings held at New York, Feb., 1939; Cleveland, April, 1939; Chicago, Oct., 1939 and New York, Feb., 1940. There are 22 papers in addition to Dr. C. H. Herty's Howe Memorial Lecture entitled " Slag Control " and a report of a round-table discussion on Experimental Methods in the Study of Steelmaking, at which methods for the analysis of gases and inclusions, slag-metal reactions, gas-metal reactions, temperature measurement of molten iron and steel, methods for the study of slag composition and characteristics, and the application of these to practical open-hearth problems were dealt with. Abstracts of all the papers contained in the volume are to be found in the *Bulletin of the Iron and Steel Institute*. A complete list of the papers is as follows : Slag Control, by C. H. Herty, jun. (Howe Memorial Lecture); Effect of the Solution-Loss Reactions on Blast-Furnace Efficiency, by P. V. Martin; Effect of the Volume and Properties of Bosh and Hearth Slag on Quality of Iron, by G. E. Steudel; Desulphurisation of Pig Iron with Calcium Carbide, by C. E. Wood, E. P. Barrett and W. F. Holbrook; Reduction of Iron Ores under Pressure by Carbon Monoxide, by M. Tenenbaum and T. L. Joseph; Experimental Methods in the Study of Steelmaking, Round Table; Slag-Metal Relationships in the Basic Open-Hearth Furnace, by K. L. Fetters and J. Chipman; Equilibria in Liquid Iron with Carbon and Silicon, by L. S. Darken; The Solubility of Nitrogen in

Molten Iron-Silicon Alloys, by J. C. Vaughan, jun. and J. Chipman; Formation of Inclusions in Steel Castings, by W. Crafts, J. J. Egan and W. D. Forgeng; Heat Capacity of Iron Carbide from 68° to 298° K. and the Thermodynamic Properties of Iron Carbide, by H. Seltz, H. J. McDonald and C. Wells; Rate of Diffusion of Carbon in Austenite in Plain Carbon, in Nickel and in Manganese Steels, by C. Wells and R. F. Mehl; Crystallography of Austenite Decomposition, by A. B. Greninger and A. R. Troiano; Study of Lattice Distortion in Plastically Deformed Alpha Iron, by N. P. Goss; Crystal Orientation in Silicon-Iron Sheet, by J. T. Burwell; Some Observations on the Recrystallisation of an Iron-Nickel Alloy, by G. Sachs and J. Spretnak; Magnetic Analyses of Transformations in a Cold-worked 18/8 Alloy, by R. Buehl, H. Hollomon and J. Wulff; The Nature of Passivity in Stainless Steels and Other Alloys, III.–Time-Potential Data for Cr–Ni and Cr–Ni–Mo Steels, by H. H. Uhlig; Pitting of Stainless Steels, by H. H. Uhlig; Effects of Low-Temperature Heat Treatment on Elastic Properties of Cold-Rolled Austenitic Stainless Steels, by R. Franks and W. O. Binder; Effects of Temperature of Pre-Treatment on Creep Characteristics of 18/8 Stainless Steel at 600° to 800° C., by C. R. Austin and C. H. Samans; A New Instrument for the Magnetic Determination of Carbon in a Steel Bath, by H. K. Work and H. T. Clark; Tensile Strength and Composition of Hot-Rolled Plain Carbon Steels, by C. F. Quest and T. S. Washburn; Precipitation-Hardening of a Complex Copper Steel, by J. W. Halley.

BURN, D. L. "*The Economic History of Steelmaking*, 1867–1939. *A Study in Competition.*" 8vo. Pp. 548. Cambridge, 1940: The University Press. (Price £1 7*s.* 6*d.*)

A modern war cannot be conducted without ample supplies of steel, oil and foodstuffs. Steel is the foundation upon which all industries producing ships, armaments and munitions are built. A year ago we hoped and expected that Germany's war effort would suffer through lack of steel, but by the occupation of Norway, Belgium, Luxemburg and Lorraine she secured ample supplies of the raw material and a large increase in iron- and steel-producing plants. Since then we, rather than the Germans, have been the sufferers, though the relative scarcity is more an anxiety than a menace.

When the war is over the steel industry will probably become a political problem of the first importance. We are not likely to allow our strategic position to be weakened by the decline of the industry under the pressure of foreign competition. Steel production will be regarded as a "key" industry, to be fostered at all costs. An island that does not maintain a large and progressive steel industry cannot hope to defend itself successfully against a highly industrialised enemy country.

It is possible that the post-war world demand for steel will be so great as to secure the prosperity of the British industry for several years. During the war of 1914–18 the world's producing capacity increased to such an extent that a substantial reduction was generally regarded as inevitable to restore the balances; but the 'twenties proved to be a decade of even greater expansion in the United States of America, to meet the growing needs of the building and motor-vehicle industries. History may repeat itself. The rebuilding of Great Britain and of other countries will mean a large and continuing demand for structural steel. Nor is that the only market likely to expand.

But a growing world demand for steel, though favourable to the British industry, will not itself guarantee prosperity. The expanding twenties, which fostered the growth of the American industry and witnessed the reconstruction of the German industry, proved to be an unhappy period for the British industry. In spite of several financial reorganisations, its competitive position grew steadily worse, and when the Great Depression brought tariffs to its aid it also created a new relationship with the State and its agencies. The recovery that followed added materially to our strength in the present war.

If we cannot again take the same risks, it is clear that the position of the steel industry and its relationship to the State will present a domestic problem of outstanding importance. If we are to understand and solve the problem, we must know the history of the past, particularly that of the inter-war period. Mr. D. L. Burn has performed an invaluable service by giving precisely that information which the statesman and the public require for their guidance.

It is impossible, in a short review, to do justice to the most satisfactory monograph upon the economic development of the steel industry that has yet been published in this country. To select one or two details for extended comment might convey a false impression of the character of the book, which is more than the "study in competition," indicated in the sub-title. In fact, it is a comparative study of the industry in Great Britain, Germany and the United States over a period of about seventy years; at least it gives us sufficient about our competitors to provide a standard of judgment of our own methods. Problems of location, technical structure and capital organisation are examined as bearing upon the more fundamental problems of efficiency or real cost. The reader will find, too, a careful, if not interestingly written account of traditions as represented by hereditary or family control. Such tests as the reviewer was able to apply by virtue of personal knowledge of small corners of the field of investigation inspired confidence in the accuracy of the work as a whole, although a conversation with Herr Thyssen, some years ago, suggested that important changes were taking place in the German organisation subsequent to the last described by Mr. Burn. If one may venture a criticism of a book describing the results of elaborate and highly competent research, it is that while generalisation is always to be found it has to be sought amid a great mass of detail. If, at appropriate stages, the author had boldly summarised the broad trends and conclusions, the book would have been more helpful to those who are more interested in broad issues of policy than in the technicalities of the industry.

To those likely to be responsible for the policy of the industry in days to come the most interesting chapter is the last, which gives a long and careful account of the steps taken to reconstruct the industry after the Great Depression and the introduction of import duties upon steel and steel products. The description of the part played by the May Committee, following the report of the Sankey Committee, and of the pressure exerted by the existence of "special" areas (of extreme depression) leaves little to be desired—which is high praise when we recall the controversial character of the issues. The book will immediately take its rightful place as the standard work upon the subject and be accepted, by economists, as a contribution of real importance to recent economic history. J. H. JONES.

IMPERIAL INSTITUTE. MINERAL RESOURCES DEPARTMENT. "*Chrome Ore and Chromium.*" By Robert Allen and G. F. Howling. (Reports on the Mineral Industry of the British Empire and Foreign Countries). 8vo. Pp. 118. London 1940: The Imperial Institute. (Price 2*s*. 6*d*.).

This Monograph discusses the mining and dressing of chrome ore, its utilisation, marketing and prices, and the world's production.

The main bulk of the book, however, is devoted to the occurrence of chrome ore both in the British Empire and foreign countries, no less than 43 different countries being dealt with. The British Empire is very strongly situated in regard to supplies of this strategic mineral, Southern Rhodesia and the Union of South Africa being among the world's leading producers. Just prior to the present war, about one-third of the world's output came from British countries and two-thirds was produced by companies under British control. Some measure of the rapidly increasing use of chrome ore may be gathered from the fact that the world's output rose from less than 250,000 tons to more than 1,000,000 tons during the period between the close of the war of 1914–1918 and the outbreak of the present war.

The Monograph contains a large number of statistical tables of production and trade, and concludes with a selected list of references for more detailed reading.

BIBLIOGRAPHY

Allen, R. M. "*The Microscope.*" 8vo, pp. viii + 286. London, 1940: Chapman and Hall, Ltd. (Price 15*s.*)

American Institute of Mining and Metallurgical Engineers. "*Transactions*, volume 140. *Iron and Steel Division*, 1940." 8vo, pp. 511. New York: The Institute. (Price $5.00.) [See notice, p. 79 A.]

American Society for Metals. "*Age-Hardening of Metals.*" A Symposium on Precipitation-Hardening held during the Twenty-First Annual Convention of the American Society for Metals, Chicago, October 23–27, 1939. 8vo, pp. 448. Illustrated. Cleveland, Ohio, 1939: The Society. (Price $5.00.)

Berglund, T. "*Handbuch der metallographischen Schleif-, Polier- und Ätzverfahren.*" Revised and enlarged by Antoine Meyer. pp. 300. Berlin: Julius Springer. (Price 28.80 RM.)

Brown, G., and A. L. Orford. "*The Iron and Steel Industry.*" 8vo, pp. xi + 122. London, 1940: Sir Isaac Pitman & Sons, Ltd. (Price 6*s.*)

Casimir, H. B. G. "*Magnetism and Very Low Temperatures.*" (Cambridge Physical Tracts.) 8vo, pp. viii + 94. Cambridge, 1940: University Press; New York: Macmillan Company. (Price 6*s.*)

Chamot, E. M., and C. W. Mason, "*Handbook of Chemical Microscopy. Vol. II. Chemical Methods and Inorganic Qualitative Analyses.*" Second edition. 8vo, pp. xi + 438 with 233 illustrations. New York: John Wiley and Sons, Inc.; London: Chapman and Hall, Ltd. (Price 30*s.*)

Chappell, L. "*Historic Melingriffith.*" 8vo, pp. 87. Cardiff: Priory Press, Ltd. (Price 2*s.* 6*d.*)

CLARK, G. L. "*Applied X-Rays.*" Third, revised edition. (International Series in Physics.) 8vo, pp. 675. New York: McGraw-Hill Book Co., Inc.; London: McGraw-Hill Publishing Co. (Price $6.00.)

CORNELIUS, H. "*Kupfer im technischen Eisen.*" Reine und angewandte Metallkunde in Einzeldarstellungen. Band IV, pp. 219. Illustrated. Berlin, 1940: Julius Springer. (Price 27 RM.)

DEUTSCHE GESELLSCHAFT FÜR METALLKUNDE. "*Werkstoffhandbuch Nichteisenmetalle.*" Schriftl. G. Masing, W. Wunder und H. Groeck. Abschnitte D-F: *Kupfer, Messing und Sondermessing, Bronze, und Rotguss.* pp. 140. Berlin 1940: VDI-Verlag G.m.b.H. (Price 12 RM.)

DE VRIES, L. "*French–English Science Dictionary.*" 8vo, pp. 515. New York: McGraw-Hill Cook Co., Inc. (Price $503.)

FOTOS, J. T., and R. N. SHREVE. "*Advanced Readings in Chemical and Technical German: from Practical Reference Books (Ullmann; Houben; Meyer and Jacobson; Beilstein; Gmelin; Oberhoffer; Guertler); with a Summary of Reading Difficulties, a Frequency Vocabulary List and Notes.*" 8vo, pp. xliii + 304. New York, 1940: John Wiley and Sons, Inc.; London: Chapman and Hall, Ltd. (Price 15*s.*)

HAYWARD, C. R. "*An Outline of Metallurgical Practice.*" Second, revised edition. 8vo, pp. 669. New York: D. Van Nostrand Co. (Price $7.00.)

HESSENBRUCH, W. "*Metalle und Legierungen für hohe Temperaturen.*" pp. 254. Illustrated. Berlin, 1940: Julius Springer. (Price 30 RM.)

HILL, F. T. "*The Materials of Aircraft Construction.*" Fourth edition. 8vo, pp. ix + 449. London, 1940: Sir Isaac Pitman and Sons, Ltd. (Price 20*s.*)

HUEBNER, W. "*A Thesaurus and a Co-ordination of English and German Specific and General Terms.*" pp. 405. New York, 1939: Veritas Press. (Price $7.50.)

IMPERIAL INSTITUTE. MINERAL RESOURCES DEPARTMENT. "*Chrome Ore and Chromium.*" By Robert Allen and G. F. Howling. (Reports on the Mineral Industry of the British Empire and Foreign Countries.) 8vo, pp. 118. London, 1940: The Imperial Institute. (Price 2*s.* 6*d.*) [See notice, p. 81 A.]

KURREIN, M., and F. C. LEA. "*Cutting Tools for Metal Machining.*" 8vo, pp. xiii + 219. London, 1940: Charles Griffin & Co., Ltd. (Price 16*s.*)

PANETH, F. A. "*The Origin of Meteorites: being the Halley Lecture delivered on 16 May, 1940.*" 8vo, pp. 26. Illustrated. Oxford, 1940: Clarendon Press; London: Oxford University Press. (Price 2*s.* 6*d.*)

RICHARDS, R. H., and C. E. LOCKE. "*Text-Book of Ore Dressing.*" pp. 608. New York, 1940: McGraw-Hill Book Co., Inc.; London: McGraw-Hill Publishing Co., Ltd. (Price $5.50.)

SACHS, G., and K. R. VAN HORN. "*Practical Metallurgy.*" 8vo, pp. 540, with 335 illustrations. Cleveland, Ohio, 1940: American Society for Metals. (Price $5.00.)

SEITZ, F. "*The Modern Theory of Solids.*" (International Series in Physics.) 8vo, pp. 689. New York: McGraw-Hill Book Co., Inc.; London: McGraw-Hill Publishing Co. (Price $7.00.)

SMITH, T. B. "*Analytical Processes: a Physico-Chemical Interpretation.*" 8vo, pp. viii + 470. London, 1940: Edward Arnold & Co. (Price 18*s.*)

"*Standard Metal Directory.*" 8th edition. New York, 1940: Atlas Publishing Co. (Price $10.)

STRONG, J. "*Modern Physical Laboratory Practice.*" In Collaboration with H. V. Neher, A. E. Whitford, C. H. Cartwright and R. Hayward. 8vo, pp. x + 642. Illustrated. London, 1940: Blackie & Son, Ltd (Price £1 5*s.*)

REFRACTORY MATERIALS

(Continued from pp. 39 A–41 A)

The Production of Highly Refractory Chrome-Dolomite Material and Tests Made on it under Service Conditions. P. P. Budnikov, M. S. Feygin, I. I. Susidko and E. I. Yudin. (Metallurg, 1939, No. 10–11, pp. 54–59). (In Russian). Following some remarks regarding the difficulties attaching to the production of stable dolomite-base refractories, the authors describe the production of a refractory dolomite clinker stabilised by additions of quartzite and chrome ore. The analyses of the raw material used were:

	Dolomite.	Chrome Ore.	Quartzite.
Silica %	3·41	5·89	96·32
Alumina %	0·73	16·79	0·59
Ferric oxide %	1·05	...	1·17
Ferrous oxide %	...	20·29	...
Lime %	29·48	1·92	0·38
Magnesia %	20·22	14·84	0·29
Chromic oxide %	...	40·32	...
Alkalis %	...	...	0·36
Sulphur trioxide %	...	...	0·43
Loss on ignition %	45·13	0·02	0·31

The amount of chrome ore was varied between 3% and 20% (calculated on the basis of the burnt dolomite) and that of the quartzite was 5% or more. The finely ground mixture was moistened with water (8%) and pressed into cylindrical briquettes which were fired at a maximum temperature of 1560–1580° C. for 1 hr. It was found that an addition of at least 18–20% of chrome ore was necessary to obtain a stable product. A specimen containing silica 10·98%, ferric oxide 2·31%, lime 37·97%, magnesia 35·96% and chromic oxide 6·62%, could resist temperatures of more than 1900° C. Under a load of 2 kg. per sq. cm. deformation set in at 1520°, and was complete at 1610° C. The finely ground product was found to possess cementing and hydraulic properties. For bricks, the chrome-dolomite clinker was bonded with chrome-dolomite cement and mixed with 8% of water. The air-dried bricks had the following properties: Refractoriness 1900° C.; mechanical strength (resistance to crushing) after one day 60 kg. per sq. cm., after four days 98 kg. per sq. cm., after seven days 175 kg. per sq. cm., and after twenty-eight days 270 kg. per sq. cm.; deformation under a load of 2 kg. per sq. cm. set in at 1450° C. and the brick disintegrated at 1590° C.; absorption of water 5·5–6·5; porosity 15–17% by volume; bulk density 2·64–2·67; 0·8%

shrinkage on heating for 1 hr. at 1400° C. A measure of its resistance to spalling was the fact that it could be heated to 850° C. and cooled in air twenty-eight times without failure. Tests under service conditions in an electric furnace and in the side and end walls of an open-hearth furnace are described. These demonstrated the excellent behaviour of the air-dried bricks, which, however, proved unsuitable for lining casting ladles.

A New Slag-Testing Furnace. J. C. McMullen. (Bulletin of the American Ceramic Society, 1940, vol. 19, Nov., pp. 439–442). The author describes a furnace and testing procedure for investigating the action of slag on a section of a 9-in. refractory brick at temperatures up to 1600° C. For the test in question, the slag, in the form of rods ½ in. in dia., was fed automatically into the furnace so as to melt and drip directly on to the brick. After running down the face of the brick, the slag is caught in sand on the furnace bottom. On completion of the test, the specimen, the spent slag and the sand were easily removed, leaving the furnace clean. Some advantages of this method of testing slag are: (1) It has a wide range of applicability and at the same time it is economical, rapid and easy, and the apparatus is compact; (2) it gives reproducible results; and (3) it is possible to control the conditions with respect to temperature, atmosphere, slag composition, slag viscosity, rate of flow of the slag and complete removal of the slag.

The All-Basic Open Hearth. J. H. Chesters. (Iron Age, 1940, vol. 146, Aug. 15, pp. 35–37; Aug. 22, pp. 39–41). The author reviews the progress that has been made in the development of all-basic open-hearth furnaces. Rapid progress has been made in the production of volume-stable chrome-magnesite bricks with a high thermal-shock resistance and high refractoriness under load. These bricks are still liable to " burst " owing to the formation of a solid solution of iron oxide in the chrome ore, but if adequate attention is given to details of design, in particular to the springing and suspension of the roof, all-basic furnaces can be built that will operate for long periods. The use of such furnaces will permit of higher working temperatures with consequent greater output, will reduce shut-down time, will lessen the quantity of slag, and will reduce the amount of fettling material used. The first installations in England had relatively short lives, but much better results have since been obtained, the record furnace having done over 1300 casts. Further improvements in roof design, together with the replacement of some of the chrome-magnesite bricks in all-basic furnaces by non-spalling dolomite bricks (or other basic bricks of lower price), will hasten the day when this type of construction becomes the rule rather than the exception.

Refractory Life in Electric-Arc Melting of Steel and Alloys. J. H. Chivers. (Bulletin of the American Ceramic Society, 1940, vol. 19, Nov., pp. 442–443). The author discusses the types of refractories used in acid and basic electric-arc furnaces, the construction of the lining and the furnace shell, and the effect of varying the voltage on the life of the lining. The furnaces are usually designed so that three different voltages, called the high, intermediate and low taps, can be applied. The author favours the use of low voltages with a correspondingly high amperage, as this permits of the furnace being kept on the high tap for a longer period, thus decreasing the melting time.

An Examination of Samples of Silica Brickwork Taken from a Vertical Retort and Showing Evidence of Pronounced Alkali Attack. F. H. Clews, W. Hugill, and A. T. Green. (Transactions of the British Ceramic Society, 1940, vol. 39, Nov., pp. 337–344). The authors examined the brickwork of a retort which had been used at a gasworks for seven years at moderate temperatures. During the first four years coals containing an average of 0·66% of sodium chloride had been carbonised. In the following three years they had been mixed in equal parts with another quality containing only about 0·1% of sodium chloride. The ill-effects due to the exceptionally high salt content of the coals were mainly found at the flue side of the retort wall. The authors consider the deleterious effect of sodium chloride due to three causes, *viz.*: (1) The formation of low-melting-point silicates, resulting in thinning by erosion at the zones of maximum temperature; (2) interaction of alkali salts with the brickwork at the zones of minimum temperature, causing expansion with consequent disruption of the structure; and (3) a general decrease in refractoriness of the retort brickwork.

Effect of Water Content and Mixing Time on Properties of Air-Setting Refractory Mortars Containing Sodium Silicate. R. A. Heindl and W. L. Pendergast. (Bulletin of the American Ceramic Society, 1940, vol. 19, Nov., pp. 430–434). The authors determined the relation between the strength after drying and the moisture content of the following refractory materials: (*a*) An Ohio fireclay; (*b*) a mixture of this clay and grog; and (*c*) this mixture with an addition of a solution of sodium silicate. It was found that the strengths after drying increased at first, and then decreased as water was added to change the consistency from a very stiff to a somewhat fluid state. The consistency at maximum strength was, however, far too stiff for trowelling. Mixture (*c*) increased in strength but decreased in workability as the mixing time was increased from 2 hr. to 5 hr. Some additional tests demonstrated that specimens made from mortars stored in sealed iron drums were relatively strong after storage from two to five years, whereas when the specimens were prepared from mortars stored in jute sacks lined with moisture-proof paper, their strength was weak after only three months' storage.

FUEL

(Continued from pp. 41 A–42 A)

Flame Temperature. B. Lewis and G. von Elbe. (Journal of Applied Physics, 1940, vol. 11, Nov., pp. 698–706). The authors discuss the temperatures of flames produced during the use of fuels for heating and the generation of power. They distinguish between the theoretical and the experimental flame temperature, the former corresponding to complete statistical equilibrium, the latter to equilibrium in the translational degrees of freedom. In industrial furnaces the difference between these two temperatures should be negligible. Experimental flame temperatures considerably higher than the theoretical temperatures were observed only for fast flames, such as hydrogen and coal-gas flames, in closed vessels. The authors point out that the temperature difference in these cases is due to excitation lag in the internal, presumably vibrational, degrees of freedom. They describe various methods for the determination of flame temperatures, in particular that known as the line-reversal method, as well as some methods involving the measurement of the brightness and the absorptivity in the infra-red, the use of thermocouples or a series of progressively thinner wires with extrapolation of the wire temperatures obtained to zero thickness. They consider soot-forming luminous flames, which they regard as being in complete statistical equilibrium, the temperature difference between soot particles and the surrounding gas being negligible, and, in conclusion review the methods applicable to the determination of the temperature and emissivity of such flames.

Some Modern Applications of Boilers in Steel Mills. D. N. Mauger. (Iron and Steel Engineer, 1940, vol. 17, Oct., pp. 40–48). The author reviews the developments in steam-boiler design and practice which have taken place at American steelworks in recent years. He illustrates and discusses a number of boilers of very modern design and analyses thirty-four proposals for boiler plant submitted by the iron and steel industry during the last three years. This analysis shows that about one-half were in favour of low-pressure plant. The choice of fuels is also of interest; the number of proposals in favour of each fuel were: Pulverised coal alone, 12; pulverised coal and blast-furnace gas, 5; pulverised coal and coke-oven gas, 2; blast-furnace gas alone, 3; coal or coke breeze on chain-grate stokers, 11; and natural gas, 1.

The Curran-Knowles Process. R. F. Haanel, E. J. Burrough and R. A. Strong. (Coke and Smokeless-Fuel Age, 1940, vol. 2, Oct., pp. 222–225; Nov., pp. 242–244, 248). The authors describe the Curran-Knowles plant at West Frankfurt, Illinois, and some tests made there with Canadian coals. Some particulars of the plant and process have already been given in a paper entitled "Advantages of Coal Carbonization as Exemplified in the Curran-Knowles Process" by M. D. Curran. (*See* Journ. I. and S.I., 1940, No. I., p. 175 A).

Changing the Fuel of Industrial Furnaces. M. Tigerschiöld. (Jernkontorets Annaler, 1940, vol. 124, No. 9, pp. 483–510). (In Swedish). The author, after reviewing the fuel situation created in Sweden by the present war, discusses the properties of Sweden's indigenous fuels, namely wood, wood tar and peat. He describes next a number of gas generators for use with wood, with special reference to one designed by A. Olsson. In conclusion he gives some practical examples of how furnaces have been adapted for heating with a fuel different from that for which they were originally intended.

Gas Mixing and Distribution at Great Lakes Steel Corporation. W. H. Collison and F. C. Frye. (Iron and Steel Engineer, 1940, vol. 17, Oct., pp. 18–25, 38). As a result of expansion of and improvements to the plant of the Hanna Furnace Division of the Great Lakes Steel Corporation more coke-oven and blast-furnace gas has become available. In the present paper the authors describe the provision made for cleaning, mixing and distributing these gases.

PRODUCTION OF IRON

(Continued from pp. 42 A–45 A)

Smelting of Jehol Titaniferous Iron Ore. F. Kakiuchi. (Tetsu to Hagane, 1940, vol. 26, Feb. 25, pp. 89–94). (In Japanese). The author describes some experiments on the reduction of Jehol titaniferous iron ore which contains about 50% of iron, 13–14% of titanium dioxide and 0·3–0·56% of vanadium pentoxide.

Reduction of Purple Ore. Y. Fujii and S. Taniguchi. (Tetsu to Hagane, 1940, vol. 26, Jan. 25, pp. 1–13). (In Japanese). The authors describe some experiments on methods of reducing a purple ore containing 60% of iron, 10% of silica, 1–2% of phosphorus and 0·1–0·25% of copper. From the results obtained they considered that sponge iron and lumps could be produced from the ore on an industrial scale in a rotary kiln, a tunnel kiln or a reverberatory furnace, using carbon monoxide and solid carbon at very high temperatures.

The Charcoal Iron Industry of Powys Land. A. S. Davies. (Reprinted from Montgomeryshire Collections, 1939). A descriptive account is given of the iron industry of Powys Land in Wales in the period from about 1670 to 1840. Several of the forges and processes are described.

Henry Cort's Bicentenary. H. W. Dickinson. (Newcomen Society, Nov. 13, 1940). The author presents an account of the life and work of Henry Cort (1740–1800), who greatly improved the arts of refining iron by puddling with mineral coal and of rolling metals in grooved rolls. Cort's forge and mills were at Fontley, near Gosport, Hampshire, and most of his work was under contract for the Portsmouth Dockyard.

Prehistoric and Primitive Iron Smelting. E. W. Hulme. (Newcomen Society: Engineering, 1940, vol. 150, Dec. 20, pp. 498–500). In this, the second part of his paper (*see* Journ. I. and S.I., 1938, No. II., p. 202 A), the author describes the crucible processes of iron smelting as applied in the Far East from 1200 B.C., dealing separately with the processes used in China, India, Ceylon and Japan. He comes to the conclusion that, on the whole, these old processes may be described as more flexible than those of contemporary Europe.

FOUNDRY PRACTICE

(Continued from pp. 45 A–48 A)

Medium Manganese Steels in the Foundry. H. Taylor. (Metallurgia, 1940, vol. 23, Nov., pp. 1–2). The author recommends the use of pearlitic medium-manganese steels containing 1–3% of manganese for castings, pointing out that their cost, in comparison with other low-alloy steels, is low, and that the mechanical properties of the former bear favourable comparison with those of the latter. He gives some indication as to the effect of various methods of heat treatment on these steels.

Cast Steel Gears. W. J. Phillips and T. D. West. (Steel, 1940, vol. 107, Oct. 21, pp. 48–50). The authors describe the correct moulding technique with regard to the position of the risers when casting gear blanks of low-alloy steel.

Modern Core Ovens. H. M. Lane. (Iron Age, 1940, vol. 146, Oct. 17, pp. 42–45). The author describes recent developments in the design of core-drying ovens and makes some practical suggestions for the improvement of existing ovens by installing forced circulation and gas-mixing devices.

Some Points in Casting Production. E. Longden. (Institute of British Foundrymen: Foundry Trade Journal, 1940, vol. 63, Dec. 12, pp. 379, 382; Dec. 19, pp. 397–400). In the first part of his paper, the author discusses in an elementary way the effects of carbon, silicon, phosphorus and manganese on the properties of cast iron. In the second part, he describes, with numerous illustrations, the method he adopted for the casting of three large boring bars 43 ft. long and 15 in. in dia., 45 ft. long and 18 in. in dia., and 47 ft. long and 22 in. in dia., respectively.

Foundry Problems. B. Hird. (Foundry Trade Journal, 1940, vol. 63, Nov. 28, pp. 349–350). The author describes some difficulties which arose in making three different iron castings because of occluded gases, and the moulding technique which was developed to prevent this from happening. In the first casting, that of a three-way filter pipe 6 ft. long, the gases emanated from the chills after the mould had been used several times. This trouble was dealt with by using new chills after every fourth casting. The second instance was that of an iron head cast on a 4-in.-dia. steel gudgeon pin; in this case the gases came from the steel pin. The third casting was that of a pug-mill roll with a steel shaft through the centre; in this case also the trouble was due to gases coming from the steel shaft. The author also describes some experiments relating to gases evolved from heated iron, from which he draws the following conclusions: (1) Gas is given off by cast iron in the molten state and during the solidification period down to a definite temperature not yet established; (2) when iron or steel is heated up to a certain temperature gas is also given off; (3) the amount of gas and the time during which it is evolved are controlled by the relative volume of the chill or insert to that of the surrounding molten metal.

PRODUCTION OF STEEL

(Continued from pp. 48 A–50 A)

Oxygen Samples from Open-Hearth Bath. K. C. McCutcheon and L. J. Rautio. (Transactions of the American Institute of Mining and Metallurgical Engineers, Iron and Steel Division, 1940, vol. 140, pp. 133–135). The authors describe and illustrate a sampling spoon for taking 1-kg. samples of steel from a furnace for making oxygen determinations. This spoon is a small jug-shaped thick-walled casting, split down the middle and held together by a steel loop attached to a handle. They also describe the sampling procedure by which no slag is allowed to enter the spoon and contaminate the steel. This method has been found to give reliable and reproducible results.

Rapid Method of Correlation Applicable to Study of Steelmaking Reactions. K. Fetters. (Transactions of the American Institute of Mining and Metallurgical Engineers, Iron and Steel Division, 1940, vol. 140, pp. 166–169). The author explains a method of statistical analysis and its application for the study of the degree of correlation which exists between any two of a number of variables in the process of steelmaking. As an example, he shows the relation between the ferrous-oxide content and the reciprocal of the carbon content of a melt of steel, as calculated from a large number of tests.

Rustless Doubles Capacity. (Iron Age, 1940, vol. 146, Nov. 14, pp. 53–55). A brief illustrated description is given of the works of the Rustless Iron and Steel Corporation, Baltimore, where a large extension programme has recently been completed. This concern manufactures stainless steel strip, bars and wire from ingots from its own electric furnaces which comprise three 12-ton and two 16-ton furnaces with a capacity of 75,000 tons per annum.

A Method of Melting Steel in a Basic Electric-Arc Furnace. K. Matsuyama, T. Iki and K. Muramoto. (Tetsu to Hagane, 1940, vol. 26, Aug. 25, pp. 597–608). (In Japanese). The authors describe a melting technique, using a basic electric-arc furnace, the object of which is to reduce the tendency to flake formation in the ingot.

The Application of Electric Heat to Industrial Furnaces. A. E. Pickles. (Australasian Engineer, 1940, vol. 40, Oct. 8, pp. 10–12). The author reviews the application of electric heating to metallurgical furnaces, dealing with the subject under the following headings: Indirect, direct-series and conducting-hearth arc furnaces; the control of arc-furnaces; low-frequency and high-frequency induction furnaces; resistance furnaces; muffle, direct-radiation, convection and conduction furnaces; controlled atmosphere and conveyor furnaces; and the control and application of resistance furnaces.

Actual Metallographic Problems in the Production and Treatment of Steel. A. Hultgren. (Jernkontorets Annaler, 1940, vol. 124, No. 8, pp. 323–338). (In Swedish). The author reviews a number of metallurgical problems which arise during the production, treatment and testing of steel. He refers to some Swedish experiments carried out mainly with 1·10% carbon steel from open-hearth and electric furnaces with a view to determining the slag inclusions at different stages of the process. For this purpose small samples were taken from the melt and cooled in small moulds; these were then sectioned and examined under the microscope. These tests have shown that, with the ordinary melting practice for this steel, there are practically no inclusions in the steel in the furnace towards the end of the process, and that even the deoxidation products rise to the top of the bath in a short time. Soon after tapping into the ladle many silicate inclusions will, however, be found; most of these rise to the surface after a few minutes' holding in the ladle. It is too soon to judge whether this occurs under other conditions, but the inference is that the conditions under which the metal is teemed into the ladle and moulds, and especially the degree of oxidation during teeming, have an important bearing on the occurrence of inclusions in steel. With regard to surface defects in ingots, these can have the character of wrinkles, frozen splashes, cracks, holes, slag inclusions and segregation. Wrinkles occur after teeming too slowly or at too low a temperature, and cracks form after teeming rapidly or at too high a temperature. A reaction between the carbon in the steel and scale in the mould is a possible cause of surface blowholes. Slag inclusions and holes near the surface may be caused by reactions between the steel and the mould surface or with the air. A crack which forms in the early stages of cooling has a tendency to be filled up with mother-liquor which causes segregation, and it thus becomes more difficult for it to be welded up during subsequent rolling or forging. In the machining of soft steel there is a tendency for long turnings to be formed, leaving a rough surface, but, if 0·1–0·3% of sulphur is present together with a high manganese content, the process of the formation of the turnings is changed and short chips will be produced. An addition of selenium to stainless steel has a similar effect. The author refers also to the recently introduced high-speed steel, containing about 8·5% of molybdenum and 2% of tungsten, which is interesting from the heat-treatment point of view in that it should not be hardened at just under its melting point, but at about 1225° C. There is now a tendency to prolong the tempering time in the treatment of high-speed steels to 1 hr. or even more, as this is said to give increased strength. The repeated tempering of high-speed steel, which transforms the retained austenite, has also been found advantageous. The nitriding of hardened and tempered high-speed steel in a cyanide bath at about 550° C. is said to increase the life of certain tools of this steel. In discussing heat-treatment practice the author refers to a new type of salt-bath furnace heated by electrodes arranged in pairs; in this, electro-magnetic forces are applied in order to cause a rapid circulation of the molten salt and thus keep it at an even temperature. In conclusion reference is made to recent improvements in laboratory equipment and technique. An extensive bibliography is appended.

The Influence of Turbulence upon the Structure and Properties of Steel Ingots. L. Northcott. (Iron and Steel Institute, 1941, this Journal, Section I.). Seven steel ingots were cast by different methods which were selected as offering different conditions of turbulence of the molten metal in the mould. The methods employed involved: (1) Bottom-casting, (2) top-casting with a single stream down the centre of the mould, (3) the use of a multi-hole tundish, (4) a sand mould with a single stream, (5) a single stream near one side of the mould, (6) top-casting with a single stream, stirring with a poker after casting, and (7) casting in a sloping mould. The macrostructure, chemical segregation and mechanical properties of the ingots were studied. Further information on the influence of the casting method upon turbulence was obtained from a number of small composite non-ferrous alloy ingots, in the preparation of which a red alloy was poured first and was immediately followed by a white alloy of similar density and melting point. Knowledge of the distribution of the stream in the mould was then obtained by examining the distribution of the differently coloured alloys in the ingot. Differences in the structures of the steel ingots are discussed from the point of view of the influence of turbulence, and theories on the mechanism of the solidification of steel are put forward.

FORGING, STAMPING AND DRAWING

(Continued from pp. 9 A–10 A)

8-in. Upsetting Forging Press. (Engineering, 1940, vol. 150, Dec. 13, pp. 467–468). An illustrated description is given of a heavy forging press, which, although primarily designed for general purposes, is used largely in the United States for forging cylinders of internal-combustion engines. The process is stated to be more economical than machining from seamless steel tubes or extruding in a hydraulic process. The press is of the crankshaft-operated type and is driven, through multiple V-belts, by a motor of 150 h.p. It is known as the 8-in. size and accommodates dies 46 in. high, which are separated transversely to permit the work to be moved up or down to the next stage; thus as many as four progressive operations can be carried out. The forgings produced, generally made from a nitriding or Nitralloy steel, range from 4½ to 6½ in. in internal dia. and from 25 to 125 lb. in weight.

Production Control in the Forge Shop. B. Nelson. (Heat Treating and Forging, 1940, vol. 26, Sept., pp. 442–445). The author describes a system of supervision for the control of material and operations in a forging shop where a number of steam hammers are in operation.

Manufacture of Heavy Ordnance and Armor Plate. R. D. Galloway. (Metals Treatment Society of Victoria, Australia: Heat Treating and Forging, 1940, vol. 26, Apr., pp. 168–172; May, pp. 231–233). The author first classifies the steels used for ordnance manufacture into three broad groups which cover a range of carbon content of 0·10–0·45%. He then describes the sequence of manufacturing processes from the production of the steel to the final heat treatment of a large forging. In the second part he describes the sequence of operations in the production of armour plate of 4% nickel steel.

Manufacture of Heavy Crankshafts and Torsion of the Bearing Journal. O. Hara. (Tetsu to Hagane, 1940, vol. 26, Aug. 25, pp. 617–629). (In Japanese). The author discusses the forging of heavy crankshafts from ingots of 20 tons and of even greater weight, and suggests a method of eliminating surface defects. In the concluding part of his paper he describes a torsion-fatigue test on a crankshaft journal 300 mm. in dia.

Shell Forgings. (Iron Age, 1940, vol. 146, Oct. 10, pp. 57–60). An illustrated description is given of the Baldwin-Omes horizontal shell-forging machine. For shell forgings up to 5 in. in dia. the drawing machine is built on, and combined with, the piercing machine, whereas for larger sizes the drawing machine is a separate unit. Square-section billets are used in this machine and the rate of production is very much higher than that with hydraulic presses. The number of rejections due to eccentricity is very low, amounting to only 1–1½% with the machines for 6-in. shells.

Bolts and Nuts. F. B. Jacobs. (Steel, 1940, vol. 107, Oct. 21, pp. 52–55, 72). The author describes some of the conveyor systems used for moving material from one department to another in a large American bolt and nut factory.

Wire Die Manufacture. (Wire Industry, 1940, vol. 7, Nov., pp. 459–460, 464, 468). A description is given of the tools and processes used for boring and polishing diamonds and hard metals in the manufacture of wire-drawing dies.

Die Inspection in the Wire Industry. C. L. Mantell. (Wire and Wire Products, 1940, vol. 15, Sept., pp. 457–459). The author draws attention to the fact that very little information is available on the design of wire-drawing dies, the methods of measuring their hardness and curvature, and their maintenance. He suggests that wire-drawers should institute a system of regular inspection of their dies and keep die record cards on which particulars of their performance and causes of failure should be entered.

ROLLING-MILL PRACTICE

(Continued from pp. 50 A–52 A)

Actual Problems in Swedish Rolling Technique. P. G. Ekman. (Jernkontorets Annaler, 1940, vol. 124, No. 8, pp. 360–376). (In Swedish). The author reviews some problems confronting Swedish rolling-mill engineers and describes some advancements in rolling-mill plan , equipment and technique. He deals in turn with mill-stand design, reheating furnaces, drives, the Steckel hot-strip mill, reconditioning worn rolls, removal of scale by jets of water at high pressure, bearings, the changing of rolls, roll trains, coilers, stitchers and shears.

Influence of Roll Passes on the Rolling-Mill Capacity. K. Sonoda. (Tetsu to Hagane, 1940, vol. 26, June 25, pp. 436–450). (In Japanese). The author discusses methods of increasing the capacity of rolling mills with particular reference to the relation between the changes in roll contour and the number of passes required to produce a given section.

Cold-Rolled Strip Steel. T. B. Montgomery. (Steel, 1940, vol. 107, Oct. 21, pp. 56–61, 75). The author discusses the theory and practice of the application of backward and forward tension in the cold-rolling of steel strip.

Marking Hot Steel. V. L. Staley. (Steel, 1940, vol. 107, Nov. 4, pp. 64–66). The author refers to the need for marking coils of steel strip while still at high temperatures and states that there are now available two types of marking compounds, one of which can be applied to steel in the temperature range 150–900° F., and the other when the temperature range is 350–1400° F. Each type can be obtained in one of seven colours.

HEAT TREATMENT

(Continued from pp. 52 A–59 A)

Cementation Equilibrium and Cementation Reaction in CO-Gas Current. O. Madono. (Tetsu to Hagane, 1939, vol. 25, Sept. 25, pp. 734–744). (In Japanese). The author discusses the theory of the carburisation of iron under the action of carbon monoxide. He shows that the carbide Fe_3C cannot be produced in iron in a current of carbon monoxide because the iron acts as a catalyst for the dissociation of this gas. Fe_3C might, however, be formed at temperatures below 500° C., as the dissociation velocity is small at comparatively low temperatures. The carburisation action of carbon monoxide at low temperatures becomes less as the velocity of the gas decreases, for a high velocity is required to sweep away any carbon dioxide produced on the surface of the metal.

Practical Gas Carburizing. J. F. Wyzalek and M. H. Folkner. (Metal Progress, 1940, vol. 38, Sept., pp. 261–268). The authors review the development in the United States of gas-carburising furnaces, and describe and illustrate several modern continuous plants.

Salt-Bath Heat Treating. W. F. Sherman. (Iron Age, 1940, vol. 146, Nov. 14, pp. 43–47). The author describes a number of salt-bath hardening plants used for the hardening of large numbers of small parts such as clutch rings and valve pusher rods at the works of American automobile manufacturers.

A Study of the Salt-Bath Quenching. S. Abe. (Tetsu to Hagane, 1939, vol. 25, Dec. 25, pp. 1065–1073). (In Japanese). The author reports on an investigation of the salt-bath heat treatment of 0·5–1·3% carbon steels and steels containing the following elements : (*a*) Carbon 0·7% and chromium 1·0–3·0% ; (*b*) carbon 0·7% and nickel 1·0–3·0% ; (*c*) carbon 0·3–0·7%, nickel 3·0% and chromium 0·5% ; and (*d*) carbon 0·3–0·7%, nickel 3·0%, chromium 0·5% and molybdenum 0·5%. His general conclusion is that this form of heat treatment is very effective when applied to 0·7–0·9% carbon steels, and to steels (*a*), (*c*) and (*d*), but that it has little effect on steels containing nickel and about 0·7% of carbon.

The Tocco Process for Hardening Steel Surfaces. (Engineering, 1940, vol. 150, Nov. 29, pp. 426–427). The essential features of the Tocco process of hardening, in which the heat is applied by means of a high-frequency current, have been described previously (*see*

p. 12 A). In the present article a more detailed description is given and some illustrations of the equipment used for hardening small crankshafts are presented.

Instruments for Control and Analysis of Controlled Atmospheres. E. E. Slowter and B. W. Gonser. (Metal Progress, 1940, vol. 38, Oct., pp. 555–566, 569). The authors consider the control of furnace atmospheres, dividing the subject into four sections. In the first section they very briefly discuss the complete chemical analysis of the atmosphere. In the second section the partial analysis by chemical means is considered, *i.e.*, the determination of carbon dioxide, oxygen and hydrogen sulphide; and the instruments developed by the Mine Safety Appliances Co. for the determination of oxygen and hydrogen sulphide are described. In the third section the authors deal with the determination of the physical properties of controlled atmospheres and describe some of the apparatus used in practice for the determination of the specific gravity, the thermal conductivity, the calorific value and the water content. In the last section some methods are discussed by which the atmosphere is examined by actually trying it on the material to be heat-treated. This part also includes descriptions of instruments for recording the visual and sub-surface effects of furnace atmospheres on the steel under treatment as well as their effect on special steels.

Equilibria for Gas-Steel Reactions. (Metal Progress, 1940, vol. 38, Oct., p. 594). Four groups of graphs relating to heat-treatment conditions are presented. These demonstrate the relation between furnace-atmosphere composition, temperature and the carbon content of the steel under conditions of equilibrium in the steel-gas reactions.

Some Surface Studies on Treated High-Speed Steel. J. G. Morrison. (American Society for Metals, Oct., 1940, Preprint No. 19). The author reports on an investigation of the hardness, the scaling and the distribution of carbon in the surface layer to a depth of about 0·012 in. in 18/4/1 tungsten-chromium-vanadium high-speed steel after various forms of heat treatment. Reducing and oxidising atmospheres were used, the former containing 10% of carbon monoxide and the latter 2·5% of oxygen. From the data obtained the author reached the following conclusions: (1) With the same holding time at 2350° F., the absorption of carbon was greater in the oxidising atmosphere than in the reducing atmosphere; (2) under the conditions of this investigation the carbon content at the surface of this steel increases rapidly to 0·95% when heat-treated at 2350° F., after this, the rate of carbon absorption decreases, or the carbon content at the surface diminishes; (3) the extent of the scaling, as measured by the decrease in diameter of the specimens, appears to be dependent on the rate at which the carbon content at the surface reaches 0·95%; (4) 18/4/1/ high-speed steels containing 0·60%, 0·70% and 0·93% of carbon when coated with borax prior to heat treatment at 2350° F. in different atmospheres all show some absorption of carbon at the surface under the conditions of this investigation; and (5) the hardness after heat treatment, is greater near the surface than at a depth of 0·0154 in. The results of hardness tests are given in tables and the changes in carbon content with increasing depth are shown in a series of curves.

The Influence of the Tempering of High-Speed Steel on the Efficiency of Cutting Tools. V. Ya. Dubovoy. (Metallurg, 1939, No. 10–11, pp. 99–117). (In Russian). In order to determine the optimum conditions of tempering high-speed steel of the 18/4/1 type, and in particular to settle the controversy regarding the relative effectiveness of single and repeated tempering, an extensive investigation was carried out, in which some specimens quenched from 1290° C. mainly in oil, were tempered repeatedly (from two to twelve times for 30 min.) and others only once (with a holding time of 30 min. to 12 hr.) at temperatures ranging from 100° to 700° C. The investigation included dilatometric, magnetic and electrical resistance tests, X-ray and microscopic examinations and determinations of the hardness, impact strength and cutting efficiency. Some cutting tools were also tested after stepped quenching. Consideration of the numerous results, which are presented diagrammatically, led the author to the following conclusions: (1) After quenching in oil from 1280° to 1290° C. the steel contains 25–35% of residual austenite. (2) Low-temperature tempering (50–400° C.) causes practically no transformation of the austenite, the main processes being of a qualitative nature such as transformation of the α-martensite into β-martensite and stress relief. (3) The higher the tempering temperature and the longer the holding time, the higher is the temperature at which the martensite transformation will set in on cooling and the more complete the transformation. (4) Complete transformation of the austenite can be achieved by either single or repeated tempering at the same temperature, the total holding time necessary being shorter when repeated tempering is used. (5) As regards cutting efficiency, repeated tempering does not appear to have any superiority over single tempering. Either form of tempering should, for maximum cutting efficiency, effect complete, or almost complete, transformation of the residual austenite and reduction of internal stresses. (6) After quenching in oil from 1290° C., the best results were obtained by tempering at 525° C. for 8 hr. (7) Stepped quenching with prolonged holding at 600–650° C. resulted in a marked deterioration of the cutting efficiency. (8) The maximum decrease in impact resistance occurs after high-temperature tempering at 550–650° C. and is connected with the transformation of the residual austenite into martensite and perhaps also with the precipitation of carbides from the austenite and martensite grains along the grain boundaries.

Keeping Them Straight. A. Fletcher. (Machine Shop Magazine,

1940, vol. 1, Nov., pp. 90–93). The author describes the plant in the modern heat-treatment department of a firm manufacturing automobile shafts and axles. The problem of keeping the shafts perfectly straight throughout the treatment has been solved by having all the furnaces and tanks designed so as to take the shafts in the vertical position, and by attaching each batch of shafts to a perforated carrier plate from which they remain suspended in the vertical position throughout the whole of the treatment.

Observations on the Tarnishing of Stainless Steels on Heating in Vacuo. V. C. F. Holm. (American Society for Metals, Oct., 1940, Preprint No. 54). The author observed the formation of temper colours when heating specimens of 18/8 chromium-nickel steel and 14% chromium steel in the range of 700–800° C., under pressures of about 10^{-4} mm. of mercury. Specimens of ordinary iron remained perfectly bright under the same conditions. Alloy steels containing high percentages of nickel, or tungsten and chromium steels with less than 5% of chromium, were also not discoloured. The colours developed on the stainless steels were similar in appearance to temper colours that form on iron and steel when heated in air, and they were eliminated by heating to higher temperatures (1050–1250° C.) at the same low pressure. The author points out that his observations are in agreement with the fact that a lower partial pressure of oxygen in the atmosphere is required while bright-annealing stainless steels than while bright-annealing iron or ordinary steels.

Centrifugal Quenching. (Machine Shop, 1940, vol. 1, Dec., pp. 90–93). It is pointed out that distortion of gear wheels, sprocket wheels, flat cams, &c., during quenching is avoided when the quenching medium is applied uniformly to the entire circumference and progressively from the circumference towards the centre, because, although the part under treatment may contract, the contraction will be uniform and the quenched part will be both round and flat. A centrifugal quenching machine of American design which enables quenching to be carried out in this way is described briefly.

WELDING AND CUTTING

(Continued from pp. 59 A–60 A)

The Practical Application of Resistance Welding. J. M. Sinclair. (Machine Shop Magazine, 1940, vol. 1, Sept., pp. 56–66; Oct., pp. 57–62; Nov., pp. 48–52). After explaining the principles of electrical-resistance welding, the author describes in detail its application in industry by the four processes of spot, seam, flash-butt and projection welding. The first three of these processes are well known, whilst the last is a recent development. Where an assembly calls for a concentrated number, or cluster, of spot welds, it is advisable to carry out the group of welds simultaneously. This can be done by "dimpling," *i.e.*, the raising of projections, on one or both of the plates to be joined. These small projections ultimately form the welds and are pressed flat during the operation. It is possible to make up to sixteen welds in one operation by this process. The types of joint and the machines used are shown in numerous illustrations, and the author also presents data respecting current strength, welding time and dimensions of projections for various gauges of sheet. In the second and third parts the author describes in detail the principles of flash-butt welding and seam welding and some of the machines for automatic welding by these processes.

Welding of High-Pressure Boilers. J. S. Train. (Australasian Engineer, 1940, vol. 40, Oct. 8, pp. 16–17). The author reviews the development of the electric welding of high-pressure vessels and briefly describes the present-day practice, giving particular consideration to the electrodes used. In conclusion he briefly discusses the examination of the welds, dealing in separate paragraphs with tensile tests (both across the joint and of the pure weld metal); bend tests; impact tests; determination of the density of the weld metal; and macro- and micro-examination.

A Summary of Reports of Investigations on Selected Types of High-Tensile Steels. L. Reeve. (Transactions of the Institute of Welding, 1940, vol. 3, Oct., pp. 177–200). In May, 1938, the Sub-Committee of the Welding Research Council on the Weldability of High-Tensile Structural Steels undertook a comprehensive programme of work in order to determine the correct procedure for the successful welding of high-tensile steels, and, if possible, to submit recommendations for high-tensile steels with improved weldability. In the present summary, details are given of the tests carried out by the sub-committee to date, together with some of the general conclusions derived from them. The tests carried out included chemical analyses of the eleven selected steels, mechanical tests on welded and unwelded specimens, standard Reeve crack tests using different sizes of fillet welds and including hardness measurements at sections cut through the test weld, quench-hardening tests on tapered specimens of each steel, oxy-acetylene cutting tests, and corrosion tests in conjunction with the Corrosion Committee of the Iron and Steel Institute. The results of testing the butt welds were not altogether conclusive as the electrodes used proved to be below standard in some respects, but it is concluded that butt welds between high-tensile steel plates, even with the type of electrode used, will develop approximatey the same strength as the unwelded plate, except for the upper strength range of steels. Particular attention must be paid to the technique in making the sealing runs of single-V butt welds, as it has been shown that low impact values and relatively high hardness in the locality of the sealing run may develop if the run is too small. In order to prevent this, it is necessary to use either a large sealing run or to preheat the material. Alternatively, the weld may be deposited in a double V, which eliminates the sealing run and also minimises the distortion. The cracking tests and measurements of hardness in fillet welds on high-tensile steels have demonstrated that maximum hardness and liability to cracking at the weld junction are related to the carbon content of the steels, although other alloying elements have a considerable influence on the average hardness. The quench-hardening test also bears out this general relationship. Cracking in a restrained joint may be avoided if the maximum hardness does not exceed 400 $\mathrm{HD}/_{10}$ (diamond pyramid hardness). In general, the hardness is also influenced by the size of fillet deposited with the first run, an increase in size reducing the value of the mean and maximum hardness and decreasing the cracking tendency. The Sub-Committee has confirmed the advantages to be gained in freedom from cracking by the use of certain types of special soft electrodes for the welding of high-tensile steels. In the light of the test data, only six of the eleven types of steel examined can be regarded as fully weldable, and of these six only two have a tensile strength exceeding 33 tons per sq. in. None of the steels complying with the requirements of British Standard Specification 548 (*i.e.*, 37–43 tons tensile) can be regarded as fully weldable on the basis of the tests described, although they can be welded successfully when suitable precautions are taken. The criterion of weldability would appear to be that, in a fillet test of the Reeve type, the mean weld-hardness of normal size fillets should not exceed 350 $\mathrm{HD}/_{10}$. Whether it will be possible to produce steels in accordance with the above specification with a hardenability within this limit is doubtful, but the most promising lines of attack would appear to be to reduce the carbon content of the steels and to compensate for the consequent loss in mechanical strength by the addition of other alloying elements.

A Guide to the Selection and Welding of Low Alloy Structural Steels. J. Dearden and H. O'Neill. (Transactions of the Institute of Welding, 1940, vol. 3, Oct., pp. 203–213). In this discussion of the welding of low-alloy structural steels the authors develop, from actual tests on fillet welds, empirical expressions by means of which the yield strength, maximum stress and welding-hardenability of these steels can be calculated from the chemical composition. These expressions also enable compositions to be selected which will give the best weldability for given tensile properties. As a result of their investigations the authors find that certain plain manganese steels, or manganese-molybdenum steels, are satisfactory if a yield point of 21 tons per sq. in. is required. If a yield point of 27 tons per sq. in. has to be provided for, a manganese-vanadium, a manganese-nickel or a manganese-molybdenum composition is suggested. The effects of various sizes of fillet, types of joint and thicknesses of plate and electrodes are evaluated and discussed, and attention is drawn to precautions which must be observed in welding low-alloy steels.

The Repair of Cast Iron Parts. (Welding Industry, 1940, vol. 8, Dec., pp. 295–296). A number of examples of the repair by welding of heavy cast iron machinery are illustrated and described. The examples include shearing machines and locomotive and Diesel engine cylinders.

A Defect in Mild Steel Welds Due to Non-Metallic Inclusions in the Parent Metal. L. E. Benson. (Transactions of the Institute of Welding, 1940, vol. 3, Oct., pp. 215–218). The author describes an investigation of the causes of a peculiar type of defect in a butt-welded ¾-in. steel plate. The defect was seen in the fracture of the plate in a standard reverse bend test. The abnormal feature was the presence of numerous round marks, 0·01–0·1 in. across, within which were pronounced concentric ripple marks similar to those in the crater left by an arc. The evidence obtained made it clear that these marks are discontinuities in the weld metal caused by the melting of the ends of abnormally large slag streaks in the original plate, possibly with the evolution of some gas. This is a striking example of how a defective weld can result solely from unsatisfactory parent plate.

British Standard Specification for Oxy-Acetylene Welding in Mild Steel. (British Standards Institution, No. 693, Revised Oct., 1940). The present specification is a revision of that issued in 1936. Since that time the question of tests on welds has been reviewed and provision is now made for more simple tests which are considered to be sufficient to ensure sound welding.

PROPERTIES AND TESTS

(Continued from pp. 62 A–69 A)

Automatic Speed Control for Tension and Compression Testing Machines. R. K. Bernhard. (ASTM Bulletin, 1940, Oct., pp. 31–34). In the first part of his paper the author considers the theory of loading in systems of direct, lever and hydraulic loading in tensile-testing machines, and the effects of changes in the mass and rigidity of the machine and the speed at which the load is increased. In the second part he describes a fully automatic control unit for tensile-testing machines with which it is possible to maintain (*a*) a constant load, (*b*) a constant motion of the crosshead of the machine at low and high testing speeds, (*c*) a constant rate of change of load, and (*d*) a constant rate of change of strain. The device consists essentially of two photo-electric cells mounted close together on the rim of a disc which is rotated by a variable-speed drive of a motor connected to the loading system of the machine. The two cells act as inertia-free and frictionless relays. By means of suitable circuits one cell starts the motor and the other reverses it. In order to use the device to maintain a constant load, a pressure guage connected to the hydraulic system of the machine is mounted on the centre of the disc. A small mirror is fixed to the spindle of the gauge needle, and this is so placed relative to a source of light that the reflected beam falls between the two cells when the gauge is registering the desired pressure. A small movement of the gauge needle is thus sufficient to deflect the beam on to one of the cells, thus starting up the motor in the required direction. In order to maintain a constant motion of the crosshead, the pressure gauge in the above description is replaced by an ammeter connected to a Wheatstone bridge circuit, which also incorporates a resistance wire fixed to the base of the machine with a sliding contact connected with the crosshead. The mirror is mounted on the ammeter needle spindle and the cells operate as previously described. To maintain a constant rate of change of strain a dial deflectometer is placed at the centre of the disc. The arrangement for maintaining a constant rate of change of load is similar to that for maintaining a constant load. It is claimed that this control device is adaptable to any type of testing machine, that it is not too expensive to be used for commercial testing, and that all the components can be mounted on a control board which is easily transportable to any testing machine in the laboratory.

Changes of Structure at the Yield Point of Mild Steel. S. Tanaka and K. Takamiya. (Nippon Kinzoku Gakkai-Si, 1940, vol. 4, June, pp. 185–187). (In Japanese). The authors describe their investigation, by means of the back-reflection X-ray technique, of the changes which occur in the structure of mild steel at the yield point.

Stress Analysis by Three-Dimensional Photo-Elastic Methods. D. C. Drucker and R. D. Mindlin. (Journal of Applied Physics, 1940, vol. 11, Nov., pp. 724–732).

Quantitative Evaluation of Distortion in Silicon Steel and in Aluminium. G. L. Clark and W. M. Shafer. (American Society for Metals, Oct., 1940, Preprint No. 22). The authors report on an investigation of the relation between measured stresses in silicon steel and in aluminium and the elongation of the Laue spots as observed in photograms. A special apparatus is described in which strips of silicon steel made up of a single crystal were bent in such a way that the radius of curvature could be measured and X-ray patterns could be obtained in this position. The elongation of the Laue spots was measured with the aid of Leonhardt curves and this was correlated directly to the curvature of the specimen. The stress applied was then calculated. Under the conditions governing the experiments, radial asterism was observed when single crystals of silicon steel were bent under stresses below the elastic limit;

when the stress was greater than the elastic limit a part of the rotation or bending of the planes was elastic. Equivalent amounts of cold-work on this steel, when expressed in terms of Nadai's theory of octahedral shear, whether by rolling or by elongation under tension, produced a similar effect as revealed by the X-ray patterns.

Measurements and Treatments of the Internal Friction of Carbon Steel. T. Endo. (Nippon Kinzoku Gakkai-Si, 1940, vol. 4, Feb., pp. 59–64). (In Japanese). The author describes an investigation of the internal friction of some carbon steels and discusses the results obtained. He found that the internal friction of some carbon steels decreases with the amplitude of the vibrations, and that this decrease is more rapid at smaller amplitudes. The internal friction at larger amplitudes is, however, nearly constant, and is subject to the laws proposed by Kimball and Lovell.

The Fatigue and Bending Properties of Cold-Drawn Steel Wire. H. J. Godfrey. (American Society for Metals, Oct., 1940, Preprint No. 37). The author studied the fatigue and bending properties of cold-drawn carbon-steel wire. He developed special fatigue and bend-testing machines for this purpose, which are described in the paper. The bend-testing machine was so designed that the severity of the test could be controlled and the true bending properties of the wire determined. The investigation led to the following conclusions : (1) The fatigue-test results are influenced by the carbon content of the material, and the fatigue limit of a low-carbon material should, therefore, be determined by a fatigue test of at least 10,000,000 cycles ; for higher-carbon steels the length of the test may be considerably less. (2) Normal cold-working by means of wire-drawing increases the fatigue limit of steel wire in proportion to its tensile strength. The ratio between the fatigue limit and the tensile strength (called the fatigue ratio) decreases, however, if the cold-working exceeds a critical amount. This ratio decreases also with an increase in the carbon content. (3) The fatigue properties of wire are improved by polishing its surface. (4) Decarburisation on the surface of the wire deteriorates its fatigue properties. The depth of decarburisation does not seem to have any effect on the fatigue ratio, however. (5) Electro-galvanising does not lower the fatigue ratio of a decarburised wire. (6) The bending-fatigue properties of hot-galvanised bridge wire are approximately the same whether cold-drawn or heat-treated. (7) The bending properties of patented steel wire improve with cold-work up to a critical amount ; beyond this point the bending properties deteriorate. (8) Both the bending-fatigue test and the bend test should be considered when deciding upon the desired amount of cold-drawing. (9) The amount of orientation of the metal crystals, as shown by X-ray photograms, is more obvious when cold-working is excessive, and the maximum amount of orientation takes place in the core of the wire.

Screw Threads—The Effect of Method of Manufacture on the Fatigue Strength. A. M. Smith. (Iron Age, 1940, vol. 146, Aug. 22, pp. 23–28). The author describes an investigation of the influence of the manner of manufacturing a screwthread on the tensile strength, the endurance limit and the ratio of these two values of the threaded material. An ordinary steel containing 0·22–0·26% of carbon was used for the tests. The three methods of producing the thread which were investigated were: (1) Cutting the thread in a lathe; (2) rolling the thread on single-extruded wire; (3) rolling the thread on double-extruded wire. The conclusions reached are as follows: (*a*) single-extruding and rolling the thread increase the tensile strength of the original material by about 12%; (*b*) double-extruding and rolling the thread increase the tensile strength of the original material by about 28·5%; (*c*) cut-threaded specimens have an endurance limit equal to about 30% of their tensile strength; (*d*) single-extruded and roll-threaded specimens have an endurance limit equal to about 33% of their tensile strength, and equal to about 37% of the tensile strength of the original material; and (*e*) double-extruded and roll-threaded specimens have an endurance limit of about 33% of their tensile strength, and about 42·6% of the tensile strength of the original material.

The Recovery of Fatigue Due to Annealing. F. Oshiba. (Nippon Kinzoku Gakkai-Si, 1940, vol. 4, Jan., pp. 13–20). (In Japanese). The author reports on the effect of the annealing time and temperature on the degree of recovery from fatigue of a 0·2% carbon steel and Flodin iron. On annealing specimens of the steel in hydrogen, maximum recovery was obtained after a definite period of annealing which depended on the temperature, and a hyperbolic relation was established between the annealing time and temperature. On annealing Flodin iron in a vacuum, the degree of recovery from fatigue became constant after a definite annealing time which depended on the temperature, and a hyperbolic relation was again established between the minimum time required to attain a constant degree of recovery and the temperature.

Hardness Conversion for Hardened Steel. H. Scott and T. H. Gray. (Metal Progress, 1940, vol. 38, Oct., p. 428). The authors present a hardness conversion table applicable to steel up to its maximum hardness, irrespective of composition and structure, and to any other material having an elastic modulus of about 30×10^6. The table covers diamond-pyramid hardness, five Rockwell scales, scleroscope, monotron and Brinell hardness (for tungsten-carbide, Hultgren and steel balls).

Investigation of Steel 9*Kh*. P. Orlets and T. Sergievskaya. (Stal, 1939, No. 10–11, pp. 64–66). (In Russian). The investigation of steel 9*Kh* (containing carbon 0·87%, silicon 0·32%, manganese 0·23%, sulphur 0·015%, phosphorus 0·016%, chromium 1·51% and nickel 0·48%) had for its object the determination of the optimum oil-quenching temperature, and a study of the effect of this temperature and the holding time on the hardenability and hardness, as well as the effect of the subsequent tempering temperature on the mechanical properties. Maximum surface hardness was obtained by quenching from 850° C., the holding time at the above temperature being that recommended in the A.S.S.T. (now A.S.M.) standards. For maximum ductility, the quenching temperature should be lowered to 830° C. and the holding time trebled. The hardness remained unchanged by tempering at up to 150° C. and it then fell five Rockwell C units on tempering at 200° C., subsequently remaining unchanged when the tempering temperature was increased up to 300° C., above which the hardness again decreased. For improved ductility tempering should be conducted at 300° C.

Quantitative Measurement of Strain Hardness in Austenitic Manganese Steel. D. Niconoff. (American Society for Metals, Oct., 1940, Preprint No. 15). The author examined the strain-hardening properties of an austenitic manganese steel (manganese 12·5%) by subjecting specimens to impacts of different intensities and by measuring the hardness at the surface and at different depths. The maximum strain-hardness of approximately Rockwell C 50 was not found at the surface, but at a short distance below it; with increasing depth the strain-hardness effect gradually diminished. Repeated impacts only slightly increased this maximum value. It was observed that increasing the number of impacts on the already hardened surface slightly decreased the hardness of the outer layer, whilst it increased the depth of the strain-hardening. In order to produce the maximum work-hardening effect on a flat surface of this steel with a single blow the impact value had to be very high. For example, to produce a surface hardness of Rockwell C 50, a stress concentration of 35,000 ft.-lb. per sq. in. was required.

Further Notes of Precipitation Hardening in the Heavy Alloys. W. P. Sykes. (American Society for Metals, Oct., 1940, Preprint No. 25). In continuation of previous investigations (*see* Journ. I. and S.I., 1940, No. I., p. 41 A) the author studied further aspects in connection with the precipitation-hardening of binary iron-tungsten and iron-molybdenum alloys, *viz.*: (*a*) The hardness changes at elevated temperatures accompanying precipitation and the attainment of structural equilibrium at these temperatures; (*b*) the behaviour of single-phase as compared with two-phase alloys during precipitation-hardening at intermediate temperatures; and (*c*) the effect on precipitation-hardening characteristics of the rate of cooling after solution treatment. The alloys studied contained 16·0–23·9% of tungsten and 8·9–14·1% of molybdenum, respectively. The investigation led to the following conclusions: (1) The precipitation hardening-characteristics of a two-phase alloy seem to be identical with those of the solid-solution phase. (2) No evidence was obtained which indicated that the presence of a second phase, either highly dispersed or in massive form, introduces sufficient additional strain during the quench to alter the rate of precipitation-hardening. (3) Water-quenching after the solution treatment leads to the formation of solid solutions which are measurably harder than those obtained by air-cooling and which, in addition, harden more rapidly during subsequent ageing.

Steel Mill Gear Tooth Wear and Failure. J. H. Jones. (Iron and Steel Engineer, 1940, vol. 17, Oct., pp. 50–52). The author discusses from a practical viewpoint the following types of gear-tooth wear and failure: (1) Abrasion or common wear; (2) pitting; (3) scoring; (4) surface flow; and (5) tooth breakage. He points out some ways of obtaining the maximum service from gears of current design.

Abrasion of Steel in Vacuum, in Hydrogen Gas, and in Nitrogen Gas. S. Saito and N. Yamamoto. (Nippon Kinzoku Gakkai-Si, 1940, vol. 4, Jan., pp. 26–35). (In Japanese). The authors report on a series of tests in which the rates of wear of a 0·7% carbon steel in a vacuum, in air, in hydrogen and in nitrogen were compared. The tests were made with an Amsler abrasion-testing machine. It was established that the wear in nitrogen and in hydrogen is several times greater than that in air. As the pressure decreases and the degree of oxidation decreases, the loss in weight by wear gradually increases, the maximum loss being reached at approximately 0·1 mm. of mercury. This maximum loss in weight in a vacuum is greater than that in either hydrogen or nitrogen, and is very much greater than that in air. When, however, the vacuum is greater than 0·1 mm. of mercury, a sudden change in the rate of wear occurs and it becomes much less than that in air.

Friction and Surface Finish. (Proceedings of the Special Summer Conferences on Friction and Surface Finish. Massachusetts Institute of Technology, June 5, 6 and 7, 1940). The Department of Mechanical Engineering and the Department of Physical Metallurgy of the Massachusetts Institute of Technology organised a symposium on friction and surface finish in order to bring before engineers a summary of the present knowledge of this subject. The symposium was held at the above Institute in June, 1940. The present publication contains all of the papers which were presented together with the written discussions. The titles and authors of the papers are as follows:

Some General Aspects of Rubbing Surfaces, by A. F. Underwood.
The Metallurgy of Surface Finish, by J. Wulff.
The Preparation of Smooth Surfaces, by D. A. Wallace.
Description and Observation of Metal Surfaces, by S. Way.
Surface Friction of Clean Metals, by H. Ernst and M. E. Merchant.
Boundary Lubrication, by G. B. Karelitz.
On the Mechanism of Boundary Lubrication, by O. Beeck, J. W. Givens, A. E. Smith and E. C. Williams.
Thin Film Lubrication, by F. C. Linn.
Mechanisms of Wear: Their Relation to Laboratory Testing and Service, by R. W. Dayton.
How Should Engineers Describe a Surface? O. R. Schurig.

Thermal Changes in Iron and Steel. E. Griffiths. (Sheffield Metallurgical Association: Iron and Coal Trades Review, 1940, vol. 141, Nov. 29, p. 551). A brief account is given of the work carried out in the Physics Department of the National Physical Laboratory on determinations of the specific heats, heats of transformation, coefficients of expansion and thermal conductivities of carbon steels, nickel steels, manganese steels, nickel-chromium steels and high-speed steels. This work was undertaken for the Thermal Treatment Sub-Committee of the Alloy Steels Research Committee. A detailed account of it will be found in Section IX. of the Second Report of the Alloy Steels Research Committee (a Joint Committee of the Iron and Steel Institute and the British Iron and Steel Federation).

The Development of Alloys for Use at Temperatures above 1000 Degrees Fahr. E. R. Parker. (American Society for Metals, Oct., 1940, Preprint No. 2). The author discusses some of the factors which affect the resistance of metals and alloys to deformation at high temperatures. He points out that whilst the thermal activity of atoms is a factor which need not be considered with regard to deformation at low temperatures, it is of great importance at high temperatures. In examining the properties of specimens under stress at temperatures above 1000° F., the author has found that alloys containing precipitates of the type Fe_2W, Fe_3Mo_2, Fe_3Nb_2, Fe_2Ta and Fe_2Ti possess greater resistance to deformation than alloys which depend on carbides for their strength. He reports on a series of tensile tests at room temperature and at 1100° F. on specially prepared alloys containing the metals indicated above and on the relation of the recrystallisation temperature to the results obtained. The results of this work supported the hypothesis that a dispersed phase of an intermetallic compound which did not agglomerate rapidly at 1100–1200° F. has a great stabilising influence and that alloys containing such compounds have a high creep resistance.

Sharp Transition from Brittle to Ductile in Hot Ingot Iron. M. Charlton. (Metal Progress, 1940, vol. 38, Sept., pp. 287–290). The author reports on a series of tests the object of which was to determine the critical temperature at which a marked change in the ductility of hot ingot iron takes place. Metal from four heats of ingot iron containing 0·03% of carbon were tested; two of these contained about 0·25% of copper. Specimens 24 in. long were used on which an 8-in. gauge length was turned to 0·800 in. in dia. These were placed in the centre of an electric-resistance furnace and the protruding ends were clamped in a tensile-testing machine. The temperature at the centre of the specimen was measured with a thermocouple. Tests were run at intervals of 100° F. in the range 1400–2400° F., and at 10° intervals in the range 1900–2000° F. Specimens from the four heats all increased in brittleness with increasing temperature, but there were slight differences in the end of the brittle temperature range, the critical temperatures being

1940°, 1910°, 1930° and 1910° F. The tensile strength increased slightly as the temperature rose from 1600° to 1900° F., whilst the elongation and reduction of area gradually decreased. A sharp increase in the elongation and reduction of area marked the termination of the brittle range, and the tensile strength decreased gradually from this point with increasing temperature. It appeared that the presence of copper had no influence on the results. All four heats contained approximately the same amount of sulphur, but two of them contained slightly more manganese than the other two, and it was the former which had the lowest critical temperature.

On the Adsorption and Absorption of Nitrogen by Iron at High Temperature. I. Hayashi. (Tetsu to Hagane, 1940, vol. 26, Feb. 25, pp. 101–122). (In Japanese). The author presents the results of his investigation of the adsorption and absorption of nitrogen by pure iron powder. The gas was passed over the powder, which was held at different temperatures in the range 500–1350° C. and then rapidly cooled, and the amount of nitrogen in the metal was determined by chemical analysis.

Properties of Low-Nickel Steels at Low Temperatures. V. Chernyak and V. Yanchevskiy. (Stal, 1939, No. 10–11, pp. 66–67). (In Russian). The authors investigated the effect of cooling to low temperatures on the mechanical properties of three chromium-nickel steels of the following analyses :

	EI 180	*EI* 100	*EI* 130
Carbon %	0·07	0·19	0·08
Silicon %	0·04	0·21	0·66
Manganese %	...	9·58	16·35
Phosphorus %	0·011	...	0·033
Chromium %	16·8	13·8	12·00
Nickel %	1·22	4·32	0·35

Tensile, hardness and impact tests were made at + 20°, −80° and −182° C. Repeated cooling to low temperatures appeared to have no effect on the mechanical properties of the steels as measured subsequently at room temperature. The impact strengths determined at −183° C. were 82%, 90% and 37% below the results obtained at room temperature for steels *EI* 130, *EI* 180, and *EI* 100, respectively.

Vanadium in Cast Iron. E. Piwowarsky. (Giesserei : Foundry Trade Journal, 1940, vol. 63, Nov. 28, pp. 345–346 ; Dec. 5, pp. 363–364 ; Dec. 12, pp. 383–384). The author summarises the available information on the effects of vanadium on the properties of cast iron. This element is generally added in amounts varying from 0·08% to 0·35%. It is a very powerful promoter of carbide formation, and this influence appears to be the more pronounced the lower the carbon and silicon contents of the iron, the harder the melting furnace is driven and the thinner the walls of the casting. Additions of up to about 0·3% of vanadium alter the structure of cast iron only slightly, the pearlite tending to take a sorbitic form. Above

0·3% of vanadium, free carbides occur; these have a spherical form, tend to increase segregation and impair the casting properties. The temperature of the commencement of the formation of temper carbon is raised by about 30–40° C. for each 0·1% of vanadium. The mechanical properties of commercial grades of cast iron are appreciably improved by the addition of vanadium, the tensile strength being increased by about 1·3–1·9 tons per sq. in. for each 0·1% of vanadium. With very high-grade iron (with tensile strength above about 20·3 tons per sq. in.) vanadium has only a slight beneficial effect; if the tensile strength exceeds about 22·8 tons per sq. in., additions of this element may even have an adverse effect. The addition of vanadium frees castings from gas and increases their density. Many of the effects of vanadium in cast iron can be explained by the fact that vanadium carbides, which are only slightly, if at all, soluble in molten iron, remove carbon from the material, thus displacing the entire iron-carbon diagram to the right. Both wear and resistance to heat, especially in the 400–500° C. range, appear to be somewhat improved by vanadium additions. In chilled cast iron, such as cast-iron rolls, the hardness of the surface is apparently improved to a lesser extent than is its cleanness and freedom from pores. The influence of vanadium on the resistance to scale formation has not yet been fully elucidated, but the author considers that no special advantages are to be expected. Vanadium in nitrided cast iron appears to have a similar effect to that of chromium. In general, vanadium appears to be able to replace the element chromium rather than molybdenum in cast iron. The author also presents a table of the analyses of a number of vanadium cast irons and their applications in accordance with American practice.

A Practical Study on High-Chromium Cast Iron. I. Naito. (Tetsu to Hagane, 1940, vol. 26, Feb. 25, pp. 71–75). (In Japanese). The author describes tests on a series of high-chromium cast irons, which had as their object the determination of the optimum compositions for good heat-resisting and wear-resisting properties. As a result of these tests he recommends the following compositions for heat and wear resistance respectively: (*a*) Carbon 2·0–2·4%, chromium 22–26%, silicon 0·6–0·8%, manganese 1·0–2·0% and aluminium 0·5–1·0%; (*b*) carbon 1·8–2·2%, silicon 0·4–0·8%, manganese 0·4–0·8% and chromium 17–22%.

Ladle Additions to Cast Iron. R. Schneidewind. (Metal Progress, 1940, vol. 38, Oct., pp. 374–375, 377). The author discusses the effect of various ladle additions in the manufacture of alloy cast irons, pointing out that the aim of modern ladle treatment is to select the alloys added so as to balance those having a graphitising effect with those which act as carbide stabilisers. In conclusion he tabulates data on the effects of some ladle additions on the mechanical properties of a cast iron containing 3·10% of carbon, 2·12% of silicon and 0·69% of manganese.

On the Effect of Addition of Special Elements (Ni, Cr, W, Mo, V) upon the Mechanical Properties of Special Steel. S. Tamaki. (Tetsu to Hagane, 1940, vol. 26, June 25, pp. 450–455). (In Japanese).

Equivalent British and S.A.E. Nickel Alloy Steels. (The Mond Nickel Company, Limited, 1940). This brochure has been published by the Mond Nickel Co., Ltd., to give information on some of the S.A.E. nickel steels which might be of assistance to British users. It contains general particulars concerning the different types of steels, followed by charts showing how the mechanical properties are affected by heat treatment. In explanation of the S.A.E. numbering system it is stated that the first of the four digits indicates to which class the steel belongs; the second digit, in the case of alloy steels, generally shows the approximate percentage of the predominating element, and the last two digits indicate the average carbon content in hundredths of 1%.

On Alloy Steels for the Coal Liquefaction and other High-Temperature and High-Pressure Chemical Industries. M. Kinugawa. (Tetsu to Hagane, 1940, vol. 26, Aug. 25, pp. 609–616). (In Japanese). The author discusses the suitability of a number of alloy steels for the manufacture of plant to operate at high temperatures and pressures, *e.g.*, for ammonia or methanol synthesis.

Properties of the Principal Cr-Fe and Cr-Ni-Fe Alloys. (Metal Progress, 1940, vol. 38, Oct., pp. 464–465). Two tables are reproduced, the first of which contains data for the chemical composition and the more important physical and mechanical properties of 2%, 5%, 9%, 12%, 17% and 27% chromium steels and of cutlery material containing 13–14% of chromium. The second table gives corresponding information on 18/8, 18/12, 25/12, 25/20 and 18/26 chromium-nickel steels.

Chromium Steels from 2 to 16% Chromium. H. D. Newell. (Metal Progress, 1940, vol. 38, Oct., pp. 384–385). The author briefly reviews the properties of commercial steels containing between 2% and 16% of chromium, pointing out that a standardisation of their composition is desirable. He calls particular attention to a highly creep-resistant steel, containing 2·25% of chromium and 1% of molybdenum, and to a recently introduced highly corrosion-resistant steel, containing 7% of chromium, 0·50% of molybdenum and 1% max. of silicon.

New Developments in Stainless Steels. (Machinery, 1940, vol. 57, Dec. 19, pp. 330–331). Some recent developments in the composition and production of stainless steels are briefly reviewed, under the following headings: The Pluramelt process; carbon-molybdenum steel plate; etching medium for nickel-chromium steel; influence of annealing; cypritic steels; stainless steels with non-metallic backing; a new ageing treatment (heating to about 200° C. for about 24 hr.); electrolytic polishing; and new alloys.

The Effect of Molybdenum and Columbium on the Structure, Physical Properties and Corrosion Resistance of Austenitic Stainless

Steels. R. Franks, W. O. Binder and C. R. Bishop. (American Society for Metals, Oct., 1940, Preprint No. 17). The authors report on an investigation of the influence of additions of molybdenum with and without niobium on the properties of nickel-chromium steels. In the steels examined, the chromium content ranged from 16% to 25%, the molybdenum from 0·90% to 3·25%, the niobium from 0·40% to 1·02%, and the nickel from 8% to 22%. The difficulty in the hot-working of some of these steels is associated with the formation of a σ-phase, and it was found that, in the steels containing 18–19% of chromium and 12–14% of nickel, the addition of 1·50–2·25% of molybdenum prevented this phase from being formed, and the alloy of this composition had satisfactory hot-working properties, high toughness and excellent resistance to corrosion when attacked by either oxidising or reducing agents. The steels containing 1·50–2·25% of molybdenum were subject to intergranular corrosion and the highest resistance to this form of corrosion was obtained when the niobium content was ten times the carbon content; in many cases, however, an addition of niobium to the extent of six times the carbon content was sufficient. The alloys containing these elements in this ratio of 6 : 1 could be stress-relieved without loss of general resistance to corrosion and were practically immune from intergranular corrosion, and they are therefore suitable for fabricating into vessels and other articles which must be given stress-relieving treatment.

Influence of Heating and Clenching Conditions on Mechanical Properties and Microstructures of High-Tensile Steel Rivets. G. Mima. (Tetsu to Hagane, 1939, vol. 25, Sept. 25, pp. 721–733). (In Japanese). The author describes some of the effects of the operations of rivet-making and rivet-closing on the mechanical properties and microstructure of high-tensile steel rivets.

War Emergency British Standard Specification for Steel Tubes and Tubulars, Light Weight and Heavy Weight Qualities. (Revised Weights). (British Standards Institution, No. 789A–1940). This specification provides for the replacement of the three qualities of tube laid down in British Standard Specification No. 789, namely, gas, water and steam qualities by two grades designated respectively " Light Weight " and " Heavy Weight."

METALLOGRAPHY AND CONSTITUTION

(Continued from pp. 69 A–76 A)

Exposure Chart for Radiography of Steel. H. R. Isenburger. (Metal Progress, 1940, vol. 38, Oct., p. 553). The author explains the use of diagrams designed to give information on the exposure time required, under certain experimental conditions, for the radiographic examination of steel specimens of various thickness.

Non-Destructive Production Test for Steel Tubing. (Steel, 1940, vol. 107, Oct. 21, pp. 38–40, 75). A description is given of a non-destructive testing machine for the rapid detection of faults in steel tubes. The machine is a development of the Sperry fault-detecting car for rails. The machine consists of a stand on which are mounted sets of energising and detecting coils with amplifiers, drive rolls, controls and auxiliary equipment, so designed that a medium-frequency current is induced in a circular direction in the tube. The machine will test tubes from ½ in. to 2½ in. in dia. and from 7 to 22 gauge. By suitable calibration it is possible to adjust the detector so that defects below a specified size are not indicated.

The Examination of Metals by Ultrasonics. A. Behr. (Metallurgia, 1940, vol. 23, Nov., pp. 7–11). The author gives an account of the development, during the last decade, of the use of ultrasonics for the examination of metals, referring, in particular, to Russian publications on this subject. After an introduction dealing in an elementary way with the properties of ultrasonics and the methods of their production, he describes at some length the arrangement of apparatus incorporating various types of defectoscopes in which use is made of ultrasonics for the examination of metals. In conclusion he briefly discusses the advantages and disadvantages of the ultrasonic method of testing, as compared with radiographic methods.

The Tracer Method of Measuring Surface Irregularities. E. J. Abott. (American Society for Metals, Oct., 1940, Preprint No. 55). The author points out that the microscope and the profile tracer are two instruments which supplement each other for measuring surface irregularities, the former being valuable for vertical observations of closely spaced irregularities and the latter for profile observations of widely spaced irregularities. He briefly outlines the technical details of the tracer method, and states that several profile recorders have been built during the last decade. They all have in common a fine tracer point which is drawn over the surface to be measured. The motions of this point are magnified and recorded in various ways in order to obtain a record of the profile of the surface. He discusses the correlation of profile tracer records and microscopic observations, comparing the results obtained by both methods when examining the surfaces of ground and polished steel specimens.

Method for Identification of Inclusions in Iron and Steel. (Metal Progress, 1940, vol. 38, Oct., p. 376). A flow sheet is reproduced representing a method for the identification of inclusions in iron and steel, which was originated by Campbell and Comstock (*see* Proceedings of the American Society for Testing Materials, 1923, vol. 23, Part 1, pp. 521–522) and modified by C. R. Wohrmann, M. Scheil and M. Bayertz.

Rating of Inclusions (" Dirt Chart "). (Metal Progress, 1940, vol. 38, Oct., p. 378). Two charts, called " dirt charts," are reproduced which were developed by G. W. Walker for the rating of silicate and oxide inclusions respectively in steels used in automobile manufacture.

On the Macrostructures of Alloy Steel Forgings. S. Nishigori. (Tetsu to Hagane, 1940, vol. 26, June 25, pp. 429–435). (In Japanese). The author discusses the macrostructure of low-alloy steel forgings, such as gear blanks, crankshafts and connecting rods, with particular reference to the influence of the manner and temperature of the heating prior to forging.

Standard Grain Sizes for Steels. (Metal Progress, 1940, vol. 38, Oct., p. 380). Eight micrographs are reproduced representing standard grain sizes, at 100 diameters, of steels which were carburised and slowly cooled to develop a cementite network. The following ranges of grain size per sq. in. are represented : (1) Up to $1\frac{1}{2}$, (2) $1\frac{1}{2}$ to 3, (3) 3 to 6, (4) 6 to 12, (5) 12 to 24, (6) 24 to 48, (7) 48 to 96, and (8) more than 96 grains per sq. in.

On the Austenitic Grain Size of Carbon Steel. S. Yanagisawa and M. Yamashita. (Tetsu to Hagane, 1939, vol. 25, Dec. 25, pp. 1027–1034). (In Japanese). The authors report the results of tests on high-carbon steel sheets, the object of which was to ascertain the relative merits of acid and basic open-hearth steels and the influence of the grain size on the mechanical properties. The two kinds of steel had approximately the same tensile strength, but the basic steel had a lower impact strength than the acid steel; this was, however, due to differences in grain size.

Influence of Silicon and Aluminium Additions on the Constitutional Diagram of 4/6 Chromium-Molybdenum Steels. C. L. Clark and M. A. Bredig. (American Society for Metals, Oct., 1940, Preprint No. 34). The authors studied the effects of silicon and aluminium additions on the constitutional diagram of 4/6 chromium-molybdenum steel. They used two series of steels with varying aluminium and silicon contents, respectively. The first series contained 0·52–0·59% of aluminium and the silicon content varied between 0·25% and 5·00%, whereas in the second series, containing between 1·24% and 1·34% of silicon, the aluminium content varied between 0·002% and 1·51%. Both series of steels contained carbon 0·12–0·15%, manganese 0·33–0·38%, phosphorus 0·010–0·018%, sulphur 0·014–0·019%, chromium 4·86–5·23% and molybdenum 0·52–0·59%. The authors determined the critical ranges of the steels by heating and cooling curves and carried out hardness determinations on specimens quenched from 1400° C. In addition many of the quenched specimens were subjected to a metallographic examination. The investigation indicated that in steels containing about 0·55% of aluminium the inner γ loop is closed at approximately 1·40% of silicon, while in the presence of 1·25% of silicon, the same result is accomplished by 0·65% of aluminium. The additive effect of these two elements in eliminating the α–γ transformation is thus not the same as that known for carbon-free alloys. The results also showed that, in the presence of 0·12–0·14% of carbon, the mixed $\alpha + \gamma$ field is several times as wide as the homogeneous γ area at the temperature of maximum γ solubility. The authors also found that the mixed $\alpha + \gamma$ field passes directly into a partially liquid state on heating, not into a homogeneous α condition.

Kinetics and Reaction Products of the Isothermal Transformation of a 6 per cent Tungsten–6 per cent Molybdenum High-Speed Steel. J. L. Ham, R. M. Parke and A. J. Herzig. (American Society for Metals, Oct., 1940, Preprint No. 18). By means of dilatometric tests, metallographic examinations and hardness determinations the authors investigated the isothermal reaction rates and constructed the S cooling curve (according to Davenport and Bain) for a high-speed steel containing carbon 0·80%, chromium 4·07%, tungsten 5·70%, vanadium 1·65% and molybdenum 6·09%. The curve discloses that the austenite of this steel transforms very slowly, if at all, between 595° and 370° C. when quenched from a solution temperature just below the solidus. The rate of transformation is not rapid at any temperature between 815° C. and room temperature, but is faster at room temperature than at any other investigated. It would therefore be expected that this steel would undergo transformation at close to room temperature when light sections are quenched in oil and consequently might develop quenching cracks. That such cracks do not occur frequently may be attributed to the retention of considerable amounts of austenite. Between room temperature and 370° C., the austenite of this steel is converted to an acicular product, whilst between 595° and 815° C. the product is spheroidal. The heat treatment of this steel is discussed in relation to the information revealed by the S-curve.

Dilatometric Studies in the Transformation of Austenite in a Molybdenum Cast Iron. D. B. Oakley and J. F. Oesterle. (American Society for Metals, Oct., 1940, Preprint No. 11). The authors report the results of an investigation on the transformation of austenite in two cast irons both containing 2·9% of carbon and 2·1% of silicon, one of which contained 0·5% of molybdenum. Each specimen was heated in an electric furnace to 1525° F. for about 15 min. and then plunged into a salt-bath, and the expansion was measured with a dilatometer as the cooling proceeded. A large number of tests were made with salt-baths held at temperatures varying from 1000° to 400° F. in stages of 100° F. From the time-expansion curves obtained the rate of transformation could be determined. The results showed that the addition of 0·5% of molybdenum had little effect on the rate of transformation except at the lowest temperature studied, *i.e.*, 400° F. At the lower temperatures the curves show periods of no expansion after some primary expansion has taken place. This may be evidence of the two-stage reaction observed by Davenport and Bain. This type of reaction occurred in the molybdenum iron at temperatures up to 500° F., whereas the iron with no molybdenum showed it up to 600° F. A number of micrographs of the specimens after expansion are reproduced.

Equilibrium Relations in the Solid State of the Iron-Cobalt System. W. C. Ellis and E. S. Greiner. (American Society for Metals, Oct., 1940, Preprint No. 16). The authors have investigated the α–γ transformation in iron-cobalt alloys by thermal and X-ray diffraction methods. In the present paper they compare their results with those of earlier investigators and present an equilibrium diagram of the iron-cobalt system in the solid state. The addition of increasing quantities of cobalt at first raises the A_3 point until a maximum is reached at about 45% of cobalt; further additions lower the transformation temperature which, at about 80% of cobalt, rapidly approaches room temperature. An extended two-phase region between 76·5% and 88·5% cobalt was established at 600° C. An order-disorder transition occurs in the α-phase in the region of 50% of cobalt. The critical temperature of order is at about 700° C. depending upon the composition. The ordered arrangement has the cesium-chloride structure. The lattice constants of the α-phase deviate widely from a linear function of the cobalt content. The first additions of cobalt increase the cell size to a maximum at approximately 20% of cobalt and further additions result in a contraction in the cell size to the limit of the α-phase.

CORROSION OF IRON AND STEEL

(Continued from pp. 76 A–78 A)

Corrosion of Metals and Alloys by Flue Gases.—Appendix I. L. Shnidman and J. S. Yeaw. (Proceedings of the American Gas Association, Twenty-first Annual Convention, 1939, pp. 542–552). This paper constitutes an appendix to a report of the work of the Gas Conditioning Committee of the American Gas Association which is studying the advisability of removing organic sulphur from gaseous fuels. The authors describe an investigation of the rates of corrosion of various metals and alloys in flue gases containing up to 0·4 g. of sulphur per 100 cu. ft. They compare these rates with those obtained previously with gases containing higher concentrations of sulphur. The results showed that 18/8 stainless steel was far superior to any of the metals or alloys tested. The materials tested were, in decreasing order of resistance, 18/8 stainless steel, zinc, 16% chromium steel, a number of copper alloys, aluminium, lead and iron.

Report on Corrosion of Metals and Alloys by Flue Gases.—Appendix II. E. J. Murphy. (Proceedings of the American Gas Association, Twenty-first Annual Convention, 1939, pp. 553–556). The author reports on an investigation of the comparative rates of corrosion of various metals and alloys in flue gases containing, and free from, sulphur compounds. For the sulphur-free gas a synthetic mixture containing approximately 5% of carbon dioxide, 12·5% of oxygen and 82·5% of nitrogen was used. In both tests samples of stainless steel were unaffected by the gases. In the synthetic mixture galvanised iron was corroded to the greatest extent, followed in order by Wilder metal, copper, and lead-coated copper. In the sulphur-bearing gases Wilder metal suffered the greatest corrosion, followed by galvanised iron, copper and lead-coated copper. It was evident that the presence of sulphur markedly increased the rate of corrosion.

The Results of Some Corrosion Tests of Metals and Alloys in the By-Product Coke Industry. O. B. J. Fraser and G. L. Cox. (Proceedings of the American Gas Association, Twenty-first Annual Convention, 1939, pp. 605–614). The authors present and discuss the results of a number of corrosion tests in processes associated with the coke-oven by-product industry. A general consideration of the data obtained indicates that Inconel and stainless steels of the chromium-nickel and chromium-nickel-molybdenum types are suitable for equipment for the distillation of crude benzol and coal tar, for resisting the attack of ammonium thiocyanate, and for processes in which sulphur dioxide is present.

The Acid-Resistivity of Iron Alloys (5th Report). The Acid-Resistivity of Fe-Cr-Co and Fe-Cr-Cu Alloys. T. Murakami and T. Sato. (Nippon Kinzoku Gakkai-Si, 1940, vol. 4, Mar., pp. 65–68). (In Japanese). The authors discuss the results obtained in a series of corrosion tests in 10% aqueous solutions of nitric, hydrochloric and sulphuric acids at 25° C. on the following groups of alloys: (1) Iron-chromium-cobalt alloys containing 12% and 18·5% of chromium with cobalt in the range 4·5–50·0%; (2) a ternary iron alloy containing chromium 20·93%, cobalt 50·48% and carbon 0·06%; and (3) a number of iron-chromium-copper alloys containing 6%, 13%, 18% and 24% of chromium with copper varying in the range 0·27–2·06%.

The Acid-Resistivity of Iron Alloys (6th Report). The Acid-Resistivity of Fe-Cr-Mn Alloys. T. Murakami and T. Sato. (Nippon Kinzoku Gakkai-Si, 1940, vol. 4, June, pp. 160–162). (In Japanese). The authors present the results obtained in their investigation of the rates of corrosion in nitric, hydrochloric and sulphuric acids of a series of iron-chromium-manganese alloys containing 6%, 14%, 18·5% and 26% of chromium with manganese in the range 1·82–12·45% and carbon in the range 0·03–0·25%.

ORES—MINING AND TREATMENT

(Continued from p. 39 A)

The Preparation of Ironstone. J. B. Bannister and G. D. Elliot. (Lincolnshire Iron and Steel Institute, Nov. 19, 1940). The authors describe in detail the quarrying and preparation of the Lincolnshire and Northamptonshire ores smelted in the blast-furnaces of the Appleby-Frodingham Steel Co., Ltd., with particular reference to blending by the Robins-Messiter system of bedding, and to sintering in Greenawalt and Dwight-Lloyd plants.

Sink-and-Float Separation Applied Successfully on the Mesabi. G. J. Holt. (Engineering and Mining Journal, 1940, vol. 141, Sept., pp. 33–38). The author describes and illustrates the ore-concentration plant at Cooley, Minnesota, where ores from the Mesabi range are treated. The plant operates on the Butler Brothers' system. The principal feature of this is the double-cone separator in which the medium is a suspension of crushed ferro-silicon (iron 83%, silicon 14–15%) in water. This suspension has a specific gravity of 6·7–7·0. The medium is cleaned by passing it through a direct-current magnetising coil which magnetises the ferro-silicon particles so that they adhere to each other to form larger particles, a factor which increases the rate of settling. The plant described has a capacity of about 350 tons per hr.

Equipment for the Purification of Effluent from Concentration Plants. S. Mörtsell. (Jernkontorets Annaler, 1940, vol. 124, No. 8, pp. 393–458). (In Swedish). The author discusses the theory and practice of the purification of the water used in ore-concentration plants. After reviewing the nature and properties of the dissolved and undissolved impurities, the author devotes the major part of his paper to the treatment of slimes, quoting the results obtained by many, particularly American, investigators as well as some of his own work. In the latter part of the paper he describes and discusses plant and equipment for the thickening, centrifuging and filtering of slimes.

FUEL

(Continued from pp. 88 A–89 A)

Steel Mill Boiler Plants. J. W. Vicary. (Steel, 1940, vol. 107, Dec. 16, pp. 64–66). The author gives a brief illustrated description of the high-pressure boiler plant of the Weirton Steel Co., in West Virginia, which was placed in operation in 1936 and which, on account of the very satisfactory results obtained, will be extended in the near future.

Power Generation Expansion at Bethlehem's Lebanon Plant. H. U. Johns. (Iron and Steel Engineer, 1940, vol., 17, Dec., pp. 52–57). The author presents data relating to the size, capacity and operating costs of the boilers of the power station at the Lebanon plant of the Bethlehem Steel Company, and describes a steam-generating unit that has recently been added. This unit consists of a three-drum bent-tube boiler with a heating surface of 12,500 sq. ft. designed for a maximum working pressure of 725 lb. per sq. in.

The Calorific Value of Carbon in Coal : The Dulong Relationship. R. A. Mott and C. E. Spooner. (Fuel in Science and Practice, 1940, vol., 19, Nov., pp. 226–231 ; Dec., pp. 242–249). The authors point out that in all considerations of the thermal reactions in a boiler, gas producer or blast-furnace the heat produced in the reaction :

$$C + O_2 = CO_2 + Q \text{ calories}$$

is of major importance, but that the values for Q used by different authorities vary. The range of values used leads to many misconceptions and errors, and in the present paper the authors show how values for the calorific value of carbon in coal or coke can be selected to provide a more accurate basis for theoretical considerations of the combustion of coal and coke. By analysing the United States Bureau of Mines data for the calorific value of laboratory cokes made at different temperatures from a series of twenty-eight coals varying considerably in rank, and applying suitable corrections, the authors show that the calorific value of carbon in bituminous coals does not vary and may be taken as 8000 ± 10 cal. per g.

Barvoys Washery at a Midland Colliery. (Iron and Coal Trades Review, 1941, vol. 142, Jan. 10, pp. 31–32). An illustrated description is given of a Barvoys coal-washing plant recently put in operation at an English colliery. The coal in question contains a hard band of low specific gravity and dull appearance about 1 in. thick. The plant has a through-put of 100 tons per hr., and the scheme of operation provides for three-product separation of the raw coal in a single Barvoys bath, filled with a medium of 1·30 sp. gr., and provided with middlings extraction equipment. The heavy medium for the separation consists of a mixture of clay, barytes and water, and the system provides for the circulation of liquids comprising heavy medium, diluted medium, clarified water and fresh water for the final spraying of the washed products.

Washed Coal as Applied to Metallurgical Practice. I. M. McLennan. (Bureau of Steel Manufacturers of Australia, 1940, vol. 1, Sept., pp. 3–49). After a general review of the purpose of coal washing and of the principles on which ash removal is based, the author briefly describes the development of the Chance, Barvoys, Tromp and Staatsmijnen-Loess processes. From this general consideration of the subject he turns to the particular problem of cleaning Australian coals and describes some tests made in 1939 with an experimental plant to ascertain whether the regular coking coals used by the Broken Hill Proprietary Co. were amenable to cleaning by dense media methods. It was found that a suitable stable separating medium could be prepared from blast-furnace flue dust which is removed from the gas in a dry cyclone-type collector. It is a very cheap medium as it is virtually an unused by-product of the Australian iron smelting industry. This flue dust was used without any pretreatment, it being simply made into a slurry with water, and it was also unnecessary to add any clay, as the bath soon picked up sufficient clay material from the raw coal. The author is of the opinion that this medium would be equally successful in a large-scale commercial installation. He presents data and discusses the results obtained in these experiments, and in conclusion describes the Barvoys, Tromp and Staatsmijnen-Loess processes in detail.

PRODUCTION OF IRON

(Continued from pp. 89 A–90 A)

Recent European Developments in Pig-Iron Manufacture. N. L. Evans. (Institute of British Foundrymen: Foundry Trade Journal, 1941, vol. 64, Jan. 9, pp. 19–22). The author describes briefly the experiments which were made both in England and on the Continent during the last ten years in order to develop the manufacture of pig iron from lean ores. Most of this work relates to desulphurising the iron with sodium carbonate after it leaves the blast-furnace. The author mentions the application of desulphurisation in the recently revived Armstrong Whitworth process for the production of high-carbon iron from remelted scrap, this iron being intended as a substitute for pig iron in the manufacture of steel. In discussing ladle linings, the author points out that the efficiency of sodium carbonate as a desulphurising reagent is impaired when it is contaminated with silica, and that such contamination invariably occurs when an acid-lined vessel is used, or when a siliceous furnace slag is allowed to enter the ladle. The author is at present investigating tar-dolomite linings for ladles, and has already succeeded in reducing the silica content of the soda slag, thus bringing about a marked improvement in the degree of desulphurisation of the iron.

Use of Sinter in Blast-Furnace Burdens. J. H. Slater. (American Institute of Mining and Metallurgical Engineers, Technical Publication No. 1263: Metals Technology, 1940, vol. 7, Dec.). The author describes some tests of the benefits accruing from adding sinter to the charge of blast-furnaces at the works of the Republic Steel Corporation. The sinter was prepared from a mixture of

40% blast-furnace flue dust and 60% Mesabi ore fines under ⅜ in., and it had the following approximate analysis: iron 57·80%, silica 12·10%, phosphorus 0·096%, manganese 0·98%, alumina 2·01%, lime 2·08% and magnesia 0·20%. Two furnaces of identical dimensions producing about 550 tons of iron per day were used and three tests each of 30 days' duration were run. In two of the tests one furnace was run without sinter and in the third with 35% of sinter in the burden. In the two tests using sinter there was a considerable increase in the amount of iron produced and a reduction in the coke consumption per ton of iron. An additional benefit was the greater smoothness of furnace operation with an improved uniformity in the quality of the iron. Further tests have shown that increasing the proportion of sinter in the burden to more than 50% is not advantageous, because the charge becomes so open that intimate gas-solid contact is not attained.

Limitations of Powder Metallurgy. E. S. Patch. (Iron Age, 1940, vol. 146, Dec. 19, pp. 31–34). The author considers the present position in the development of the manufacturing process known as powder metallurgy. He deals particularly with the production of gears and then discusses the following factors which affect the continued development of the process: (1) The metal powder does not flow freely so that pieces with re-entrant angles cannot be formed; (2) tolerances in length must be slightly more liberal than those necessary for conventional methods of finishing; (3) a tolerance for concentricity of about 0·003 in. must be allowed; (4) a part can only be made profitably by powder metallurgy when, in producing it by machining a casting, the cost of machining is high in proportion to the cost of the material; and (5) there must be a market for a sufficient number of parts to justify the high cost of the special dies and tools.

FOUNDRY PRACTICE

(Continued from pp. 90 A–91 A)

The Foundryman and the Metallurgist. F. Dunleavy. (Institute of British Foundrymen: Foundry Trade Journal, 1941, vol. 64, Jan. 9, pp. 24–26; Jan. 16, p. 41). The author discusses the need for co-operation between the foundryman and the metallurgist and cites a number of cases in which a metallurgist has been able to detect the causes of changes in the behaviour of a cupola when there was no apparent change in the cupola practice.

Some Recent Developments in Iron and Steel Castings. R. C. Good. (Pittsburgh Foundrymen's Association: Foundry, 1940, vol. 68, Oct., pp. 46–48, 110–112; Nov., pp. 42–44, 103–105). The author reviews recent developments in foundry technique for the improvement in quality of grey iron and steel castings. He reproduces micrographs which show that by using a selected charge of steel, cast iron and briquetted alloys, a fair degree of strength and ductility can be obtained in the iron produced. A reduction of the depth of chill and a softening of light sections can be accomplished by adding a graphitiser composed essentially of ferro-silicon and graphite. To prevent the formation of soft spots composed of eutectic ferrite-graphite an addition of 1% of a chromium-manganese-silicon-zirconium alloy has been found to be effective. The production of a high-duty cast iron without resorting to the addition of substantial amounts of alloys or to heat treatment is not difficult if a charge low in silicon and low in total carbon is melted and deoxidised in the ladle, but the alloy added to the ladle must be balanced, *i.e.*, its composition must be such that it has the correct degrees of fusibility and solubility. Turning to steel castings, the author considers the influence of alloying elements on the solubility of gases in steel. He also presents a series of diagrams which show how adding increasing quantities of deoxidisers affects the size and distribution of the inclusions. In conclusion he summarises the influence of additions of calcium alloys, manganese, zirconium, vanadium, silicon and aluminium on the properties of cast steel.

Marks Progress in Melting Gray Iron. D. J. Reese. (Foundry, 1940, vol. 68, Nov., pp. 45, 98–102). The author discusses recent advancements in cupola melting practice at a grey iron foundry, dealing in turn with the duplex process, accuracy in weighing, size of coke, charging steel turnings, methods of analysis, frequency of tapping, and preheating and drying the blast.

The Production of High-Silicon Acid-Resisting Castings. F. Marsden. (Foundry Trade Journal, 1941, vol. 64, Jan. 23, pp. 51, 64). The author states that, owing to their friability, high-silicon acid-resisting cast irons have to be cast under special precautions, and he gives, in brief paragraphs, some information on the following processes which have to be adapted specially to the casting of the above alloys, *viz.*, melting, moulding, core making, casting operation, stripping the casting after teeming, dressing and machining.

PRODUCTION OF STEEL

(Continued from pp. 91 A–93 A)

The Present and Future in Open-Hearth Steelmaking. J. O. Griggs. (Blast Furnace and Steel Plant, 1940, vol. 28, Sept., pp. 879–884; Oct., pp. 981–986, 995). The author discusses recent developments in open-hearth practice. The use of auxiliary slag pockets is recommended as this helps to keep the main pocket free of slag—a practice which has three outstanding advantages. First, by keeping the slag from coming in contact with the silica walls, the slag-pocket arches are not subjected to the cutting effect of the slag. Secondly, the average rebuilding time can be reduced by at least 48 hr., because the slag is easily removed without disturbing the brick-work in the main pocket. Thirdly, rebuilding costs will be lower because of the increased life of the main pocket. After comparing the merits of fixed and tilting open-hearth furnaces, the author points out that, although the cost of constructing the former is only about 40% of that of the latter, the latter offer many advantages in operation and are more flexible in the sense that they can easily be switched over from working on charges high in scrap to " all pig " charges. The author gives some examples of simple methods of determining the correct composition of a charge to meet certain requirements, and gives a table showing the amount of aluminium per ton of metal which should be added in the ladle to correct under-deoxidised steels containing different percentages of iron oxide. The design of ingot moulds, furnace ports, doors and auxiliary equipment is also discussed.

Transformers for Electric Furnaces. (Engineering, 1940, vol. 150, Dec. 27, pp. 504–505). The design and auxiliary equipment of modern transformers for electric steel furnaces and heat-treatment furnaces are discussed and illustrated.

Coreless Induction Furnace Melting Losses. (Iron and Coal Trades Review, 1941, vol. 142, Jan. 10, pp. 25–26; Foundry Trade Journal, 1941, vol. 64, Jan. 23, pp. 59, 66). An abridged English translation is presented of Weitzer's paper on the losses due to oxidation of the elements carbon, manganese, chromium, nickel, cobalt, tungsten, molybdenum, tantalum, niobium, zirconium and titanium in the manufacture of alloy steels in acid coreless induction furnaces. This paper appeared in Stahl und Eisen, 1939, vol. 59, Dec. 21, pp. 1353–1356. (*See* Journ. I. and S.I., 1940, No. I., p. 184 A).

Study of the Direct Steel Manufacture from Iron Sand. M. Sano. (Tetsu to Hagane, 1940, vol. 26, Sept. 25, pp. 685–688). (In Japanese). The author describes a direct method of manufacturing vanadium steel from a low-phosphorus vanadium-bearing iron sand.

On Steel Made from Sponge Iron and Loop. I. Taniyama. (Tetsu to Hagane, 1940, vol. 26, July 25, pp. 511–520). (In Japanese). The author discusses methods of producing steel from sponge iron and from iron blooms (loop), and the reasons for the difference in the qualities of steel produced from these raw materials as compared with that produced from a charge containing steel scrap.

FORGING, STAMPING AND DRAWING

(Continued from pp. 94 A–95 A)

The Maintenance of Hard Metal Drawing Dies. (Wire Industry, 1940, vol. 7, Dec., pp. 529–530). The use of wet and dry lubricants for wire-drawing and the cooling and polishing of wire-drawing dies are discussed. It is recommended that die-polishing machines which can be set to produce the desired contour be employed, as the most skilful polisher cannot accurately produce the correct shape by hand.

Reactive Wire Drawing. H. A. Stringfellow. (Wire and Wire Products, 1940, vol. 15, Oct., pp. 527–538, 635). The author develops formulæ and explains methods for the calculation of the work done in reducing wire when a back pull is applied to the wire entering the die.

Lime for Wire Drawing. D. E. Washburn. (Wire and Wire Products, 1940, vol. 15, Oct., pp. 575–577). The author discusses the different types of lime used as a lubricant in wire-drawing. He shows that the chemical composition of a lime has little effect, except for the magnesia content, on the type of coating produced. In theory, magnesia has a higher neutralising value on any acid carried over from the pickling bath than has calcium oxide, and the preference for lime containing no dolomite for wire-drawing is due rather to its physical characteristics than to its chemical composition.

ROLLING-MILL PRACTICE

(Continued from pp. 95 A–96 A)

Bakelised Bearings for Rolling Mills. S. Uchikawa. (Tetsu to Hagane, 1940, vol. 26, July 25, pp. 526–536). (In Japanese). The author discusses some experience gained at a Japanese steelworks in the use of roll-neck bearings of a synthetic resin material known as "Nittelite." Great advantages are claimed for this material as compared with bearings of babbit metal and other alloys. The setting-up and the lubrication of Nittelite bearings are described.

Controlling the Temper of Tinplate. T. B. Montgomery. (Blast Furnace and Steel Plant, 1940, vol. 28, Oct., pp. 991–995). The author describes the electrical system of control for regulating the roll pressures and tension when cold-rolling steel strip in order to impart to it the necessary temper and finish for tinplating.

Ward-Leonard Control for Strip Mill Auxiliary Drives. E. S. Murrah and H. W. Poole. (Iron and Steel Engineer, 1940, vol. 17, Dec., pp. 35–46). The authors discuss the advantages and limitations of Ward-Leonard control for motors driving roll trains and other auxiliary equipment for strip mills, and consider some actual examples of such installations.

HEAT TREATMENT

(Continued from pp. 96 A–99 A)

Heat-Treating Troubles and Their Correction. R. B. Seger. (Machinist, 1940, vol. 84, Dec. 21, pp. 837–839). The author presents a table in which are enumerated side by side the causes and the methods of correcting the more common defects occurring during heat treatment. The defects are classified as follows: (1) Cracking; (2) warpage; (3) soft spots; (4) change in size; and (5) spalling.

Flame-Hardening. R. H. Zeilman. (Welding Journal, 1940, vol. 19, Oct., pp. 746–749). The author describes the application of flame-hardening to many of the parts used in the construction of mechanical excavating machinery, particularly to pinions and dog-clutch faces.

Flame-Hardening Cast Iron Bearing Rings. J. L. Foster. (Iron Age, 1940, vol. 146, Dec. 19, pp. 44–45). The author describes the equipment developed for flame-hardening the faces of cast-iron thrust rings for bearings and the structure of the metal in the hardened area.

Flame and Induction Hardening. (Automobile Engineer, 1940, vol. 30, Dec., pp. 398–400). Descriptions are given of the latest developments in the Shorter process of flame-hardening and the Tocco process of induction-hardening (*see* Journ. I. and S.I., 1940, No. II., p. 148 A and this volume, p. 54 A). Data are also presented concerning the types of steel to which these treatments can be applied.

The Production Flame-Hardening of Machine Parts. J. Erler and P. H. Tomlinson. (Welding Journal, 1940, vol. 19, Oct., pp. 705–709). The authors discuss some of the advantages to be gained by flame-hardening forged, rolled and cast steel and Meehanite castings, and point out certain precautions which should be taken in each case. The following ways of either reducing or entirely eliminating distortion are mentioned: (1) Keeping the part cool by submerging the greater portion of it in water, or by spraying water on certain sections; (2) machining the part in such a way that, after hardening, the distortion will pull it into the desired shape; (3) bending the part in the opposite direction to the curve of distortion; and (4) increasing the mass in the sections to be flame-hardened so as to give greater stiffness, and, after hardening, removing the extra material from the back or soft side. The authors found that the best operators for flame-hardening machines were young men who had worked in a laboratory doing routine analyses for a few years, for this training made them conscious of changes which might occur as a result of minute variations in operation or procedure.

Recent Heat-Treatment Furnace Installations. (Metallurgia, 1940, vol. 23, Dec., pp. 57–63). An illustrated review is given of recent heat-treatment furnace installations, and particular consideration is given to the production and control of protective atmospheres. Among the furnaces and installations considered are a continuous sheet normalising furnace fired by clean gas on the Wellman-Chantraine principle; a bright-annealing furnace comprising four hearths and one gas-fired furnace, and another bright-annealing furnace for wire and strip in coil form; an electrically heated pit-type furnace for aero-engine parts; an installation for the heat treatment of high-speed steel drills and reamers; an installation for the continuous scale-free hardening and tempering of bolts; a vertical furnace for the general hardening of automobile parts; and a duplex nitriding furnace.

Electric Furnaces. (Automobile Engineer, 1940, vol. 30, Dec., pp. 390–394). Descriptions and numerous illustrations are given of some modern types of batch and continuous electric furnaces for the heat treatment of automobile parts and high-speed steel tools.

Concentrator of Eddy Currents for Zonal Heating of Steel Parts. G. Babat and M. Losinsky. (Journal of Applied Physics, 1940, vol. 11, Dec., pp. 816–823). The authors have developed a device called a concentrator of eddy currents, which makes it possible for the magnetic field of a multi-turn coil to be transformed and directed so that the magnetic flux is concentrated on that portion of a steel part which is to be heat-treated. The concentrator consists of a massive copper bushing with a slot. Eddy currents flow on the surface facing the multi-turn coil on the slot sides, and close themselves on the crest which faces the piece to be treated. Since this crest is considerably less in height than the surface facing the heater coil, the density of the eddy currents will be greatest on the crest, and the magnetic field of greatest intensity will therefore be in the interstice between the crest of the concentrator and the piece.

Electric Patenting, Tempering and Annealing of Steel Wire. J. P. Zur. (Wire and Wire Products, 1940, vol. 15, Oct., pp. 582–586). After discussing some disadvantages of the conventional methods of patenting, tempering and annealing steel wire and strip, the author describes a new technique of electric direct-resistance continuous heat treatment in which the electric current is applied directly to the moving strand of wire or strip and the material is brought up to the desired temperature by virtue of its own electrical resistance. The advantages claimed for this method are: (1) The temperature is practically uniform throughout the cross-section of the wire; (2) the temperature can be easily controlled by varying the voltage; (3) the material attains the desired temperature in a few seconds; (4) the wire or strip can be passed through the apparatus at much higher speeds than through a conventional type of furnace; and (5) the heat losses are exceedingly small, an efficiency of over 90% being claimed.

Hardening Small Parts Uniformly. R. Trautschold. (Steel, 1940, vol. 107, Dec. 23, pp. 56–59). The author describes the equipment of a heat-treatment shop which is designed for the hardening of large quantities of small parts such as the components of roller chains. The principal feature of this equipment consists of a number of gas-fired rotating retorts into which the parts are charged in batches. This form of furnace ensures the uniform heating of all parts and has proved to be very efficient, as there are no massive boxes or trays which, in other furnaces, absorb a large proportion of the heat.

Steel Hardening. J. L. Burns. (Iron Age, 1940, vol. 146, Dec. 12, pp. 55–60). The author presents a series of curves for predicting the hardenability of a steel bar of given diameter when its composition is known. In these curves quantitative relationships are established by plotting the area below the hardness-penetration curve against a factor calculated from the chemical composition. Three groups of curves are given for fine-grained steels, one each for low-, medium- and high-carbon steels, and additional curves are reproduced which show what must be the composition of an alloy steel bar of given diameter in order that it may be completely hardenable throughout its section.

WELDING AND CUTTING

(Continued from pp. 99 A–102 A)

Changes in the Shape of Spherical Spot-Welding Electrodes. W. F. Hess and R. A. Wyant. (Welding Journal, 1940, vol. 19, Oct., pp. 345-S–350-S). The authors report on an investigation of the influence of the shape of the electrode, the mechanical pressure, the area of contact, the current density and some other factors on the properties of spot welds in steel sheet 0·036 in. thick. The conclusions reached are as follows: (1) Flat areas begin to form on dome-shaped copper electrodes as soon as welding is commenced; after about 200 welds the diameters of these flat areas increase slowly with the number of welds made, and, under proper conditions, it is probable that this diameter reaches a limit at which it remains constant. (2) It is advisable to degrease the sheets before welding, because the welding of oily sheets causes a black deposit to form on the electrodes which is detrimental to the quality of the weld. (3) Higher mechanical pressure can be applied to dome-shaped tips than to flat tips without extruding metal between the sheets. (4) When molten metal is extruded from welds made with dome-shaped tips, large cavities are left irrespective of the pressure applied. (5) The current density increases much more rapidly with increasing mechanical pressure than with increasing diameter of weld. (6) The mechanical pressure has a more important influence on the size of the weld than any other factor.

The Arc Torch—A Widely Adaptable Heat Source. F. W. Scott. (Steel, 1940, vol. 107, Dec. 23, pp. 42–46). The author describes a hand-operated form of torch for arc welding. This tool has two arms which carry a pair of carbon electrodes, the tips of which are maintained sufficiently close together for a continuous arc to be maintained. The weld metal is added from a welding rod as in oxy-acetylene welding. With this torch there is no gas pressure behind the flame to force the molten metal away from the point of impact. This technique is recommended for welding cast iron and non-ferrous metals, and for brazing, but not for welding steel.

Oxy-Acetylene Welding of Stainless Steels. L. Sanderson. (Metallurgia, 1940, vol. 23, Dec., pp. 55–56). The author points out that the growing employment of stainless steels in aircraft, and particularly in the modern seaplane, has involved a considerable use of welding, and he states that oxy-acetylene welding is quite suitable for welding these steels. It is emphasised, however, that the welding technique must be adapted to the special properties of these steels, and information is given regarding the most favourable temperatures of welding, the oxygen/acetylene ratio, the welding rods, and annealing after welding. He also considers briefly the technique to be employed in welding stainless steel sheets of various thicknesses, and castings of different design. In conclusion he points out some special precautions to be taken in the welding of 12–16% chromium, 17–19% chromium and 18/8 nickel-chromium steels.

Methods for Reclaiming Spindles, Crabs and Coupling Boxes. D. B. Rice. (Blast Furnace and Steel Plant, 1940, vol. 28, Oct., pp. 979–980). The author describes the welding procedure adopted at an American rolling mill for building up the worn surfaces of wobblers and coupling boxes. When the worn parts consist of 0·45–0·55% carbon steel, the author recommends preheating the part to 400–500° F., then building it up to within $\frac{1}{2}$ in. of the required contour with carbon-steel electrodes of similar analysis and completing the last $\frac{1}{2}$-in. layer with electrodes containing 3·5–4% of nickel, 14% of manganese and 0·75–0·90% of carbon. If the material to be repaired is a 12–14% manganese steel, the preheating temperature should not exceed 300° F. and the whole of the deposit should be made with nickel-manganese steel electrodes.

Resistance Flash Welding of Strip in Steel Mills. J. H. Cooper. (Welding Journal, 1940, vol. 19, Oct., pp. 721–729). The author describes the advantages of using flash-butt welding machines for welding together the ends of the coils when rolling steel strip. He illustrates a number of these machines and other machines for trimming the weld, and, in conclusion, discusses the metallurgical characteristics and mechanical properties of the joints.

Welding and Flame Cutting Wrought Iron. H. Lawrence. (Steel, 1940, vol. 107, Dec. 9, pp. 84–87). The author describes some of the precautions necessary to produce sound welds in wrought iron by the oxy-acetylene and electric-arc methods. In order to work the silicates out of the iron it is necessary to maintain a quiescent pool of liquid weld metal; this can be done by welding at a slower rate than with steel, thus allowing time for the slag to float to the top. For heavy sections it is good practice to anneal the welds and this can be done at 700–800° F., as compared with 1100–1250° F. for steel.

Flame-Gouging Welds. H. Lawrence. (Steel, 1940, vol. 107, Dec. 16, pp. 54–56, 88). The author enumerates the advantages of flame-gouging for the removal of weld metal and recommends the process as a substitute for chipping. He briefly describes the technique of flame-gouging and gives some practical advice to operators.

Cutting Steel with Oxy-Propane. W. T. Tiffin. (Iron Age, 1940, vol. 146, Dec. 12, pp. 61–64). The author describes tests of the cutting speeds and gas consumptions in the cutting of steel plate with an oxy-propane torch. He presents data showing the results obtained and the cost per foot under American conditions for cutting plate of various thicknesses from ¼ in. to 2½ in.

MACHINING

(Continued from p. 61 A)

On the Study of Cutting Efficiency and the Form of Cutting Edge of High-Speed Steel Tools. I. T. Kikuta and S. Koshiba. (Nippon Kinzoku Gakkai-Si, 1940, vol. 4, July, pp. 203–209). (In Japanese). The authors carried out cutting tests with three kinds of tool steel on an annealed chromium steel, an oil-quenched and tempered nickel-chromium steel and on an austenitic stainless steel in order to determine how various factors affected the cutting efficiency. The three tool steels used were of the following analyses :

	(1).	(2).	(3).
Carbon. %	0·72	0·75	0·83
Chromium. % . . .	4·23	4·20	4·22
Tungsten. % . . .	17·76	19·47	20·49
Vanadium. % . . .	0·94	1·68	2·24
Cobalt. %	...	5·70	11·60

Their conclusions are as follows: (1) The cutting efficiency increases with the tool size, provided that each size is properly hardened; (2) changes in the cutting speed have almost no effect on the force exerted on the edge of the tool; (3) the cutting force increases directly with increased depth of cut; (4) the cutting efficiency decreases rapidly with increasing width of cut; and (5) under the conditions of this investigation, decreasing the feed on the same width of cut increases the cutting efficiency.

On the Study of Cutting Efficiency and the Form of Cutting Edge of High-Speed Steel Tools. II. T. Kikuta and S. Koshiba. (Nippon Kinzoku Gakkai-Si, 1940, vol. 4, Aug., pp. 262–267). (In Japanese). The authors report on their investigation of the effect of tool contour on the cutting efficiency of two types of high-speed steel when machining an annealed chromium steel, a quenched and tempered nickel-chromium steel and an austenitic stainless steel. The tool steels tested were of the 18/4/1 tungsten-chromium vanadium and 5–6% cobalt types. They found that the highest cutting efficiency was attained with a cutting angle of 60° for machining the chromium steel, 70° for the nickel-chromium steel and 50° for the stainless steel.

Carbides for Cutting Steel. P. M. McKenna. (Machinist, 1941, vol. 84, Jan. 18, pp. 917–919). The author discusses the design of tools tipped with a double carbide of titanium and tungsten for various high-speed machining operations on steel. He deals in particular with shapes designed to break up turnings into short coils.

Efficient Control and Disposal of Swarf and Scrap Metal. D. F. Galloway. (Journal of the Institution of Production Engineers, 1940, vol. 19, Dec., pp. 476–515). The author describes methods and equipment for the economic control, disposal and recovery of scrap metal in the form of swarf accruing from various machining operations. Chip control, de-swarfing of machines, recovery of cutting oil, and the breaking, sorting, sintering, baling and magnetic handling of swarf are among the processes mentioned. Special attention is given to the efficient recovery of aluminium scrap.

PROPERTIES AND TESTS

(Continued from pp. 102 A–112 A)

Studies on the Cold-Drawn Steel Bars (Report 1): An X-Ray Study of Residual Strains in Cold-Drawn Steel Bars. K. Takase and T. Watari. (Tetsu to Hagane, 1940, vol. 26, Oct. 25, pp. 737–744). (In Japanese). The authors report on an X-ray investigation, by the back-reflection method, of the causes of the changes in the mechanical properties of cold-drawn steel bars brought about by the drawing process.

Cost and Procedure Control by Use of Polarized Light. E. W. P. Smith. (Welding Journal, 1940, vol. 19, Oct., pp. 733–737). The author explains the fundamental principles of polarised light and its application for studying stress distribution. He shows how, by the study of models of welded joints in polarised light, the parts where no stresses exist are revealed to the designer, who can thus effect important economies in welding by specifying the correct thicknesses of the weld metal to be deposited.

On Crack Formation in Steam Boiler Material. G. Wållgren. (Teknisk Tidskrift, 1940, vol. 70, Oct. 19, Mekanik, pp. 109–113). (In Swedish). The author discusses the causes of the formation of cracks in boiler steel with particular reference to the investigation of the failure of the shell plate of the upper dome of a marine boiler. The author differentiates between fatigue cracks without corrosion, intercrystalline cracks caused by caustic embrittlement, and corrosion-fatigue cracks. Specimens of the boiler plate taken from positions close to, and away from, the cracks and etched by Fry's method revealed slip lines at angles of about 45° to the surface of the plate. The cracks, which were examined under high magnification, were, however, found to be at an angle of about 90° to the surface. The cracks passed partly along grain boundaries and partly through the grains. In discussing whether stress is the primary cause of crack formation in boiler plate, the author states that this is the case only when the slip lines are so marked as to have been caused by stresses up to the elastic limit of the steel. He is of the opinion that corrosion-fatigue cracks can also form and grow in regions where there are no slip lines but where there are high stresses less than the elastic limit. A steel can be cold-worked without any slip lines appearing and such a region is very sensitive to corrosion. Another point which supports the theory that slip lines are not the primary cause of corrosion cracks is the fact that those parts of a boiler which are cold-worked during erection to the extent that slip lines occur, but which are not heavily stressed when the boiler is at pressure, have no tendency to the formation of corrosion cracks. The author's general conclusion is that stresses play the most important part in the formation of corrosion cracks in boiler steel so that their prevention is a question of design.

Fatigue Tests of Structural Steel Flats and Bars Cut from Butt-Welded Boiler Plate. F. C. Lea and J. G. Whitman. (Proceedings of the Institution of Mechanical Engineers, 1941, vol. 144, Jan., pp. 132–139). The authors describe an investigation of the fatigue strength of butt-welded flats and bars cut from steel boiler plate, the objects of which were: (*a*) To discover whether the resistance of the welded plates to repeated stresses depends on the size of the plate, or, in other words, to see if there is any " scale effect "; and (*b*) to compare the results obtained from welded plates with those from plates with holes drilled in them. The results demonstrated that for specimens from 1 in. to 3 in. in thickness prepared in three different works no scale effect was indicated. The tests made on drilled plates confirmed the results of earlier work and showed that drilling a hole in a plate reduces the fatigue-range load very considerably. It was also found from tests on thick welded plate that, with properly controlled welding, fatigue ranges for the welds as high as those for the black plate can be obtained. From these results the authors conclude that there seems to be no reason, from the point of view of fatigue, why the working stresses in the welds of welded pressure vessels should not be as high as those at present used in riveted vessels, as well-made welded joints may be assumed to have an efficiency of 100%.

Shot-Blasting and Its Effect on Fatigue Life. F. P. Zimmerli. (American Society for Metals, Oct., 1940, Preprint No. 51). The author discusses a method of surface finishing called shot-blasting. This is a further development of cloud-burst hardening and consists of propelling small steel balls against the surface of the object to be treated. This is commercially accomplished by two types of machines. In the older of these air is used as the propelling agent, as in the common sand-blast equipment; in the more recent type of machine the shot is thrown, by centrifugal force, from a rapidly rotating wheel having radial blades. The author reports on fatigue tests carried out on small springs which had been subjected to shot-blasting and describes at some length the fatigue-testing machine designed for this purpose. Data obtained for springs of a variety of compositions, in particular for valve spring wire, indicated increases in fatigue values exceeding 40% for specimens subjected to shot-blasting. The beneficial effects of shot-blasting are, however, eliminated by excessive heating.

Wear-Resistant Coatings of Diesel Cylinder Liners. J. E. Jackson. (S.A.E. Journal, 1941, vol. 48, Jan., pp. 28–32). The author points out that newly assembled Diesel engines require running-in before being placed into service, and that the problem of " scuffing " during the run-in may be largely overcome by treating the honed liners before assembly with a concentrated aqueous solution of sodium hydroxide in the presence of a small amount of sulphur. This treatment, called by the author a " controlled pitting process," ensures running-in without scoring or seizure owing to (1) the removal of undesirable components from the surface, (2) the deposition of certain reaction products on the surface, and (3) some change in the surface structure. The coating deposited consists of ferrous oxide and sulphide tightly adhering to the unetched underlying iron. It causes the surface to become matt, so that the spreading of lubricating oil is facilitated and some of it can be retained in the porous inner structure of the coating. Several micrographs of the coating, before and after running in, are reproduced, and also some showing the honed and run-in surfaces of the untreated liner.

On the Magnetic Hysteresis Loop by Alternating Current. Y. Ishihara. (Nippon Kinzoku Gakkai-Si, 1940, vol. 4, Aug., pp. 228–239). (In Japanese). The author describes his investigation of the hysteresis of annealed and water-quenched steels of different carbon content, using an apparatus of his own design.

The Adiabatic Temperature Changes Accompanying the Magnetization of Ferromagnetic Materials in Low and Moderate Fields. L. F. Bates and J. C. Weston. (Proceedings of the Physical Society, 1941, vol. 53, Jan., pp. 5–34). The authors have developed a new and relatively simple method for the measurement of the heat changes which accompany magnetisation processes in fields of the order of a few hundred oersteds. They measure the adiabatic changes in the temperature of a rod which occur when it is taken step-by-step through a hysteresis cycle. The temperature-measuring system, *i.e.*, a series of thermocouples the "hot" junctions of which are directly attached to the rod, is capable of detecting changes in temperature of the order of 10^{-6}° C. Experiments are described which were made with pure annealed and hard-drawn nickel and with three nickel-iron alloys, strained and unstrained, containing 49%, 42% and 36% of nickel, respectively. The observed temperature changes, plotted as functions of the intensity of magnetisation and of the applied field, are represented in numerous graphs. The results, which provide a very effective proof of Warburg's law, are said to make possible a detailed analysis of the energy changes accompanying magnetisation. They indicate that Becker's views on the magnetisation processes in ordinary ferromagnetic materials are essentially correct, but they do not support Preisach's conception of the formation of demagnetisation nuclei in the case of extremely soft ferromagnetic materials.

Properties of Steels as a Basis for Design for High Temperature Service. Part I. Carbon Steels. H. J. Tapsell and A. E. Johnson. (Proceedings of the Institution of Mechanical Engineers, 1941, vol. 144, Jan., pp. 97–106). The authors discuss the influence of stress, temperature and time on the behaviour of 0·15–0·50% carbon steels. The data presented, which provide a basis for design purposes, are derived from investigations carried out at the National Physical Laboratory, and consist of the results of tensile and creep tests at temperatures up to 1000° F. In a final summary of the creep stresses of carbon steels in relation to design stresses, the authors consider that all good quality carbon steels in the range 0·15–0·5% carbon, in the hot-rolled or normalised conditions, may, for practical purposes, be considered to have similar creep properties between 400° and 510° C., but that, in view of their different hot-tensile properties, account should be taken of the limits of proportionality up to 426° or 455° C. If a steel is in a highly spheroidised state it is considerably weaker than when normalised or air-cooled after hot-rolling, and the use of such a steel should be avoided. Slight cold-work is not likely to alter the properties of a steel sufficiently to put it outside the range of steels of average quality, but heavy cold-work would render the steel subject to more rapid spheroidisation during service, especially in the higher working temperature range.

Alloys for Corrosive Services. (Chemical Age, 1941, vol. 44, Jan. 4, pp. 11–12). Some information is given respecting the properties of an austenitic cast iron known as "Audcoloy" which contains nickel, copper and silicon. This alloy has excellent corrosion resistance and is recommended for certain machinery parts in contact with sulphuric acid, hot or cold caustic soda and fatty acids.

Modern Steels to Combat High Temperatures. C. L. Clark. (Mining and Metallurgy, 1940, vol. 21, Nov., pp. 521–523). The author explains the function of alloying elements, separately and in combination, in heat-resisting steels. He discusses in particular the low-alloy steels containing chromium, molybdenum and silicon and their application in high-temperature boiler plant and oil refineries.

Development of Low-Alloyed High-Speed Steels. D. W. Rudorff. (Metallurgia, 1940, vol. 23, Dec., pp. 43–46). Basing his information on a report by H. A. Minkevitch and O. S. Ivanov (Metallurg, 1940, No. 1, pp. 31–46), the author reviews recent developments in high-speed steels in Russia, where the aim of the research has been not only to improve the service life of the tools, but, mainly, to modify their compositions so as to enable greater use to be made of the alloying elements readily available. In this way a number of high-speed steels have been developed by alloying high-chromium steels with relatively small amounts of tungsten, molybdenum and vanadium. The physical and mechanical properties of these steels, in dependence on their composition, are dealt with at some length.

Some Developments in Alloy Steel Production. J. H. G. Monypenny. (Metallurgia, 1940, vol. 23, Dec., pp. 37–39). The author considers the factors which have contributed to the development of the alloy-steel industry, referring, in particular, to the production of alloy steels which combine high strength with toughness and are relatively free from non-metallic inclusions. By a comparison of some specifications issued during the war period 1914–18 with more recent ones, he demonstrates the progress made in meeting the more exacting demands on alloy steel. He also directs attention to the striking development in the commercial use of stainless and heat-resisting steels, which he considers is due partly to their improved machinability. In conclusion he suggests a wider application of "lower"-chromium/"higher"-nickel steels, which would result in less low-carbon ferro-chromium being required in their production—a saving particularly desirable in war time.

Developments in Tool Steels. J. P. Gill. (Canadian Metals and Metallurgical Industries, 1940, vol. 3, Oct., pp. 252–256;

Nov., pp. 285–289). The author reviews the progress made in the last twenty years in the production and heat treatment of tool steels and discusses the properties and applications of many of the types of steel which are now available. Particular reference is made to a graphitic steel developed by Timken; this contains graphitic carbon 0·65–0·70%, combined carbon 0·80–0·85%, manganese 0·40% and silicon 0·80–0·90%. When the material is hardened the graphitic carbon remains in the graphitic state, so that the steel has something of the appearance of cast iron and is in some degree self-lubricating.

Study of Case-Hardening Steels Without Nickel. S. Ishida and S. Higashimura. (Tetsu to Hagane, 1940, vol. 26, July 25, pp. 521–525). (In Japanese). The authors compare the properties of case-hardening nickel-chromium steels with those of chromium-manganese and chromium-manganese-molybdenum steels. They find that the mechanical properties of the last two alloy steels are equal, or even superior to those of the nickel-chromium steels, and, as nickel has to be imported by Japan, they recommend the increased application of the former alloys for certain purposes.

Influence of Chemical Composition on the Hot-Working Properties and Surface Characteristics of Killed Steels. G. Soler. (American Institute of Mining and Metallurgical Engineers, Technical Publication No. 1262 : Metals Technology, 1940, vol. 7, Dec.). The author discusses the relation between the chemical composition of killed steels and (*a*) the structure of the steel as cast, and (*b*) the hot-working properties and surface characteristics of the steel. He gives examples of the effects of changes in the composition by reference to illustrations of internal and external defects in solid-drawn tubes and ingots of a number of alloy steels.

METALLOGRAPHY AND CONSTITUTION

(Continued from pp. 112 A–116 A)

The Scientific Method in Metallurgy. S. L. Hoyt. (Fifteenth Edward De Mille Campbell Memorial Lecture : Transactions of the American Society for Metals, 1940, vol. 28, Dec., pp. 757–796). The author reviews the succession of theories respecting the behaviour of elements and metals since the time of Aristotle, touching on the work of Galileo, Black, Guldberg and Waage, Nernst, Le Chatelier, Gibbs, Roozeboom, Roberts-Austen, Beilby and Rosenhain, Hume-Rothery, and Bragg. In tracing the advancement of knowledge of the properties and structure of metals, the author shows how theories carefully deduced from the results of scientific experiments have contributed far more to this advancement in the last fifty years than did reasoning and assumption without experiment in the 1500 years after Aristotle.

The Polishing of Cast-Iron Micro-Specimens and the Metallography of Graphite Flakes. H. Morrogh. (Iron and Steel Institute, 1941, this Journal, Section I.). A polishing technique has been developed for grey cast irons, whereby the graphite flakes can be obtained perfectly preserved and smoothly polished. The polishing medium used is either Diamantine or magnesium oxide. The method owes its success to the correct application of a repeated polishing and etching operation.

With specimens prepared in this manner, it is possible to examine the internal structure of the graphite flakes and temper-carbon nodules. Secondary graphite is shown to be deposited either on the existing eutectic graphite flakes or in a Widmanstätten pattern. Graphite flakes show complicated internal structures, which are illustrated micrographically using polarised light. " Inclusions " are also shown intimately associated with graphite flakes.

Electrolytic Polishing of Metals. (Metal Progress, 1940, vol. 38, Oct., p. 554). A table, compiled by G. E. Pellissier, H. Markus and R. F. Mehl, is reproduced which gives information on approved conditions regarding selection of solution, current density, voltage, temperature and time for the electrolytic polishing of all types of carbon steels, 3% silicon steel, austenitic steels, Armco iron, white cast iron, silicon iron and numerous non-ferrous metals and alloys. A list of references to the literature is added.

Radiographs in the Modern Manner. R. C. Woods. (Iron Age, 1940, vol. 146, Dec. 5, pp. 35–39). The author explains the purpose of a layer of fluorescent calcium-tungstate crystals in conjunction with photographic emulsion for X-ray films and discusses the efforts which have been made to prevent the scatter of rays and the consequent lack of definition in the radiograph. He describes a recently developed technique, using lead foil screens 0·005 in. thick, in which the casette is loaded with a lead screen between the object and the film and a second lead screen to back up the calcium-tungstate screen behind the film. Some radiographs of castings taken with and without lead screens are reproduced.

Sub-Boundary Structures of Recrystallized Iron. N. P. Goss. (American Institute of Mining and Metallurgical Engineers, Technical Publication No. 1236: Metals Technology, 1940, vol. 7, Dec.). The author reports on an X-ray investigation of the effects of cold-reduction on the structure of strip hot-rolled from ingot iron containing carbon 0·04%, manganese 0·017%, phosphorus 0·006% and sulphur 0·020%. The hot-rolled strip, 0·076 in. thick, was cold-rolled down to various thicknesses in the range 0·004–0·022 in., annealed and strained by further slight cold-reduction. Specimens were heat-treated at temperatures above and below the A_3 point. X-ray Laue diagrams of the specimens were then taken. Very little distortion was observed in the Laue diagrams of the strip heat-treated at below the A_3 point, the Laue spots being sharply defined, but the Laue spots of the strip annealed at above the A_3 point exhibited asterism and were diffused, showing that the grains were distorted. The mechanism of the distortion was also examined by obtaining Laue diagrams of a large single crystal of α-iron, about 0·010 in. thick, before and after a reduction of less than 1% by cold-rolling. The diagrams revealed that lattice blocks were displaced relative to each other by rotation, the displaced blocks being fairly large and the degree of rotation surprisingly large. From this the author deduces that deformation does not proceed by a uniform displacement upon all possible slip planes, but that large rotational displacements take place between very large lattice fragments of the order of 10^{-4} cm. This evidence is believed to support the theory of the existence of a sub-boundary structure within the grain, and to give additional proof that plastic deformation is not homogeneous.

Application of X-Rays to the Study of Alloys. H. Lipson. (Nature, 1940, vol. 146, Dec., 21 pp. 798–801). The author reviews recent knowledge concerning alloy systems which has been revealed by X-ray investigations. It had been known for some time that it was difficult to produce equilibrium in certain iron-rich alloys, many of the physical properties showing hysteresis in their variation with temperature. With normal heat treatments, X-ray photographs showed lines that were very blurred, but it was found that by heat treatment at low temperatures for lengthy periods much sharper lines were obtained. This gave a valuable indication of the approach to equilibrium, and two diagrams based solely on X-ray data have recently been published. Although in their equilibrium state certain alloys (*e.g.*, iron-nickel alloys) contain both body-centred and face-centred cubic phases, they can be maintained as single-phase alloys with a face-centred structure by quenching from high temperatures. If such a single-phase alloy is placed in liquid air, its structure changes over completely to the body-centred cubic form. This does not happen if the alloy is in the duplex state. The explanation of the fact that alloys which are so reluctant to change their structures at temperatures as high as 400° C. should be able to change so completely at − 200° C., lies in the difference between transformations which require migration of atoms and those which do not.

The deduction of the crystal structure of an alloy from its composition has been carried out by applying the Hume-Rothery rule, which points out that there is a general connection between crystal structure and concentration of valency electrons. By " valency electrons " is meant those electrons in the outer shell of an atom which are not firmly bound to the nucleus; it is these electrons which give to the atoms their peculiar metallic properties. The transition elements, such as iron, which have incomplete inner shells, may take extra electrons into those shells and thus cancel out the effect of their own valency electrons. Thus, over large ranges of composition, these elements behave as though they were nulvalent.

Neutron Study of Order in Iron-Nickel Alloys. F. C. Nix, H. G. Beyer and J. R. Dunning. (Physical Review, 1940, vol. 58, Dec. 15, pp. 1031–1034). The authors studied the order in iron-nickel alloys with the help of neutron transmission measurements, the neutron transmission being the greater the higher the state of order. They found that the difference in neutron transmission between fully annealed and quenched alloys plotted against the nickel content displays a broad peak at about Ni_3Fe and falls to infinitely small values near 35% (atomic) of nickel and for pure nickel. The substitution of 2·3% (atomic) of molybdenum or 4·1% (atomic) of chromium for iron in the annealed alloys containing 78% (atomic) of nickel caused a decrease of 15·6% and 21·2% respectively, in the neutron transmission. The disorder produced by the cold-working of an annealed binary alloy containing 75% (atomic) of nickel was indicated by a decrease in neutron transmission of 20·6%. In conclusion the authors discuss the value of neutron transmission measurements in the study of order-disorder phenomena in alloys.

The Equilibrium Diagram of the Iron and Silicon System. A. Osawa and T. Murata. (Nippon Kinzoku Gakkai-Si, 1940, vol. 4, Aug., pp. 228–239). (In Japanese). The authors report on their investigation by dilatometric, thermal, magnetic and X-ray methods of the iron-silicon system up to 65% of silicon.

CORROSION OF IRON AND STEEL

(Continued from pp. 116 A–117 A)

Measurement of Electrode Potentials and Polarization in Soil-Corrosion Cells. R. B. Darnielle. (Journal of Research of the National Bureau of Standards, 1940, vol. 25, Oct., pp. 421–433). The author discusses means of measuring the potentials at the electrodes in soil-corrosion cells and describes the Hickling method in detail. Methods involving the interruption of the current passing through the cell are only satisfactory when the period of interruption is sufficiently short to prevent error due to depolarisation. The Hickling method for measuring the potential of polarised electrodes utilizes an electronic interrupter and an electronic potentiometer, by means of which potentials can be measured a very short time after the current has been interrupted. The author found this method to be accurate within about 0·01 V. in the conditions under which he carried out his tests.

Caustic Embrittlement in Steam Boiler Material. F. Odqvist. (Teknisk Tidskrift, 1940, vol. 70, Oct. 19, Mekanik, pp. 103–109). (In Swedish). After a brief review of the literature on the causes of caustic embrittlement, the author describes his investigation of the causes of the repeated failure of some of the rivets in the drum of a water-tube boiler at a Swedish works. The boiler had a heating surface of 335 sq. m. and operated at a maximum pressure of 275 lb. per sq. in., but the conditions were such that the pressure fell from 275 to 180 lb. per sq. in. three times every hour. He examined several of the rivets under the microscope and tested their mechanical properties. From the results obtained he came to the following conclusions: (1) Most of the cracks originated from cavities in the neck, but in a few cases there were cracks in the shank; (2) the propagation of the cracks was independent of the type of microstructure of the rivets; (3) tensile and hardness tests proved that the mechanical properties of the steel between two cracks had not deteriorated in any way; (4) some chemical action in the nature of an alkali attack on the rivets had taken place; (5) the structure of the steel at the cracks was of intercrystalline character in those parts of the rivets where individual grains could be observed; and (6) the high humus content of the feed-water may have been a contributory cause of the failure of the rivets.

ANALYSIS

(Continued from pp. 78 A–79 A)

Improvements in the Accuracy of the Vacuum-Fusion Method for the Determination of Oxygen in Steel. S. Marshall and J. Chipman. (Transactions of the American Institute of Mining and Metallurgical Engineers, Iron and Steel Division, 1940, vol. 140, pp. 127–131). An investigation was made of the effect of changes in the pumping speed and of the addition of tin to the crucible on the accuracy of oxygen determinations in the analysis of steel by the vacuum-fusion method, especially with regard to samples of high manganese content. The results indicated that the error introduced by metallic films of vaporised metals may be reduced by the addition of tin in the manner described below, and by increasing the pumping speed. The method of carrying out an analysis using tin is as follows: After the crucible has been sufficiently baked out at about 2000° C. the temperature is lowered to 1600° C. and a " blank " is collected and analysed. The temperature is then lowered to 1300° C. and 10–20 g. of tin and an equal amount of low-manganese iron are dropped into the crucible. The temperature is then raised to 1600° C. and the evolved gas is pumped away and discarded. During this time the walls of the furnace are being coated with a film of tin which is vaporised from the melt. When the furnace pressure has returned to its initial value, indicating that the gas from the iron-tin alloy has been removed, the first sample may be dropped into the crucible and the gas evacuated. When the evacuation is complete and the collected gas has been transferred to the analytical train, another sample can be treated and the gas stored in the reservoir. This procedure practically doubles the number of samples that can be analysed in one day. The original addition of tin is usually sufficient to maintain a vapour of tin in the furnace for the duration of any one day's run.

Rapid Determination of Oxygen in the Molten Steel. S. Tawara and N. Sato. (Tetsu to Hagane, 1940, vol. 26, Sept. 25, pp. 693–698). (In Japanese). The authors describe a method of determining the oxygen content of molten steel. This method is a modification of Herty's process and a determination can be made in 15 min.

On the Form of Nitrogen in Cr, Ni–Cr, and Ni–Cr–Mo Steels. I. Hayashi and M. Ebisuda. (Tetsu to Hagane, 1939, vol. 25, Dec. 25, pp. 1035–1042). (In Japanese). The authors describe a method of determining the nitrogen in alloy steels containing chromium, nickel and molybdenum and discuss the results obtained when it was applied to study the effect of different heat treatments on the form in which the nitride is present in the steel.

Determination of Nitrogen in Iron and Steel by the Kjeldahl Method. I. T. Somiya. (Tetsu to Hagane, 1940, vol. 26, Jan. 25, pp. 43–46). (In Japanese). The author describes the apparatus used for the distillation of the ammonia formed when determining the nitrogen content of iron or steel by the Kjeldahl method.

A Comparative Investigation of Methods of Determining the Nitrogen Content in Steel and in Alloys. G. Phragmén and R. Treje. (Jernkontorets Annaler, 1940, vol. 124, No. 9, pp. 511–531). (In Swedish). It has been observed that considerable differences frequently occur in the nitrogen determinations of the same steel when carried out in different laboratories. The authors have therefore investigated the accuracy of a number of different methods, and in the present paper they survey the literature on the subject and report on some of the results obtained in their laboratory work. Determinations can easily be made with unalloyed steels by dissolving the sample in a dilute acid which converts the nitrogen into an ammonium salt; but with alloy steels difficulties arise owing to nitrogen being retained in the undissolved residue. A number of methods of treating the residue were examined. These included filtering it on asbestos, drying at about 100° C., and dissolving the residue either in sulphuric acid with a sulphate addition, or in perchloric acid with an addition of sulphuric acid, or alternatively by oxidation with sodium peroxide by Klinger's method. The vacuum-fusion method was also examined. The authors found that in most cases the results were in fairly good agreement. In their opinion the method using sulphuric acid with a sulphate addition is a simple and reliable one. A comprehensive table, in which the nitrogen determinations for a large number of alloy steels carried out by six different methods are compared, is presented, and a bibliography with 82 references is appended.

On Hydrogen in Steel. S. Kobayashi. (Tetsu to Hagane, 1939, vol. 25, Sept. 25, pp. 745–771). (In Japanese). The author studies the vacuum-fusion method for the determination of hydrogen in iron and steel, and discusses the results obtained when applying it to determine the variations in the hydrogen content of the melt during steelmaking operations. He also examines the relation between the amounts of hydrogen in the steel and in the slag.

On the Determination of the Total Sulphur in Steel by the Combustion Method. Y. Kanamori. (Tetsu to Hagane, 1940, vol. 26, Aug. 25, pp. 630–636). (In Japanese). The author describes a combustion method of determining the sulphur in iron and in alloy steels. In this method the sample is burnt in oxygen and the sulphur compounds are collected by passing the evolved gases through hydrogen peroxide ; after removal of any carbon dioxide, the solution is titrated with sodium hydroxide.

On the Determination of Silicate Inclusions in Steel. S. Kobayashi, Y. Kanamori and K. Kosiya. (Tetsu to Hagane, 1939, vol. 25, Dec. 25, pp. 1074–1080). (In Japanese). The authors describe a modification of Dickenson's method of determining the percentage of silicate in non-metallic inclusions in steel. Dickenson's method was described in Journ. I. and S.I., 1926, No. I., p. 177.

The Determination of Non-Metallic Inclusions in Steel. I. Araki. (Tetsu to Hagane, 1940, vol. 26, Jan. 25, pp. 14–19). (In Japanese). The author compares a number of methods of determining the amount of silica and alumina in steel. For silica determinations the method he proposes is to dissolve the sample in nitric acid and treat the residue with hydrofluoric and sulphuric acids, and he puts forward an oxime method for alumina determinations. He also discusses the results obtained by applying these methods to samples taken from a basic electric-arc furnace.

On the Introduction of the Quantitative Spectrum Analysis of Iron and Steel in Work's Practice. I. Kadokawa and K. Nagata. (Tetsu to Hagane, 1939, vol. 25, Dec. 25, pp. 1043–1052). (In Japanese).

Spectroscopic Analysis. (Automobile Engineer, 1940, vol. 30, Nov., p. 358). A description is given of a simplified form of spectroscope called the " Spekker Steeloscope." (*See* p. 79 A).

Quantitative Spectrographic Analysis of Steels. S. Vigo. (A.S.T.M. Bulletin, 1940, Dec., pp. 17–22). The author describes the spectrographic apparatus used at the laboratory of the Watertown Arsenal, Massachusetts, and some of the work done with it. The spectrograph is a diffraction grating instrument and dispersion is obtained by a Johns Hopkins concave speculum grating with a curvature of 3 m. radius, ruled with 15,000 lines per in. over a distance of 4 in. and height of 2 in. Three orders of spectra are available. The first with a dispersion of 5·2 Å per mm., is used for quantitative and for general qualitative analysis. The second order (2·6 Å per mm.) is applied in the qualitative analysis of complex alloys. The need for the third order (1·7 Å per mm.) has not arisen. Spectrographic analysis at this arsenal is based on standard steels and the author describes how these are prepared. The mean relative errors in routine analyses of some of the elements were determined and found to vary between 3·52% and 8·38%, and the mean relative deviation in reproducibility varied between 1·70% and 4·96%.

BOOK NOTICE

(Continued from pp. 79 A–82 A)

WILSON, W. KER. "*Practical Solution of Torsional Vibration Problems With Examples from Marine, Electrical, Aeronautical, and Automobile Engineering Practice.*" Second Edition. Volume 1. 8vo. Pp. xx + 731. Illustrated. London, 1940. Chapman & Hall, Ltd. (Price £2 2*s*.)

The importance of vibration in the case of installations having petrol or heavy oil engines as a prime mover is now well known, and in the past many expensive lessons have been learnt by bitter experience.

This book contains just the practical information which is required by the designer so that vibration may be obviated in the running range in the design stage. Geared systems are dealt with by the author at some length and methods are indicated for the correct phasing of engines working on the same main wheel. Flexible couplings as well as torsionally stiff and axially flexible couplings are dealt with. These matters are regarded by the writer as highly important, because it is likely that at the end of this war much more attention will be paid to the problem of the geared marine oil-engine installation. In this connection a careful study of the author's sections on geared systems and flexible couplings is recommended.

The book contains much useful information on pendulum dampers—one well-known builder of marine oil engines has adopted this form of damper—and it seems probable that in the future it will be used more frequently in marine installations.

It is noted that the author maintains a semi-empirical method for the estimation of vibration stresses at resonant speeds, and the writer agrees that such methods are the only practical solution of the problem. Chapter VI. contains some very valuable information on the determination of stresses at non-resonant speed, and indicates methods of phasing and the effect of the order of firing of the various cylinders.

This book in re-edited form has now become what may be termed a standard reference book, and should be available to the draughtsmen and designers responsible for transmission systems in which the prime mover delivers a periodic fluctuating torque.

S. F. DOREY.

BIBLIOGRAPHY

(Continued from pp. 82 A–84 A)

AMERICAN CHEMICAL SOCIETY. DIVISION OF GAS AND FUEL CHEMISTRY. "*Symposium on Furnace Atmospheres for Metallurgical Purposes.*" 4to. Easton, Pa., 1940: The Society.

AMERICAN SOCIETY FOR TESTING MATERIALS. "1940 *Supplement to A.S.T.M Standards Including Tentative Standards.*" 8vo. Part I. Metals. Pp. xiii + 478: Part II. Nonmetallic Materials—Constructional. Pp. xiii + 348: Part III. Nonmetallic Materials—General. Pp. xiii + 574.. Philadelphia: The Society.

BEAUMONT, R. A. "*Mechanical Testing of Metallic Materials With Special Reference to Proof Stress.*" (Aeronautical Engineering Series Ground Engineers.) 8vo, pp. viii + 120. Illustrated. London, 1940: Sir Isaac Pitman & Sons, Ltd. (Price 6*s.*)

BURTON, E. F., H. G. SMITH and J. O. WILHELM. "*Phenomena at the Temperature of Liquid Helium.*" 8vo, pp. xi + 362. New York, 1940: Reinhold Publishing Corporation.

BURNHAM, T. H., and G. O. HOSKINS. "*Engineering Economics (Book I), Works Organisation and Management.*" London, 1940: Sir Isaac Pitman and Sons, Ltd. (Price 10*s.*)

DEPARTMENT OF SCIENTIFIC AND INDUSTRIAL RESEARCH. Fuel Research. Physical and Chemical Survey of the National Coal Resources, No. 44: "*Methods of Analysis of Coal and Coke.*" 8vo, pp. vi + 85. London, 1940: H.M. Stationery Office. (Price 2*s.*)

DICKASON, A. "*The Geometry of Sheet Metal Work.*" 8vo, pp. 335. London, 1940: Sir Isaac Pitman and Sons, Ltd. (Price 12*s.* 6*d.*)

MATTHEWS, F. J. "*Works Boiler Plant.*" 8vo, pp. 184. London, 1940: Hutchinson & Co., Ltd. (Price 10*s.* 6*d.*)

PARTINGTON, J. R. "*Chemical Thermodynamics: a Modern Introduction to General Thermodynamics and its Applications to Chemistry.*" Third edition. 8vo, pp. x + 230. London, 1940: Constable & Co., Ltd. (Price 14*s.*)

ROYAL AERONAUTICAL SOCIETY. "*International Index to Aeronautical Technical Reports*, 1939." 8vo, pp. 191. London, 1940: Sir Isaac Pitman and Sons, Ltd. (Price 7*s.* 6*d.*)

SHUMARD, F. W. "*A Primer of Time Study.*" First edition. 8vo, pp. xii + 519. New York and London, 1940: McGraw-Hill Book Company, Inc. (Price 35*s.*)

WILSON, W. KER. "*Practical Solution of Torsional Vibration Problems With Examples from Marine, Electrical, Aeronautical, and Automobile Engineering Practice.*" Second Edition. Volume I. 8vo, pp. xx + 731, Illustrated. London, 1940: Chapman & Hall, Ltd. (Price £2 2*s.*) [See notice, p. 142 A.]

REFRACTORY MATERIALS

(Continued from pp. 85 A–87 A)

The Resistance of Magnesite Bottoms of 185-Ton Open-Hearth Furnaces. V. Dement'ev. (Stal, 1939, No. 12, pp. 27–31). (In Russian). The author has found that the percentage of magnesia in the magnesite is a measure of the quality of a fused-on magnesite bottom to an open-hearth furnace, and experience at the Magnitogorsk Works has shown that 75% of magnesia is the optimum proportion. Further analyses showed that the main change in the composition of the furnace bottom during the period of its service was a reduction of the magnesia content. In this paper he deals with the causes of the reduction of the stability of the bottoms. These are: (1) *Diffusion of admixtures into the fused-on bottom.* This in turn is conditioned by the difference in the compositions of the bottom, the metal and the slag, as well as by the temperature and porosity of the bottom. (2) *The composition of the bottom and the duration of its burning-in.* Ample time for burning-in should be allowed. (3) *Deposits adhering to the bottom.* These are caused mainly by the practice of charging lime directly on to the bottom of the furnace. The best method is to charge a layer of clean fine iron scrap, which will protect the bottom from the lime and ore, and will also damp out shocks during charging. If iron scrap is not available, a layer of ore is the next best thing. (4) *Chemical activity of the bottom.* This is partly the consequence of (1), (2) and (3). The impurities which find their way into the bottom, in particular ferrous oxide and manganous oxide, undergo repeated reduction and oxidation during the various stages of the open-hearth process. Statistical investigations have shown that the oxidising action on the bottom during the time the furnace is standing empty between heats is directly connected with the time lost in repairing bottoms. The oxidising period during the heat has a like effect. (5) *Basicity of the slag.* The maximum stability of the bottom is obtained with a lime/silica ratio in the slag of 2·6. (6) *Carbon content of the metal.* The higher the carbon content, the lower is the melting point and the higher is the stability of the bottom.

Effects upon Furnace Refractories of Protective Gases for High Carbon Steels. J. H. Loux. (American Chemical Society, Symposium on Furnace Atmospheres for Metallurgical Purposes, Sept., 1940: Industrial and Engineering Chemistry, Industrial Edition, 1941, vol. 33, Jan., pp. 42–46). The author discusses the problem of finding a suitable refractory brick which will not be disintegrated by the action of the protective atmospheres used in the heat treatment of high-carbon steels. (*See* p. 40 A).

FUEL

(Continued from pp. 118 A–120 A)

The Influence of Mixing of Gas and Air on the Speed of Combustion. W. Trinks. (Industrial Heating, 1940, vol. 7, Oct., pp. 928–936). The author presents an abridged account of the conclusions of Rummel and of Schwiedessen relating to methods of mixing fuel gas and air and to the effects of different methods on the rate of combustion (*see* Journ. I. and S.I., 1937, No. II., pp. 124 A, 166 A, 255 A, and 1938, No. I., p. 54 A). He also discusses these conclusions, pointing out that if the mixing of the gas and air is perfect, the flame will be very short and intense, but if the mixing is not perfect, the rest of the mixing must take place within the furnace and the length of the flame will then depend not only on the burner design, but on the manner in which the gases impinge on the furnace walls and roof and on the charge.

Shrinkage of Coke. H. S. Auvil, J. D. Davis and J. T. McCartney. (United States Bureau of Mines, 1940, Nov., Report of Investigations No. 3539). The Bureau of Mines has investigated the shrinkage of coke by methods which eliminated the effect of the swelling in the plastic range and the results obtained have not all been published. As other investigators have disagreed with some of the previous conclusions of the Bureau (*see* Journ. I. and S.I., 1940, No. I., p. 174 A), the present report covers the entire investigation and makes known some hitherto unpublished data. In particular, studies were made of the " after-shrinkage " of coke, *i.e.*, the shrinkage obtained on reheating a newly formed coke in an inert atmosphere to the maximum temperature of a coke-oven. It was found that the range of shrinkage on reheating cokes made at 500° C. to 900° C. was 23·8% to 27·2% (on a volume basis), from which it was concluded that the after-shrinkage of all cokes from commercial coking coals is virtually the same. In fact, the maximum variation found was hardly large enough to affect appreciably the net expansion or the pushing properties of the coke.

The Production of Dry-Quenched Coke. V. Bazanishvil'. (Stal, 1939, No. 12, pp. 18–20). (In Russian). Dry-quenched coke was first produced in Russia in 1936 at the Kirov Works. The coke is quenched by a stream of neutral gas which is circulated in a closed circuit, giving up to boilers the heat taken up from the coke. The article presents a summary of one month's determinations of the properties of dry-quenched coke as compared with those of wet-quenched coke prepared at the same time from the same coal. The screen analysis showed the dry-quenched coke to be superior and more uniform. It was also better as regards mechanical strength and abrasion resistance and had less cracks. The ash and sulphur contents were the same for both types of coke, but the dry-quenched coke contained less moisture and less volatile matter. Very brief production details are given. On the whole, it is concluded that dry-quenched coke is to be preferred for metallurgical purposes.

PRODUCTION OF IRON

(Continued from pp. 120 A–121 A)

An Electrically Operated Clay Gun. A. F. Morgan. (Iron and Steel, 1940, vol. 14, Oct., pp. 12–13). An illustrated description is given of an electrically driven clay gun for plugging the tap-hole of blast-furnaces. (*See* Journ. I. and S.I., 1940, No. II., p. 178 A).

The Production of Charcoal Pig Iron in the Urals. I. Sokolov. (Stal, 1939, No. 12, pp. 14–17). (In Russian). The present output of pig iron in the Urals amounts to about 430,000 tons per annum. This includes about 150,000 tons of iron which is worked up by the acid open-hearth process, about 50,000 tons of nickel-chromium foundry iron, the remainder being mainly worked up by the basic open-hearth process, with an insignificant amount for foundry work. One ton of iron requires about 7 cu. m. of charcoal, *i.e.*, 945–980 kg. The author summarises the various properties which make charcoal iron superior to coke iron. These properties are inherent or hereditary in the irons and their origin must be sought in differences in the conditions obtaining during the production of the two types of iron. Owing to the greater reactivity of the charcoal, reduction of the ore in the blast-furnace is more complete at a lower temperature. There is little or no solution of ferrous oxide in the reduced iron and there are reasons to believe that the content of other oxide inclusions would be lower in the charcoal iron. This fact would determine the structure on solidification. Owing to the lower temperature and the different composition of the slag, the gas content of the charcoal iron would also be less than that of coke iron. In the concluding section, brief reference is made to ore deposits in the Urals from which ores suitable for the production of charcoal irons low in sulphur and phosphorus, and of alloy irons containing titanium and nickel-chromium, could be obtained. An estimate of the timber resources leads to the conclusion that, with the necessary organisation, enough charcoal could be made available to render possible an output of not less than 1·5 million tons of charcoal iron per annum.

Break-Through of the Hearth of the No. 4 Blast Furnace at the Magnitogorsk Works. P. Natarov. (Stal, 1940, No. 1, pp. 9–12). (In Russian). A break-through of the hearth which occurred in one of the blast-furnaces at the Magnitogorsk Works about 400–500 mm. below the iron notch is briefly described. The causes of the break-through and the repairs, which took sixteen days, are dealt with and some suggestions are made with a view to the prevention of similar accidents.

Effects of Scrap in the Blast-Furnace Burden. C. L. T. Edwards. (American Institute of Mining and Metallurgical Engineers, Technical Publication No. 1270; Metals Technology, 1941, vol. 8, Jan.). The author discusses, from the technical and economic points of view, the effects of the addition of scrap to the blast-furnace burden and briefly reports on the experience gained by the Bethlehem Steel Co. in this respect. He comes to the conclusion that there is no fundamental difference between pig irons produced from ore and from scrap and that differences sometimes occurring are only due to the nature of the materials involved, *i.e.*, variables which are inherent in ore mixtures just as much as in mixtures that include scrap.

Recent European Developments in Pig-Iron Manufacture. N. L. Evans. (Institute of British Foundrymen: Iron and Coal Trades Review, 1941, vol. 142, Feb. 14, p. 217; Feb. 21, pp. 245–246, Feb. 28, p. 268). The author's paper on English and Continental experiments during the last ten years on the desulphurisation of pig iron in the ladle with soda ash is reproduced. (*See* p. 120 A).

An Investigation of the Comparative Reducibility of Swedish Red, Black and Bog Ores. J. Petren. (Jernkontorets Annaler, 1940, vol. 124, No. 11, pp. 589–599). (In Swedish). The author reports the results of a laboratory investigation of the reducibility of a number of Swedish iron ores. The results are presented in a number of comprehensive tables. From these it can be said that the bog ores have about the same reducibility as the hematite ores, but they could perhaps be fully reduced at a slightly lower temperature. The three hematite ores tested behaved in a similar manner and the rate of reduction was unaffected by the particle size; the reduction began earlier and at a much lower temperature than was the case with the magnetites. The tests on the six different magnetites produced widely differing results, showing that the reducibility of these ores varied considerably and was related to their structure and grain size.

Blast-Furnace Slag Aggregates in Building and Road Construction. T. W. Parker. (Chemistry and Industry, 1941, vol. 60, Feb. 1, pp. 59–63). The author reports on an investigation on the uses of blast-furnace slag in building construction, which, for the past few years, has been carried out at the Building Research Station, Garston, Herts., for the Blastfurnace Committee of the Iron and Steel Industrial Research Council. In the present paper he reviews only the use of the material as heavy aggregate in macadam and concrete, and gives no consideration to foamed slag used as a light-weight aggregate. In the first section he deals with the chemical composition and petrography of blast-furnace slag, pointing out that most of the slag used in Great Britain may be regarded as a synthetic mineral composed chiefly of melilite and calcium silicates or anorthite, together with calcium sulphide and minor compounds. In the second section he reviews various methods employed for the preparation of slag for the market, and states that more uniformity of product would no doubt be obtained if suitable specifications for slag were available. He enumerates the properties it would be necessary to specify. In the third section he briefly considers the use of slag in road construction. Experimental work on this problem has not yet been carried out in the laboratory, and his information is based on experience gained in the survey of slag sources and on observations made by officials of slag-producing works and local road authorities. Finally, in the fourth section, he summarises the results of a comprehensive series of tests on the strength and stability of concretes made from slags from various sources and on the influence of slag on steel reinforcements, and reports very briefly on the examination of some samples of concretes containing blast-furnace slag in which failures or troubles had arisen.

FOUNDRY PRACTICE

(Continued from pp. 121 A–122 A)

Desulphurizing Cupola Cast Iron from the Practical Angle. W. Levi. (Transactions of the American Foundrymen's Association, 1941, vol. 48, Mar., pp. 623–636). The author describes the desulphurising procedure adopted at an American foundry making castings for ploughs and all sizes of pipes. There are six large cupolas with fore-hearths at this foundry and a high proportion of scrap (over 83%) is used in the charge. The iron is desulphurised by additions of fused soda ash. The fore-hearth is preheated to at least 2400° F. before any iron is tapped into it. After sufficient iron has been run into the fore-hearth to cover the hole leading to the spout, a predetermined amount of soda ash is added, the fore-hearth is filled up with more iron, and the metal is tapped into a thoroughly preheated transfer ladle. When this is done, sufficient soda ash is immediately added to the fore-hearth to treat an amount of iron equal to that which has been removed. The fore-hearth must be completely refilled before any further ladles are filled from it. After several ladles of iron have been drawn off and an equal number of additions of soda ash have been made, a layer of slag will have formed on the surface. This slag is skimmed off by allowing the level of the iron in the fore-hearth to rise to just below the slag spout. More efficient desulphurisation is obtained when there is a thin layer of slag over the iron. At this foundry the sulphur contents of the iron before and after the soda ash treatment average 0·138% and 0·093%, respectively; 8 lb. of soda ash per ton of iron are added and the refining action takes about 17 min. Data on the cost of the treatment are also presented.

Effect of Manganese on Second-Stage Graphitization. D. P. Forbes, P. A. Paulson and G. K. Minert. (Transactions of the American Foundrymen's Association, 1941, vol. 48, Mar., pp. 574–586). A "pearlitic malleable" iron differs from ordinary malleable iron in that a portion of the carbon (usually 0·40% to 0·80%) is retained as combined carbon in the finished product. White cast iron can be graphitised by subjecting it to two cycles of heat treatment which the present authors refer to as "first-stage" and "second-stage" graphitisation. The effect of the manganese content on first-stage graphitisation has been previously reported (*see* Journ. I. and S.I., 1939, No. I., p. 121 A). In this paper the authors report on an investigation of the effect of manganese as a retarder of graphitisation and carbide stabiliser in the second stage. Three sets of test bars were prepared containing total carbon 2·28%, sulphur 0·09%, phosphorus 0·14% and silicon about 1%, and the manganese content was varied by adding ferro-manganese to the ladle so that the three sets contained 0·45%, 0·91% and 1·28% of manganese respectively. For the first stage, all the bars were heated up to 1720° F. in 18 hr., held at this temperature for 30 hr., and cooled to room temperature in still air in about 1 hr. For the second stage, three bars from each set were assembled in four groups, reheated to 1300° F., and removed from the furnace at intervals of 45 min., 3 hr., 12 hr. and 48 hr., respectively. The combined carbon of the specimens was determined and microscopical examinations were made. It was found that the greater the manganese content, the greater was the retardation of the graphitisation rate in the second stage at constant subcritical temperatures. The logarithm of the time required to graphitise a given percentage of combined carbon at subcritical temperatures was approximately proportional to the percentage of manganese present. In general, the practical advantages of increasing the manganese in pearlitic malleable iron are: (1) Graphitisation is retarded, with the result that minor variations in temperature, time and composition will not cause large variations in the tensile properties of the castings; (2) the rapid loss of combined carbon while spheroidisation of the cementite is proceeding will be prevented; and (3) it enables some combined carbon to be retained in metal subjected to a heat treatment which would completely graphitise ordinary malleable iron.

Inverse Chill in Malleable Iron. E. Touceda. (Transactions of the American Foundrymen's Association, 1941, vol. 48, Mar., pp. 449–461). The author reports on an investigation of a number of test-pieces of malleable iron the object of which was to ascertain the cause of an unsatisfactory condition often called "inverse chill." The fracture of specimens reveals this condition when the outer layer of metal is dark and the core light in appearance. The theory put forward by the author is that if the ferrite grains are ductile they will elongate and cause the fracture to have innumerable tiny pits; the ridges between them will cast shadows and cause this region to appear dark. On the other hand, should the ferrite be brittle, or if there are particles of free cementite distributed in the matrix, the fracture will be smoother, with less shadow, and therefore lighter in appearance.

Chaplets and the Steel Casting. H. F. Taylor and E. A. Rominski. (Transactions of the American Foundrymen's Association, 1941, vol. 48, Mar., pp. 481–513). The authors investigated the influence of a number of factors affecting the degree of fusion obtained between the metal of a chaplet and the metal of a casting. For this purpose a series of similar castings was made using chaplets with plain, threaded and specially shaped stems, and the effects of copper, tin, cadmium, silver and aluminium coatings and of impregnating the chaplet metal with silicon were examined. Sections were cut in two directions across the finished casting so that the fusion line could be studied under the microscope. The conclusions reached were as follows: (1) A threaded stem on a chaplet does not necessarily improve the fusion; in some cases this shape is detrimental, because dust and moisture may accumulate at the bottom of the threads and cause gases to form. Threaded stems key the chaplet to the casting by fusion along the ridges of the thread, but without complete fusion at the root the casting may fail in service. (2) The storage and handling of chaplets should receive greater attention with a view to increasing the cleanliness. (3) A heavily-coated steel chaplet fuses readily, but does not always give sufficient support to the core. The ideal is to have a thin alloy case with a core of steel that does not melt so easily. (4) Low-carbon steel is a good material for chaplets for steel castings. (5) Fusion is aided by the migration of carbon from the cast metal into the low-carbon steel of the chaplet. (6) Silicon-impregnated chaplets fuse readily, but grain-coarsening takes place in a small area round the chaplet. (7) The results obtained with chaplets dipped in tin were not consistent and no general conclusion could be drawn. (8) Copper- and nickel-plated chaplets are satisfactory when they are properly prepared and kept clean. (9) Silver-plated chaplets are very satisfactory, but silver-cadmium and cadmium coatings are unsatisfactory because of the high vapour pressure of cadmium. (10) Chaplets sprayed with aluminium proved to be good in this investigation, but further study of them is required. (11) The stream-lining of chaplets assists the gases to escape; this also is a subject requiring further study.

Recent Experiments with Gray Iron Synthetic Molding Sand. F. Holtby and H. F. Scobie. (Transactions of the American Foundrymen's Association, 1941, vol. 48, Mar., pp. 465–476). The authors report on an investigation of the properties of some synthetic sand mixtures for making grey-iron castings using as a base sand from the St. Peter sandstone formation in the Mid-Western States of America, in particular that found in Minnesota. Mixtures were also made up with additions of the fines which are removed as waste material from the moulding sand in steel foundries. The authors are of the opinion that the above sandstone formation offers an almost inexhaustible supply of practically pure silica sand which can be used as a base for mixtures for both moulds and cores, and that the fines from steel foundries can be blended with other sands of similar composition to control the permeability of the mixture.

French Views on Moulding Sands. (Foundry Trade Journal, 1941, vol. 64, Jan. 30, pp. 71–73). The effects of drying, milling and variations in moisture content on the permeability of moulding sands are discussed.

One Way of Making a Loam Mould. J. Potter. (Institute of British Foundrymen: Foundry Trade Journal, 1941, vol. 64, Feb. 6, pp. 85–86). The author describes the method which he favours for preparing a loam mould for casting a Diesel engine silencer.

Production of Automotive Castings. H. S. Austin. (Metal Progress, 1940, vol. 38, Dec., pp. 775–780). After discussing how the structure of cast iron varies with different carbon and silicon contents and the influence of silicon on the combined carbon during heat treatment, the author describes some recent improvements in cupola practice. He shows in particular how the cupola practice is adjusted for the manufacture of iron for piston rings and other automobile parts.

Influence of the Mold on Shrinkage in Ferrous Castings. H. L. Womochel and C. C. Sigerfoos. (Transactions of the American Foundrymen's Association, 1941, vol. 48, Mar., pp. 591–618). The authors studied the effects of the hardness of the mould and of the compressive strength, permeability, moisture content and composition of the moulding sand on the degree of pipe in cast round bars of grey iron and steel. Their general conclusions were: (1) A high moisture content, a high compressive strength of the green sand, soft ramming and an uneven distribution of the grains of different sizes all tend to promote shrinkage defects in grey-iron castings; (2) the presence of sea coal in green sand decreases the amount of piping in risers in grey-iron castings; (3) small grey-iron castings made in baked oil-sand moulds have no shrinkage defects; and (4) small steel castings are not so susceptible as grey-iron castings to changes in the moulding conditions. It is pointed out that the influence of the compressive strength of green sand as stated in conclusion (1) applies only under the conditions of casting described by the authors.

PRODUCTION OF STEEL

(Continued from pp. 122 A–123 A)

Works of the Chamber of Mines Steel Products. W. Wallace. (Journal of the South African Institution of Engineers, 1940, vol. 39, Dec., pp. 101–110). After an introduction in which information is given on the capital of the Chamber of Mines Steel Products Co., its yearly output during the last 24 years, the number of employees, &c., the author deals, in separate paragraphs, with the following subjects: Historical development of the company; raw materials; the foundry; the ball-mill plant; the die press plant; and process losses. In conclusion he reproduces a flow sheet showing the monthly requirements of raw ma erials, the output, and the intermediate products and stages of production.

Rustless Iron and Steel Corporation Increases Product of Plant. (Blast Furnace and Steel Plant, 1940, vol. 28, Dec., pp. 1160–1162). A brief description is given of the recently completed extensions to the works of the Rustless Iron and Steel Corporation in Baltimore. (*See* p. 92 A).

Plant Expansion of the Rustless Iron and Steel Corp. E. F. Cone. (Metals and Alloys, 1940, vol. 12, Dec., pp. 769–777). The author briefly describes the extension of the Rustless Iron and Steel Corporation of Baltimore, which was designed in 1935 and completed in 1940. Numerous photographs are reproduced.

A Method of Rapid Dephosphorisation of Bessemer Steel. G. M. Yocom. (American Institute of Mining and Metallurgical Engineers, Technical Publication No. 1265; Metals Technology, 1941, vol. 8, Jan.). After an introduction, in which the development of the acid and basic Bessemer processes in America and Europe are reviewed at some length, the author describes the method applied for the dephosphorisation of acid Bessemer steel at the Benwood plant of the Wheeling Steel Corporation, Benwood, West Virginia, where 250,000 tons of low-phosphorus steel were produced during the last three years. The dephosphorising mixture used consists of 50% of calsifer (impure lime), 30% of dried mill scale and 20% of dried flux. The analysis of this mixture when melted together to form a slag is as follows: CaO 48%, Fe_2O_3 28%, SiO_2 7% and Al_2O_3 7%. The mixture is added in the cold state to the stream of metal as it is being poured from the converter into the ladle, and the operation is completed in about 30 sec. The phosphorus content is reduced in that short time from 0·95–0·100% to 0·020–0·040%. The author gives information on the following subjects of importance in connection with this method of dephosphorising: Properties of low-phosphorus and normal-phosphorus iron; low-phosphorus Bessemer steel; iron and slag control for dephosphorisation; iron specification for dephosphorisation; blowing practice for dephosphorised heats; separation of slag and metal; simultaneous dephosphorisation and ferro-manganese addition; ladle linings; preparation and properties of the dephosphorising mixture; fluxes; ladle reaction; ladle slags; physical and hot-working properties, and welding and threading quality of dephosphorised steel. In conclusion the author enumerates the advantages and disadvantages of the dephosphorising method.

The Mechanics of the Gas Stream in the Bath of a Bessemer Converter. I. Kazantsev. (Stal, 1940, No. 1, pp. 16–18). (In Russian). The author obtains mathematically a dimensionless expression for the ratio W_x/W_a where W_x is the axial velocity of the air stream in the medium at a distance x from the nozzle, and W_a is the axial velocity at the mouth of the nozzle. Some numerical results for the expression were obtained experimentally by measuring the velocities of air streams in water and mercury, the air stream being blown in at the bottom of the container holding the liquid. In the case of mercury, evidence was obtained showing that a large number of droplets of metal were carried by the air stream inside the liquid. The conditions under which the air stream will carry liquid away with it above the surface of the liquid are also considered. The conclusions are that in a converter the amount of metal thrown out by the air stream will be less the smaller the diameter of the nozzle and the deeper the bath. The degree of utilisation of the oxygen of the blast in the converter, and consequently the speed of the process, will be greater the smaller the value of W_x/W_a at the surface of the bath. It is suggested that a suitable value of the ratio of the depth of the metal to the diameter of the nozzles would be 50–60 instead of the value of 25–30 used at present. This would permit the speed of the blast at the ends of the nozzles to be reduced from 300 to 250 m. per sec. without impairing the efficiency of the reaction of the oxygen with the metal.

Notes on Acid Open Hearth Practice with Reference to the Manufacture of Tyre Steel. E. C. Houston. (Journal of the West of Scotland Iron and Steel Institute, 1940, vol. 48, Oct., pp. 3–10). The furnace and casting procedures in the manufacture of steel tyres from acid open-hearth steel are described and discussed. A high standard of ingot surface is required, for the dimensions of the bloom or " cheese " after forging do not include sufficient machining allowances for the removal of surface seams. In order to obtain the standard required, the ingots are cast uphill in groups, the moulds having been cleaned and coated with a light bitumastic liquid. A moderate treatment with aluminium to suppress the slagging action partially is recommended. When teeming clusters of two and of four ingots at the same rate of rise in inches per minute, the metal in the first case is twice as long in passing through the central feeder and is consequently cooled more, so that to get the metal into the moulds at the same temperature in both cases, a faster rate of rise is required in the smaller cluster. It is generally conceded that the round ingot is the most difficult to cast without cracking. If the following expression is taken as a measure of the departure from roundness of an ingot :

$$\frac{\text{Difference in the diameters of the circumscribed and inscribed circles}}{\text{Diameter of the inscribed circle}},$$

it has been found that the degree of freedom from cracking decreases as this ratio decreases. The melting and refining practice is described in detail in the latter part of the paper.

Stakhanov Working at the New Open-Hearth Shop at the Dzerzhinskiy Works. M. Orman and L. Roytburd. (Stal, 1939, No. 12, pp. 21–27). (In Russian). The melting shop at the Dzerzhinskiy works has four basic furnaces each with a hearth area of 50·74 sq. m. The timing of the various stages of the open-hearth process is discussed. A heat of axle steel takes 9 to 10½ hr., depending on the gas supply. The life of the furnaces has been improved by certain structural alterations and efficient upkeep. Reference is made to certain improvements in the working conditions in the shop and to the reorganisation of the wage system, which is based on an appraisal of results. At the time the shop changed over to operation to a time schedule, a control and despatch system was organised. Some data characterising the way in which the planned schedule was carried out in the shop during 1939 are given.

The Melting of Steel *KhNM* in a Basic Open-Hearth Furnace. A. Madyanov and A. Denisov. (Stal, 1940, No. 1, pp. 19–21). (In Russian). The melting of chromium-nickel, ball-bearing and other high-grade steels as usually done in acid open-hearth or electric furnaces takes rather a long time and the process can be carried out more rapidly by conducting it, under appropriate conditions, in a basic open-hearth furnace. The necessary conditions and some of the experience gained from two experimental heats are discussed. The process involves diffusional deoxidation, by means of which the need for adding ferro-alloys to deoxidise the steel with the consequent unavoidable and undesirable formation of non-metallic inclusions can be avoided. For efficient diffusional deoxidation in the practice under discussion, the slag composition had to be adjusted to give a fluid slag. When the carbon content had been reduced to within 0·10–0·15% of the final value, the slag was treated with a deoxidising mixture (ferro-silicon, quick-lime and coke). This reduced the rate of elimination of the carbon. Another more active deoxidising mixture, containing more ferro-silicon and coke, was then thrown on to the slag. Boiling was thereby stopped. Preheated ferro-chromium was added with further portions of the deoxidising mixture, and the bath was stirred. The addition of the deoxidising mixture is shown by the metal-composition/time curves to cause a slight increase in the phosphorus content of the metal, which, however, did not exceed the permissible limits. The mechanical properties and inclusion rating (sulphides and oxides) of the cast steel from the basic open-hearth furnace were found to be essentially the same as those of electric furnace steel.

The Chromium Reduction Process with Diffusional Deoxidation in Acid Electric Furnaces (the Zaporozhstal' Method). Yu. Shul'te. (Iron and Steel, 1940, vol. 14, Oct., pp. 2–6). An English translation is presented of a paper on the theory and practice of remelting a charge consisting entirely of chromium-nickel or chromium-molybdenum structural steel scrap with carbon 0·25% min. and chromium 1·6% max. This paper appeared in Russian in Stal, 1939, No. 8, pp. 18–22 and an English abstract was published in Journ. I. and S.I., 1940, No. I., p. 281 A.

Production of Chromium-Molybdenum-Aluminium Steel by Melting Works Scrap in Basic Electric Furnaces without Oxidation. V. Gol'dman. (Stal, 1940, No. 1, pp. 27–31). (In Russian). The author describes a method of producing chromium-molybdenum-aluminium steel in basic electric furnaces utilising the works' own scrap containing carbon 0·35–0·42%, chromium 1·4%, molybdenum 0·3–0·5% and aluminium 1·0%. Preliminary experiments in which the scrap was first melted in an acid electric furnace and then mixed with a charge in a basic electric furnace, proved unsuccessful. In the method developed, the scrap was mixed with 30% of low-carbon iron or mild steel and melted down in a basic electric furnace, no lime being added. A fluid slag comprising 1·5–2% of the melt was formed by oxidation of the constituents of the charge and by some attack on the refractory lining; this slag contained lime 10–35%, silica 15–39%, magnesia 9–29%, alumina 25–42%, ferrous oxide 1·5–7%, manganous oxide 3–12% and chromic oxide 2–7%. The composition of the metal after melting was: Carbon 0·30–0·40%, manganese 0·20–0·35%, chromium 0·80–1·10%, molybdenum 0·25–0·40%, aluminium 0·15–0·30% and silicon 0·05–0·25%. The slag was removed, but a little was left to protect the metal, and deoxidation was effected by adding either coke or, better still, charcoal and lime mixed with small amounts of ferro-silicon. The refining slags contained lime 48–65%, silica 12–26%, magnesia 10–30%, alumina 3–15%, ferrous oxide 0·3–1·2%, manganous oxide 0·12–0·42%, and chromic oxide 0–0·8%. The alumina content favoured rapid deoxidation of the refining slag. The necessary amounts of ferro-chromium and ferro-molybdenum were added at the beginning of the refining period. The steel was tapped into a ladle to which the required amount of aluminium had been added. The results of the examination of the microstructure and of mechanical tests are discussed; these results were very uniform and in some cases the steel obtained from charges which included the works own scrap was superior to that obtained from charges with no scrap, and it satisfied the technical specifications. In subsequent experimental heats it was shown that the self-forming slag (*i.e.*, slag formed without the addition of fluxing agents) could be used for refining and deoxidising purposes. By eliminating the need for forming a second refining slag, the time for remelting was shortened by 1 hr., or 16%. In conclusion, the results of microscopical examinations of the cast and of the rolled steel are discussed. The number of non-metallic inclusions in the steel refined under a second slag was about the same as in that refined under a self-forming slag, but sulphide inclusions predominated in the latter. In the rolled steel the distribution and shape of the non-metallic inclusions caused less ferrite banding in the steel produced with works scrap in the charge than in that for which no scrap was used.

The Problem of the Large Ingot at the Kuznetskiy Metallurgical Works. I. Demko. (Stal, 1939, No. 12, pp. 39–45). (In Russian). For a number of technical reasons the sizes of the ingots cast at the Kuznetskiy Works are limited to a minimum of 6·33 tons with wide-end-up moulds with feeder-heads, and 3·15 tons with wide-end-down moulds without feeder heads. An extensive study of the problem of obtaining sound ingots of such large size had to be made. The large amount of evidence as to the structure of ingots obtained under different conditions is summarised and explanations of the observed facts are advanced as a preliminary to a practical solution. In wide-end-down ingots piping and porosity were very marked and did not weld up during rolling, even though the access of air to the surfaces concerned had been prevented. This, it was shown, was due to the presence of a coating of a slag of oxides originating from the molten metal and not to a residue of the furnace slag. On changing over to wide-end-up moulds, a number of difficulties were encountered in connection with the behaviour of the mould bottoms and with refractory feeding-head scrap getting into the soaking pits. While piping could be eliminated, shrinkage porosity continued to be troublesome. Sectioned ingots exhibited the following four zones: (1) A contaminated feeder-head zone; (2) a zone of sound metal; (3) an unsound zone showing segregation and shrinkage pores; and (4) a zone of sound metal at the bottom amounting to 50–60% of the whole. It is pointed out, incidentally, that the ingots obtained were not worse than those claimed as satisfactory by Gathmann (*see* Blast Furnace and Steel Plant, 1937, vol. 25, Feb., p. 204), but the requirements and the method of inspection (by making sulphur prints of finely ground surfaces) were more stringent at the Soviet works. A mechanism which would explain the above structural features of the ingots is suggested for the solidification process. This involves essentially the formation of the zone of columnar crystals growing from the mould walls into the interior and the precipitation from the liquid phase of austenite solid-solution crystals. These sink to the bottom of the square ingots, forming a heap having the shape of a truncated pyramid. The thickness of the zone of columnar crystals increases with the distance from the bottom of the ingot, reaching a maximum. The formation of the intermediate, third zone of unsoundness is explained by movement and constraint imposed on the residual melt by the two structural formations referred to. The effect on the above mechanism of a change in pouring temperature is indicated. The observations made on a large number of ingots and their interpretation lead the author to make a number of suggestions regarding mould design.

The Axial Heterogeneity in Ingots of Carbon Steel. F. Khablak. (Stal, 1939, No. 10–11, pp. 68–74). (In Russian). Large white spots were noticed near the centre in sulphur prints of cross-sections of 75-mm. square ingots of 0·55–0·65% carbon steel. In the absence of any data in the literature regarding the nature of these spots an investigation was carried out which showed that they corresponded to regions low in segregating constituents—in particular sulphur and carbon and, to a lesser extent, phosphorus. It was further established that the white spots occurred only along the axis of the rolled product and might be situated at various distances from the top end. Observations on different steels showed that the more pronounced the minor segregation and the higher the carbon content, the larger and more distinct were the white spots in the sulphur prints. Heat treatment had no effect on the macrostructure of the spots. The mechanical properties of the steel in the region of the spots were about the same as those of steel from other parts of the ingot, the impact strength in some cases being actually higher, whilst the ductility (as measured by the elongation and reduction of area) was generally lower. The latter fact may, however, be connected with specific structural defects in the axial region. Axial heterogeneity was studied in a number of experimental, bottom-poured ingots from two heats of 0·47% and 0·58% carbon steel respectively. Ingots were poured in both wide-end-up and wide-end-down moulds. Ingots were sectioned and their axial heterogeneity was examined, whilst others were examined after rolling down to 65 mm. square. In sections cut from ingots poured in wide-end-down moulds white spots were found to extend from the top for three-fifths of its length. In ingots from wide-end-up moulds no white spots could be found. The experiments also confirmed the superiority of wide-end-up moulds from the point of view of the depth of the shrinkage cavity. Segregation was also studied by making chemical analyses of samples from different parts of the ingot. In general it was found to be slight. It is suggested that the formation of the white spots in deeply piped ingots may be accounted for by the residual melt accumulating near the bottom of the cavity, leaving the dendrites in the upper part more free from segregates and more ready to weld up during rolling. Some decarburisation of these dendrites may also occur in closed shrinkage cavities by the action of hydrogen evolved in such cavities.

Designs Device for Producing Partial Vacuum in Ingot Moulds. (Steel, 1941, vol. 108, Jan. 20, p. 82). An illustrated description is given of a device for assisting the removal of gases from ingot moulds as they are being filled. The device consists of a flat steel box attached to the underneath of the ladle, so constructed that it fits tightly round the edges of the upper end of the ingot mould. A tube, rectangular in section and protected by firebricks, passes horizontally through the box; one end of this tube is connected to a small blower and the exhaust end is trumpet-shaped. A venturi nozzle is constructed through the lower wall of this tube to form a passage for the gases which are drawn out of the mould by the action of the air blown through the tube.

FORGING, STAMPING AND DRAWING

(Continued from p. 124 A)

Some Typical Shell Forging Methods. F. G. Schranz. (Steel, 1941, vol. 108, Jan. 13, pp. 48–52). The author describes some of the latest types of hydraulic machines for piercing and forging billets to make shells and presents some data on the quantity of water and the pumping capacity required to operate a given number of these machines.

Shell Forging by the Upset Method. W. W. Criley. (Iron Age, 1941, vol. 147, Jan. 16, pp. 25–28). A description is given of the upset forging machine and process developed by an American company for producing shells up to 5 in. in dia. In this process the shell is made from the billet in five operations, the last four of which pierce the billet without increasing its length. It is stated that a gang of six or seven men operating one of these machines can produce an average of 70 shells 75 mm. in dia. per hr. The interior of the shell in the as-forged state does not require any machining.

Notes on Cold Trimming Drop Forgings. J. R. Thain. (Heat Treating and Forging, 1940, vol. 26, Nov., pp. 531–533). The author discusses some factors affecting the pressure required for the trimming of drop forgings in the cold state. The most important factor is the clearance or the distance between the punch and the trimmer blades when the punch is entering the die centrally. Opinions differ as to the correct amount of clearance; some maintain that the total clearance should be 10% of the flash thickness, for mild steel, 12% for medium carbon steels and 5% for brass. Another view is that the clearance should be related to the ductility of the metal as shown by the reduction of area in a tensile test, or to the percentage reduction in thickness before failure in shear.

Die Design. W. G. Kifer. (Iron Age, 1941, vol. 147, Jan. 9, pp. 37–39). The author defines a number of terms used in die manufacture and makes some recommendations on the design of dies. He also gives some particulars of a special die steel which has been developed. This is known as "Graphitic Steel." This material contains 1·50% of total carbon and three types for three different applications are made. These are "Graph-Sil" which can be water-quenched, "Graph-Mo" for oil-quenching and "Graph-Tung" for making coins.

Now Tubing is Shaped, Strengthened, Hardened in One Operation by New Forming Process. (Steel, 1940, vol. 107, Dec. 30, pp. 40–43, 56). An illustrated description is given of a new type of tube-forming machine. The machine consists essentially of three heads mounted on a long horizontal bed. One head revolves the tube to be processed; the second head holds two forming wheels which bear on the outside of the tube at diametrically opposite points; the third is hydraulically controlled and is used to exert pressure or tension on the tube according to whether it is desired to increase or decrease the wall thickness. The machine illustrated is capable of forming a tube 20 ft. long × 8 in. in dia. Such a tube can, for example, be reduced to 2 in. in dia. with the wall increased in thickness from No. 8 gauge to $\frac{5}{16}$ in. Several changes in the diameter and wall thickness can be made along the length of the same tube within the limits of cold-work of the material.

Tube Shaping. D. James. (Iron Age, 1940, vol. 146, Dec. 26, pp. 31–34). The author describes the development of the Dewey process of tube forming. This is the process described in the preceding abstract.

The Origin of Gauges for Wire, Sheets and Strip. H. W. Dickinson and H. Rogers. (Newcomen Society: Engineering, 1941, vol. 151, Jan. 24, p. 75; Jan. 31, pp. 85–86). In the first part of their paper the authors state that wire-drawing in the true sense of the word, *i.e.*, drawing a rod of metal through successively smaller holes in a tempered steel plate, called a "wortle," dates back certainly to the fourteenth century, and probably earlier. They refer to some old descriptions of the wortle and review its development to the modern gauge. They also deal briefly with the introduction of an adjustable thickness gauge known in the mill as a "catchem," which became necessary as thinner and thinner sheets were required. In the second part of their paper the authors review the attempts to standardise gauges made in Great Britain, France and the United States during the nineteenth century.

ROLLING-MILL PRACTICE

(Continued from pp. 124 A–125 A)

Roll More Tons. A. E. Lendl. (Iron and Steel, 1941, vol. 14, Jan., pp. 146–150). The author emphasises the bad effect of side work on the lateral surfaces of the grooves in rolls used for rolling angles. He considers formulæ which the roll designer can use to calculate the free lateral spread from which the amount of preventable lateral spread can be deduced. Some experimental work is described which proved that Ekelund's formula produced the best results, and an instance is quoted where the adoption of a design obtained by this formula led to eight times as many tons of angles being rolled between two roll dressings.

Medium Sheet Mill at the Zaporozhstal Works. F. Panasenko. (Stal, 1939, No. 12, pp. 32–39). (In Russian). The new sheet mill at the Zaporozhstal Works consists of one two-high stand and one three-high Lauth stand placed 28·6 m. apart in tandem. A description is given of the slab-preheating furnaces and of the roller conveyors connecting the various units of the mill. The mechanism for counter-balancing the upper roll of the three-high mill is referred to as a special feature. In the two-high mill the slabs are cross-rolled—the length of the slab forming the width of the sheet. Reduction is from 100–250 mm. to 20–50 mm. in 15–19 passes. Details are given of the loads on the motor (the nominal current rating is 4250 amp. with permissible overloads of 210–220%). In the three-high mill reduction is from 20–50 mm. to 6–25 mm. The width of sheet rolled is from 1·4 to 2·0 m. The maximum length of sheets is 16 m. In conclusion, some details are given regarding the quality of the output for several types of steels rolled.

Continental Steel Corporation Puts New Sheet Mill on Production. R. J. Leckrone. (Blast Furnace and Steel Plant, 1940, vol. 28, Dec., pp. 1189–1190). The author describes and illustrates a three-high sheet mill recently installed at the works of the Continental Steel Corporation, Kokomo, Indiana. This mill will take sheet bars up to $\frac{5}{8}$ in. in thickness and roll them down to 0·109 in. in five passes. The roll speed is about 30 r.p.m. and only the bottom roll is power-driven.

Bethlehem Enlarges Continuous Pipe Mill at Sparrows Point. (Industrial Heating, 1940, vol. 7, Dec., pp. 1186–1188). A brief illustrated description is given of the tube mill and skelp heating furnace recently installed at the Maryland works of the Bethlehem Steel Co. (*See* Journ. I. and S.I., 1940, No. II., p. 146 A).

PYROMETRY

(Continued from p. 11 A)

A High Sensitivity Radiation Pyrometer. N. E. Dobbins, K. V. Gee and W. J. Rees. (Refractories Journal, 1941, vol. 17, Jan., pp. 11–14). This is a reproduction of a paper which appeared in the Transactions of the British Ceramic Society, 1940, vol. 39, Aug., pp. 253–257, describing a photo-electric pyrometer. (*See* p. 11 A).

Temperature Measurement. I. Samuels. (Metals Treatment Society of New South Wales: Australasian Engineer, 1940, vol. 40, No. 7, pp. 85–89). The author explains the principles applied for the measurement of temperatures in liquid-expansion thermometers, mercury-in-steel and vapour-pressure thermometers, electrical-resistance and thermo-electric pyrometers, and optical and radiation pyrometers. Several diagrams of connections and of methods of mounting thermocouple sheaths are presented, and the author concludes with some notes on the installation of industrial pyrometers.

HEAT TREATMENT

(Continued from pp. 125 A–127 A)

Saturation of Iron-Carbon Alloys with Beryllium. I. E. Kontorovich and M. Ya. L'vovskiy. (Vestnik Metallopromyshlennosti, 1939, No. 12, pp. 64–70). (In Russian). The surfaces of a number of plain carbon steels (carbon 0·02–1·68%) and of three cast irons (total carbon 2·81%, 3·58% and 4·05%) were cemented with beryllium by heating specimens at 1000° C. for 10 hr. in a powder containing 96% of beryllium to which had been added 5–10% of beryllium oxide. The layers formed were examined microscopically and by X-rays, and their hardness and chemical stability were studied. Surface layers on high-carbon iron alloys developed a hardness up to 2000 Vickers units. These layers consisted of (*a*) an outer zone of extremely hard columnar crystals of beryllium carbide (Be_2C) orientated perpendicularly to the surface, and (*b*) a neighbouring zone consisting of a saturated solution of beryllium in α-iron and of particles of iron beryllide. This zone was up to 1 mm. thick. The thickness and hardness of the first zone increased with increasing carbon content of the iron or steel. The formation of the first zone interfered with the formation of the second zone. Unfortunately, from the point of view of practical applications of this process of surface hardening, it was found that the surface zone consisting of beryllium carbide was slowly decomposed by atmospheric moisture.

Production Methods for Case-Hardening. A. J. G. Smith. (Metallurgia, 1941, vol. 23, Jan., pp. 89–90). The author describes an American plant in which the whole process of case-hardening is continuous and automatic. For carrying the work through the complete cycle light grid trays are employed, and these are moved through the furnace by means of hydraulic cylinders operating at a pressure of 500 lb. per sq. in. Mechanically controlled oil valves and interlocking pilot valves, together with interlocking switches, ensure that no fresh operation in the cycle can be undertaken until the preceding one is completed. The time cycle is controlled by an electrically operated device which permits of any desired time interval being set and maintained. Either town gas or natural gas may be employed as a carburising medium as well as for heating.

Recent Developments in Gas Carburizing. The Hypercarb Process. A. Darrah. (American Chemical Society, Symposium on Furnace Atmospheres for Metallurgical Purposes, Sept., 1940). **Gas Carburizing by the Hypercarb Process.** A. Darrah. (Industrial and Engineering Chemistry, Industrial Edition, 1941, vol. 33, Jan., pp. 54–59). The author describes the development of the gas-carburising process and the particular form of treatment called the " Hypercarb Process." The principal feature of this process is that the hydrocarbon gases are specially treated before they come into contact with the steel. The purpose of this treatment is, first, to activate the atmosphere, and, secondly, to remove the excess of carbon which would otherwise form a deposit on the steel. In the plant described, town gas, or straight natural gas, is preheated to about 1700° F. by passing it over a number of internally heated radiant tubes and allowing it to rise into the heat-treatment chamber, through which it passes in the opposite direction to that of the charge.

Protective Tinning in the Nitriding Process. (Tin and Its Uses, 1941, Jan., pp. 8–9). A brief description is given of one of the less common applications of the tinning process. In this instance the coating is applied to those surfaces of an article which are to be left unhardened in a subsequent nitriding treatment. The surfaces which are to be nitrided are left untinned by coating them with a suitable varnish or wax before they are put in the bath of molten tin alloy.

Local Hardening. D. McPherson. (Machine Shop Magazine, 1941, vol. 2, Feb., pp. 94–99). The author discusses some of the disadvantages of the cyaniding process. Some cyanide hardened tools have a dull or an irregular mottled appearance, which is usually a matter of appearance that does not affect the cutting properties. This the author has found to be intimately associated with the presence of a nickel impurity in the bath. Experiments were conducted by applying an external e.m.f. to pieces of tool steel immersed in a central position in a cyanide bath. Three pieces of brightly polished high-speed steel were given identical treatments, during which one was made strongly anodic, the second strongly cathodic, and the third kept neutral. Each was found to be discoloured, the second one being the most marked. The first was uniformly discoloured with a pleasing grey matt finish. The precise explanation of this is now being sought. A cyanide bath which has " aged " gives the best performance. The ageing is simply the oxidation of some of the cyanide, and the process can be accelerated by raising the temperature of the bath to about 850–900° C., but it must be allowed to cool before use to about 570° C. which is the optimum temperature for cyaniding. In conclusion, some examples are given of the increased performance of high-speed steel tools obtained as a result of cyaniding.

Producing Annealing Atmospheres from the Products of Combustion of Gaseous Fuels. A. G. Hotchkiss. (American Chemical Society, Symposium on Furnace Atmospheres for Metallurgical Purposes, Sept., 1940). **Annealing Atmospheres from the Combustion Products of Gaseous Fuels.** A. G. Hotchkiss. (Industrial and Engineering Chemistry, Industrial Edition, 1941, vol. 33, Jan., pp. 32–38). The author traces briefly the development of equipment for producing annealing-furnace atmospheres from the products of combustion of gaseous fuels. He describes and illustrates a gas converter with a capacity of 2000 cu. ft. per hr. suitable for town gas, natural gas, propane, butane and cracked ammonia. This converter is equipped with indicating flow-meters for gas and air, an automatic fuel-air proportioning mixer, burner, water-cooled combustion chamber, condenser, water separator and condensate trap. He presents combustion data for various gaseous fuels and explains methods of calculating the correct atmosphere compositions for various purposes.

Chemical Equilibrium and the Control of Furnace Atmospheres—A Review of Equilibrium Data. J. B. Austin and M. J. Day. (American Chemical Society, Symposium on Furnace Atmospheres for Metallurgical Purposes, Sept., 1940: Industrial and Engineering Chemistry, Industrial Edition, 1941, vol. 33, Jan., pp. 23–31). The concept of equilibrium and the properties of the equilibrium constant, particularly its significance in relation to the control of conditional atmospheres in metallurgical furnaces, are described and illustrated by typical reactions for iron and steel. The equilibrium constants for the reactions:

$$\begin{aligned}
Fe + H_2O &= FeO + H_2\\
Fe + CO_2 &= FeO + CO\\
CO + H_2O &= CO_2 + H_2\\
CO_2 + C &= 2CO\\
CH_4 &= C + 2H_2\\
Fe_3C + CO_2 &= 3Fe + 2CO\\
Fe_3C + 2H_2 &= CH_4 + 3Fe
\end{aligned}$$

are critically reviewed, and selected values are given.

Protective Atmospheres for Hardening Steel. J. R. Gier. (American Chemical Society, Symposium on Furnace Atmospheres for Metallurgical Purposes, Sept., 1940: Industrial and Engineering Chemistry, Industrial Edition, 1941, vol. 33, Jan., pp. 38–41). The author explains that the properties required of a heat-treatment furnace atmosphere are that it shall not form scale or damage the surface of the steel under treatment and that it shall not affect the carbon content of the surface layer of the steel. He discusses the composition of and methods of producing gases that comply with these requirements with particular reference to " endogas controlled atmosphere," which is sufficiently low in price to be used in large continuous furnaces as well as in small tool-room furnaces. This atmosphere can also be used for the bright-hardening of steel of any carbon content in any temperature range. It is produced by the endothermic reaction of air with hydrocarbon gases in a special generator containing an electrically heated catalyst; with this technique, the completeness of the reactions obtained produces a mixture the components of which are in chemical equilibrium and are present in amounts which can be accurately controlled. This condition enables the carbon pressure of the gas mixture to be readily adjusted to balance the carbon in any steel. The gas is formed in a single-stage process and is delivered directly from the generator to the furnace; no intermediate drying or scrubbing to remove carbon dioxide is necessary. An account is given of some laboratory experiments from which it was evident that the desired control of the carbon pressure of the atmosphere could be attained by regulating the proportions of gas and air in the feed of a suitable generator. "Endogas" atmospheres consist mainly of carbon monoxide, hydrogen and nitrogen with traces of steam, carbon dioxide and methane. A table is presented of the compositions of four "Endogas" atmospheres which are in equilibrium at 1700° F. with steel containing 0·20%, 0·64%, 0·94% and 1·34% of carbon respectively.

City Gas for Special Atmospheres. C. R. Cline and C. G. Segeler. (American Chemical Society, Symposium on Furnace Atmospheres for Metallurgical Purposes, Sept., 1940: Industrial and Engineering Chemistry, Industrial Edition, 1941, vol. 33, Jan., pp. 46–54). The authors review the properties of the following types of protective atmospheres for heat treatment: (*a*) Purified partially burned gases; (*b*) cracked gas; (*c*) cracked liquid ammonia; (*d*) pure nitrogen; (*e*) natural gas treated with steam and air with subsequent removal of carbon dioxide and water; (*f*) cracked methanol; and (*g*) charcoal producer gas. They also describe how natural gas and town gas can be treated to produce suitable furnace atmospheres, and discuss the effects of variations in the quantity of air on the composition of the products of combustion when using different gaseous fuels.

Modern Electric Heat-Treatment Furnaces. F. W. Haywood. (Journal of the Birmingham Metallurgical Society, 1940, vol. 20, Dec., pp. 84–101). After an introduction, in which he outlines the development of electric heat-treatment furnaces during the last 25 years, the author briefly describes the furnaces used nowadays, under the following headings: Batch-type furnaces, forced-air-circulation furnaces, continuous furnaces and miscellaneous furnaces. A special chapter is devoted to bright-annealing in which, before reviewing the electric furnaces using controlled atmospheres, the author deals with the production and composition of the more frequently employed atmospheres.

Heat Treatment of Tool Steels. J. English. (Iron Age, 1940, vol. 146, Dec. 26, p. 35). Some practical hints on the heat treatment of tool steels are given. One of these is a simple method of estimating the amount of oxygen in a furnace atmosphere when this is between 1% and 5%. A small block of wood not exceeding ¾ in. square is placed on the hearth, the door is closed and the way it burns is observed. If the wood smokes and chars without visible flame, the oxygen content is below 1·5%. If the block smokes and then shows intermittent flashes of pale blue flame then 1½–2½% of oxygen is present. When the flame from the block is about half blue and half yellow, the oxygen is between 2½% and 4%. At about 5% of oxygen or more the block will burn steadily with a yellow flame and the residual charcoal will glow all over.

Heat Treatment of Alloy Steels. J. M. Robertson. (Iron and Coal Trades Review, 1941, vol. 142, Jan. 31, pp. 117–118). After a brief outline of the theories put forward to account for flake and hair-line crack formation in steel, the author dwells in particular on the hydrogen theory and compares this with the thermal-stress theory. It has been calculated that 0·001% of hydrogen remaining in steel is sufficient to produce rupture at 200° C. The hydrogen theory provides an explanation of the observation that once a bloom has been made non-susceptible by slow cooling, flakes cannot be produced by later forging or cooling treatment. The thermal-stress theory would not explain this. The hydrogen theory, on the other hand, does not seem to account for the way in which the susceptibility to flake formation varies from one type of steel to another. There is no evidence that alloy steels contain more hydrogen than carbon steels, nor that the hydrogen content tends to increase with the alloy content. The relation between the hydrogen which diffuses out to the surface of the metal and that which diffuses into cavities is not clear. It is supposed that, during slow cooling, the hydrogen escapes from the metal at the surface, but to reach the surface it must pass the cavities, and it would therefore be easier for it to diffuse into the cavities than to go to the surface. The hydrogen theory appears to be more satisfactory and has been more generally accepted than any other theory relating to flakes. The thermal-stress theory has been regarded as a rival to the hydrogen theory, but there is no reason why the two should not be combined. When this is done, it is seen that practically every aspect of the phenomena associated with flakes can be explained. The author deals next with hardening cracks and explains why these cracks are not produced when a steel part is first immersed in the quenching medium, but actually develop when the interior cools to near room temperature and the martensite change is approaching completion. Alloy steels are less liable to develop hardening cracks than carbon steels. One reason for this is that the general run of alloy steels contains less carbon than the carbon steels that are heat-treated commercially; this is so because it is possible to suppress the normal changes in alloy steels of low carbon content much more easily than those in plain carbon steels of similar carbon content.

WELDING AND CUTTING

(Continued from pp. 127 A–129 A)

Oxy-Acetylene Welding of Carbon-Molybdenum Pipe. E. R. Seabloom. (Welding Engineer, 1940, vol. 25, July, pp. 23–26). The author discusses the suitability of molybdenum steel for high-pressure and high-temperature applications and compares the temperature-expansion curves for a carbon steel, a molybdenum steel and a chromium-molybdenum steel. He also describes a suitable technique for oxy-acetylene welding and stress-relieving thick-walled tubes of molybdenum steel.

Welding of Carbon-Molybdenum Piping for High Temperature-High Pressure Service. R. W. Emerson. (Welding Journal, 1941, vol. 19, Oct., pp. 366-S–376-S). The author reports on an investigation of the welding characteristics of a low-carbon, 0·5% molybdenum steel, which is the material generally used in the United States for the tubes of boilers operating at 750–1000° F., and pressures of up to 1800 lb. per sq. in. The results obtained led to the following conclusions : (1) Hardness readings and microstructure examinations on a single-bead weld made under conditions of severe cooling indicated that this steel had only a slight weld-hardening tendency; (2) the heat treatment of small specimens in a laboratory confirmed conclusion (1), for, in order to cause severe hardening, it was necessary to quench the specimens in water from 1925° F. Quenching in water from as high a temperature as 1650° F. failed to produce excessive hardening; (3) of preheating and stress-relieving had a readily noticeable effect on the elongation before failure in free bending, but such treatment had little, if any, effect on the tensile-test results; (4) both hardness tests and microscopic examination of a multi-pass weld in a tube 6⅝ in. in outside dia. by 0·432 in. thick which was neither preheated nor stress-relieved indicated a structure similar to that of air-cooled heat-treated specimens; and (5) the design of joints in high-pressure tubes of this steel must be such that proper fusion is obtained in the root of the weld, and the angle of the V groove must be sufficiently large to prevent excessive undercutting of the walls.

Atomic Hydrogen Welding in Aircraft Production. R. Smallman-Tew. (Transactions of the Institute of Welding, 1941, vol. 4, Jan., pp. 22–28). The author discusses the factors affecting the application of atomic-hydrogen welding in the production of aircraft. There are at present certain forgings which the aircraft manufacturer obtains from a stamping firm, which is in turn dependent on the steel supplier, and a third party may even be concerned in the machining of the forging. If such a component could be made by the welding of pressed blanks, such a process would be of great value in war time. In the manufacture of a prototype aircraft the cost of making forging dies to produce only a few components is a very high charge, so that, if welding can be applied instead of forging, considerable economies could be effected. The theory and practice of atomic-hydrogen welding are described in the present paper and its advantages as compared with oxy-acetylene and arc welding processes for aircraft components are pointed out. The question whether a part is to be designed as a welded component, or as a replica of the forging it is to replace, is dealt with. In conclusion, a detailed description of the various stages in the production of spar wing root attachment fittings is given.

The Arc Welding of High Tensile Alloy Steels. (Transactions of the Institute of Welding, 1941, vol. 4, Jan., pp. 3–21). This report covers the work carried out for the *R*13 Sub-Committee appointed in 1938 by the *R*1 Committee on the Weldability of Ferrous Metals appointed by the Institute of Welding. The *R*13 Sub-Committee has investigated problems arising in the welding of high-tensile steels of medium and high alloy content. The report is published in three parts, abstracts of which will be found below.

The Arc Welding of High Tensile Alloy Steels. Part I. The Cracking Problem with Special Reference to Thermal Characteristics. E. C. Rollason. (Transactions of the Institute of Welding, 1941, vol. 4, Jan., pp. 3–9). In the first part of this paper the author discusses the general behaviour of alloy steels on cooling and how this affects the tendency to form cracks under the thermal conditions prevailing in welding. He then describes experiments from which thermal data respecting two medium alloy steels were obtained. The steels in question contained (*a*) 5·88% of chromium, 0·35% of molybdenum and 0·16% of carbon, and (*b*) 3·62% of nickel, 0·95% of chromium, 0·34% of molybdenum and 0·35% of carbon. It was found that the brittle constituents, in which cracks usually occur on welding, formed in the range 200–140° C. in steel (*b*), and in the range 350–250° C. in steel (*a*). A number of welding tests using these two steels are described, and the benefit of preheating is illustrated by the results obtained. In conclusion, the design of joints, welding technique and modifications of the composition of the steel to be welded are discussed.

The Arc Welding of High Tensile Alloy Steels. Part II. Base Metal Cracking. E. C. Rollason and A. H. Cottrell. (Transactions of the Institute of Welding, 1941, vol. 4, Jan., pp. 9–16). The authors carried out a number of welding tests on a nickel-chromium-molybdenum steel and a 6% chromium, 0·35% molybdenum steel, using high-tensile and special 18/8 austenitic steel electrodes. The conclusions reached from the data obtained were as follows : (1) The cracking temperature is not identical with the temperature of the martensite formation, the former usually being about 40–70° C., whilst the latter is in the region 270–210° C. The transverse stress required to produce cracking is of the order of 50 tons per sq. in.

(2) All welds creep under the contraction stresses and this creep finishes at about 250° C. Preheating reduces the effect of this creep, but if austenitic steel electrodes are used there is much less creep than usual. (3) Preheating reduces the cooling rate of the weld region and this alone is sufficient to prevent cracking. It is thought that this delayed cooling causes the martensite to be tempered and toughened without loss in hardness. (4) Cracks in the base metal are usually preceded by small cracks near the weld interface and this is the region most susceptible to cracking. (5) Two possible reasons why no cracks form when austenitic steel electrodes are used are: (*a*) Cracking is prevented by localised creeping and stress-relieving in the weld at relatively low temperatures; and (*b*) the production of suitable austenitic material in the region of the martensite adjacent to the weld toughens it sufficiently to prevent cracking. (6) Poor root penetration of the weld metal is liable to create a notch effect which can initiate a crack.

Arc Welding of High Tensile Alloy Steels. Part III. The Effect of Delayed Cooling on Properties of Martensite, and a Magnetic Test for Determining the Transformation Temperature. A. H. Cottrell, K. Winterton and P. D. Crowther. (Transactions of the Institute of Welding, 1941, vol. 4, Jan., pp. 17–21). The authors describe a series of tests on specimens of three types of steel the object of which was to ascertain whether delayed cooling makes a steel more resistant to cracking by increasing its toughness. Three types of steel were tested, the compositions of which were:

	A.	*B.*	*C.*
Carbon. %	0·35	0·32	0·16
Nickel. %	3·62	3·38	...
Chromium. %	0·95	0·65	5·88
Molybdenum. %	0·34	0·26	0·35

The specimens were heated to 1000° C. and cooled in a stream of cold air from a fan. This cooling was interrupted at the particular temperature under consideration by immersing the specimens for a given time in an oil-bath held at that temperature, and then quenching them in water. After cleaning, the specimens were tested in a Hounsfield tensometer arranged for notched-bar bending tests, and the load was increased to fracture. The tests on steel *B* showed that the delayed-cooling treatment improved the toughness of the martensitic structure and imparted to it a capacity for plastic deformation. Much improvement was obtained within the first 10 min. of delayed cooling, and a temperature of 150° C. appeared to give the best results for tempering times of this order. Cracking was induced very considerably in steels *B* and *C* when delayed-cooling treatments at 100° C. for 10 to 20 min. were applied, and cracking was entirely prevented in the case of steel *B* when the delay was for 10 min. at 150° C.

The authors also describe a magnetic method of testing which they devised with the object of studying under actual welding conditions the martensite transformation in the metal adjacent to the weld. These tests revealed that the martensitic changes in the transformed zone of the base plate occurred at temperatures between 270° C. and 210° C., and that the weld-metal transformation occur at 515° C. to 500° C. with the shielded arc electrode.

Fundamentals of Resistance Welding. R. S. Pelton. (Welding Journal, 1941, vol. 19, Oct., pp. 426-S–432-S). The author describes the theory and practice of automatic resistance spot-welding, dealing in particular with methods of measuring the welding current, the effects of varying the mechanical pressure on the electrodes and the effects of different methods of cleaning the surfaces to be welded on the quality of the weld.

Welding 70,000 Tensile Steel. W. G. Theisinger. (Iron Age, 1941, vol. 147, Jan. 16, pp. 35, 71). The author explains some of the requirements of the boiler codes of the American Society for Testing Materials and the American Society of Mechanical Engineers, and makes some recommendations regarding the procedure for welding steel with a tensile strength of 70,000 lb. per sq. in. so that the strength of the joint will comply with these codes.

Review of Welding Procedures for Creep-Resistant Steels. H. Thompson. (Welding Engineer, 1940, vol. 25, Sept., pp. 13–15, 34). After explaining the terms " creep " and " creep strength," the author discusses the welding of creep-resisting molybdenum steels. He points out that preheating is necessary before welding molybdenum steels and that 200° F. is a sufficiently high preheating temperature when the steel is under 1 in. in thickness, whilst 400° F. need not be exceeded for greater thicknesses. When a molybdenum steel is being used for a pressure vessel, it should be borne in mind that this steel is not so ductile as a carbon boiler steel, so that greater power would be required to drive the bending rolls and chipping would take longer; on the other hand there would be no difference in the cost of deseaming or cutting with the oxy-acetylene torch.

Review of Recent Developments in Pressure Vessel Welding. R. Lattice. (Welding Engineer, 1940, vol. 25, July, pp. 17–21). After pointing out some of the differences between the API-ASME and the ASME codes for the welding of pressure vessels, the author compares some of the advantages and limitations of hand and automatic welding. He describes three different techniques for welding flanges on to pressure vessels, and, in conclusion, discusses some of the advantages of deseaming with an oxy-acetylene torch instead of using pneumatic chisels; not the least of these advantages is the absence of noise.

Cast Iron Electrodes for Welding Gray Cast Iron. G. S. Schaller. (Welding Journal, 1940, vol. 19, Oct., pp. 395-S–401-S). The author discusses the results obtained in the course of a laboratory investigation of the effects of different electrode coatings and current strengths on the quality of the welds produced when welding grey cast iron. Tests made with low and high heat inputs showed that, with the former, the inclusion of gas in the weld metal was much more sensitive to the composition of the coating, because, with more heat, concentrated gases and slag were more rapidly liberated and fewer inclusions were found. With a high heat input a more distinct line of fusion was noted and microscopic studies showed that free graphite was plentiful in this region. With less heat, on the other hand, the fusion zone was narrower, whilst the change in size and shape of the graphite flakes was more abrupt. The quality of the welds produced using many different compositions of coating is shown in a comprehensive table. The investigation demonstrated that coated cast-iron electrodes which will produce sound machinable welds can be developed.

Defects in Weld Metal and Hydrogen in Steel. C. A. Zapffe and C. E. Sims. (Welding Journal, 1941, vol. 19, Oct., pp. 377-S–394-S). The authors first discuss the purely metallurgical aspects of hydrogen in steel and then describe many experiments which demonstrate the relation of hydrogen to defects in welds, as well as the influence of temperature, time and thickness of metal on the rate of hydrogen removal. The general conclusions reached by the authors were: (1) The effects of hydrogen on weld metal are insufficiently understood by welders, and they are very important, because hydrogen is frequently a constituent of the atmosphere surrounding a weld while it is being made; (2) hydrogen dissolves in steel only in the atomic state, but on coming out of solution it collects in molecular form under tremendous pressure at discontinuities in the steel; (3) the solubility of hydrogen in steel decreases several thousand-fold as the temperature falls from the melting point to room temperature. Below 400° C. the amount in true solution becomes immeasurably small. The solubility varies with the square root of the pressure at constant temperature; (4) the much lower solubility of the hydrogen in the cooler metal surrounding a freshly deposited layer of weld metal causes gas to be evolved, which may reach the surface or may be trapped by the freezing of the weld metal; (5) a fracture generally has a shiny surface in the region of occluded hydrogen; (6) specimens welded in the presence of moisture show a spotty fracture due to oxide inclusions, whereas, in the absence of moisture, the fracture shows extensive embrittled areas that form a fibrous structure when the hydrogen is removed; (7) the principal source of hydrogen contained in welds is either a reducing atmosphere or moisture present during welding; (8) the most important sources of moisture in welding are the electrode coating and the atmosphere; ideal welding therefore calls for previously dried welding electrodes and a dehumidified atmosphere; (9) the absence of ferrous oxide in killed steel tends to increase the hydrogen solubility: (10) non-metallic impurities such as sulphur, phosphorus and carbon assist in the retention of absorbed hydrogen, *e.g.*, hydrogen may react with carbon to form methane; (11) some alloying elements in steel restrict the diffusion of hydrogen and thus cause occlusion under pressure; (12) the presence of nickel increases the solubility of hydrogen in steel; (13) the rate of diffusion of hydrogen increases logarithmically with increasing temperature, but there is an optimum temperature for removing the gas, because, at elevated temperatures, the solubility increases so much that increasing quantities of the gas dissolve instead of being expelled. In the present investigation this optimum temperature was found to be about 600° C.; (14) the time required for removing hydrogen increases with decreasing temperature and may be limited for economic reasons; (15) because it is often undesirable, or impossible, to anneal welded structures at temperatures as high as 600° C., the possibility of annealing for longer periods at lower temperatures is important. At temperatures from 400° C. down to about 75° C., hydrogen can be satisfactorily removed in a matter of hours, and, even at room temperature, a sufficient quantity of the gas will be released in the course of some weeks to confer good strength and ductility upon a weld; and (16) peening assists hydrogen to escape from a weld.

A Comparison of Tests for Weldability of Twenty Low-Carbon Steels. C. E. Jackson and G. G. Luther. (Welding Journal, 1940, vol. 19, Oct., pp. 351-S–363-S). The authors study the relative merits of various methods of determining the weldability of steel. The tests were made on twenty different commercial steels ranging in carbon content from 0·10% to 0·54% with approximately similar manganese, silicon, sulphur and phosphorus contents. The following eight weldability tests were used: (1) The single-V groove weld test suggested by R. D. Williams; (2) the T bend test as used by the American Navy Department; (3) the V-notched slow-bend test of Cornelius and Fahsel; (4) the V-notched impact test of Jackson and Rominski (*see* Journ I. and S.I., 1940, No. I., p. 27 A); (5) the longitudinal-bead slow-bend test suggested by Harter, Hodge and Schoessow (*see* Journ. I. and S.I., 1939, No. II., p. 342 A); (6) a weld-hardness test discussed by French and Armstrong (*see* Journ. I. and S.I., 1940, No. I., p. 151 A); (7) the weld-quench test proposed by Bruckner (*see* Journ. I. and S.I., 1938, No. II., p. 160 A); and (8) the cold-rolling test proposed by Dowdell and Hughes (*see* p. 172 A). Each of these tests is described in detail and the data obtained in their application are presented in tables and graphs. Bend tests in accordance with test (1) on the whole range of carbon steels revealed a marked change at about 0·28–0·35% of carbon, for above this range the angle of bend obtained at the maximum load of the apparatus decreased markedly. Test procedure (2) showed a similar relationship between the angle of bend at maximum load and the carbon content. The quality of both the base metal and the welded joint is revealed by this test. The results of procedure (4) demonstrated that when high impact values were obtained the results of tests (2) and (3) would also be good. Plotting the angle of bend at maximum load and the elongation obtained by procedure (5) against the carbon content indicated a definite deterioration in these two properties at about 0·35% of carbon, which continued with increasing carbon content. As to test (6), the ease with which a hardness survey can be made has led to it being widely adopted in order to give a rough estimate of the weldability of steel. In procedure (7) specimens are heated to 1350–1365° C., and double-quenched in salt baths at 530° and 310° C. and then subjected to Charpy impact tests; in this investigation no correlation was established between the results obtained by this and by the other procedures. In test (8) the cold-rolling capacity of samples of welded plate was determined, but here also no relationship was established between the carbon content and the degree of cold-reduction required to produce failure in the weld metal or the base metal.

Metallurgical Changes at Welded Joints and the Weldability of Steels. R. H. Aborn. (Welding Journal, 1941, vol. 19, Oct., pp. 414-S–426-S). The author discusses the metallurgical changes which occur during the process of welding and how these changes can be controlled. He compares the transformation taking place on continuous cooling with that occurring at constant temperature and explains a method of estimating the safe maximum cooling rate from the S-curve. In the concluding part of his paper he considers the weldability of steels, dividing them into the three following groups: (1) Carbon steels with carbon under 0·30% and low-alloy steels with carbon under 0·15% which are readily welded and require neither preheating nor stress-relieving; (2) steels with carbon in the range 0·25–0·50% and low-alloy steels with 0·15–0·30% of carbon which are weldable with care and for which preheating and stress-relieving treatment are desirable; and (3) steels with more than 0·50% of carbon, low-alloy steels with more than 0·30% of carbon and steels containing more than 3% of an alloying element which are difficult to weld and for which preheating and stress-relieving are necessary.

Cold Rolling Testing of Welds. T. P. Hughes and R. L. Dowdell. (Welding Journal, 1940, vol. 19, Oct., pp. 364-S–365-S). The authors describe a test procedure the object of which is to determine the effect of carbon in the higher-carbon steels on the weldability of the steel and the effect of the welding heat on the malleability of the weld metal and the base metal. For this test specimens of ½-in. plate are welded by making a single V joint and are then ground to give a smooth surface. Tehea re then cold-rolled in a small mill with the weld transverse thy to rolling direction, and a reduction of 0·010 in. is applied with each pass until failure occurs. The number of passes required to produce cracks and rupture is noted. The authors claim the following advantages for this test when applied to high-carbon and alloy steels: (1) The test-pieces are easily prepared; (2) the plasticity of the weld metal and the base metal is readily shown and insufficient weld penetration is revealed in a few minutes; and (3) the test is more suitable for this class of steel than ordinary tension and bend tests.

Hard-Facing Technique. T. B. Jefferson. (Welding Engineer, 1939, vol. 24, Dec., pp. 37–39; 1940, vol. 25, Jan., pp. 32–34; Feb., pp. 27–29; Mar., pp. 23–25, 38; Apr., pp. 29–32; June, pp. 28–31). In this series of articles the author discusses the technique of "hard-facing," which is the deposition by welding or other means of a hard abrasion-resisting material on mild steel. In Part I. he classifies the hard-facing materials into three groups, as follows: (1) Ferrous alloys; (2) non-ferrous alloys; and (3) diamond substitutes. In Part II. he describes tests which demonstrated the different degrees of hardness obtained by different methods of deposition. In Part III. the oxy-acetylene process of deposition is discussed with particular reference to the effect of the adjustment of the flame on the hardness of the metal deposited, whilst in Part IV. the electric welding process is dealt with. In Part V. the wear resistance of different thicknesses of deposited metal and the relative merits of the single-bead and multiple-bead techniques of welding are considered. In Part VI., which concludes the series, a simple wear-testing device for determining the wear resistance of specimens of weld metal used for hard-facing is described. The results obtained with this device show that the Brinell hardness number of the specimen is not necessarily an index of its wear resistance.

Hard-Facing Permits Combination of Toughness, Hardness. G. Z. Griswold. (Machine Design, 1940, vol. 12, Dec., pp. 60–63). The author describes and illustrates a number of examples of the successful application in industry of hard-facing, or the deposition of wear-resisting alloys by the oxy-acetylene process.

Hardfaced Parts. (Automobile Engineer, 1941, vol. 31, Jan., pp. 13–16). Some particulars are given of the special alloys used for building up worn parts of internal combustion engines such as exhaust valves, valve seats, valve rockers and pistons. The technique of Stelliting is described in detail.

CLEANING AND PICKLING OF METALS

(Continued from pp. 19 A–21 A)

Alkaline Cleansers for Cleaning Metal Components. P. Mabb. (Metallurgia, 1941, vol. 23, Jan., p. 80). The author outlines the mechanism of the degreasing of metals and alloys with alkaline cleansers and discusses suitable compositions of such cleansers.

Diffusion Coatings on Metals. F. N. Rhines. (American Society for Metals, Oct., 1940, Preprint No. 46). The author designates as diffusion coatings all coatings in the production of which diffusion phenomena are involved, *i.e.*, those produced by so-called cementation processes. He discusses the general principles governing the operation of these processes and points out that the diffusion may occur between two solids, *e.g.*, in cladding, between a solid and a liquid as in tinning, and between a solid and a gas as in carburising. He describes and illustrates the simple principles derived from the phase rule which govern the structure of diffusion coatings in binary systems, and also gives some data for more complicated ternary systems. He considers the factors determining the structure of the coatings formed, dealing separately with those formed by solid, liquid and gaseous diffusion. He discusses, in particular, the rate of formation of diffusion layers and the physical properties of the coatings. Finally, the author presents tabular data on the various systems which have been employed in the formation of cemented coatings on iron and steel and on some non-ferrous metals. A bibliography on the subject with 271 references is appended.

Chrome Hardening Can Increase the Useful Life of Machine Parts Says " Works Chemist." (Sheet Metal Industries, 1940, vol. 14, Dec., pp. 1295–1298). A review is presented of the theory and practice of chromium-plating, the characteristics of the deposit and the many industrial applications of chromium-plated steel. Particular attention is drawn to the reduction of friction in pressing operations by using chromium-plated dies.

Notes on the Spot Test for Thickness of Chromium Coatings. W. Blum and W. A. Olson. (Proceedings of the American Electroplaters' Society, 1940, June, pp. 25–28). The authors discuss the effects of changes in the temperature and in the acid concentration on the results obtained by the spot test for determining the thickness of very thin chromium coatings on steel. The basis of this test is the time required for one drop of concentrated hydrochloric acid to dissolve the chromium. They give the test procedure which they recommend for determining coatings 0·00005 in. or less in thickness.

An Electrolytic Chromium Plate Thickness Tester. S. Anderson and R. W. Manuel. (Electrochemical Society, Oct., 1940, Preprint No. 3). The apparatus designed by the authors for the determination of the thickness of chromium-plate is based on the principle that the ampère-seconds required for the anodic dissolution of a given small area of chromium-plating are proportional to the thickness of the plating. The method thus corresponds to that suggested by Britton for the examination of zinc coatings on wire (*see* Journ. I. and S.I., 1936, No. I., p. 224 A). The areas stripped in the test are 4·8 or 2·4 mm. in dia. and the electrolyte used is a mixture of trisodium phosphate and sodium sulphate, the concentration of each of the two salts being $1N$. The moment an area has been completely stripped the potential of the metal in this area changes sharply. Full data obtained under various experimental conditions are presented in tables.

Disposal of Waste Liquors from Chromium Plating. C. R. Hoover and J. W. Masselli. (Industrial and Engineering Chemistry, Industrial Edition, 1941, vol. 33, Jan., pp. 131–134). The authors report on the results of an investigation of the cost and efficiency of various methods of treating the waste liquors from chromium-plating plants. The methods studied included: (*a*) the reduction of hexavalent chromium followed by the precipitation of hydrated chromic oxide; (*b*) the direct chemical precipitation of hexavalent chromium; and (*c*) miscellaneous processes involving precipitation, coagulation and electrolysis. A number of reducing and precipitating agents were studied; of these the most promising were sulphur dioxide, sodium sulphide, sodium sulphite, sodium bisulphite, barium sulphide, ferrous sulphate, iron and steel scrap, zinc, lead acetate and barium hydroxide. After preliminary tests, three effluent-purification processes, using sodium sulphide, barium sulphide and scrap iron respectively, and one chromium-recovery process using sulphur dioxide, were more thoroughly examined. All these reagents precipitated the chromium completely and no difficulties arose from the formation of undissociated chromium complexes. The small amounts of copper, nickel, cadmium and zinc sometimes present in spent liquors containing chromium were for the most part removed in the sulphide and scrap-iron processes, and to a less extent by sulphur dioxide. The quantity of cyanide was reduced by all the processes, the reduction amounting to 50% when either iron scrap or sulphur dioxide was used. These two methods can readily be adapted for the purification of waste liquors from the chromium-plating of steel. A neutral effluent was obtained with all four processes. The special characteristics of these four methods of treatment are discussed in detail. The authors' general conclusion is that the most practical method for recovering a saleable chromium compound is a modification of the sulphur-dioxide process in which a hydrated chromium oxide of a high degree of purity is precipitated with soda ash or caustic soda. The precipitate is dried and roasted at 500° C., when it forms the dense green β-chromium-trioxide which is saleable as a green pigment. In conclusion, a brief description is given of a plant now under construction for the treatment of 40,000 to 100,000 gal. of liquor in 24 hr.

Influence of Basis Metal on the Plate. A. Bregman. (Iron Age, 1940, vol. 146, Dec. 19, pp. 39–43). The author reviews the literature on the effects of cleaning and polishing a plate or sheet on the quality of metallic coatings deposited on it by the electrolytic process. He deals mainly with nickel coatings.

Adhesion of Nickel Deposits. E. J. Roehl. (Iron Age, 1940, vol. 146, Sept. 26, pp. 17–20; Oct. 3, pp. 30–33). The author describes a refinement of Ollard's method of testing the adhesion of nickel coatings on steel, cast iron and some non-ferrous metals. The method consists of nickel plating a heavy deposit on the end of a cylindrical specimen of the steel (or other metal) which is 1 in. in dia. × 1½ in. long; a hole ⅜ in. in dia. is drilled through the centre of the deposited metal into the end of the specimen; the coated end of the specimen is then carefully machined so that the nickel/steel interface formed a ring exactly 1 in. in outside dia. × ⅜ in. in inside dia. The test-piece is then supported in a suitable die and a load is applied by means of a rod in a tensile-testing machine until the nickel ring is forced off. From the value of the load and the area of contact, the adhesive value can be calculated. In the second part of the paper the author describes a number of tests in which the adhesion values of nickel on steel and cast iron were determined and the influence of different methods of cleaning and coating the specimen was studied. He also discusses how variations in the coating thickness and the diameter of the drilled hole affect the values obtained. The results are presented in nine tables.

On the Hot-Dip Zinc Coatings of Welded Gas Pipes. Parts I. and II. Y. Ogino. (Tetsu to Hagane, 1940, vol. 26, Jan. 25, pp. 20–42; Feb. 25, pp. 76–88). (In Japanese). The author reports the results of a series of tests on galvanised butt-welded gas pipes ¾ in. in dia. The pipes were galvanised by the hot-dip method, and the efficiency of the coating was examined by tensile, compression, bending, hardness and corrosion tests.

Zinc Coatings—Unit Operations, Costs and Properties. J. L. Bray and F. R. Morral. (American Society for Metals, Oct., 1940, Preprint No. 45). The authors describe in an elementary way the various methods of producing zinc coatings on steel by hot dipping and electrogalvanising, and discuss the properties of the coatings produced by the various processes, in particular with regard to composition and corrosion resistance. They briefly consider the equipment required for the methods discussed, and compare the costs involved.

The Electrogalvanizing of Wire. E. H. Lyons. (Electrochemical Society, Oct., 1940, Preprint No. 2). The author reviews recent developments in the electrogalvanising of steel wire, giving data on production costs and describing plants operating the Tainton and Meaker processes. He considers, in particular, the difficulties experienced in producing wire surfaces capable of receiving adherent deposits and discusses modern methods for cleaning the wire surfaces before coating. He points out that ordinary pickling is ineffective and suggests that this might be due to the fact that, owing to the heat and pressure during drawing, the organic compounds used to facilitate the drawing process are transformed into tarry substances insoluble in pickling liquor. He also discusses the deleterious effect of copper in the steel wire which reduces the hydrogen overvoltage and thus causes blisters. Finally, he compares the properties of hot-galvanised and electrogalvanised wires, and outlines the methods for testing the coating thickness, recommending in particular the hydrochloric-acid/antimony-chloride strip test.

New Bethanizing Unit. (Steel, 1940, vol. 107, Dec. 23, pp. 60–66). A description is given of one of the departments of the wire division of the Bethlehem Steel Company. The process known as " bethanizing " is carried out in this department. This is an electrolytic process of coating the wire with zinc, using a zinc-sulphate solution, with the wire itself as the cathode and electrodes of a silver-lead alloy as anodes. The wire passes continuously through the solution and the thickness of the coating can be delicately adjusted in the range 0·4–2·4 oz. per sq. ft. by regulating the speed of passage of the wire through the bath.

Standard Methods of Determining the Thickness of Zinc Coatings. D. S. Abramson. (Zavodskaya Laboratoriya, 1940, No. 4, pp. 390–398). (In Russian). The author presents a critical review of methods described in the literature for the determination of the average and local thickness of zinc coatings. In addition some of the author's results obtained in checking Preece's immersion method, which is regarded as unsuitable, and the dropping and constant-flow methods of determining local thickness, are given. The constant-flow method is regarded as the most reliable. The apparatus used for the method is described. The solution employed contained 70 g. per litre of ammonium nitrate, 70 ml. per litre of $1N$ hydrochloric acid and 7 g. per litre of copper sulphate. This solution can be employed for coatings irrespective of the method of preparing the surface of the base metal. The accuracy of the method is $\pm$ 10%. For works use the volumetric constant-flow method is simpler and will give an accuracy of $\pm$ 16%.

Electrical Equipment for Strip Processing Lines. G. A. Caldwell. (Iron and Steel Engineer, 1940, vol. 17, Oct., pp. 26–38). The author describes in detail the various operations and equipment for cleaning and coating steel to make tinplate and galvanised strip by continuous methods, with particular reference to the electric motors and their control. Numerous illustrations and diagrams of the machinery at a modern American strip mill are presented.

Study of Sodium Stannate Tin Plating Solution. F. Bauch. (Proceedings of the American Electroplaters' Society, 1940, June, pp. 112–118). The author reports on a study of sodium stannate tinning solutions which was made to collect data concerning the difficulties which occur in the electro-deposition of tin and to find the optimum operating conditions. His conclusions were as follows : (1) Bivalent tin ions in a sodium stannate plating solution cause large, porous and blistered deposits. (2) The anodic behaviour of tin in sodium stannate solutions is more uniform when the free sodium-hydroxide content of the bath is above 1 oz. per gal. (3) A hydroxide film forms on the tin anode in an alkaline tinning solution when the current is cut off; unless this film is removed by an initial increase in the current density, the tin will be dissolved as stannite. This high current density is required for a short period only, and it is desirable to reduce the current later in order to obtain a high anode efficiency. Unless the source of electrical energy is designed for an overload, it must be of greater capacity than that required for actual operation. (4) The current density at the anode must be maintained within a certain range. If the anode is coated with a yellowish-green film and the gassing at the anode is slight, the optimum current density is being applied. It is difficult to maintain this condition when the quantity of goods to be plated varies greatly; full tank loads should therefore be used, and, with plating machines, dummy cathodes can be used to make up a full load. (5) The formation of bivalent tin ions is accompanied by a low potential across the solution. (6) An excess of oxidising agents in the bath lowers the cathode efficiency and the load may be attacked, so that oxidising agents should be diluted and added to the solution with vigorous stirring.

Electro-Tinning—Good and Bad. W. H. Tait. (Tin and Its Uses, 1941, Jan., pp. 6–7). The author discusses some causes of unsatisfactory tin coatings which were produced in a tin-chloride/caustic-soda bath. He states that the alkaline stannate and the acid sulphate baths, for which simple tests and controls have been worked out, are more reliable. He also gives some information on the necessary thickness of tin coatings for certain purposes.

Hot-Tinning. C. E. Homer. (Tin Research Institute, Dec., 1940, Publication No. 102). A comprehensive account is presented of the tinning of steel, iron, copper, brass and bronze by the hot-dipping process. The descriptions are confined to the tinning of fabricated articles whilst the manufacture of tinplate and the continuous tinning of strip and wire are not included.

Plating and Painting of Aircraft Cylinders. H. E. Linsley. (Iron Age, 1940, vol. 146, Dec. 12, pp. 43–47). The author describes the plant and processes which have been developed by the Wright Aeronautical Corporation for degreasing, tinning, rinsing, drying, painting and baking aircraft engine cylinders. The tin coating is applied to the cylinder barrel in the rough-machined state and it is then removed from the bore prior to nitriding. A high degree of mechanisation has been achieved by suspending the cylinder barrels from special carriers attached to pulleys on an overhead conveyor system. The barrels pass through the tinning equipment at the rate of about 30 per hr., and through the paint tanks and baking ovens at the rate of 25 per hr.

Protective Films on Tinplate by Chemical Treatment. R. Kerr. (Journal of the Society of Chemical Industry, 1940, vol. 59, Dec., pp. 259–265). The author describes a process of producing a sulphur-resisting coating on tinplate used for packing sulphur-containing foodstuffs. The process is a chemical one requiring less complicated equipment than electrolytic processes. With this chemical method a protective film is produced on tinplate by immersion in a hot chromic acid solution after preliminary degreasing; alternatively, a hot alkaline phosphate-chromate solution which simultaneously degreases and films the tinplate can be used. The films produced also inhibit rusting of the tinplate at discontinuities in the tin coating. The results of tests on filmed and unfilmed cans in which meat and vegetable products had been stored for various periods are presented. Canning trials with fruit packs showed that no advantage accrued from the chromate film.

Streaky Tin Coatings on Steel. (Tin and Its Uses, 1941, Jan., p. 9). It is stated that the causes of rippled and streaky coatings on steel articles tinned by the hot-dip process are usually traceable to the inadequacy of the degreasing, pickling or fluxing treatments. Some suggestions to remedy this trouble include annealing or normalising the steel at 550° C., or a drastic pickling under careful control for 20 to 40 sec. in dilute nitric acid (25% by volume).

History of Mill-Addition Opacifiers in Vitreous Enamelling. H. D. Prior. (Bulletin of the American Ceramic Society, 1940, vol. 19, Oct., pp. 379–383). The author reviews the development of the use of opacifiers in the preparation of vitreous enamel. In 1850, tin oxide was added to the melt to promote opacity and sometimes as much as 37% of this was added. Much work was done between 1900 and 1910 on enamel compositions to which 8–12% of tin oxide was added to the mill, then, as normal opaque frits appeared, the amount added was reduced to about 6% and substitutes for tin oxide became available; these substitutes included antimony-base and zirconium-base compounds. At the present time, with a combination of the technique of fine-grinding and the use of super-opaque frits, mill additions as low as 2% are quite satisfactory. A recent development is the use of cerium compounds as mill-added opacifiers.

Properties and Characteristics of Enamelling Iron. J. C. Eckel and J. A. Eckel. (Bulletin of the American Ceramic Society, 1940, vol. 19, Nov., pp. 419–423). After summarising the characteristic properties of continuously-rolled iron sheet and the effects of various forms of heat treatment on it, the authors discuss how the properties at various stages of production influence subsequent operations in the enamelling shop. The authors' general conclusions are as follows: (1) After cold-reduction, normalising, pickling and cold-rolling, enamelling-quality iron sheet has good drawing and enamelling properties; (2) enamelling iron shows a poor response to box-annealing after cold- or hot-reduction; if the annealing temperature exceeds about 1250° F., "rosettes" are usually formed which will cause enamelling defects; (3) the surface of the sheet should be of correct roughness to facilitate proper drainage of the enamel slip without impairing the drawing properties; and (4) enamelling iron must be free from non-metallic inclusions, which have a harmful effect on its drawing, welding, pickling and enamelling properties.

The Metalastik Process. (Machinery, 1941, vol. 57, Jan. 30, pp. 492–493). Brief particulars are given of a process of bonding rubber to steel and other metals and several applications of the process in general engineering are described and illustrated. The applications include vibration-absorbing mountings, flexible couplings and metal-rubber bushes.

Concrete Coating for Underground Piping. J. H. T. McGee. (Journal of the American Water Works Association, 1940, vol. 32, Oct., pp. 1723–1731). The author describes the difficulties which were experienced in Everglades—a town built on tidal marshes—and in Hollywood, Florida, owing to failure of the underground pipe lines embedded in particularly corrosive soils. These difficulties were overcome by coating the pipes with concrete with a cement/sand ratio of 1 : 2½ to 1 : 4, depending on the condition of the underlying pipe. The cement used was a modified Portland cement containing not more than 8% of tricalcium aluminate. The author describes at some length the method by which the coating was applied to the damaged pipe lines. In conclusion, he deals very briefly with the manufacture of cement-coated pipes as developed by the American Cast Iron Pipe Company.

Flame Pretreatment of Structural Steel Surfaces for Painting. J. G. Magrath. (American Society for Metals, Oct., 1940, Preprint No. 50). The author emphasises the importance of producing a clean and completely dry surface before applying paint to structural steel, and recommends for this purpose the treatment of the surface with an oxy-acetylene flame prior to application of the primer coating. He outlines the procedure and describes the apparatus employed. In conclusion, he briefly discusses the use of the process for burning off old paint and reconditioning the surface of structures before repainting them.

PROPERTIES AND TESTS

(Continued from pp. 130 A–135 A)

Hot Tensile Testing with Miniature Test Pieces. G. T. Harris. (Engineering, 1941, vol. 151, Feb. 7, p. 101). The author describes the apparatus used for making tensile tests at temperatures in the range 600–800° C. on very small steel test-pieces. The specimens were 1·2 in. long overall, with the parallel portion 0·632 in. long by 0·1785 in. in dia. and they were pulled in a standard worm-gear, long-base, Hounsfield tensometer. The electric furnace for heating consisted of a coil of wire wound round a steel tube 5 in. long by 1 in. in dia. Temperature measurements were made with three platinum/platinum-rhodium thermocouples. The furnace tube was inserted in a hole drilled through an insulating brick, 6 in. by 4 in. by 3½ in., which afforded excellent thermal insulation with the minimum of constructional difficulty. Some of the test results are presented and these are consistent and reliable except with regard to the elongation. In view, however, of the small quantity of material required and the low cost of preparation of the test-pieces and equipment, the fact that reliable figures for maximum stress and percentage reduction of area can be obtained by this method suggests that it might be more widely used with success.

Flakes and Cooling Cracks in Forgings. F. B. Foley. (Metals and Alloys, 1940, vol. 12, Oct., pp. 442–445). The author discusses some of the causes of flakes and cooling cracks in steel and means of preventing their formation. A coarse-grained structure makes steel liable to produce cooling cracks, but such cracks in the ingot are not of great importance in the manufacture of forgings because they readily weld up during the forging operation. The significance of hydrogen is dealt with, but the author does not believe the presence of hydrogen to be the sole cause of flakes.

Flakes in Forgings. (Metallurgist, 1941, vol. 13, Feb., pp. 4–6). The causes of circular white areas (called flakes) on the fractures of tensile test-pieces from steel forgings are reviewed, with particular reference to recent papers by Foley (*see* preceding abstract) and by Zapffe and Sims (*see* p. 25 A).

The Fatigue Strength of Steels at Low Temperatures. I. V. Kudryavtsev and V. S. Chernyak. (Vestnik Metallopromyshlennosti, 1939, No. 12, pp. 40–44). (In Russian). Earlier work on the low-temperature fatigue strengths of steels carried out at temperatures down to $-40°$ C. is briefly reviewed. The authors' work was carried out at $+20°$, $-75°$ and $-183°$ C. The steels used were: (1) *E*-8 containing carbon 0·34%, manganese 0·40%, silicon 0·27%, sulphur 0·008%, phosphorus 0·037%, chromium 1·30% and nickel 3·28%; and (2) *St*-3 containing carbon 0·15%, manganese 0·51%, silicon 0·09% and phosphorus 0·061%. Electric welds in the construction of liquid-air apparatus of steel *St*-3 were also tested. The two steels exhibited marked increases in tensile strength, yield point and fatigue strength as the temperature was lowered.

The Hardness Testing of Micro-Specimens, Micro-Constituents, and Minerals. H. O'Neill. (Metallurgia, 1941, vol. 23, Jan., pp. 71–74). The author gives some reasons why the term "micro hardness testing," which is in common use, is not very adequate and suggests the use of the term "hardness testing by micro-indentation." He reviews the types of apparatus used for this test and some of the results obtained.

The Surface Temperature of Sliding Surfaces under Conditions of Boundary Lubrication. L. V. Elin and M. D. Krylov. (Vestnik Metallopromyshlennosti, 1939, No. 12, pp. 33–39). (In Russian).

The authors studied the changes in the temperature of sliding surfaces in relation to the coefficient of friction and the pressure, using a modification of Bowden and Ridler's apparatus with which the temperature could be measured at 0·1 mm. from the surface. With this apparatus it was also possible to check the continuity of the film of the lubricant used by measuring its electrical resistance. They found that the coefficient of friction tended to decrease as the speed of sliding was increased. It was observed that the temperature varied inversely with the coefficient of friction, not directly as established by Bowden and Ridler.

The Creep Properties of a 0·31 per cent Carbon, 0·54 per cent Molybdenum Cast Steel. H. J. Tapsell and L. E. Prosser. (Proceedings of the Institution of Mechanical Engineers, 1940, vol. 144, pp. 91–96). The authors describe an investigation of the creep properties of a cast molybdenum steel, of a type suitable for turbine castings, which was undertaken to obtain data for the purposes of designing for high-temperature service. The steel tested contained carbon 0·31%, silicon 0·29%, sulphur 0·035%, phosphorus 0·035%, manganese 0·55% and molybdenum 0·54%. Complete creep curves from tests lasting, in some cases, over 10,000 hr. are shown, and the estimated stress-temperature relationships for 0·1% and 0·3% creep in 100,000 hr., and for 0·1% creep in 10,000 hr., are recorded.

On Heat Resistant and High Tensile Alloy Cast Iron. T. Saito. (Kyoto Imperial University, Transactions of the Mining and Metallurgical Alumni Association, 1940, vol. 10, Sept. 25, pp. 297–302). (In Japanese). The author reports on an investigation of the properties at 600°, 700° and 800° C. of a number of low-alloy cast irons prepared in a 3-ton electric furnace.

Applications of Cast Iron in Modern Automobile Construction. E. C. Toghill and R. V. Dowle. (Proceedings of the Institution of Automobile Engineers, 1939–40, vol. 34, pp. 253–281). The authors review briefly the manufacture and properties of cast iron with the object of bringing to the notice of the automobile engineer and designer some of the irons with special properties for specific applications. They explain the metallurgical terms used in connection with the structure, and the effect of added elements on the structure, the physical properties and on the founding of grey iron. A section is also devoted to inoculated and malleable cast irons. The application of cast irons in automobile construction is dealt with in the concluding part. (An abridged version of this paper appeared in Foundry Trade Journal, 1940, vol. 62, May 23, pp. 383–384; May 30, pp. 403–404, 406).

Alternative British and American Alloy Steels. (Iron Age, 1940, vol. 146, Dec. 26, pp. 40–41). Two tables are presented in which are shown side by side the designations of a large number of British steels to British Standard and Air Ministry requirements together with the nearest equivalent S.A.E. number. In the first table approximately similar British and American steels are compared, and the second table gives the full chemical composition of nickel steels to British and American standard specifications.

Special and Alloy Steels. W. H. Hatfield. (Sheffield Society of Engineers: Iron and Coal Trades Review, 1941, vol. 142, Feb. 7, pp. 189–190: Foundry Trade Journal, 1941, vol. 64, Feb. 13, pp. 105, 114). In reviewing the economic and metallurgical aspects of the manufacture of special and alloy steels during the present war, the author observes that although the special elements required to impart certain properties to steel are still available in sufficient quantity, strict economy is necessary and the policy to adopt is to produce steel with the properties required for a given application under the most economic and convenient circumstances. He points out the fallacy of estimating the cost of an alloy steel by simply adding the cost of the quantity of the alloying element without considering the additional processing charges incurred for such items as special annealing and slow cooling. In conclusion, the author refers to the progress made since 1918 by the introduction of free-cutting steel, creep-resisting steel and magnet steel.

Heat Resisting Steels. H. E. Arblaster. (Metals Treatment Society of Victoria: Australasian Engineer, 1940, vol. 40, Nov. 7, pp. 101–104, 108–116). In this discussion of heat resistance, the properties of heat-resisting steels and the methods of determining the suitability of a steel for a given application, the author first considers the nature of oxide films, the factors affecting their growth and their mechanical properties. He then devotes the major portion of his paper to the properties of the following classes of heat-resisting steel: (*a*) Straight chromium steels containing 12–30% of chromium; (*b*) silicon-chromium steels; (*c*) austenitic chromium-nickel steels with and without additions; (*d*) ferrous alloys containing aluminium; (*e*) chromium-manganese steels; and (*f*) alloys high in nickel and chromium, *e.g.*, 60% of nickel and 20% of chromium. The influence of the carbon content on the mechanical properties of a number of steels at high temperatures is dealt with in the concluding part of the paper.

Metals and Alloys in the Dairy Industry. H. A. Trebler. (Metals and Alloys, 1940, vol. 12, Dec., pp. 769–777). The author points out that ten years ago about 90% of all equipment for direct contact with dairy products was made of copper, with or without a tin coating, and that the same equipment, when replaced or newly constructed, is to-day to a great extent made of stainless steel or Inconel. He deals under separate headings with equipment made of the following materials: Iron and steel; low-alloy steels; 18/8 chromium-nickel stainless steel; stainless steel castings; Inconel and chromium-iron alloys; clad metal; copper; copper-nickel alloys; nickel; and aluminium.

Moly Steels Combine High Physicals, Workability. W. G. Patton. (Machine Design, 1941, vol. 13, Jan., pp. 50–53, 106.) The author discusses some applications of molybdenum steels in the petroleum and automobile industries and presents four comprehensive tables showing the composition, appropriate heat treatment, mechanical properties and applications of a large number of cast irons and steels containing molybdenum.

Molybdenum in Iron and Steel. T. N. Parker. (Mechanical Engineering, 1940, vol. 62, Nov., pp. 793–799). The author reviews the development and properties of molybdenum-bearing cast irons and steels. He refers particularly to the replacement by molybdenum of a large proportion of the tungsten in high-speed steels and discusses the influence of molybdenum on the shape of the S-curves for molybdenum steels.

The Formation and Properties of Martensite on the Surface of Rope Wire. E. M. Trent. (Iron and Steel Institute, 1941, this Journal, Section I.). The occurrence of thin layers of martensite on the worn surfaces of wires in wire ropes used for mining is described. These layers are very easily cracked, and the cracks lead to a rapid failure of the wires through fatigue. This produces a very dangerous form of deterioration, particularly in mining haulage ropes.

Under conditions of friction the rope may frequently seize or weld locally on to the object against which it is rubbing, and a thin layer of the object may be torn away and remain adherent to the wire surface.

The metallurgical structures of these thin layers are described. Martensitic surfaces similar to those found in service were reproduced on wires in the laboratory by a number of methods, such as striking the wire a glancing blow with a hardened steel tool or rubbing the wire with a steel tool under heavy pressure. These layers are examined metallurgically and their effect on the mechanical properties of the wire is examined briefly. The cause of a certain type of corrosion pitting known as "chain pitting," which occurs in service on the worn surfaces of wires in haulage ropes, is traced to a localisation of corrosion at cracks in a martensitic surface.

Spring Materials. A. M. Wahl. (Machine Design, 1940, vol. 12, Oct., pp. 46–49, 94–96). Comprehensive tables and graphs are presented of the properties of helical and flat springs of a wide range of carbon and alloy steels. The properties shown include the hardness, modulus of rupture in torsion, elastic limit, ultimate tensile strength and the fatigue limit.

METALLOGRAPHY AND CONSTITUTION

(Continued from pp. 135 A–138 A)

How to Electrolytically Polish Metals for Metallographic Examination. F. Keller. (Iron Age, 1941, vol. 147, Jan. 9, pp. 23–26). The author discusses the advantages and limitations of the electrolytic method of polishing specimens of metals and alloys for metallographic examination. The advantages include the following: (1) Large areas can be prepared for examination at low or high magnification; (2) the element of skill is eliminated; (3) the metal on the polished surface does not flow; (4) all scratches are removed; and (5) the final polishing can be done quickly and economically. The disadvantages are that the method is unsatisfactory for heterogeneous alloys, that different electrolytes and current densities are required for different alloys, and that the surface must be smooth and uniformly active before the electrolytic process is applied.

Electrical Detection of Flaws in Metal. H. C. Knerr. (Metals and Alloys, 1940, vol. 12, Oct., pp. 464–469). An electrical non-destructive method of detecting flaws in metal tubes is described. The principle upon which this method of testing is based is that a discontinuity in the wall of a tube which is passed through a set of energising coils carrying an alternating current will cause changes in the value of the current induced in the tube wall. By suitable electronic apparatus such changes in the induced current can be indicated and recorded. In addition, automatic switches can be incorporated in the apparatus which, when a fault is detected, stop the rolls feeding the tube through the testing machine. The detection of short cracks and pinholes requires a modified apparatus; in such case the effect of a small flaw upon the total circumferential current is small, but there is an area surrounding the flaw from which the current is almost completely deflected. Thus, by placing a conductor of short length in close inductive relation to the area immediately adjacent to the flaw, a relatively large change is produced in the induced current in this conductor. In the application of this principle the tube is passed between pairs of D-shaped coils, called "tangent coils," because it is easier to detect differences in the inductive effects in two similar coils than it is to detect changes in a single coil. Testing machines applying the above principles capable of testing tubes up to 5 in. in dia. have been designed, and, with smaller diameters, the tubing can be tested at a rate of about 50 ft. per min.

Electrical Detection of Flaws in Metallic Tubing. (Metallurgia, 1941, vol. 23, Jan., pp. 79–80). The possibilities of flaw detection in tubes by non-destructive electrical methods are discussed with especial reference to H. C. Knerr's method, (*see* preceding abstract).

Methods of Determining Austenite Grain Size in Steels. S. A. El'got. (Zavodskaya Laboratoriya, 1940, No. 4, pp. 428–436). (In Russian). A method of determining the grain size of steel should not cause any change in the grain size during the determination, but it should enable studies to be made of the influence on it of temperature and the time for which the steel is heated, and, finally, it should be universally applicable to all compositions of carbon steels. The methods which were tested experimentally on medium-carbon steels (the McQuaid-Ehn method, normalisation, annealing, partial quenching, and etching of quenched specimens) were all found to be unsatisfactory in one way or another and unable to satisfy all the above conditions. The McQuaid-Ehn method, in particular, suffered from a number of disadvantages, whilst some of the other methods did not give sufficiently definite results. Experiments with high-temperature oxidation or etching as a means of revealing austenite grain boundaries were shown actually to reveal the grain size as it existed during the first part of the heating period. Normally, the much larger grain size obtained at the maximum temperature remained undetected. It was found that in order to avoid this, the polished section should be heated to the required temperature for the required period of time in a non-oxidising and non-etching medium. Molten alkali carbonates are particularly suitable for this purpose. The melting point of the bath used should be only 50–100° C. below the temperature at which it is used to avoid dissociation. The specimens should then be transferred to the chloride etching salt bath for 2–5 min., or to a furnace for 10–20 min. The temperature of the etching reagent or of the furnace should be slightly below that to which the specimens had been heated. Etched specimens are quenched in paraffin oil in which all salt residues must be completely removed. After washing with cold water and drying, they are ready for examination. A black network corresponding to the grain size at the maximum temperature will be seen. Oxidised specimens are cooled in water, polished and etched with alcoholic picric acid.

Some Characteristics of Metal Surfaces. E. A. Smith. (Machinery, 1941, vol. 57, Feb. 13, pp. 541–544). The author illustrates and explains the crystallographic systems of symmetry of some of the common metals and their oxides and the significance of these systems in relation to the fatigue properties of the metal. Reference is made to the Beilby layer and the discovery of Finch that this layer on the surface of metals can take up, by solid solution at room temperature, crystalline materials which may lie on it, whilst a crystalline surface cannot do this. It is also possible to roll colloidal graphite into the Beilby layer formed on mild steel, and the possibility of cold-working bearing surfaces by buffing in order to impart anti-friction qualities is now being investigated.

Causes of the Formation of "Naphthalene" Type Fracture in High-Speed Steel. N. M. Lapotyshkin. (Vestnik Metallopromyshlennosti, 1939, No. 12, pp. 71–72). (In Russian). It was shown that, contrary to the suggestion of a previous investigator, work-hardening of high-speed steel resulting from various methods of deformation and machining followed by heating to a temperature within the α-lattice range, did not produce the characteristic coarse-grained "naphthalene" type of fracture. This type of fracture was obtained only after repeated quenching.

Reinforced Cast Iron. N. M. Levanov. (Vestnik Metallopromyshlennosti, 1939, No. 12, pp. 11–18). (In Russian). In order to strengthen long cast-iron parts they may be reinforced with mild-steel bars (carbon 0·10–0·25%), the cast iron being poured round them at a temperature of 1250–1380° C. As a result of the diffusion of carbon into the mild steel core it has been found in a microsection that the original structure of the cast iron near the core was transformed into one with finer graphite lamellæ. At a distance of 1·5–1·7 mm. from the core the structure was sorbitic. At a distance of 0·22 mm. from the core the graphite disappeared completely, the structure being that of eutectoid steel. In a tensile test the cast iron was the first to fracture, this being followed by the fracture of the steel core at a higher stress. In compression and compression-torsion tests reinforced cast-iron specimens behaved as a homogeneous material and the same thing was observed in transverse bending tests on cast-iron beams with multiple steel reinforcing cores. In the latter tests the stress-strain diagram was found to be linear up to 80–85% of the ultimate stress, the strain being completely elastic up to 60–65% of the ultimate stress. The use of reinforced cast iron for metallurgical equipment and structural elements is considered.

Iron from Heaven. R. F. Mehl and G. Derge. (Metal Progress, 1940, vol. 38, Dec., pp. 799–804). The authors review some of the beliefs which have been held since the earliest times regarding the origin and composition of meteorites, and then trace the sequence of discoveries which have been made concerning their origin and composition by scientific methods during the last 100 years. Meteorites are classified into three groups, known as siderites, siderolites and aerolites. Paneth measured the helium content of meteorites, and from this calculated that they originated at about the same time as the solar system. The iron meteorites (siderites) consist mainly of iron-nickel alloys, with about 8% of nickel. In 1863, Rose succeeded in cleaving a meteorite along the planes of the lamellæ and produced a nearly perfect octahedron, but why the meteorite had this structure and how this structure was formed remained a mystery until the science of physical metallurgy was developed in the present century. The Widmanstätten figure in meteorites consists of striations of kamacite (the α-iron solid solution) in the form of plates—which are formed by the slow transformation of the γ phase in the iron-nickel system, stable at higher temperatures—lodging preferentially parallel to the octahedral planes of the γ crystal. Between these plates of kamacite lie layers of taenite, which are the undecomposed residue of the parent γ phase. In 1930, and the succeeding years, the reason for the orientation relationship shown in part by Young was demonstrated, namely, that one crystal phase forms from another by a series of lattice movements which ordinarily can be described as a series of shearing operations. Studies of this subject have given metallurgists a picture of the atomic-crystallographic mechanism by which one phase may be born of another, a picture which has an important bearing on the age-hardening process and on the processes which lead to the formation of martensite-like structures.

The Pearlite Interval in Gray Cast Iron. A. Boyles. (Transactions of the American Foundrymen's Association, 1941, vol. 48, Mar., pp. 531–569). The author studied the pearlite transformation range in both the stable and metastable conditions of the iron-carbon-silicon system. To do this he selected two irons, one containing total carbon 3·03% and silicon 2·34%, and the other, total carbon 2·93% and silicon 2·19%; he heat-treated these in such a way as to arrest the transformation at various points. He presents micrographs which show that, in these two irons, ferrite, austenite and graphite exist in equilibrium in the temperature range 1450–1550° F. He also formed the following conclusions: (1) The rate of formation of both ferrite and pearlite was accelerated at sub-critical temperatures. (2) Under identical conditions of heat treatment, an iron containing fine graphite flakes showed more ferrite than one containing large graphite flakes. (3) When specimens were heated to 1600° F., quenched in molten lead standing at a constant sub-critical temperature, held for various lengths of time and finally quenched in water, the rate of formation of both pearlite and ferrite was accelerated at subcritical temperatures; when similar experiments were made with an initial temperature of 1800° F., the rate of formation of free ferrite was still accelerated at subcritical temperatures, but less ferrite was formed. (4) In small castings quenched at various stages of cooling in the mould, it was observed that ferrite began to form along the graphite flakes prior to the formation of pearlite; it continued to develop during the transformation period, but no additional ferrite appeared after transformation was complete. (5) Silicon not only promotes graphitisation, but also provides a mechanism for the formation of free ferrite by causing the α, γ and graphite phases to exist in equilibrium over a range of temperature well above that at which pearlite forms.

Rate of Diffusion of Nickel in Gamma Iron in Low-Carbon and High-Carbon Nickel Steels. C. Wells and R. F. Mehl. (American Institute of Mining and Metallurgical Engineers, Technical Publication No. 1281; Metals Technology, 1941, vol. 8, Jan.). The authors have determined, with an accuracy of 20%, the diffusion rates of nickel in γ-iron between 1050° and 1450° C. They examined alloys containing 0–100% of nickel and 0·01–1·23% of carbon. From the data obtained from 4% and 16% nickel steels containing respectively 0·03% and 0·6% of carbon, they determined graphically the values of the activation energy. Values for higher nickel concentrations were calculated by the Dushman-Langmuir equation. They found that the rate of diffusion of nickel is increased by several orders of magnitude as the nickel content is increased from close to zero to nearly 100%. At constant nickel concentration the diffusion rate increases by more than 300% as the carbon content is raised from zero to 1·5%. Impurities in amounts usually present in commercial steels proved to have no appreciable effect on the rate of diffusion of nickel. The authors determined the diffusion equations at nickel concentrations of 4% and 16% and carbon concentrations of 0·03% and 0·6%, and they developed two empirical equations from which the values of the diffusion rate can be calculated at nickel contents between zero and 20% and at carbon contents between zero and 1·5% respectively.

Influence of Transformation on the Oxidation of Steels. T. Ikesima. (Research Reports of the Sumitomo Metal Industries, Ltd., 1940, vol. 4, Aug., pp. 103–110). (In Japanese). Using test-pieces 50 × 30 × 2 mm., the author heated specimens of carbon steel, chromium-molybdenum steel and nickel-chromium steel to 900° C. in an electric furnace, and then determined their gain in weight during heating by means of a special balance described in the paper. He found that the gain-in-weight/temperature curves changed their direction suddenly at the transformation point, which indicated severe oxidation at that point. A corresponding result was obtained when cooling the specimens from 900° C. This anomaly was also clearly observed when examining specimens of air-hardening steels, the transformation points of which lie at relatively low temperatures. The author explains the phenomenon observed as follows: At the transformation temperature the metal changes its volume abruptly, whilst the oxide film on its surface, formed during the preceding heating, cannot deform without cracking, unless it is very thin or of high plasticity. If cracks occur in the film, however, oxygen can reach the underlying metal, which results in severe oxidation.

Influence of Austenite Grain Size upon Isothermal Transformation. Behavior of S.A.E. 4140 Steel. E. S. Davenport, R. A. Grange and R. J. Hafstein. (American Institute of Mining and Metallurgical Engineers, Technical Publication No. 1276; Metals Technology, 1941, vol. 8, Jan.). The authors studied the isothermal transformation of austenite in a commercial quality of steel S.A.E. 4140, fine- and coarse-grained specimens of which were prepared. by the appropriate selection of the austenising temperature. The investigation led to the following results: (1) Increasing the austenite grain size from 7–8 to 2–3 on the A.S.T.M. scale greatly retards the transformation of the austenite at temperatures of about 565° C. and above, at which soft lamellar structures form. At lower temperatures, where the structures are acicular, grain size has no appreciable effect upon the speed of transformation. (2) The effect of the grain size on the rate of transformation to pro-eutectoid ferrite and lamellar products varies with the temperature, being greatest just below the Ae_1 temperature and decreasing in intensity as the transformation temperature (480° C.) is approached. (3) Coarse-grained austenite tends to reject less free ferrite than fine-grained austenite at temperatures of about 650° C. (4) Grain size does not appreciably affect the hardness of the transformation products, except at temperatures at which there is a considerable difference in the amounts of free ferrite rejected by coarse- and fine-grained austenite.

An X-Ray Investigation of the Aluminium-Cobalt-Iron System. O. S. Edwards. (Journal of the Institute of Metals, 1941, vol. 67, Feb., pp. 67–77). The author reports on an investigation by means of X-rays of alloys of the aluminium-cobalt-iron system with less than 50% (atomic) of aluminium quenched from 800° C. The constitutional diagram obtained shows: (1) a small face-centred cubic p ase field α, including pure cobalt; (2) an extensive body-centredhcubic phase field β, including pure iron; and (3) a two-phase field between them. Examination of the main superlattice in the β-phase field by photometry of the films showed that, for three component alloys with less than 50% (atomic) of aluminium, the cobalt atoms always " oppose " the aluminium atoms as far as possible in the body-centred cubic lattice. The conditions for the appearance of an unexplained second face-centred cubic phase α′ in the two-phase region were examined. The determination of the tie-lines in this region was also attempted; it was complicated by spacing variations in the two-phase regions of the binary alloys, and by the presence of the α′ phase.

CORROSION OF IRON AND STEEL

(Continued from pp. 138 A–139 A)

Corrosion Control with Threshold Treatment. G. B. Hatch and O. Rice. (Industrial and Engineering Chemistry, Industrial Edition, 1940, vol. 32, Dec., pp. 1572–1579). Threshold treatment is a process by which slight supersaturation of water with calcium carbonate is stabilised by the addition of sodium hexametaphosphate in concentrations of 0·5–5 parts per million. This treatment was also found by the authors to inhibit to a marked extent the corrosive effect of such water on iron and steel. They report on their experiments on the passage of treated water through black-iron pipe and, in a specially designed apparatus, through a column of steel wool. For the quantitative evaluation of the effect of the treatment they used the decrease in oxygen concentration after the passage of the water through the column of steel wool as a measure of the corrosion. They give data on the effect of the hexametaphosphate concentration, the rate of flow of the water, the presence of previously formed rust and the *p*H value of the water, upon the rate of corrosion at normal tap-water temperatures. The authors think that the inhibitive effect of the threshold treatment upon the corrosion is due to the adsorption of hexametaphosphate, or a complex thereof, upon the metal or metal-oxide surface.

The Acid Corrosion of Steel. G. H. Damon. (Industrial and Engineering Chemistry, Industrial Edition, 1941, vol. 33, Jan., pp. 67–69). The author reports on some experiments which were carried out to ascertain the influence of the carbon content of steel and the concentration of acid on the rate of corrosion of steel in sulphuric acid. Thirteen different concentrations of acid ranging from 1*N* to 35·5*N* were used in the tests. In all cases the maximum rate of corrosion occurred at concentrations between 11*N* and 14*N*, the steels high in carbon corroding much more rapidly than those low in carbon. Steels with carbon in the range 0·06–0·37% had the lowest rates of corrosion irrespective of the acid concentration. In sulphuric acid with a concentration of 17*N* or more the evolution of hydrogen from the specimens was at first rapid, but it stopped almost completely after a few minutes; this behaviour was unaffected by the carbon content. This was found to be due to the formation of a film of ferrous sulphate.

A Method of Determining the Corrosion-Resistance of Iron and Steel. S. Johansson. (Jernkontorets Annaler, 1940, vol. 124, No. 11, pp. 629–631). (In Swedish). After a brief review of the corrosion-resistance tests of Palmær, Brennert and Sjövall, and their limitations, the author puts forward a very sensitive method of his own, in the development of which the following objects were kept in view: (1) The dissolved iron was to be kept in solution so that there would be no "complications" such as rust formation; (2) the determination of the dissolved iron at regular intervals without interfering with the course of the corrosion must be possible; and (3) the time required for the test was to be reasonably short. In the first tests sodium-chloride solutions of varying strength were used with additions of 0·02% of calcium ferricyanide and 1% of gum arabic. The ferrous ions dissolved from the specimen with the calcium ferricyanides formed Turnbull's blue (ferrous-ferricyanide) and the gum served as a protective colloid to hold the double salt in solution. It was found possible by this method to hold 1–2 mg. of iron in clear solution in 50 c.c. of liquid. The amount of iron could be determined by colorimetry at suitable time intervals. Accurate and reproducible results were obtained by this method with a large number of steels, but it had nevertheless two disadvantages. With high concentrations of dissolved iron some oxidation to tervalent iron occurred, causing the determinations to be too low, and no relationship could be established which governed the concentration at which this oxidation began. Secondly, the small addition of calcium ferricyanide appeared to have a slight accelerating effect on the corrosion, especially with the high-alloy steels. The method was therefore modified so that the dissolved iron would be fixed in bivalent form by the addition of an organic compound, either dipyridyl or phenanthroline; both of these form exceedingly stable, cherry-red, complex salts with ferrous ions. These coloured solutions are very suitable for colorimetry, as the colour is quite strong, even in very dilute solutions. Even a few millionths of a gramme of iron in 50 c.c. will cause an appreciable colour. The two compounds are of about equal value in this respect, and, using either one or the other, no oxidation to tervalent iron takes place; there is also no need to add gum arabic. Colorimetric determinations are made with a photo-electric cell and as little as 10^{-5} g. of iron in 50 c.c. can be determined without difficulty. With specimens of ordinary steel 100–200 sq. mm. in area, an iron determination can be made after only a few minutes; certain stainless steels with an area of 1000–2000 sq. mm. require a few hours, and steel with a very high corrosion resistance requires a longer time. An addition of 1·5 c.c. of a 1% aqueous solution of phenanthroline will fix at least 1 mg. of iron as a complex.

Atmospheric Exposure Tests on Copper-Bearing and Other Irons and Steels in the United States. E. S. Taylerson. (Iron and Steel Institute, 1941, this Journal, Section I.). The results of atmospheric exposure tests of twelve different irons and steels for a period of five years at three locations in the United States of America are reported. These include three steels, containing 0·03%, 0·2% and 0·5% of copper, respectively, tested by the (British) Corrosion Committee, six steels ranging in copper content from very low to 0·5%, a copper-bearing wrought iron, and two low-alloy steels. The last nine materials were of American origin and were selected to illustrate the large difference in corrosion rate that can be obtained owing to variation in analysis. The locations included an industrial district on marine marshes, an inland industrial district and a rural district. The results illustrate the great influence of copper and the even greater protective value of higher percentages of alloying elements.

The comparative pollution of the atmosphere at these three locations was evaluated by exposure of the Corrosion Committee's standard pollution samples for a period of two years.

Corrosion Testing of Electrodeposited Metal. N. J. Gebert. (Monthly Review of the American Electroplaters' Society, 1940, vol. 27, Oct., pp. 755–760). In discussing routine methods of testing nickel-, copper- and chromium-plating on iron and steel, the author considers the relative merits of a salt spray test and a rapid routine method of testing in use at the works with which he is connected. For the latter test the corrosive medium is made up of 15 g. of agar, 6 g. of potassium ferricyanide, 60 g. of sodium chloride, 1000 c.c. of distilled water and 250 c.c. of alcohol. This solution is kept in a small bottle which is maintained at about 120° F. by keeping it in a larger dish of water on a small electric hot-plate. A syphon is attached to the bottle and a small oxygen cylinder is used to blow the solution on to the material to be tested. The test is made by examining the number of blue spots which appear within two minutes on the sprayed surface. If any piece shows more than the stipulated number of blue spots, the plater and the laboratory are immediately notified, so that corrective measures can be applied without delay.

The Relative Effect of Various Polishing Finishes on Corrosion Resistance of Electro-Plated Nickel Deposits on Steel. (Proceedings of the American Electroplaters' Society, 1940, June, pp. 144–145). The results are presented of an investigation of the effects of six different procedures for the mechanical polishing, with an artificial alumina abrasive, of cold-rolled steel on the corrosion resistance of a subsequently applied nickel coating. The variations in procedure were introduced by using various rag and sheepskin polishing wheels in different sequence.

BOOK NOTICES.

(Continued from p. 142 A)

British Cast Iron Research Association. Special Publication No. 7. "*The Sampling and Chemical Analysis of Cast Ferrous Metals.*" Revised and enlarged by E. Taylor-Austin. 8vo. pp. 140. Illustrated. Birmingham, 1941: The Association. (Price 15*s*. 0*d*.)

This work, originally published in 1929, has been out of print for some time, and the third edition, re-written and greatly enlarged, is now issued in response to demand, in spite of difficulties due to war-time conditions.

It represents the most comprehensive collection of methods available at the present time for the sampling and chemical examination of cast ferrous metals, excluding steel, though the majority of the methods are applicable to this material. It covers the determination of all the elements commonly found in white and grey cast iron, pig iron and malleable cast iron. Methods are also described for ascertaining the amounts of less common elements present in these materials; twenty-three elements are dealt with in all. A complete new section dealing with the examination of ferro-alloys has been added in the new edition.

Many of the procedures described in the earlier editions have undergone considerable modification, and much new material covering recent analytical developments has been incorporated in the text. Alternative methods are given in many instances and their comparative accuracy and limitations are critically discussed. The basic chemical principles underlying each procedure are separately treated, in order to assist the reader's understanding of any particular process.

General laboratory technique is described and recommended analytical tolerances are given, and notes are included on the care and maintenance of platinum apparatus.

"*Factory Training Manual.*" Being a Practical Textbook for Use in the Factory and Workshop in Connection with the Ministry of Labour Scheme for Training Skilled and Semi-Skilled Operatives. By a Group of Engineers. Edited by Reginald Pugh. With a Foreword by The Rt. Hon. Ernest Bevin, P.C. 8vo. Pp. xii + 286. Illustrated. Bath, 1941: Management Publications Trust, Ltd. (Price 5*s.* 0*d.*)

The vast expansion of the engineering industry to meet the national emergency and the increased demand for craftsmen has emphasised the need of a training scheme to enable men and women to take an efficient part in the production of materials vital to the successful termination of the war. This handbook is intended primarily for the use of the training instructor in the factory and workshop and has been prepared by a group of engineers employed by the British Thomson-Houston Co., Ltd., each one of whom is a specialist in his branch of engineering. Part I. of the handbook deals with the general features of an engineering works, Part II. with the machines, Part III. with assembly and inspection, Part IV. with finishing, fabrication and welding, and Part V. with the human factor. It is well illustrated and should be a valuable aid to those engaged in the training of operatives in the engineering industry.

MINERAL RESOURCES

(Continued from p. 38 A)

Magnetite Deposits near Daltonganj, Palamau District, with a Note on Electric Smelting. K. K. Sen Gupta and J. Sen Gupta. (Quarterly Journal of the Geological, Mining and Metallurgical Society of India, 1939, vol. 11, Dec., pp. 143–148). The authors give an account of the geology of a magnetite deposit which occurs as a midrib of a range of hills 9 miles south-west of Daltonganj on the East Indian Railway. Some analyses of samples of the ore revealed a total iron content of 65–68%, but as the main ore body is a mixture, this percentage should not be taken as representative. It is estimated that about 4,000,000 tons of ore are available. The authors give some reasons for recommending the smelting of this ore in electric furnaces.

Lignite in the United States. I. Lavine. (Fuel in Science and Practice, 1941, vol. 20, Jan., pp. 14–19; Feb., pp. 31–38; Mar., pp. 48–51). The author describes the occurrence, geology and properties of lignite in the United States. The United States possesses nearly 950,000 million tons of lignite, which is almost entirely confined to two large fields, one in the northern Great Plains province and the other in the southern Gulf province. Some lignite has been found also in California, but the deposits have not been developed commercially.

ORES—MINING AND TREATMENT

(Continued from p. 118 A)

The Separation of Silica from the Ore before the Blast-Furnace Process. W. Luyken. (Stahl und Eisen, 1941, vol. 61, Jan. 30, pp. 97–100). The author reviews and compares the efficiency of a number of processes which have been tried out in Germany for beneficiating the lean German iron ores by the removal of surplus silica. In the author's opinion the Krupp-Renn process (*see* Journ. I. and S.I., 1940, No. I., p. 74 A) is better than the wet, the dry-magnetic and the magnetising-roasting processes when judged by the percentage removal of silica, but the latter processes can be worked with a much higher throughput of ore in a given time.

The Preparation for Smelting of Ores from the Bakal Deposits. V. Kulibin. (Stal, 1940, No. 1, pp. 12–15). (In Russian). The Bakal deposits supply, or will supply, the various iron works using the charcoal blast-furnace process in the Urals. The material mined comprises mainly dense brown ironstone, siderite and various powdered ores. The quality of these ores, which in many cases are low in sulphur and phosphorus, is referred to. Both the brown ironstone ores and the siderites have to be roasted, and the use of continuous shaft furnaces, possibly burning local peat or coal-fired, in place of the present wood-fired batch-type furnaces, is envisaged. The powdered ores have to be agglomerated, for which purpose consideration of the various available and suggested methods leads the author to the conclusion that the Dwight-Lloyd sintering process will be the best. Suggestions are also made for methods of dealing with and utilising the very large accumulations of ore fines which are screened off.

Mining and Beneficiation of Appalachian Manganese Ores. E. Newton. (United States Bureau of Mines, 1941, Jan., Information Circular 7145). The author describes the simple methods of mining and concentrating manganese ores in eastern Tennessee and compares these methods with those used elsewhere in the Appalachian region. The report contains general information on manganese in order to give a clearer understanding of the methods and costs of mining, and discusses some of the difficulties met with in these operations.

REFRACTORY MATERIALS

(Continued from p. 144 A)

The Influence of Zinc Oxide on the Corrosiveness of Checker Slags. J. H. Chesters and T. W. Howie. (Transactions of the British Ceramic Society, 1941, vol. 40, Feb., pp. 33–39). The authors have examined the causes of a number of examples of excessive corrosion of checker bricks in basic open-hearth furnaces. Analyses of the dust or slag removed from these checker bricks showed that large amounts of zinc oxide and smaller amounts of lead were present. From the results of a series of laboratory experiments they constructed melting-point diagrams of ternary mixtures of: (*a*) Silica brick, " synthetic checker-brick dust " and zinc oxide; and (*b*) china clay, " synthetic checker-brick dust " and zinc oxide. From these diagrams it can be deduced that the addition of a typical checker-brick dust to silica brick does not cause such a rapid fall in the melting point as it does when added to the corresponding china-clay mixture. This was confirmed by works trials, and the use of high-alumina bricks for the upper portion of checkerwork is therefore recommended if high concentrations of zinc oxide are expected.

The Determination of the Thermal Conductivity of Refractory Materials. E. Griffiths and A. R. Challoner. (Transactions of the British Ceramic Society, 1941, vol. 40, Feb., pp. 40–53). The authors describe two types of apparatus for determining the thermal conductivity of refractory materials in the 400–1600° C. temperature range. In one of these a specimen 18 in. square is used; it is built up of standard-size bricks, and the heat transmitted through it is measured by a water-flow calorimeter. The maximum temperature for this apparatus is about 1000° C. In the other apparatus the specimen is in the form of a disc 8 in. in dia. and 1 in. in thickness; this disc forms the top of a muffle furnace. A block of steel of known conductivity rests on the specimen, and the heat transmitted is calculated from the temperature gradient in and the known thermal conductivity of the steel. The maximum hot-face temperature attainable with this apparatus is about 1600° C. Measurements made on fireclay and silica refractories in the two forms of apparatus have given results in good agreement.

Steel Plant Refractories. J. H. Chesters. (Iron Age, 1941, vol. 147, Feb. 6, pp. 33–36; Feb. 13, pp. 47–51). The author describes methods of determining the following properties of refractory bricks used in steelmaking: (*a*) Porosity and bulk density; (*b*) specific gravity; (*c*) cold-crushing strength; (*d*) permeability to air; (*e*) Seger-cone equivalent; (*f*) refractoriness-under-load; (*g*) thermal shock resistance; (*h*) resistance to slag; and (*i*) the properties revealed by X-rays. He compares some of the British methods of testing with those laid down in the Manual of the A.S.T.M. Standards on Refractory Materials (1937).

Constructional Refractory Materials Containing Chromium Ore. K. Konopicky. (Stahl und Eisen, 1941, vol. 61, Jan. 16, pp. 53–59). The author reviews the literature on the production and properties of chrome refractory bricks and discusses the effects of the burning and the slag reactions on the properties of the chromium-ore aggregates used in their manufacture. He traces the development of different compositions of brick and their applications with reference to diagrams of the quaternary system magnesia-silica-lime-(chromium oxide + iron oxide + alumina) and discusses the causes of the failure of chrome-magnesite bricks in open-hearth furnaces. A bibliography with 379 references is appended.

Memorandum on Blast-Furnace Refractories. L. A. Smith. (Blast Furnace and Steel Plant, 1941, vol. 29, Jan., pp. 63–65). The author suggests some of the directions in which the development of blast-furnace refractories may be expected to proceed in the next two or three years. The expected improvements include the use of denser bricks made from selected Missouri flint clays for 20–30 ft. of the wall above the mantle in order to prevent disintegration in this area. He suggests that less shrinkage, disintegration and hot spots would occur in the area up to 10–15 ft. from the bottom if the lining were built right up against the shell plating.

Refractory Requirements for the Electric Smelting of Iron Ores. C. Hart. (Bulletin of the American Ceramic Society, 1941, vol. 20, Feb., pp. 53–56). The author describes the high-shaft, low-shaft and pit-type electric furnaces used in Norway, Sweden and Finland for smelting iron ore, with particular reference to their refractory linings.

FUEL

(Continued from pp. 145 A–146 A)

Steam Generation in Steel Mills. H. J. Kerr. (Blast Furnace and Steel Plant, 1941, vol. 29, Jan., pp. 112–115, 128). After reviewing the improvements which have taken place in the design and efficiency of steam boilers for rolling-mill power-stations, the author considers four factors which are delaying the introduction of very high pressures, *i.e.*, from 1500 to 2200 lb. per sq. in., in these boilers. These factors are: (1) The proper circulation of the steam and the difficulty of separating the water from the steam; (2) the difficulty of removing silica from the feed-water; (3) the supply of suitable steel for boilers and equipment to operate at these high pressures and temperatures; and (4) changes in the technique of operation.

Waste-Heat Boilers. J. B. Crane. (Combustion, 1941, vol. 12, Jan., pp. 27–31). The author discusses types of boilers adapted to the utilisation of the heat of industrial exhaust or waste gases. He points out certain fundamental differences between these boilers and those fired with fuel, and presents data illustrating the economy of waste-heat recovery as achieved by the former.

Low-Temperature Carbonisation and Its Ultimate Development. G. Cellan-Jones. (Coke and Smokeless-Fuel Age, 1941, vol. 3, Feb., pp. 33–37). The author gives an illustrated description of the Gibbons/Cellan-Jones low-temperature coke-oven, which has been designed mainly for the production of domestic fuel. He reviews the results so far obtained with this type of oven, and considers the possibilities of its development after the present war.

PRODUCTION OF IRON

(Continued from pp. 146 A–148 A)

Blast-Furnace Rehabilitation. J. H. Slater. (Iron and Steel Engineer, 1941, vol. 18, Feb., pp. 38–42). The author discusses some of the factors involved in the reconstruction of blast-furnaces with a view to increasing the production. These factors include the capacity of the foundations, the contour of the interior, the facilities for charging ore and coke, the number of tuyeres, instrumentation, the capacity of the stoves and the cleaning of blast-furnace gas.

Things New in Blast-Furnace Charging. G. Fox. (Steel, 1941, vol. 108, Feb. 10, pp. 70–73). The author describes some of the control facilities which are now being applied to blast-furnace charging equipment. One of these is a new and simplified type of revolving distributor with a control such that the angle of rotation of the distributor is not changed after a fixed number of small-bell discharges, but is changed in response to, and in step with, the discharges of the large bell; this means that all the skip-loads placed on the large bell for one discharge have a given degree of rotation. Another automatic control provides for the discharge of a predetermined quantity of water into selected skips. An improvement has been made in the type of test-rods used for determining the height of the burden in a blast-furnace. This new testing device is in the form of a conical weight suspended from a semi-flexible cable; the cable is sufficiently rigid to support its own weight and sufficiently flexible to be drawn taut by the weight attached to it. It has a smooth surface, and therefore makes a good fit in the stuffing-box where it passes through the furnace top.

The Manufacture of Basic Pig Iron and Basic Steel. J. A. Thornton. (Journal of the Institution of Production Engineers, 1941, vol. 20, Feb., pp. 33–37). A brief outline of the coke-oven, blast-furnace and basic open-hearth processes is presented.

The Physical Behaviour of the Ores and Additions in the Blast-Furnace. F. Hartmann. (Stahl und Eisen, 1940, vol. 60, Nov. 14, pp. 1021–1027). The author presents and discusses the results of a large number of both full-scale and laboratory tests the object of which was to obtain information on the movement and physical behaviour of the various materials which make up a blast-furnace burden. The materials tested were Gällivare ore, Freiburg ore, minette, lime, open-hearth slag, mill-cinder and coke. Laboratory tests on the ores showed that they behaved very differently with increasing temperature. Diffusion within the individual pieces of ore appeared to have no practical significance. At between 500° and 700° C. the ore split up as a result of the reaction between the carbon monoxide and the iron oxide; this of course reduced the size of the pieces. Except in the case of minette, the ores were not broken up by the evolution of carbon dioxide. The temperature at which the ores fritted and sintered was in some cases as much as 300° C. higher than in others, and it was also dependent on the nature of the gases present; the fritting and sintering temperatures therefore affect the rate at which the burden passes down the blast-furnace. The temperatures at which the ores became soft enough to change shape were also different for each ore. While being heated up to the melting point the lumps of ore changed in volume, and this also affected the space occupied by the ore in the blast-furnace and its rate of descent. The volume changes for the different ores tested varied between a growth of 3% and a contraction of 18%. The melting points of the ores were very different, and some of them flowed freely immediately after melting, whilst others required considerable additional heating before they ran easily; this of course influences the rate at which the molten metal reaches the hearth. Additional factors which affect the viscosity of the molten materials include: (*a*) The reactions between the particles left after the reduction of the iron oxide; and (*b*) the gases given off by the different components of the burden.

Relation Between the Temperature of the Slag and the Composition of the Pig Iron in the Blast-Furnace Process. N. Yakubtsiner. (Stal, 1940, No. 1, pp. 1–9). (In Russian). A brief description is first given of the construction of the tungsten-graphite thermocouple used for measuring blast-furnace slag temperatures. The experimental results were obtained in 1938 with No. 3 blast-furnace at the Zaporozhstal works producing converter iron, and furnaces No. 1 and No. 2 at the Azovstal works producing foundry and converter iron. The silicon content of the metal from No. 3 blast-furnace varied between 0·3% and 1·5%. The slag was comparatively acidic with a lime/silica ratio of about 1·0–1·1. The slag temperatures, which were usually some 50–60° C. higher than those of the metal, varied between 1320° and 1510° C. The silicon content of the iron was found to increase with the temperature of the slag by about 0·1% for every 20° C. The above slag temperature range corresponded to a decrease in the sulphur content from 0·11% to 0·03%. Provided that certain conditions remained constant, the manganese content of the iron increased with the slag temperature, but the relation was not as definite as in the case of silicon. With the Azovstal furnaces, essentially the same but much less definite relations were found to exist between the temperature of the slag and the silicon, sulphur and manganese contents of the iron. As was to be expected, similar relations were established between the temperature of the iron and its composition. Certain deviations from the above observations resulted from, in particular, changes in the volume of the blast and its temperature. The temperature measurements of the upper and lower slag, the variation in temperature and composition of the iron in the course of a tapping, and changes in the composition of the iron resulting from interruption of the tapping are described and discussed. It was also found that there was a direct relation between the compositions of the iron and slag tapped from the furnace and the temperatures measured at the axis of the hearth of No. 3 furnace. Some practical deductions are made from the data obtained.

Ferro Alloys. F. R. Kemmer. (B.H.P. Review, 1940, vol. 18, Dec., pp. 1–3). The author gives some details of the recent extensions to the Newcastle Steel Works of The Broken Hill Proprietary Co., Ltd., which comprise electric furnaces and equipment for the manufacture of ferro-manganese, ferro-silicon, ferro-chromium, some carbon-free alloys and tungsten powder.

FOUNDRY PRACTICE

(Continued from pp. 148 A–151 A)

Malleable Iron. H. F. Davis. (American Society for Metals: Canadian Metals and Metallurgical Industries, 1940, vol. 3, Dec., pp. 304–307). The author describes in detail the construction and operation of reverberatory furnaces and the casting and heat-treatment processes in the manufacture of black-heart malleable iron.

The Testing of Foundry Sands. H. A. Stephens. (Society of Chemical Industry of Victoria, 1940, vol. 40, Apr.–May, pp. 260–266). The author describes methods of determining the moisture content, permeability, degree of fineness, sintering point and strength of moulding sands.

Fundamental Theory of the Molding Sand Composition for Steel Castings. M. Yoshida. (Tetsu to Hagane, 1940, vol. 26, Nov. 25, pp. 788–795). (In Japanese). The author considers the factors affecting the selection of a suitable clay and mixing liquid to form a binding material for moulding sand for steel castings.

The Randupson Process. W. Parker. (Institute of British Foundrymen: Foundry Trade Journal, 1941, vol. 64, Mar. 20, p. 194). The author briefly describes the materials and technique employed in the Randupson process of moulding and discusses some of its advantages and disadvantages. The advantages include: (1) The great saving in moulding time, which may vary from 30% to 75%, according to the type of casting; (2) the ramming of the cement-sand mixture is easy and the mixture flows well; (3) reinforcement of the mould with straight iron rods can be done quickly, and it is unnecessary to have perfect ramming round the rods; and (4) after removing the pattern, no smoothing of the mould surface is necessary. The disadvantages include: (1) The inability to make the mould and cast in one day; (2) the life of the sand-cement mixture is only 3–5 hr., so that all the mixed sand must be used within that period; and (3) in the green state the sand has little or no bond, and the moulding method must therefore be arranged so that there will be no necessity to lift the mould before the mixture has had time to set.

Vertical Casting of Iron Pipe. C. H. Vivian. (Foundry Trade Journal, 1941, vol. 64, Feb. 27, pp. 141–142). The author gives a detailed description of the method of preparing the moulds and cores for the vertical casting of large diameter iron pipes at the Phillipsburg foundry of the Warren Foundry and Pipe Association.

Producing Cast-Iron Pipe Centrifugally. E. F. Cone. (Metals and Alloys, 1941, vol. 13, Feb., pp. 155–161). After a brief introduction, in which he outlines the development of the centrifugal casting of pipes, the author reproduces numerous photographs showing how this process is carried out at the United States Pipe and Foundry Co., Burlington, New Jersey.

Continuous Casting of Metals. (Steel, 1941, vol. 108, Feb. 17, pp. 80–83). A description is given of the Goss method for the continuous casting of steel. The equipment for this method of casting consists of a feeder ladle with a heavy refractory lining surrounded by an induction coil to keep the steel molten. This ladle is fixed over a water-cooled, die-casting unit, the upper portion of which is lined with refractory bricks and the lower portion with a copper lining surrounded by the water-circulation system. The cooling is adjusted so that a thin skin forms by solidification as the metal enters the chamber, and this skin becomes thicker when the copper lining is reached. There are ports at intervals down the sides of the chamber through which graphite or some other suitable lubricant is injected. By the time the metal reaches the lower open end of the casting unit it is sufficiently solid to be pulled out by rolls. After passing through these rolls it is cut off into suitable lengths for hot-rolling. The dimensions of the equipment and the speed of casting are not stated.

PRODUCTION OF STEEL

(Continued from pp. 152 A–158 A)

The Colorado Fuel and Iron Corporation Has Served the West for Sixty Years. C. Longenecker. (Blast Furnace and Steel Plant, 1941, vol. 29, Jan., pp. 78–98). The author presents a detailed and fully illustrated description of the plant at the iron and steel works of The Colorado Fuel and Iron Corporation at Pueblo, Colorado, where the first blast-furnace was put in operation in 1881. The plant now includes 192 coke-ovens, 3 blast-furnaces, 16 open-hearth furnaces, rolling-mills, a wire mill, a cast-iron pipe foundry, bolt and nut shop, and forge, welding, pattern and engineering shops.

The Further Development of Hydraulic Converter Tilting Mechanism. F. W. Körver. (Stahl und Eisen, 1940, vol. 60, Nov. 14, pp. 1037–1038). The author gives a detailed description, with diagrams, of a newly developed hydraulic apparatus, to operate at about 200 atm., for tilting 60-ton Bessemer converters.

The Utilisation of Bessemer Steel for Certain Rolled Sections and Parts. S. Loshchilov and B. Fastovskiy. (Stal, 1940, No. 1, pp. 22–26). (In Russian). In the introduction it is pointed out that, although there are large deposits of low-phosphorus ores in the U.S.S.R. suitable for working up into Bessemer steel, the latter has so far been utilised only for the production of rails. The present investigation was designed to study the possibilities of rolling Bessemer steel into various sections, as it is pointed out that this steel is superior in many respects to open-hearth steel. In this, the first part of the article, characteristics relating to some experimental heats of Bessemer steel produced at the Petrovskiy and the Dzerzhinskiy works are described. In both cases the metal blown was characterised by a high silicon content (1·8–2·8%). The deoxidation practice is described and the effect of final deoxidation with aluminium in the ladle on the oxygen content is shown by analyses from a number of heats. The metal was top-poured into the ingot moulds. The pouring temperatures and rates are given and the appearance and composition of the ingots are described. The surface of the ingots was improved by placing a sheet-iron sleeve on the bottom of the moulds, whilst at the Dzerzhinskiy works two or three buckets of saw-dust were also added.

Design and Operation of Regenerators for Open-Hearth Furnaces. F. H. Loftus. (Blast Furnace and Steel Plant, 1940, vol. 28, Nov., pp. 1078–1082, 1086; Dec., pp. 1167–1169, 1181; 1941, vol. 29, Feb., pp. 197–199). The author discusses the factors affecting the efficiency of regenerators for open-hearth furnaces. He presents drawings and tables of 36 different designs of checker construction and their characteristics, and in a discussion of these gives examples of some of the heat-transfer calculations involved in the design of regenerators. Descriptions are also given of the Loftus checker installation for single-pass open-hearth regenerators and of a special design of regenerator which provides for the cleaning of the waste gas before it passes through the checker chamber.

Open-Hearth Practice Substantially Advanced in 1940. W. J. Reagan. (Blast Furnace and Steel Plant, 1941, vol. 29, Jan., pp. 40–45). In reviewing the improvements in open-hearth practice which have taken place in the United States during 1940, the author refers particularly to the increased output per furnace which has been achieved by improved design and improved metallurgical control. The improvements in design include the incorporation of auxiliary slag pockets and auxiliary checkers, increasing the regenerator capacity and the increased use of insulation. More economical working has been achieved by the installation of devices for automatically controlling the gas pressure in the furnace, the air-gas ratio and the reversals of the gas flow. The use of " synthetic scrap " is also of interest; this practice can be applied at plants where surplus hot metal and a Bessemer converter are available. The hot metal is blown until the carbon is at about 0·10%, and it is then cast into synthetic scrap ingots In conclusion the author refers to the great increase in the output of alloy steel, and mentions a new method of making chromium steel. This consists of adding a mixture containing about 50% of metallic chromium to the ladle; the other ingredients include a substance which causes an exothermic reaction, and the heat thus produced offsets the chilling effect of adding the mixture to the ladle. About 95% of the chromium in the mixture passes into the steel.

Open-Hearth Problems and Improvements in 1940. H. M. Griffith. (Blast Furnace and Steel Plant, 1941, vol. 29, Jan., pp. 58–60). The author presents a summary of some of the improvements in open-hearth design and practice which have taken place in the United States in 1940. An innovation in furnace-bottom preparation is the use of a double-burnt dolomite aggregate which forms a very dense mass when rammed in the dry state. It is claimed that with this bottom the furnace can be brought up to temperature and charged without any burning in, and that in the first heat it will absorb practically no steel at all.

Developments in the Iron and Steel Industry during 1940. W. H. Burr. (Iron and Steel Engineer, 1941, vol. 18, Jan., pp. 54–72). The author reviews the various phases in the development of the iron and steel industry of the United States which have taken place during 1940.

Process Regulations for Increasing the Capacity of the Iron and Steel Industry. I. R. Risser. (Stahl und Eisen, 1940, vol. 60, Nov. 28, pp. 1069–1075). The author describes some of the measures adopted by the process-management of the August-Thyssen-Hütte A.-G. in order to improve the quality and quantity of their steel production. These measures included improving the method of weighing the charges for the steel furnaces, a new roll design for rolling rails, the adoption of new methods of reporting daily output and idle time of plant, and the use of a new form of time-sheet.

Influence of Various Factors on the Working Efficiency of Basic Open-Hearth Furnaces. F. Horiuchi. (Tetsu to Hagane, 1940, vol. 26, Nov. 25, pp. 771–776). (In Japanese). The author studied the effect on the efficiency of the basic open-hearth process of variations in the pig-iron/scrap-metal ratio, the composition of the pig iron and of the air/gas ratio of the incoming gases.

Equilibria of Liquid Iron and Slags of the System CaO–MgO–FeO–SiO_2. K. L. Fetters and J. Chipman. (American Institute of Mining and Metallurgical Engineers, Technical Publication No. 1316 : Metals Technology, 1941, vol. 8, Feb.). The authors present the results of an experimental study of the equilibrium between slag and molten iron at temperatures used in steelmaking. In order to facilitate the interpretation of the results they chose the simplest mixture that would be comparable to open-hearth slag, *viz.*, combinations of CaO, MgO, SiO_2, FeO and Fe_2O_3 in a great variety of proportions. In the course of their investigation they made 22 heats from which 170 slag samples and a still larger number of metal samples were examined after heating at various temperatures and for various periods. The analytical results obtained are listed in an extensive table. The oxygen content of the iron in equilibrium with the slags studied was found to depend on the temperature, the relationship corresponding to that known for simple iron-oxide slags. At equilibrium, the ferric oxide of the slag was found not to be completely reduced to ferrous oxide. Further, the authors determined the solubility of magnesia in molten iron oxide in equilibrium with the metals in the range 1540–1800° C. The solubility of magnesia in $CaO–FeO–SiO_2$ slags at 1600° C. is represented diagrammatically. The distribution of sulphur between slag and metal proved to be only slightly affected by the temperature of heating, but to be dependent to a great extent on the composition of the slag. In conclusion, the authors state that their results indicate the existence of stable orthosilicates and monoferrites in the molten slag.

1,200-Ton Hot-Metal Mixer. R. L. Knight. (B.H.P. Review, 1940, vol. 18, Dec., pp. 12–13). The author describes a hot-metal mixer which has recently been constructed at the Port Kembla Works of Australian Iron and Steel, Ltd. This tilting mixer has a capacity of 1200 tons of molten iron and is almost cylindrical in shape. It is lined with high-alumina refractory bricks which are bonded with a cement with a diaspore clay base. The tilting mechanism is driven by two 115-h.p. electric motors. The safety devices include a compressed-air engine which automatically returns the mixer to the normal position should there be any failure in the current supply to the tilting motors.

Modern Arc-Furnace Design and Practice. H. F. Walther. (Iron and Steel Engineer, 1941, vol. 18, Jan., pp. 22–32). After a brief outline of the history of the development of electric furnaces for steelmaking, the author describes the plant and practice at the works of the Timken Roller Bearing Co., Canton, Ohio. This plant includes two 35-ton, one 60-ton and one 100-ton electric furnaces. The 100-ton furnace is elliptical in plan, the assembled roof weighs 30 tons, it has six 18-in. graphite electrodes and power is supplied from two 10,000-kVA. transformers. On an average, the 18-in. lining lasts 86 heats and the 12-in. roof 43 heats. One of the 35-ton furnaces is a cone-shaped Héroult furnace with an all-welded shell 15 ft. 8 in. in dia. at the hearth and 18 ft. in dia. at the top. The object in this design was to improve the life of the refractory lining, and it is claimed that as many as 275 heats have been obtained with one lining and 100 heats with one roof. The general trend of electric-furnace design is towards still higher power input, but there is no general rule for the calculation of the power requirement, as this depends mainly on the type of scrap to be melted and the depth of the bath. The largest graphite electrode being produced at the present time is 20 in. in dia. with a capacity in the range 12,000–15,000 kVA. In the author's opinion there are practical considerations which put a limit on the size of a cylindrical furnace, and the maximum dimensions would be about 20 ft. in dia. × 11 ft. 6 in. deep, and 6 ft. from the door-sill to the top of the shell; such a furnace would produce about 75 tons per heat. The author explains a formula for calculating the power efficiency of electric furnaces. Applying this formula to 10-, 30-, 35-, 60- and 100-ton furnaces, their power efficiencies are found to be 50%, 60·2%, 61·7%, 65% and 66·9%, respectively. With regard to the life of linings, silica-brick linings have proved to be the most practical and economical for large furnaces, as the use of the much dearer magnesite brick is only justified in small furnaces where exceedingly high temperatures are required. The insulation of electric-furnace roofs is not practicable.

The Development of the Electric-Arc Furnace as a Large-Capacity Furnace and Its Metallurgical Application. W. Rohland. (Stahl und Eisen, 1941, vol. 61, Jan. 2, pp. 2–12). Commencing with a historical survey of the development in Germany of the electric furnace for steelmaking from the first 30-cwt. furnace which was put in commission at Remsheid in 1906, the author goes on to compare the characteristics of this furnace with those of modern 15- and 30-ton furnaces. He describes in detail the design of modern skip-charging and removable-top electric-arc furnaces, and discusses the cost and efficiency of operating furnaces of different capacities. From this it is seen that 6–15-ton furnaces are most suitable for making tool steel and special steels of high quality, while 15–30-ton furnaces are most efficient for structural and ball-bearing steels. For the duplex process furnaces of 40–45 tons are most suitable. The author discusses next the metallurgical processes involved, dealing with possible developments of the Perrin and similar processes and the application of low-frequency currents for accelerating the reactions between the steel and slag. With regard to these possibilities a brief description is given of a Siemens-Halske electric furnace which incorporates the electric arc for heating and the application of low-frequency current to produce a stirring action. Some preliminary experiments in desulphurising and dephosphorising pig iron have already been carried out. These showed that, for a given slag with a quiet bath, the sulphur content was reduced by only 0·007–0·0035% of the original content every 10 min. until 0·0014% was reached, whereas, with the application of low-frequency current, the rate of desulphurisation was increased to 0·025–0·03% in 10 min. Clear indications as to how the dephosphorisation of steel could be accelerated were not obtained in the experiments with pig iron as the high carbon content introduced complications. Using pig iron in the above furnace with a quiet bath, no dephosphorisation at all took place, but, irrespective of the carbon content, it commenced instantly on switching on the low-frequency current, and in 30–35 min. the phosphorus content was reduced by 35–55%. Owing to war-time conditions these experiments could not be pursued further, and the author suggests that investigations with a view to developing the combined electric-arc and low-frequency technique should be undertaken by the Kaiser-Wilhelm-Institut für Eisenforschung at some future date.

Steel Plant Maintenance Shops. T. R. Moxley. (Iron and Steel Engineer, 1941, vol. 18, Jan., pp. 75–82). The author discusses the organisation of maintenance shops in a steelworks. He deals with the financial aspect, and the location, size, lighting and heating of the buildings and then describes the equipment of the shops for machining, forging, heat treating, boiler-making, founding, welding, pipe-making and pattern-making.

The Use of the Normal Curve in Problems of the Steel Industry. H. Ziebolz. (Iron and Steel Engineer, 1941, vol. 18, Jan., pp. 50–52). The author explains the use of the "normal curve" (frequency curve) as a method of presenting statistics, and illustrates this by examples of its application to problems in the steel industry.

REHEATING FURNACES

(Continued from p. 9 A)

Soaking-Pit Control. M. J. Boho. (Iron and Steel Engineer, 1941, vol. 18, Jan., pp. 34–42). The author studies the progress made during the last five years in methods of soaking-pit control. He presents the problem of delivering perfectly heated ingots to the rolling-mill and describes how the heat input, the fuel/air ratio and the furnace pressure of soaking-pits can be measured and controlled, giving details of a number of control systems in actual operation.

Billet-Heating Furnace Developments. (Metallurgia, 1941, vol. 23, Feb., pp. 123–126). Illustrated descriptions are given of a number of modern furnaces for the reheating of billets of steel and of non-ferrous metals. A special type of gas burner is described which is designed to give instantaneous and complete combustion of all gases, whether clean or dirty and tarry. The burner can be used with any type of industrial gas, and it will also work with air preheated up to 400° C.

The Heating of Billets for Shell Forging. W. Trinks. (American Society of Mechanical Engineers: Industrial Heating, 1940, vol. 7, Dec., pp. 1155–1162). **Recommendations for Heating Billets for Shell Forging.** W. Trinks. (Steel, 1941, vol. 108, Jan. 20, pp. 54–57). The author examines the advantages and limitations of batch, continuous and rotating-hearth furnaces for the heating of round and square billets for forging shells.

FORGING, STAMPING AND DRAWING

(Continued from pp. 158 A–159 A)

Casting, Cutting Off and Heating Billets for Forging High-Explosive Shell. A. F. Macconochie. (Steel, 1941, vol. 108, Feb. 10, pp. 54–60). The author describes the preliminary stages in the manufacture of shell forgings as practised by the National Steel Car Corporation, Ltd., at Hamilton, Ontario. At these works the billets are cut from the long square bars with an oxy-acetylene cutting machine, and are then charged to the forging furnaces. This company has a number of rectangular and two rotating-hearth furnaces. The former are economical, in that they can be arranged with regenerative heating, whilst the latter offer the advantages of continuous charging and unloading. The present trend is towards using the rotating-hearth type.

3·7-In. Shell Forgings. (Machinist, 1941, vol. 84, Mar. 15, pp. 1080–1086). An illustrated description is given of the Stewart-Lloyd process of forging shells. After the square billets have been cut off in the required lengths they are heated to a forging temperature of about 2080° F. and transferred to a vertical piercing machine which produces a " bottle "; the bottle is then drawn down to the required outside diameter by being pushed on a mandrel through five sets of reducing rolls, each set consisting of three rolls mounted in a head with their axes at 60° to each other.

Shaping Steel to Form High-Explosive Shell. A. F. Macconochie. (Steel, 1941, vol. 108, Feb. 17, pp. 58–62, 91). The author describes some methods of pressing and rolling billets to make shells. The machines described include a Southwark vertical hydraulic press for piercing the blank, a Manfred-Weiss horizontal machine with five sets of rolls for reducing the pierced blank, an Assel shell-rolling mill and a Witter mill.

Forging Practice at the Sedalia Shop of Missouri Pacific Railroad. C. Cleveland. (Heat Treating and Forging, 1940, vol. 26, Dec., pp. 581–586). The author describes the reheating, forging and hardening equipment and processes at the Sedalia workshops of the Missouri Pacific Railroad. The manufacture and repair of leaf springs and the forging of box saddles from scrapped wagon axles are described in detail.

Forging the Stainless Steels. E. R. Johnson. (Metal Progress, 1941, vol. 39, Jan., pp. 54–59, 118). In the first part of his paper the author discusses the properties of the following four groups of stainless steels: (1) The martensitic steels containing 10–14% of chromium; (2) the ferritic steels containing 14–18% of chromium, or 23–30% of chromium with carbon 0·35% max.; (3) the austenitic steels; and (4) the high-chromium (22–26%) austenitic steels. He deals in particular with their hot-working characteristics, and in the concluding part describes briefly some of the rolling, forging and heat-treatment processes applied to the different groups.

Inspection Control in the Forge Shop. J. Mueller. (Heat Treating and Forging, 1940, vol. 26, Oct., pp. 477–480). The author discusses the organisation of inspection procedure for a shop making large numbers of medium and small forgings, and reproduces some examples of suitable inspection report forms.

New " Marshall Richards " Fine Wire-Drawing Machine. (Wire Industry, 1941, vol. 8, Feb., pp. 69–70). A description and illustration are presented of a new type of wire-drawing machine which will draw from a maximum inlet size of 0·036 in. in dia. to a finishing size in the range 0·010–0·001 in. in dia., and will accommodate a maximum of sixteen dies to suit any elongation required. The machine is designed so that the heavy end of the wire can be drawn on cones flooded with liquor and the fine end on dry cones in the usual manner, but if it is required to draw from an inlet size of 0·004 in. in dia. or less, the liquor spray can be cut off from the cones and confined to the dies only.

ROLLING-MILL PRACTICE

(Continued from pp. 159 A–160 A)

Manufacture and Composition of Grease Lubricants. T. Lennox. (Iron and Steel Engineer, 1941, vol. 18, Feb., pp. 29–34). The author describes the raw materials, equipment and processes used in the manufacture of lubricating grease.

The Rolling of Heavy and Medium Plates. E. Howahr. (Stahl und Eisen, 1941, vol. 61, Jan. 23, pp. 73–83; Jan. 30, pp. 100–107). The author presents a detailed account of the plate mills at the works of Marrel at Rive de Gier and of the Dortmund-Hoerder Hüttenverein at Dortmund. These descriptions illustrate the development of the four-high stand for rolling both thick and medium plates.

The Tandem Mill at the Port Kembla Works of Lysaght's Newcastle Works Pty., Ltd. (B.H.P. Review, 1940, vol. 18, Dec., pp. 18–19). An illustrated description is given of the sheet mill at the Port Kembla Works of Lysaght's Newcastle Works Proprietary, Ltd. This mill consists of seven two-high stands in tandem, each stand being independently driven by a 400-h.p. motor.

Follansbee Steel Corp. Expands Cold Reduction Facilities. T. J. Ess. (Iron and Steel Engineer, 1941, vol. 18, Feb., pp. 55–66). The author gives a detailed description with numerous illustrations of the sheet mills at the Follansbee, West Virginia, and the Toronto, Ohio, plants of the Follansbee Steel Corporation, where extensions have recently been completed.

Follansbee Modernizes Production Facilities. T. C. Campbell. (Iron Age, 1941, vol. 147, Feb. 13, pp. 53–55). The author describes and illustrates some of the new sheet-mill equipment which has recently been installed at the works of the Follansbee Steel Corporation at Follansbee, West Virginia. This equipment includes two four-high cold reversing mills, one four-high temper mill, two cutting and shearing lines, tinning plant and two bell-type gas-fired annealing furnaces.

Rolling-Mill Design Assists the National Effort in the Production of Sheet and Strip Metal. C. E. Davies. (Sheet Metal Industries,

1941, vol. 15, Mar., pp. 320–322). The author discusses some improvements in rolling-mill design which have been instrumental in considerably increasing the production of steel sheet and strip in Great Britain. Particular reference is made to the advantages of the four-high mill stand.

Properties of Flat-Rolled Products Improved by Change in Type of Mill. F. G. White. (Blast Furnace and Steel Plant, 1941, vol. 29, Jan., pp. 51–53, 57). The author presents an outline of the development of the continuous rolling of wide steel strip. He points out that, owing to the metallurgical requirements of this process, the product is unsuitable for such a large variety of applications as sheets rolled separately and heat-treated in packs. Many of the difficulties experienced by users have been overcome by the steel mill employing metallurgist-salesmen who are able to understand these difficulties and trace the cause.

Contribution on the Investigation of the Slip of Friction-Roll Drives. O. Meurer. (Stahl und Eisen, 1941, vol. 61, Jan. 2, pp. 13–16). The author describes some of the difficulties which arise in the roll of steel strip in a two-high stand when one roll is power-driven and the other is rotated by friction in contact with the strip. He describes a mechanical slip-measuring device with which he investigated the factors affecting the amount of slip of the idle roll of a small mill-stand when rolling strip 80–100 mm. wide down to 1 mm. It was found that the type and condition of the roll bearings were an important factor. The use of roller bearings or the constant and careful maintenance of synthetic-resin bearings lubricated with an emulsion considerably reduced the amount of slip. A heavy reduction and a soft strip of small cross-section were factors which reduced the slip. The rolling speed had practically no effect.

Demagnetizing and *IR* Drop Compensation for Cold Mills. G. E. Stoltz. (Iron and Steel Engineer, 1941, vol. 18, Feb., pp. 17–24). In the cold-rolling of steel strip, the amount of reduction depends on both the roll pressure and the tension on the strip. The latter is a function of the difference between the driving speeds of the mill-stands, and special attention has therefore to be paid to methods of controlling the speed of the mill motors. In the present paper the author describes methods of controlling the following factors which affect the variations in speed of a direct-current motor with change in the load: (1) The voltage drop caused by the internal resistance of the armature circuit; (2) the occurrence of a magnetic flux across the armature; and (3) the reduction in magnetisation of the main shunt field due to over-compensation of the commutating pole windings.

Design and Operation—Continuous Butt Weld Pipe-Mills. J. H. Loux and E. T. Trebilcock. (Iron and Steel Engineer, 1941, vol. 18, Feb., pp. 44–52). The authors give a detailed description of the Fretz-Moon process of manufacturing butt-welded steel tubes. In the years 1938, 1939 and 1940 ten tube-mills operating this process have been erected in the United States; the authors give details of the capacity of these mills and draw attention to the improvements which have been incorporated from time to time.

HEAT TREATMENT

(Continued from pp. 161 A–165 A)

Improved Facilities for the Manufacture of Heat Treatment Equipment. (Metallurgia, 1941, vol. 23, Feb., pp. 93–94). A brief description is given of the plant and testing equipment at the new works of Wild-Barfield Electric Furnaces, Ltd., at Holloway, where electric furnaces for the heat treatment of metals are manufactured.

Rationalisation of the Technology of Heating and Developments of Heat-Treatment Furnaces. N. Minkevich. (Stal, 1940, No. 1, pp. 32–39). (In Russian). The author presents a critical survey of the theory of heat-treatment furnaces and present trends in their development. Stakhanovite methods of improving and speeding-up furnace outputs include mainly an intensification of the heat supply and heating rates during the first part of the heating-up stage, increases in furnace charges and suitable modifications or amplifications of the method of heating. The concluding part of the article includes some illustrated descriptions of present-day heat-treatment furnaces in the United States.

Special Heat-Treating Furnace for Military and Other Parts. H. C. Knerr. (Metals and Alloys, 1941, vol. 13, Jan., pp. 49–50). The author describes a new furnace for the heat treatment of small metal parts. The furnace embodies an inverted metal retort, the heating space of which is about 4 ft. high and 10 in. in dia. The retort is closed by a sealed door at the bottom, and heat is applied externally by means of a gaseous fuel, so that the products of combustion do not come in contact with the heat-treated material. A rod of heat-resisting alloy passes through a stuffing-box in the upper end of the retort and it can be moved up and down by means of a winch and cable. The parts to be heat-treated may be attached to a fixture on the lower end of the rod and thereby pulled up into the furnace, where they remain freely suspended during heating. When ready for quenching, the rod is lowered so that the charge moves directly out of the bottom of the furnace into the quenching tank beneath. The atmosphere of the furnace consists of completely dissociated anhydrous ammonia. The furnace can be used at up to 1950° F., so that it is suitable for nearly all ferrous and non-ferrous alloys, with the exception of high-speed steel. The furnace is also suitable for bright-annealing, and in this case the materials are allowed to remain in the protective atmosphere until they are cold.

Small Furnace with Panel Burners. C. B. Williams. (Metal Progress, 1941, vol. 39, Jan., pp. 51–53). The author gives a detailed description of a batch furnace with a hearth 36 in. × 48 in. for the heat treatment of alloy steels. The furnace is fired with a natural-gas/air mixture through "radiant combustion" burners 9 in. × 18 in. in area placed in the walls.

Electrical Resistance Furnaces for High Temperatures. P. Schwarzkopf. (Metals and Alloys, 1941, vol. 13, Jan., pp. 45–49). After reviewing the different types of electrical resistance furnaces for high temperatures, with particular reference to the heating elements used in them, the author deals at some length with Stratit heating elements. These elements consist of rods or wires of molybdenum or tungsten, which are built into vacuum-tight ceramic tubes. The rods or wires are made by the usual procedure, *i.e.*, by pressing and sintering the powdered metal. The ceramic tubes must be made in such a manner that their internal surfaces do not react with the metallic conductor, and it was found advisable to use pure alumina or beryllium oxide for the inner layer nearest to the metal. For the outer layer sillimanite proved to be very suitable. In conclusion the author states that the cost of Stratit elements, and of their operation, compares favourably with any other element used in practice. It is expected that these elements will shortly be available to industrial users.

The Deep-Cementation of Chromium-Molybdenum- and Chromium-Manganese Case-Hardening Steels in Salt Baths. H. Diergarten. (Stahl und Eisen, 1940, vol. 60, Nov. 14, pp. 1027–1035). In the case-hardening of chromium-molybdenum and chromium-manganese steels to depths of 1·2–2·0 mm. it was observed that large round carbide particles were formed, which led to grinding difficulties and to a deterioration in the strength of the steel. In the present paper the author describes an investigation of the case-hardening properties of two types of 0·2% carbon steels, one containing manganese 0·8–1·2%, chromium 1·0–1·4% and molybdenum 0·2–0·3%, and the other, manganese 1·1–1·5% and chromium 0·8–1·5%. From the results obtained the author found that a second treatment for 20 min. at 900–930° C. in a bath of a neutral salt brought the coarse carbide into solution and that it did not reform with the subsequent hardening treatment. The surface hardness and the strength of the core also remained unaffected by this double salt-bath treatment. One difficulty did, however, arise, for, after treatment in this manner, it was found that internal cracks were formed which, on closer investigation, proved to be flakes. The occurrence of these flakes was found to be due to the steel having absorbed hydrogen from the salt-baths. It is therefore essential to see that the salt-baths are free from both hydrogen and water. Another way of preventing the formation of flakes is to cool the steel slowly (but not so slowly as to permit of a cementite network forming), or to quench it in a salt bath at 220–230° C. instead of in water or in oil.

The Case-Hardening of Stainless Chromium Steels. R. Weihrich. (Stahl und Eisen, 1941, vol. 61, Jan. 23, pp. 83–84). The author discusses the problem of case-hardening stainless steel containing about 15% of chromium. Attempts to case-harden this steel in the ordinary commercial compounds at temperatures under 1000° C. have not been successful, because the case produced was not uniform and the corrosion resistance of the metal was in part destroyed. This steel can, however, be case-hardened by holding for 6 hr. at a temperature of 920–950° C. in freshly prepared charcoal which has been activated by heating to 800–900° C. with the exclusion of air; after the case-hardening, the steel should be quenched in oil from about 1020° C. with or without an intermediate anneal. A surface hardness of Rockwell C 57–59 can be produced in this way, and the corrosion resistance remains unimpaired. The case-hardened stainless steels are suitable for the dies used in the manufacture of synthetic-resin articles.

Recent Investigations of the Dry Cyaniding Process. D. W. Rudorff. (Metallurgia, 1941, vol. 23, Feb., pp. 99–102). The author surveys some of the results obtained in investigations of the properties of the case produced by the dry-cyaniding process, which consists of passing a carburising gas and ammonia over the heated steel parts (*see* Journ. I. and S.I., 1938, No. II., p. 361 A). Experimental work by Prosvirin (*see* Journ. I. and S.I., 1939, No. II., p. 150 A), using the three following methods of supplying the gases: (*a*) Separate admission to the muffle through individual supply connections; (*b*) joint admission of the gases in the mixed state; and (*c*) alternating admission of the two gases, showed that the degree of hardness obtained was identical in all three cases, but method (*b*) produced the greatest depth of case. Examination with the microscope and chemical analyses demonstrated that, with this treatment, the nitrogen penetrated the steel to a considerably greater depth than the depth of case revealed under the microscope. Another series of tests to determine the influence of the time of treatment on the depth of case obtained, showed that the depth increased linearly with increasing time, whereas with liquid cyaniding the rate of penetration does not remain constant, but decreases. The explanation given for this is that, at temperatures below the Ac_3 point, the diffusing power of nitrogen is considerably greater than that of carbon, so that the surface first becomes saturated with nitrogen, and this leads, as the treatment continues, to a depression of the Ac_3 point, and the α-iron changes to γ-iron; in this state the conditions for the penetration of carbon are greatly improved, and it is this increase in the depth of the pearlitic case, not the increase in total case depth, which becomes visible under the microscope.

On the Cause of the Formation of Abnormal Structure in Carburized Steels. K. Iwase and M. Homma. (Nippon Kinzoku Gakkai-Si, 1940, vol. 4, Nov., pp. 351–362). (In Japanese). The authors discuss two theories which have been advanced to account for the formation of the abnormal structure and soft spots frequently found in carburised hypereutectoid steels. These are: (1) The oxygen theory, which is based on the fact that the structure of steel becomes abnormal when it is carburised with an agent containing oxygen, whilst when a hydrocarbon agent is used the structure is normal; and (2) the high-purity iron theory, recently developed in Germany, which is based on the fact that very pure carbonyl or electrolytic iron has an abnormal structure after carburisation.

Flame-Hardening Equipment. (Metallurgia, 1941, vol. 23, Feb., pp. 127–128). Descriptions and illustrations are given of a number of flame-hardening machines for surface-hardening by the Shorter process. (*See* Journ. I. and S.I., 1940, No. II., p. 148 A).

WELDING AND CUTTING

(Continued from pp. 166 A–173 A)

The Electro-Magnetic Stirring Action in a Spotweld. A. M. Unger, H. A. Matis and W. A. Knocke. (Welding Journal, 1941, vol. 20, Jan., pp. 42–47). The authors examine the properties of spot welds in dissimilar metals made at different temperatures. The metals chosen were Cor-ten, a low-alloy high-tensile steel, and 18/8 stainless steel, because these etched in different ways, and an examination of a section through a spot weld would reveal to what extent one metal had migrated into the other. It is well known that the flow of current through molten metal causes a stirring action in accordance with Fleming's left-hand rule. By examining sections of spot welds made with different current durations the successive stages of migration of the stainless steel into the Cor-ten could be studied. In welds made with sufficient current strength and duration of flow to melt both metals, these were well mixed, as was shown not only by the etching, but also by the fact that the chemical composition of the mixture represented the average of the analyses of the two separate alloys. Welds of this kind were hard, brittle and definitely martensitic in character. A further series of tests was made with welds in which the timing was adjusted so as to bring the alloys up to the plastic stage; this was done by arranging two brief cooling periods by means of an interruptor in the circuit. Because of the different thermal conductivities of the alloys, it was possible by this technique to bring the stainless steel up to the plastic condition before the Cor-ten. Hardness tests and examination of the microstructure of spot welds made in this way proved that no migration of metal and only the customary incerase in hardness had taken place. In discussing the results of this investigation the authors state that it proved the necessity of considering what type of alloy would result from the welding of two dissimilar metals or alloys, and if the mixture were a brittle alloy, a technique must be developed which enabled the weld to be made in the plastic stage.

A Defect in Mild-Steel Welds Due to Non-Metallic Inclusions in the Parent Metal. L. E. Benson. (Iron and Steel, 1941, vol. 14, Feb., pp. 181–184). The author reports on an investigation of the causes of peculiar crater-like defects appearing in the fracture when testing welded specimens. This paper appeared previously in Transactions of the Institute of Welding, 1940, vol. 3, Oct., pp. 215–218. (*See* p. 102 A).

British Standard Specification for Metal Arc Welding as Applied to Tubular Steel Structural Members. (British Standards Institution No. 938–1941). British Standard No. 538 provides for the welding of mild steel in general building construction, and excludes the welding of tubular steel sections; the present standard specification fills the gap which was thus left and gives details of the design and procedure for making the following butt-welded joints in tubes up to 6 in. in dia.: square, single-V, single-U, single-J and single-bevel.

MACHINING

(Continued from pp. 129 A–130 A)

Surface Quality of an S.A.E. 3140 Steel. O. W. Boston and W. W. Gilbert. (Mechanical Engineering, 1940, vol. 62, Nov., pp. 785–789). The authors report on a series of machining tests, the object of which was to determine the influence of the cutting speed, the cooling agent and the heat treatment on the machined surface of specimens of steel *S.A.E.* 3140. The steel tested contained carbon 0·42%, silicon 0·17%, nickel 1·41%, chromium 0·61%, sulphur 0·021%, phosphorus 0·017% and manganese 0·69%. Their general conclusions were: (1) At low cutting speeds the surface quality of the above steel in a normalised and annealed state, machined without a cooling agent, was poor, but it improved greatly as the speed was increased; (2) the different cooling agents used had little effect on the finished surface as long as the cutting speed was high, but at low cutting speeds the surface produced, using a sulphurised mineral oil, was better than that produced without cooling; (3) the optimum cutting speed for the several structures tested was lowest on the hardest steel and highest on the softest steel; and (4) at cutting speeds above the optimum, changes in the speed, the cooling agent or the structure had little effect on the quality of the surface.

Tests Tell the True Story of Machinability. J. P. Walsted. (Steel, 1941, vol. 108, Feb. 17, pp. 76–78). The author describes a simple and practical method of testing the machinability of cast iron. The machinability index is based on the time taken to drill a ¼-in.-dia. hole in inches per minute, and great care is taken to standardise the conditions. The conditions which the author has found to be suitable are: (*a*) Drill speed: 490 r.p.m.; (*b*) pressure on the drill point: 60 lb.; (*c*) point angle: 118°; (*d*) the drills are ground on a new machine with the same setting by the same operator; and (*e*) the time taken to drill a hole ½ in. in depth is taken with a stop-watch. The results of some tests on new iron and on some 25-year-old castings are presented.

Anti-Aircraft Gun Barrels. (Machinist, 1941, vol. 84, Mar. 15, pp. 1070–1079). A detailed and illustrated description is given of the turning, boring and milling procedure adopted by a Canadian engineering works for the manufacture of Bofors anti-aircraft gun barrels from forgings.

PROPERTIES AND TESTS

(Continued from pp. 182 A–186 A)

The Strength Features of the Tension Test. F. B. Seely. (Proceedings of the American Society for Testing Materials, 1940, vol. 40, pp. 535–550). The author explains and discusses the nature of the movement which takes place within polycrystalline aggregates during tensile tests. He differentiates, with particular reference to steel, between plastic deformation accompanied by strain-hardening and failure by brittle fracture as a result of the cohesive strength of the material being exceeded. Reference is made to the influence of the state of stress, the rate of increasing strain, strain-hardening and temperature changes on these two types of fracture, and an outline of a testing procedure for steel is given, the object of which is to demonstrate how these factors can change the mode of failure from plastic yielding to brittle fracture.

Speed in Tension Testing and Its Influence on Yield Point Values. L. H. Fry. (Proceedings of the American Society for Testing Materials, 1940, vol. 40, pp. 625–636). The author analyses the results of over 300 tensile tests made at various rates of extension and on different types of testing machines to determine how this rate affects the yield-point value. His conclusions from this analysis were: (1) Yield-point values determined under normal conditions of routine testing show a scatter considerably greater than that attributable to variations in the quality of the material; such variations are undoubtedly due to variations in stress concentrations during testing; (2) yield-point values tend to increase with increasing rate of extension—this increase is related directly to the rate of extension and only indirectly to the rate of travel of the cross-head of the machine; (3) generally speaking, the yield-point value increases in a straight line with the logarithm of the rate of extension; (4) There is some indication that for the same increase in the rate of extension, the rise in the yield-point values is less rapid with specimens of 8-in. gauge length than with those of 2-in. gauge length; (5) only a small fraction of the cross-head travel is revealed as strain in the specimen—the proportion of strain to cross-head travel varies during a test as the load is increased, but is unaffected by the rate of cross-head travel.

The Limited Significance of the Ductility Features of the Tensile Test. H. W. Gillet. (Proceedings of the American Society for Testing Materials, 1940, vol. 40, pp. 551–575). The author discusses the relative importance of the elongation, reduction-of-area, yield point and ultimate strength of a steel when assessing its usefulness for a particular application.

An Investigation of the Effect of Rate of Strain on the Results of Tension Tests of Metals. P. G. Jones and H. F. Moore. (Proceedings of the American Society for Testing Materials, 1940, vol. 40, pp. 610–624). The authors describe an investigation of the effects of variations in the rate of strain on the values obtained for the tensile strength of four ferrous and ten non-ferrous metals. The ferrous metals tested were: (*a*) A 0·45% carbon steel; (*b*) a 0·20% carbon steel; (*c*) 18/8 stainless steel; and (*d*) a steel containing 0·41% of copper and 1·01% of chromium. The apparatus used comprised a 100-ton Riehlé screw-pattern testing machine with an electric resistance extensometer incorporating an oscillograph, which recorded the extension with a high degree of sensitivity. It was found that, for steel specimens, doubling the rate of extension did not raise the tensile strength values by more than 1%, and the elongation and reduction of area by more than 3%. In the case of stainless steel, doubling the rate of extension increased the yield point and the tensile strength by 1·8% and 0·5%, respectively, and decreased the elongation on 2 in. and the percentage reduction of area by 3·7% and 2·3%, respectively.

On the Serrated Elongation in Different Metals. T. Sutoki. (Nippon Kinzoku Gakkai-Si, 1940, vol. 4, Oct., pp. 315–324). (In Japanese). The author gives an account of his investigation of the elongation of carbon and alloy steels, and of single crystals of iron, duralumin, brass, copper, nickel, aluminium and zinc at temperatures ranging from that of liquid nitrogen up to red heat, with special reference to the fluctuations in the stress-strain diagram for steel between the yield point and fracture.

Mechanical Properties of Gray Cast Iron in Torsion. J. O. Draffin, W. L. Collins and C. H. Casberg. (Proceedings of the American Society for Testing Materials, 1940, vol. 40, pp. 840–846). The authors investigated the mechanical properties of hollow grey cast-iron cylinders with outside diameters of 3½ in. and 4½ in. and wall thicknesses ranging from 0·10 in. to 0·70 in. The results corroborated those obtained in earlier torsional tests on small solid and hollow cylinders. The graph of the modulus of rupture against the tensile strength gave a fairly smooth curve, the modulus changing from 1·36 times the tensile strength for a solid cylinder to 0·73 times the tensile strength for a hollow cylinder with a wall-thickness/outside-radius ratio of 0·057.

Drill-Steel Shanks and Striking-Face Failures. F. R. Anderson. (Engineering and Mining Journal, 1941, vol. 142, Jan., pp. 40–43). The author describes and illustrates a number of different types of failure which frequently occur at the striking face of power-driven hammers and rock-drill shanks. From an examination of these it is seen that, to obtain the maximum service life of these parts, it is essential that the maximum possible area of contact is maintained, and that the hardness of the drill shank is less than that of the piston or tappet which delivers the blow.

Fatigue Tests on Zinc-Coated Steel Wire. D. G. Watt. (Proceedings of the American Society for Testing Materials, 1940, vol. 40, pp. 717–729). The author describes an investigation of the fatigue properties of galvanised-steel wire of three types with tensile strengths of 98–110, 76–85 and 49–54 tons per sq. in., respectively. He came to the following conclusions from the data obtained: (1) Within the limits of tensile strength investigated, *i.e.*, 49 to 110 tons per sq. in., the endurance limits of the zinc-coated wire increased with the tensile strength, but the endurance ratios diminished; (2) hot-dip galvanising reduced the endurance limit of the patented wire by 15·4–28·4%; (3) the coating thickness had no perceptible influence on the endurance limit; (4) reduction by drawing, within the range of wire sizes studied, increased the endurance limit of the wire in both the patented and the galvanised states; (5) steel wires of all three types, galvanised by electrolytic processes, had somewhat higher endurance ratios than wires galvanised by hot-dip methods; and (6) in the fatigue testing of hot-dip galvanised wire, closely spaced hair-line cracks developed, but no evidence of such cracks could be found in electrolytically coated wire tested in the same manner.

Rockwell Hardness of Cylindrical Specimens. W. E. Ingerson. (Bell Telephone System Technical Publications, 1940, Monograph B-1229). A theoretical method for correcting the apparent Rockwell hardness numbers obtained on cylindrical specimens using a ball penetrator is developed. The deviation of apparent from true hardness is found to vary approximately linearly with the depth of impression for a given specimen diameter, and this deviation is calculated for various specimen diameters between ⅛ in. and 1 in.

Analysis of Rockwell Hardness Data. R. L. Peek, jun., and W. E. Ingerson. (Bell Telephone System Technical Publications, 1940, Monograph B-1230). The authors analyse Rockwell hardness data in order to correlate readings made with different loads and ball diameters and to simplify the relation between hardness and tensile-strength measurements. They conclude that, for the Rockwell hardness test on homogeneous materials of the types studied, the relation between the difference in penetration (h), the major and minor loads (W and W_0, respectively), and tensile strengths for ball diameter D, is given by the equation

$$\frac{h}{D} = \left(\frac{C(W - W_0)}{SD^2}\right)^{1/m}$$

where C and m are constants of the material. The quantity $m - 1$ is a measure of the work-hardening properties of the material similar to the quantity $n - 2$ in the Meyer analysis of the Brinell test. They also observed that indentation hardness readings are affected by the thickness of the specimen when the penetration is half the thickness or more. Evidence of this effect is given by a departure from a straight-line relation in the plot on logarithmic paper of $(W - W_0)/D^2$ against h/D.

Optical Protractor and Brinell Microscope. (Engineering, 1941, vol. 151, Mar. 14, p. 210). Descriptions are given of two new types of instrument which are now available for inspection and hardness testing. The first instrument is an optical protractor for determining angles in finished work or for setting work to a required angle. It is claimed that unskilled operators can quickly learn to make readings and settings to a degree of accuracy comparable with those obtained by a 5-in. sine bar in the hands of a skilled operator. The second instrument is a microscope for measuring the diameter of Brinell hardness impressions. The magnification is 20 diameters. The field covered is approximately 8·2 mm., and the scale of the eye-piece is calibrated to read the diameter of the impression in tenths of a millimetre.

The End-Quench Test for Evaluating Heats of Steel. A. P. Terrile and P. R. Brucker. (Metal Progress, 1941, vol. 39, Jan., pp. 37–42). The authors describe how the end-quench hardenability test of Jominy and Boegehold (*see* Journ. I. and S. I., 1938, No. II., p. 378 A) has been applied by the Pittsburgh Crucible Steel Co. to provide a means of evaluating the hardenability of the centre of bars of water- and oil-hardening S.A.E. steels of various diameters. The hardenability numbers adopted by this company represent the distance from the end of a Jominy specimen, expressed in sixteenths of an inch, at which the hardness is Rockwell C 50. From the data obtained by many tests, curves have been prepared which demonstrate the relation between this hardenability number and the diameter of round steel bars which have an equivalent cooling rate at the centre when quenched in water, or oil, or when cooled in air. A further series of experiments on steels tempered at various temperatures was made to show the reduction in hardness after tempering at points with a hardness of Rockwell C 50 in the as-quenched state. The authors also give some practical examples of the application of the relationships thus established to the control of the steelmaking process, with the object of ensuring the production of a steel with the desired hardenability.

Investigation of Abrasion of Carbon Steels for Gear Material. S. Saito and N. Yamamoto. (Nippon Kinzoku Gakkai-Si, 1940, vol. 4, Sept., pp. 297–304). (In Japanese). The authors report on an investigation of the abrasion resistance of a number of steels with from 0·2% to 0·7% of carbon, the object of which was to study the influence of the carbon content and the heat treatment. When the heat treatment was regulated to give the specimen a Shore hardness of 60, it was observed that the low-carbon steels had superior tensile and impact properties to the high-carbon steels.

Study of the Effects of Variables on the Creep-Resistance of Steels. H. C. Cross and J. G. Lowther. (Proceedings of the American Society for Testing Materials, 1940, vol. 40, pp. 125–153). The authors report on some research work carried out at the Battelle Memorial Institute to study the relative creep-resistance of fine and coarse-grained iron and steels and to determine how the deoxidation practice applied in the steelmaking process influenced the creep strength. The materials tested were: (*a*) a very pure iron; (*b*) a very pure iron with an addition of 0·5% of silicon; (*c*) a rimmed steel with 0·06–0·08% of carbon; (*d*) a steel similar to (*c*) but killed with silicon; (*e*) a rimmed steel with carbon 0·16–0·26%; and (*f*) four killed steels with carbon 0·14–0·17%. The fine- and coarse-grained structures were produced by appropriate heat treatments. The creep tests were made at 850° F. The heat treatment which produced coarse grains in materials (*a*) and (*b*) improved the creep-resistance. The silicon-killed steels had consistently better creep-resistance than the rimmed steel, and this applied to both fine- and coarse-grained material. With regard to the steels containing about 0·15% of carbon, in the fine-grained condition, the silicon-killed steel and the silicon-aluminium-killed steel to which 1 lb. of aluminium per ton had been added, had creep-resistance properties far superior to those of similar steel killed with 2 lb. of aluminium per ton either with or without silicon. All the steels, which, in their heat treatment were cooled in air from a temperature producing coarse grains, had better creep-resistance than those cooled in air from temperatures producing fine grains, the greatest improvement being observed in the aluminium-killed steels.

Influence of the Oxygen Content on the Overheating Sensitivity and the Structural Abnormality of Steel. R. Ziegler. (Stahl und Eisen, 1941, vol. 61, Jan. 9, pp. 43–44). The author describes some experiments undertaken to test the sensitivity to overheating of specimens of basic electric-furnace steel killed with different amounts of aluminium. The results showed that the annealing temperature at which grain-coarsening began decreased with increasing manganese content as well as with increasing carbon content, but it increased with increasing silicon. The sensitivity to overheating increased with decreasing oxygen content. The highest temperature at which grain-coarsening began was obtained when the amount of aluminium added was 0·025–0·25%.

On Heat-Resistant and High-Tensile Alloy Cast Iron. T. Saito. (Transactions of the Mining and Metallurgical Alumni Association, Kyoto Imperial University, 1940, vol. 10, Nov. 30, pp. 311–326). (In Japanese). The author reports on an investigation of the properties of heat-resisting high-tensile cast irons containing 2·5–3·5% of carbon with additions of one or more of the following elements: silicon, nickel, copper, tungsten, molybdenum, manganese, vanadium, chromium, phosphorus and sulphur.

Acid-Resisting Cast Iron—The User's Point of View. P. Parrish. (Institution of Chemical Engineers: Iron and Coal Trades Review, 1941, vol. 142, Mar. 14, pp. 321–322; Mar. 21, pp. 347–348). The author discusses the manufacture of the large cast-iron pots used in the preparation of sulphuric acid and the properties of high-silicon cast irons. In designing these pots large flat surfaces should be avoided, and the thickness should be kept as uniform as possible, all large flanges should be webbed, and slots rather than bolt holes should be provided. A competent inspector who can detect faulty castings with a hammer test should be employed by the user to examine all castings. A diameter of 7 ft. 9 in. appears to be the limit of size for cast-iron pots in Great Britain. In discussing high-silicon irons, the author refers to the work of Wüst and Petersen, who observed the regular decrease in the carbon content of iron and the equally regular rise in the freezing point with successive increments of silicon. It seems to be accepted that in grey iron the silicon present causes the carbide constituent to break down into iron and graphite, leaving the phosphide structure free; the phosphorus has no appreciable effect on the condition of the carbon, but it reduces the solubility, and thus acts indirectly as a hardener. It is suggested that further research on two questions is required; these are: (1) Does the addition of nickel to high-silicon iron decrease its resistance to corrosion?; and (2) as silicon-iron castings are not brittle when red-hot, but only when cooled, can any steps be taken to avoid undue brittleness in the cold state?

Pure Iron-Phosphorus Alloys. G. T. Motok. (Iron Age, 1940, vol. 146, Oct. 31, pp. 27–32). The author describes an investigation of the properties of pure iron when alloyed with up to 1% of phosphorus. The alloys were prepared and melted in an atmosphere of helium, which was passed through molten lithium to remove any nitrogen; the apparatus for this is described in detail. The results of Charpy impact tests showed that the impact values were so low, even after grain-refining treatment, that no conclusions could be drawn as to the effect of phosphorus, heat treatment, or degree of cold-working on the impact value. On the other hand, the changes in hardness after various treatments could readily be determined, and it was noted that the hardness increased fairly uniformly with increasing phosphorus content. No perceptible age-hardening occurred on quenching with subsequent ageing at room temperature, at 200° C., or at 450° C., either with or without cold-working.

The conclusion appeared, therefore, to be justified that, in the absence of other elements such as carbon and nitrogen, phosphorus in amounts up to 1% did not give rise to ageing effects. The embrittlement which was caused by the phosphorus did not appear to be due to precipitation-hardening, but to phosphorus dissolved in the ferrite. Examination with the microscope showed that, on quenching the iron-phosphorus alloys from the $\alpha + \gamma$ field, a two-phase structure was obtained. This duplex structure was not found after cooling in the furnace, but traces of it remained after cooling in air. It did not appear, however, that this comparative sluggishness of the iron-phosphorus alloys in attaining a homogeneous structure on cooling from above the critical temperature could play a very important part in the mechanism of embrittlement, since the embrittlement still took place even after very slow cooling. The solid-solubility of phosphorus in iron at room temperature was found to be not less than 1%. The fracture of the specimens showed some evidence of being intercrystalline, but it was doubtful whether the brittleness was due to weakness at the crystal boundaries. Phosphorus appeared neither to accelerate nor to retard the excessive grain growth produced by cold-work with subsequent annealing at 650° C. With up to 0·7% of phosphorus the depth of carbon penetration on case-hardening was unaffected, but with a higher percentage of phosphorus less penetration was observed.

A Comparison of Different Case-Hardening Alloy Steels. H. Schrader and F. Brühl. (Stahl und Eisen, 1940, vol. 60, Nov. 21, pp. 1051–1058). The authors discuss the results obtained in tensile and impact tests on square test-pieces of quenched and of case-hardened low-alloy steels. The specimens tested were of the steels known in Germany as *EC*80, *ECMo*80, *EC*100 and *ECMo*200, the analyses of which are :

	*EC*80	*ECMo*80	*EC*100	*ECMo*100	*ECMo*200
Carbon. % . .	0·14–0·19	0·14–0·19	0·18–0·23	0·18–0·23	0·17–0·23
Silicon. % . .	0·35 max.	0·35 max.	0·35 max.	0·35 max.	0·35 max.
Manganese. % .	1·1–1·4	1·1–1·4	1·2–1·5	1·2–1·5	1·3–1·5
Chromium. % .	0·8–1·1	0·8–1·1	1·2–1·5	1·2–1·5	1·7–2·0
Molybdenum. % .	...	0·15–0·25	...	0·15–0·25	0·15–0·25

It was observed that small specimens of steels *EC*80 and *ECMo*80 in the case-hardened state had granular fractures in the core. Steel *EC*100 differed from *ECMo*100 in that the fractures of specimens up to 30 mm. square were fibrous, and the sections had to be larger than this to obtain a granular appearance. It was possible to obtain tough fibrous fractures of steel *ECMo*200 with specimens up to 60 mm. square. The strength of the cores of large specimens of steels *EC*100 and *ECMo*100 in the quenched state decreased rapidly with increasing size up to 250 mm. square. Chromium-molybdenum steels such as steel *ECMo*200 and steels high in chromium and manganese, on the other hand, maintained a high strength when the cross-section was large, but their core-hardness was sometimes on the high side. A similar hardness penetration and a core strength up to 105 kg. per sq. mm. were attained with large-section specimens of chromium-nickel steels with additions of tungsten or molybdenum, whilst small specimens of these steels did not exhibit core-strengths above the safety limit. The impact strength of chromium case-hardening steels with more than 2% of chromium and no nickel depended to a great extent on the tensile strength. In the case of steel *ECMo* 200 the optimum tensile strength was 150 kg. per sq. mm. With steels *ECMo* 100 and *EC* 100 the notch-sensitivity remained practically unchanged over a large range of tensile strength. To test the possibility of using these steels for gears, special specimens with a tooth-shaped projection were subjected to heavy impacts. These tests demonstrated that with the surface in good condition both the chromium-molybdenum and the chromium-manganese steels were just as good as chromium-nickel steels.

The three types of steel tested had a tendency to form too much carbide at the case-core boundary, and it was found that this danger could be reduced to a great extent by mixing a proportion of used powder in the case-hardening compound.

On a Proprietary Stainless Steel. Y. Okura.. (Tetsu to Hagane, 1940, vol. 26, Nov. 25, pp. 802–812). (In Japanese). The author reports on an investigation of the structure and the mechanical and welding properties of specimens of a proprietary brand of stainless steel which varied in composition within the following limits: Carbon up to 0·25%, silicon 0·1–2·0%, copper 0·05–4·5%, nickel 1–7%, chromium 16–30% and tungsten 0·2–8·0% with additions of up to 1% of vanadium, cobalt, aluminium or titanium. From the results obtained it was concluded that steel containing chromium 18%, nickel 7% and tungsten 5–8% is suitable for applications where resistance to the attack of ferric-chloride solution or sea-water plus hydrogen peroxide is required.

Stainless Steels for Aircraft. O. Fraser, jun. (Steel, 1941, vol. 108, Jan. 27, pp. 52–53, 61). The author elaborates some of the difficulties with which the producer of stainless steel has to contend in meeting the increasingly exacting requirements of the user, especially those of the aircraft industry. Some of the ways in which production could be facilitated with advantage to both producer and buyer are: (1) As far as possible the ordering of very small quantities of steel to special analyses should be avoided; (2) 18/8 stainless steel stabilised with titanium is about 2*d*. per lb. cheaper than that stabilised with niobium, and, as tests show no discernible difference in the properties of the two steels, it appears to be unnecessary for customers to insist on the latter quality; (3) the buyer should specify the particular purpose for which the steel will be used; and (4) buyers should realise that stretching or rolling annealed stainless steel is bound to work-harden it, and that to order stainless-steel sheets fully soft and flat imposes conditions impossible of fulfilment.

A New Die Steel. L. Sanderson. (Metallurgia, 1941, vol. 23, Jan., p. 80). The author discusses a recently developed die steel containing 0·97% of carbon, 5·05% of chromium, 0·18% of vanadium and 1·07% of molybdenum. This steel is said to enable a 40% saving in die cost on short-run dies to be made and to give excellent results in service, so that, although originally designed for dies only, it has since been applied to a wide range of parts, such as cams, clutch parts, ways on lathes and grinders, plugs, gauges, &c. He gives data for the Rockwell C hardness of the steel in dependence on the heat treatment. He points out that the high hardness is due to the molybdenum content, and that vanadium is added to provide a wide range of heat treatment and to prevent excessive grain growth.

Low-Alloy High-Speed Steels. A. Gulyaev and K. Osipov. (Stal, 1939, No. 12, pp. 47–54). (In Russian). The authors consider that high-speed steels should contain alloying elements to form carbides which dissociate with difficulty. The high-speed steel substitutes used in Russia (*e.g.*, steels *EI*116, *EI*172, *EI*173 and *EI*184), in which chromium is the chief alloying constituent, are not of this type. Consideration of previous work led the authors to the choice of the following compositions :

	(1)	(2)	(3)	(4)	(5)
Carbon. % . .	1·01	1·19	1·56	1·3	1·35
Vanadium. % .	3·18	4·56	6·68	4·88	4·77
Tungsten. % .	3·13	3·34	2·84	...	3·03
Molybdenum. % .	3·50	3·35	3·20	3·13	...
Chromium. % .	4·3	4·4	4·4	4·6	4·5

In these steels the carbon content is based on the vanadium content. The above alloys were prepared in an H.F. furnace, cast, forged into billets, annealed at 900° C. and cooled in the furnace. Curves from data obtained with a dilatometer showed that the Ac_1 points of the steels were between 820° and 890° C. S-curves for the steels showed that they all possessed minimum austenite stability within the range 720–770° C. Maximum cooling rates on annealing and critical cooling rates for quenching were also obtained from the S-curves. Investigation of the optimum quenching temperatures showed that the permissible range was wide. In all the molybdenum-bearing steels the amount of residual austenite after quenching from 1260–1280° C. was about 20%, *i.e.*, even less than in tungsten high-speed steel quenched from the same temperatures. A first tempering treatment rapidly reduced the austenite content to 5–10%, except in steel (5). Optimum tempering conditions were studied with reference to the hardness and microstructure obtained. Cutting tests on axle carbon steels, in which the steels were compared with the chromium-bearing substitutes and with tungsten high-speed steel, demonstrated that the new steels were very much superior to the substitutes, and that they were in fact equal, and in some cases, even superior to, ordinary high-speed steels. Compositions (1), (3) and (4) were the best. Data regarding the recommended practice of forging, annealing, quenching and tempering these types of tool steel are summarised.

The Properties of Hardenable Manganese Steels Containing Other Alloy Additions. H. Cornelius. (Stahl und Eisen, 1940, vol. 60, Nov. 28, pp. 1075–1083). The author reports on an investigation of the properties of a number of hardenable steels used in the German aircraft industry. The composition of the steels tested was within the following limits: carbon 0·27–0·42%, silicon 0·26–0·47%, manganese 0·46–2·1%, chromium 0–2·1%, nickel 0–2·1%, molybdenum 0–0·29%, and vanadium 0–0·35%. The steels were made in a small high-frequency furnace, and specimens 18 mm. in dia. were examined under the microscope and tested mechanically. It was found that a small addition of vanadium to steels containing more than 1% of manganese was necessary to make them sufficiently non-sensitive to overheating. Any difference in hardenability on cooling in air could be reduced by a subsequent tempering. The relation between the tensile strength, elastic limit, elongation, reduction of area and impact strength are shown by several series of curves. These indicate that, among other characteristics, the low-alloy steels containing manganese as the important element have a higher elastic limit within the practical limits of tensile strength than the chromium-molybdenum steel. The alloy steels with no molybdenum had the lowest impact strength. An addition of vanadium also lowered the brittleness of tempered manganese steels. The author's general conclusion was that the steels containing 1·0–1·7% of manganese are suitable substitutes for the chromium-molybdenum steels.

Study on the Experimental Manufacture of High-Tensile Structural Steel Sheets. I. Kohira, K. Moridera and G. Maeda. (Tetsu to Hagane, 1940, vol. 26, Nov. 25, pp. 777–787). (In Japanese). The authors investigated the mechanical properties of steel plates rolled from four types of high-tensile steel. These steels contained: (*a*) carbon 0·18%, manganese 1·65% and copper 0·246%; (*b*) carbon 0·18%, manganese 1·09%, chromium 0·66% and copper 0·246%; (*c*) carbon 0·18%, manganese 1·21%, molybdenum 0·28% and copper 0·254%; and (*d*) carbon 0·25%, manganese 1·31%, and copper 0·320%. No difficulty was experienced in the melting and rolling of the steels, and the differences in the mechanical properties in the as-rolled condition were negligible. The tensile strength was 55–65 kg. per sq. mm., the yield point 40–45 kg. per sq. mm. and the elongation 20% for all of the four types.

METALLOGRAPHY AND CONSTITUTION

(Continued from pp. 186 A–192 A)

The Use of Plane Polarised Light and Sensitive Tint Illumination in the Analysis of the Microstructures of Steel. B. L. McCarthy. (Mordica Memorial Lecture: Wire and Wire Products, 1941, vol.

16, Jan., pp. 17–26). The author explains the theory and practical application of a method of examining the microstructure of steel which involves the use of polarised light and colour-photography. An American microscope is now available which makes provision for the application of polarised light and sensitive-tint illumination. In this instrument a modification of the ordinary Nicol prism is used as a vertical illuminator, and the arrangement is such that this prism acts not only as a vertical illuminator, but also as both polariser and analyser. A carbon arc forms the source of light, and from this the light passes through the vertical polarising illuminator perpendicular to the specimen. At the cemented interface of the prism the ordinary rays are totally reflected against the prism side, where they are absorbed, whilst the extraordinary ray, travelling in a refractive index matching the index of the cement, passes on to the specimen. This polarised beam, if reflected unchanged from the specimen, passes back along the original path, through the specimen back to the source of light. If, however, a portion of the surface of the specimen depolarises the light, such ordinary rays are reflected at the cemented interface of the prism on to a silvered mirror, and thence to the eye-piece. When the sample is viewed through the eye-piece, only that portion with depolarising properties is visible. By placing a quarter-wave plate in the beam emerging from the polariser, the plane-polarised light is converted to circularly polarised light, which, as it is symmetrical to the axis of the ray, acts in the same manner as ordinary unpolarised light. This provision for ordinary light increases the light efficiency to about four times that obtainable when a plane-glass reflector is employed. The author reproduces a number of micrographs in black-and-white and in colour, and discusses the differences revealed by the two processes. He points out that, because of the accuracy with which sensitive-tint illumination reveals areas of weak depolarisation, the determination of isotropic and anisotropic areas is greatly facilitated. The use of polarised light in the study of preferred orientation in cold-worked steel has been suggested, but the author considers that, while it may be useful in the study of the effect of cold-heading, or where the reduction is extremely light, there is little to be gained by using it on heavily worked steel wire. His own examination of a cold-worked low-carbon steel wire with the instrument described revealed that the basic purple colour persisted after cold-working, which was evidence that the surface remained isotropic.

A Time-Saving Adaptation for Photomicrography. J. A. Quense and W. M. Dehn. (Industrial and Engineering Chemistry, Analytical Edition, 1941, vol. 13, Jan., p. 68). A brief description is given of a viewing tube and holder for use on the microscope of a microphotographic apparatus. The purpose of this is to obviate the removal and replacement of the camera when the microscope is used with a Nicol analyser for selecting a representative field for examination.

A Holder for Metallographic Specimens during Polishing. G. R. Kauffman. (Metals and Alloys, 1941, vol. 13, Feb., pp. 171–172). The author gives an illustrated description of a device to hold metallographic specimens on a rotating polishing disc at a constant pressure. This holder is inexpensive and can easily be fitted to almost any type of polishing machine. It consists of an aluminium ring 2 in. in dia. and 2 in. high. This is held on a rod which in turn is clamped to a stand so that the ring can easily be fixed in any desired position relative to the disc. In practice, it is clamped so that its lower edge is ⅛ in. above the disc. The specimen is placed on the disc within the ring, and the holder thus retains it in any one position for as long as desired at a constant pressure, which is simply that exerted by its own weight.

The Electrolytic Polishing of Stainless Steels. H. H. Uhlig. (Transactions of the Electrochemical Society, 1940, vol. 78, pp. 265–272). The author describes the electrolytic polishing of 18/8 stainless steel, using a glycerine/phosphoric-acid mixture as the electrolyte. He determined the optimum polishing conditions with the aid of a photo-electric spectrophotometer, and found that a maximum degree of polish was obtained with an electrolyte made up of approximately 42% of phosphoric acid, 47% of glycerine and 11% of water by weight, held at 100° C. or higher, with a current density at the anode of at least 0·1 amp. per sq. in. Good solutions could also be made by adding other organic substances to phosphoric acid. Such substances consist, in general, of high-boiling-point materials containing one or more hydroxyl groups which are soluble in phosphoric acid.

Dendritic Structure. C. H. Desch. (Sheffield Metallurgical Association: Foundry Trade Journal, 1940, vol. 63, Nov. 28, p. 351; Dec. 5, pp. 365–366). The author's paper on the influence of dendritic structure on the heterogeneity of alloys is reproduced (*see* p. 71 A).

The Problem of the Primary Crystallisation of Steel: Undercooling and Nucleus Formation in the Liquid State. P. Bardenheuer and R. Bleckmann. (Stahl und Eisen, 1941, vol. 61, Jan. 16, pp. 49–53). The authors studied the factors affecting the degree to which steel can be undercooled and describe a method of determining how undercooling affects the formation of nuclei. They found that steel with less than 0·01% of carbon could be undercooled as much as 250° C., and steel with 0·4% of carbon about 205° C., provided that it was entirely surrounded by slag and not in contact with the crucible. Spontaneous crystallisation would only take place with a high degree of undercooling. Their experiments demonstrated that it would be impossible to teem steel in the undercooled state. Further investigations were made based on the theory that, if an addition of another element can be made to steel already in the undercooled state without increasing the degree of undercooling, this addition would introduce foreign nuclei into the metal. The effects of a large number of elements added to a low-carbon stee on the nucleus-formation were studied by applying the above theory, and it was found that aluminium, beryllium, boron, calcium-aluminium, calcium-aluminium-silicon, titanium, vanadium and zirconium completely stopped the undercooling, in most cases an addition of only 0·1% being sufficient. On the other hand, manganese, silicon, calcium-silicon, chromium, cobalt, molybdenum, nickel, niobium-tantalum, phosphorus, sulphur, tungsten and nitrogen did not prevent undercooling, and therefore did not lead to nucleus-formation in the liquid steel. Hydrogen and oxygen fell between these two groups, for increasing quantities of these gases decreased the undercooling capacity of the steel more and more without inhibiting it altogether. The authors also carried out tests to ascertain whether the nuclei of foreign bodies could be destroyed by superheating; these showed that the nuclei remained after superheating. Superheating to from 20° to 190° C. above the melting point had also no effect on the undercooling of a nucleus-free melt; but a slight superheating of up to 20° C. above the melting point inhibited undercooling.

Isothermal Transformation of Austenite in Grey Cast Iron. C. R. Hilliker and M. Cohen. (Iron Age, 1941, vol. 147, Feb. 13, pp. 43–46). The authors studied the isothermal transformation of the austenite in two grey cast irons, one of which was unalloyed, whilst the other contained 2·03% of nickel. They discuss the results obtained and present the transformation curves, which show that, as with steel, the presence of nickel decreases the rate of transformation. The hardness of the isothermal-transformation products of both irons did not increase continuously with decreasing transformation temperature, for it was observed that the products formed in the 950–850° F. range were slightly but definitely softer than those formed at transformation temperatures above and below this range.

The Equilibrium Diagram of the Al–Fe–Ti System and the Segregation of Fe and Ti. H. Nishimura and E. Matsumoto. (Nippon Kinzoku Gakkai-Si, 1940, vol. 4, Oct., pp. 339–343). (In Japanese). The authors studied the aluminium corner of the aluminium-iron-titanium equilibrium diagram by thermal analysis and determined the limits of solubility of titanium and iron in liquid aluminium at various temperatures.

CORROSION OF IRON AND STEEL

(Continued from pp. 192 A–195 A)

Vagaries of Corrosion Testing. F. A. Rohrman. (Monthly Review of the American Electroplaters' Society, 1941, vol. 28, Jan., pp. 27–34). The author discusses the principles to be followed in the planning of corrosion tests on specimens of sheet metal.

Corrosion of Metals by Non-Aqueous Solutions. The Action of Ethyl Alcohol on Metals. L. G. Gindin, R. S. Ambarzumian and E. P. Belchikova. (Comptes Rendus (Doklady) de l'Académie des Sciences de l'U.R.S.S., 1940, vol. 29, Oct. 10, pp. 44–47). This is the introductory paper to a projected series on the corrosion of metals in non-aqueous electrolytes, mainly alcohols and alcoholic solutions. The authors review the literature on the effect of ethyl alcohol on metals, and they describe the procedure they adopted for the purification of the ethyl alcohol used for their investigation. In the series of experiments described in the present paper, the tests were carried out in sealed glass tubes in an apparatus which is illustrated, and tables are given of the results obtained after keeping samples of magnesium, aluminium, zinc and steel (carbon 0·28%, manganese 0·51% and silicon 0·22%) for 210 days under absolute alcohol and for 150 days under 99·7% alcohol. Only magnesium proved to be slightly corroded by ethyl alcohol under the experimental conditions described, whereas the three other materials examined were not affected at all.

Corrosion of Metals by Non-Aqueous Solutions. R. S. Ambarzumian, L. G. Gindin and E. P. Belchikova. (Comptes Rendus (Doklady) de l'Académie des Sciences de l'U.R.S.S., 1940, vol. 29, Oct. 20, pp. 91–94). The authors studied the influence of carbon dioxide on the action of ethyl alcohol on magnesium, aluminium and steel. They used alcohol saturated with carbon dioxide, and the experimental procedure was as described in the first paper of the series (*see* preceding abstract). They found that in the presence of carbon dioxide, steel and aluminium are also not corroded by ethyl alcohol, whereas the corrosion of magnesium is considerably enhanced.

Effect of Chemical Composition and Microstructure on the Galvanic Corrosion of Cast Iron. Y. Saito. (Tetsu to Hagane, 1940, vol. 26, Nov. 25, pp. 795–801). (In Japanese). The author has studied the galvanic corrosion of pairs of cast-iron specimens of different composition when immersed in contact in a 3% sodium-chloride solution. His tests revealed that increasing the total carbon and the silicon increased the solution pressure by their graphitising effect, whilst manganese, chromium, nickel and copper decreased the solution pressure. For this reason it is pointed out that more attention should be paid to the composition of cast-iron marine-engine components which are in contact with each other.

Controlling Factors in Galvanic Corrosion. W. A. Wesley. (Proceedings of the American Society for Testing Materials, 1940, vol. 40, pp. 690–702). The author presents data which demonstrate the effects of variations in the cathode area and resistance of the galvanic circuit on the currents produced in corrosion tests on iron-copper couples in a neutral sodium-chloride solution. Anode and cathode polarisation curves are given for iron and copper respectively, and a simple equation is derived from the data which relates the current, the anode and cathode areas and the resistance.

Some Corrosion Tests in a Railway Tunnel. S. C. Britton. (Journal of the Institution of Civil Engineers, 1941, vol. 16, Mar., pp. 65–72). The author reports on a series of corrosion tests made by exposing samples of non-ferrous metals and alloys; mild steel protected by paints, by metal coatings, or by thick wrappings; cast iron; stainless steels; and a few non-metallic materials, in the ventilating shaft of a steam-operated railway tunnel for periods ranging up to 3½ years. Copper, although superficially corroded, became covered with a protective coating of a corrosion product, and proved to be the best metallic material; copper alloys did well, but were not so durable as copper alone. Aluminium and stainless steel suffered severe local attack, and 14-S.W.G. sheets were perforated at many points within three years. Mild steel of the same gauge was entirely destroyed within three years, but almost complete protection was afforded to it by a thick rubber coating. Other thick coatings, such as bitumen reinforced by asbestos and cotton lint impregnated with petroleum jelly, also gave fair protection. Metal coatings of aluminium and lead sprayed on to mild steel, and other aluminium coatings applied by a cementation process, failed quickly. Well-chosen paint coatings had as long a life as these metal coats, but allowed rusting to commence after, at most, a year. A plain cast iron suffered considerable "graphitisation," *i.e.*, conversion of the metal into a mixture of rust, graphite and residual iron. A special cast iron with finely dispersed graphite flakes was less badly attacked, and a highly alloyed austenitic cast iron suffered only shallow localised graphitisation. It was evident that painting and metal spraying would have given relatively better performances if conditions had not been so continuously wet, and stainless steels might have justified their name had there been no soot.

Some Observations on the Potentials of Metals and Alloys in Sea-Water. F. L. LaQue and G. L. Cox. (Proceedings of the American Society for Testing Materials, 1940, vol. 40, pp. 670–686). The authors review the literature on the potentials of metals and alloys in sea-water and similar chloride solutions, and describe a particular method of determining the potentials of a number of metals and alloys using specimens in the form of short lengths of pipe through which sea-water is circulated for a test period of six months. The results obtained are discussed and from them a galvanic series is set up in which the relative positions of about forty common metals and alloys can be observed.

A Review of Data on the Relationship of Corrosivity of Water to its Chemical Analysis. V. V. Kendall. (Proceedings of the American Society for Testing Materials, 1940, vol. 40, pp. 1317–1322). The author reviews some of the attempts to classify waters in accordance with their corroding power, with particular reference to Langelier's mathematical evaluation of the ability of a water to precipitate calcium carbonate. He also discusses the results obtained in pitting tests in hot-water pipes in seventeen cities in the United States; from these it was evident that the duration of the test must be taken into account in assigning a corrosion value to a particular water, or in comparing the corrosiveness of several waters.

Measuring the Scale-Forming and Corrosive Tendencies of Water by Short-Time Tests. E. P. Partridge and G. B. Hatch. (Proceedings of the American Society for Testing Materials, 1940, vol. 40, pp. 1329–1339). The authors review a number of short-time tests for determining the corrosiveness of water. They refer in particular to the marble test of Tillmans and Heublin for determining the degree to which a water is saturated with calcium carbonate and to Enslow's continuous-flow test, in which the water is passed slowly through a column packed with calcium carbonate and the difference between the *p*H values of the inflowing and outflowing water is measured. They also present curves showing the value of sodium hexametaphosphate as a corrosion inhibitor.

The Effect of Gas Pressure on the Passivity of Iron. R. S. Crog and H. Hunt. (Electrochemical Society, Apr., 1941, Preprint 79–9). The authors describe an attempt to determine the effect of decreased pressure on the passivity of iron in chromic acid solutions. The passivity or activity of the iron was measured by the potential of an iron/chromic-acid-solution half-cell with an electrode of $Hg/Hg_2CrO_4/(1{\cdot}0$ mol. $H_2CrO_4)$ as the reference half-cell. The iron was found to be active in chromic acid solutions containing moderate concentrations of halide ions at the reduced pressures used in the experiments. Iron wires heated to redness in a vacuum (0·01 mm. of mercury) were found to be passive in chromic acid solutions. From this it appeared improbable that an adsorbed film of oxygen could cause the passivity. The apparatus used for the experiments is described and illustrated.

Intercrystalline Cracking in Boiler Plates. (A Report from the National Physical Laboratory: Iron and Steel Institute, 1941, this Journal, Section I.). Previous work having shown that in the absence of corrosive attack no cracks are formed in boiler-plate steel specimens kept under tension for 5 years, even when concentrations of stress are present, various combinations of stress and exposure to caustic solutions have been investigated. Part I., by C. H. Desch, is an introductory survey of the subject. In Part II., C. H. M. Jenkins and F. Adcock describe experiments in which specimens, with or without notches or drilled holes, were kept under tension in concentrated sodium hydroxide solution at 225° C. An electrically heated pressure vessel was used, with special devices for maintaining constant temperature and stress, and for indicating the onset of cracking. The typical form of intercrystalline crack was not obtained, but in regions of concentrated stress breakdown was caused by the growth of non-metallic inclusions. Heavily cold-worked steel resisted better than annealed material. The black magnetic oxide formed had the composition Fe_3O_3.

In Part III., by F. Adcock and A. J. Cook, experiments under pressure at temperatures up to 470° C. are described. Using small pressure bombs, intercrystalline cracking was found in boiler-plate steel immersed in a solution of pure sodium hydroxide at 310° C. Decarburisation occurred, the carbon being removed as methane. In other experiments the steel specimens and the solution were enclosed in a steel pressure vessel lined with silver. Under these conditions, no intercrystalline cracks were formed in annealed material at 410° C., but if the steel were cold-worked, cracks similar to those produced by hydrogen at high temperatures were formed. With highly concentrated caustic solutions intercrystalline cracks penetrating from the surface became filled with oxide. Experiments with powdered-silver filters showed that masses of oxide could be precipitated at a distance from the specimen, and this may contribute to the cracking of boilers, by sealing cavities and allowing a pressure of hydrogen to be built up. Pure iron developed cracks of the oxide type. Sodium sulphate in solution did not inhibit cracking. Some alloys of iron were also examined.

In Part IV., by F. Adcock and C. H. M. Jenkins, observations of strain-etching in acid open-hearth boiler-plate steel are described. Such markings are usually found only in basic steel. They were not produced in plates which had been deformed cold, but were found in material which had been bent at 100° C., and also in a rolled plate which had been presumably finished at a low temperature. They coincided with the stress lines found by magnetic testing, and with the directions of cracking in a marine boiler plate which had developed corrosion cracks in service.

In Part V., H. J. Gough and H. V. Pollard describe the behaviour of specimens of boiler-plate steel and of both riveted and welded joints in the same material, when subjected to slow cycles of alternating bending stress while immersed in a boiling solution of sodium hydroxide. A machine taking plates 2 ft. 3 in. long, $\frac{3}{4}$ in. thick, and up to 12 in. wide (for riveted joints) was constructed. When failure occurred, it was due to corrosion-fatigue and the cracks were transcrystalline. The typical caustic cracking was thus not obtained, but the cracks observed closely resembled certain defects found in boilers as the result of service.

Boiler Embrittlement Protection. C. W. Rice. (Combustion, 1941, vol. 12, Jan., pp. 37–39). The author reviews the present knowledge regarding the causes and prevention of caustic embrittlement. He offers explanations for certain apparently conflicting experiences and shows how, under different circumstances, some salts may either accelerate or prevent embrittlement.

A Corrosion-Fatigue Test to Determine the Protective Qualities of Metallic Platings. J. N. Kenyon. (Proceedings of the American

Society for Testing Materials, 1940, vol. 40, pp. 705–714). The author describes a corrosion-fatigue test for determining the degree of protection afforded by metallic coatings on steel wire. The wire tested was 0·037 in. in dia., and was a sample of that used to reinforce the beading of rubber tyres. For this purpose a stress-reversal machine was used in which a length of the wire was bent to arc curvature and rotated by an electric motor, a 7·5-in. portion of the arc dipping into a bath of oil, distilled water or salt solution (*see* Journ. I. and S. I., 1935, No. II., p. 486). The fatigue limits for 10^7 cycles of stress of uncoated and copper-zinc-coated wire tested in a non-corrosive oil were practically identical. In distilled water, copper-zinc-plated wire was much superior to unplated wire, and was also superior to bronze-plated wire. Bronze-coated specimens tested in oil after storage for 48 hr. in distilled water lost about 65% of their original tensile strength after 10^7 cycles of stress and this was found to be due to embrittlement, but this embrittlement did not occur in specimens that had been galvanised prior to bronze-plating.

ANALYSIS

(Continued from pp. 139 A–142 A)

The Sampling and Chemical Analysis of Cast Ferrous Metals. E. Taylor-Austin. (British Cast Iron Research Association, Special Publication No. 7, March, 1941). This publication is the third edition of the first report on recommended methods for the sampling and analysis of cast ferrous metals and alloys issued by the British Cast Iron Research Association Tests and Specifications Sub-Committee. It is in three parts, the first dealing with plain, the second with alloyed materials, and the third with ferro-alloys which find an increasing use in foundry practice. (*See* p. 195 A).

Electrolytic Determination of Copper in Steel. H. A. Frediani and C. H. Hale. (Industrial and Engineering Chemistry, Analytical Edition, 1940, vol. 12, Dec. 15, pp. 736–737). The authors review some methods for the electrolytic determination of copper in steel, and briefly report on their own experiments which yielded satisfactory results on Bureau of Standards samples representing a wide range of copper concentrations. They used a special cell, a sketch of which is shown, which enabled the temperature of the electrolyte to be controlled, and they added either hydrofluoric or phosphoric acid in order to suppress the effect of ferric iron on deposited copper.

The Use of Sodium Thiosulphate instead of Sodium Arsenite in the Determination of Manganese by the Persulphate Method. N. V. Udovenko and E. V. Smekh. (Zavodskaya Laboratoriya, 1940, No. 4, pp. 398–400). (In Russian). The results obtained by the persulphate method of determining manganese were unchanged when sodium thiosulphate was used for the final titration instead of sodium arsenite. Thiosulphate can also be used for the potentiometric titration. Sodium thiosulphate solution is to be preferred to sodium arsenite, as it is non-poisonous and more stable, particularly in warm weather.

Photo-Electric Method of Determining Nickel in Steels and Cast Irons. V. E. Mal'tsev and T. P. Temirenko. (Zavodskaya Laboratoriya, 1940, No. 4, pp. 386–390). (In Russian). In a preliminary investigation a study was made of the optimum conditions for the determination of nickel by the colorimetric method involving the oxidation of the dimethylglyoxime complex with bromine water in ammoniacal solution as suggested by Feigl. In the course of the investigation the use of sodium hydroxide in place of ammonia was found to be preferable. A spectro-photometric curve which was obtained showed that the oxidised nickel dimethylglyoxime compound has a maximum absorption in the range of 4300–5000 Å. In analysing irons and steels the iron is best suppressed by the addition of sodium potassium tartarate. The following elements, in the amounts stated, caused errors which, at most, were within the limits of experimental error: Titanium 2%, molybdenum 3%, vanadium 2%, chromium 10%, copper 2% and cobalt 1·5%. The procedure is as follows: 0·1 g. of the sample is dissolved by warming with 10 ml. of (1 : 3) nitric acid in a 100-ml. graduated flask. The solution is cooled, made up to the mark, well mixed and filtered if necessary to remove graphite. 5 ml. of the solution are transferred to a 100-ml. graduated flask and 10 ml. of a 20% solution of potassium sodium tartarate, 10 ml. of bromine water (1 ml. of bromine per litre), 3 ml. of a 1% alcoholic solution of dimethylglyoxime and 5 ml. of a 5% solution of sodium hydroxide are added, the solution is made up to the mark, mixed and the intensity of the coloration is determined. The final nickel determination is obtained from a calibration curve. The time required is 13 min. for steel and 18 min. for iron, and the results are accurate within 2–4%.

A Note on the Determination of the Total Sulphur in Iron and Steel by the Combustion and Volumetric Methods. (Vita's Method). I. K. Yamada, S. Mitui and S. Kiriyama. (Research Reports of the Sumitomo Metal Industries, Ltd., 1940, vol. 4, Aug., pp. 83–103). (In Japanese). With a view to overcoming the difficulties experienced in the determination of sulphur in iron and steel by combustion, the authors studied the absorption and acidity of pure SO_2, $SO_2 + SO_3$, CO_2 and P_2O_5. They also endeavoured to find the most favourable temperature of combustion, a suitable rate of oxygen flow, and a method which would render the admixture of oxides unnecessary.

Third Report of the Oxygen Sub-Committee of the Committee on the Heterogeneity of Steel Ingots. (Iron and Steel Institute, 1941, this Journal, Section I). Since the publication of the Second Report of the Oxygen Sub-Committee, which appeared as Section VI. of the Eighth Report on the Heterogeneity of Steel Ingots, the work of the co-operating members has been concerned mainly with a more detailed study of existing methods of determining the oxygen, nitrogen and hydrogen in steel, rather than with developing new ones, and the additional knowledge presented in this, the Third Report, is a definite advance on previous conceptions of those methods. The present Report is published in eleven sections. Section I. is introductory in character, and gives a list of the personnel of the Sub-Committee and some remarks on the scope of the present report. Section II. is in three parts. In Part A, H. A. Sloman reviews the work of the Oxygen Sub-Committee on the vacuum fusion method and compares the forms of apparatus in use in the co-operating laboratories; a note on the solid solubility of oxygen in high-purity iron is included. In Part B, by T. Swinden, W. W. Stevenson and G. E. Speight, some typical results obtained by the fractional vacuum fusion procedure for several types of steel are presented and compared, in some instances with results obtained by the alcoholic-iodine method. In Part C, by W. C. Newell, the author states that a change in the type of vacuum pump when determining the rate of evolution of hydrogen from steel improved the efficiency of the apparatus previously described (*see* Journ. I. and S. I., 1940, No. I., p. 243 P). The aluminium reduction method of determining oxygen is dealt with in Section III., which is also in three parts. In Part A, N. Gray and M. C. Sanders describe recent developments in the design and manipulation of the apparatus for this method using an atmosphere of hydrogen. In Part B, W. W. Stevenson and G. E. Speight give a description of the aluminium reduction method as operated at the Central Research Department of The United Steel Companies, Ltd. In Part C, E. Taylor-Austin gives an account of the application of the aluminium reduction method for the determination of total oxygen to a series of pig-iron samples. The chlorine method is dealt with in Section IV. In Part A of this section, by E. W. Colbeck, S. W. Craven and W. Murray, details of the present procedure for the determination of non-metallic inclusions in steel by the chlorine method are given. In Part B, the same investigators review the applicability and utility of the chlorine method. Section V. is concerned with the alcoholic-iodine method. T. E. Rooney discusses in Part A the present position, limitations and possibilities of this method of determining the oxides in low- and medium-carbon steels. W. W. Stevenson and G. E. Speight describe in Part B the conditions necessary for the successful operation of the standard alcoholic-iodine method for determining the oxide inclusions in iron and steel, as well as the development and operation of a simplified method. Section VI. deals with the aqueous-iodine method of determining oxide inclusions. In Part A, by E. Taylor-Austin, a modification of this method of determining non-metallic inclusions in pig iron and cast iron is described; the method is one which has been successfully applied to a series of more than forty typical pig irons. J. G.

Pearce discusses in Part B the present position of the determination of oxide inclusions in pig iron and cast iron. W. W. Stevenson explains in Section VII. that experimental work is proceeding on the suitability of the hydrogen reduction method for determining that part of the oxygen content of steel which corresponds to the low-temperature fraction of the fractional vacuum fusion method, but that difficulty has been met with in the selection of a suitable refractory material for the furnace tube; experiments with transparent silica are now being made. In Section VIII., G. E. Speight discusses the improvements in the analysis of non-metallic residues which have been effected since the publication of the Second Report of the Oxygen Sub-Committee, and the progress made towards complete colorimetric determination. In Section IX., accounts are presented of the results of oxygen determinations by various methods. Part A, by T. Swinden and W. W. Stevenson, is a report of an examination of the oxygen content of a basic Bessemer rimming steel by the following methods: (*a*) Microscopical examination; (*b*) chlorine extraction; (*c*) alcoholic iodine extraction; and (*d*) vacuum fusion determination by both the total and the fractional procedure. In Part B, W. R. Maddocks submits a note on the influence of carbon on the alcoholic iodine method for the determination of oxygen in steel. Section X. consists of a general summary of the present report by T. Swinden. The work in hand and future programme of investigations are discussed in Section XI. The outstanding feature of the present Report as regards total oxygen is the increased confidence which is being obtained in the aluminium reduction method; further work on the possible limitations and on improvements of this method form a definite feature of the forward programme.

A Microphotometer for Spectrochemical Analysis. E. M. Thorndike. (Industrial and Engineering Chemistry, Analytical Edition, 1941, vol. 13, Jan., pp. 66–67). A description is given of a microphotometer for measuring the density of spectrum lines. The apparatus is easily assembled from standard equipment, parts of which are available in most laboratories. A microscope forms a convenient foundation for the apparatus. The photographic plate to be measured is mounted on a mechanical stage and is illuminated by a single-filament automobile headlamp bulb. An enlarged image of the spectral line is directed on to a horizontal slit in front of a photo-electric cell by a prism mounted on a strip of metal which is hinged near one end and supported by a micrometer screw at the other. The cell is connected to a galvanometer the scale of which is mounted over the microscope. In measuring a spectrogram, the image of the spectral line is brought near to the slit by adjusting the mechanical stage, and the slit is turned until it is parallel to the image of the line. The image of the line is then centered on the slit approximately by adjusting the mechanical stage, and precisely by adjusting the micrometer screw until the minimum galvanometer deflection is obtained. The density of the line is computed from this deflection.

The Iron Content and the Angle of Polarization of Chilled Surfaces of Open Hearth Slag. M. Tenenbaum and T. L. Joseph. (Blast Furnace and Steel Plant, 1940, vol. 28, Dec., pp. 1157–1159). The authors describe some experiments undertaken with the object of determining the relation between the refractive index of an open-hearth slag and the iron content of the slag. The refractive index of an isotropic substance is equal to the tangent of the polarisation angle. In the present investigation the angles of polarisation of the samples of slag were measured with a modified Fuess goniometer. The slag samples were taken from industrial furnaces during the refining periods of typical basic open-hearth heats and were chilled rapidly by pouring them into cast-iron moulds. A portion of each sample was used for chemical analysis and pieces ¼ in. to ½ in. square from the lower surface were polished for examination with polarised light. It became apparent that the total iron content of the slag expressed as mols of ferrous oxide was more significant than the percentage values when determining the angle of polarisation. In calculating the mols of ferrous oxide in the slag, each mol of ferric oxide was taken to be equivalent to two mols of ferrous oxide, and the molar value of the former was added to that of the latter, which was calculated from the ferrous-oxide content obtained by chemical analysis. A table of the iron contents and the angles of polarisation of a large number of slag samples is presented, together with a curve demonstrating the relation between these two factors. It was evident that the angle of polarisation increases with increasing iron content expressed as mols of ferrous oxide.

Determination of Organic Sulphur in Combustible Gas. F. M. Rogers and R. F. Baldaste. (Industrial and Engineering Chemistry, Analytical Edition, 1940, vol. 12, Dec. 15, pp. 724–725). The authors have developed a relatively rapid method for the determination of organic sulphur in fuel gas, which is suitable for the determination of less than 13 mg. of sulphur per 100 cu. ft. of gas. The gas is burned with purified air in a special burner, illustrated in the paper. The sulphur dioxide resulting from the combustion of the gas is absorbed in sodium hypobromite solution, and the sulphate formed is determined "turbidimetrically" according to Sheen, Kohler and Ross (*see* Industrial and Engineering Chemistry, Analytical Edition, 1935, vol. 7, p. 262).

Determination of Sulphur in Coal and Coke. H. L. Brunjes and M. J. Manning. (Industrial and Engineering Chemistry, Analytical Edition, 1940, vol. 12, Dec. 15, pp. 718–720). The authors discuss the three methods for the determination of sulphur in coal and coke which have been accepted as standard by the A.S.T.M., *viz.*, the Eschka, the peroxide-fusion and the bomb-washing methods. They also describe the modified Eschka method they have developed, in which the ignition period can be shortened to approximately 50–60 min. They applied this modified Eschka method in some three thousand tests made on coals ranging from anthracite to lignite and containing 0·40–15·00% of sulphur. Some of the results obtained are reproduced in a table and compared with those obtained with the standard Eschka and peroxide-fusion methods.

Method of Analyzing Chrome Ore and Chrome-Ore Refractory Products. A. J. Boyle, D. F. Musser and O. S. Keim. (Bulletin of the American Ceramic Society, 1941, vol. 20, Jan., pp. 7–9). The authors have developed a comparatively rapid method for the analysis of chrome-ore and chrome-ore refractories, using four separate samples for the determination of chromium, silicon, iron, and iron and aluminium, respectively. The samples are in all cases fused with sodium peroxide. The solvents used and the materials of the crucibles are as follows:

Determination.	Crucible.	Solvent.
Cr_2O_3	Iron	Water.
SiO_2	Iron	Perchloric acid (50%)
$Fe_2O_3 + Al_2O_3$ . . .	Silica	Perchloric acid (50%)
Fe_2O_3	Nickel	Perchloric acid (50%)

Calcium and magnesium are determined in the filtrate of the $Fe_2O_3 + Al_2O_3$ precipitate. The authors point out that the chromate present does not interfere with the precipitation of magnesium as $MgNH_4PO_4$.

BIBLIOGRAPHY

(Continued from p. 143 A)

British Cast Iron Research Association. Special Publication No. 7. "*The Sampling and Chemical Analysis of Cast Ferrous Metals.*" Revised and enlarged by E. Taylor-Austin. 8vo, pp. 140. Illustrated. Birmingham 1941: The Association. (Price 15*s*.) [See Notice, p. 195 A.]

Department of Scientific and Industrial Research. "*Physical and Chemical Survey of the National Coal Resources, No. 52: The Leicestershire and South Derbyshire Coalfield. South Derbyshire Area—The Kilburn Seam.*" 8vo, pp. iv + 58. London, 1940: H.M. Stationery Office. (Price 2*s*.)

"*Factory Training Manual.*" Being a Practical Textbook for Use in the Factory and Workshop in Connection with the Ministry of Labour Scheme for Training Skilled and Semi-Skilled Operatives. By a Group of Engineers. Edited by Reginald Pugh. With a Foreword by The Rt. Hon. Ernest Bevin, P.C. 8vo, pp. xii + 286. Illustrated. Bath, 1941: Management Publications Trust, Ltd. (Price 5*s*.) [See Notice, p. 196 A.]

" *Kingzett's Chemical Encyclopædia.*" A Digest of Chemistry and its Industrial Applications. Revised and Edited by R. K. Strong. With Foreword by Sir Gilbert T. Morgan, O.B.E., F.R.S. Sixth Edition. 8vo, pp. x + 1088. London, 1940: Baillière, Tindall and Cox. (Price 45*s*.)

LEAGUE OF NATIONS. ECONOMIC INTELLIGENCE SERVICE. " 1939/40 *Statistical Year-Book of the League of Nations.*" 8vo, pp. 285. Genève, 1940. (Price 12*s*. 6*d*.)

MANTELL, C. L. " *Industrial Electrochemistry.*" Second Edition. (Chemical Engineering Series.) 8vo, pp. x + 656. Illustrated. London, 1940: McGraw-Hill Publishing Co., Ltd. (Price 38*s*.)

MASSACHUSETTS INSTITUTE OF TECHNOLOGY. " *Proceedings of the Special Summer Conferences on Friction and Surface Finish.*" Held at Massachusetts Institute of Technology, Cambridge, Massachusetts, June 5–7, 1940. Publication made possible by the generosity of the Chrysler Corporation. 8vo, pp. 244. Illustrated. Cambridge, Mass., 1940.

" *Mineral Industry. Its Statistics, Technology and Trade During* 1939." Edited by G. A. Roush. Vol. 48. 8vo, pp. xliv + 761. New York and London, 1940: McGraw-Hill Book Co., Inc. (Price 70*s*.)

MOND NICKEL CO., LTD. " *Equivalent British and S.A.E. Nickel Alloy Steels.*" 8vo, pp. 15. London: Mond Nickel Co., Ltd.

SCHOELLER, W. L., and A. R. POWELL. " *The Analysis of Minerals and Ores of the Rarer Elements.*" Second Edition, Entirely Rewritten. 8vo, pp. 323. London, 1940: Charles Griffin and Co., Ltd. (Price 18*s*.)

STUTZER, O. " *Geology of Coal.*" Translated and revised by A. C. Noé. 8vo, pp. xiii + 461. Chicago, 1940: University of Chicago Press; Cambridge: University Press. (Price 30*s*.)

SWIGERT, A. M., jun. " *The Story of Superfinish.*" 8vo, pp. 672. Illustrated. Detroit, Mich., 1940: Lynn Publishing Co.

UNITED STATES BUREAU OF MINES. " *Minerals Yearbook* 1940." 8vo, pp. x + 1514. Washington, 1940: Government Printer. (Price $2.)

VOGEL, A. I., " *A Text-book of Qualitative Chemical Analysis.*" Second Edition. 8vo, pp. xi + 486. London, 1941: Longmans, Green & Co. (Price 10*s*. 6*d*.)

WAY, R. B., and N. D. GREEN, " *Welding and Flame Cutting.*" 8vo, pp. 64. London, 1940: Percivall Marshall & Co., Ltd. (Price 2*s*.)

SUBJECT INDEX.

[References to the papers read before the Institute are indicated by the word *Paper* following the page number. The letter *P.* denotes a reference in Section I. of the Journal which contains the reports of the proceedings of meetings, the papers read, and the discussions thereon. The letter *A.* denotes a reference to the section dealing with abstracts.

Indexing of Alloy Steels and Other Alloys. In the indexing of alloy steels, carbon and iron are ignored and the alloying elements contained in the steel are arranged in alphabetical order; for example, all references to nickel-chromium-molybdenum steel will be found under the heading chromium-molybdenum-nickel steel. In the indexing of other alloys, carbon and iron, when present, are included in the title; iron, when present, is always mentioned first and the other elements follow in alphabetical order, carbon being in all cases mentioned last. Examples: "Iron-silicon-carbon alloys" and "iron-chromium-nickel-carbon alloys."]

NAME INDEX.

PRINTED IN GREAT BRITAIN BY RICHARD CLAY AND COMPANY, LTD.,
BUNGAY, SUFFOLK.

Lightning Source UK Ltd.
Milton Keynes UK
UKHW021447090119
335262UK00011B/971/P